Medicinal and Aromatic Plants

Also of interest

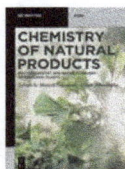

Chemistry of Natural Products.
Phytochemistry and Pharmacognosy of Medicinal Plants
Mayuri Napagoda, Lalith Jayasinghe (Eds.), 2022
ISBN 978-3-11-059589-5
e-ISBN 978-3-11-059594-9

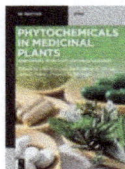

Phytochemicals in Medicinal Plants.
Biodiversity, Bioactivity and Drug Discovery
Charu Arora, Dakeshwar Kumar Verma, Jeenat Aslam,
Pramod Kumar Mahish (Eds.), 2023
ISBN 978-3-11-079176-1
e-ISBN 978-3-11-079189-1

Essential Oils.
Sources, Production and Applications
Rajendra Chandra Padalia, Dakeshwar Kumar Verma,
Charu Arora and Pramod Kumar Mahish (Eds.), 2023
ISBN 978-3-11-079159-4
e-ISBN 978-3-11-079160-0

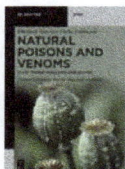

Natural Poisons and Venoms.
Plant Toxins: Alkaloids and Lectins
Eberhard Teuscher, Ulrike Lindequist, 2024
ISBN 978-3-11-112740-8
e-ISBN 978-3-11-113621-9

Medicinal and Aromatic Plants

Pharmaceutical, Food, and Cosmetic Applications

Edited by
Elyor Berdimurodov, Esra Uçar, and Burak Tuzun

DE GRUYTER

Editors
Dr. Elyor Berdimurodov
Faculty of Chemistry
National University of Uzbekistan
Tashkent, 100034
Uzbekistan
elyor170690@gmail.com

Prof. Esra Uçar
Medicinal and Aromatic Plant Department
Technical Sciences
Vocational School of Sivas
Sivas Cumhuriyet University
58140 Sivas, Turkey
eucar@cumhuriyet.edu.tr

Prof. Dr. Burak Tuzun
Associate Professor
Medicinal and Aromatic Plant Department
Technical Sciences
Vocational School of Sivas
Sivas Cumhuriyet University
58140 Sivas, Turkey
theburaktuzun@yahoo.com

ISBN 978-3-11-146866-2
e-ISBN (PDF) 978-3-11-146971-3
e-ISBN (EPUB) 978-3-11-147038-2

Library of Congress Control Number: 2025934050

Bibliographic information published by the Deutsche Nationalbibliothek
The Deutsche Nationalbibliothek lists this publication in the Deutsche Nationalbibliografie;
detailed bibliographic data are available on the internet at http://dnb.dnb.de.

© 2025 Walter de Gruyter GmbH, Berlin/Boston, Genthiner Straße 13, 10785 Berlin
Cover image: sefa ozel/E+/Getty Images
Typesetting: Integra Software Services Pvt. Ltd.

www.degruyterbrill.com
Questions about General Product Safety Regulation:
productsafety@degruyterbrill.com

Contents

Part I: **Introduction**

Gamze Tüzün, Burak Tüzün*, Dilara Ülger Özbek,
and Elyor Berdimurodov

Chapter 1
The importance of medicinal and aromatic plants for living things

Abstract: Medicinal and aromatic plants have had an important place in both traditional and modern medicine throughout history, and have become an indispensable resource in various fields such as health, cosmetics, and food industries. This book focuses on the basic properties, biological activities, and versatile usage areas of medicinal and aromatic plants. The book explains the historical and scientific basis of medicinal and aromatic plants, and details the chemical structures and pharmacological effects of alkaloids, flavonoids, terpenoids, and other active compounds. It also covers developments in the field of phytochemistry and the discovery of active compounds obtained from these plants, and their analysis methods. The role of aromatic plants in daily life is discussed through essential oils and aromatherapy applications, and their effects in nutrition, beauty, and health. In addition, methods and legal regulations developed to protect and ensure sustainable use of natural resources are discussed with their local and international dimensions. The challenges faced by medicinal and aromatic plants, such as climate change, genetic innovations, and biotechnological approaches, and future opportunities constitute one of the sections of the book that offers a forward-looking vision. In addition, the importance of medicinal and aromatic plants in the local and global context is emphasized, focusing on successful conservation projects in different regions, and the integration of traditional knowledge and modern science.

Keywords: medicinal and aromatic plants, sustainable use, phytochemistry, biodiversity conservation, traditional knowledge integration

*Corresponding author: Burak Tüzün, Plant and Animal Production Department, Technical Sciences
Vocational School of Sivas, Sivas Cumhuriyet University, 58140 Sivas, Turkey,
e-mail: theburaktuzun@yahoo.com, https://orcid.org/0000-0002-0420-2043
Gamze Tüzün, Department of Chemistry, Faculty of Science, Cumhuriyet University, 58140 Sivas, Turkey
Dilara Ülger Özbek, Advanced Technology Research and Application Centre, Sivas Cumhuriyet
University, 58140 Sivas, Turkey
Elyor Berdimurodov, Chemical and Materials Engineering, New Uzbekistan University, 54 Mustaqillik
Ave, Tashkent 100007, Uzbekistan; Faculty of Chemistry, National University of Uzbekistan, Tashkent
100034, Uzbekistan

https://doi.org/10.1515/9783111469713-001

1.1 Introduction to medicinal and aromatic plants

1.1.1 Historical background

Medicinal and aromatic plants (MAPs) are one of the oldest natural resources used in health and treatment processes in human history. The first traces of their use dates back to prehistoric times. Archaeological evidence shows that Neanderthals used plants with medicinal properties in their diets and treatment practices. For example, fossil remains found in El Sidrón Cave in Spain revealed that Neanderthals used chamomile and yarrow for herbal treatment [1].

In Mesopotamia, herbal treatment methods are recorded on clay tablets dating back to 3,000 BC. These tablets include recipes from plants such as garlic, fennel, and thyme. Ancient Egypt was also one of the pioneer civilizations in the use of medicinal plants. The Ebers Papyrus, dating back to 1550 BC, contains approximately 850 herbal prescriptions. Aloe Vera, myrrh, and garlic are among the plants frequently used by Egyptians in treating infections and healing wounds [2].

Ayurvedic medicine, developed in India, has provided comprehensive information on the systematic use of medicinal plants. Texts such as the Charaka Samhita and Sushruta Samhita have detailed the health effects of plants such as neem, turmeric, and ashwagandha [3]. During the same period, traditional Chinese medicine (TCM) in China used plants such as ginseng, ephedra, and licorice to balance yin-yang and treat diseases.

Ancient Greek and Roman civilizations brought herbal medical practices to the Western world. Hippocrates (460–370 BC) emphasized the role of plants in disease treatment and laid the foundations of modern medicine. Dioscorides' De Materia Medica described approximately 600 plants and it was used as a medical reference source in Europe for centuries [4]. During the Islamic Golden Age, Ibn Sina's book Al-Qanun fi't-Tıb systematically organized information on medicinal plants and influenced both Eastern and Western civilizations [5].

In other parts of the world, indigenous communities have also discovered the healing properties of plants in their environment. From the Amazon rainforest to the African savannas, indigenous peoples have used MAPs to treat illness for thousands of years. This traditional knowledge from the past forms the basis of many studies that form the basis of modern pharmaceutical chemistry research today [6–9]. Today, researchers have studied the medicinal potential of chemicals found in plant extracts of many plants for the treatment of important diseases [10–15].

1.1.1.1 Historical background

MAPs are one of the oldest natural resources used in health and treatment processes in human history. The first traces of use date back to prehistoric times. Archaeological evidence shows that Neanderthals used plants with medicinal properties in their diets and treatment practices. For example, fossil remains found in El Sidrón Cave in Spain revealed that Neanderthals used chamomile and yarrow for herbal treatment [16].

In Mesopotamia, herbal treatment methods are recorded on clay tablets dating back to 3,000 BC. These tablets include recipes from plants such as garlic, fennel, and thyme. Ancient Egypt was also one of the pioneer civilizations in the use of medicinal plants. The Ebers Papyrus, dating back to 1,550 BC, contains approximately 850 herbal prescriptions. Aloe vera, myrrh and garlic are among the plants frequently used by Egyptians in treating infections and healing wounds [17].

Ayurvedic medicine, developed in India, has provided comprehensive information on the systematic use of medicinal plants. Texts such as the Charaka Samhita and Sushruta Samhita have detailed the health effects of plants such as neem, turmeric, and ashwagandha [18]. During the same period, traditional Chinese medicine (TCM) in China used plants such as ginseng, ephedra, and licorice to balance yin-yang and treat diseases.

Ancient Greek and Roman civilizations brought herbal medical practices to the Western world. Hippocrates (460–370 BC) emphasized the role of plants in disease treatment and laid the foundations of modern medicine. Dioscorides' De Materia Medica described approximately 600 plants and was used as a medical reference source in Europe for centuries [19]. During the Islamic Golden Age, Ibn Sina's book Al-Qanun fi't-Tıb systematically organized information on medicinal plants and influenced both Eastern and Western civilizations [20].

In other parts of the world, indigenous communities have also discovered the healing properties of plants in their environment. From the Amazon rainforest to the African savannah, indigenous peoples have used MAPs to treat illnesses for thousands of years. This traditional knowledge has formed the basis of modern pharmaceutical research today.

1.1.2 Traditional and modern uses

1.1.2.1 Traditional uses

MAPs have been one of the most important components of traditional medicine. Herbal treatments have often been applied in the form of infusions, decoctions, pastes, and oils. For example, in India, holy basil (tulsi) has been used to treat respiratory diseases. In China, ginseng has been consumed as an energy booster and immune booster [21, 22].

In Africa, rooibos tea has been used both as a dietary supplement and as a medicine for its antioxidant properties. Native American communities have used aloe vera and wormwood to heal wounds. Aromatic plants have also been widely used in religious rituals. Lavender has been used as incense due to its relaxing effects, and plants such as mint and eucalyptus have been used to clear the respiratory tract [23].

1.1.2.2 Modern uses

Modern science, recognizing the value of traditional knowledge, has widely used active ingredients obtained from MAPs in the pharmaceutical, cosmetic, and food industries. Aspirin, derived from salicylic acid obtained from willow bark, is an example of the adaptation of this herbal knowledge to modern medicine [24].

The active ingredient of turmeric, curcumin, is being investigated in the treatment of conditions such as cancer, arthritis, and Alzheimer's disease due to its anti-inflammatory and antioxidant properties. Lavender oil is used as an effective natural solution in the treatment of anxiety and insomnia [25, 26].

Aromatic plants are widely used in the food industry as natural sweeteners and preservatives. Essential oils obtained from plants such as thyme and rosemary are used to prevent microbial spoilage in foods. The cosmetic industry prefers essential oils obtained from plants such as lavender, chamomile, and sandalwood in skin care products [27]. An example of a table summarizing the modern applications of MAPs is presented in Table 1.1.

Table 1.1: The modern applications of medicinal and aromatic plants.

Plant name	Active compounds	Modern application
Lavender	Linalool and linalyl acetate	Aromatherapy, stress relief, and wound healing
Turmeric	Curcumin	Anti-inflammatory, anticancer, and antioxidant
Peppermint	Menthol and menthone	Irritable bowel syndrome treatment, analgesic, and antispasmodic
Rosemary	Rosmarinic acid and carnosic acid	Cognitive function improvement and food preservative
Ginger	Gingerol and shogaol	Antiemetic, anti-inflammatory, and antioxidant
Aloe vera	Polysaccharides and anthraquinones	Skin care, wound healing, and digestive health

1.1.2.3 Sustainability and future perspectives

Sustainable use of MAPs is important for both ecosystem protection and human health. Climate change, habitat loss, and overharvesting threaten the natural populations of these plants. Modern biotechnology holds promise for producing the active ingredients of these plants in laboratory conditions and contributing to the protection of natural resources [28, 29].

1.2 Alkaloids, flavonoids, terpenoids, and other active compounds

Bioactive compounds found in MAPs are molecules that support the defense mechanisms of plants and provide various pharmacological benefits for humans. Among these compounds, alkaloids, flavonoids, terpenoids, and other active phytochemicals stand out in Figure 1.1.

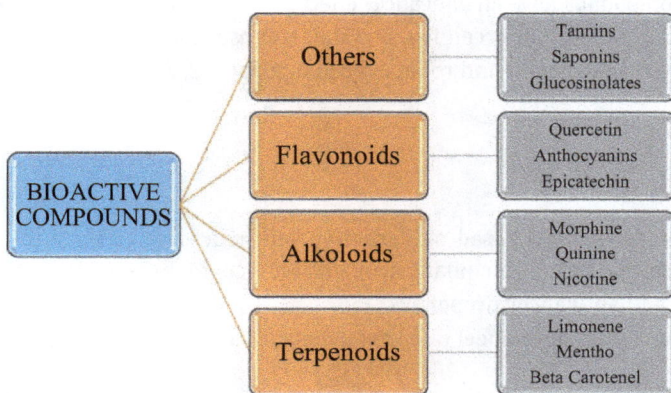

Figure 1.1: Bioactive compounds in medicinal and aromatic plants.

1.2.1 Alkaloids

Alkaloids are nitrogen-containing heterocyclic compounds and are widely found in plants. Known for their medicinal effects, alkaloids have been used throughout history as analgesics, and antimalarial and anticancer agents:

1. Morphine: Obtained from the poppy plant (*Papaver somniferum*), it is used as a powerful analgesic to control post-surgical pain.
2. Quinine: This alkaloid, isolated from the *Cinchona* tree (*Cinchona* spp.), is one of the first effective drugs used in the treatment of malaria.
3. Nicotine: Found in the tobacco plant (*Nicotiana tabacum*), nicotine has stimulating effects on the central nervous system, and has been carefully studied due to its addictive properties [30, 31].

1.2.2 Flavonoids

Flavonoids are phenolic compounds that support the color, taste, and defense mechanisms of plants. They are known for their antioxidant properties and offer positive effects on health:

1. Quercetin: Found in onions, apples, and grapes, this compound has strong antioxidant properties and can help prevent cardiovascular diseases.
2. Anthocyanins: Found in red-purple fruits such as blueberries, blackberries, and strawberries, these compounds have an anticancer effect by scavenging free radicals.
3. Epicatechin: Found abundantly in green tea, it stands out especially for its effect in protecting cardiovascular health and reducing cell damage [32, 33].

1.2.3 Terpenoids

Terpenoids are isoprene derivatives based on carbohydrate structures and are the main components of essential oils. Their pharmacological effects include antimicrobial, anti-inflammatory, and anticancer properties.

1. Limonene: Limonene, found in the peel of citrus fruits, has antimicrobial and anticancer effects.
2. Menthol: Menthol, obtained from the mint plant, is used for skin irritation and respiratory tract disorders due to its cooling and analgesic effects.
3. Beta-carotene: This terpenoid, found in carrots, sweet potatoes, and tomatoes, is a precursor of vitamin A and supports vision health [34, 35].

1.2.4 Other active compounds

1. Tannins: These compounds, found in tea, coffee, and oak bark, have astringent properties and help in wound healing.
2. Saponins: They have soap-like properties and have the potential to strengthen the immune system and prevent cancer.
3. Glucosinolates: These compounds, found in cabbage, broccoli, and cauliflower, reduce the risk of cancer by activating detoxification enzymes [36].

1.3 Chemical structures and pharmacological effects

1.3.1 Chemical structures and effects of alkaloids

The chemical structures of alkaloids generally contain nitrogen-containing heterocyclic rings. This structure allows them to be effective on the central nervous system:
1. Morphine: It has a purine-based structure and shows analgesic effects by activating opioid receptors.
2. Quinine: It has a quinoline ring structure and prevents the proliferation of the Plasmodium parasite.
3. Nicotine: Nicotine, which has pyridine and pyrrole rings, shows addictive effects by stimulating nicotinic acetylcholine receptors [37, 38].

1.3.2 Chemical structures and effects of flavonoids

Flavonoids consist of a three-ring (C6-C3-C6) structure and carry phenolic hydroxyl groups. This structure determines their antioxidant properties:
1. Quercetin: It is a powerful antioxidant with a 3-OH group and reduces inflammation [39].
2. Anthocyanins: They contain glycosidic bonds and reduce oxidative stress by scavenging free radicals.
3. Epicatechin: It belongs to the flavan-3-ol class and increases the elasticity of blood vessels [40].

1.3.3 Chemical structures and effects of terpenoids

Terpenoids are derived from isoprene units and their different chain lengths determine their pharmacological properties:
1. Limonene: It belongs to the monoterpene class and has an antimicrobial effect by disrupting microbial cell membranes.
2. Menthol: It has a cyclic monoterpene structure and shows its cooling effect by activating TRPM8 receptors.
3. Beta-carotene: It has a tetraterpene structure and contributes to the synthesis of retinol (vitamin A) [41–43].

1.3.4 Structures and effects of other compounds

1. Tannins: It has a polyphenolic structure and prevents the growth of microorganisms by binding to proteins [44].
2. Saponins: It has a glycosidic structure and strengthens the immune system by stabilizing the cell membrane [45].
3. Glucosinolates: It activates detoxification enzymes with its sulfur-containing structure and prevents the growth of cancer cells [46].

1.4 Phytochemistry: active compounds in medicinal and aromatic plants

1.4.1 Alkaloids, flavonoids, terpenoids, and other active compounds

MAPs are known for their positive effects on health, thanks to the various phytochemicals they contain. The biological activities of these compounds depend on the role they play in the plant's metabolism and their chemical structures. The most frequently researched and used phytochemicals are alkaloids, flavonoids, terpenoids, and other bioactive compounds [47]:

1. Alkaloids: Alkaloids are nitrogen-containing heterocyclic compounds and generally play an important role in the defense system of plants.
2. Morphine: Obtained from the poppy plant (*Papaver somniferum*), this alkaloid is used as a strong painkiller. It binds to opioid receptors and affects the central nervous system.
3. Quinine: This compound, isolated from the *Cinchona* tree (*Cinchona* spp.), is used in the treatment of malaria. It has a parasitic effect with its quinoline ring
4. Nicotine: Nicotine, found in the tobacco plant (*Nicotiana tabacum*), has stimulating effects on the central nervous system, but it is addictive.
5. Capsaicin: This alkaloid, found in hot peppers, is widely used in pharmacology for its pain-relieving effect.

1.4.2 Flavonoids

Flavonoids are polyphenol compounds with strong antioxidant properties due to their phenolic structure. Flavonoids are naturally occurring polyphenolic compounds that are widely distributed in plants. Chemically, they are benzopyrone derivatives with a C6-C3-C6 carbon skeleton. Flavonoids play an important role in determining the properties of plants such as color, odor, and taste, and are also part of protective mechanisms against environmental stresses [48, 49]:

1. Quercetin: This flavonoid, found in apples, onions, and grapes, protects against cardiovascular diseases by neutralizing free radicals.
2. Anthocyanins: Found in foods such as blueberries, blackberries, and red cabbage, it offers cell damage-preventing effects.
3. Epicatechin: This flavonoid, found in abundance in green tea, increases the flexibility of blood vessels and supports heart health.

1.4.3 Terpenoids

Terpenoids are the main components of volatile oils derived from isoprene units. These compounds are known for their antimicrobial, anticancer, and anti-inflammatory effects [50, 51]:

1. Limonene: This monoterpene found in the peel of citrus fruits has antimicrobial effects and is widely used in cosmetic products.
2. Menthol: Menthol obtained from the mint plant provides cooling and analgesic effects.
3. Beta-carotene: This tetraterpene, a precursor to vitamin A, is found in carrots, sweet potatoes, and spinach.

1.4.4 Other active compounds

1. Tannins: Tannins found in tea and coffee have antioxidant properties and support cell renewal.
2. Saponins: These compounds strengthen the immune system and have been associated with anticancer effects.
3. Glucosinolates: These compounds, found in vegetables such as cauliflower, broccoli, and cabbage, protect against cancer by activating detoxification enzymes [46].

1.5 Chemical structures and pharmacological effects

1.5.1 Chemical structures and effects of alkaloids

The chemical structure of alkaloids generally contains heterocyclic rings with nitrogen atoms. This structure plays a decisive role in their pharmacological effects [52, 53]:

1. Morphine: It contains methoxy groups attached to the phenanthrene ring and shows its pain-relieving effect by interacting with opioid receptors. The molecular structure of morphine is given in Figure 1.2.

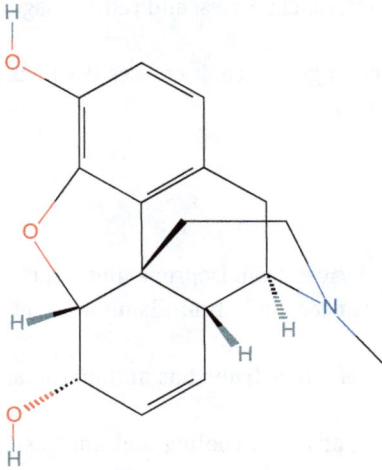

Figure 1.2: Molecular structure of morphine.

2. Quinine: Quinine, which carries a quinoline ring, is an alkaloid effective in the treatment of parasitic infections. The molecular structure of quinine is given in Figure 1.3.

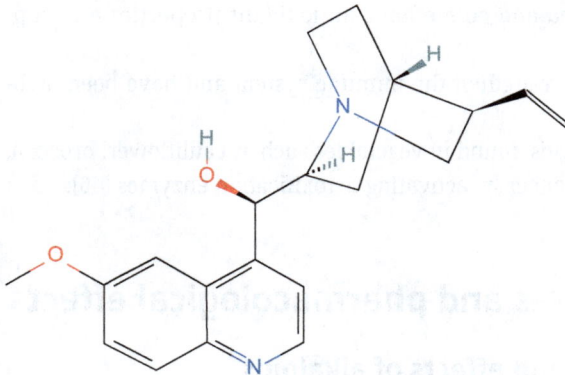

Figure 1.3: Molecular structure of quinine.

3. Nicotine: It has pyridine and pyrrole rings; this structure allows it to bind to nicotinic acetylcholine receptors. The molecular structure of nicotine is given in Figure 1.4.

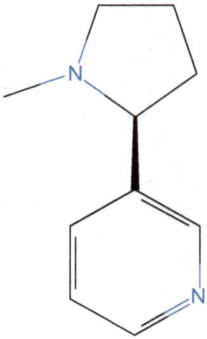

Figure 1.4: Molecular structure of nicotine.

1.5.2 Chemical structures and effects of flavonoids

Flavonoids are polyphenolic compounds built on the C6-C3-C6 skeleton. This structure has strong antioxidant properties [54, 55]:

1. Quercetin: The presence of 3-OH and 5-OH groups increases the radical scavenging capacity. The molecular structure of quercetin is given in Figure 1.5.

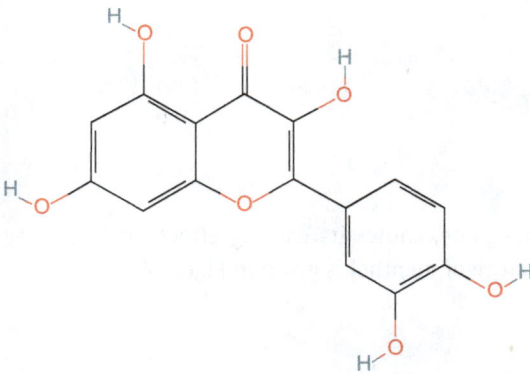

Figure 1.5: Molecular structure of quercetin.

2. Anthocyanins: They contain sugar molecules connected by glycosidic bonds, which increases water solubility and optimizes bioavailability. The molecular structure of anthocyanıns is given in Figure 1.6.

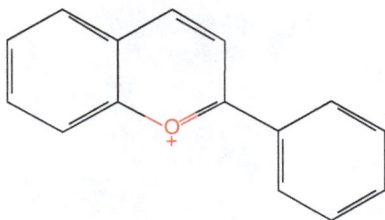

Figure 1.6: Molecular structure of anthocyanıns.

1.5.3 Chemical structures and effects of terpenoids

Terpenoids consist of isoprene units and their pharmacological activities vary depending on their structure [41, 43, 56]:

1. Limonene: It has a monoterpene structure containing double bonds and has antimicrobial and anticancer properties. The molecular structure of limonene is given in Figure 1.7.

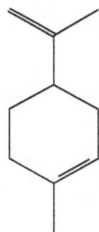

Figure 1.7: Molecular structure of limonene.

2. Menthol: Menthol, a cyclic monoterpene, shows its cooling effect by activating TRPM8 receptors. The molecular structure of menthol is given in Figure 1.8.

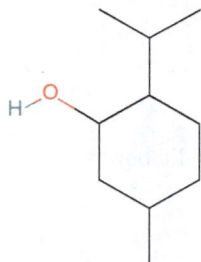

Figure 1.8: Molecular structure of menthol.

3. Beta-carotene: Its structure, containing conjugated double bonds, plays a leading role in the production of vitamin A and supports eye health. The molecular structure of beta-carotene is given in Figure 1.9.

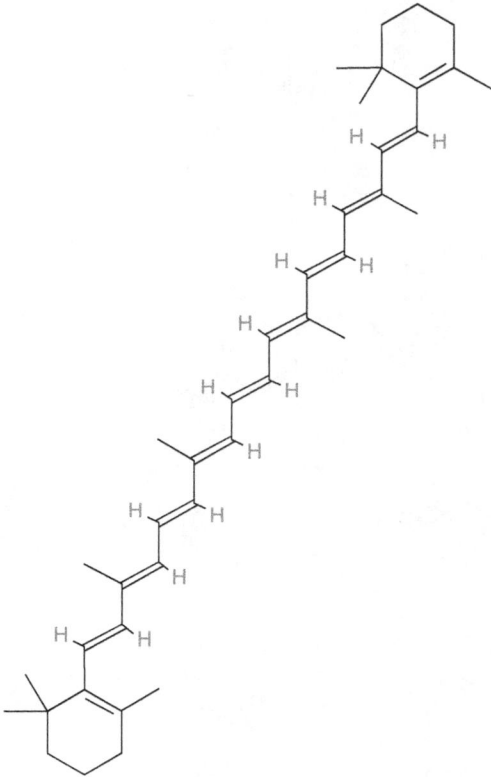

Figure 1.9: Molecular structure of beta-carotene.

1.5.4 Structures and effects of other compounds

1. Tannins: It has a polyphenolic structure and prevents the growth of microorganisms by binding with proteins. The molecular structure of tannic acid is given in Figure 1.10.
2. Saponins: It contains triterpene structures connected by glycosidic bonds and stabilizes the cell membrane.
3. Glucosinolates: Their sulfur-containing structures activate detoxification enzymes and neutralize toxins.

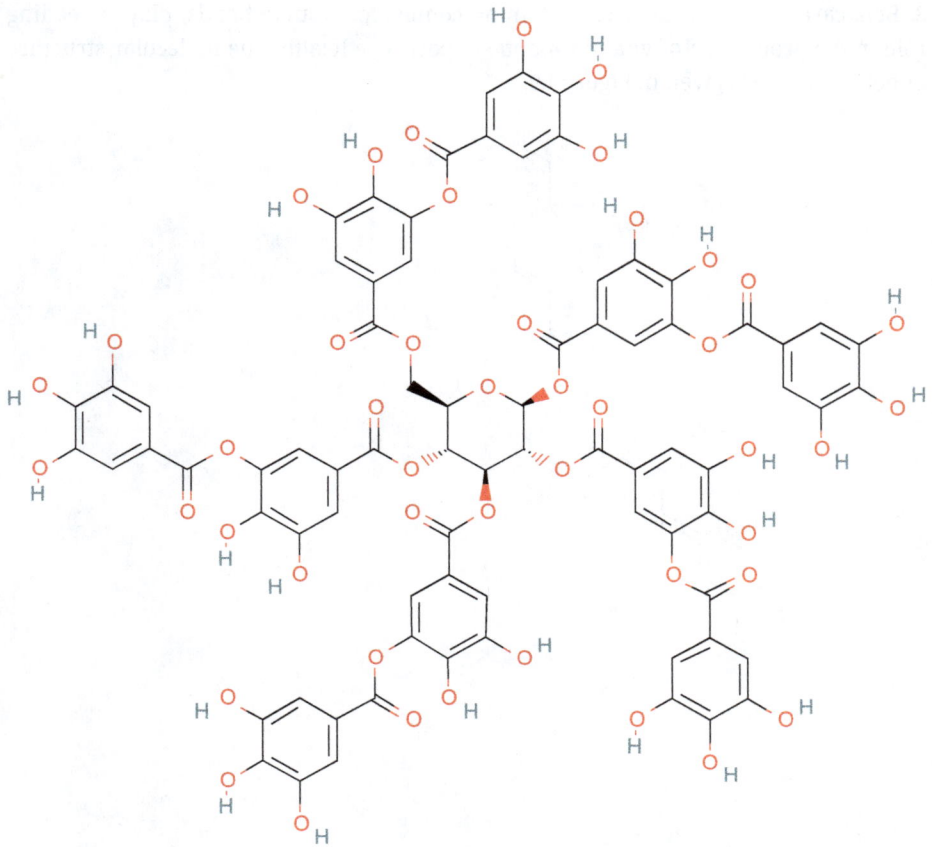

Figure 1.10: Molecular structure of tannic acid.

1.6 Aromatic plants in everyday life

Aromatic plants have been a part of our cultural heritage from the past to the present and have played an indispensable role in human life in the fields of medicine, cosmetics, and food [57]. Thanks to the volatile oils and other active ingredients they contain, these plants not only emit pleasant scents, but also offer many benefits in different areas such as health, beauty, and taste. In particular, the importance of essential oils and aromatherapy and applications in the cosmetics and food industry allow us to better understand the effects of aromatic plants in our daily lives [58, 59].

1.6.1 The importance of essential oils and aromatherapy

Essential oils are extremely dense and effective natural compounds that carry the essence of aromatic plants [60]. They are usually obtained from the leaves, flowers, barks, roots, or resins of plants by methods such as steam distillation, cold pressing, or solvent extraction [61]. These oils contain chemical components that plants produce to protect themselves from pests, to communicate with other living things, or to adapt to environmental stresses.

Aromatherapy is a natural treatment method in which essential oils are used to support physical, mental, and emotional balance [62]. In aromatherapy, essential oils are usually used by inhalation, topical application as massage oil, or by mixing them into the air through diffusers. Table 1.2 lists the many applications for essential oils used in aromatherapy. The main benefits of aromatherapy are:
1. Stress and anxiety management: Essential oils such as lavender, bergamot, and ylang-ylang are known for their calming effects. Inhaling these oils reduces stress hormones and creates a sense of relaxation. Aromatherapy, especially used during meditation or yoga, can increase mental calmness.
2. Improving sleep quality: Lavender oil is a common solution for individuals with sleep problems. Putting a few drops of lavender oil on your pillow in the evening or spreading it into the air through a diffuser creates an environment for deep and peaceful sleep.
3. Strengthening the immune system: Oils such as thyme, tea tree, and eucalyptus protect the body against infections, thanks to their antimicrobial properties. Inhaling or topically applying these oils, especially during cold and flu season, supports the immune system.
4. Pain- and muscle-relaxing effects: Peppermint, rosemary, and ginger oils offer effective natural solutions for relieving problems such as muscle pain and headaches. Peppermint oil provides relief when used by massaging the temples, especially in migraine attacks.
5. Mental vitality and concentration: Lemon, orange, and rosemary oils reduce mental fatigue and increase focus. These oils have a productivity-increasing effect when used in the work environment.

Table 1.2: Areas of use of essential oils used in aromatherapy [61].

Essential oil	Usage areas	Action mechanism
Lavender oil	Stress and anxiety reducer, sleep regulator, headache reliever, and skin care	Relaxing and calming properties
Peppermint oil	Energy booster, concentration enhancer, headache reliever, and muscle relaxant	Cooling and invigorating effect
Ylang-Ylang oil	Mood stabilizer, blood pressure lowering, and aphrodisiac	Calming and relaxing properties

Table 1.2 (continued)

Essential oil	Usage areas	Action mechanism
Lemon oil	Energy booster, offers mental clarity, and natural cleanser	Refreshing and purifying properties
Eucalyptus oil	Respiratory tract opener, nasal congestion reliever, and mental relief	Refreshing and cleansing effect
Bergamot oil	Stress and anxiety reducer, and mood enhancer	Mild stimulating and balancing properties
Tea tree oil	Antiseptic and relieves skin problems (acne and infection)	Antimicrobial and healing properties
Sandalwood oil	Meditation support, relaxing, and skin moisturizer	Grounding and calming properties
Rose oil	Emotional balancer, aphrodisiac, and skin care	Soothing and refreshing
Rosemary oil	Memory enhancer, mental clarity enhancer, and muscle pain reliever	Stimulating and circulatory properties

Aromatherapy creates positive effects not only on physical health but also on emotional balance. In today's fast-paced living conditions, these natural solutions offered by aromatherapy play an important role in improving the quality of life of individuals.

1.6.2 Applications in the cosmetics and food industry

Volatile oils and other active ingredients obtained from aromatic plants have a wide range of uses in the cosmetics and food industries [63, 64]. These ingredients are in great demand in modern formulations because they are both natural and effective [65]. The areas of use of the essential oil of the aromatic plant in cosmetics are shown in Figure 1.11:

1   Cosmetics industry: Aromatic plants are widely used in skin and hair care products, perfumes, and spa therapies. The main advantages of essential oils in the cosmetic field are as follows:

2   Skin care: Essential oils such as lavender, tea tree, and rose oil have antioxidant and anti-inflammatory properties. These oils are effective in treating acne, maintaining the moisture balance of the skin, and reducing the signs of aging.

3   Natural fragrances: As an alternative to synthetic perfumes, essential oils offer pleasant and natural scents. For example, oils such as jasmine and sandalwood are among the main components of luxury perfumes.

4   Hair care: Rosemary and mint oils nourish the scalp, reduce hair loss, and promote hair growth. They are also effective in eliminating dandruff problems.

Figure 1.11: Uses of essential oil of aromatic plant in cosmetics.

In spa therapies, the use of essential oils in massage and steam applications provides both a relaxing experience and supports skin health.

1.6.3 Food industry

Aromatic plants are indispensable in the food industry as flavor and aroma enhancers. Extracts obtained from plants such as thyme, rosemary, mint, and basil add characteristic flavors to dishes and drinks while also offering protective properties [66]. The use of MAPs in the food sector is shown in Figure 1.12:

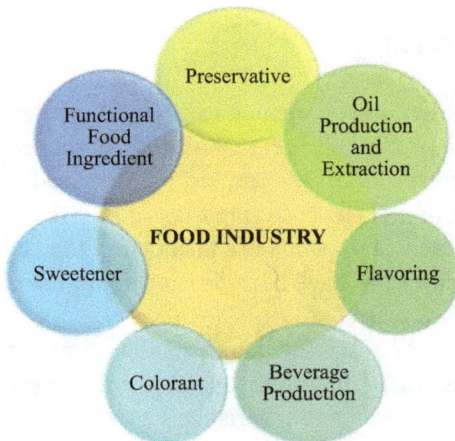

Figure 1.12: Use of medicinal and aromatic plants in the food sector.

1. Natural preservatives: Rosemary extract delays the oxidation of fats in foods, extending their shelf life. These extracts, used as an alternative to synthetic preservatives, are both healthy and environmentally friendly.
2. Taste and aroma additives: Ingredients obtained from aromatic plants such as lemongrass, mint, and vanilla enrich the flavor profile of chocolate, ice cream, candy, and beverages.
3. Functional foods: Some ingredients obtained from aromatic plants can enhance the health benefits of foods. For example, the antioxidant properties of plants such as turmeric and ginger are widely used in functional foods.

Essential oils obtained from aromatic plants offer sustainable solutions in both the cosmetic and food industries with their naturalness and versatile uses.

1.7 Protection and sustainable use of medicinal and aromatic plants

MAPs are natural resources that people have used for thousands of years to improve their health, beauty, and quality of life [67]. These plants play an important role in a wide range of areas, from modern medicines to traditional treatment methods, and from cosmetic products to food additives. However, the high demand for these plants and the various pressures that ecosystems face put many MAP species at risk of extinction. The protection of these endangered species and the development of sustainable harvesting methods not only ensure the continuity of these plants but also contribute to the preservation of ecosystem balances [68].

1.7.1 Protection of endangered species

MAPs are very important both because of their direct economic value and their ecological role [69]. However, factors such as over-collection, habitat loss, climate change, and agricultural expansion threaten many plant species. The Red List prepared by the International Union for Conservation of Nature (IUCN) is an important guide on endangered species and it is seen that many medicinal plants are included in this list [70]. For this reason, various conservation strategies have been developed to protect these species.

Conservation strategies generally include both in situ (in natural environment) and ex situ (outside natural environment) methods [71]. In situ conservation involves protecting the natural habitats of threatened plants. This method supports the survival and reproduction of plants in their natural ecosystems. National parks, biosphere reserves, and special protection areas are effective tools for in situ conservation. Protecting plants in these areas also helps protect other elements of biodiversity.

Ex situ conservation aims to protect threatened plants in environments outside their natural habitats [72]. Seed banks, botanical gardens, and genetic material storage projects are the most common methods of ex situ conservation. For example, the seeds of a plant can be saved and reintroduced to nature in the coming years. Similarly, the cultivation and production of these species in botanical gardens makes significant contributions to the preservation of genetic diversity.

The role of local communities is also critical in the protection of endangered species. Traditional knowledge helps us understand the importance of these plants in the natural ecosystem [69]. Actively involving local people in conservation projects ensures both sustainable use of these species and in awareness raising. Education and awareness-raising programs support communities to participate more effectively in conservation efforts.

1.7.2 Sustainable harvesting methods

Intensive collection of MAPs threatens the natural populations of many species [73]. Uncontrolled harvesting not only causes plant populations to decline, but also disrupts ecosystem balances. Therefore, the development and implementation of sustainable harvesting methods are of great importance for the protection of these valuable resources [74].

Sustainable harvesting methods involve collection processes that respect the regenerative capacity of plants without harming nature. For this, it is necessary to first understand the biology and ecology of plants. When deciding which part (e.g., leaves, roots, flowers) to use and the amount to be harvested, the plant's growth cycle and population density should be taken into account.

The following basic principles should be observed during the harvesting process:
a. Only a certain part of the plants should be taken and the rest should be left in their natural habitat.
b. Harvesting should not be done during certain periods so as to allow the population to regenerate.
c. Harvested areas should be rehabilitated to support the continuity of the natural cycle.

Timing is a critical factor in sustainable harvesting. For example, if seeds are used, plants should be allowed to complete their reproductive cycle. Otherwise, natural population renewal may not be possible. Similarly, in plants where roots are used, it is necessary to leave a portion of each plant's root rather than removing the entire root, for the continuity of the population.

Modern technologies also support sustainable harvesting methods. Tools such as remote sensing and geographic information systems allow plant populations to be monitored and harvest plans to be optimized. These technologies also provide more accurate data on the geographical distribution and ecological needs of plants.

The role of local people in this process includes both the correct application of collection methods and the support to sustainability [74]. Traditional methods and knowledge can be combined with modern scientific approaches to achieve more effective results. The economic benefits provided by local communities also increase interest in these processes [75]. However, it is of great importance that such benefits are organized in a way that they do not harm nature.

1.8 Conservation and sustainable use of medicinal and aromatic plants

1.8.1 Protection of endangered species

MAPs are among the important resources for human health and quality of life [67]. Used for thousands of years in traditional medicine, religious rituals, cosmetics, and the food industry, these plants are one of the most valuable gifts that nature has offered to humanity [76]. However, in recent years, due to high demand for these plants, habitat destruction, and climate change, many species have become endangered. The protection of endangered plant species is of critical importance not only for the continuity of these valuable natural resources, but also for the preservation of ecosystem balance and maintenance of biodiversity [69].

1.8.1.1 Threats to endangered species

The main reasons why MAPs are under threat include:

1. Over-collection: Excessive and uncontrolled collection, especially for commercial purposes, seriously reduces the natural populations of many plant species. Incorrect practices during the collection of roots, leaves, flowers, or seeds of plants prevent plants from completing their natural cycles and make it difficult for them to regenerate [71].
2. Habitat loss and destruction: Human activities such as agriculture, urbanization, industrial expansion, and mining cause the natural habitats of MAPs to disappear. Habitat loss reduces the survival chances of these species and seriously disrupts ecosystem balances [77].
3. Climate change: Global warming and other climatic changes negatively affect the living conditions of many plant species. In particular, temperature increases, changes in precipitation regimes, and extreme weather events seriously threaten the distribution and population density of MAPs [78].
4. Unconscious agriculture and chemical use: Pesticides and fertilizers used in agricultural activities can have toxic effects on many plant species. In addition, mono-

culture agricultural practices reduce biodiversity and cause the destruction of natural vegetation [79].

5. Loss of traditional knowledge: Traditional knowledge accumulated by local communities over generations is rapidly disappearing with modernization and urbanization. Traditional knowledge is vital to ensuring the correct use and preservation of MAPs.

1.8.2 Conservation strategies

Various strategies are implemented both locally and globally to protect endangered plant species [67, 69, 80]. These strategies aim to protect natural populations of plants, ensure their sustainable use, and increase biodiversity.

1.8.2.1 Protection of natural habitats (in situ conservation)

In situ conservation refers to the protection of plants in their natural habitats. This method ensures the continuity of natural processes by preserving the relationships of plants with other organisms in the ecosystem [69, 80]:

1. Protected areas: National parks, biosphere reserves, and special protection zones provide safe living spaces for endangered plants. In these areas, human intervention is limited and the sustainability of natural populations is ensured.

    Habitat restoration: Reforestation of degraded areas and restoration of natural vegetation create new living spaces for endangered plants.

2. Conservation outside the natural habitat (ex situ conservation): Ex situ conservation ensures the protection of endangered plants outside their natural habitats. This approach preserves the genetic material of plants and enables their reintroduction to nature in the future.

3. Seed banks: Seeds can be preserved for long periods of time by storing them in low temperature and humidity conditions. This method is one of the most effective ways to preserve the genetic diversity of plants.

4. Botanical gardens: In botanical gardens, endangered plants are grown and protected for scientific purposes. These gardens are also an important tool for raising public awareness.

5. Gene banks: The genetic material of plants is stored for scientific research and breeding studies.

1.8.3 Participation of local communities

Local communities play a critical role in the conservation of MAPs. Traditional knowledge helps us understand the importance of plants in natural ecosystems. Involving local people in conservation projects encourages both their economic benefits and their support for these projects.

1.8.3.1 Education and awareness

Raising awareness throughout society is vital for the conservation of MAPs. Education programs should aim to explain the importance of these plants, the threats they face, and how individuals can contribute to this process. In particular, raising environmental awareness among younger generations will support long-term conservation efforts.

1.8.3.2 International collaborations

The conservation of MAPs should be supported not only by local efforts but also by international collaborations. International agreements such as the United Nations Convention on Biological Diversity (CBD) promote the conservation of biodiversity and the sustainable use of natural resources. In addition, CITES (Convention on Trade in Endangered Species) regulates international trade in endangered plants.

1.8.3.3 Sustainable harvesting and trade

Sustainable harvesting of MAPs plays an important role in the conservation of these species. During harvest, the natural regenerative capacity of plants should be respected and commercial activities should be organized in a way that they do not disrupt the ecosystem. Sustainable trade certificates and organic farming practices are among the important tools in this area.

1.8.4 Sustainable harvesting methods

The protection of natural resources and the preservation of a healthy environment for future generations have made the concept of sustainability more important than ever. The increasing demand, especially for the use of MAPs, necessitates the proper management of these resources. Sustainable harvesting methods are of vital importance in preventing the depletion of natural populations, maintaining the balance of ecosystems, and sustaining biodiversity. These methods aim to ensure that plant resources not only meet today's needs but also ensure that future generations can benefit from these resources.

1.8.4.1 The importance of sustainable harvesting

Sustainable harvesting methods aim to carry out the collection process without harming the natural regeneration capacity of plants. MAPs are widely used in many sectors such as the pharmaceutical, cosmetic, and food industries. However, uncontrolled and excessive collection of these plants causes habitat destruction and many species face the risk of extinction. Sustainable harvesting reduces these risks and ensures the long-term usability of plant resources.

1.8.4.2 Sustainable harvesting principles

1. Respect for the regeneration capacity of natural populations: Harvesting should be done by taking into account the growth cycles and renewal rates of plants. For example, if a plant whose seeds or roots are collected is completely uprooted before it has had the opportunity to leave enough seeds, it becomes impossible to renew the population.
2. Control of harvest frequency: Plants in the same area should not be harvested at frequent intervals. After the plant is collected, a certain period of time should be provided for the population to renew itself.
3. Collect only as much as needed: Unnecessary and excessive harvesting can lead to the depletion of plant resources. The amount harvested should be enough to meet the demand, and no more should be collected.
4. Protection of habitat: Care should be taken not to damage the habitat during the plant collection process. In particular, situations such as the use of vehicles, compaction of the soil, or damage to other plants should be prevented.
5. Benefiting from the knowledge and experience of local communities: Local communities have deep knowledge of the natural life cycles and methods of use of plants. This knowledge can guide the development of sustainable harvesting methods.

1.8.4.3 Sustainable harvesting techniques

1. Selective harvesting: Selective harvesting involves collecting only certain plants or plant parts. For example, collecting only the leaves or flowers of a plant instead of uprooting it completely increases the plant's chances of survival.

2. Rotational harvesting: After harvesting in one area, a sufficient period of time should be given before harvesting in the same area again. This allows the plants to renew their populations and maintains the balance of the habitat.
3. Harvest timing: The optimal harvest time should be determined by considering the life cycle of the plants. For example, if a plant whose seeds are used, is harvested before its seeds mature, its regenerative capacity may be seriously damaged.
4. Use of traditional methods: Traditional harvesting methods developed by local communities generally take into account environmental sustainability. Combining these methods with modern practices can provide more effective results.
5. Minimum intervention techniques: Methods that minimize environmental damage during the harvest of plant resources should be preferred. For example, collecting only the surface leaves instead of uprooting can help preserve soil structure.

1.8.4.4 Monitoring and evaluating the harvesting process

Continuous monitoring and evaluation are necessary for sustainable harvesting methods to be successful [81, 82]. The amount of plants harvested, population density, and the habitat status should be checked regularly. This process may include the following steps:
1. Keeping harvest records: Recording the type and amount of plants harvested and the area harvested is important for understanding population dynamics.
2. Habitat assessment: The effects of harvesting on habitat should be regularly examined.
3. Scientific research: Collecting scientific data on the regeneration rates of plants and their sensitivity to environmental factors contributes to the development of sustainable practices.

1.8.4.5 The economic dimension of sustainable harvesting

Sustainable harvesting methods not only provide environmental benefits, but also offer economic advantages. These methods, which prevent the depletion of natural resources, enable the continued commercial use of MAPs in the long-term. In addition, sustainable harvesting practices can be supported by labels such as organic certification and sustainable trade, which can increase the value of products.

1.8.4.6 International approaches and legal regulations

The protection and sustainable use of MAPs are of critical importance in ensuring the continuity of ecosystems, social well-being, and biodiversity on a global scale [69].

These plants are the basis not only for the health and cosmetic industries, but also for traditional medicine and the livelihoods of local peoples. However, many plant species are under threat due to factors such as overuse of natural resources, habitat loss, and climate change [83]. This situation requires the development of effective protection and sustainable use strategies through international cooperation and legal regulations.

1.8.4.6.1 International approaches

Various international organizations and initiatives play an active role in the protection of MAPs. The United Nations (UN) and its affiliated organizations are at the forefront of these. In particular, the United Nations Convention on Biological Diversity (CBD) provides a global framework for the protection of biodiversity, its sustainable use, and the equitable sharing of benefits from genetic resources. The CBD protects the rights of local communities, especially those with traditional knowledge, and encourages these communities to manage natural resources sustainably.

The International Union for Conservation of Nature (IUCN) is another important actor supporting efforts to protect MAPs. The IUCN Red List identifies threatened species and sets priorities for their protection. Conservation projects are developed, especially for critically endangered species, and recommendations are made to protect their natural habitats.

In addition, FAO (Food and Agriculture Organization) emphasizes the importance of MAPs within the framework of agricultural biodiversity and sustainable agricultural practices. FAO provides guidance for the sustainable management of natural stocks of these plants and the enhancement of the economic benefits of biological resources.

Another important initiative is CITES (Convention on International Trade in Endangered Species of Wild Fauna and Flora). This convention aims to protect the natural populations of these species by regulating international trade in endangered MAP species. CITES prevents uncontrolled collection of plants, especially those with high commercial value, and develops sustainable trade policies.

1.8.4.6.2 Legal regulations

Various legal regulations are implemented at national and international levels for the protection and sustainable use of MAPs. These regulations generally focus on areas such as the protection of biodiversity, management of genetic resources, and protection of the rights of local communities.

The European Union (EU) has developed comprehensive regulations for the protection of MAPs within the scope of its biodiversity policies. The EU Habitat Directive aims to protect natural habitats and wild flora and fauna. This directive ensures the protection of the natural habitats of many MAP species and the sustainable use of these species. In addition, the EU's Organic Farming Regulations encourage the production of these plants using organic and sustainable methods.

The Nagoya Protocol provides an international framework for access to genetic resources and the fair sharing of the benefits obtained from these resources. The protocol aims to protect the rights of local communities in the use of MAPs and to ensure the sustainable management of biodiversity.

Many countries have national laws regulating trade in MAPs. For example, countries rich in biodiversity, such as India, China, and Brazil, have developed specific laws that regulate the trade and use of these plants. These laws generally control the collection, processing, and export of plants, while also protecting the rights of local communities.

1.8.4.6.3 Protection of local communities and traditional knowledge

Local communities play a critical role in the conservation and sustainable use of MAPs. Traditional knowledge stems from the centuries-old experiences of these communities and their way of life in harmony with nature. However, with the rise of modern biotechnology and pharmacological research, this knowledge is being used for commercial purposes, and the rights of local communities are being neglected.

The Nagoya Protocol and the United Nations Declaration on the Rights of Indigenous Peoples aim to protect the knowledge and resources of local communities. These regulations require that the benefits derived from biological resources be shared fairly with local communities. In addition, many countries have developed national policies that recognize the rights of local communities over these plants.

1.8.5 Many countries are protecting biodiversity

Biodiversity plays a vital role in maintaining ecological balance and providing resources necessary for human life. MAPs have been used in traditional medicine, cosmetics, and food industry for centuries. However, factors such as overexploitation, habitat destruction, and climate change have caused a serious decrease in the populations of these plants. Aware of these problems, many countries are taking steps toward the protection and sustainable use of MAPs by developing various policies, strategies, and programs to protect biodiversity.

1.8.5.1 Global conservation efforts

Many countries around the world have adopted comprehensive strategies to protect their unique biodiversity, especially MAPs. The basis of these efforts is the United Nations Convention on Biological Diversity (CBD). Signed by 196 countries, this convention promotes the protection of biodiversity, the sustainable use of its components, and the equitable sharing of benefits from genetic resources. Many countries aim to secure their natural resources by integrating the goals of CBD into their national policies.

For example, India, known for its rich biodiversity and traditional knowledge systems, aims to protect biological resources and promote their sustainable use through the Biological Diversity Act of 2002. This act has established Biodiversity Management Committees at the local level to document biological resources and related traditional knowledge, and regulates their use for research and commercial purposes. India also supports the protection of MAPs through the National Medicinal Plants Board (NMPB), promotes agriculture, and carries out conservation projects in natural areas.

In South America, Brazil has taken important steps, especially in protecting the Amazon rainforest. Brazil has implemented strict regulations to prevent overexploitation of natural resources and has established structures such as the Genetic Heritage Management Council to ensure the sustainable use of genetic resources and traditional knowledge.

China has also prioritized the protection of MAPs through the Chinese Medicine Resources Protection and Development Plan, which focuses on the sustainable harvesting and cultivation of medicinal plants used in TCM. Many nature reserves and botanical gardens have been established across the country, protecting endangered species.

1.8.5.2 Protected areas and conservation in natural habitats

One of the most effective methods for protecting biodiversity is the establishment of protected areas such as national parks, wildlife sanctuaries, and biosphere reserves. These areas are safe habitats that prevent overharvesting, by allowing MAPs to reproduce naturally.

Countries such as Kenya, South Africa, and Tanzania have designated large land areas as protected areas, ensuring the protection of both flora and fauna. In Europe, the Natura 2000 Network, established under the EU Habitats Directive, protects endangered species and habitats throughout the European Union. This network covers areas with many MAP species.

In addition to protected areas, in situ conservation initiatives focus on protecting MAPs in their natural habitats. Countries such as Australia and Canada are implementing projects to restore degraded ecosystems and reintroduce native plant species to the wild. These efforts not only increase biodiversity but also support the ecological functions of MAPs.

1.8.5.3 Ex situ conservation and gene banks

Many countries have invested in ex situ conservation methods to protect endangered MAPs. Botanical gardens, seed banks, and tissue culture laboratories play an important role in preserving genetic diversity.

The Millennium Seed Bank Partnership in the United Kingdom has collected and preserved seeds of rare and threatened MAPs from around the world. Similarly, Ethiopia's Biodiversity Conservation Institute manages a gene bank that stores seeds of medicinal plants vital to traditional medicine. These efforts ensure that genetic resources are available for future research, breeding, and reintroduction activities.

1.9 Challenges and future prospects

1.9.1 Impacts of climate change

Climate change is considered one of the greatest environmental challenges in human history, and its effects on biodiversity and natural resources are becoming increasingly evident [69]. MAPs, in particular, constitute a group of plant species that are highly sensitive to the effects of climate change. The genetic diversity, adaptation capacities, and changes in the ecological conditions specific to their habitats of these plants bring about serious problems that threaten their sustainable use and conservation.

Climate change includes many factors such as temperature increase, changes in precipitation patterns, drought, severe weather events, and deterioration in soil quality [84]. These changes directly affect the growing conditions of MAP species. For example, drought and water stress limit the growth, photosynthetic capacity and secondary metabolite production of plants. Secondary metabolites are the components that are generally responsible for their medicinal and aromatic properties. Therefore, climate change can negatively affect the quantity and quality of these components, which can lead to both economic losses and difficulties in product supply in the health sector.

Habitat loss and ecosystem degradation put many MAP species at risk of extinction [85]. For example, endemic plants growing in high altitude regions may be forced to migrate to higher altitudes due to climate change. However, this may exceed the adaptation capacity of these species and cause the species to become extinct. In addition, the disruption of biotic relationships such as plant–pollinator interactions in ecosystems may negatively affect the reproductive success and population dynamics of plants.

Agricultural practices are also affected by climate change, which can make it difficult to grow MAPs. Limited water and nutrient resources in traditional agricultural systems limit farmers' efforts to increase their production in a sustainable way. This can cause both economic losses for local communities and disruptions in global supply chains.

1.9.2 Genetic and biotechnological approaches

Genetic and biotechnological approaches offer great potential to mitigate the negative effects of climate change on MAP [86]. These technologies provide innovative solutions to improve plant genetics, resistance, and biochemical properties [87]. Genetic engineering and biotechnological techniques play an important role in the protection and sustainable use of MAP species.

1.9.2.1 Protection of genetic diversity and breeding studies

Genetic diversity is a fundamental element that enables plants to adapt to changing environmental conditions. Molecular marker technologies can be used to analyze the genetic diversity and population structure of MAP species. This information helps develop strategies for the protection of threatened species. In addition, combining traditional breeding studies with biotechnological methods enables the development of new plant varieties resistant to stress factors such as drought, salinity, and disease.

1.9.2.2 Genomic and transcriptomic approaches

Next-generation sequencing technologies allow detailed examination of the genomic and transcriptomic profiles of MAP species. These analyses help us understand the genetic pathways that control plant stress responses and secondary metabolite production. Thus, desired traits can be increased by manipulating target genes. For example, to increase the production of secondary metabolites, biosynthetic pathways can be regulated or gene transfer methods can be used.

1.9.2.3 Culture tissue techniques

Culture tissue techniques allow rapid and efficient propagation of endangered TAB species. These techniques allow plants to be produced in a laboratory environment without ecological pressures, while preserving their genetic characteristics. At the same time, these methods can support the commercial production of rare plant species and reduce the pressure on natural populations.

1.9.2.4 CRISPR/Cas9 technology

Gene editing technologies can be used to modify the genetic makeup of TAB species in a targeted manner. Tools such as CRISPR/Cas9 can be used to increase the resistance of plants to stress factors or to optimize secondary metabolite production pathways. This technology provides faster and more precise results compared to traditional genetic methods.

1.9.2.5 Metabolic engineering and synthetic biology

Metabolic engineering techniques can be used to optimize the secondary metabolite production pathways of TAB species. Synthetic biology approaches can increase the production of commercially valuable compounds by reprogramming the natural biosynthetic pathways of plants. For example, the production of active ingredients used in the pharmaceutical industry by microorganisms can contribute to the conservation of natural resources.

1.9.2.6 Bioinformatics and data analysis

Bioinformatics tools play a critical role in the analysis of genomic, transcriptomic, and metabolomic data. These data help us understand the adaptation mechanisms and metabolic profiles of TAB species. Additionally, bioinformatics approaches provide a powerful platform for targeting genetic modifications and predicting potential side effects.

1.10 Case studies and regional practices

1.10.1 Successful projects in specific regions

Successful projects have been implemented in various regions around the world for the conservation and sustainable use of MAP. These projects support the economic development of local communities, while ensuring the sustainability of natural resources. The success of such projects is possible through the integrated application of local knowledge, ecosystem-oriented approaches and scientific research.

1.10.1.1 India: Ayurveda and biodiversity conservation projects

India, with its rich biodiversity and traditional medicinal knowledge systems, is home to successful projects for the conservation of MAPs. The Government of India, through the National Medicinal Plants Board, is implementing various programs to protect medicinal plant reserves and promote sustainable harvesting methods. These projects aim not only at the conservation of endemic species but also at the economic development of farmers and communities. For example, organic cultivation and marketing of plants used in Ayurveda has both increased the incomes of local farmers and met the demand for these products on a global scale.

1.10.1.2 Brazil: sustainable collection projects in the Amazon forest

Brazil has developed various projects to protect the biodiversity of the Amazon forest and to utilize these resources sustainably. For example, the "ProNatura Project" is an initiative that promotes the sustainable collection and processing of MAPs. Within the scope of this project, local communities have been trained in the collection, processing, and sale of plants in international markets. At the same time, legal regulations have been developed against overexploitation and protocols have been established to ensure the use of these plants without harming nature.

1.10.1.3 Turkey: protection and production of endemic plants

Turkey is one of the countries with the richest plant diversity in the world and is of great importance, especially in terms of endemic species. Within the scope of the "National Action Plan for Medicinal and Aromatic Plants," the protection of medicinal plants and their cultivation as cultivated plants are encouraged. For example, projects to grow plants such as lavender, thyme, and sage in various regions of Anatolia have both protected natural resources and revitalized the local economy. In addition, universities and research institutes have contributed to the development of new products in the health sector by investigating the active ingredients of these plants.

1.10.1.4 Africa: integration of local knowledge with modern practices

Many successful projects are being carried out on the African continent where traditional medicinal plants are integrated with modern science. Initiatives such as the "Traditional Healers' Cooperative," especially in South Africa, have protected the knowledge of traditional healers and integrated this knowledge into modern health systems. The traditional knowledge of local people about medicinal plants has been included in modern drug development processes after undergoing scientific validation processes.

1.10.2 The bridge between traditional knowledge and modern science

MAPs have held an important place in traditional knowledge systems for centuries and have been used as a source of healing in many cultures. However, modern science offers a unique platform for understanding the potential benefits of these plants more deeply and in developing new therapeutic products. The bridge between traditional knowledge and modern science plays a vital role in the preservation and sustainable use of these plants.

1.10.2.1 Documentation and protection of traditional knowledge

Traditional knowledge is often transmitted through oral culture, which increases the risk of its loss. Therefore, documenting the knowledge of local communities about medicinal plants is one of the most effective ways to ensure that this knowledge is passed on to future generations. For example, projects supported by UNESCO have recorded the knowledge of local communities in written form and made this information accessible for scientific research.

1.10.2.2 Scientific validation and application

Modern science offers various methods to test the validity and effectiveness of traditional knowledge. Laboratory studies analyze the chemical components of traditionally used plants to determine their pharmacological effects. For example, studies on some plants used in Africa have proven that these plants have antibacterial, antiviral, and anticancer properties. This process allows traditional knowledge to be integrated into modern medical practices.

1.10.2.3 Education and awareness

Another important way to bridge the gap between traditional knowledge and modern science is to increase education and awareness. Universities, research centers, and civil society organizations organize educational programs on MAPs, bringing together both traditional knowledge holders and scientists. Such collaborations encourage the two different knowledge systems to work together and provide mutual benefits.

1.10.2.4 Patents and intellectual property rights

The use of traditional knowledge systems by modern science raises the issue of intellectual property rights. Scientific projects using traditional knowledge must provide material and moral benefits to local communities. For this reason, many countries have developed legal regulations for the patenting of products based on traditional knowledge and the protection of the rights of local communities.

1.10.2.5 Public and private sector collaboration

Public and private sector collaboration provides an effective model for combining traditional knowledge with modern science. For example, pharmaceutical companies are developing collaborations to integrate traditional knowledge obtained from local communities into modern drug development processes. Such projects both stimulate scientific innovation and contribute to the economic development of local communities.

1.11 Conclusions

MAPs have been an integral part of human life in many areas such as health, beauty, food, and many others from the past to the present. These plants, which are the healing treasures of nature, have not only been a part of traditional treatment methods, but have also been integrated with modern science and technology to offer more effective and innovative solutions. MAPs have been examined over a wide range, from their chemical structures to their pharmacological effects, and from their roles in daily life to sustainable use strategies. By detailing the biological effects of active ingredients such as alkaloids, flavonoids, and terpenoids, it sheds light on the interactions of these molecules with pharmacological targets. Studies on phytochemistry and identification of active compounds reveal both the scientific and industrial potential of these plants. The role of aromatic plants in daily life, their effects on health and well-being, and the variety and value of their contributions to human life have been discussed in detail.

The protection and sustainable use of natural resources are of vital importance in order to provide long-term benefits of these plants. Therefore, the protection of endangered species, the adoption of sustainable harvesting methods, and the implementation of international legal regulations stand out as indispensable elements for the balance of natural ecosystems and the preservation of biodiversity. Finally, the effects of climate change and environmental stress factors on these plants increase the challenges and risks that may be encountered in the future. However, genetic and biotechnological approaches hold promise for providing innovative solutions to these challenges. Regional success stories and the blending of traditional knowledge with modern science provide inspiration for effective methods and collaboration models that can be applied in this field.

References

[1]   Hardy, K., Buckley, S., Collins, M. J., Estalrrich, A., Brothwell, D., Copeland, L. . . . and Rosas, A. (2012). Neanderthal medics? Evidence for food, cooking, and medicinal plants entrapped in dental calculus. Naturwissenschaften, 99, 617–626.

[2]   Nunn, J. F. (2002). Ancient Egyptian Medicine, University of Oklahoma Press.

[3]   Patwardhan, B., Vaidya, A. D. and Chorghade, M. (2004). Ayurveda and natural products drug discovery. Current Science, 86(6), 789–799.

[4]   Ozdogan, E. (2019). De materia medica: Where art and scientific principles pome together. Marmara Medical Journal, 32(2), 94–96.

[5]   Ghaffari, F., Taheri, M., Meyari, A., Karimi, Y. and Naseri, M. (2022). Avicenna and clinical experiences in Canon of Medicine. Journal of Medicine and Life, 15(2), 168.

[6]   Kapancık, S., Çelik, M. S., Demiralp, M., Ünal, K., Çetinkaya, S. and Tüzün, B. (2024). Chemical composition, cytotoxicity, and molecular docking analyses of Thuja orientalis extracts. Journal of Molecular Structure, 1318, 139279.

[7]   Poustforoosh, A., Faramarz, S., Negahdaripour, M., Tüzün, B. and Hashemipour, H. (2024). Investigation on the mechanisms by which the herbal remedies induce anti-prostate cancer activity: Uncovering the most practical natural compound. Journal of Biomolecular Structure and Dynamics, 42(7), 3349–3362.

[8]   Erdogan, M. K., Gundogdu, R., Yapar, Y., Gecibesler, I. H., Kirici, M., Behcet, L. . . . and Taslimi, P. (2023). In vitro anticancer, antioxidant and enzyme inhibitory potentials of endemic Cephalaria elazigensis var. purpurea with in silico studies. Journal of Biomolecular Structure and Dynamics, 41(21), 11832–11844.

[9]   İnanir, M., Uçar, E., Tüzün, B., Eruygur, N., Ataş, M. and Akpulat, H. A. (2024). The pharmacological properties of Gypsophila eriocalyx: The endemic medicinal plant of northern central Turkey. International Journal of Biological Macromolecules, 266, 130943.

[10]   Erdogan, M. K., Gundogdu, R., Yapar, Y., Gecibesler, I. H., Kirici, M., Behcet, L. . . . and Taslimi, P. (2022). The evaluation of anticancer, antioxidant, antidiabetic and anticholinergic potentials of endemic Rhabdosciadium microcalycinum supported by molecular docking study. ChemistrySelect, 7(17), e202200400.

[11]   Tüzün, B., Sayin, K. and Ataseven, H. (2022). Could Momordica charantia Be effective in the treatment of COVID19?. Cumhuriyet Science Journal, 43(2), 211–220.

[12]   Rbaa, M., Galai, M., Dagdag, O., Guo, L., Tüzün, B., Berdimurodov, E. . . . and Lakhrissi, B. (2022). Development process for eco-friendly corrosion inhibitors. In: Eco-Friendly Corrosion Inhibitors, Elsevier, Cambridge. 27–42.

[13]   Eruygur, N., Uçar, E., Tüzün, B., Ataş, M., İnanır, M., Demirbaş, A. . . . and Uskutoğlu, T. (2024). Evaluation of antioxidant, antimicrobial, enzyme inhibition activity, and cell viability capacity of Hypericum heterophyllum vent., an endemic species in Turkey's Flora. Journal of Molecular Structure, 1307, 137908.

[14]   Saraç, H., Demirbaş, A. and Tüzün, B. (2023). Could Zingiber officinale plant be effective against Omicron BA. 2.75 of SARS-CoV-2?. Turkish Computational and Theoretical Chemistry, 7(3), 42–56.

[15] Tüzün, B. (2024). Evaluation of cytotoxicity, chemical composition, antioxidant potential, apoptosis relationship, molecular docking, and MM-GBSA analysis of rumex crispus leaf extracts. Journal of Molecular Structure, 140791.

[16] Wang, J. F., Wei, D. Q. and Chou, K. C. (2008). Drug candidates from traditional Chinese medicines. Current Topics in Medicinal Chemistry, 8(18), 1656–1665.

[17] Singh, R. (2015). Medicinal plants: A review. Journal of Plant Sciences, 8(2), 50–55.

[18] Balick, M. J. and Cox, P. A. (2020). Plants, People, and Culture: The Science of Ethnobotany, Garland Science, New York.

[19] Van, J. R. (1971). Inhibition of prostaglandin synthesis as a mechanism of action for aspirin-like drugs. Nature New Biology, 231, 323–328.

[20] Noorafshan, A. and Ashkani-Esfahani, S. (2013). A review of therapeutic effects of curcumin. Current Pharmaceutical Design, 19(11), 2032–2046.

[21] Koulivand, P. H., Khaleghi Ghadiri, M. and Gorji, A. (2013). Lavender and the nervous system. Evidence-Based Complementary and Alternative Medicine, 2013(1), 681304.

[22] Burt, S. (2004). Essential oils: Their antibacterial properties and potential applications in foods – A review. International Journal of Food Microbiology, 94(3), 223–253.

[23] Schippmann, U., Leaman, D. J. and Cunningham, A. B. (2002). Impact of cultivation and gathering of medicinal plants on biodiversity: Global trends and issues. Biodiversity and the Ecosystem Approach in Agriculture, Forestry and Fisheries.

[24] Cragg, G. M. and Newman, D. J. (2013). Natural products: A continuing source of novel drug leads. Biochimica Et Biophysica Acta (Bba)-general Subjects, 1830(6), 3670–3695.

[25] Heinrich, M. (2013). Ethnopharmacology and drug discovery. Comprehensive Natural Products II: Chemistry and Biology, Development & Modification of Bioactivity, 3, 351–381.

[26] Dewick, P. M. (2002). Medicinal natural products: A biosynthetic approach.

[27] Tapas, A. R., Sakarkar, D. M. and Kakde, R. B. (2008). Flavonoids as nutraceuticals: A review. Tropical Journal of Pharmaceutical Research, 7(3), 1089–1099.

[28] Middleton, E., Kandaswami, C. and Theoharides, T. C. (2000). The effects of plant flavonoids on mammalian cells: Implications for inflammation, heart disease, and cancer. Pharmacological Reviews, 52(4), 673–751.

[29] Thoppil, R. J. and Bishayee, A. (2011). Terpenoids as potential chemopreventive and therapeutic agents in liver cancer. World Journal of Hepatology, 3(9), 228.

[30] Gershenzon, J. and Dudareva, N. (2007). The function of terpene natural products in the natural world. Nature Chemical Biology, 3(7), 408–414.

[31] Robins, R. J. (1998). The biosynthesis of alkaloids in root cultures. In: Margaret F. Roberts & Michael Wink (eds) Alkaloids: Biochemistry, Ecology, and Medicinal Applications, Springer US, Boston, MA, 199–218.

[32] Gershenzon, J. (1999). Alkaloids: Biochemistry, ecology, and medicinal applications. Crop Science, 39(4), 1251–1251.

[33] Crozier, A., Clifford, M. N. and Ashihara, H. (2006). Plant secondary metabolites. In: Alan Crozier, Michael N. Clifford & Hiroshi Ashihara (eds) Occurrence, Structure and Role in the Human Diet, Blackwell-Publishers.

[34] Harborne, J. B. and Willians, C. (2000). Advances in flavonoid research since 1992. Phytochemical Oxford, 55(6), 481–504.

[35] Harrewijn, P., Van Oosten, A. M. and Piron, P. G. (2001). Natural Terpenoids as Messengers: A Multidisciplinary Study of Their Production, Biological Functions, and Practical Applications, Springer Science & Business Media, London.

[36] Wagner, K. H. and Elmadfa, I. (2003). Biological relevance of terpenoids: Overview focusing on mono-, di-and tetraterpenes. Annals of Nutrition and Metabolism, 47(3–4), 95–106.

[37] Virk-Baker, M. K., Nagy, T. R. and Barnes, S. (2010). Role of phytoestrogens in cancer therapy. Planta Medica, 76(11), 1132–1142.

[38] Orouji, N., Asl, S. K., Taghipour, Z., Habtemariam, S., Nabavi, S. M. and Rahimi, R. (2023). Glucosinolates in cancer prevention and treatment: Experimental and clinical evidence. Medical Oncology, 40(12), 344, 1–21.

[39] Kelly, G. S. (2011). Quercetin. Alternative Medicine Review, 16(2), 172–194.

[40] Abdulkhaleq, L. A., Assi, M. A., Noor, M. H. M., Abdullah, R., Saad, M. Z. and Taufiq-Yap, Y. H. (2017). Therapeutic uses of epicatechin in diabetes and cancer. Veterinary World, 10(8), 869.

[41] Christianson, D. W. (2006). Structural biology and chemistry of the terpenoid cyclases. Chemical Reviews, 106(8), 3412–3442.

[42] Christianson, D. W. (2017). Structural and chemical biology of terpenoid cyclases. Chemical Reviews, 117(17), 11570–11648.

[43] Griffin, S. G., Wyllie, S. G., Markham, J. L. and Leach, D. N. (1999). The role of structure and molecular properties of terpenoids in determining their antimicrobial activity. Flavour and Fragrance Journal, 14(5), 322–332.

[44] Chung, K. T., Wong, T. Y., Wei, C. I., Huang, Y. W. and Lin, Y. (1998). Tannins and human health: A review. Critical Reviews in Food Science and Nutrition, 38(6), 421–464.

[45] Cheok, C. Y., Salman, H. A. K. and Sulaiman, R. (2014). Extraction and quantification of saponins: A review. Food Research International, 59, 16–40.

[46] Halkier, B. A. and Gershenzon, J. (2006). Biology and biochemistry of glucosinolates. Annual Review of Plant Biology, 57(1), 303–333.

[47] Roaa, M. H. (2020). A review article: The importance of the major groups of plants secondary metabolism phenols, alkaloids, and terpenes. International Journal for Research in Applied Sciences and Biotechnology (IJRASB), 7(5), 354–358.

[48] Panche, A. N., Diwan, A. D. and Chandra, S. R. (2016). Flavonoids: An overview. Journal of Nutritional Science, 5, e47.

[49] Harborne, J. B. and Mabry, T. J. (2013). The flavonoids: Advances in research.

[50] Graßmann, J. (2005). Terpenoids as plant antioxidants. Vitamins & Hormones, 72, 505–535.

[51] Cheng, A. X., Lou, Y. G., Mao, Y. B., Lu, S., Wang, L. J. and Chen, X. Y. (2007). Plant terpenoids: Biosynthesis and ecological functions. Journal of Integrative Plant Biology, 49(2), 179–186.

[52] Aniszewski, T. (2015). Alkaloids: Chemistry, Biology, Ecology, and Applications, Elsevier, Helsinki, Finland.

[53] Rajput, A., Sharma, R. and Bharti, R. (2022). Pharmacological activities and toxicities of alkaloids on human health. Materials Today: Proceedings, 48, 1407–1415.

[54] Jeong, J. M., Choi, C. H., Kang, S. K., Lee, I. H., Lee, J. Y. and Jung, H. (2007). Antioxidant and chemosensitizing effects of flavonoids with hydroxy and/or methoxy groups and structure-activity relationship. Journal of Pharmacy & Pharmaceutical Sciences, 10(4), 537–546.

[55] Karak, P. (2019). Biological activities of flavonoids: An overview. International Journal of Pharmaceutical Sciences and Research, 10(4), 1567–1574.

[56] Siddiqui, T., Sharma, V., Khan, M. U. and Gupta, K. (2024). Terpenoids in essential oils: Chemistry, classification, and potential impact on human health and industry. Phytomedicine Plus, 100549, 100549.

[57] Bhattacharya, S., Saha, T., Das, P., Sarkar, S., Koley, S., Bhattacharjee, S. . . . and Chakraborty, A. J. (2022). Aromatic plants: Role and uses in human prosperity and sustainability. Journal of Pharmaceutical Innovation, 11(5), 341–348.

[58] Solomou, A. D., Martinos, K., Skoufogianni, E. and Danalatos, N. G. (2016). Medicinal and aromatic plants diversity in Greece and their future prospects: A review. Agricultural Science, 4(1), 9–21.

[59]   Maknea, K. I., Asănică, A., Fabian, C., Peticilă, A., Tzortzi, J. N. and Popescu, D. (2022). The use of co-cultivation of aromatic, medicinal plants and vegetables in sustainable urban horticulture. AgroLife Scientific Journal, 11(1).

[60]   Lis-Balchin, M. (1997). Essential oils and 'aromatherapy': Their modern role in healing. Journal of the Royal Society of Health, 117(5), 324–329.

[61]   Ali, B., Al-Wabel, N. A., Shams, S., Ahamad, A., Khan, S. A. and Anwar, F. (2015). Essential oils used in aromatherapy: A systemic review. Asian Pacific Journal of Tropical Biomedicine, 5(8), 601–611.

[62]   Vora, L. K., Gholap, A. D., Hatvate, N. T., Naren, P., Khan, S., Chavda, V. P. . . . and Khatri, D. K. (2024). Essential oils for clinical aromatherapy: A comprehensive review. Journal of Ethnopharmacology, 330, 118180.

[63]   Lubbe, A. and Verpoorte, R. (2011). Cultivation of medicinal and aromatic plants for specialty industrial materials. Industrial Crops and Products, 34(1), 785–801.

[64]   Dikme, T. G. (2023). Use of medicinal and aromatic plants in food. The Eurasian Clinical and Analytical Medicine, 11(1), 6–10.

[65]   Vartak, A., Sonawane, S., Alim, H., Patel, N., Hamrouni, L., Khan, J. and Ali, A. (2022). Medicinal and aromatic plants in the cosmetics industry. In: Ákos Máthé & Irfan Ali Khan (eds) Medicinal and Aromatic Plants of India, Vol. 1, Springer International Publishing, Cham, 341–364.

[66]   Inoue, M., Hayashi, S. and Craker, L. E. (2019). Role of medicinal and aromatic plants: Past, present, and future. In: Perveen, S. & Al-Taweel, A. (eds) Pharmacognosy-medicinal Plants, Vol. 13, intechopen, 1.

[67]   Barata, A. M., Rocha, F., Lopes, V. and Carvalho, A. M. (2016). Conservation and sustainable uses of medicinal and aromatic plants genetic resources on the worldwide for human welfare. Industrial Crops and Products, 88, 8–11.

[68]   Grigoriadou, K., Krigas, N., Lazari, D. and Maloupa, E. (2020). Sustainable use of Mediterranean medicinal-aromatic plants. In: Feed Additives, Academic Press, 57–74.

[69]   Padulosi, S., Leaman, D. and Quek, P. (2002). Challenges and opportunities in enhancing the conservation and use of medicinal and aromatic plants. Journal of Herbs, Spices & Medicinal Plants, 9(4), 243–267.

[70]   Chandra, L. D. (2016). Bio-diversity and conservation of medicinal and aromatic plants. Advances in Plants & Agriculture Research, 5(4), 00186.

[71]   Lange, D. (2002, August). Medicinal and aromatic plants: Trade, production, and management of botanical resources. In: XXVI International Horticultural Congress: The Future for Medicinal and Aromatic Plants, Vol. 629, ISHS Acta Horticulturae, 177–197.

[72]   Hassanpouraghdam, M. B., Ghorbani, H., Esmaeilpour, M., Alford, M. H., Strzemski, M. and Dresler, S. (2022). Diversity and distribution patterns of endemic medicinal and aromatic plants of Iran: Implications for conservation and habitat management. International Journal of Environmental Research and Public Health, 19(3), 1552.

[73]   Schippmann, U. W. E., Leaman, D. and Cunningham, A. B. (2006). A comparison of cultivation and wild collection of medicinal and aromatic plants under sustainability aspects. In: Bogers, Robert J, Craker, Lyle E. & Lange, Dagmar (eds) Medicinal and Aromatic Plants, Springer, Dordrecht, 75–95.

[74]   Sharma, N. and Kala, C. P. (2018). Harvesting and management of medicinal and aromatic plants in the Himalaya. Journal of Applied Research on Medicinal and Aromatic Plants, 8, 1–9.

[75]   Ur-Rahman, I., Sher, H. and Bussmann, R. W. (Eds.) (2019). Reference Guide on High Value Medicinal and Aromatic Plants–sustainable Management and Cultivation Practices, University of Swat, Pakistan.

[76]   Shafi, A., Hassan, F., Zahoor, I., Majeed, U. and Khanday, F. A. (2021). Biodiversity, management and sustainable use of medicinal and aromatic plant resources. Medicinal and Aromatic Plants: Healthcare and Industrial Applications, 85–111.

[77] Ahmad, J., Malik, A. A. and Shakya, L. (2013). Urban development: A threat to wild species of medicinal and aromatic plants. Middle East Journal of Scientific Research, 13, 947–951.

[78] Cavaliere, C. (2009). The effects of climate change on medicinal and aromatic plants. Herbal Gram, 81, 44–57.

[79] Şenkal, B. C. (2020). The role of secondary metabolites obtained from medicinal and aromatic plants in our lives. ISPEC Journal of Agricultural Sciences, 4(4), 1071–1079.

[80] Okigbo, R. N., Eme, U. E. and Ogbogu, S. (2008). Biodiversity and conservation of medicinal and aromatic plants in Africa. Biotechnology and Molecular Biology Reviews, 3(6), 127–134.

[81] Omogbadegun, Z. O. (2013). Medicinal and aromatic plants' productivity and sustainability monitoring framework. European Journal of Medicinal Plants, Accepted 9 March 2013, in press.

[82] Marcelino, S., Hamdane, S., Gaspar, P. D. and Paço, A. (2023). Sustainable agricultural practices for the production of medicinal and aromatic plants: Evidence and recommendations. Sustainability, 15(19), 14095.

[83] Parvin, S., Reza, A., Das, S., Miah, M. M. U. and Karim, S. (2023). Potential role and international trade of medicinal and aromatic plants in the world. European Journal of Agriculture and Food Sciences, 5(5), 89–99.

[84] Dajic-Stevanovic, Z. and Pljevljakusic, D. (2015). Challenges and decision making in cultivation of medicinal and aromatic plants. In: Medicinal and Aromatic Plants of the World: Scientific, Production, Commercial and Utilization Aspects, Spinger, Budapest, Hungary, 145–164.

[85] Baričevič, D., Máthé, Á. and Bartol, T. (2015). Conservation of wild crafted medicinal and aromatic plants and their habitats. In: Medicinal and Aromatic Plants of the World: Scientific, Production, Commercial and Utilization Aspects, Spinger, Budapest, Hungary, 131–144.

[86] Kumar, N. (Ed.) (2018). Biotechnological Approaches for Medicinal and Aromatic Plants: Conservation, In: Genetic Improvement and Utilization, Springer, Panchanpur, Gaya, Bihar, India.

[87] Kumar, J. and Gupta, P. K. (2008). Molecular approaches for improvement of medicinal and aromatic plants. Plant Biotechnology Reports, 2, 93–112.

İrem Yıldız Özbaş*, Severina Pacifico, and Emre Özbaş

Chapter 2
Methods of obtaining drugs from medicinal and aromatic plants

Abstract: Many of the therapeutic agents we use today are plant-based, highlighting the importance of medicinal and aromatic plants in drug discovery and the development of new drugs. The complex nature of plants and the challenges in effectively and efficiently separating and purifying plant-based compounds have historically directed scientists toward synthetic sources for drug development. However, recent advancements in technology and modern trends in drug discovery that integrate natural product development processes indicate that plants will continue to play a significant role in this field in the future. Overcoming the challenges encountered at every stage of drug development from plants requires an integrated interdisciplinary approach that leverages both traditional methods and technological advancements. A fast and efficient selection of source plants utilizing ethnopharmacological approaches, high-throughput screening (HTS), and virtual screening methods, coupled with well-designed and bioactivity-guided extraction and isolation methods, advanced structural characterization techniques, and the transfer of bioactive compounds in preclinical and clinical studies, are essential steps for successful plant-based drug development. New analytical technologies, combinatorial chemistry, computational and screening methods including artificial intelligence, molecular modeling, omics technologies for studying interactions between bioactive molecules and their targets, and the design of biological models represent modern approaches that have demonstrated significant success in obtaining drugs from medicinal and aromatic plants. These methods also hold substantial potential for future research. This section provides an overview of the stages involved in obtaining drugs from medicinal and aromatic plants and the methods used at each stage. It discusses traditional and modern methodologies in drug development, covering source plant selection, bioactive compound extraction, isolation, characterization, biological assays, clinical trials, new analog development, and optimization processes. Finally, the chapter evaluates future expectations for plant-based drug development in light of emerging trends.

---

*Corresponding author: İrem Yıldız Özbaş, Pharmacy Services Department, Aşık Veysel Vocational School of Şarkışla, Sivas Cumhuriyet University, 58140 Sivas, Turkey, e-mail: irem@cumhuriyet.edu.tr, https://orcid.org/0000-0002-9657-9074
Severina Pacifico, Department of Environmental Biological and Pharmaceutical Sciences and Technologies, University of Campania "Luigi Vanvitelli", via Vivaldi 43, 81100 Caserta, Italy
Emre Özbaş, Medicinal and Aromatic Plants Programme, Plant and Animal Production Department, Technical Sciences Vocational School of Sivas, Sivas Cumhuriyet University, 58140 Sivas, Turkey

https://doi.org/10.1515/9783111469713-002

Keywords: medicinal and aromatic plants, bioactive compounds, natural products drug discovery, plant-based drug development

2.1 Introduction

Plants and herbal products have been used in various ways against diseases since the beginning of human existence. Plants and plant-based mixtures, known for their lower potential to cause undesirable effects than synthetic drugs, ease of access, and rich composition of active compounds, persist as a common choice among people today, much like in the past. The number of plants used in treatments has steadily increased, and recent studies have begun to establish the scientific foundations for the role of plants in medicine [1].

Despite advancements in technology and accumulated knowledge, many diseases worldwide still lack effective treatments, and for some, no curative drugs are currently available. Thus, the discovery of new drugs remains a critical necessity. Moreover, developing more effective and safer alternatives to existing medications for treatable diseases is of paramount importance. Nature offers an immense reservoir for drug discovery, containing countless compounds yet to be explored. Harnessing the potential of natural products is essential in the course of identifying and developing new medicines [2].

Medicinal plants are rich resources that can be used to develop new drugs. Nearly half of today's medicines are plant-based. A significant portion of new drugs and active pharmaceutical ingredients are either directly derived from plants or inspired by the chemical structures of plant compounds [3, 4]. Naturally derived molecules from plants that are physiologically bioactive can be used directly in treatment, while some are also utilized as precursor molecules in the chemical synthesis or semisynthesis of drugs [5]. The different therapeutic applications of natural compounds are shown in Figure 2.1 [6, 7]. Medicinal and aromatic plants, along with the natural products derived from them and the drugs developed from these sources, exhibit a wide pharmacological spectrum of effects. They demonstrate therapeutic properties in the treatment of infections, cancers, gastrointestinal disorders, respiratory, digestive, cardiovascular diseases, and chronic conditions such as diabetes. Many of these plants share common properties, including antioxidant, antimicrobial, and immunomodulatory effects, making them valuable for prophylactic use as well. Among the most widely used drugs derived from natural products globally are numerous examples of antibiotics, antifungals, cancer chemotherapeutics, cholesterol-lowering agents, antihypertensives, and immunosuppressants [7].

Plants are organisms that display remarkable adaptation to their habitats. They have developed a wide range of strategies to defend against potential threats and attacks from their environment. To defend against predators and environmental fac-

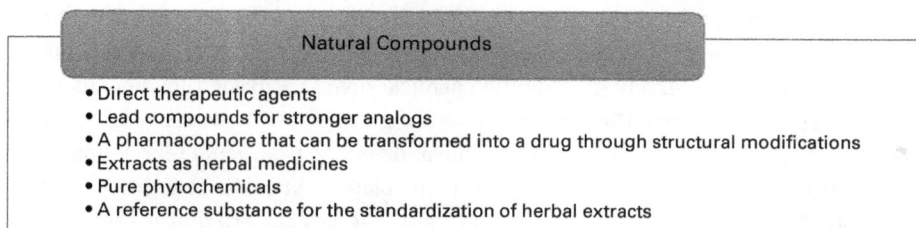

Figure 2.1: The various therapeutic applications of natural compounds.

tors, they produce secondary metabolites, such as toxins, pigments, and aromatic compounds. Phytochemical compounds exhibit a broad spectrum of polarities and are found in highly complex matrices. These metabolites are diverse, complex in structure, and abundant, including alkaloids, flavonoids, glycosides, terpenoids, lipids, waxes, peptides, and phenolics (Figure 2.2) Secondary metabolites are also considered bioactive substances, meaning they exhibit biological activity. Bioactive substances can be defined as substances that have pharmacological or toxic effects on both humans and animals. These bioactive components are increasingly being studied by scientists with great interest for their potential in innovative treatments [5, 8–10].

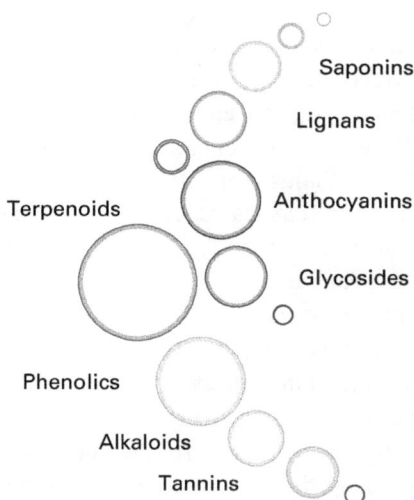

Figure 2.2: Plant secondary metabolites.

Drug discovery from medicinal and aromatic plants offers several advantages over synthetic molecules. The most important of these is the diversity of compounds with much more complex structures compared to synthetic molecules. Plants and other natural sources perform chemical transformations using various enzymes. This enables highly specific structural changes to occur at particular sites in a stereospecific manner, resulting in the formation of a complex molecule. Plant-derived molecules, many of which remain undiscovered, largely provide the chemical diversity needed for new drug research. Through their chemical diversity, plants demonstrate a wide range of pharmacological effects [5, 11]. With their phytochemical diversity, traditional foundations, and bioactivity of metabolites, medicinal and aromatic plants will continue to serve as sources of potential drug raw materials in every era. The key factors that make plants a potential source for drug discovery are illustrated in Figure 2.3.

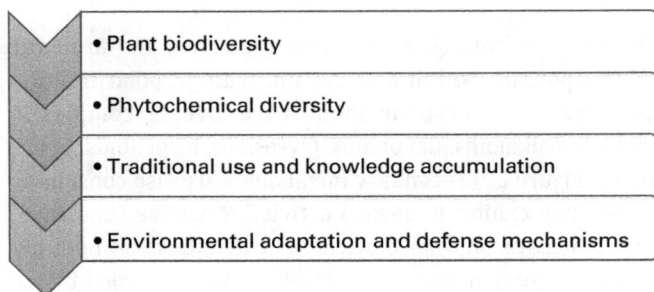

- Plant biodiversity

- Phytochemical diversity

- Traditional use and knowledge accumulation

- Environmental adaptation and defense mechanisms

Figure 2.3: The factors that highlight the potential of plants for development as drugs.

Libraries of natural compounds, which cover a wide chemical space, are significantly richer than libraries of synthetic molecules. However, this also brings certain challenges. The rich phytochemical content of plants enhances their medicinal effects and synergistic potential but complicates the identification of the component or components responsible for the effect. Determining the mechanism of action of the compounds is also challenging and quite time-consuming. In some cases, isolating individual components that exhibit synergistic effects can reduce the efficacy of the natural compound. In these conditions, combination studies can be conducted by considering the synergistic effects of therapeutic candidate components [9, 12].

To determine the chemical composition of bioactive components in plants and evaluate their pharmacological properties, these compounds must first be isolated, purified, and characterized from the plant. Drug discovery from natural products fundamentally involves the processes of screening, isolation, characterization, and optimization. The technical barriers encountered in each of these processes highlight the challenges of discovering drugs from plants compared to synthetics. The structural complexity of plant-derived products makes it challenging to determine structure-activity relationships and optimize chemical structures. In conclusion, obtaining new drugs from natural compounds requires a variety of innovative approaches [9, 12]. With advancements in technology, new strategies have been developed for drug discovery from natural products and recent technical developments have significantly overcome the obstacles faced in drug extraction from natural sources. Technologies such as several engineering methods, genome mining, advanced analytical tools, bio-informatics and artificial intelligence (AI), and increasing detection power and sensitivity in analytical methods have made it easier to research plants and plant-derived products [5, 13].

This section aims to provide an overview of the stages involved in obtaining medicines from medicinal and aromatic plants, as well as the methods used at each stage. This section will comprehensively address the methods and techniques used in source plant selection, extraction, isolation, characterization, and optimization of active components with the application of advanced drug development approaches to plant-based natural products. Additionally, recent trends and future prospects in plant-derived drug development will also be discussed in this section.

2.2 An overview of the history of plant-based medicines

The majority of the data on medicinal and aromatic plants that has reached us today has been empirically obtained through trial and error, based entirely on observation. This knowledge dates back approximately 5,000 years in India, China, and Egypt and at least 2,500 years in Greece and Central Asia [14]. Although there was no documentation in the beginning, word-of-mouth communication provided a simple way for community members to share knowledge. Since writing and other recoding techniques made it possible for communities to preserve information about the therapeutic properties of plants, many people have been treated with plant-based extracts documented in these records [2]. The oldest written records of the clinical use of plants originate from India and China. Over time, these clinical records accumulated and transformed into the global pharmacopoeias of ancient civilizations, such as Egypt's Ebers Papyrus, Greece's De Materia Medica, and China's Shen Nong Ben Cao. These sources documented various plants and formulations used as medicines. This ancient wisdom and legacy of experience have served as an inspiration for modern drug discoveries [15]. Therefore, plant extracts and mixtures have been applied over the centuries to remedy various ailments, leading to the development of medications for microbial organisms and cancer [16].

Initially, plants or plant parts were used in their raw form, but over time, tinctures, poultices, powders, or teas derived from them began to be used in treatments. Since these are often in the form of extracts containing multiple components as mixtures, information on which compounds are responsible for the healing effects is either very limited or entirely absent [17]. However, as time progressed, the discovery of the therapeutic effects of plants sparked increased interest in research aimed at isolating the active compounds responsible for these effects. From the nineteenth century onwards, with the advancements in chemistry, a period began in which active components were isolated from plants. The discovery of drugs from plants and herbal products has accelerated with the identification of bioactive compound groups and the detection of pharmacological activity of natural products through preclinical and clinical studies [18].

Scientists' interest in researching medicinal plants led to the emergence of the first drugs. The discovery of plant-based medicines created a revolution in medicine [9]. The German pharmacist Friedrich Sertürner isolated the alkaloid morphine from the *Papaver somniferum* L. plant in 1805. Morphine was the first active compound to be isolated from a plant and marked a turning point in drug discovery from plants [19]. Isolation of morphine is also the beginning of natural product chemistry. The isolation of morphine was followed by the isolation of quinine in 1820, caffeine in 1821, nicotine in 1828, atropine in 1831, and digitalin in 1868 [13]. Aspirin, digoxin, pilocarpine, cocaine, codeine, paclitaxel, tetracycline, artemisinin, doxorubicin, and cyclosporine are some of the plant-derived active ingredients still used as medicines today [5, 9]. Some plant-derived active compounds and their therapeutic uses are shown in the Table 2.1.

Following the isolation of natural products, studies to elucidate and characterize their structures began. Structural determination studies of natural molecules accelerated in the 1940s with the introduction of physical tools by Robert Burns Woodward. A new era in drug discovery from natural products has begun with the elucidation of the structures of active compounds isolated from plants, allowing their chemical synthesis and enabling modifications to their structure to alter their efficacy and side effect profiles. Robert Burns Woodward pioneered the total synthesis of natural products by synthesizing bioactive compounds from natural sources, such as quinine, cholesterol, cortisone, chlorophyll, and reserpine [13]. The period from the 1950s to the 1960s was considered the Golden Age of drug discovery from natural products [32]. Before the emergence of high-throughput screening (HTS) and the post-genomic era, more than 80% of all drug active ingredients were either entirely natural products or derived from natural sources, including semisynthetic analogs. It has been noted that these naturally sourced compounds and their by-products served as inspiration for the advancement of a majority of pharmaceutical compounds [6].

Today, the approach of isolating and evaluating individual components from plants has shifted to examining and formulating potential therapeutic components by utilizing libraries of natural compounds [9].

2.3 Methods for drug discovery and development from plants

The production of drugs from medicinal and aromatic plants requires a multifaceted and meticulous scientific research process. There are numerous stages involved in the discovery of a bioactive component from a medicinal plant and its subsequent transition to clinical application as a drug. Different research methods are used in each of these stages.

Table 2.1: Some plant-derived active compounds and their resources, chemical structures, and therapeutic uses.

Natural compound	Plant	Chemical structure	Therapeutic use	References
Paclitaxel	Taxus brevifolia		Ovarian and breast cancer	[20]
Artemisinin	Artemisia annua		Antimalarial Chemotherapeutic	[21]

(continued)

Table 2.1 (continued)

Natural compound	Plant	Chemical structure	Therapeutic use	References
Silymarin	*Silybum marinum*		Antihepatotoxic	[22]
Morphine	*Papaver somniferum*		Analgesic	[19]
Quinine	*Cinchona officinalis*		Antimalarial	[23]
Caffeine	*Coffea arabica*		Psychoactive	[24]

Nicotine Nicotiana tabacum Psychoactive [25]

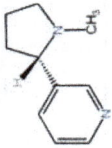

Vincristine Vinca rosea Chemotherapeutic [26]

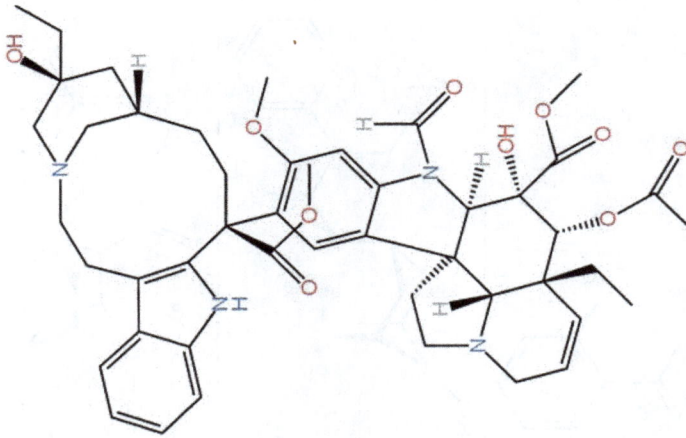

(continued)

Table 2.1 (continued)

Natural compound	Plant	Chemical structure	Therapeutic use	References
Aspirin	*Salix alba*		Analgesic Anti-inflammatory Antipyretic	[27]
Atropine	*Atropa belladona*		Anticholinergic Spasmolytic	[28]
Cocaine	*Erythroxylum coca*		Anesthetic	[29]

Codeine

Papaver somniferum

Analgesic

[30]

Colchicine

Colchicum autumnale

Gout treatment

[31]

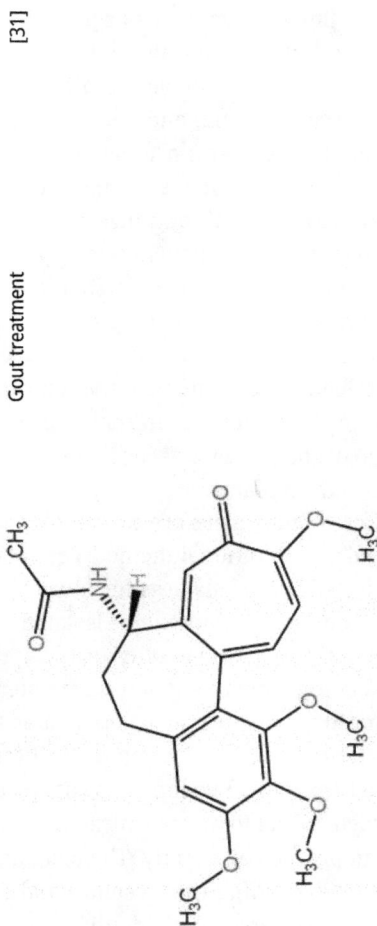

The process of drug development from plants begins with the selection and collection of plant materials. The collected materials undergo extraction using appropriate methods. The plant extracts are then divided into fractions to isolate bioactive compounds. The separation, quantification, and structural determination of the desired components from the obtained extracts are performed using chromatographic and spectroscopic techniques. Once the active compound is fully identified, structure-activity relationships are examined to optimize bioactivity, pharmacokinetics, and other pharmacological parameters, enhancing its applicability as a drug. Analogs of the lead compound can be synthesized to further the drug development process through structural modifications, semisynthesis, or total synthesis routes [33]. The evaluation of biological activity can be conducted by screening natural compound libraries prior to plant selection or through active extracts or isolated and purified compounds obtained after plant selection. In some cases, bioactivity assessment is integrated into every stage of the process. Accordingly, various approaches exist in the drug development process from plants.

In drug discovery from plants, there are two approaches: traditional and modern, utilizing different methods. In traditional methods, extracts obtained from plants and plant materials are subjected to various tests for bioactivity. Extracts that show activity are fractionated, and the active compound is isolated. Here, the extraction and isolation processes can be guided by bioactivity tests, or they may proceed independently of bioactivity, isolating components for subsequent bioactivity evaluation. In modern methods, advanced robotic technologies such as HTS are used to rapidly test hundreds of molecules found in natural compound libraries. Using this approach, a lead compound required for drug development can be quickly identified, allowing subsequent processes such as isolation, structural analyses and modifications, bioassays, and clinical trials to proceed efficiently. The drug discovery process from plants can be divided into stages, including the selection and collection of plants or plant materials, extraction, isolation, structural identification, bioassays, clinical studies, and optimization (Figure 2.4) [34].

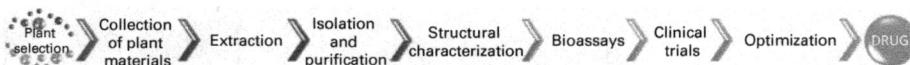

Plant selection 〉 Collection of plant materials 〉 Extraction 〉 Isolation and purification 〉 Structural characterization 〉 Bioassays 〉 Clinical trials 〉 Optimization 〉 DRUG

Figure 2.4: The main processes in the production of drugs from plants.

A comprehensive, cross-disciplinary approach that utilizes technological progress is crucial for enhancing the drug development process from medicinal and aromatic plants. In this process, the integration of bioinformatics, HTS technologies, genomics and metabolomics, efficient extraction and suitable isolation procedures, and structural elucidation tools will make the process significantly more efficient. Various bioactivity assay methods are also employed to assess the pharmacological suitability of phytochemical components [2, 7].

The discovery of new therapeutic compounds from plants is made possible through the collaboration of various fields: botany, taxonomy, ethnobotany, and plant ecology in the selection, collection, and species identification of plants; pharmacognosy and chemistry in the extraction, analysis, and isolation of raw materials; and molecular biology, biochemistry, microbiology, pharmacology, and toxicology in the evaluation of therapeutic efficacy (Figure 2.5).

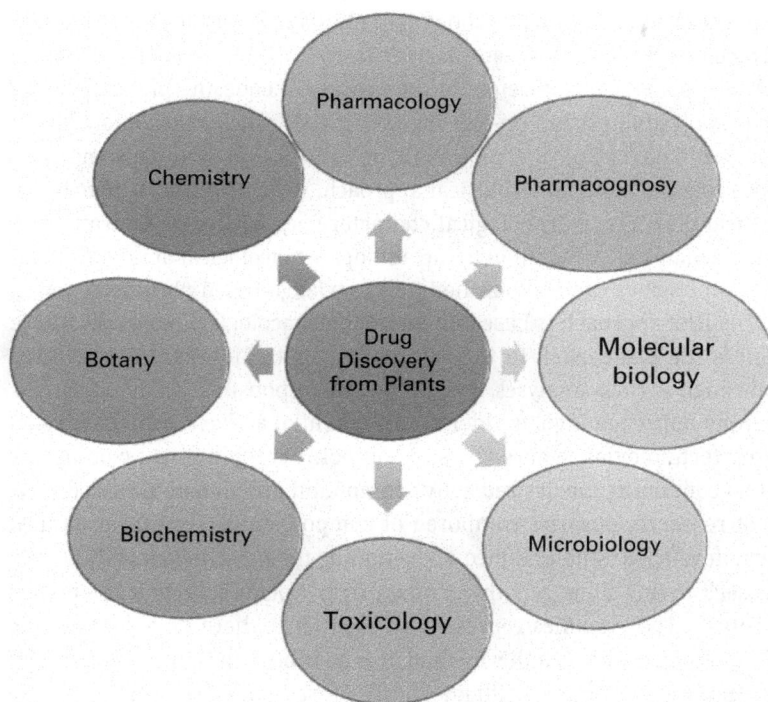

Figure 2.5: Interdisciplinary collaboration in drug discovery from plants.

2.3.1 Plant selection

The first step in discovering new drugs from plant sources is identifying and collecting plant species with potential therapeutic effects. The successful discovery and development of drugs heavily rely on the effective and efficient selection of plants that align with the research objectives [11].

Plants can be selected based on knowledge or randomly without prior information and subjected to screening for potential therapeutic effects. However, whether selected based on knowledge or randomly, they remain extremely valuable resources for the pharmaceutical industry [35].

Selecting plant species for drug development based on a well-designed, knowledge-based strategy increases the success rate compared to random plant selection [7]. In knowledge-based plant selection, records of traditional uses and natural compound libraries should be thoroughly investigated [6]. For any natural product to be used as a drug, it must first be identified as having a potential therapeutic effect on a symptom or disease. Following this, it must undergo preclinical and clinical trials to establish its efficacy and safety. This process is long and challenging. Proceeding without an appropriate and detailed screening and well-founded hypothesis may result in wasted time and inefficient outcomes in bioactivity tests [36].

Knowledge-based plant selection refers to the process of choosing plants for drug discovery or other applications based on existing information, such as traditional uses, ethnopharmacological data, chemical composition, or biological activity. In some studies, plants are also selected using an ecological approach, which considers their biodiversity and chemical diversity, morphological characteristics, phylogenetic traits, and chemical defense mechanisms, in addition to the ethnopharmacological approach [11].

Two primary approaches can be identified for knowledge-based plant selection in drug discovery. The first approach is based on ethnopharmacological research, where the starting point is the plant itself. In this approach, a potential plant is identified first, and then, through various analyses, the compound responsible for the activity is obtained. The second approach involves identifying potential active compounds using modern screening technologies, where the starting point is the natural compound. The plant or plants containing the desired active compound are then determined [35].

Over years of research, libraries composed of compounds derived from natural sources have been developed. Studies aimed at examining the bioactivities of these natural compounds seek to link findings either to the original plant source or to another potential plant source. These studies represent an approach where plant selection is based on specific molecules. Through this method, it is possible to discover new biological activities from natural compounds with known structures and origins [11].

In recent years, HTS technologies that accelerate the selection process of plants promising therapeutic effects have been developed. Since these technologies allow the testing of very small sample quantities, it facilitates the screening of natural products that are difficult to isolate, purify, and synthesize. With the integration of AI into the drug discovery process from plants, the library of natural compounds has expanded, and the number of potential target molecules with therapeutic effects has increased [13]. The application of genomics, transcriptomics, proteomics, and metabolomics in the evaluation of natural molecules with drug potential has enabled the discovery of new therapeutic candidates through more effective and advanced screening methods. Various contemporary techniques, such as database mining, virtual screening, natural product libraries, and molecular modeling, are also applied in the transformation of natural compounds into pharmaceutical agents [7, 9]. The main methods used in plant selection are shown in Figure 2.6.

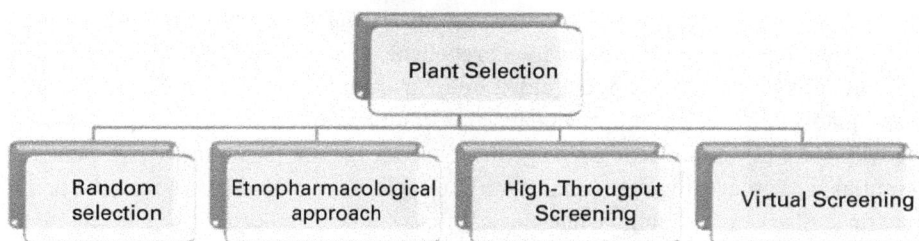

Figure 2.6: Selecting plants for new drug discovery and development.

2.3.1.1 Random plant selection

Random selection of medicinal and aromatic plants is a method of source plant research where ethnobotanical or ethnopharmacological knowledge plays almost no role. In this approach, plants are typically screened regionally for activity or sometimes for specific target secondary metabolites (flavonoids, alkaloids, terpenoids, etc.) [7]. The plant-based screening and research efforts by the United States National Cancer Institute (NCI) for cancer therapeutic agents serve as an example of this approach. The NCI and the Central Drug Research Institute (CDRI) screened approximately 35,000 plant species for anticancer activity between 1960 and 1980 [37]. Paclitaxel and camptothecin are notable results of this screening process and are now used in chemotherapy [38].

The large number and complex chemical composition of plants limit the approach of randomly collecting plants in the drug discovery process. After random selection, the processes become more complex and time-consuming, making it more challenging to identify plant-based bioactive compounds and understand their mechanisms of action. Additionally, conducting trials with numerous plants to identify bioactive compounds that exhibit the desired efficacy can be a highly costly endeavor [11, 35].

2.3.1.2 Plant selection based on ethnopharmacology and traditional uses

The knowledge accumulated from centuries of plant use by humans has provided valuable guidance for numerous scientific studies. The ways in which plants can be used for various diseases, and the specific applications, have been passed down from generation to generation, first through oral traditions and later through written documents and records. The scientific investigation of the effects of plants with traditional uses, as well as the identification of their phytochemical contents and compounds potentially responsible for these effects, is extremely important for their preparation and application as medicinal products.

The decision to investigate a specific plant species for drug discovery is often based on its traditional use as a medicine, insect repellent, or for a cultural purpose. The analysis of plants or plant extracts is guided by their traditional uses [11]. The traditional use of plants in medicine can provide insights into their efficacy and safety. Collaborating with local communities to leverage their knowledge and experience with plants is essential for researching medicinal plants suitable for treatment and for advancing new drug discovery. Ethnopharmacological field studies conducted prior to the plant selection stage will serve as a guide for this purpose. As a result of these field studies, the creation of regional ethnopharmacological databases will provide an invaluable resource for future research [5]. However, traditional medicinal plants cannot proceed to the clinical usage phase unless the knowledge obtained from traditional uses is validated through a long and labor-intensive series of analyses. Only 1 in 10,000 of the tested compounds from plants with traditional uses has successfully made it to the path of becoming a drug over an approximate 10-year timeframe [35].

The investigation of bioactive compounds in traditional herbal products is the focus of ethnopharmacology. The fundamental principle of ethnopharmacology is to investigate traditional medicines and their uses regionally, by combining them with field observations. Most of the natural products that have been developed into drugs currently have ethnomedical uses [18]. Drug discovery studies based on a plant's known activity according to traditional uses are still frequently applied by scientists today. Drugs developed from plants such as *Rauwolfia serpentina* and *Digitalis purpurea*, as well as morphine derived from *Papaver somniferum*, and berberine derived from *Berberis aristata*, are examples of drug discovery through ethnopharmacological approaches. Many important drug compounds, such as artemisinin, quinine, atropine, taxol, and aspirin, demonstrate the successful use of knowledge-based strategies with natural products on the path to drug development [6, 13].

Advancements in isolation and characterization techniques, the development of specific chemoinformatics methods, the rise in bioassay techniques, and HTS technologies have established systematic methodologies that bridge traditional ethnopharmacology and modern drug discovery. Ethnopharmacological studies remain a vital starting point for drug discovery, just as they were historically. However, in the past, these studies would first identify a plant of interest, followed by the isolation of its active components. In contrast, modern approaches typically identify the active compound initially, and then use existing ethnopharmacological data to pinpoint plants that contain these active ingredients. In plant-based drug development using ethnopharmacological data, the first step is to determine whether extracts obtained from plants exhibit efficacy against a specific disease, independent of their structure-activity relationship [35]. There are numerous scientific publications reporting positive activity in various tests for plant extracts selected using ethnopharmacological criteria. Once the activity is confirmed, the next step involves isolating and identifying the active compound. The isolated active compound becomes a verified potential drug candidate or can be developed into a new drug through further studies. However, combining this approach with other technologies can significantly increase the chances of success [11].

2.3.1.3 Plant selection by HTS technologies

For many years, natural products with potential as drug candidates have been screened. HTS is a widely used technological method for this purpose [39].

Over years of research, libraries consisting of compounds derived from natural sources have been created. Studies that aim to investigate the bioactivities of these natural compounds and link the findings either to the original plant source or to another potential plant source focus on the selection of plants based on specific molecules. Through this method, it is possible to discover new biological activities from natural compounds with known structures and sources [11].

The emergence of databases for natural compounds, the increase in computational power, and the application of new technologies such as AI have enabled the development and application of computational methods for identifying new drug-like compounds and their derivatives. Advancements in analytical and fractionation techniques used for the identification, isolation, and purification of natural compounds have made the screening of natural compounds using HTS more compatible and efficient. This facilitates the screening of thousands of compounds for their therapeutic effectiveness and safety in a short time and at significantly lower costs [40].

In the past, creating natural product libraries was a challenging, complex, and slow process. However, with the emergence of new technologies, this process has become faster and more efficient, allowing natural compounds to be seamlessly integrated into modern screening technologies [34].

HTS is an advanced technology that plays a critical role in the drug discovery process. Essentially, it is a method supported by robots, detectors, and software that enables the rapid and efficient screening and testing of chemical compounds. The primary function of HTS is to accelerate drug development by identifying the potential interaction of chemical compounds with biological targets. By allowing the simultaneous screening of large volumes of compounds, this method enables the analysis of up to 100,000 compounds per day in modern applications [41].

HTS technologies are also crucial in screening plant-derived compounds. Plants are a rich source of compounds with therapeutic potential. However, traditional methods for studying plants often yield results over extended periods. To address this issue, high-throughput pharmacological screening (HTPS) can be applied to crude plant extracts. Using the "differential smart screening" method, the biological activities of compounds within crude extracts are measured, allowing plants that exhibit the desired activity to be prioritized for further investigation [42].

Since the 2000s, automated HTS has become a central focus in drug discovery. This innovation positioned combinatorial chemistry as a preferred method for developing drug candidates suitable for HTS leading numerous drug manufacturers to move away from extract libraries developed through traditional screening. This shift was driven by the perception that extract-based screening often led to the rediscovery of known compounds. Additionally, the structural complexity of natural products was seen as a significant challenge, requiring costly and time-intensive processes like total synthesis and derivatization. Consequently, natural product-based drug discovery was considered impractical, especially when compounded by supply chain issues and long development timelines. In contrast, HTS technologies employ combinatorial chemistry to rapidly generate large compound libraries, accelerating the identification of potential drug candidates. Over the past two decades, traditional natural product chemistry has been mostly supplanted by drug discovery that targets specific molecular pathways, which focuses on efficiently obtaining "hits" from these extensive combinatorial libraries. Combinatorial chemistry has significantly transformed the discovery of new chemical entities with biological activity, enabling the efficient development of structural analogues [43].

The success of the HTS process depends on the diversity of compound libraries and the quality of screening assays. A successful HTS operation requires careful selection of biological activity tests. Through accurate assay methods, undesirable biological activities can be filtered out, leading to better outcomes [44]. As a result of numerous studies conducted on plants to date, libraries containing hundreds or even thousands of natural compounds have been established [35]. This has accelerated the discovery of drug candidate compounds by screening the natural compound chemical space using advanced methods, including HTS technologies.

The isolation of the first natural protein tyrosine phosphatase 1B (PTP1B) inhibitor from *Broussonetia papyrifera* has been a successful example of the application of HTS technology in drug discovery [13].

2.3.1.4 Plant selection through virtual screening

Virtual screening is a method that can be applied to both combinatorial chemistry-generated molecular libraries and natural compound libraries. The primary goal of this approach is to select potential drug candidate molecules from large libraries in a more specific and reduced manner. Virtual screening focuses on two main approaches: ligand-based and structure-based screening. Ligand-based virtual screening selects potential compounds for further testing based on the structural and activity data of known bioactive compounds. In contrast, structure-based virtual screening utilizes the three-dimensional structure of a compound and techniques such as molecular docking to determine the compound's optimal position and orientation within a binding site, thereby predicting its potential bioactivity. When sufficient structural and activity data for a molecule are available, both techniques can be used together to achieve more successful results. Virtual screening has been shown to outperform HTS in certain cases [11].

2.3.1.5 Phytochemical databases

Bioactive phytochemicals possess the ability to bind to molecular targets or receptors associated with specific diseases or physiological conditions. This characteristic of plant-derived compounds makes them suitable for use in drug design through virtual screening methods. Due to the high drug development potential of phytochemicals in the field of computational drug design, database management systems are essential specifically for these compounds. Phytochemical databases are digital resources that systematically store chemical, biological, and pharmacological information about compounds derived from plant sources. A comprehensive database containing information about medicinal plants and their components serves as a valuable resource for researchers working on drug development from medicinal plants. Such databases should include detailed information about the chemical structures, pharmacological properties, and physicochemical characteristics of plant-derived compounds (Figure 2.7) Examples of phytochemical natural product databases include CVDHD, KNAPSACK, Nutrichem, Phytochemica, TCMID, TCM@Taiwan, TCM-Mesh, MAPS, and Phytochemdb. While all these databases generally provide basic information, some are also suitable for virtual screening. However, there is a need to expand these resources to encompass both phytochemical and pharmacological data comprehensively. By doing so, phytochemical databases can become even more effective tools for in silico drug design, playing a larger role in the discovery of drugs from medicinal plants [45].

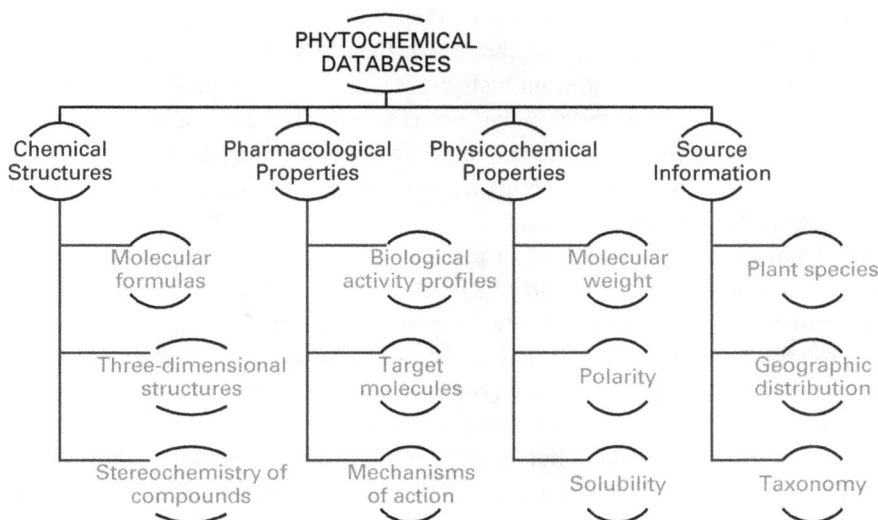

Figure 2.7: Basic information included in phytochemical databases.

2.3.2 Collection and identification of selected plants and pretreatment of plant materials

The step following the selection of candidate plants for drug discovery and development is the identification of the plant species and the collection of the necessary plant materials. The accurate identification of the plant species is of vital importance in the subsequent processes. For species identification of the collected plant materials, various analyses, including macroscopic, microscopic, and instrumental techniques, are conducted in collaboration with the fields of botany and taxonomy. Advanced methods such as chemometrics, immunoassays, and DNA fingerprinting can also be used for this purpose [7]. The morphological and anatomical characteristics of the plant material should be determined by experts, knowledgeable and experienced individuals in the field, particularly taxonomists [33].

The first step in correctly identifying selected medicinal and aromatic plants is to determine the plant's botanical origin and identify its species name in binomial nomenclature. At this stage, organoleptic properties such as color, smell, taste, shape, size, fracture characteristics, surface, and textural features should first be examined macroscopically. Subsequently, specific structural and anatomical characteristics at the tissue and cellular levels should be evaluated microscopically. In addition to these methods, DNA barcoding is a reliable technique that provides secure information, identifying plant species and quality assessment of medicinal and aromatic plants. Botanical species can be performed using DNA barcoding, where a short region of the plant's DNA sequence is used as a genetic marker [7].

For the extraction, isolation, and characterization of a bioactive compound from plants, it is essential to have sufficient biological material. This requires the collection of an adequate amount of material in the correct manner [12]. Legal and ethical regulations must be followed when collecting plants and plant materials. A sample of the species-verified reference material must be recorded and preserved in a herbarium with a designated accession number [33].

The chemical makeup of medicinal plants is extremely complicated and can be influenced by a variety of factors, including soil composition, growth and storage conditions, genetic makeup, harvest timing, processing techniques, and more [46]. Preserving the biomolecules in medicinal plants is crucial for all processes involving these plants. Therefore, after plant materials are collected, they undergo certain pretreatment processes before the extraction stage. Sample pretreatment is an important component of the sample preparation process in modern analytical methods. It is also the most error-prone stage during analysis, and the pretreatment procedures applied to plant materials significantly affect the phytochemicals in the final extract. The primary goal of pretreatment is to isolate target metabolites from the matrix and to enhance the selectivity, accuracy, reliability, reproducibility, and determinability of the analysis. The proper preparation of plant material is one of the most important factors that enhance extraction efficiency [47, 48]. An effective plant material prepara-

tion method should allow for the efficient and comprehensive isolation of both volatile and nonvolatile, alongside polar and nonpolar substances, irrespective of the compound's location within the matrix, its classification, or the existence of other factors.
The method should also be durable and sensitive, for example, resistant to high temperatures. A well-chosen material pretreatment method enhances the accessibility of
phytoactive compounds and simultaneously facilitates their extraction [49].

Plant material to be processed is typically subjected to pretreatment steps such as
drying, lyophilization, crushing, grinding, homogenization, or steam distillation.
These pre-extraction processes enhance the active surface area, significantly improving extraction kinetics and, consequently, the yield of targeted metabolites [50, 51].

Drying

It is possible to extract plant samples from either fresh or dried plant material, including leaves, bark, roots, fruits, and flowers [48]. However, the analysis of plants and
herbal products is typically conducted on dried materials. This approach allows the
determination of component ratios on a dry mass basis [50].

Another reason for preferring dried materials in experimental studies is that
fresh materials tend to deteriorate more quickly over time, whereas dried materials
are more stable [48]. To prevent microbial and/or enzymatic degradation of the material, water activity must be eliminated through drying, freezing, or lyophilization [51].
The primary goal of the drying process is to prevent metabolic activities that could
lead to alterations in the chemical composition of the plant. This is achieved by reducing the water content in the plant material, which is essential for the proper functioning of plant enzymes. Thus, the drying process helps eliminate issues related to the
high water content in the material. The absence of water, coupled with high drying
temperatures, helps inhibit enzymes that might degrade the active compounds. Additionally, effective drying reduces the microbial load in the end product. It also substantially decreases the mass and volume of the material, leading to decreased expen-
ditures on packaging, transportation, and warehousing [49].

The drying process can be carried out under natural or artificial conditions and
in various ways. The drying technique and temperature depend on the type and features of the components contained in the plant. The drying of natural products is typically performed in hot air or nitrogen-flow ventilated ovens. In the presence of volatile components, low-temperature drying is preferred. Drying with high heat can lead
to the depletion of these components and may also trigger the degradation of compounds in essential oils [52]. The drying process can lead to unpredictable degradation
of the phytochemical content of the plant, depending on the method used and the
characteristics of the plant components. Therefore, the appropriate drying method
should be selected after evaluating all necessary parameters [51]. The most commonly
used drying methods for plant materials are shown in Figure 2.8 [53].

One of the oldest drying methods used for plant-based raw materials is open-air
drying. While it can be carried out under sunlight or in the shade, drying under direct

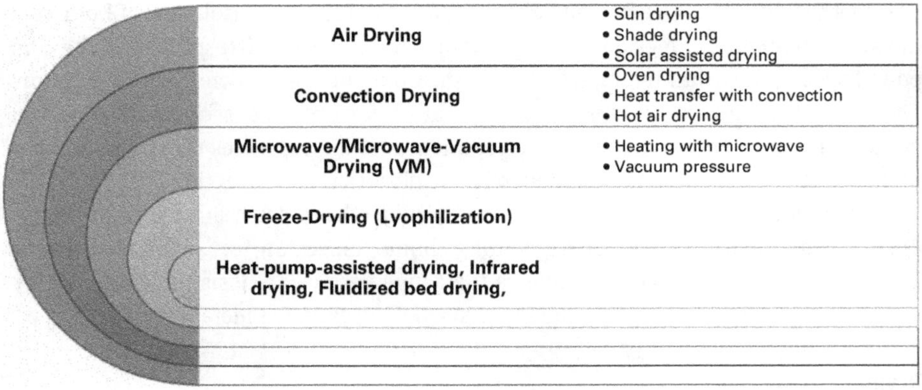

Air Drying	• Sun drying • Shade drying • Solar assisted drying
Convection Drying	• Oven drying • Heat transfer with convection • Hot air drying
Microwave/Microwave-Vacuum Drying (VM)	• Heating with microwave • Vacuum pressure
Freeze-Drying (Lyophilization)	
Heat-pump-assisted drying, Infrared drying, Fluidized bed drying,	

Figure 2.8: Drying methods for plant materials.

sunlight often leads to issues such as the degradation of the material's aroma and color. In the shade-drying method, the material is left to dry in an open or semi-open space without direct exposure to sunlight. Drying is carried out under natural conditions, benefiting from air circulation. Shade drying allows for better preservation of volatile compounds as well as the aroma and color of the material. However, all open-air drying methods expose the material to environmental contaminants. Additionally, due to the long drying times and the inability to control parameters such as temperature and pressure, this method has lost its importance in modern applications. Due to the long drying times associated with sun and shade drying methods, a commonly used alternative is hot air drying, also known as oven drying. Under artificial conditions, both temperature and pressure can be regulated as needed, depending on the characteristics of the material. For this purpose, ventilated chambers heated using various methods are preferred. This technique utilizes convection for heat transfer and allows precise control over key parameters such as temperature, air circulation speed, and drying duration. These adjustable features make it an efficient option for drying plants and herbs while ensuring consistency and quality [53].

Due to the various limitations of traditional drying techniques, new and modern methods have been developed for the drying of plant materials. Compared to traditional methods, these techniques offer improved preservation of bioactive compounds as well as enhancements in the physical and chemical characteristics and organoleptic features of the dried products. Advanced drying techniques such as freeze-drying, microwave drying, infrared drying, spray drying, and supercritical drying are particularly noteworthy. The most suitable drying techniques and conditions should be selected by considering the differences in the properties of the materials to be dried and/or the target plant components of interest. Each drying method should be evaluated for its advantages and disadvantages in terms of drying kinetics and the quality of the final dried products [54].

2.3.2.1 Comminution and homogenization

Comminution and homogenization are important steps in preparing raw plant material for extraction and subsequent comprehensive analyses.

The particle size of plant materials affects extraction efficiency, making the fragmentation of materials a critical pretreatment process. When particle size is reduced, there is greater interaction between the samples and the extraction solvents at the surface level. Powdered samples, with their smaller and more uniform particle size, enhance surface contact with the solvents, whereas ground samples result in larger, irregularly sized particles. For extraction to be effective and efficient, the solvent must interact with the target analytes as much as possible [48]. Comminution and grinding increase the rate at which the solvent penetrates the solid material. However, in some cases, very fine grinding and pulverization may not be suitable. Extremely fine grinding can cause the solids to compact and form a mass during extraction, obstructing the free flow of the solvent. Additionally, in materials with a cellular structure, grinding may lead to cell rupture, which can result in the extraction of unwanted components [55, 56].

An appropriate comminution technique can be selected based on the texture and hardness of the plant material.

The following points should be considered during the comminution and grinding of materials:

1. Essential oils are sensitive to temperature, so any increase in temperature should be avoided when working with raw materials containing these compounds. To prevent the loss of essential oils, the material should be ground in small quantities.
2. Roots, tough stems, fruits, and seeds are initially chopped by hand or machine and then processed into smaller pieces using different mechanical grinding mills
3. Manual cutting is a simple and equipment-free method for comminution. However, since it produces pieces of varying sizes, sieving the cut material is recommended to ensure uniformity [50].

The next step after the comminution process is the homogenization of the material. The mixing process is essential both during the initial preparation of the material and in subsequent steps to ensure the homogeneity of the material [48]. For this purpose, various manual or mechanical homogenizers can be used. Sample homogenization can also be accomplished via enzymatic lysis (often hydrolysis), freezing in a way that ruptures the cellular wall or membrane, high-energy ultrasound vibrations, and other nonmechanical physicochemical processes. One important point to consider during the homogenization process is to prevent thermal degradation of the material due to excessive heating [50].

2.3.3 Extraction

Parts such as the roots, stems, leaves, flowers, seeds, or even the entire plant can be used fresh or dried for therapeutic purposes. Medicinal plants contain numerous active compounds that can exhibit different physiological effects in various parts of the plant. To identify and isolate these active compounds, they must first be separated from the plant through the extraction procedure [8].

The extraction of plant materials is the most critical step before the isolation and purification of plant components. Plant constituents naturally exist in a complex matrix, and their physical and chemical characteristics vary significantly. The properties of the target molecule, believed to be responsible for the pharmacological activity or considered a drug candidate, must be well understood. To obtain it in pure form, it must be carefully separated from the rest of the plant. This is only achievable through an appropriate extraction method and optimized extraction parameters [57]. Extraction involves isolating the medicinally active components of plants using selective solvents and standardized methods. The distribution of a compound between two immiscible phases, which permits their subsequent separation and recovery of the extracted compound, is the fundamental principle of extraction [56]. Utilizing variations in the mixture of components' physical or chemical characteristics is the foundation of extraction techniques. Particle or molecular size and shape, density, solubility, and electrostatic charge are some of the more often used characteristics in separation procedures. Some operations include more than one of these qualities. Nonetheless, the majority of the processes are physical in nature [55]. The extraction processes of secondary metabolites in plants are related to the solubility differences of compounds in a solvent mixture. During this process, the solvents penetrate the plant material and dissolve components with similar polarity, thereby separating them from others [58]. Separating the soluble plant metabolites from the insoluble cellular marc (residue) is the aim of all extraction processes. Crude extracts obtained through these methods typically consist of a complex mixture of various plant metabolites, including alkaloids, glycosides, phenolics, terpenoids, and flavonoids [48].

The extraction step is one of the initial stages in obtaining an active compound from plant material. Further separation, identification, and characterization of bioactive compounds can only be achieved after performing an appropriate extraction process. Therefore, the extraction process and techniques can significantly influence the outcomes. In the process of identifying and isolating an active compound from a plant, several critical steps must be carefully considered, such as accurately identifying the plant, accounting for potential transformations throughout the process of pretreatment and extracting the material, followed by the elimination of known compounds at the initial stage of fractionation [33].

The efficiency and outcomes of the extraction process are influenced by numerous factors. The most significant among these are the matrix properties of the plant material in which the components are embedded, the type of solvent used, temperature, pressure, and time (Figure 2.9) [8, 59].

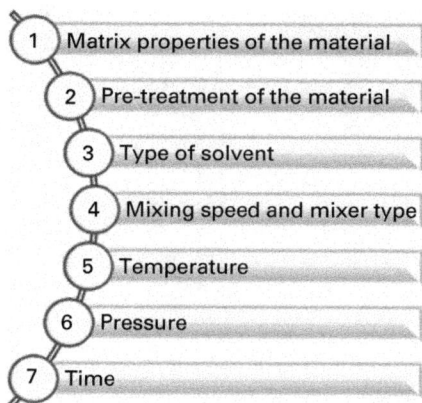

1 Matrix properties of the material

2 Pre-treatment of the material

3 Type of solvent

4 Mixing speed and mixer type

5 Temperature

6 Pressure

7 Time

Figure 2.9: The factors affecting extraction efficiency.

Plant metabolites are typically found as complex mixtures containing numerous substances with varying degrees of polarity and hydrophobicity. The main categories of these substances in plant materials include low-polar compounds (e.g., waxes, terpenoids), semipolar compounds (e.g., lipids, phenolic compounds, low-polar alkaloids), and high-polar compounds (e.g., polar glycosides, polar alkaloids, saccharides, peptides, and proteins) [50]. Based on the differing polarities and structural characteristics of metabolites, one of the most crucial parameters in the extraction process is the choice of solvent. Solvent selection should consider the extraction method, the part of the plant to be used, the target active metabolites, and the intended use of the obtained extract. The polarity of the solvent is also an important criterion to reach the target components in the extraction process. A wide range of solvents is available, from low polarity to high polarity, where polar solvents are used for polar components, and nonpolar solvents are used for components with lower polarity. Generally the selectivity of extraction can be enhanced by using solvent mixtures instead of a single solvent. The solvents commonly used in extraction processes and their polarities are shown in Figure 2.10 [60]. Beyond polarity, many other factors influence solvent selection. The most significant of these include the solvent's safety, selectivity, toxicity, and environmental impact [33, 61]. Besides the type of solvent, other parameters influencing extraction efficiency include the ratio of the sample to the solvent, the temperature of the extraction environment, and the physicochemical features of the material [62].

To prevent oxidative damage in plant materials and to preserve the biological activities and other properties of the components extracted from plants, extraction parameters such as pH, temperature, and time must be carefully adjusted [51]. Tempera-

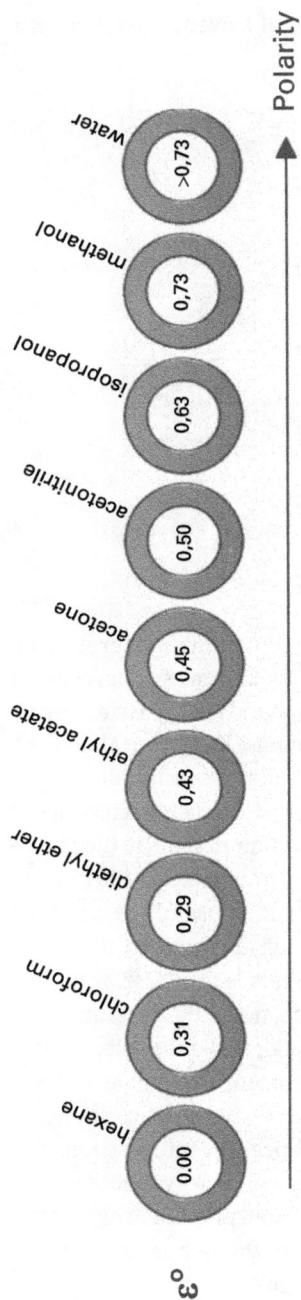

Figure 2.10: Solvent polarity and eluotropic strength (ε°).

ture is a crucial factor in the extraction of solid materials. High temperatures enhance the solubility of components in solvents, allowing for higher extraction yields. Raising the temperature decreases the viscosity and surface tension of the solvent, resulting in improved diffusion efficiency. However, elevated temperatures may also lead to solvent losses, the extraction of some unwanted components, and most importantly, damage to certain sensitive compounds in the plant material. Therefore, the proper temperature should be selected based on the characteristics of the plant material being processed [51, 55].

Although it is possible to extract all components from medicinal and aromatic plants using various extraction methods, it is not always feasible to determine the effect of each individual component in an extract. Some components may be present in quantities too small to detect their activity, while others in the extract may mask the effects of certain compounds. Pre-fractionation and the application of novel extraction techniques are two strategies that can be used to accomplish this. It has been demonstrated that employing these strategies improves the quality of hit leads for medication development [2].

Numerous factors related to the plant itself also influence the extractability of its components. The properties of the matrix in which the components are embedded (which vary depending on the plant's botanical and anatomical origin and the part used) can become the most critical criteria in selecting the extraction method [51].

The growing interest in natural bioactive compounds has increased the demand for more advanced extraction methods. At the small manufacturing enterprise (SME) or small research setting levels, traditional techniques like maceration and Soxhlet extraction are frequently employed. These traditional extraction techniques, which are widely used and rely on simple equipment, require long processing times, high energy consumption, and large amounts of solvents. Due to these disadvantages, traditional methods have been increasingly replaced by modern and innovative techniques. Significant progress has been achieved in the processing of medicinal plants, including the use of contemporary extraction techniques like supercritical fluid extraction (SFE), ultrasound-assisted extraction (UAE), and microwave-assisted extraction (MAE), which are intended to boost output at a reduced cost. These advanced methods achieve significantly higher yields in the recovery of bioactive compounds while greatly reducing the need for large quantities of raw materials Additionally, changes to the techniques are always being created. With so many different approaches available, choosing the best extraction technique requires careful consideration. The main methods used for the extraction of plant samples are summarized in Figure 2.11 [48, 62].

2.3.3.1 Conventional extraction techniques

Several traditional extraction methods can be used to extract bioactive chemicals from plant sources. The majority of these methods rely on the extraction capabilities

CONVENTIONAL EXTRACTION TECHNIQUES	ADVANCED EXTRACTION TECHNIQUES
Maceration	Ultrasound-assisted extraction (UAE)
Infusion	Pulsed-electric field extraction (PEF)
Decoction	Microwave assisted extraction (MAE)
Percolation	Supercritical fluid extraction (SFE)
Soxhlet extraction	Pressurized liquid extraction (PLE)
Distillation	Enzyme assisted extraction (EAE)
	Solid-phase micro extraction (SPME)

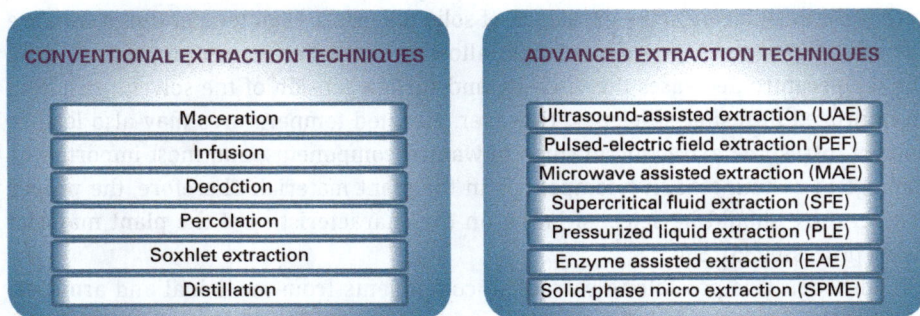

Figure 2.11: Conventional and advanced extraction methods.

of the various solvents being applied, as well as the use of heat and/or mixing. In all these methods, the extraction process is performed by treating the material with a solvent at high thermal condition and/or with agitation [63].

Maceration, infusion, decoction, percolation, and Soxhlet extraction are traditional extraction methods applied to medicinal and aromatic plants. The most significant disadvantages of these methods are the long processing times and the excessive use of organic solvents. Decoction and hydrodistillation methods, on the other hand, use water as the solvent. Traditional extraction methods are based on solid-liquid (matrix-solvent) extraction, where phytochemical components are extracted from their matrix using various solvents depending on their solubility properties. In these methods, the solvents penetrate the solid plant materials and dissolve the compounds with similar polarity. Applying a solvent with suitable polarity in combination with a compatible extraction method is critically important, depending on the target compounds [64]. The most commonly used solvents, based on the type and polarity of the compounds intended for extraction from plants, are shown in Figure 2.12 [64, 65].

Compared to modern techniques, conventional extraction methods have two main disadvantages: they require higher temperatures and take longer, which can lead to the degradation of certain components in plants. Despite these drawbacks, conventional methods continue to be widely used due to the easy availability of extraction equipment and their lower cost compared to advanced alternatives [66]. The advantages and disadvantages of traditional extraction methods are shown in Table 2.2.

2.3.3.2 Maceration

Maceration is a simple and widely used extraction method. This technique is based on leaving crushed or powdered plant materials in contact with a solvent at room temperature. Over a period of two to three days, frequent stirring ensures adequate diffusion of the solvent into the plant sample. As the cell walls of the plant weaken and

Water	Ethanol	Methanol	Dichloromethanol	Chloroform	Ether	Acetone
□ Tannins	□ Flavonols	□ Polyphenols	□ Terpenoids	□ Flavonoids	□ Alkaloids	□ Flavonols
□ Anthocyanins	□ Polyphenols	□ Flavones		□ Terpenoids	□ Terpenoids	
□ Terpenoids	□ Alkaloids	□ Anthocyanins			□ Coumarins	
□ Saponins	□ Terpenoids	□ Terpenoids			□ Fatty acids	
□ Lectins	□ Tannins	□ Tannins				
	□ Sterols	□ Lactonens				
	□ Flavonoids	□ Saponins				

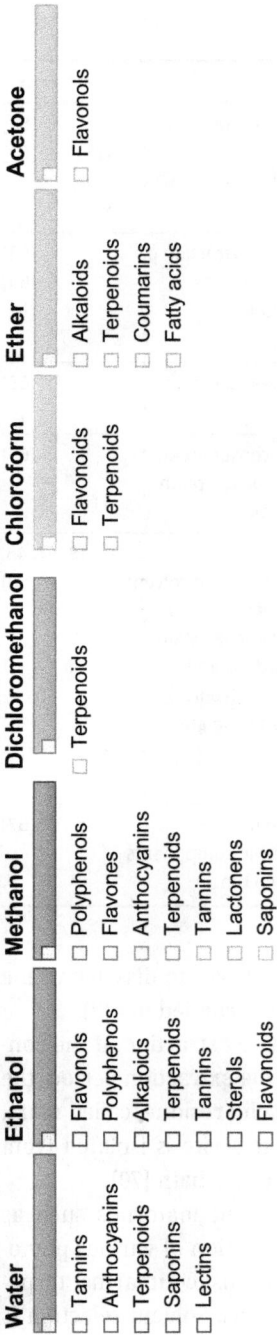

Figure 2.12: Commonly used solvents for the extraction of secondary metabolites.

Table 2.2: The advantages and disadvantages of traditional extraction methods.

Method	Advantages	Disadvantages	References
Maceration	Low-cost and simple equipment Ease of application	Limited to heat-resistant components Long extraction time Low productivity	[58] [67]
Decoction	Efficient for water-soluble bioactive compounds Avoiding the degradation of stable compounds	Not suitable for the extraction of heat sensitive constituents Long extraction time Energy-intensive	[68] [69]
Infusion	Simple Accessible	Not suitable for heat-sensitive compounds	[68]
Percolation	Highly efficient	Excessive solvent consumption Excessive energy consumption Long extraction time	[69]
Soxhlet extraction	Low cost Continuous contact with the solvent No filtration required after the process Simple equipment and simple method Suitable for the extraction of large amounts of material	Long extraction time Requires large amounts of solvent Difficulty in automation Limitations in solvent selection Exposure to hazardous and flammable organic solvents Unsuitability for shaking and stirring	[8, 47, 48, 50]
Hydrodistillation and steam distillation	Low cost	Low extraction yield Partial loss of volatile components Lengthy processing times	[67]

break down, the phytochemical components within the plant begin to dissolve in the solvent. At the end of the extraction stage, a filtration process is carried out [8].

During this process, periodic shaking is crucial for effective extraction. If the container is a bottle, occasional shaking is recommended. After the extraction period, the liquid extract, known as the miscella, is separated from the solid residue, called marc, using methods such as filtration or decantation. Then, the miscella is isolated from the menstruum by evaporating the solvent using an oven or water bath [70].

Maceration can be used to extract coarse powdered plant materials such as leaves, bark, or root bark [71]. This method allows for the extraction of various phytochemicals, including polyphenols, flavonoids, alkaloids, tannins, coumarins, terpenoids, polypeptides, glycosides, steroids, quinones, and saponins. Solvent selection is crucial in determining the bioactive compounds to be extracted. For example, ethanol effectively extracts glycosides, alkaloids, and carbohydrates, while water is suitable for terpenoids, alkaloids, glycosides, and carbohydrates. On the other hand, methanol is effective in extracting phenolic compounds, flavonoids, tannins, glycosides, and amino acids [69].

This simple solid-liquid extraction method is notably advantageous for extracting thermolabile components [72]. However, it has limitations such as long extraction times and relatively low efficiency [67].

2.3.3.3 Infusion

Infusions are preparations in the form of dilute solutions containing easily soluble components of raw plant materials. Fresh infusions are typically obtained by soaking solid materials in cold or hot water for a short period of time [73].

The basic principle of the infusion method involves moistening raw materials, cut into appropriately sized pieces, with a small amount of water for about 15 min. The concentrated infusion is then diluted with water up to 10 times its volume. Modified filtration or maceration processes may be used in the preparation of concentrated infusions. After dilution with water, concentrated infusions are similar to fresh infusions in terms of strength and aromatic properties. Infusions are prone to fungal and bacterial growth [74].

This method is a convenient way to isolate heat-stable compounds from plants. It is simple and accessible because it does not require expensive equipment or highly skilled practitioners. However, a significant disadvantage of this technique is that heat-sensitive plant compounds may degrade during the process, making them unsuitable for extraction [68].

2.3.3.4 Decoction

This method is based on boiling dry or wet plant parts with water for a certain period. Woody plant materials such as roots and bark are processed with this method to extract heat-resistant components, resulting in a higher yield of water-soluble compounds [70].The preparation involves heating the required amount of herbs with water for 30 min until approximately 50% of the water evaporates. The vessel must remain closed during the heating process to prevent the loss of essential volatile compounds [75].

It is suitable for extracting hard and fibrous plant parts such as fruits, roots, and shells that carry active ingredients and are stable under high heat [68]. After boiling, water used as a solvent is removed with the help of a vacuum evaporator, leaving behind a concentrated extract referred to as "quath" or "kwath." [76].

This traditional decoction method ensures that the water-soluble bioactive compounds are efficiently extracted from plant materials while avoiding the degradation of stable compounds. However, the process is not suitable for thermolabile or volatile compounds, which may be lost during heating [77, 78].

Finally, although the decoction method is a widely used technique for the extraction of plant compounds, it has several limitations. The most obvious disadvantage of this method is that it is costly in terms of energy and time due to the long-term boiling and high temperature requirements. In addition, it can reduce the extraction effectiveness by causing the decomposition of heat-sensitive components and the evaporation of volatile components. The method also leads to the extraction of unwanted water-soluble substances, which negatively affects the purity of the product obtained. The difficulty of standardization is another important problem of the decoction method; the composition of the product can differ based on the quality of the plant material and the process parameters. While the voluminous extracts resulting from the use of high amounts of water require additional concentration processes, the need for more raw materials for hard-textured plants can increase the cost. These limitations prevent the decoction method from being preferred in all cases and encourage the use of alternative extraction techniques [69].

2.3.3.5 Percolation

Percolation is an effective and widely used method for extracting active components from plant materials, offering a more controlled extraction process compared to maceration. The term is derived from the Latin word *percolo*, meaning "to flow through," and the process involves gradually passing a solvent drop by drop through a solid material. The percolation technique is simple in terms of equipment and easy to perform. In this technique, the powdered sample is tightly filled into a tank called a percolator, moistened with the solvent, and then continuously infused with the extraction solvent while the extract is simultaneously collected. Common solvents include ethanol, water, or hydro-alcoholic mixtures, and the process continues until the eluate becomes colorless. After extraction, the residual plant material is pressed to recover the absorbed solvent. The recovered solvent is then combined with the collected extract, and evaporation is used to produce a concentrated extract [79].

Since percolation involves the continuous addition of fresh solvent to a saturated solution, it is both efficient and effective [80]. The technique is suitable for extracting components that are unstable under thermal conditions. Additionally, it preserves the quality and concentration of the final product while achieving high extraction efficiency. However, disadvantages include high solvent consumption, long extraction times, and increased energy requirements during subsequent concentration processes [81, 82].

2.3.3.6 Hydrodistillation and steam distillation

These two methods are commonly used for extracting essential oils from plant materials, based on the principle of separating components according to differences in their physical properties. In hydrodistillation, plant samples are placed in a closed container with an appropriate amount of water. The mixture is then boiled or subjected to direct steam. The steam carries the extracted essential oil to a condenser, where it cools down and forms a liquid mixture. This process also results in the creation of a by-product called "hydrosol," a water component containing part of the plant's essence [83].

Steam distillation is a suitable extraction technique for temperature-sensitive materials such as oils, resins, hydrocarbons, and other compounds that are water-insoluble and can be separated at their respective boiling points. It has been used for many years for the extraction of essential oils from plants. The process involves distilling a component or mixture of components at temperatures significantly lower than their specific boiling points. Fresh or dried plant material is placed in the steel chamber of the apparatus, and the generated steam passes through the plant material, penetrating its cells, softening them, and facilitating the volatilization of the essential oil. Once released, small droplets of oil are formed and mix with the steam (carrier), passing into a cooling system. The mixture condenses there, forming a liquid mixture where the oil phase is typically at the top. The less dense oil is easily separated from the water [84].

Hydrodistillation and steam distillation are traditional extraction methods for isolating essential oils. The primary advantage of these methods is their low cost. However, they also have disadvantages, including low extraction yield, partial loss of volatile components, lengthy processing times, and the potential degradation of some components. Despite these drawbacks, these two methods remain the most commonly preferred techniques for essential oil isolation [85].

2.3.3.7 Soxhlet extraction

The Soxhlet extractor, developed by German chemist Franz Ritter Von Soxhlet in 1879, has remained a popular apparatus for many years and is widely used today for the extraction of natural source compounds. It has also served as a reference model for newly developed extraction techniques [59].

For the extraction process, the finely ground dry sample is first placed into the extraction chamber of the Soxhlet apparatus. This chamber typically consists of a porous bag or "thimble" made of filter paper or cellulose. The extraction solvent is then heated above its boiling point in the distillation flask. Vapors from the boiling solvent move into the condenser, where they condense and drip back onto the sample. Once the solvent reaches the siphon level, the siphon empties the solution back into the distillation flask. The solutes dissolved in the solvent are transferred to the bulk liquid in the flask. As the solvent flows back into the solid plant material bed, the dissolved metabolites are retained in the distillation flask. This allows the hot solvent to circulate through the material multiple times. Since only pure solvent is vaporized, fresh solvent is used in every cycle, while the extracted metabolites remain in the solvent flask. This process is repeated until the extraction is complete [47, 59]. Therefore, it is a time-consuming process that requires multiple cycles to complete the extraction. Prolonged exposure of bioactive compounds to high temperatures can result in the degradation of thermolabile components, reducing their quality. The efficiency of Soxhlet extraction depends on various factors such as the average particle size of the material, extraction time, and the choice of solvents, whether polar or nonpolar [86].

Soxhlet extraction is a straightforward and practical technique that allows for an endless cycle of extraction using a new solvent until all of the solute in the raw material has been extracted [58]. In fact, the primary characteristic of Soxhlet system is the gradual recycling of the extracting solvent, which prevents the solvent from potentially settling during the maceration step, which is traditionally seen upon simple contact between the solvent and the sample matrix, and displaces transfer equilibrium to ensure a high extraction yield [56].

Soxhlet extraction can be applied to solid and semisolid plant materials, but it is primarily used for extracting components from solid samples. A dry, finely divided solid is the acceptable sample for Soxhlet extraction. A number of variables, including temperature, solvent-sample ratio, and agitation speed, must be taken into account. The extraction solvents are usually pure organic solvents or their mixtures, and high purity is required for these solvents. However, this increases exposure to toxic organic solvents and their environmental impact [47].

Although the Soxhlet extraction method has drawbacks, such as prolonged processing times and significant solvent usage, it is still frequently utilized for plant material extraction because of its simplicity [56]. In recent years modern versions of Soxhlet extractors, including pressurized, automated, ultrasound-assisted, and microwave-assisted versions, have also been developed [47].

2.3.3.8 Advanced extraction techniques

There are several methods available for extracting plant materials. Due to the long processing times, high solvent consumption, and low yields associated with traditional extraction techniques, the use of modern extraction methods has increased in recent years. Innovative and more environmentally friendly advanced extraction techniques, which minimize the use of synthetic and organic chemicals, have been developed to replace traditional methods. Most of these techniques use mechanisms such as heating and ultrasonic vibrations to break down cell walls more rapidly, enhancing the solubility of desired active compounds and improving extraction efficiency [87].

Modern methods are highly automated, allowing for the simultaneous control of multiple parameters. By selecting the most suitable technique, both sample and solvent consumption can be reduced. These methods enable efficient extraction in a shorter time, often providing extracts with higher yield and quality compared to conventional methods. One common feature of these techniques is their ability to operate at high temperatures and pressures. Many modern methods are much more suitable for the extraction of heat-sensitive and volatile compounds compared to traditional methods [88].

The most prominent modern extraction methods include microwave-assisted extraction, pressurized liquid extraction, supercritical fluid extraction, and ultrasonic-assisted extraction. These techniques are suitable for industrial-scale extraction of active compounds from plants. Additionally, the use of green technology, which combines the use of green solvents such as deep eutectic solvents and ionic liquids, offers a good alternative for the extraction of natural compounds, achieving higher yields with less solvent and energy consumption [89]. A brief comparison of the advantages and disadvantages of advanced extraction techniques is presented in Table 2.3.

Table 2.3: The advantages and disadvantages of advanced extraction techniques.

Method	Advantages	Disadvantages	References
Ultrasound-assisted extraction	Reduced reaction/preparation times Minimal material consumption Effective and economical solvent use Increased sample throughput Short extraction time High efficiency	Decline of extraction of power with time High cost Nonselective Heat can damage thermal labile compound	[88, 90]
Pulsed-electric field extraction	Short extraction time High efficiency Low energy Less solvent Low extraction temperature	Free radical formation Expensive equipment Efficiency dependent on the conductivity of the environment	[91, 92]
Microwave-assisted extraction	Short extraction time Ease of use High efficiency Less amount of solvent Automation of the instrument Easily coupled with other analytical methods Low energy consumption	It can damage heat-sensitive compounds Limited penetration depth Uneven heating in complex matrices Equipment and maintenance cost	[8, 93]

Table 2.3 (continued)

Method	Advantages	Disadvantages	References
Supercritical extraction	Short extraction time High selectivity Suitable for thermally unstable compounds Better diffusivity Requires less sample and solvent Environmentally friendly Ability to operate at low temperatures Minimal waste production Ease of automation Possibility of on-line coupling with separation and detection techniques	Limited to low-polarity compounds Low extraction yields Expensive	[94, 95]
Pressurized liquid extraction	High efficiency and extraction yield Green technology Automation Reducing time and solvent consumption Protection sensitive compounds Selectivity	High instrument cost	[96, 97]
Enzyme-assisted Extraction	Sustainable and eco-friendly High efficiency Low energy consumption Simple recovery with reduced solvent usage	Slow process Difficulty in achieving optimal conditions Expensive	[76, 98, 99]
Solid-phase microextraction	Ease of use Low cost High efficiency Rapidity Being solvent-free A lack of requirement for special equipment Improved sensitivity Automation, miniaturization High-throughput performance Online coupling with various analytical instruments	Fiber breakage Sample carry-over problems pH instability	[100, 101]

2.3.3.9 Ultrasound-assisted extraction

Ultrasound-assisted Extraction (UAE) is a technique that utilizes high-frequency sound waves for the extraction of target compounds. UAE is considered an environmentally friendly technology due to its ability to reduce the need for organic solvents.

The significant increase in the production of targeted plant-based molecules makes UAE a highly efficient method [102].

In ultrasound-assisted extraction, also known as sound wave-assisted liquid extraction, sound waves are emitted into the medium through devices that generate ultrasonic sound, enabling the extraction process. This method typically relies on sound waves produced by devices operating at frequencies ranging from 20–50 kHz, which disrupt the structure of the cell walls in the sample, thereby facilitating the penetration of the solvent into the plant material [69].

Several parameters influence the extraction efficiency in UAE. These include the frequency and intensity of ultrasonic waves, the type of solvent, extraction time, and temperature. These factors can be optimized based on the characteristics of the material and components to be extracted, ensuring that target compounds are not degraded or subjected to thermal damage, thereby maximizing extraction efficiency. UAE can be performed in two ways: bath extraction and probe (horn) extraction. In bath extraction, the sample container holding the plant material is immersed in a liquid medium (usually water) or placed in a bath directly exposed to ultrasonic waves. In probe extraction, ultrasonic horns are applied directly to the sample. Both methods generate cavitation through vibrations, breaking down cellular barriers and facilitating the extraction of desired compounds [90].

In UAE, solvents such as ethylene glycol, water, ionic liquids, and its oligomers, glycerol, or other solvents derived from biomass can be used [103]. This method enables the extraction process to be carried out with lower energy consumption, shorter durations, and at lower temperatures. Additionally, it requires fewer instruments and smaller solvent volumes, making it an environmentally friendly approach [104].

2.3.3.10 Pulsed-electric field extraction

In extraction processes, nonthermal technological methods are gaining increasing importance as alternatives to thermal treatments. Pulsed electric field (PEF) is a technique that uses moderate to high electric fields to reduce the damage caused by traditional heating methods to plant materials. PEF extraction enhances mass transfer by disrupting the matrix in which the components are embedded within the plant material. PEF technology is a promising alternative to many other extraction methods, as it allows the extraction of plant components without affecting their activities. The use of PEF for extraction has increased extraction yield, shortened processing time, prevented the decomposition of temperature-sensitive materials due to the absence of thermal treatment, reduced energy costs, and eliminated negative environmental impacts. Recently, this technology has also been employed to stimulate the biosynthesis of metabolites in plants beyond its application in extraction [91, 105].

The fundamental principle of PEF-assisted extraction is based on placing plant material between two metal electrodes and exposing it to repetitive short pulses of moderate electric fields and low energy input. This process induces permeabilization of plant cell membranes through pore formation, a phenomenon known as electroporation or electropermeabilization. This technique shows great potential for the selective recovery of target intracellular compounds. One of the primary reasons why purer extracts can be obtained with PEF is its selective effect on the cytoplasmic membrane, allowing the targeted discharge of intracellular components without disrupting the overall cell structure. As a result, the need for additional purification steps is reduced. For this reason, PEF is applied to plant tissues as a pretreatment that facilitates extraction [106].

PEF technology is a nonthermal, minimally invasive, and environmentally friendly technique. Due to its ability to enhance mass transfer of intracellular components through electroporation, PEF has found its place in plant extraction processes. The effects of PEF on cells are illustrated in Figure 2.13. The electric field applied disrupts the cell's lipid bilayer membrane, alters its permeability, and facilitates contact between the solvent and target compounds. This results in an increase in extraction efficiency. Consequently, it reduces the solvent temperature and concentration required for extraction. Lower temperatures help preserve the structural and bioactive properties of heat-sensitive compounds during extraction [92].

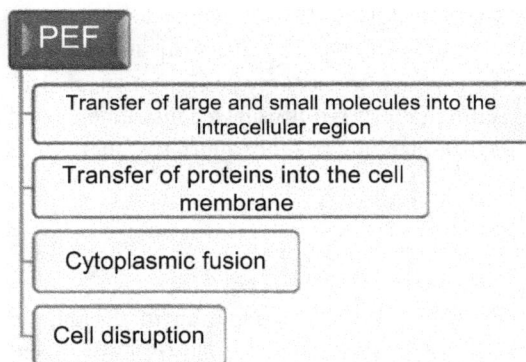

Figure 2.13: Effects of PEF on cells.

The efficiency of electroporation is generally improved by increasing crucial factors like the strength of the electric field, treatment duration, and application temperature. Additionally, the ease, speed, and scalability of adapting PEF to industrial tools make it a versatile technology for integration with other methods. However, PEF parameters must be tailored to each species, considering their structures, sizes, and other factors that influence extraction efficiency [107].

2.3.3.11 Microwave-assisted extraction

Microwave-assisted extraction is a technique in which the sample is extracted by applying microwave energy in an appropriate solvent. In this method, high-frequency microwave energy as a type of electromagnetic waves with wavelengths ranging from 1 mm to 30 cm is used. Microwave energy heats the sample from the inside out, providing simultaneous and homogeneous heating. Since microwave energy accelerates heating, it also speeds up the extraction process and allows the use of less solvent. This type of extraction enables rapid and effective extraction of the materials. Key parameters impacting extraction efficiency include the choice of solvent, operating temperature, microwave power, exposure time, as well as the properties of the plant material, its matrix, and particle size [108].

The basic principle of the microwave-assisted extraction method involves heating intracellular water, leading to the breakdown of plant cells and allowing the solvent to penetrate the plant matrix, resulting in the transfer of components into the solvent. Microwaves disrupt hydrogen bonding in organic molecules and induce dipole rotation. This causes ions with increased kinetic energy to continuously move and change direction. The disruption of hydrogen bonds also enhances the ability of solvents to penetrate the plant matrix [109].

Microwave radiation's most significant characteristic is its interaction exclusively with the dipoles of polar or polarizable substances (solvents and samples). The heat generated through microwaves is transferred via conduction on the surface of these materials. Since energy transfer occurs solely through dielectric absorption, nonpolar liquids exhibit very weak heating. Microwave-assisted extraction (MAE) is selectively applied using solvents with high dielectric constants and polar substances. MAE is suitable for certain secondary metabolites, such as phenolic acids and some flavonoids, but not for those sensitive to thermal degradation, such as anthocyanins and certain tannins [110]. In a closed MAE system, it is possible to reach temperatures 2–3 times the boiling points of certain solvents (such as acetone, acetone-hexane, dichloromethane-acetone). This significantly increases the extraction efficiency of components from the plant matrix. Solvents like water, methanol, and ethanol have high microwave absorption capacities and can rapidly increase in temperature, thereby reducing the processing time [93].

2.3.3.12 Supercritical extraction

Supercritical fluid extraction (SFE) is an environmentally friendly extraction technique. Its key feature is the use of supercritical fluids as solvents. These fluids operate above their critical temperature and pressure, showing physicochemical features that exhibit a balance between gas-like and liquid-like behaviors (Figure 2.14) [111]. The low viscosity and high diffusivity of supercritical fluids allow the solvent to penetrate the material more effectively, significantly reducing extraction time. Among supercritical fluids, carbon dioxide (CO_2) is the most widely used solvent as an excellent alternative to organic solvents. CO_2 is an efficient solvent for an extensive variety of components, and its inert, nontoxic, safe, and recyclable nature makes it ideal for SFE. Moreover, its low cost and easy availability further enhance its appeal as a solvent for this method. SFE can be performed at low temperatures, preserving the biological activity of heat-sensitive compounds. The ability to reuse the solvent minimizes waste generation, contributing to economic and environmental sustainability [94]. Some of the solvents that can be utilized as supercritical fluids, apart from carbon dioxide, include hydrocarbons such as pentane, butane, and hexane; aromatic solvents like benzene and toluene; alcohols (methanol, ethanol, isopropanol, n-butyl alcohol); and gases such as ethylene and propane and water [112].

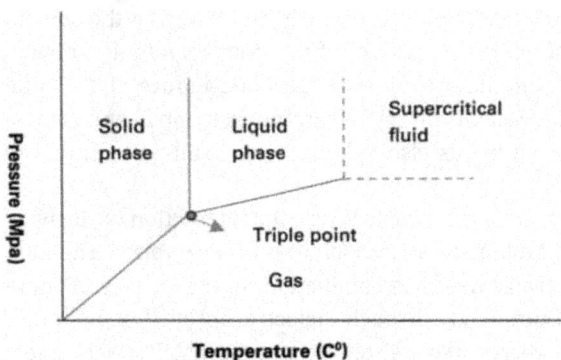

Figure 2.14: The supercritical fluid region.

The main drawback of supercritical CO_2 is its low polarity, which limits its use. Nonpolar or moderately polar substances are easily dissolved by CO_2, a nonpolar molecule. As a result, SFE is primarily used for the extraction of nonpolar or moderately polar compounds such as lipids, essential oils, and carotenoids. To overcome this limitation, CO_2 can be combined with polar organic solvents (modifiers) for the extraction of polar compounds. Small amounts of cosolvents such as ethanol or water can be used to enhance the solubility of polar compounds. These cosolvents, being more polar than CO_2, increase the polarity of the supercritical mixture, significantly improving extraction efficiency [113].

There are three approaches to the implementation of supercritical extraction: static, dynamic, and a combination of both modes. In the static mode, the supercritical solvent is allowed to contact the plant matrix for a specific period. In the dynamic SFE mode, fresh supercritical solvent is continuously introduced over the plant material. Consequently, the flow rate of the supercritical fluid is directly proportional to extraction efficiency. When both modes are combined, the process begins with static extraction for certain duration, followed by a transition to the dynamic mode. This combination positively influences extraction efficiency [95].

SFE allows for shorter operation times, versatile applications, and safer, greener experiments through the use of a low-viscosity fluid for the extraction of plant-based components. Another key advantage of SFE is its ability to establish a direct online connection with a chromatographic method (e.g., GC, HPLC), enabling the immediate measurement of components after extraction [114].

2.3.3.13 Pressurized liquid extraction

Pressurized liquid extraction (PLE), first introduced in 1995 as accelerated solvent extraction (ASE), is also referred to by several other names, including pressurized solvent extraction, enhanced solvent extraction, and superheated liquid extraction (SHLE). When water is used as the extraction solvent, the technique is known as pressurized hot water extraction (PHWE) [97].

PLE is an automated and rapid extraction method that utilizes liquid solvents at high temperatures and pressures, enabling more efficient extraction of components from solid and semisolid plant matrices. Increasing temperature and pressure significantly enhances the extraction performance compared to traditional methods [115]. In this technique, solvents are maintained in a liquid state above their boiling points under high pressure. Using solvents at temperatures above their atmospheric boiling points provides several advantages, including improved solubility, enhanced diffusion, and better mass transfer mechanisms, which facilitate the extraction of target plant components. Additionally, under high temperatures and pressures, the viscosity and surface tension of solvents are reduced, allowing for deeper penetration into the solid matrix. This accelerates solvent penetration and the overall extraction process, thereby increasing the efficiency of bioactive compound extraction from plant materials [116].

PLE can generally be performed in static mode, dynamic mode, or a combination of both modes. In the dynamic mode of the extraction procedure, fresh solvent is continuously pumped through the sample, resulting in a constant shift in equilibrium and an increase in the mass transfer rate. However, its main disadvantages include the requirement for larger solvent volumes and the necessity of a concentration step to dilute the components in the extract before chromatographic analysis of the target compounds. Extraction efficiency in dynamic mode is equal to or higher than in static mode, and the extraction time is generally similar in both modes. However, incorporating a pre-extraction step in static mode before transitioning to dynamic mode can reduce the overall extraction time [117].

Widely used for years, PLE provides numerous benefits compared to traditional extraction methods, such as faster extraction times, lower solvent usage, reduced costs, and the ability to scale the extraction process easily to industrial levels [118].

2.3.3.14 Enzyme-assisted extraction

The fundamental principle of enzyme-assisted extraction is the catalytic action of enzymes to hydrolyze and break down plant cell walls. This process enables the intracellular components to be released under optimal experimental conditions. Initially, the plant cell wall binds to the active site of the enzyme. The substrate-enzyme interaction between the cell wall and the enzyme induces a conformational change in the enzyme upon binding. This structural alteration in the enzyme leads to the breakdown of the bonds within the cell wall, releasing the active components from the plant cells. This method significantly preserves the bioactive potential of the extracted compounds [98].

Enzyme-assisted extraction (EAE) is particularly necessary for phytochemicals bound within the lignin-polysaccharide network of certain plants. These phytochemicals are stabilized by hydrophobic interactions, such as hydrogen bonds and van der Waals forces, making their separation highly challenging. In such cases, phytochemicals are often dispersed in the cytoplasm and cannot be extracted through standard solvent-based methods. To address this issue, specific enzymes are employed. These enzymes hydrolyze structures like cellulose and lipids, facilitating the release of bound phytochemicals. Enzyme treatment is employed as a pretreatment to degrade cell walls, enhancing the extraction efficiency of phytochemicals [76].

EAE is performed using two primary approaches: enzyme-assisted aqueous extraction (EAAE) and enzyme-assisted cold pressing (EACP). EAAE is typically applied for the extraction of oils and other lipophilic components from seeds, while EACP is commonly used to hydrolyze seed cell walls with enzymes, thereby enhancing extraction efficiency [99]. The parameters affecting the EAE procedure are listed in Figure 2.15 [99, 119].

2.3.3.15 Solid-phase microextraction

Solid-phase extraction (SPE) operates on a principle similar to liquid-liquid extraction (LLE), involving the partitioning of dissolved substances between two phases. However, in SPE, one phase is a liquid while the other is a solid (sorbent). The stationary phases used in solid-phase extractions are the same type as those used in liquid chromatography columns. The stationary phase is housed in a glass or plastic column. Commercial SPE cartridges are designed in the form of injectors, are single-use, and have a capacity of approximately 1–10 mL. Solid-phase extraction is often used as a sample preparation step to clean the sample before performing chromatographic or other analytical methods for determining the quantities of components in the sample [120]. Solid-phase extraction was initially employed as a purification method before HPLC or GC analysis. However, its application has expanded and it is now commonly used for the rapid fractionation of crude plant extracts or for transferring purified

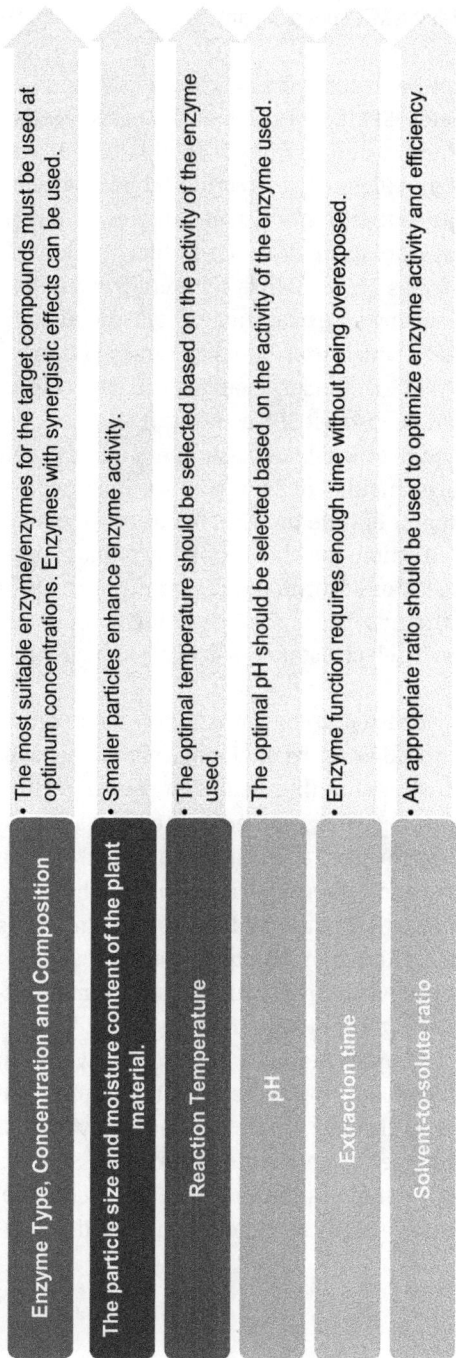

Enzyme Type, Concentration and Composition
- The most suitable enzyme/enzymes for the target compounds must be used at optimum concentrations. Enzymes with synergistic effects can be used.

The particle size and moisture content of the plant material.
- Smaller particles enhance enzyme activity.

Reaction Temperature
- The optimal temperature should be selected based on the activity of the enzyme used.

pH
- The optimal pH should be selected based on the activity of the enzyme used.

Extraction time
- Enzyme function requires enough time without being overexposed.

Solvent-to-solute ratio
- An appropriate ratio should be used to optimize enzyme activity and efficiency.

Figure 2.15: Key parameters in enzyme-assisted extraction (EAE).

compounds from HPLC separations to capillary NMR spectroscopy for structural identification [121].

Solid-phase microextraction (SPME) can be described as a modified version of SPE and is widely applied across various fields. SPME is a robust solid-based extraction technique developed in the 1990s by Authur and Pawliszyn [122]. SPME differs from solid-phase extraction in several ways. In SPE, analytes require liquid-phase extraction after sorption. SPE is limited to the extraction of liquid samples, whereas SPME is a technique that facilitates chromatographic analyses of solutions from challenging matrices in either liquid or gaseous states [123]. SPME is a solvent-free method aimed at increasing sensitivity by directly injecting all extracted analytes into analytical instruments, in contrast to traditional SPE, which uses larger volumes of extraction phases. An important advantage of the SPME method is its ability to avoid macromolecules and particulate organic materials that obstruct SPE columns [124].

The target analytes, which can be gas, liquid, or solid, are desorbent and analyzed when a sample is frequently exposed to trace amounts of an extractant immobilized on a solid substrate for a predetermined amount of time in SPME. To increase the extraction efficiency and selectivity of microextraction, the choice of adsorption materials is important [125]. The amounts and speeds of adsorption are significantly influenced by the analytes' interactions with the adsorbent surface, which can occur through hydrogen bonding, π-π, dipole-dipole, electrostatic, or hydrophobic/hydrophilic interactions [126].

In this technique, the substances to be extracted are adsorbed onto high boiling point polymers coated on the surface of a silica fiber as the stationary phase. The substances absorbed or adsorbed by the fiber coatings are thermally desorbed in a chromatography injection port after the extraction device is exposed to the head space of the sample or sample solution. Because it only involves a few steps and small sample sizes, the fiber SPME method is simple to use and offers superior cleanup. The partition equilibrium of analytes between the extraction phase and the sample matrix is the foundation of SPME, which produces quantitative or semiquantitative results. In recent years, different analytical tools have been combined with alternative microextraction devices to develop various SPME-related technologies [100].

The integration of SPME with chromatography methods consists of two main steps. The first involves the absorption of dissolved substances from the sample matrix onto the adsorbent, while the second step transfers the absorbed analytes to a chromatography inlet system through either thermal or liquid desorption. SPME is gaining increasing attention as a green and versatile sample preparation technique. Combining SPME with an automated sampler significantly enhances the speed and efficiency of the process [127].

2.3.3.16 Bioassay-guided fractionation of plant extracts

Bioassay-guided fractionation involves testing an extract for activity, separating it chemically, and then testing the resulting fractions for activity. This procedure is iterative; until one or more active molecules are isolated, the most active fraction can be separated and its fractions examined again. With bioassay-guided fractionation, drug interaction analysis is done in reverse; rather than making and testing a mixture, a naturally occurring mixture is separated and tested in order to identify any interactions that may be present. Data gathered from bioassay-guided fractionation can be utilized to measure synergy in bioactive extracts without the need for further experiments and help guide go/no-go choices by treating natural substance extracts as combinations of their fractions [128]. The general steps for bioassay-guided fractionation and the identification of bioactive chemicals are shown in Figure 2.16 [129, 130].

When biologically active molecules are isolated using chromatographic separation techniques combined with bioassay-guided fractionation of plant extracts, the fact that subfractions obtained through fractionation exhibit more drug-like properties compared to crude extracts can lead to a focus on more promising compounds for drug discovery. Performing chemical analysis after the active fraction is isolated also accelerates the process. Fractionation strategies focused on biological activity instead of a specific group of components have gained importance in drug development processes involving medicinal and aromatic plants [131].

Common challenges encountered in fractionation under bioanalysis guidance include the potential loss of bioactive biological activity during the fractionation process or failure resulting from the procedure. Key reasons for this include the degradation of bioactive components during the process, the presence of components at very low concentrations, and the bioactivity being due to the synergistic effects of multiple components. To avoid these issues, it is crucial to identify the target bioactive compounds early in the purification procedure. However, in recent years, the re-isolation of previously identified plant molecules as a result of fractionation under biological assay guidance has been encountered. To prevent this, a pre-evaluation step known as "dereplication" is used [132].

2.3.4 Isolation and purification

Modern research requires the isolation of individual components from plant extracts and their evaluation as potential drugs, as opposed to the traditional medicine, which uses whole plant extracts for treatment. Both approaches – using whole extracts and purifying individual components – have their own advantages and disadvantages. In some cases, it has been observed that the herbal extract obtained from the whole plan tor specific parts of the plant, or a mixture of different plant extracts without isolating the herbal components provides better therapeutic efficacy. In many cases,

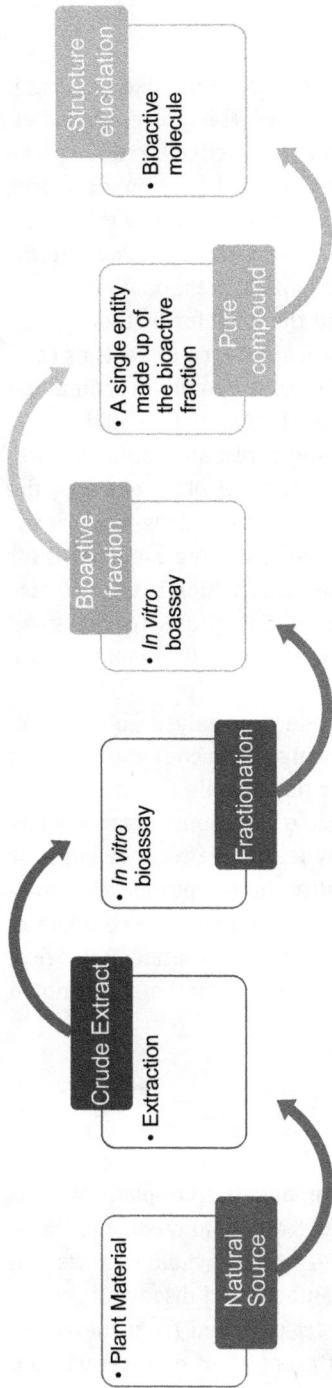

Figure 2.16: The general steps for bioassay-guided fractionation of natural sources.

the isolation of the "active compound" has resulted in the compound becoming completely inactive. The generally better therapeutic effects observed when using whole extracts instead of isolated compounds without any purification steps can be attributed to the potential synergistic effects of the active components in plants or the specific properties of the matrix in which they naturally occur in the plant [57].

A lead compound (bioactive pure compound), if found within a mixture of other compounds derived from a plant source and is intended to be developed as a stand-alone drug, must be isolated and purified. The process of isolating and purifying bioactive compounds from plants is a highly challenging and complex endeavor. The physical or chemical differences between each particular natural product determine the separation. The ease of isolation and purification is tightly associated with the structure stability and concentration of the compound within the material [133].

The separation and isolation of components from a plant extract is typically performed using chromatographic techniques, followed by the characterization of the isolated compounds, which is generally carried out using spectroscopic techniques [134]. Within this framework, chromatographic separation methods are used iteratively to produce fractions enriched with specific components or groups of components, ultimately leading to the isolation of single compounds. Spectroscopic methods allow the analysis of extracts, fractions, or single compounds and provide insights into the chemical character and structural properties of the compounds [135].

In recent years, significant advancements have been made in the field of natural compound isolation. New methods combining chromatographic and spectroscopic or spectrometric techniques aim to elucidate the structures of known or novel compounds without requiring isolation. There has been an increasing trend toward isolation techniques based on pharmacological or biological activity. Bioassay-guided isolation strategies enable the correlation of chemical profiles of extracts and fractions with activity data from micro-scale in vitro biological assays. This approach has significantly shortened the time required to identify bioactive compounds [136].

Chromatographic techniques are among the most important bioanalytical methods used in the analysis of natural product sources. By utilizing these methods, different and complex components in a complex plant extract can be separated, identified, and purified both qualitatively and quantitatively [137]. This section will focus on chromatographic techniques commonly used in isolation and purification.

2.3.4.1 Thin-layer chromatography (TLC) and high-performance thin-layer chromatography (HPTLC)

Thin-layer chromatography (TLC) is a chromatographic technique that is simple to prepare and apply, versatile, sensitive, and highly efficient. It is generally used to separate nonvolatile compounds from plant materials [138].

TLC is a chromatographic technique that utilizes the adsorption mechanism to separate a component from a mixture. Separation, as in all chromatographic methods, relies on the interaction between the compounds in the mixture and the stationary phase. The basic principle of the method involves a multistep distribution process that includes the target compounds, solvents, or mixtures of solvents (mobile phase or eluent), and the stationary phase (adsorbent). It is applicable for the separation of low molecular weight compounds. The stationary phase typically consists of materials such as silica gel, sephadex, aluminum oxides, or cellulose, cast at an appropriate thickness onto glass, plastic, or aluminum plates. The compounds in the extract migrate on the plate based on their solubility in the mobile phase. Each separated compound is identified by calculating the retention factor (Rf), which is the ratio of the distance traveled by the compound to the distance traveled by the mobile phase, and is then compared with known compounds. The method's key advantages include its time efficiency and stability against acidic solvents [70].

(TLC is among the earliest developed chromatographic techniques. However, with the development of devices, automation, and the advancement of new adsorbents and supports, it is still widely used today. HPTLC is an advanced form of TLC that uses higher-performance adsorbents. The HPTLC technique is a standardized method that can be used for the qualitative and quantitative analysis of components in plant samples. At the same time, HPTLC is a technique recognized by the European Pharmacopoeia and widely used for quality control and analysis of herbal medicines and their preparations in compliance with GMP standards. The TLC method is a versatile, sensitive, and high-efficiency technique with simple sample preparation and application. With TLC/HPTLC, chromatographic fingerprint analyses can be performed by quickly comparing a range of plant components with references [46]. Compared to TLC, it offers higher accuracy and reproducibility due to improved separation efficiency and detection limits. The use of high-resolution sorbents with specific particle sizes and chemically modified plates, combined with compatibility with various instruments and detectors, enables rapid quantification of phytochemicals and chromatographic fingerprint analyses [138–140].

2.3.4.2 High-performance liquid chromatography (HPLC) and ultra-performance liquid chromatography (UHPLC)

Liquid chromatography is widely regarded as the most popular method for herbal fingerprinting due to its numerous advantages, including broad applicability, high resolution, excellent selectivity, sensitivity, reproducibility, and the capability for full automation [46].

High-performance liquid chromatography or high-pressure liquid chromatography (HPLC) is a modern, powerful, and versatile chromatographic separation technique routinely used to separate, identify, and quantify components from complex mixtures. HPLC analysis serves as a cornerstone of phytochemical studies, especially in the characterization and isolation of components from crude plant extracts and in obtaining their chemical profiles or fingerprints [141].

HPLC operates based on chromatographic separation principles, where analytes are separated according to their differential interactions with the stationary phase and the mobile phase. The separation mechanism in HPLC relies on the differences in the affinities of the compounds to the stationary and mobile phases. The analytes eluting from the HPLC column are detected by various detectors, and their signals are recorded by a data system [142]. Typically, the stationary phase inside the column engages with the molecules of interest through mechanisms dependent on the type of separation method employed. A liquid solvent or a mixture of solvents serves as the mobile phase, facilitating the movement of analytes through the column. The differences in interactions between analytes and the stationary phase result in varying retention times, enabling the separation of components in the mixture. HPLC is particularly suitable for the analysis of nonvolatile and thermally unstable plant metabolites. Parameters such as the type and properties of the solvent, column temperature, and flow rate significantly influence the separation efficiency of HPLC. By controlling these parameters, it is possible to achieve high-resolution and efficient separation of complex extracts [143].

Ultra-high-performance liquid chromatography (UHPLC) systems are advanced techniques that operate at significantly higher pressures compared to HPLC and utilize packing columns with particles smaller than 2 μm. This method is widely applied in various areas of plant analysis, including the chemical profiling of plant components, fingerprinting, dereplication, and metabolomics. Beyond identifying compounds, one of the fundamental applications of both HPLC and UHPLC is the dereplication process, which involves recognizing known metabolites in extracts and is conducted during the early stages of the fractionation process to expedite analysis [144]. Compared to HPLC, UHPLC offers numerous advantages, including operating at high flow rates, significantly reducing analysis time, providing highly efficient separation with excellent reproducibility, enhanced sensitivity, and lower solvent consumption than other analytical methods. This method provides fast and sophisticated chromatographic separation with reduced analysis time, while also ensuring exceptional precision and selectivity, which helps in the precise and dependable identification of compound structures across various samples [145].

2.3.4.3 Gas chromatography (GC)

Gas chromatography (GC) is a method capable of performing both qualitative and quantitative determination of target analytes. It is a chromatographic technique in which the mobile phase is a gas and the stationary phase is a liquid. In this method, the chromatography column contains a liquid stationary phase adsorbed onto the surface of an inert solid. The migration rate of the compounds to be analyzed relies on their distribution within the gas phase. In GC, volatile or volatilizable compounds are vaporized and injected into the chromatographic column, where they are carried along the column by the flow of the gaseous mobile phase and detected using various detectors [146].

Gas chromatography/mass spectrometry (GC/MS) is considered the gold standard for comprehensive qualitative and quantitative analysis of volatile organic compounds found in natural products [147]. In GC-MS analysis, compounds are first injected into the gas chromatograph, where they are separated based on their volatility. The separated compounds then enter the mass spectrometer, where they are bombarded with electrons and fragmented into ions. These ions are detected by the system's detectors, allowing for the analysis to be performed [148]. Given that substances suitable for evaluation by GC – characterized by low molecular weight, medium or low polarity, and concentrations in the ppb-ppm range – also meet the requirements for mass spectrometry (MS), the combination of GC and MS forms a highly advantageous and synergistic method. Furthermore, both analytical processes occur in the same aggregation state, which is the vapor phase [149].

2.3.4.4 Column chromatography (CC)

Column chromatography (CC) is a technique used for the isolation of bioactive compounds identified in plants and for the separation of metabolites in various plant extracts. Additionally, it is a preferred method for the removal of impurities and purification of biological mixtures. This method can be utilized to separate and purify both solid and liquid samples. The basic principle of CC is based on the separation of compounds by adsorption onto a stationary phase placed inside a narrow column, with the help of a liquid mobile phase. Compounds are adsorbed by the stationary phase to varying degrees depending on their chemical structural properties, and elution occurs in this manner [150].

CC, based on the principle of adsorption, is commonly used in the initial separation stage of plant extracts. The main reasons for its widespread use include the simplicity of the technique, the high capacity of the process, and the low cost of adsorbents like silica gel and macroporous resins. Since the separation process primarily depends on the adsorption affinities of natural compounds to the surface of the adsorbents, it is crucial to carefully select the adsorbent (stationary phase) and the mobile phase to ensure efficient separation, high recovery of target compounds, and to prevent irreversible adsorption of target compounds onto the adsorbents [151]. The use of silica gel as an adsorbent is suitable for the separation of most phytochemical compounds. Alumina can be used for alkaloids, steroids, and terpenoids, which are alkaline or neutral lipophilic components. Activated carbon is suitable for hydrophilic components such as amino acids, carbohydrates, and some glycosides. Polyamide, based on the formation of various hydrogen bonds, can primarily be used for the separation of phenols, quinones, flavonoids, anthraquinones, tannins, and others [152].

2.3.4.5 Ion exchange chromatography (IEC)

Ion exchange chromatography (IEC) is a widely used fractionation method that enables the separation of ions and ionizable molecules based on differences in their electrostatic properties [153]. Among all LC techniques, it is one of the most widely used and versatile due to its large sample-handling capacity, broad applicability (especially to proteins and enzymes), moderate cost, powerful resolving ability, ability to perform simultaneous quantification, and ease of scalability and automation [154, 155].

IEC can be applied in both solid-gas and solid-liquid systems. Ion exchangers are used as the stationary phase in ion chromatography. Different types of ion exchangers are utilized based on their polarity, chemical and physical resistance, particle size distribution, internal and specific surface area, density, porosity, and pore radius distribution [156]. Ion-exchange resin might capture and release the charged molecules by altering the mobile phase's ionic strength (e.g., changing pH or salt concentration) [157]. The type of stationary phase, detection method, and eluent type are the most important factors affecting the separation quality [158].

2.3.5 Elucidation of the chemical structure

The process of identifying and characterizing phytochemicals is still greatly challenged by the fact that plant extracts typically consist of a mixture of several bioactive compounds with varying polarities [159]. The complex chemical composition of herbal samples is represented by distinctive profiles and patterns, called fingerprints, which can be developed through multiple approaches, such as chromatographic and spectroscopic techniques [46].

The structure of a purified active compound obtained through extraction and isolation methods can be determined using various spectroscopic techniques. Nuclear magnetic resonance (NMR) spectroscopy is frequently employed for the structural determination of natural products, especially for unknown compounds, as it offers significant advantages. Materials analyzed by NMR can be recovered after analysis. Other commonly used structural elucidation methods include mass spectrometry (MS) for determining molecular weight and infrared (IR) spectroscopy for identifying functional groups [160].

Combined instrumental analysis methods are used to profile the structural composition of the numerous and complex secondary metabolites found in plants. In the study of the effects of phytochemical compounds and the quality control of herbal medicines, "hyphenated techniques," which integrate sensitive and rapid analytical methods with online spectroscopic techniques to simultaneously provide both structural and activity information, are widely used. These methods are highly successful in the rapid online identification of known components, preventing dereplication, and ensuring the standardization or quality control of a complex extract [161].

2.3.5.1 Nuclear magnetic resonance (NMR)

NMR spectroscopy is a technique applicable for identifying target bioactive metabolites from complex plant extracts. The primary advantages of NMR spectrometers include their ability to perform measurements without requiring any prior sample preparation or preprocessing. Moreover, its noninvasive nature, rapid operation, and high sensitivity make it a highly preferred method. However, due to the significant equipment costs, NMR is predominantly used for the structural elucidation of previously uncharacterized compounds rather than known ones [162].

The sample preparation for NMR is quite simple, as it does not require detailed pretreatment or fractionation. It is also a highly reproducible method. NMR provides quantitative and detailed information about the structure of metabolites. However, its main disadvantages are its low sensitivity and the fact that it generally profiles only the major components. Additionally, NMR is not very useful for detecting trace components, as it can only detect compounds at concentrations as low as 0.1% [163].

The NMR technique has never lost its importance as it is used not only for elucidating chemical structures but also for structural studies of biomolecules in three dimensions, identifying reaction mechanisms, and ligand binding screening in drug discovery [164].

2.3.5.2 Mass spectrometry (MS) and high-resolution mass spectrometry (HRMS)

Mass spectrometry (MS) is a spectroscopic technique that generates ions from atoms or molecules in the gas phase and measures their mass-to-charge (m/z) ratios. Mass spectrometers differentiate ions with different mass-to-charge ratios using static, pulsed, or periodically changing electric and/or magnetic fields. The main applications of mass spectrometers include determining molecular mass, elemental and isotopic compositions, structural elucidation, and quantification [165].

The most prominent separation techniques commonly combined with mass spectrometry (MS) are HPLC, GC, and capillary electrophoresis (CE). GC-MS is the most frequently employed technique among these for the separation and analysis of mixtures that contain volatile organic compounds or those that can be made volatile, along with thermally stable components. GC-MS is an inexpensive and highly sensitive method. However, its applicability is limited compared to other methods, as it is only suitable for the analysis of volatile compounds and relatively lower molecular weight components [166]. The combination of MS with GC is a suitable method for the structural determination of target components present in small quantities within complex mixtures. When necessary, the derivatization of target compounds and the GC-MS analysis of the resulting derivatives can provide valuable information about the parent compound. Another important consideration is that the removal of oxygen-containing functional groups prior to analysis may be beneficial [167].

In phytochemical analyses, the main applications of MS-MS spectra include the partial determination of sugar sequences in glycosides, the detection of characteristic losses, such as in prenylated compounds, the fragmentation of flavonoids to determine the positions of substituents on the A or B rings, and the differentiation of isomers. However, for the structural elucidation of completely unknown compounds, MS-MS alone may be insufficient; therefore, it must be combined with other complementary methods [168].

High-resolution mass spectrometry (HRMS) can be used for both qualitative and quantitative determination of metabolite profiles [169]. HRMS analysis is generally characterized by high selectivity. Among mass spectrometers, HRMS has features such as high mass precision, superior resolution, fast scanning capabilities, and excellent sensitivity [145]. Since HRMS-based analytical techniques offer fast and precise solutions for the characterization of secondary metabolites found in highly complex matrices and the detailed elucidation of their structures, they have become widely preferred techniques today [170].

The presence of a wide variety of isomers and their derivatives in plant extracts complicates the overall plant matrix, leading to the co-elution of isomeric species with the same mass. This makes it difficult to accurately identify and quantify the targeted components. Traditional precise techniques such as LC-MS and HPLC-MS also face these challenges. Some of these limitations can be overcome by using multistage analyzers (MS/MS) or at least one high-resolution, specially equipped instrument. In MS/MS or HRMS analyses, narrowing the mass range analyzed reduces background signal noise from interfering ions detected by the detector, thus increasing sensitivity [171]. The UHPLC-HRMS technique combines the high sensitivity and excellent separation capability of UHPLC with HRMS. UHPLC-HRMS is an advanced analytical technique used for the separation and identification of components in complex mixtures [166].

2.3.5.3 Infrared spectroscopy (IR) and Fourier transform infrared spectroscopy (FTIR)

Infrared (IR) spectroscopy is a technique that provides information about the presence of functional groups such as hydroxyl, primary and secondary amines, carbonyl groups, alkenes, and arenes. The fingerprint region of the infrared spectrum is particularly useful for identifying and distinguishing substances due to its unique and highly distinctive absorption patterns [172].

For the characterization of chemicals or chemical bonds (functional groups) contained in an unknown combination of plant extract, Fourier Transform Infrared Spectroscopy (FTIR) has proven to be a very helpful approach. Furthermore, FTIR spectra of pure substances are typically sufficiently distinct that they resemble a chemical "fingerprint." For the majority of common plant chemicals, comparing an unknown compound's spectrum to a library of known compounds can help identify them [159]. For FTIR analysis, both liquid and solid materials can be used. A drop of liquid sample can be placed between the plates to form a thin film. Solid materials, on the other hand, can be milled with potassium bromide and pressed into a thin pellet for analysis [173].

2.3.5.4 UV-visible spectroscopy

UV-visible spectroscopy is a fast and easy-to-apply analytical technique based on measuring the absorption or transmittance of light. The wavelength range in which UV-visible spectroscopy operates is between 200 nm and 800 nm. Ultraviolet-visible (UV-Vis) spectrophotometers primarily consist of a light source that passes light through a sample and a detector on the opposite side that records the transmitted light. Covalently bonded unsaturated compounds capable of absorbing light at specific wavelengths known as chromophores, exhibit electronic transition energy differences that match the energy of UV-visible light. Covalently bonded saturated groups that influence the absorption of chromophores but do not themselves absorb UV-Vis electromagnetic radiation are known as auxochromes. When UV-Vis radiation strikes chromophores, ground-state electrons are excited to higher energy states. Auxochromes act as electron donors and, while they do not change color themselves, they influence the color of chromophores. Since water and alcohols do not absorb light in the UV-Vis range, they are transparent and therefore serve as suitable media for UV-visible spectroscopy. Chromophores have characteristic absorption bands; however, changes such as the addition of another compound to the medium or an increase in temperature can alter energy levels and, consequently, absorption intensity [174].

UV-Vis spectroscopy is a versatile and powerful analytical technique. One of its key advantages is its ability to simultaneously measure the electronic transitions of organic molecules (primarily through $n \to \pi^*$ and $\pi \to \pi^*$ transitions) and transition metal oxides or ions (via $d\text{-}d$ and charge transfer transitions). However, absorption bands in the UV-Vis range are often broad and may overlap, making the interpretation of results occasionally challenging. Combining UV-Vis spectroscopy with other analytical methods provides more comprehensive insights. Examples of such complementary techniques include X-ray absorption spectroscopy and diffraction, vibrational spectroscopy, and magnetic resonance [175].

UV-visible spectroscopy is a widely used method for the qualitative and quantitative analysis of phytochemicals, elucidation of the structure of plant components, and examination of the potency, quality, and purity of compounds due to its suitability for small-scale studies, simplicity, and low equipment cost [176].

2.3.6 Evaluation of therapeutic efficacy with bioassays

The physiological activities exhibited by natural products stem from their capacity to engage with different cellular targets [177]. In the process of developing drugs from plant-based sources, it is essential to thoroughly evaluate these interactions. Biological activity studies focus on determining both the therapeutic effects and the potency of these effects. To conduct an appropriate biological experiment, steps such as literature review, evaluation of the stability of the sample product or compound, performance testing in biosystems, and determination of 50% effective concentration (EC50) or cytotoxic concentrations (IC50) must be carried out [178].

Testing biological activity after purification ensures more reliable results by eliminating the matrix surrounding the compound in the extract and removing unwanted interactions [136]. Additionally, bioactivity assays can be conducted at multiple stages of the drug development process. In the preliminary screening phase, plant samples, extracts, various fractions of the extracts, or even libraries of pure compounds may undergo biological analyses to detect bioactivity potential. Conducting these analyses in a HTS format can increase efficiency while reducing costs and time. In later stages, these tests can assist fractionation processes, such as purification or bioactivity-guided approaches, in isolating and identifying bioactive compounds [179].

Conducting bioactivity assays in a stepwise manner is an appropriate approach. The first phase emphasizes high capacity and low cost, and if a positive effect is detected in this screening, the process advances to the second phase of biological activity testing, which is more precise and accurate. In the second phase, pure compounds with potential as drug candidates should be investigated under additional models and test conditions to select candidates for clinical trials. Biological experiments can be classified based on the target used. These primarily include lower organisms (bacteria, fungi, insects, lower plants, etc.), live cells under tissue culture conditions (e.g., cancer cells), animal or human-derived tissues, animals, isolated organs of vertebrates, and isolated subcellular systems (e.g., enzymes, receptors, etc.) [180].

Biological analyses in the drug development process are generally classified into in vitro and in vivo experiments conducted prior to preclinical and clinical research. For drug discovery from medicinal and aromatic plants, it is essential to select appropriate biological assays to evaluate the activity and potency of bioactive compounds (hits or leads) against the relevant disease. In vitro research focuses on primary activity, specificity, cellular toxicity, and physiologically significant activity of compounds using various assays [181]. These analyses include:
– Comparative screening
– Interaction studies

- Bioactivity-guided fractionation
- Biological characterization
- Stability studies
- Investigation of mechanisms of action

While in vitro researches are indispensable for the primary evaluation of natural products, they are limited by the absence of pharmacokinetic data, lack of direct compatibility with in vivo/clinical doses, and the narrow scope of physiological mechanisms represented in the assay systems [182].

In vivo bioassays aim to evaluate the complete effects of bioactive compounds in disease models and assess toxicity and safety on a cellular or organismal basis. These experiments involve the use of animal models, isolated systems (e.g., enzymes, receptors), or isolated organ preparations to investigate biological activity, toxicity, pharmacokinetics, and pharmacodynamics. However, ethical concerns regarding animal use and the physiological and metabolic differences between animals and humans can limit the applicability of the data obtained from these studies [183].

In vitro and in vivo tests can be used to conduct toxicity studies and perform toxicological evaluations of compounds. While in vitro studies are often sufficient to examine direct effects on cell proliferation and phenotype, in vivo studies provide more detailed qualitative and quantitative assessments of toxicological effects. Since the effects and toxicity of many drugs vary by species, selecting appropriate animal models for toxicity studies is crucial. Almost all in vivo studies evaluating pharmacological and toxicological effects, including the mode of action, are conducted to generate fundamental data for the proposed use of the product in subsequent clinical trials [184].

2.3.7 Preclinical and clinical researches

The discovery of a bioactive compound is the first step toward its development as a drug. The subsequent steps involve preclinical and clinical studies. A plant-derived molecule discovered through in vitro and in silico experiments demonstrating bioactivity must be designed according to the characteristics of the target cell or mechanism in the body to exhibit the desired effect in vivo [36].

Drug discovery from medicinal and aromatic plants is essentially a stepwise optimization process of a pharmacologically active lead compound. For this, it is first necessary to investigate the bioactivity of the plant-derived molecule. However, this is not sufficient for the molecule to be clinically applicable. Although the therapeutic effects of a medicinal plant or its components have been demonstrated through bioactivity tests, the molecule cannot be converted into a drug if there are deficiencies in preclinical and clinical research. Especially in vivo studies are needed to determine parameters such as effect profile, bioavailability, side effects, and toxicity. Numerous studies are conducted daily on the pharmacological activities of plant extracts and isolated plant components. However, most of these studies have not resulted in drug production due to the lack of clinical trials.

The purpose of preclinical studies in drug discovery is to present one or more clinical candidate molecules that are effective, safe, and possess drug-like properties, supported by sufficient data demonstrating biological activity at a disease-related target. These molecules are then selected for further clinical trials. Drug discovery programs generally aim to generate data through the collaboration of chemistry, biology, toxicology, and pharmacology, evaluating compounds at different doses and across multiple experiments. Theoretically, discovering a bioactive molecule is of no practical use if the molecule cannot be tolerated by the target or fails to produce a therapeutic effect when administered to humans. Compounds intended for drug development must have appropriate pharmacokinetic parameters, including dose, speed, extent and duration of reaching the site of action, and binding to the relevant target. Thus, comprehensive preclinical and clinical studies must be conducted for drug candidate compounds [185].

Clinical trials are conducted on either patients or healthy volunteers, depending on the phase being carried out. Their primary goal is to provide data on the safety and efficacy of drugs. Clinical trials are conducted in accordance with a protocol designed by the investigator or sponsor. The study begins once the conditions regarding the number and selection criteria of participants, dosage information, study duration, parameters, and data analysis procedures are established. The phases, general characteristics, and analyses involved in preclinical and clinical studies are presented in Figure 2.17 [184].

For traditionally used medicinal plants and herbal products, an exception exists under the European Medicines Agency (EMA) regulations regarding clinical trials. According to this regulation, herbal medicines with documented traditional use for at least 15 years within the EU and 30 years in other countries do not require clinical trials to demonstrate their safety and efficacy for their traditionally indicated uses [186].

2.3.8 Structural modifications and developing new analogues

Chemical modifications of natural products play a significant role in drug development, as altering the chemical structures of these compounds can lead to the discovery of new therapeutic options and improve existing treatments. Such small modifications in the synthesis of natural compounds can enhance their biological activity, improve pharmacokinetic profiles, and result in more successful clinical applications [187].

Despite their intricate chemical structures, plant secondary metabolites demonstrate enhanced drug-likeness compared to synthetic molecules. However, some natural products, while demonstrating therapeutic efficacy, may have poor oral bioavailability, excessive side effects, or suboptimal activity. To address these limitations, functional group modifications are often required to produce drugs that are more sol-

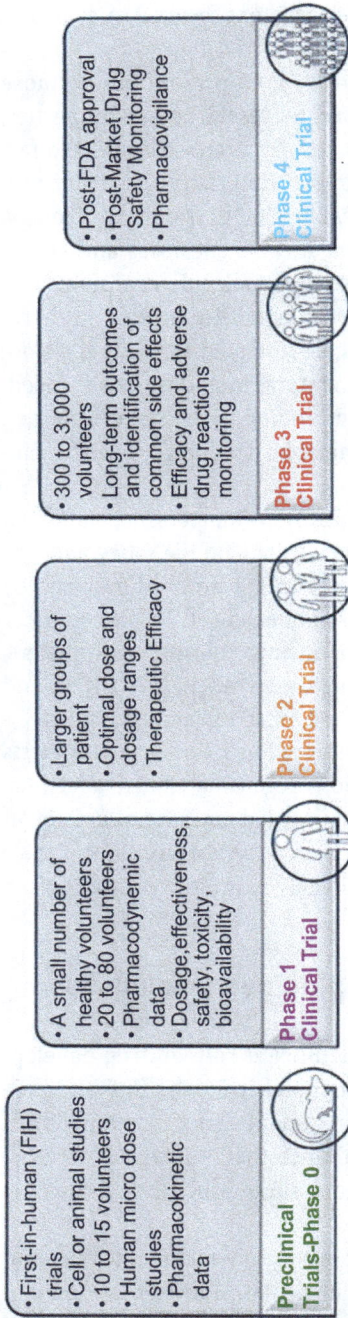

Figure 2.17: Preclinical and clinical studies.

Preclinical Trials-Phase 0
- First-in-human (FIH) trials
- Cell or animal studies
- 10 to 15 volunteers
- Human micro dose studies
- Pharmacokinetic data

Phase 1 Clinical Trial
- A small number of healthy volunteers
- 20 to 80 volunteers
- Pharmacodynemic data
- Dosage, effectiveness, safety, toxicity, bioavailability

Phase 2 Clinical Trial
- Larger groups of patient
- Optimal dose and dosage ranges
- Therapeutic Efficacy

Phase 3 Clinical Trial
- 300 to 3,000 volunteers
- Long-term outcomes and identification of common side effects
- Efficacy and adverse drug reactions monitoring

Phase 4 Clinical Trial
- Post-FDA approval
- Post-Market Drug Safety Monitoring
- Pharmacovigilance

uble, better absorbed and distributed, more selective, and less toxic. Addressing these challenges necessitates the chemical production of natural product analogs, which serve as lead compounds for the creation of more potent drugs. Structural modifications are implemented through strategies such as the genomes to natural products (GNP) platform to enhance activity [7, 13].

The extraction of natural compounds from their original sources is constrained by the availability of source plants. Additionally, the extraction process is often complex and inefficient, and acquiring the large amounts of raw materials needed for drug approval and distribution is typically expensive. Consequently, once plant-derived drugs are discovered, synthetic production often becomes the preferred method. Structural information is crucial for the synthesis of natural product analogs. Understanding the configuration of stereogenic centers is essential for designing an effective synthetic strategy [33]. Advances in NMR spectroscopy, HPLC, microfluidic systems, and algorithmic developments have been effectively applied in medicinal chemistry, enabling the synthesis of numerous natural compound analogs. Computational chemistry tools have further contributed to drug discovery by facilitating the development of structural analogs from natural molecules [2, 151, 188].

An excellent example of clinical application through structural modifications is the transformation of the natural product vinblastine into vinorelbine by adding methyl groups and an oxygen atom. These changes improved the pharmacokinetic properties of the compound and enhanced its efficacy in cancer treatment. Specifically, the addition of oxygen increased the compound's bioavailability and facilitated better targeting to cellular regions [189].

Prodrug and isomerism strategies are other prominent approaches in structural modifications. The term "prodrug" was introduced by Adrien Albert in 1958. Prodrugs are biologically inactive derivatives that can be converted into pharmacologically active drug molecules. These designs typically include functional groups such as esters, amides, phosphates, carbonates, or carbamates, which can be enzymatically or chemically cleaved in the body. The prodrug strategy addresses physical barriers related to solubility, bioavailability, chemical instability, and therapeutic effects at target sites, aiming to optimize absorption, distribution, metabolism, excretion, and toxicity (ADMET) processes to enhance therapeutic efficacy [190].

Isomerism has been one of the most groundbreaking results in clinical research for improving the pharmacokinetics and efficacy of compounds used as drugs. Findings from studies on isomerism play a critical role in discovering new drugs and improving the bioavailability of existing ones. Most drugs currently in use have undergone chiral switching, transitioning from racemic mixtures to one of their isomers [191]. By definition, isomers are molecules with the same atomic composition but different bonding arrangements or spatial orientations of atoms, meaning they are distinct substances sharing the same molecular formula. Isomerism leads to different therapeutic applications; for instance, quinine exhibits antimalarial activity, while quinidine has antiarrhythmic properties [192].

2.4 The use of omics technologies in drug discovery and development

Today, studies utilizing omics approaches are increasingly preferred, with the most common omics technologies including genomics, transcriptomics, proteomics, and metabolomics [193]. The application of omics studies to phytotherapy is referred to as "phytomics" [194].

The therapeutic effects of plants are often not attributed to a single compound but rather to the combinatorial effects of the components within the extract. Therefore, focusing solely on a single isolated compound in drug development studies from plants may not always be the most accurate approach. Since many diseases are already treated with pharmacological combinations, a combination strategy should be taken into consideration instead. The effects of these combinations on genes and proteins involved in various cellular processes should be thoroughly investigated using available "-omics" platforms. Such an approach can capture an effect that operates through synergistic mechanisms on multiple targets within a physiological system, rather than searching for a specific molecule targeting a single objective. Integrating technologies such as genomics, transcriptomics, proteomics, metabolomics/metabonomics, automation, and computational strategies into the research process will enable the development of a systems biology approach, paving the way for more efficient and innovative drug designs [188].

Recent technological advancements in genomics, proteomics, and metabolomics have introduced significant innovations across various scientific research fields, sparking great interest and excitement among scientists. Genomics aims to study genetic information, proteomics focuses on proteins, and metabolomics involves the qualitative and quantitative analysis of all low-molecular-weight metabolites within a cell or organism and their dynamics in biological systems [193]. These new platforms, referred to as "-omics" technologies, are high-throughput systems capable of simultaneously detecting tens of thousands of genes and proteins. They enable detailed analysis and comprehensive characterization of biological systems. These technologies have the potential to correlate complex mixtures with intricate effects in the form of gene/protein expression profiles, providing evidence of the efficacy of phytochemical components and defining their activity profiles [195].

2.4.1 Genomics

When the therapeutic properties of plants were first discovered, humanity was far from the rigor of scientific evidence, the principles of philosophical and experimental methodologies, and the ability to identify bioactive molecules through advanced genomic technologies that we possess today. Plant genomics has emerged as a transformative approach, enabling the comprehensive exploration and analysis of vast botanical diversity and the intricate biochemical repertoire it encompasses. This progress has been supported by rapid advancements and decreasing costs in genetic sequencing technologies [196]. Recent developments in genomic techniques, such as DNA barcoding and other innovative methods, have established accurate identification criteria for plants and other natural product sources. These techniques provide much faster and more accurate identification compared to the commonly used morphological and other traditional methods [197].

Genomics can serve various roles in the target identification process during drug discovery from natural products. Specifically, genomic analysis can identify targets such as cellular signaling pathways and enzymes metabolizing specific compounds. Genome-based methods, including sequencing and transcriptomic studies, have enabled the evaluation of many systems for compound targeting [198].

Genome-wide association studies in humans have identified thousands of genetic polymorphisms associated with diseases. The complex interactions of genes within the human genome underpin conditions such as diabetes, autoimmune diseases, cancer, and neurological disorders. Functional genomics is an innovative field aimed at elucidating the relationship between genotype and phenotype. Leveraging genetic editing tools and large datasets, it allows for deeper exploration of gene functions and biological interactions. It also plays a vital role in revealing disorder mechanisms and discovering new drug targets. Integrating functional genomic approaches into drug development pipelines is expected to accelerate the creation of innovative and effective therapies [199].

2.4.2 Metabolomics

Metabolomics is a nonselective, universally applicable, comprehensive, and simultaneous analytical method used for the identification and quantification of metabolites in biological samples. Metabolomics provides meaningful and useful data for large-scale analysis of primary and secondary metabolites in plant extracts, holistic interpretation of results, and the monitoring and evaluation of cellular function or systems biology. This research field aims to profile metabolites and detect differences between them. In metabolomics, various analytical strategies are used to determine the phytochemical composition of a specific plant extract or matrix [200].

Advancements in instrumental analysis methods, such as chromatography and spectroscopy, have accelerated progress in metabolomic technologies. In recent years, omics technologies, including metabolomics, have become widely used in the research of plant-based drugs. Through metabolomic methods, secondary metabolites found in medicinal and aromatic plants can be accurately and comprehensively analyzed. This, in turn, speeds up the processes of identification and characterization of these metabolites. In addition, metabolomics is a method that can be used to comprehend the mechanisms of action of plant components at the molecular level [12]. Although metabolomics does not aid in the initial selection of plants, it provides a rapid analysis of the active components in the selected plant extract. The concentrations and properties of phytochemical components can be determined through metabolomics, and by utilizing various statistical analysis methods, these components can be studied in relation to their physiological activities. The aim of metabolomic studies is to evaluate and identify bioactive components within a complex plant extract before isolating the relevant compound. The strength of the metabolomic approach in plant-based drug discovery lies in its ability to identify all components contributing to the plant's therapeutic effects and, consequently, account for potential synergistic interactions [11].

Metabonomics primarily focuses on analyzing how living systems respond metabolically to biological stimuli or genetic changes on a global and dynamic scale. In recent years, the term has evolved beyond its traditional usage and is now often associated with a systems biology-driven approach. This perspective examines the functional changes and disruptions within biological systems triggered by pharmacological effects, providing a thorough understanding of both the natural product and its impact on the organism. Metabolomic profiling of natural products using technologies such as ultra-performance liquid chromatography–quadrupole TOF MS (UPLC–MS) enables the identification of components responsible for therapeutic effects in plants. Metabolomic and metabonomic profiling conducted using NMR, MS, and UPLC also provides insights into the pharmacodynamic, pharmacokinetic, and toxicological properties of natural products [188].

2.4.3 Proteomics

The proteome represents the cumulative composition of all proteins expressed in a cell, tissue, or organism. Proteomics, on the other hand, is the scientific field that studies the flow of information through pathways and networks to understand the functional relationships of proteins. Proteomics involves the detailed analysis of proteins in a sample, covering protein mapping and characterization, as well as the study of their associated structures and functions. However, proteomics is highly complex due to the vast scope of the analyzed domain (over 100,000 proteins) and the challenges in detecting rare proteins. Nonetheless, the dynamic responsiveness of the proteome to both genetic and environmental variations makes it a highly promising area for biomarker discovery. The potential of proteins to be widely affected in disease conditions highlights their role in the diversity of disease biomarkers discovered to date, enabled by proteomic technologies. Therefore, proteomic data analysis provides a multifaceted perspective for understanding disease mechanisms and developing new therapeutic strategies [201].

Advancements in MS-based proteomics have significantly contributed to unraveling biological systems, understanding disease mechanisms, and establishing links between genotype and phenotype. While traditional approaches often focused on a limited number of proteins, innovations in mass spectrometry and related technologies now allow for the holistic examination of biological systems as unified entities. These technological advancements have also driven the development of new bioinformatics analysis methods, enabling deeper molecular-level insights in proteomics. Although many bioinformatics techniques commonly used in proteomics are adapted from other omics fields, the unique nature of proteomic data necessitates the development of specialized analytical strategies [202].

2.5 Future scope

The drug discovery process has historically been long and labor-intensive, requiring significant effort in sourcing. This process begins with identifying a lead compound for a specific drug target, followed by rigorous optimization and preclinical studies in animal models to assess efficacy and toxicity. These initial steps typically take 5–6 years, and only 1 out of every 5,000 lead compounds advances to human trials. Moreover, obtaining final approval from regulatory bodies, such as the U.S. Food and Drug Administration (FDA), adds to the timeline, resulting in an average time of 12 years and high costs for a drug to reach the market [203].

AI has emerged as a transformative force in the pharmaceutical industry, bringing the ability to simplify drug development and lower related costs. Experts in the field predict that AI-powered drug discovery methods can shorten the process, and drugs developed by AI could reach the market in a relatively short time [204].

If AI advancements are applied to the plant-based drug discovery process, they could enable the rapid identification of drug candidates, offering time savings compared to traditional methods. Using an AI-supported platform, a molecule for obsessive-compulsive disorder was developed in just one year, marking it as the first AI-invented molecule to enter human trials. Integrating AI with databases containing plant-derived chemical compounds offers a promising path for innovation in drug discovery. Rich in bioactive compounds, traditional medicine can benefit from AI's analytical capabilities to identify and optimize potential therapeutic agents. By leveraging machine learning algorithms, researchers can predict compound efficacy and toxicity, thus accelerating the discovery of new treatments. Successfully combining AI with traditional medicine could bridge gaps in current drug development practices. AI-supported platforms can analyze large datasets to reveal relationships between compounds and therapeutic effects, offering a more efficient alternative to traditional trial-and-error methods. This approach not only accelerates drug development but also has the ability to enhance the treatment of rare and neglected illnesses [205].

2.6 Conclusion

Medicinal and aromatic plants have been used for therapeutic purposes for centuries. Today, many approved drugs are derived from plants. Since plants host a vast phytochemical diversity compared to synthetic molecules, they are an immense source for new drug development. Over the years, research on plants has accumulated enough knowledge to form natural product libraries. However, the existence of plants worldwide whose effects and contents remain unexplored makes them an almost unlimited resource.

New drug discovery and development, whether based on synthetic or natural compounds, is a complex, difficult, and lengthy process. Additionally, the complex nature of plants and the intricate chemical structure of their components make the drug development process from plants even more challenging. The process of drug discovery from plants generally consists of identifying the source plant, collecting plant material, performing preliminary treatments, extraction, bioactivity studies, isolation and purification, biological experiments, clinical trials, and optimization. At each stage, various obstacles are encountered, depending on the plant's characteristics and the methods used. If these obstacles are not overcome, even if the plant's therapeutic properties are discovered, they cannot be translated into clinical applications. Despite the development of various innovative methods and technologies to overcome these challenges, some issues still persist. Among these issues, the most critical are the lack of standardization, the inability to isolate and purify pure chemical compounds, insufficient elucidation of biological mechanisms, and the limited success of transitioning to controlled clinical trials with very few molecules. To facilitate, accelerate, and ultimately succeed in drug discovery and development from plants, innovative drug design methods are essential.

Identifying the source plant is one of the most crucial steps in the drug discovery procedure from medicinal and aromatic plants, as it will influence the success of subsequent stages. When identifying candidate plants or plant compounds, the utilization of high-throughput screening (HTS) technologies and virtual screening methods not only increases success but can also significantly speed up the process. Additionally, systematic ethnopharmacological studies will contribute to the plant identification process.

In some diseases and conditions, isolated single compounds from plants do not reflect the therapeutic effect exhibited by the plant. This is due to the synergistic effects exhibited by the components in the plant's complex structure. To overcome this issue, combinatorial chemical approaches should be used when evaluating drug candidate compounds obtained from plants. The innovative technologies of the computational molecular design era, along with new analytical and computational techniques, have opened new horizons in the process from plants to drugs.

New analytical technologies and combinatorial chemistry, including artificial intelligence (AI) and new computational and screening methods, omics technologies for studying interactions between bioactive molecules and targets, microfluidics for designing biological models, and innovative drug design, are methods that have the potential for great success in drug discovery from medicinal and aromatic plants and possess unique potential in future studies. Considering that natural products are unique and rich sources for drug discovery, the integration of advanced methods will inevitably lead to an increase in plant-based drugs in the future.

References

[1] Vaou, N., Stavropoulou, E., Voidarou, C., Tsigalou, C., and Bezirtzoglou, E. (2021 Sep 27). Towards advances in medicinal plant antimicrobial activity: A review study on challenges and future perspectives. Microorganisms, 9(10), 2041, doi: 10.3390/microorganisms9102041.

[2] Dzobo, K. (2022). The role of natural products as sources of therapeutic agents for innovative drug discovery. Comprehensive Pharmacology, 2(2), 2–10.

[3] Harvey, A.L. (2008). Natural products in drug discovery. Drug Discovery Today: Technologies, 13(19–20), 894–901, doi: 10.1016/j.drudis.2008.07.004.

[4] Cragg, G.M. and Newman, D.J. (2013). Natural products: A continuing source of novel drug leads. Biochimica Et Biophysica Acta, 1830(6), 3670–3695, doi: 10.1016/j.bbagen.2013.02.008.

[5] Chaachouay, N. and Zidane, L. 2024. Plant-derived natural products: A source for drug discovery and development. Drugs. Drug Candidates, 3(2), 184–207.

[6] Katiyar, C., Gupta, A., Kanjilal, S., and Katiyar, S. (2012). Drug discovery from plant sources: An integrated approach. AYU, 33(1), 10–19, doi: 10.4103/0974-8520.100295.

[7] Najmi, A., Javed, S.A., Al Bratty, M., and Alhazmi HA. 2022. Modern approaches in the discovery and development of plant-based natural products and their analogues as potential therapeutic agents. Molecules, 27(1). 349, doi: 10.3390/molecules27010349.

[8] Ghenabzia, I., Hemmami, H., Amor, I.B., et al. (2023). Different methods of extraction of bioactive compounds and their effect on biological activity: A review. International Journal of Secondary Metabolite, 10(4), 469–494, doi: 10.21448/ijsm.1236890.

[9] Mandal, S., Kar, N.R., Jain, A.V., and Yadav, P. (2024). Natural products as sources of drug discovery: Exploration, optimisation, and translation into clinical practice. African Journal of Biological Sciences, 6(9), 2487–2504.

[10] Salamon, I. (2024). Medicinal, aromatic, and spice plants: Biodiversity, phytochemistry, bioactivity, and their processing innovation. Horticulturae, 10(1), 280, doi: 10.3390/horticulturae10010280.

[11] Schwikkard, S.L. and Mulholland, D.A. (2021). Useful methods for targeted plant selection in the discovery of potential new drug candidates. Phytochemistry Reviews, 20(3), 441–452, doi: 10.1007/s11101-021-09714-4.

[12] Atanasov, A.G., Zotchev, S.B., Dirsch, V.M., and Supuran, C.T. (2021). Natural products in drug discovery: Advances and opportunities. Nature Reviews Drug Discovery, 20(3), 200–216, doi: 10.1038/s41573-020-00114-z.

[13] Zhang, L., Song, J., Kong, L., et al. (2020). The strategies and techniques of drug discovery from natural products. Pharmacology & Therapeutics, 216, 107686. doi: 10.1016/j.pharmthera.2020.107686.

[14] Ang-Lee, M.K., Moss, J., and Yuan, C.S. (2001). Herbal medicines and perioperative care. JAMA, 286(2), 208–216, 10.1001/jama.286.2.208.

[15] Pan, S.Y., Litscher, G., Gao, S.H., et al. (2014). Historical perspective of traditional indigenous medical practices: The current renaissance and conservation of herbal resources. Evidence-Based Complementary and Alternative Medicine, 2014, 525340, doi: 10.1155/2014/525340.

[16] Priya, S. and Satheeshkumar, P.K. (2020). Natural products from plants: Recent developments in phytochemicals, phytopharmaceuticals, and plant-based nutraceuticals as anticancer agents. In: Prakash, B. (ed) Functional and Preservative Properties of Phytochemicals, Academic Press, Cambridge, MA, USA, 145–163.

[17] Kiyohara, H., Matsumoto, T., and Yamada, H. (2004). Combination effects of herbs in a multi-herbal formula: Expression of Juzen-taiho-to's immunomodulatory activity on the intestinal immune system. Evidence-Based Complementary and Alternative Medicine, 1(1), 83–91, 10.1093/ecam/neh005.

[18] Süntar, İ. (2020). Importance of ethnopharmacological studies in drug discovery: Role of medicinal plants. Phytochemistry Reviews, 19(5), 1199–1209, doi: 10.1007/s11101-020-09692-1.

[19] Tavlı, Ö.F. (2022). Papaver somniferum L. In: Kümüşer, Y. (ed) Novel Drug Targets with Traditional Herbal Medicines, Springer, Berlin/Heidelberg, Germany, 479–490.

[20] Stepp, J.R. (2004). The role of weeds as sources of pharmaceuticals. Journal of Ethnopharmacology, 92(1), 163–166, doi: 10.1016/j.jep.2004.03.002.

[21] Posadino, A.M., Giordo, R., Pintus, G., et al. (2023). Medicinal and mechanistic overview of artemisinin in the treatment of human diseases. Biomedicine and Pharmacotherapy, 163, 114866, doi: 10.1016/j.biopha.2023.114866.

[22] Post-White, J., Ladas, E.J., and Kelly, K.M. (2007). Advances in the use of milk thistle (Silybum marianum). Integrative Cancer Therapies, 6(2), 104–109, doi: 10.1177/1534735407301632.

[23] Eyal, S. (2018 Nov 23). The fever tree: From malaria to neurological diseases. Toxins (Basel). 10(12), 491, doi: 10.3390/toxins10120491, PMID: 30477182; PMCID: PMC6316520.

[24] Islam, M.T., Tabrez, S., Jabir, N.R., Ali, M., Kamal, M.A., Da Silva Araujo, L., Oliveira Santos JV, D., Mata AMOF, D., De Aguiar, R.P.S., and De Carvalho Melo Cavalcante, A.A. (2018). An insight into the therapeutic potential of major coffee components. Current Drug Metabolism, 19(6), 544–556, doi: 10.2174/1389200219666180302154551.

[25] Sansone, L., Milani, F., Fabrizi, R., Belli, M., Cristina, M., Zagà, V., De Iure, A., Cicconi, L., Bonassi, S., and Russo, P. (2023 Sep 26). Nicotine: From discovery to biological effects. International Journal of Molecular Sciences, 24(19), 14570, doi: 10.3390/ijms241914570.

[26] Dhyani, P., Quispe, C., Sharma, E., et al. (2022). Anticancer potential of alkaloids: A key emphasis to colchicine, vinblastine, vincristine, vindesine, vinorelbine and vincamine. Cancer Cell International, 22, 206, doi: 10.1186/s12935-022-02624-9.

[27] Lin, C.R., Tsai, S.H.L., Wang, C., et al. (2023). Willow bark (Salix spp.) used for pain relief in arthritis: A meta-analysis of randomized controlled trials. Life (Basel), 13(10), 2058, doi: 10.3390/life13102058.

[28] MK, D. and Hollman, A. (2002 Sep). Atropa belladonna. Heart, 88(3), 215, doi: 10.1136/heart.88.3.215-a.

[29] Restrepo, D.A., Saenz, E., Jara-Muñoz, O.A., Calixto-Botía, I.F., Rodríguez-Suárez, S., Zuleta, P., et al. (2019). Erythroxylum in focus: An interdisciplinary review of an overlooked genus. Molecules, 24(20), 3788, doi: 10.3390/molecules24203788.

[30] Chen, J.H., Lin, I.H., Hsueh, T.Y., Dalley, J.W., and Tsai, T.H. (2022). Pharmacokinetics and transplacental transfer of codeine and codeine metabolites from. Papaver somniferum L. Journal of Ethnopharmacology, 298, 115623, doi: 10.1016/j.jep.2022.115623.

[31] Foumani, A.E., Irani, S., Shokoohinia, Y., and Mostafaie, A. (2022). Colchicine of Colchicum autumnale, a traditional anti-inflammatory medicine, induces apoptosis by activation of apoptotic genes and proteins expression in human breast (MCF-7) and mouse breast (4T1) cell lines. Cell Journal. 24(11), 647–656, doi: 10.22074/cellj.2022.8290.

[32] Shen, B. (2015). A new golden age of natural products drug discovery. Cell, 163(6), 1297–1300, doi: 10.1016/j.cell.2015.11.031.

[33] Ahmad Dar, A., Sangwan, P.L., and Kumar, A. (2020). Chromatography: An important tool for drug discovery. Journal Separation Science, 43(1), 105–119, doi: 10.1002/jssc.201900835.

[34] Jamshidi-Kia, F., Lorigooini, Z., and Amini-Khoei, H. (2018). Medicinal plants: Past history and future perspective. Journal of Herbmed Pharmacology, 7(1), 1–7.

[35] Pirintsos, S., Panagiotopoulos, A., Bariotakis, M., et al. 2022. From traditional ethnopharmacology to modern natural drug discovery: A methodology discussion and specific examples. Molecules, 27(13), 4060, doi: 10.3390/molecules27134060.

[36] Diniz, G., Veysanoğlu, Ş., İnal, S., Araç, B., Düzenli, N., and Karakayalı EM. (2024). The mysterious travel of drugs from nature to pharmacy. *Forbes*. The Journal of Medical Sciences, 5(1), 1–8.

[37] Fabricant, D.S. and Farnsworth, N.R. (2001). The value of plants used in traditional medicine for drug discovery. Environmental Health Perspectives, 109(Suppl 1), 69–75, doi: 10.1289/ehp.01109s169.

[38] Oberlies, N.H. and Kroll, D.J. (2004). Camptothecin and Taxol: Historic achievements in natural products research. Journal of Natural Products, 67(2), 129–135.

[39] Mishra, K.P., Ganju, L., Sairam, M., Banerjee, P.K., and Sawhney, R.C. (2008). A review of high throughput technology for the screening of natural products. Biomedicine and Pharmacotherapy, 62(2), 94–98.

[40] Nasim, N. and Sandeep, I.S. (2022). Mohanty S. Plant-derived natural products for drug discovery: Current approaches and prospects. Nucleus, 65(5), 399–411, doi: 10.1007/s13237-022-00388-1.

[41] Szymański, P., Markowicz, M., and Mikiciuk-Olasik, E. (2012). Adaptation of high-throughput screening in drug discovery-toxicological screening tests. International Journal of Molecular Sciences, 13(1), 427–452, doi: 10.3390/ijms13010427.

[42] Littleton, J., Rogers, T., and Falcone, D. (2005). Novel approaches to plant drug discovery based on high throughput pharmacological screening and genetic manipulation. Life Science, 78(5), 467–475, doi: 10.1016/j.lfs.2005.09.003.

[43] Dias, D.A., Urban, S., and Roessner, U. (2012). A historical overview of natural products in drug discovery. Metabolites, 2(2), 303–336, doi: 10.3390/metabo2020303.

[44] Janzen, W.P. and Bernasconi, P. (2009). High throughput screening: Methods and protocols, second edition. Preface. Methods in Molecular Biology, 565, v–vii, doi: 10.1007/978-1-60327-258-2.

[45] Mahmud, S., Paul, G.K., Biswas, S., et al. (2022). Phytochemdb: A platform for virtual screening and computer-aided drug designing. Database (Oxford), 2022, baac021, doi: 10.1093/database/baac021.

[46] Barzdina, A., Paulausks, A., Bandere, D., and Brangule, A. (2022). The potential use of herbal fingerprints by means of HPLC and TLC for characterization and identification of herbal extracts and the distinction of Latvian native medicinal plants. Molecules, 27(8), 2555, doi: 10.3390/molecules27082555.

[47] Büyüktuncel, E. (2012). Gelişmiş ekstraksiyon teknikleri I. Hacettepe University Journal of the Faculty of Pharmacy, 32(2), 209–242.

[48] Azwanida, N.N. (2015). A review on the extraction methods use in medicinal plants, principle, strength and limitation. Medicinal and Aromatic Plants, 4(3), 196, doi: 10.4172/2167-0412.1000196.

[49] Krakowska-Sieprawska, A., Kiełbasa, A., Rafińska, K., Ligor, M., and Buszewski, B. (2022). Modern methods of pre-treatment of plant material for the extraction of bioactive compounds. Molecules, 27(3), 730, doi: 10.3390/molecules27030730.

[50] Romanik, G., Gilgenast, E.A., Przyjazny, A., and Kamiński, M. (2007). Techniques of preparing plant material for chromatographic separation and analysis. Journal of Biochemical and Biophysical Methods, 70(2), 253–261, doi: 10.1016/j.jbbm.2007.02.008.

[51] Gil-Martín, E., Forbes-Hernández, T., Romero, A., et al. (2022). Influence of the extraction method on the recovery of bioactive phenolic compounds from food industry by-products. Food Chemistry, 378, 131918, doi: 10.1016/j.foodchem.2021.131918.

[52] Thamkaew, G., Sjöholm, I., and Galindo, F.G. (2020). A review of drying methods for improving the quality of dried herbs. Critical Reviews in Food Science and Nutrition, 61(11), 1763–1786, doi: 10.1080/10408398.2020.1765309.

[53] Nurhaslina, C.R., Bacho, S.A., and Mustapa, A.N. (2022). Review on drying methods for herbal plants. Materials Today: Proceedings, 63, 122–139, doi: 10.1016/j.matpr.2022.02.002.

[54] Belwal, T., Cravotto, C., Prieto, M.A., Venskutonis, P.R., Daglia, M., Devkota, H.P., and Cravotto, G. (2022). Effects of different drying techniques on the quality and bioactive compounds of plant-based products: A critical review on current trends. Drying Technology, 40(8), 1539–1561, doi: 10.1080/07373937.2022.2068028.

[55] Mohdaly, A. and Smetanska, I. (2010). Methods for the extraction of metabolites from plant tissues. In: Sadler, M. (ed) Handbook of Nutritional Biochemistry: Genomics, Metabolomics and Food Supply, Nova Science Publishers, New York, 123–164.

[56] Camel, V. (2014). Extraction methodologies: General introduction. Encyclopedia of Analytical Chemistry, 1–26, doi: 10.1002/9780470027318.a1018.pub2.

[57] Mtewa, A.G., Deyno, S., Kasali, F.M., Annu, D., and Sesazi, E. (2018). General extraction, isolation and characterization techniques in drug discovery: A review. International Journal of Basic & Applied Research, 38(1), 10–24.

[58] Ngaha Njila, M.I., Mahdi, E., Massoma Lembe, D., et al. (May 22–24, 2017). Review on extraction and isolation of plant secondary metabolites. In: Rahman, M.A., Yingthawornsuk, T., & Ramli, M.F. (eds) Proceedings of the 7th International Conference on Agricultural, Chemical, Biological and Environmental Sciences (ACBES-2017), Kuala Lumpur, Malaysia, Vol. 2017, 67–70. doi: 10.15242/IIE.C0517024.

[59] Azmir, J., Zaidul, I.S.M., Rahman, M.M., et al. (2013). Techniques for extraction of bioactive compounds from plant materials: A review. Journal of Food Engineering, 117(4), 426–436, doi: 10.1016/j.jfoodeng.2013.01.014.

[60] Badawy, M.E.I., El-Nouby, M.A.M., Kimani, P.K., Lim, L.W., and Rabea, E.I. (2022). A review of the modern principles and applications of solid-phase extraction techniques in chromatographic analysis. Analytical Science, 38(12), 1457–1487, doi: 10.1007/s44211-022-00190-8.

[61] Shikov, A.N., Mikhailovskaya, I.Y., Narkevich, I.A., and Pozharitskaya, O.N. (2022). Methods of extraction of medicinal plants. In: Mukherjee, P.K. (ed) Evidence-Based Validation of Herbal Medicine: Translational Research on Botanicals, Elsevier, Amsterdam, 769–794. doi: 10.1016/B978-0-323-85542-6.00029-9.

[62] Fierascu, R.C., Fierascu, I., Ortan, A., Georgiev, M.I., and Sieniawska, E. (2020). Innovative approaches for recovery of phytoconstituents from medicinal/aromatic plants and biotechnological production. Molecules, 25(2), 309, doi: 10.3390/molecules25020309.

[63] Jha, A.K. and Sit, N. (2022). Extraction of bioactive compounds from plant materials using combination of various novel methods: A review. Trends in Food Science and Technology, 119, 580, doi: 10.1016/j.tifs.2021.11.019.

[64] Rasul, M.G. (2018). Conventional extraction methods use in medicinal plants, their advantages and disadvantages. International Journal of Basic Sciences and Applied Computing, 2(6), 10–14.

[65] Gligor, O., Mocan, A., Moldovan, C., Locatelli, M., Crişan, G., and Icfr, F. (2019). Enzyme-assisted extractions of polyphenols – A comprehensive review. Trends in Food Science and Technology, 88, 302–315, doi: 10.1016/j.tifs.2019.03.029.

[66] Osorio-Tobón, J.F. (2020). Recent advances and comparisons of conventional and alternative extraction techniques of phenolic compounds. Journal of Food Science and Technology, 57(12), 4299–4315, doi: 10.1007/s13197-020-04433-2.

[67] Handa, S.S., Khanuja, S.P.S., Longo, G., and Rakesh, D.D. (2008). Extraction Technologies for Medicinal and Aromatic Plants, 1st ed, United Nations Industrial Development Organization and the International Centre for Science and High Technology, Italy.

[68] Mondal, S., Das, M., Debnath, S., Sarkar, B.K., and Babu, G. (2024). An overview of extraction, isolation, and characterization techniques of phytocompounds from medicinal plants. Natural Products Research, 1–23, doi: 10.1080/14786419.2024.2426059.

[69] Verep, D., Ateş, S., and Karaoğul, E. (2023). A review of extraction methods for obtaining bioactive compounds in plant-based raw materials. J Bartin Fac For, 25(3), 492–513, doi: 10.24011/barofd.1303285.

[70] Abubakar, A.R. and Haque, M. (2020). Preparation of medicinal plants: Basic extraction and fractionation procedures for experimental purposes. Journal of Pharmacy and Bioallied Sciences, 12(1), 8.

[71] Ingle, K.P., Deshmukh, A.G., Padole, D.A., Dudhare, M.S., Moharil, M.P., and Khelurkar, V.C. (2017). Phytochemicals: Extraction methods, identification, and detection of bioactive compounds from plant extracts. Journal of Pharmacognosy and Phytochemistry, 6, 32–36.

[72] Pandey, A. and Tripathi, S. (2014). Concept of standardization, extraction, and pre-phytochemical screening strategies for herbal drug. Journal of Pharmacognosy and Phytochemistry, 2(5), 115–119.

[73] Bimakr, M. (2010). Comparison of different extraction methods for the extraction of major bioactive flavonoid compounds from spearmint (Mentha spicata L.) leaves. Food Bioprod Process, 89, 1–6, doi: 10.1016/j.fbp.2009.07.002.

[74] Singh, J. (2008). Maceration, percolation and infusion techniques for the extraction of medicinal and aromatic plants. In: Handa, S.S., Khanuja, S.P., Longo, G., & Rakesh, D.D. (eds) Extraction Technologies for Medicinal and Aromatic Plants, United Nations Industrial Development Organization & International Centre for Science and High Technology, Vienna., 32–35.

[75] Nagalingam, A. (2017). Drug delivery aspects of herbal medicines. In: Iwata, M. and Akutsu, T. (eds) Jpn Kampo Med Treat Common Dis Focus Inflammation, Elsevier, amsterdam, 17, 143. doi: 10.1016/B978-0-12-809398-6.00015-9.

[76] Bitwell, C., Indra, S.S., Luke, C., and Kakoma, M.K. (2023). A review of modern and conventional extraction techniques and their applications for extracting phytochemicals from plants. Science African, 19, e01585, doi: 10.1016/j.sciaf.2023.e01585.

[77] Li, S.L., Lai, S.F., Song, J.Z., Qiao, C.F., Liu, X., Zhou, Y., Cai, H., Cai, B.C., and Xu, H.X. (2010). Decocting-induced chemical transformations and global quality of Du–Shen–Tang, the decoction of ginseng evaluated by UPLC-Q-TOF-MS/MS based chemical profiling approach. Journal of Pharmaceutical and Biomedical Analysis, 53(4), 946–957, doi: 10.1016/j.jpba.2010.02.016.

[78] Hidayat, R. and Wulandari, P. (2021). Methods of extraction: Maceration, percolation, and decoction. Eureka Herba Indonesia, 2(1), 68–74, doi: 10.37275/ehi.v2i1.15.

[79] Rathi, B.S., Bodhankar, S.L., and Baheti, A.M. (2006). Evaluation of aqueous leaves extract of Moringa oleifera Linn for wound healing in albino rats. Indian Journal of Experimental Biology, 44(11), 898–901.

[80] Shrivastav, G., Prava Jyoti, T., Chandel, S., and Singh, R. (2024). Eco-friendly extraction: Innovations, principles, and comparison with traditional methods. Separation & Purification Reviews, 1–17, doi: 10.1080/15422119.2024.2381605.

[81] Wang, W.Y., Qu, H.B., and Gong, X.C. (2020). Research progress on percolation extraction process of traditional Chinese medicines. Zhongguo Zhong Yao Za Zhi= Zhongguo Zhongyao Zazhi= China Journal of Chinese Materia Medica, 45(5), 1039–1046, doi: 10.19540/j.cnki.cjcmm.20191221.305.

[82] Abdelmohsen, U.R., Sayed, A.M., and Elmaidomy, A.H. (2022). Natural products' extraction and isolation-between conventional and modern techniques. Frontiers in Natural Products, 1, 873808, doi: 10.3389/fntpr.2022.873808.

[83] Perović, A.B., Karabegović, I.T., Krstić, M.S., et al. (2024). Novel hydrodistillation and steam distillation methods of essential oil recovery from lavender: A comprehensive review. Industrial Crops and Products, 211, 118244, doi: 10.1016/j.indcrop.2024.118244.

[84] Božović, M., Navarra, A., Garzoli, S., Pepi, F., and Ragno, R. (2017). Essential oils extraction: A 24-hour steam distillation systematic methodology. Natural Products Research, 31(20), 2387–2396, doi: 10.1080/14786419.2017.1309534.

[85] Pheko-Ofitlhile, T. and Makhzoum, A. (2024). Impact of hydrodistillation and steam distillation on the yield and chemical composition of essential oils and their comparison with modern isolation techniques. Journal of Essential Oil Research, 36(2), 105–115, doi: 10.1080/10412905.2024.2320350.

[86] Daud, N.M., Putra, N.R., Jamaludin, R., Norodin, M.S.M., Sarkawi, N.S., Hamzah, N.H.S., Nasir, H.M., Zaidel, D.N.A., Yunus, M.A.C., and Salleh, L.M. (2022). Valorisation of plant seed as natural bioactive compounds by various extraction methods: A review. Trends in Food Science and Technology, 119, 204.

[87] El Maaiden, E., Bouzroud, S., Nasser, B., Moustaid, K., El Mouttaqi, A., Ibourki, M., Boukcim, H., Hirich, A., Kouisni, L., and El Kharrassi, Y. (2022). A comparative study between conventional and advanced extraction techniques: Pharmaceutical and cosmetic properties of plant extracts. Molecules, 27(7), 2074, doi: 10.3390/molecules27072074.

[88] Gupta, A., Naraniwal, M., and Kothari, V. (2012). Modern extraction methods for preparation of bioactive plant extracts. International Journal of Applied and Natural Sciences, 1(1), 8–26, Erişim tarihi: 24.11.2024 https://www.researchgate.net/publication/236229645

[89] Picot-Allain, C., Mahomoodally, M.F., Ak, G., and Zengin, G. (2021). Conventional versus green extraction techniques – A comparative perspective. Current Opinion in Food Science, 40, 144–156, doi: 10.1016/j.cofs.2021.02.009.

[90] YA, B. and Carpenter, J. VK (2024). A comprehensive review on advanced extraction techniques for retrieving bioactive components from natural sources. ACS Omega, 9(29), 31285.

[91] Ranjha, M.M.A.N., Kanwal, R., Shafique, B., et al. (2021). A critical review on pulsed electric field: A novel technology for the extraction of phytoconstituents. Molecules, 26(16), 4893, doi: 10.3390/molecules26164893.

[92] Bocker, R. and Silva, E.K. (2022). Pulsed electric field assisted extraction of natural food pigments and colorings from plant matrices. Food Chemistry, 15, 100398, doi: 10.1016/j.fochx.2022.100398.

[93] Kumar, A.P.N., Kuma, M., Jose, A., Tomer, V., Oz, E., Proestos, C., Zeng, M., Elobeid, T., S, K., and Oz, F. (2023). Major phytochemicals: Recent advances in health benefits and extraction method. Molecules, 28(2), 887, doi: 10.3390/molecules28020887.

[94] Usman, M., Nakagawa, M., and Cheng, S. (2023). Emerging trends in green extraction techniques for bioactive natural products. Processes, 11(12), 3444, doi: 10.3390/pr11123444.

[95] Quitério, E., Grosso, C., Ferraz, R., Delerue-Matos, C., and Soares, C. (2022). A critical comparison of the advanced extraction techniques applied to obtain health-promoting compounds from seaweeds. Marine Drugs, 20(11), 677, doi: 10.3390/md20110677.

[96] Zhang, H., Ren, Y., Wei, J., Ji, Y., Bai, X., Shao, Y., Li, H., Gao, R., Wu, Z., Peng, Z., and Xue, F. (2022). Optimization of the efficient extraction of organic components in atmospheric particulate matter by accelerated solvent extraction technique and its application. Atmosphere, 13(5), 818, doi: 10.3390/atmos13050818.

[97] Mustafa, A. and Turner, C. (2011). Pressurized liquid extraction as a green approach in food and herbal plants extraction: A review. Analytica Chimica Acta, 703(1), 8–18.

[98] Nadar, S.S., Rao, P., and Rathod, V.K. (2018). Enzyme-assisted extraction of biomolecules as an approach to novel extraction technology: A review. Food Research International, 108, 309–330, doi: 10.1016/j.foodres.2018.03.006.

[99] Kate, A.E., Anupama, S., Shahi, N.C., Pandey, J.P., Om, P., and Singh, T.P. (2016). Novel eco-friendly techniques for extraction of food-based lipophilic compounds from biological materials. Natural Products Chemistry and Research, 4(5), 1000231.

[100] Kataoka, H. (2021). In-tube solid-phase microextraction: Current trends and future perspectives. Journal of Chromatography A, 1636, 461787, doi: 10.1016/j.chroma.2020.461787.

[101] Kanu, A.B. (2021). Recent developments in sample preparation techniques combined with high-performance liquid chromatography: A critical review. Journal of Chromatography A, 1654, 462444, doi: 10.1016/j.chroma.2021.462444.

[102] Shen, L., Pang, S., Zhong, M., Sun, Y., Qayum, A., Liu, Y., Rashid, A., Xu, B., Liang, Q., Ma, H., and
Ren, X. (2023). A comprehensive review of ultrasonic assisted extraction (UAE) for bioactive
components: Principles, advantages, equipment, and combined technologies. Sonochemistry:
Ultrasound in Organic Chemistry, 101, 106646, doi: 10.1016/j.ultsonch.2023.106646.

[103] Lupacchini, M., Mascitti, A., Giachi, G., Tonucci, L., d'Alessandro, N., Martinez, J., and Colacino, E.
(2017). Sonochemistry in non-conventional, green solvents or solvent-free reactions. Tetrahedron,
73, 609–653.

[104] Carreira-Casais, A., Otero, P., Garcia-Perez, P., Garcia-Oliveira, P., Pereira, A.G., Carpena, M.,
Soria-Lopez, A., Simal-Gandara, J., and Prieto, M.A. (2021). Benefits and drawbacks of ultrasound-
assisted extraction for the recovery of bioactive compounds from Marine Algae. International
Journal of Environmental Research and Public Health, 18(17), 9153, doi: 10.3390/ijerph18179153.

[105] Diaconeasa, Z., Iuhas, C.I., Ayvaz, H., et al. (2023). Anthocyanins from agro-industrial food waste:
Geographical approach and methods of recovery – A review. Plants, 12(1), 74, doi: 10.3390/
plants12010074.

[106] Carpentieri, S., Jambrak, A.R., Ferrari, G., and Pataro, G. (2021). Pulsed electric field-assisted
extraction of aroma and bioactive compounds from aromatic plants and food by-products. Frontiers
in Nutrition, 8, 792203, doi: 10.3389/fnut.2021.792203.

[107] Martínez, J.M., Delso, C., Álvarez, I., and Raso, J. (2020). Pulsed Electric Field-assisted extraction of
valuable compounds from microorganisms. Comprehensive Review in Food Science Food Safety,
19(3), 530–552, doi: 10.1111/1541-4337.1251.

[108] Lama-Muñoz, A. and Contreras Md, M. (2022). Extraction systems and analytical techniques for food
phenolic compounds: A review. Foods, 11(23), 3671, doi: 10.3390/foods11223671.

[109] Akhtar, I., Javad, S., Yousaf, Z., Iqbal, S., and Jabeen, K. (2019). Microwave assisted extraction of
phytochemicals: An efficient and modern approach for botanicals and pharmaceuticals. Pakistan
Journal of Pharmaceutical Sciences, 32(1), 223–224.

[110] Patel, M., Dave, K., and Patel, P. (2021). A review on different extraction methods of plants:
Innovation from ancient to modern technology. International Journal of Biology, Pharmacy and
Allied Sciences Special Issue, 10(12), 523–524.

[111] Ahangari, H., King, J.W., Ehsani, A., and Yousefi, M. (2021). Supercritical fluid extraction of seed oils –
A short review of current trends. Trends in Food Science and Technology, 111, 250, doi: 10.1016/j.
tifs.2021.02.066.

[112] Uwineza, P.A. and Waśkiewicz, A. (2020). Recent advances in supercritical fluid extraction of natural
bioactive compounds from natural plant materials. Molecules, 25(17), 3847, doi: 10.3390/
molecules25173847.

[113] Herrero, M., Mendiola, J.A., Cifuentes, A., and Ibáñez, E. (2010). Supercritical fluid extraction: Recent
advances and applications. Journal of Chromatography A, 1217(16), 2495–251, doi: 10.1016/j.
chroma.2009.12.019.

[114] Gros, Q., Duval, J., West, C., and Lesellier, C. (2021). On-line supercritical fluid extraction-supercritical
fluid chromatography (SFE-SFC) at a glance: A coupling story. TrAC, Trends in Analytical Chemistry,
144, 116433, doi: 10.1016/j.trac.2021.116433.

[115] Mottaleb, M.A. and Sarker, S.D. (2012). Accelerated solvent extraction for natural products isolation.
In: Sarker, S. and Nahar, L. (eds) Natural Products Isolation. Methods in Molecular Biology, Humana
Press, Totowa, NJ., 864. doi: 10.1007/978-1-61779-624-1_4.

[116] Bouloumpasi, E., Skendi, A., Christaki, S., Biliaderis, C.G., and Irakli, M. (2024). Optimizing conditions
for the recovery of lignans from sesame cake using three green extraction methods: Microwave-,
ultrasound-, and accelerated-assisted solvent extraction. Industrial Crops and Products, 207(2),
117770, doi: 10.1016/j.indcrop.2023.117770.

[117] Vazquez-Roig, P. and Picó, Y. (2015). Pressurized liquid extraction of organic contaminants in environmental and food samples. TrAC, Trends in Analytical Chemistry, 71, 55–64, doi: 10.1016/j.trac.2015.04.014.

[118] Priego-Capote, F. (2013). Delgado de la torre MDP. In: Rostagno, M.A. and Prado, J.M. (eds) Natural Product Extraction: Principles and Applications, The Royal Society of Chemistry, Cambridge, UK., 157–195.

[119] Wijesinghe, W.A.J.P. and Jeon, Y.J. (2012). Enzyme-assisted extraction (EAE) of bioactive components: A useful approach for recovery of industrially important metabolites from seaweeds: A review. Fitoterapia, 83(1), 6–12, doi: 10.1016/j.fitote.2011.10.016.

[120] Zwir-Ferenc, A. and Biziuk, M. (2006). Solid phase extraction technique – Trends, opportunities and applications. Polish Journal of Environmental Studies, 15(5), 677–690.

[121] Koehn, F. and Carter, G. (2005). The evolving role of natural products in drug discovery. Nature Reviews Drug Discovery, 4(3), 206–220, doi: 10.1038/nrd1657.

[122] Khan, W.A., Arain, M.B., and Soylak, M. (2020). Nanomaterials-based solid phase extraction and solid phase microextraction for heavy metals food toxicity. Food and Chemical Toxicology: An International Journal Published for the British Industrial Biological Research Association, 145, 111704, doi: 10.1016/j.fct.2020.111704.

[123] Hinshaw, J.W. (2012). Solid-phase microextraction (SPME): A discussion. LCGC Asia Pacific, 15(4), 22–25.

[124] Peng, S., Huang, X., Huang, Y., Huang, Y., Zheng, J., Zhu, F., Xu, J., and Ouyang, G. (2021). Novel solid-phase microextraction fiber coatings: A review. Journal Separation Science, 45(1), 282–334, doi: 10.1002/jssc.202100634.

[125] Jin, H.F., Shi, Y., and Cao, J. (2024). Recent advances and applications of novel advanced materials in solid-phase microextraction for natural products. TrAC, Trends in Analytical Chemistry, 178, 117858, doi: 10.1016/j.trac.2024.117858.

[126] Zheng, J., Huang, J., Yang, Q., Ni, C., Xie, X., Shi, Y., Sun, J., Zhu, F., and Ouyang, G. (2018). Fabrications of novel solid phase microextraction fiber coatings based on new materials for high enrichment capability. TrAC, Trends in Analytical Chemistry, 108, 135–153, doi: 10.1016/j.trac.2018.09.009.

[127] Zheng, J., Kuang, Y., Zhou, S., Gong, X., and Ouyang, G. (2023). Latest improvements and expanding applications of solid-phase microextraction. Analytical Chemical, 95, 218, doi: 10.1021/acs.analchem.3c02946.

[128] Dettweiler, M., Marquez, L., Bao, M., and Quave, C.L. (2020). Quantifying synergy in the bioassay-guided fractionation of natural product extracts. PLoS One, 15(8), e0235723, doi: 10.1371/journal.pone.0235723.

[129] Rimando, A.M., Olofsdotter, M., Dayan, F.E., and Duke, S.O. (2001). Searching for rice allelochemicals: An example of bioassay-guided isolation. Journal of Agronomy, 93, 17.

[130] Kellogg, J. and Kang, S. (2020). Metabolomics, an essential tool in exploring and harnessing microbial chemical ecology. Phytobiomes Journal, 4(3), 198, doi: 10.1094/PBIOMES-04-20-0032-RVW.

[131] Malviya, N. and Malviya, S. (2017). Bioassay guided fractionation-an emerging technique influencing the isolation, identification and characterization of lead phytomolecules. International Journal of Health and Pharmaceutical, 2(5), 2, Erişim Tarihi 26.11.2024 http://escipub.com/international-journal-of-hospital-pharmacy/.

[132] Nothias, L.F., Nothias-Esposito, M., da Silva, R., Wang, M., Protsyuk, I., Zhang, Z., Sarvepalli, A., Leyssen, P., Touboul, D., Costa, J., Paolini, J., Alexandrov, T., Litaudon, M., and Dorrestein, P.C. (2018). Bioactivity-based molecular networking for the discovery of drug leads in natural product bioassay-guided fractionation. Journal of Natural Products, 81(4), 758–767, doi: 10.1021/acs.jnatprod.7b00737.

[133] Lahlou, M. (2013). The success of natural products in drug discovery. Pharmacology and Pharmacy, 4, 17–31, doi: 10.4236/pp.2013.43A003.

[134] MU, R., Abdullah, K.F., and Niaz, K. (2020). Introduction to natural products analysis. In: Silva, A.S., Nabavi, S.F., Saeedi, M., and Nabavi, S.M. (eds) Recent Advances in Natural Products Analysis, Elsevier, Amsterdam, 3–15. doi: 10.1016/B978-0-12-816455-6.00001-9.

[135] Salam, A.M., Lyles, J.T., and Quave, C.L. (2012). Methods of extraction and chemical analysis of medicinal plants. In: Albuquerque, U., De Lucena, R., Cruz da Cunha, L., and Alves, R. (eds) Ethnobiology and Ethnoecology Methods and Techniques, Springer Protocols Handbooks. Humana Press, New York, 257–283. doi: 10.1007/978-1-4939-8919-5_17.

[136] Bucar, F., Wube, A., and Schmid, M. (2013). Natural product isolation – How to get from biological material to pure compounds. Natural Product Reports, 30, 525–545.

[137] Kumari, V.B.C., Patil, S.M., Ramu, R., Shirahatti, P.S., Kumar, N., Sowmya, B.P., Egbuna, C., Uche, C.Z., and Patrick-Iwuanyanwu, K.C. (2022). Chromatographic techniques: Types, principles, and applications. In: Egbuna, C., Patrick-Iwuanyanwu, K.C., Shah, M.A., Ifemeje, J.C., and Rasul, A. (eds) Analytical Techniques in Biosciences, Academic Press, Cambridge, MA., 73–101. doi: 10.1016/B978-0-12-822654-4.00013-0.

[138] Moulishankar, A., Ganesan, P., Elumalai, M., and Lakshmanan, K. (2021). Significance of TLC and HPTLC in phytochemical screening of herbal drugs. Journal of Global Pharma Technology, 30–31, Erişim Tarihi 27.11.2024 https://www.researchgate.net/publication/348432127.

[139] Gökbulut, A. (2021). High Performance Thin Layer Chromatography (HPTLC) for the investigation of medicinal plants. Current Analytical Chemistry, 17(9), 1252, doi: 10.2174/1573411016999200602124813.

[140] Sanket, G.B. and Dighe, P.R. (2022). Analysis of Herbal Drugs by HPTLC. Asian Journal of Pharmaceutical Research and Development, 10(2), 125–128, doi: 10.22270/ajprd.v10i2.1056.

[141] Nahar, L., Onder, A., and Sarker, S.D. (2020). A review on the recent advances in HPLC, UHPLC and UPLC analyses of naturally occurring cannabinoids (2010–2019). Phytochemical Analysis: An International Journal of Plant Chemical and Biochemical Techniques, 31(5), 413–457, doi: 10.1002/pca.2906.

[142] Basharat, R., Kotra, V., Lean, Y.L., Mathews, A., and Kanakal, M.M. (2021). A mini-review on ultra performance liquid chromatography. Oriental Journal of Chemistry, 37(4), 847–857, doi: 10.13005/ojc/370411.

[143] Siddique, I. (2021). Unveiling the power of high-performance liquid Chromatography: Techniques, applications, and innovations. European Journal of Advances in Engineering and Technology, 8(9), 79–84.

[144] Eugster, P.J., Guillarme, D., Rudaz, S., Veuthey, J.L., Carrupt, P.A., and Wolfender, J.L. (2011). Ultra high pressure liquid Chromatography for crude plant extract profiling. Journal of AOAC International, 94(1), 51–70, doi: 10.1093/jaoac/94.1.51.

[145] Ma, J., Li, K., Shi, S., Li, J., Tang, S., and Liu, L.H. (2022). The application of UHPLC-HRMS for quality control of traditional Chinese medicine. Frontiers in Pharmacology, 13, 922488, doi: 10.3389/fphar.2022.922488.

[146] Rasul, M.G. (2018). Extraction, isolation, and characterization of natural products from medicinal plants. International Journal of Basic Sciences and Applied Computing, 2(6), F0076122618.

[147] Obara, K., Uenoyama, R., Obata, Y., and Miyazaki, M. (2024). Development of the gas chromatography/mass spectrometry-based aroma designer capable of modifying volatile chemical compositions in complex odors. Chemical Senses, 49, bjae007, doi: 10.1093/chemse/bjae007.

[148] Teonata, N., Wijaya, V.A., Vithaloka, V.S., and Thariq, M. (2021). An introduction to different types of gas chromatography. Jurnal Sains Dan Terapan Kimia, 15(1), 8–17, doi: 10.20527/jstk.v15i1.8621.

[149] Stashenko, E. and Martínez, J.R. (2014). Gas chromatography-mass spectrometry. In: Guo, X. (ed) Advances in Gas Chromatography, BoD – Books on Demand, Norderstedt, 1–38. doi: 10.5772/57492.

[150] Susanti, I., Pratiwi, R., Rosandi, Y., and Hasanah, A.N. (2024). separation methods of phenolic compounds from plant extract as antioxidant agents candidate. Plants, 13(7), 965, doi: 10.3390/plants13070965.

[151] Zhang, Q.W., Lin, L.G., and Ye, W.C. (2018). Techniques for extraction and isolation of natural products: A comprehensive review. Chinese Medicine, 13, 20, doi: 10.1186/s13020-018-0177-x.

[152] Feng, W., Li, M., Hao, Z., and Zhang, J. (2020). Analytical methods of isolation and identification. In: Rao, V., Mans, D., and Rao, L. (eds) Phytochemicals in Human Health, IntechOpen, London UK, 45–73. doi: 10.5772/intechopen.88122.

[153] Ngere, J.B., Ebrahimi, K.H., Williams, R., Pires, E., Walsby-Tickle, J., and McCullagh, C.S.O. (2023). Ion-exchange chromatography coupled to mass spectrometry in life science, environmental, and medical research. Analytical Chemical, 95(1), 152–166.

[154] Cummins, P.M., Rochfort, K.D., and O'Connor, B.F. (2017). Ion-exchange chromatography: Basic principles and application. In: Walls, D. and Loughran, S. eds. Protein Chromatography. Methods Mol Biol, Humana Press, New York, NY, Vol. 1485, 11, doi: 10.1007/978-1-4939-6412-3_11.

[155] Wallace, R.G. and Rochfort, K.D. (2023). Ion-exchange chromatography: Basic principles and application. In: Loughran, S.T. and Milne, J.J. (eds) Protein Chromatography, Methods Mol Biol 2699, Springer, Cham, 161–177. doi: 10.1007/978-1-0716-3362-5_9.

[156] Kammerer, J., Carle, R., and Kammerer, D.R. (2011). Adsorption and ion exchange: Basic principles and their application in food processing. Journal of Agricultural and Food Chemistry, 59(1), 22–42.

[157] Dragull, K. and Beck, J.J. (2012). Isolation of natural products by ion-exchange methods. Methods in Molecular Biology, 864, 189–219.

[158] Michalski, R. (2016). Principles and applications of ion chromatography. In: Michalski, R. (ed) Application of IC-MS and IC-ICP-MS in Environmental Research, John Wiley & Sons, Inc., Hoboken, NJ., 1. doi: 10.1002/9781119085362.ch1.

[159] Fonmboh, D.J., Ejoh, A.R., Fokunang, T.E., et al. (2020). An overview of methods of extraction, isolation and characterization of natural medicinal plant products in improved traditional medicine research. Asian. Journal of Pharmaceutical Sciences, 9(2), 1–52, doi: 10.9734/ajrimps/2020/v9i230152.

[160] McRae, J., Yang, Q., Crawford, R., and Palombo, E. (2007). Review of the methods used for isolating pharmaceutical lead compounds from traditional medicinal plants. Environist, 27, 165–174, doi: 10.1007/s10669-007-9024-9.

[161] Brusotti, G., Cesari, I., Dentamaro, A., Caccialanza, G., and Massolini, G. (2014). Isolation and characterization of bioactive compounds from plant resources: The role of analysis in the Ethnopharmacological approach. Journal of Pharmaceutical and Biomedical Analysis, 87, 218–228, doi: 10.1016/j.jpba.2013.03.007.

[162] Ivanović, M., Islamčević Razboršek, M., and Kolar, M. (2020). Innovative Extraction Techniques Using Deep Eutectic Solvents and Analytical Methods for the Isolation and Characterization of Natural Bioactive Compounds from Plant Material. Plants, 9(11), 1428, doi: 10.3390/plants9111428.

[163] Dayrit, F.M. and Dios, A.C. (2017). Chapter 5: 1H and 13C NMR for the profiling of natural product extracts: Theory and applications. In: Sharmin, E., Zafar, F. (eds) Spectroscopic Analysis – Developments and Applications. IntechOpen, London, 82. doi: 10.5772/intechopen.71040.

[164] Huang, X., Powers, R., Tymiak, A., Espina, R., and Roongta, V. (2008). Introduction to NMR and its application in metabolite structure determination. In: Zhang, D., Zhu, M., and Humphreys, W.G. (eds) Drug Metabolism in Drug Design and Development, John Wiley & Sons, Inc., Hoboken, NJ., 369–405.

[165] Gehin, C. and Holman, S.W. (2021). Advances in high-resolution mass spectrometry applied to pharmaceuticals in 2020: A whole new age of information. Analytical Science Advances, 2(3), 142–156.

[166] Ma, X. (2022). Recent advances in mass spectrometry-based structural elucidation techniques. Molecules, 27(19), 6466, doi: 10.3390/molecules27196466.

[167] Francke, W. (2010). Structure elucidation of some naturally occurring carbonyl compounds upon coupled gas chromatography/mass spectrometry and micro-reactions. Chemoecology, 20(3), 163–169, doi: 10.1007/s00049-010-0048-0.

[168] Wolfender, J.L. (2009). HPLC in natural product analysis: The detection issue. Planta Medica, 75(7), 719–734, doi: 10.1055/s-0028-1088393.

[169] Rochat, B. (2012). Quantitative/Qualitative analysis using LC–HRMS: The fundamental step forward for clinical laboratories and clinical practice. Bioanalysis, 4(14), 1709–1711.

[170] Alvarez-Rivera, G., Ballesteros-Vivas, D., Parada-Alfonso, F., Ibañez, E., and Cifuentes, A. (2019). Recent applications of high resolution mass spectrometry for the characterization of plant natural products. TrAC, Trends in Analytical Chemistry, 112, 87–101, doi: 10.1016/j.trac.2019.01.002.

[171] Bongiorno, D., Di Stefano, V., Indelicato, S., Avellone, G., and Ceraulo, L. (2023). Bio-phenols determination in olive oils: Recent mass spectrometry approaches. Mass Spectrometry Reviews, 42(4), 1462–1502, doi: 10.1002/mas.21744.

[172] Hanson, J.R. (2015). Current strategies for the elucidation of the structures of natural products. Science Program, 98(2), 177–188.

[173] Sasidharan, S., Chen, Y., Saravanan, D., Sundram, K.M., and Extraction, Y.L.L. (2011). Isolation and characterization of bioactive compounds from plants' extracts. African Journal of Traditional, Complementary and Alternative Medicines (AJTCAM), 8(1), 1–10.

[174] Rocha, F.S., Gomes, A.J., Lunardi, C.N., Kaliaguine, S., and Patience, G.S. (2018). Experimental methods in chemical engineering: Ultraviolet visible spectroscopy – UV-Vis. The Canadian Joumal of Chemical Engineering, 96(12), 2512–2517, doi: 10.1002/cjce.23344.

[175] Vogt, C., Wondergem, C.S., and Weckhuysen, B.M. (2023). Ultraviolet-Visible (UV-Vis) Spectroscopy. In: Wachs, I.E. and Bañares, M.A. (eds) Springer Handbook of Advanced Catalyst Characterization, Springer, Cham, 237–264. doi: 10.1007/978-3-031-07125-6_11.

[176] Mandru, A., Mane, J., and Mandapati, R. (2023). A review on UV-visible spectroscopy. Journal of Pharma Insights and Research, 1(2), 091–096, doi: 10.5281/zenodo.10232708.

[177] Dixon, N., Wong, L.S., Geerlings, T.H., and Micklefield, J. (2007). Cellular targets of natural products. Natural Product Reports, 24(6), 1288–1310, doi: 10.1039/B603184K.

[178] Indrayanto, G., Putra, G.S., and Suhud, F. (2021). Validation of in-vitro bioassay methods: Application in herbal drug research. In: A. Al-Majed (Ed) Profiles of Drug Substances, Excipients and Related Methodology, Elsevier, Cambridge, MA., 273–307. doi: 10.1016/bs.podrm.2020.07.005.

[179] Sabotič, J., Bayram, E., Ezra, D., Gaudêncio, S.P., Haznedaroğlu, B.Z., Janež, N., and Vasquez, M.I. (2024). A guide to the use of bioassays in exploration of natural resources. Biotechnology Advance, 71, 108307, doi: 10.1016/j.biotechadv.2024.108307.

[180] Montalvão, S.I.G., Singh, V., and Haque, S. 2014. Bioassays for bioactivity screening. In: RochaSantos, T. and Duarte, A.C. (eds) Comprehensive Analytical Chemistry, Elsevier, Amsterdam, 65, 79–114. doi: 10.1016/B978-0-444-63359-0.00005-7.

[181] Barba-Ostria, C., Carrera-Pacheco, S.E., Gonzalez-Pastor, R., Heredia-Moya, J., Mayorga-Ramos, A., Rodríguez-Pólit, C., Zúñiga-Miranda, J., Arias-Almeida, B., and Guamán, L.P. (2022). Evaluation of biological activity of natural compounds: Current trends and methods. Molecules, 27(14), 4490, doi: 10.3390/molecules27144490.

[182] Agarwal, A., D'Souza, P., Johnson, T.S., Dethe, S.M., and Chandrasekaran, C.V. (2014). Use of in vitro bioassays for assessing botanicals. Current Opinion in Biotechnology, 27, 39–44, doi: 10.1016/j.copbio.2013.08.010.

[183] Strömstedt, A.A., Felth, J., and Bohlin, L. (2014). Bioassays in natural product research: Strategies and methods in the search for anti-inflammatory and antimicrobial activity. Phytochemical Analysis:

An International Journal of Plant Chemical and Biochemical Techniques, 25(1), 13–28, doi: 10.1002/pca.2468.

[184] Deore, A.B., Dhumane, J.R., Wagh, H.V., and Sonawane, R.B. (2019). The stages of drug discovery and development process. Asian Journal of Pharmaceutical Research and Development, 7(6), 62–67, doi: 10.22270/ajprd.v7i6.616.

[185] Mohs, R.C. and Greig, N.H. (2017). Drug discovery and development: Role of basic biological research. Alzheimers Dement (N Y), 3(4), 651–657, doi: 10.1016/j.trci.2017.10.005.

[186] Routledge, P.A. (2008). The European herbal medicines directive: Could it have saved the lives of Romeo and Juliet?. Drug Safety, 31(5), 416–418, doi: 10.2165/00002018-200831050-00006.

[187] Jwh, L. and Vederas, J.C. (2009). Drug discovery and natural products: End of an era or an endless frontier?. Science, 325(5937), 161–165, doi: 10.1126/science.1168243.

[188] Thomford, N.E., Senthebane, D.A., Rowe, A., et al. (2018). Natural products for drug discovery in the 21st century: Innovations for novel drug discovery. International Journal of Molecular Sciences, 19(5), 1578, doi: 10.3390/ijms19051578.

[189] Sotelo, J.M., Branes, J., Vergara, L., and Selman, A. (2003). Vinorelbine: A review of its pharmacology and clinical efficacy in the treatment of cancer. Cancer Chemotherapy and Pharmacology, 52(6), 478–487, doi: 10.1007/s00280-003-0684-3.

[190] Rautio, J., Meanwell, N.A., Di, L., and Hageman, M.J. (2018). The expanding role of prodrugs in contemporary drug design and development. Nat. Nature Reviews Drug Discovery, 17(8), 559–587, doi: 10.1038/nrd.2018.46.

[191] Chhabra, N., ML, A., and Padmanabhan, D. (2013). A review of drug isomerism and its significance. International Journal of Applied and Basic Medical Research, 3(1), 16–18, doi: 10.4103/2229-516X.112233.

[192] Murray, R., Bender, D., Botham, K.M., Kennelly, P.J., Rodwell, V., and Weil, P.A. (2012). Harper's Illustrated Biochemistry, 29th ed, Lange Medical Books/McGraw-Hill Medical Publishing Division, New York, NY.

[193] Horgan, R.P. and Kenny, L.C. (2011). 'Omic' technologies: Genomics, transcriptomics, proteomics and metabolomics. Obstetrics and Gynaecology, 13(3), 189–195, doi: 10.1576/toag.13.3.189.27672.

[194] Sahoo, S. and Brijesh, S. (2019). Pharmacogenomic assessment of herbal drugs in affective disorders. Biomedicine and Pharmacotherapy, 109, 1148–1162, doi: 10.1016/j.biopha.2018.10.122.

[195] Ulrich-Merzenich, G., Panek, D., Zeitler, H., Vetter, H., and Wagner, H. (2009). New perspectives for synergy research with the "omic"-technologies. Phytomedicine, 16(6–7), 495–508, doi: 10.1016/j.phymed.2009.04.001.

[196] Chakraborty, P. (2018). Herbal genomics as tools for dissecting new metabolic pathways of unexplored medicinal plants and drug discovery. Biochim Open, 6, 9–16, doi: 10.1016/j.biopen.2017.12.003.

[197] Ghorbani, A., Saeedi, Y., and De Boer, H.J. (2017). Unidentifiable by morphology: DNA barcoding of plant material in local markets in Iran. PLoS One, 12(4), e0175722, doi: 10.1371/journal.pone.0175722.

[198] Barbosa, S., Carreira, S., Bailey, D., Abaitua, F., and O'Hare, P. (2015). Phosphorylation and SCF-mediated degradation regulate CREB-H transcription of metabolic targets. Molecular Biology of the Cell, 26(16), 2939–2954, doi: 10.1091/mbc.E15-04-0247.

[199] Kabadi, A.M., McDonnell, E., Frank, C.L., and Drowley, L. (2020). Applications of functional genomics for drug discovery. SLAS DISCOVERY: Advancing Life Sciences R&D, 25(8), 823–842, doi: 10.1177/2472555220902092.

[200] Wolfender, J.L., Marti, G., Thomas, A., and Bertrand, S. (2015). Current approaches and challenges for the metabolite profiling of complex natural extracts. Journal of Chromatography A, 1382, 136–164, doi: 10.1016/j.chroma.2014.10.091.

[201] Colgrave, M.L., Goswami, H., Howitt, C.A., and Tanner, G.J. (2012). What is in a beer? Proteomic characterization and relative quantification of hordein (gluten) in beer. Journal of Proteome Research 11(1), 386–396. doi: 10.1021/pr2008434.

[202] Chen, X., Wang, Y., Ma, N., et al. (2020). Target identification of natural medicine with chemical proteomics approach: Probe synthesis, target fishing and protein identification. Signal Transduction and Targeted Therapy, 5(1), 72, doi: 10.1038/s41392-020-0186-y.

[203] DiMasi, J.A., Grabowski, H.G., and Hansen, R.W. (2016). Innovation in the pharmaceutical industry. New estimates of R&D costs. Journal of Health Economics, 47, 20–33, https://doi.org/10.1016/j.jhealeco.2016.01.012.

[204] Brazil, R. (2017). Artificial intelligence: Will it change the way drugs are discovered?. The Pharmaceutical Journal, 299, 1–10, doi: 10.1002/phj.1805.

[205] Fleming, N. (2018). How artificial intelligence is changing drug discovery. Nature, 557(7707), S55–S57, doi: 10.1038/d41586-018-05267-x.

Fatemeh Ahmadi*, Maximilian Lackner, and August Starzinger

Chapter 3
Challenges encountered in growing medicinal and aromatic plants

Abstract: Medicinal and aromatic plants (MAPs) are invaluable resources, widely utilized across diverse industries, including pharmaceuticals, cosmetics, agriculture, and food production, due to their rich composition of bioactive compounds such as alkaloids, terpenoids, and phenolics. These compounds possess potent medicinal properties, including antimicrobial, antioxidant, anti-inflammatory, and anticancer activities, making them essential in the health and wellness sectors. Despite their economic and therapeutic significance, the cultivation of MAPs presents multifaceted challenges. Environmental stresses, including salinity, drought, heat, and soil pH imbalances, significantly affect their growth, metabolism, and the biosynthesis of secondary metabolites, which are critical for their bioactivity. This chapter provides a comprehensive analysis of the physiological and biochemical responses of MAPs to various abiotic stresses, highlighting the intricate mechanisms plants use to adapt and survive under unfavorable conditions. It explores the role of stress in modulating secondary metabolite production, often enhancing their concentration but at the potential cost of biomass yield. Additionally, the chapter delves into the industrial applications of MAP-derived compounds, emphasizing their importance in pharmaceuticals for drug development, cosmetics for natural formulations, and agriculture for eco-friendly pest and disease management solutions. Strategies for mitigating challenges in MAP cultivation are discussed, including integrated pest management, sustainable soil and water management practices, and the optimization of light intensity. Emerging technologies such as precision agriculture, genetic and epigenetic modifications, and advanced metabolomics approaches are presented as potential solutions to enhance MAP productivity and quality. The chapter also addresses global issues, including the impact of climate change on MAP cultivation, regulatory hurdles, and the need for equitable benefit-sharing in the use of traditional medicinal plants. By integrating traditional knowledge with modern scientific advancements, this chapter underscores the potential of MAPs to contribute to sustainable agricultural practices and produce high-quality medicinal compounds. Future perspectives emphasize the importance of interdisciplinary research to overcome the chal-

*Corresponding author: Fatemeh Ahmadi, School of Agriculture and Environment, University of Western Australia, Crawley, WA 6009, Australia, e-mail: fatemeh.ahmadi@uwa.edu.au
Maximilian Lackner, University of Applied Sciences Technikum Wien, Hoechstaedtplatz 6, 1200 Vienna, Austria
August Starzinger, Canngoo GmbH i.G., Rainerstraße 36, 5310 Mondsee, Austria

https://doi.org/10.1515/9783111469713-003

lenges in MAP cultivation, ensuring their continued role in supporting global health, environmental sustainability, and economic development.

Keywords: bioactive compounds, industrial applications, secondary metabolites, salinity stress, sustainable agriculture

3.1 Introduction

Plants possess the remarkable ability to transform water, minerals, and other soil elements into complex compounds essential for their metabolism [1]. These bioactive substances, including essential oils, alkaloids, tannins, and bitter compounds, not only benefit the plants themselves but also offer significant advantages to human health [2]. These natural compounds can enhance immune function and support various organ systems in the human body, positively impacting tissues and organs [3].

Despite rapid advancements in modern medicine, the use of natural remedies persists. Concerns about synthetic medication side effects and chemical additives have led many to prefer medicinal plants as alternatives [4]. The terms "medicinal" and "aromatic" are frequently used in conjunction to describe these herbs, generally referring to plants containing bioactive compounds with therapeutic and fragrant properties that can positively affect human and animal physiology [5].

Medicinal and aromatic plants (MAPs) serve multiple purposes, from disease prevention and health maintenance to industrial applications. Their uses span various sectors, including nutrition, cosmetics, personal care, religious rituals, and the food, pharmaceutical, and perfume industries [6]. The range of active ingredients in MAPs is extensive, and while no standardized classification exists, they can be categorized based on their families, active components, intended uses, or pharmaceutical effects [7].

3.2 Bioactive compounds

Plant metabolites are typically categorized as primary or secondary, with proteins and nucleic acids excluded from this classification. Primary metabolites, such as carbohydrates, fats, and proteins, are crucial for basic plant functions [8]. Secondary metabolites, however, are diverse compounds found in smaller quantities and are not essential for plant survival [9].

The stems and leaves of most MAPs are rich in secondary metabolites, with varied physiological activities. These compounds are generally classified into three main groups: phenolic compounds, terpenoids/terpenes, and alkaloids [10]. Common secondary metabolites in MAPs include essential oils, glycosides, steroids, saponins, flavonoids, tannins, phenols, pigments, and resins, with over 30,000 distinct types identi-

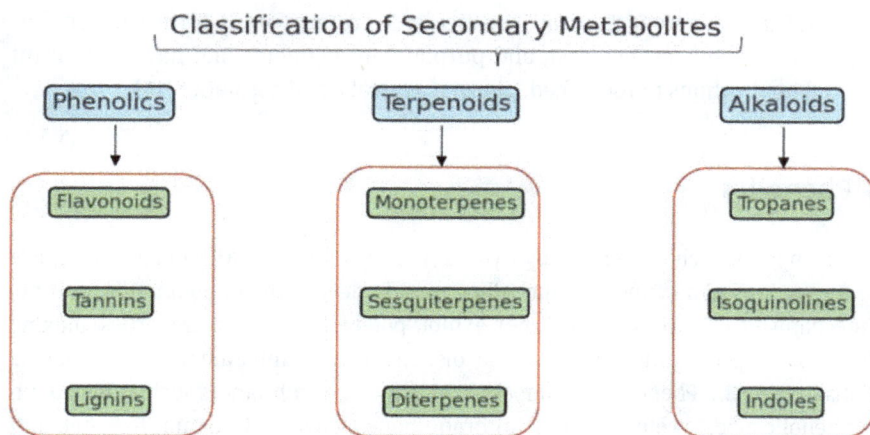

Figure 3.1: Classification of secondary metabolites in medicinal plants.

fied in plants (Figure 3.1). Additionally, MAPs may contain polyphenols, kinins, flava-nols/flavonoids, polypeptides, or their oxygenated derivatives. Some of these compounds may work synergistically, enhancing their biological effects [11].

3.2.1 Alkaloids

Alkaloids are nitrogen-containing compounds, typically with a ring structure, that exhibit alkaline properties and potent physiological effects [12]. Usually found in plants, these compounds often exist as salts with minerals or organic acids, and rarely in free form within plant cells. While alkaloids can have medicinal benefits when used correctly, misuse can lead to severe consequences [13]. Examples include morphine from Papaver somniferum, cocaine from coca plants, nicotine from tobacco, and theobromine from cacao, all of which have addictive potential.

3.2.2 Terpenoids (terpenes)

Terpenoids are fundamental components of essential oils, composed of 5-carbon isoprene units. The number of isoprene units determines their structure, ranging from monoterpenes to tetraterpenes. Monoterpenes and sesquiterpenes are prevalent in essential oils, with monoterpenoids often responsible for a plant's scent and flavor. Sesquiterpenes are known for their antimicrobial and antitumoral properties [14].

Diterpenoids are found in resins and include the plant growth hormone gibberellin. Triterpenoids, like cucurbitacins and limonoids, play crucial ecological roles in plant defense against herbivores and insects [15]. For instance, azadirachtin from *Aza-*

dirachta indica is a potent natural insecticide. Tetraterpenoids form carotenoids, which produce yellow, orange, red, and purple plant pigments and have significant commercial applications in food, feed, pharmaceuticals, and cosmetics [16].

3.2.3 Phenolics

Phenolic compounds, characterized by hydroxyl groups attached to aromatic rings, are crucial for various plant functions including growth, development, pollination, and defense mechanisms [17]. These compounds exhibit potent antioxidant properties, making them valuable in preventing cardiovascular diseases, cancer, inflammation, and cellular mutations in table 3.1. Phenolics encompass a wide range of substances such as benzoquinones, phenolic acids, acetophenones, anthraquinones, flavonoids, lignin, flavonol, and tannins [18]. Flavonoids, a subgroup containing flavones, flavonol, and anthocyanins, contribute to plant coloration, protect cells from oxidative stress, and attract pollinators [19]. Tea leaves are rich in phenolics, particularly catechins and theaflavins, while artichoke contains snaring, which promotes liver cell regeneration [20]. Silymarin, derived from thistle seeds, is utilized in liver treatments. Essential oils, complex mixtures of numerous compounds, are another important plant product. For instance, rose oil contains over a hundred constituents, with citronellol, neral, and geraniol being key components [21]. Other plants like fennel and anise are rich in menthol, thyme species contain carvacrol, lavender has linalyl acetate, mint contains menthol, and melissa has neral. Medical sage is high in thujone, while marjoram is known for its thymol content [22].

Table 3.1: Classification of secondary metabolites in medicinal and aromatic plants [161].

Class	Subcategories	Examples	Biological activities
Phenolics	Flavonoids, tannins, and lignins	Catechins (tea) and resveratrol	Antioxidant and anti-inflammatory
Terpenoids	Monoterpenes and sesquiterpenes	Limonene (citrus) and carotenoids	Antimicrobial, insecticidal, and hormonal regulation
Alkaloids	Tropanes, isoquinolines, and indoles	Morphine (poppy) and vinblastine	Analgesic, anticancer, and neurological effects

3.3 Industrial importance of biological active compounds

MAPs find applications across various industries, including pharmaceuticals, personal care, cosmetics, and organic food production [23]. Growing concerns about synthetic additives have led to increased use of MAPs, their extracts, and essential oils in pharmaceutical, food, and feed industries [24]. While synthetic medication use has risen with population growth, there is a parallel trend toward plant-based products due to concerns about medication side effects. Pharmaceutical companies are actively patenting MAP-derived products, with about 40% of medications developed in the last two decades based on natural formulations [25]. The demand for plant products in other industries is also rising. They are used in nonalcoholic beverages and confectioneries within the food industry and in perfumes, skin and hair care products, and aromatherapy in the cosmetics sector [26]. Mint exemplifies a versatile MAP with diverse applications due to its antimicrobial, antidiabetic, antioxidant, anti-inflammatory, antitumor, and pesticide properties, finding use across cosmetics, food, agriculture, and textiles [27].

3.4 Industrial use of MAPs

Bioactive compounds from plants and their bioactive compounds are utilized across various industries. In cosmetics, essential oils from lavender, mint, and rosemary provide aroma and therapeutic benefits [28]. The pharmaceutical sector leverages the antioxidant, antimicrobial, and anti-inflammatory properties of plant bioactive compounds for new medication development. The growing consumer preference for natural and organic products has made plant extracts popular in various formulations [29]. Lavender oil, for instance, is favored in cosmetics and shampoos for its antioxidant properties. Rosemary extracts are valued for their antioxidant and anti-inflammatory characteristics, showing potential in new medication formulations and as dietary supplements [30].

3.5 Essential oils

Plants produce two distinct types of oils: fixed oils and essential oils. Fixed oils consist of fatty acids and glycerol esters, while essential oils are complex mixtures of volatile organic compounds and various metabolites [31]. These essential oils contribute to the plant's distinctive taste and scent, forming its essence. They play a crucial role in the plant's immune and defense systems against environmental threats [32]. Essential oils are typically extracted from the aromatic, nonwoody parts of plants, such as flowers, leaves, fruit peels, or roots, using methods like steam distillation or hydrodistillation [33]. These volatile liquids are insoluble in water but readily dissolve in organic solvents. The global aroma and scent industry relies heavily on essential oils, which account for approximately 17% of the sector. Common sources include rose, jasmine, and mint [34]. MAPs, particularly their essential oils, exhibit a wide range of beneficial properties, including antibacterial, antiviral, antifungal, antiparasitic, and insecticidal activities. They also demonstrate hypolipidemic, antioxidant, and anti-toxigenic effects, and can help control odors and reduce ammonium and methane emissions in ruminants [35]. Essential oils find applications in various industries as additives for cosmetics, medical products, soaps, perfumes, ice creams, and disinfectants [36]. Emerging research explores their potential in preventing nutrient deficiencies, pest control, and plant growth promotion. While studies have investigated essential oils from roughly 3,000 aromatic plants, only about 300 are commercially available in table 3.2. Of these, around 50 are in high demand for industrial and commercial use, with just two dozen seeing regular, large-scale production [37].

Table 3.2: Industrial applications of bioactive compounds from medicinal plants [30].

Industry	Bioactive compounds	Example of plants	Applications
Pharmaceuticals	Alkaloids and phenolics	Poppy, willow, and sage	Pain relief, antioxidants, and antibacterial agents
Cosmetics	Essential oils and flavonoids	Lavender, calendula, and chamomile	Skincare, anti-aging, and hair care
Food and beverages	Terpenoids and phenolics	Mint, rosemary, and basil	Flavoring, preservatives, and antioxidant additives
Agriculture	Terpenes and alkaloids	Neem, tobacco, and peppermint	Pesticides, growth regulators, and soil enhancers

3.6 MAPs in the dye industry

The growing interest in natural dyes stems from the increased awareness of sustainability and environmental concerns. These natural colorants, derived from plants, insects, animals, and minerals, offer more than just aesthetic appeal [38]. Many plant-based dyes possess additional benefits such as antibacterial, antioxidant, anti-inflammatory, and UV-protective properties due to their polyphenol, flavonoid, and anthocyanin content [39]. Numerous MAPs serve as sources of natural dyes. Chlorophyll, for instance, is responsible for the ubiquitous green color in plants [40]. However, plants can produce a diverse array of colors in their flowers and leaves, ranging from white and pink to yellow and red. Even nongreen plants can harness sunlight to produce various pigments, resulting in a wide spectrum of colors [41]. The use of natural dyes in textiles has a long history that continues to this day. These dyes play a crucial role in the textile industry's efforts to reduce water pollution and promote sustainable practices in both raw materials and finished products [42]. A variety of plants yield different colors. Turmeric produces a vibrant yellow, while other yellow dyes come from *woodwax*, *Venetian sumac*, and *dyer's mignonette*. Additional yellow sources include *Adhatoda vasica*

Nees leaves, *jackfruit, Crocus sativus* L. flowers, chamomile, *Tagetes erecta* L., *Nyctanthes arbortristis* L., and *Cassia auriculata* L. seeds and flowers [43]. Red dyes are obtained from *Carthamus tinctorious* L. and *Tagetes erecta* L. flowers, while purple comes from *Galium aparine* L. roots, onion peels, and rosemary leaves and flowers. Some fruits, like *Acanthophonax trifoliatum* L. and *Garcinia mangostana* L., yield black dyes [44]. Brown dyes are derived from the leaves and bark of plants such as *Azadirachta indica* A., *Acacia catechu,* and oak trees. Tea plants can impart both color and antibacterial properties to textiles. Carotenoids in plants produce red-yellow hues, while other pigments include orellin, bixin, annatto, mordant, and lawone [45].

3.6.1 Use of MAPs in the perfumery

The perfumes sector blends natural essential oils and synthetic organic compounds to create unique scent experiences. The fragrance industry offers over 3,000 commercial products, combining science and artistry in the creation of natural, herbal, animal, and synthetic aromas. Perfume production typically involves mixing pure ethyl alcohol with animal, herbal, or synthetic essences, stabilized for consistency [46]. Natural perfume ingredients are extracted from various plant parts through distillation or other extraction methods. Flowers like jasmine, rose, lilac, narcissus, violet, and gardenia are common sources, as are citrus fruits like lemon and orange [47]. Of the approximately 1,500 known aromatic plant species, detailed information exists for about 500, with only 50 commonly used for essential oil production in perfumery. Key plant categories include:

Aromatic herbs (e.g., *lavender, melissa, sage, rosemary,* and *thyme*)
Flowers (e.g., *rose, jasmine, orange blossom,* and *narcissus*)
Citrus fruits
Grains and seeds (e.g., *anise, dill,* and *cumin*)
Balsams and resins (e.g., *camphor, myrrh,* and *galbanum*)
Barks and roots (e.g., *cinnamon, ginger,* and *vetiver*)
Forest trees (e.g., *birch, cedar, pine,* and *sandalwood*)
Other aromatic plants (e.g., *tobacco, chamomile,* and *vervain*)

Natural essences comprise various chemical compounds, including alcohols, esters, phenols, aldehydes, ketones, acids, and hydrocarbons [48]. Essential oils typically contain 20–60 different chemical constituents. Terpenoids form the basis of many natural fragrances, with different compounds contributing specific scent characteristics. For example, menthol provides a refreshing aroma, linalyl acetate offers fruity and floral notes, and carvone imparts a minty scent [49]. The unique fragrance profiles of various plants result from their specific combinations of secondary metabolites.

3.6.2 Use of MAPs in cosmetics

The cosmetics industry is experiencing a surge in demand for natural and organic products. MAPs are increasingly valued for their bioactive compounds, particularly flavonoids, which offer skin health benefits [50]. These plants also provide essential minerals that support overall bodily functions. The cosmetics sector favors plant-based ingredients in various formulations, utilizing a wide range of MAPs [51]. For example, calendula flowers are incorporated into creams, shampoos, and baby products, while comfrey leaves and roots feature medicinal ointments and hair care items [52]. Licorice root serves as a natural skin lightener, and *St. John's wort* is used to combat hair loss. Grape seed oil, rich in the antioxidant resveratrol, is a popular ingredient in antiaging products. Tea extracts, particularly from green tea, are prized for their polyphenol content, including tannins and catechins, which offer numerous skincare benefits [53].

3.6.3 Use of MAPs in plastic production

The quest for environmentally friendly alternatives to traditional plastics has led to increased interest in plant-based, biodegradable polymers. These materials offer a promising solution to plastic pollution, provided they are produced sustainably from nonfood crops or as value-added byproducts [54]. Bioplastics boast several advantages over conventional plastics, including biodegradability, renewability, recyclability, and nontoxic disposal. Recent research has explored the development of antioxidant-active packaging materials by incorporating MAP-derived antioxidants into polylactic acid matrices [55]. Another study investigated the creation of biodegradable biofilms using nanocellulose extracted from jackfruit peels, combined with plasticizers and natural fillers. These innovations demonstrate the potential for creating eco-friendly packaging materials that could replace petroleum-based plastics [56].

3.6.4 Other industrial applications

MAPs find diverse applications across various industries due to their antimicrobial, fungicidal, and bactericidal properties [57]. In food preservation, they serve as natural alternatives for meat, canned goods, and fresh produce. The animal feed industry has successfully employed MAPs to reduce reliance on synthetic antimicrobials [58]. Many plants are marketed as nutraceuticals in tablet or capsule form for daily nutritional supplementation. In landscaping, certain MAP species are valued for their aesthetic qualities and practical uses, such as forming natural hedges [59]. Research has identified potential applications in collection gardens, therapy gardens, botanical displays, and various urban green spaces. The textile industry has also embraced MAPs, particularly for their antimicrobial properties. Studies have explored microencapsulation techniques to incorporate essential oils into fabrics, resulting in long-lasting antimicrobial and antifungal effects [60]. Various plant extracts have demonstrated strong antibacterial activity, when applied to textiles. These innovations suggest promising applications in hygienic textiles for medical and food industry use, as well as in the production of naturally antibacterial clothing and home textiles [61].

3.6.5 MAPs in energy production

The growing global population, industrialization, and urbanization have led to increased fossil fuel consumption, depleting natural resources and causing environmental pollution [62]. Biofuels are emerging as a potential solution, considered the energy source of the future. The biomass byproducts from MAP industrial processes, including fruits, roots, leaves, and flowers, offer promising potential for biofuel production [63]. Utilizing waste biomass from aromatic industries can yield economic, environmental, and social benefits. While precise data on MAP waste biomass is limited, it is estimated that significant amounts are produced, as essential oil content typically comprises less than 5% of the plant material [64]. The aromatic industry is thought to generate around 200,000 tons of solid waste annually after essential oil extraction. This waste biomass, rich in polyphenols and other bioactive compounds, can be repurposed to create value-added products such as biogas, compost, biochar, biofuels, and biopesticides [65]. Plants like *jojoba, sunflower, rapeseed, madwort,* and mole bean are being explored for biofuel production. Maps may also contribute to solar energy advancements. Research has shown that pigments extracted from plants like *Malabar spinach* and *red cabbage* can absorb green light while reflecting red and blue light, suggesting potential applications in developing multicolor solar cells for agrivoltaic systems [66].

3.6.6 MAPs in agricultural applications

MAPs offer various agricultural applications, particularly in pest management. Plants naturally produce secondary metabolites like esters, ketones, and essential oils as defense mechanisms against pests and mites [67]. These compounds exhibit neurotoxicity, growth regulation, and enzyme-inhibition effects on pests. For instance, fennel extracts have demonstrated high toxicity against mosquito larvae, with terpineol and 1,8-cineol proving particularly effective against mosquito bites. Vetiver root extracts show promise as an eco-friendly insecticide against certain beetles [68]. Oregano has been suggested for agricultural pest control due to its insecticidal, antiviral, antibacterial, and antifungal properties. Industrial cannabis flowers secrete cannabinoids and terpenes that repel plant-eating insects [69]. Studies have shown cannabis essential oils to be toxic to various pests while remaining harmless to nontarget invertebrates. Intercropping MAPs with vegetables can protect crops from pests, extend storage periods, and improve quality during transport [70]. Essential oils from aromatic plants also help combat soil nematodes. This practice can alter soil composition, decreasing pH and nitrogen while enhancing organic nitrogen and water content. MAPs show potential in phytoremediation of heavy metal-contaminated soils [71]. Some species can accumulate heavy metals without transferring them to their essential oils, suggesting their suitability for cultivation on polluted lands while still producing economically viable products [72]. It is important to note that environmental factors such as salinity, temperature, light, and nutrient availability significantly influence the synthesis and accumulation of secondary plant metabolites. These factors warrant further detailed exploration to optimize MAP cultivation and utilization [73].

3.7 Salt stress

Salt stress significantly impacts the development and morphology of medicinal plants. It hinders germination by damaging the embryo or reducing soil potential, impeding water uptake. This effect has been observed in various plants, including *Ocimum basilicum*, *Eruca sativa*, and *Petroselinum hortense* [74]. The seedling stage is particularly vulnerable to salt stress. Studies have shown that salinity impairs seedling growth in plants like Thymus maroccanus by inhibiting food reserve mobilization and cell division. Similar effects have been noted in basil, chamomile, and marjoram. Salt stress also affects mature plants. In Aloe Vera, increased salinity led to decreased foliage, root growth, and dry matter, primarily due to reduced total soluble solids [75]. *Citronella java* plants exposed to high salinity showed a significant reduction in tiller numbers. Cumin's vegetative and reproductive stages were found to be highly sensitive to salt stress [76]. Growth inhibition due to salinity has been reported in numerous medicinal plants, including *Majorana hortensis*, *peppermint*, and *Matricaria recutita*. *Mentha piperita* var. *officinalis* and *Lipia citriodora* var. *verbena* exhibited reduced leaf numbers, area, and biomass under salt stress. *Milk thistle* exposed to high salinity showed decreased plant height, leaf count, and capitula number [77].

3.7.1 Nutrient

Uptake of salt stress disrupts nutrient uptake in plants by creating an ionic imbalance. The abundance of Na^+ and Cl^- ions interferes with the absorption of essential nutrients like K^+, Ca^{2+}, and NO_3^-. This imbalance affects nutrient availability, partitioning, and transport within the plant system. Studies on various medicinal plants have shown decreased levels of N, P, K^+, Ca^{2+}, and Mg^{2+} under salt stress conditions [78].

3.7.2 Productivity

Increasing salt concentrations negatively impact the productivity of medicinal plants such as fennel, cumin, and milk thistle. Salt-stressed plants typically show reduced fruit yield per plant and fewer umbels [79].

3.7.3 Photosynthesis

Photosynthesis, a crucial physiological process for plant growth and survival, is particularly vulnerable to salt stress. The stress disrupts the metabolic balance within plant cells, affecting the photosynthetic machinery. Many medicinal plants, including *Thymus vulgaris* and *Satureja hortensis*, show reduced chlorophyll content under salt stress [80]. This decrease is attributed to inhibited chlorophyll synthesis and increased degradation, leading to suppressed photosynthesis. Salt stress also negatively impacts chloroplast development and protein translation within plastids, sometimes resulting in plastid degradation, as observed in fennel [81].

3.8 Drought stress

Drought stress significantly affects the growth, development, and secondary metabolite production of medicinal plants [82]. As global climate patterns shift, understanding how water scarcity impacts these valuable species becomes increasingly important. Medicinal plants, prized for their therapeutic properties, show complex responses to drought that can both hinder growth and enhance the production of certain beneficial compounds. Under drought conditions, medicinal plants typically close their stomata to conserve water [83]. While this helps reduce water loss, it also limits carbon dioxide uptake, compromising photosynthesis and overall growth [84]. Plants often exhibit stunted growth, smaller leaves, and reduced biomass, along with visible signs like leaf rolling and wilting. Interestingly, water scarcity can trigger increased production of secondary metabolites, the compounds responsible for medicinal properties. Plants in semiarid conditions often show higher concentrations of active substances compared to those in moderate climates [85]. This increase is linked to the plant's stress response and altered metabolism. Drought causes an imbalance in the cellular redox state, leading to reactive oxygen species (ROS) accumulation [86]. To combat this, plants enhance their antioxidant defenses, producing more secondary metabolites like phenolic compounds, flavonoids, and terpenoids. These not only protect the plant but also contribute to its medicinal value. For example, drought-stressed sage plants produce higher concentrations of monoterpenes, key components of their therapeutic properties [87]. The enhancement of secondary metabolite production extends to other compounds like alkaloids. In *Cathar-*

anthus roseus, valued for its anticancer alkaloids, drought stress upregulates genes involved in alkaloid biosynthesis, increasing the concentrations of these valuable compounds [88]. However, while active substance concentration may increase under drought stress, total yield can be affected by reduced biomass production. In some cases, the concentration increase outweighs biomass reduction, resulting in higher total metabolite production per plant [89]. In others, biomass reduction may lead to an overall decrease in active compound yield. The response to drought stress is further complicated by interactions with other environmental factors like light intensity and temperature [90]. The duration and severity of drought also play crucial roles in determining its impact. Mild to moderate water deficits may stimulate secondary metabolite production without severe damage, but prolonged or severe drought can lead to irreversible harm or plant death. Understanding these effects has important implications for medicinal plant cultivation and natural medicine production [91]. Careful water management may enhance product quality without significantly compromising yield. However, implementing such strategies requires a thorough understanding of species-specific responses and optimal stress levels. Long-term drought exposure can lead to genetic and epigenetic changes, altering plant characteristics over generations. This highlights the importance of conserving diverse plant populations and their habitats, as they may harbor valuable traits for drought resistance and metabolite production [92]. As research advances, developing sustainable cultivation practices that balance plant productivity with high-quality medicinal products will be crucial. Understanding drought stress mechanisms may provide insights into improving plant-based medicine efficacy and production, contributing to the ongoing importance of medicinal plants in healthcare and traditional medicine systems worldwide [93].

3.9 Heavy metals

Heavy metals can significantly impact medicinal plants, affecting their growth and phytochemical production. While plants need certain metals for growth, excessive amounts become toxic [94]. Plants can accumulate both essential and nonessential metals, potentially leading to enzyme inhibition and oxidative stress damage to cell structures. Heavy metal contamination in medicinal plants can occur through cultivation, processing, or intentional addition for alleged medicinal purposes [95].

The increasing use of herbal drugs has raised concerns about heavy metal contamination in medicinal plants [96]. Metal accumulation in plant tissues can alter physiological and biochemical processes, affecting plant health, productivity, and the safety and efficacy of derived herbal medicines. Heavy metal stress often reduces overall biomass production in medicinal plants by interfering with essential processes like photosynthesis, respiration, and nutrient uptake [97]. For example, cadmium exposure can cause chlorosis, leaf rolling, and premature senescence, decreasing plant vigor and yield. Root systems, often the first point of contact with soil metals, may experience reduced elongation, decreased biomass, and altered architecture, further impacting water and nutrient absorption [98]. Interestingly, heavy metal stress can have complex effects on phytochemical production, sometimes stimulating secondary metabolite production as part of the plant's stress response. Moderate metal stress may increase phenolic compounds, flavonoids, and antioxidants in some species, potentially enhancing their therapeutic value [99]. This response is often associated with the plant's attempt to mitigate ROS damage caused by heavy metals disrupting cellular redox balance. However, the relationship between heavy metal stress and secondary metabolite production is not straightforward [100]. Severe or prolonged exposure can suppress important phytochemical biosynthesis due to overall plant health deterioration. The outcome depends on metal type and concentration, plant species, and genetic factors. These stress-induced changes in phytochemical profiles have significant implications for medicinal properties and therapeutic applications [101]. While some changes might enhance certain medicinal properties, they can also lead to unpredictable variations in therapeutic effects. The accumulation of heavy metals in medicinal plants raises serious safety concerns for herbal product consumers. Despite potential enhancements in beneficial compounds, the presence of toxic metals poses significant health risks. This issue is particularly problematic as many users perceive herbal medicines as inherently safe [102]. While regulatory bodies have established guidelines for acceptable heavy metal levels in medicinal plant products, global enforcement remains challenging, especially in regions with limited regulatory oversight.

Long-term heavy metal exposure can induce genetic and epigenetic changes in medicinal plants, affecting their stress adaptation and phytochemical profiles across generations [103]. This underscores the importance of protecting natural habitats from metal pollution and developing cultivation strategies in controlled, uncontaminated environments. Some medicinal plants can hyperaccumulate heavy metals, presenting challenges for medicinal use but offering potential in phytoremediation [104]. These species could serve dual purposes: cleaning contaminated soils while producing biomass for nonconsumptive applications, such as extracting specific compounds for industrial or pharmaceutical use. The complex interactions between heavy metals and medicinal plants require a multidisciplinary approach, combining plant physiology, biochemistry, pharmacology, and environmental science. Researchers are developing advanced analytical methods and molecular biology approaches to better understand these interactions [105]. Heavy metals have diverse effects on medicinal plants, generally negatively impacting growth but sometimes increasing the production of certain beneficial compounds. However, potential enhancements in medicinal properties must be balanced against health risks from metal accumulation in plant tissues [106]. As herbal medicine use grows globally, comprehensive strategies for monitoring and controlling heavy metal contamination in medicinal plants are crucial [107]. This includes implementing strict quality control measures in cultivation, harvesting, and processing, as well as thorough safety assessments of herbal products.

Ongoing research into plant responses to heavy metal stress may lead to new strategies for enhancing valuable phytochemical production, while minimizing contamination risks [108]. A balanced approach considering both the potential benefits and risks of heavy metal interactions with medicinal plants is essential for the sustainable and safe use of herbal medicines in the future.

3.10 Heat stress

Heat stress significantly impacts medicinal plants' growth, development, and phytochemical composition. As global temperatures rise due to climate change, understanding these effects becomes increasingly important [109]. When exposed to heat stress, medicinal plants often close their stomata to conserve water. While this reduces transpiration, it also limits carbon dioxide uptake, compromising photosynthesis and overall growth. This can lead to stunted growth, smaller leaves, and reduced biomass production [110]. Heat stress can alter the chemical composition of essential oils in plants like *Mentha* × *piperita* L. var. *Mitcham* and *Mentha arvensis* var. *piperascens* Malinv. ex L. H. Bailey. For instance, menthol percentages may decrease, while pulegone and menthyl acetate increase, affecting the plant's medicinal properties and commercial value [111]. Long-term heat exposure can induce genetic and epigenetic changes in plant populations, potentially altering their stress adaptation abilities and phytochemical profiles across generations. This highlights the need for habitat protection and controlled cultivation strategies. Interestingly, heat stress can sometimes stimulate secondary metabolite production as part of the plant's stress response [112]. Moderate heat stress may increase phenolic compounds, flavonoids, and antioxidants in some species. This is believed to be a defense mechanism against heat-induced oxidative stress. Plants enhance their antioxidant defense systems to combat ROS accumulation caused by high temperatures. This includes producing secondary metabolites that protect against oxidative damage and contribute to medicinal properties [113]. However, the relationship between heat stress and secondary metabolite production is complex. While moderate stress might enhance certain compounds, severe or prolonged exposure can suppress important phytochemical biosynthesis due to overall plant health deterioration. At the molecular level, heat stress triggers heat shock protein (HSP) expression, which protects cellular proteins from denaturation. Plants also activate antioxidant defense systems, increasing the production of enzymatic antioxidants like superoxide dismutase and nonenzymatic antioxidants such as ascorbic acid to scavenge ROS and protect cells from oxidative damage [114].

Heat stress significantly impacts medicinal plants' hormonal balance, affecting various physiological processes. Abscisic acid production often increases under heat stress, promoting stomatal closure to reduce water loss, but limiting CO_2 uptake and photosynthesis. Other hormones like ethylene and salicylic acid also play roles in heat stress response and can influence secondary metabolite production [115]. High temperatures can damage the photosynthetic apparatus, particularly photosystem II, reducing photosynthetic efficiency and overall plant productivity. This not only affects growth but can also influence secondary metabolite production, as many are derived from photosynthetic products. Heat stress can alter the activity of key enzymes involved in secondary metabolite biosynthesis [116]. For example, it may enhance phenylalanine ammonialyase (PAL) activity, potentially increasing concentrations of certain beneficial phytochemicals. However, effects vary, depending on the specific compound and plant species [117]. These stress-induced changes in secondary metabolite composition can have significant implications for medicinal properties and therapeutic applications. While some changes might enhance certain medicinal properties, they can also lead to unpredictable variations in therapeutic effects. The plant's response to heat stress is often influenced by other environmental factors, such as water and nutrient availability [118]. Drought stress, often accompanying heat stress, can exacerbate negative effects on plant growth and metabolism. Understanding heat stress effects on medicinal plants is crucial for cultivation and natural medicine production. Controlled environmental stress techniques may enhance product quality without significantly compromising yield. However, implementing such strategies requires a thorough understanding of species-specific responses and optimal stress levels [119]. In conclusion, heat stress effects on medicinal plants involve a complex interplay of growth impacts and potential increases in certain bioactive compounds (Figure 3.2). This presents both challenges and opportunities for cultivation in a changing climate. Ongoing research is crucial for developing sustainable practices that balance productivity with high-quality medicinal products [120]. Understanding these mechanisms may provide insights into improving plant-based medicine efficacy and production, contributing to the ongoing importance of medicinal plants in healthcare and traditional medicine systems worldwide.

3.11 Soil pH

Soil pH significantly impacts the growth, development, and phytochemical composition of medicinal plants. As a key factor affecting nutrient availability and microbial activity, soil pH plays a crucial role in determining plant health and therapeutic potential [121]. Most medicinal plants thrive in soil pH between 6.0 and 7.5, though specific species may have different preferences. Acidic soils (pH below 6.0) can lead to nutrient deficiencies, particularly phosphorus, calcium, and magnesium. This can result in stunted growth, reduced flower and seed production, and weakened root systems. Alkaline soils (pH above 7.5) may reduce the availability of micronutrients like iron, manganese, zinc, and copper, impacting various physiological processes, including photosynthesis and secondary metabolite biosynthesis [122]. Soil pH also influences rhizosphere microbial communities, affecting nutrient cycling and plant health. Acidic soils may suppress ben-

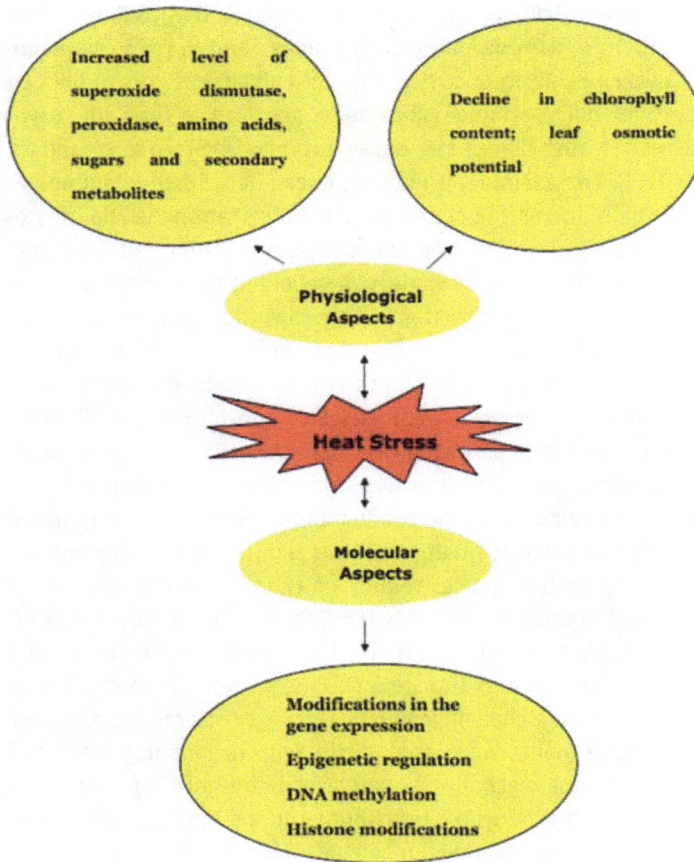

Figure 3.2: Physiological and molecular aspects of heat stress in the medicinal plants.

eficial microorganisms, while alkaline soils can favor certain microbes but also promote some pathogens [123]. These microbial community changes can indirectly impact secondary metabolite production in medicinal plants. The effect of soil pH on phytochemical compounds is complex and species-specific. Mild pH stress may stimulate the production of certain secondary metabolites as a plant stress response. For instance, slightly acidic conditions can enhance phenolic compound and flavonoid production in some species [124]. However, severe pH stress can lead to an overall decline in phytochemical content. Alkaloid biosynthesis and accumulation are particularly sensitive to soil pH changes. In some plants, like *Catharanthus roseus*, slightly acidic conditions enhance valuable alkaloid production. Essential oil composition and yield in medicinal plants also vary with soil pH [125]. For example, some mint species show higher essential oil yields in slightly acidic soils. Long-term exposure to suboptimal pH can lead to genetic and epigenetic changes in plant populations, potentially altering their adaptive strategies and phytochemical profiles over generations [126]. This highlights the importance of maintaining appropriate soil pH in natural habitats and cultivation areas to preserve the genetic diversity and medicinal properties of these plants. Managing soil pH is crucial for optimizing growth and phytochemical production in medicinal plants. Various strategies can be employed to adjust soil pH to suit the specific requirements of different medicinal plant species [127].

Soil pH management is crucial for optimizing medicinal plant cultivation. For acidic soils, liming with materials like agricultural limestone or wood ash can gradually raise pH. Careful application is necessary to avoid micronutrient deficiencies. Alkaline soils can be acidified using elemental sulfur or organic materials like peat moss [128]. Container-grown plants may benefit from acidifying fertilizers or organic acids. Regular soil testing is essential for maintaining optimal conditions. Soil pH significantly affects nutrient availability, impacting plant growth and bioactive compound synthesis. Acidic soils may reduce macronutrient availability, while increasing certain micronutrients to potentially toxic levels. Alkaline soils often decrease phosphorus and micronutrient availability [129]. Understanding these pH-dependent nutrient dynamics is vital for optimizing medicinal plant cultivation and phytochemical yield. Root architecture and rhizosphere interactions are also influenced by soil pH [130]. Acidic conditions can alter root morphology, affecting nutrient and water uptake. Soil pH impacts root exudates, shaping the rhizosphere microbiome, which in turn affects nutrient cycling and plant defense mechanisms. Innovative cultivation techniques have emerged from research on soil pH effects [131]. Controlled pH stress can stimulate the production of specific secondary metabolites, potentially enhancing the therapeutic efficacy and commercial value of certain medicinal plants. Aquatic and semiaquatic medicinal plants are also affected by pH, with water pH playing a similar role to soil pH in terrestrial systems [132]. This is crucial for cultivating and conserving aquatic medicinal species. Climate change further complicates the relationship between soil pH and medicinal plant growth. Changes in precipitation, temperature, and CO_2 levels can alter soil pH dynamics, potentially affecting medicinal plant distribution and phytochemical profiles. In conclusion, soil pH fundamentally shapes medicinal plant growth and phytochemical composition [133] (Figure 3.3.). Ongoing research is essential for developing tailored pH management strategies for different species, ensuring sustainable cultivation practices, and preserving these valuable plant resources for future generations.

3.12 Light intensity

Light intensity significantly impacts the growth, development, and phytochemical composition of medicinal plants. As photosynthetic organisms, plants depend on light for energy, and variations in light quantity and quality can substantially affect their morphology, physiology, and biochemistry [134]. Understanding these effects is crucial

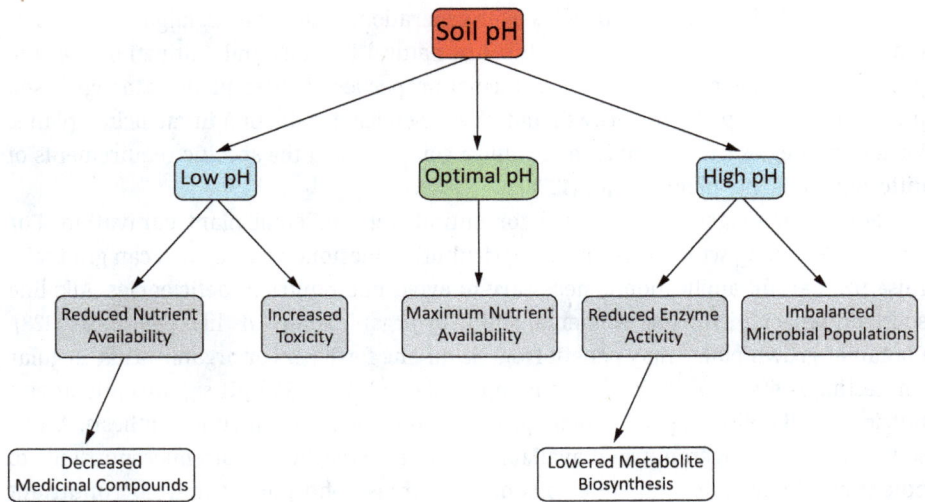

Figure 3.3: Effect of soil pH on medicinal compounds.

for optimizing medicinal plant cultivation and ensuring consistent production of high-quality phytochemicals. Plants exhibit various responses to different light intensities at multiple organizational levels. Morphologically, high light intensity often results in thicker leaves with smaller surface areas, while low light conditions produce larger, thinner leaves [135]. This adaptation optimizes light-capturing ability in different environments. For example, sage (*Salvia officinalis*) grown under partial shade developed larger leaf areas compared to those in full sunlight. Light intensity also affects stem length and plant architecture [136].

Low light conditions often lead to elongated stems and internodes (shade avoidance response), while high light intensities produce shorter, more compact growth habits [137]. This compact growth can be advantageous in medicinal plant cultivation, potentially increasing yields and simplifying management practices. Physiologically, light intensity directly impacts the photosynthetic apparatus. Chlorophyll content, crucial for photosynthesis, is particularly sensitive to light intensity changes [138]. Plants grown in lower light conditions generally have higher chlorophyll content per unit leaf area, an adaptive response to capture more light energy. However, the relationship between light intensity and photosynthetic efficiency is not linear [139]. While moderate increases in light intensity can enhance photosynthetic rates, excessively high levels can cause photoinhibition, damaging the photosynthetic apparatus and reducing growth and productivity. Light intensity also influences secondary metabolite production, which is often of primary interest in medicinal plants due to their therapeutic properties. Secondary metabolite biosynthesis is closely linked to photosynthetic activity and overall metabolic state, both heavily influenced by light intensity [140]. Higher light intensities generally promote the production of certain secondary metabolites, particularly those involved in photoprotection and antioxidant defense [141].

Flavonoids, a diverse group of phenolic compounds known for their medicinal benefits, are highly responsive to variations in light intensity. Numerous studies have indicated that medicinal plants exhibit increased flavonoid production when grown under bright light conditions [142]. For instance, research on Anoectochilus formosanus demonstrated that total flavonoid levels rose with light intensity, peaking at 60 $\mu mol/m^2/s$. This boost in flavonoid synthesis is believed to serve as a protective mechanism against excessive light and the oxidative stress it can cause [143]. Essential oils, another critical category of secondary metabolites in medicinal plants, also react to changes in light intensity. However, the specific response can differ among plant species and the compounds involved [144]. For example, higher light levels have been linked to increased essential oil content in plants like *Thymus vulgaris* (thyme) and *Matricaria chamomilla* (chamomile). Conversely, species such as *Anethum graveolens* (dill) and *Salvia officinalis* (sage) tend to produce more essential oils in shaded conditions. These variations underscore the complex and species-specific relationship between light intensity and secondary metabolite production [145]. The production of alkaloids, which are nitrogen-containing compounds with significant pharmacological effects, is also influenced by light intensity [146]. In some instances, increased light exposure correlates with higher alkaloid levels. For example, *Catharanthus roseus*, known for its anticancer alkaloids, shows enhanced production of vinblastine and vincristine under high-light conditions. However, optimal alkaloid synthesis may occur at intermediate light levels for certain species. While elevated light intensities generally boost secondary metabolite production, this relationship is not uniform across all plants and can have limitations [147]. Excessive light can induce oxidative stress and photodamage, reducing overall plant productivity and the yield of secondary metabolites. Thus, identifying the ideal light intensity for each medicinal plant species is essential for maximizing growth and phytochemical output [148].

Light intensity effects are further complicated by interactions with other environmental factors like temperature, water availability, and nutrient status. High light levels often coincide with elevated temperatures that can independently influence plant metabolism and secondary metabolite production [149]. Additionally, plants exposed to intense light may require more water; insufficient hydration can adversely affect secondary metabolite synthesis. Understanding how light intensity impacts medicinal plants has significant implications for their cultivation and natural medicine production [150]. In controlled agricultural settings such as greenhouses or indoor farms, growers can manipulate light conditions to enhance plant growth and phytochemical yields. Various lighting strategies have been developed, including artificial lighting to supplement natural sunlight [151]. LED systems are particularly popular due to their energy efficiency and precise control over light spectrum and intensity. Adjusting light exposure during different growth stages has shown potential for increasing both biomass production and secondary metabolite accumulation in medicinal plants. Some studies suggest that providing higher light intensities during specific developmental phases can stimulate phytochemical production without hindering overall growth [152]. This targeted approach allows growers to customize lighting conditions according to the needs of various medicinal plant species [153]. It is crucial to recognize that the optimal lighting for growth may not align with conditions that maximize secondary metabolite production; a balance must be struck between enhancing biomass yield and optimizing phytochemical content. The influence of light intensity on medicinal plants extends beyond immediate physiological responses; long-term exposure to varying light environments can induce genetic and epigenetic changes in plant populations [154]. These changes may alter adaptive strategies and phytochemical profiles over generations, emphasizing the importance of preserving diverse habitats to maintain the full range of medicinal properties in these species. As research progresses in this area, new technologies are being developed to deepen our understanding of how light intensity affects medicinal plants [155]. Techniques such as high-throughput phenotyping, combined with metabolomics, provide valuable insights into the intricate relationships between lighting conditions, plant physiology, and secondary metabolite production. In conclusion, light intensity plays a fundamental role in shaping the growth and phytochemical composition of medicinal plants [156]. Its effects range from influencing basic morphological traits to modulating complex biochemical pathways involved in producing important compounds. Optimizing lighting conditions requires a nuanced approach, tailored to each species' specific needs, while considering interactions with other environmental factors [157]. By effectively managing light intensity throughout the growth cycle, it is possible to enhance both yield and quality in medicinal plant cultivation, supporting sustainable production of valuable natural medicines as global demand for plant-based therapies continues to rise [158].

Long-term exposure to varying light conditions can result in genetic and epigenetic modifications within plant populations, potentially altering their adaptive strategies and phytochemical profiles across generations [159]. This is particularly significant for conserving and sustainably utilizing medicinal plant species, emphasizing the importance of safeguarding diverse habitats and genetic resources to preserve their medicinal properties. As research progresses, new technologies are emerging that enhance our understanding of how light intensity influences medicinal plants [160]. Techniques such as high-throughput phenotyping, combined with metabolomics and transcriptomics, are offering valuable insights into the intricate relationships between light conditions, plant physiology, and secondary metabolite production [161]. These advanced methods enable researchers to pinpoint key genes and metabolic pathways involved in responses to light, which could lead to new strategies for improving the medicinal qualities of plants through targeted genetic enhancements or optimized cultivation practices. In summary, light intensity is essential in shaping the growth, development, and phytochemical composition of medicinal plants. Its effects range from influencing basic morphological traits to modulating complex biochemical pathways linked to the production of important medicinal compounds [162, 163]. While general trends exist – such as increased secondary metabolite production under higher light levels – the specific responses can vary widely among different plant species and compounds. Therefore, optimizing light conditions for cultivating medicinal plants requires a detailed, species-specific approach that considers the interactions between light intensity and other environmental factors. By effectively managing light exposure throughout the growth cycle, it is possible to improve both yield and quality in medicinal plant cultivation, supporting the sustainable production of valuable natural medicines [164] (Figure 3.4). As our understanding of these complex interactions evolves, we can anticipate further innovations in cultivation practices that enhance the efficient and targeted production of therapeutic compounds to meet the rising global demand for plant-based medicines [165, 166].

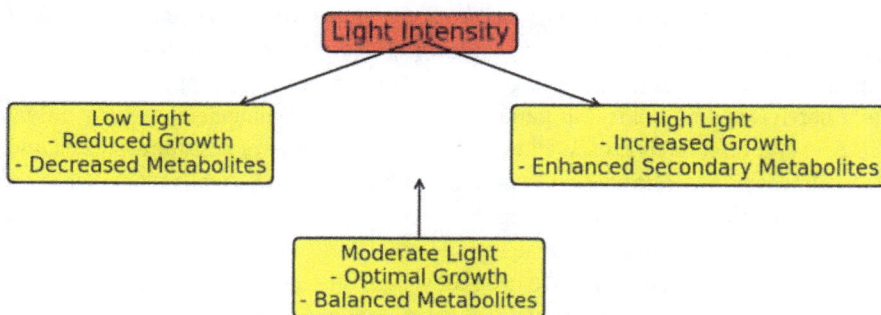

Figure 3.4: Effect of light intensity on medicinal plant growth.

3.13 Pest and disease management

Pest and disease management is a critical aspect of growing medicinal plants, as it directly impacts the quality and yield of the final product. Effective strategies must balance the need for plant protection with the requirement to maintain the purity and safety of medicinal compounds. Integrated Pest Management (IPM) forms the cornerstone of modern pest control in medicinal plant cultivation. This approach combines various techniques to minimize pest damage, while reducing reliance on chemical pesticides [167]. Cultural practices play a significant role in IPM for medicinal plants. Crop rotation helps break pest life cycles and reduces the buildup of soil-borne pathogens. Intercropping with companion plants can deter pests and attract beneficial insects. For instance, planting marigolds alongside medicinal herbs can repel nematodes and some insect pests [168]. Proper sanitation is crucial in preventing disease spread. This includes removing infected plant material, sterilizing tools, and maintaining clean growing areas. Pruning and thinning improve air circulation, reducing humidity levels that favor fungal growth. The timing of planting and harvesting can also be adjusted to avoid periods of high pest pressure or disease susceptibility [169].

Biological control methods are increasingly important in medicinal plant production. Predatory insects, such as ladybugs and lacewings, can be introduced to control aphids and other soft-bodied pests. Beneficial nematodes are effective against soil-dwelling insect larvae. Microbial pesticides, including *Bacillus thuringiensis* (Bt) preparations, offer targeted control of certain caterpillar species without leaving harmful residues. Physical barriers and traps provide nonchemical pest control options. Row covers protect young plants from insect damage and can create microclimates that accelerate growth. Sticky traps monitor pest populations and can help control flying insects. Pheromone traps disrupt the mating cycles of specific pests, reducing their populations over time [170]. Water management is crucial in disease prevention. Drip irrigation or soaker hoses deliver water directly to plant roots, keeping foliage dry and reducing the risk of fungal infections. Proper drainage prevents waterlogging, which can lead to root rot and other soil-borne diseases. Mulching helps retain soil moisture, suppress weeds, and create a barrier against soil-splash pathogens [171]. Nutrient management plays a dual role in pest and disease control. Well-nourished plants are more resistant to pests and diseases. However, overfertilization, especially with nitrogen, can make plants more attractive to certain pests and more susceptible to fungal infections. Balanced fertilization based on soil tests and plant requirements is essential [172].

When chemical interventions are necessary, selecting appropriate pesticides is crucial. For medicinal plants, products with short residual periods and low toxicity to nontarget organisms are preferred. Botanical pesticides derived from plants like neem or pyrethrum offer effective control with reduced environmental impact [167]. Sulfur and copper-based fungicides are often used in organic production systems to manage fungal diseases. Timing and application methods of pesticides are critical to their effectiveness and safety. Spot treatments targeting affected areas can minimize overall pesticide use [168]. Applying pesticides during periods of low bee activity helps protect pollinators. Rotating pesticides with different modes of action prevents the development of resistance in pest populations. Disease-resistant varieties offer a long-term solution to recurring pathogen problems. While breeding programs for medicinal plants are less extensive than for major crops, selecting cultivars adapted to local conditions can significantly reduce disease pressure. Some medicinal plants have natural resistance to specific pathogens, which can be enhanced through selective breeding. Monitoring and early detection are crucial components of effective pest and disease management. Regular scouting allows growers to identify problems before they become severe [170]. This includes inspecting plants for signs of pest damage or disease symptoms, as well as using traps and other monitoring tools to track pest populations. Early intervention often allows for the use of less intensive control measures. Climate control in a greenhouse or indoor cultivation can significantly impact pest and disease pressure. Maintaining optimal temperature and humidity levels reduces plant stress and creates an environment less favorable to many pests and pathogens. UV sterilization of irrigation water and air filtration systems can further reduce the risk of disease introduction and spread. Postharvest handling is an often overlooked aspect of pest and disease management in medicinal plants. Proper drying and storage conditions prevent the growth of molds and other microorganisms that can contaminate the final product [172]. Implementing good manufacturing practices (GMPs) throughout the processing chain ensures the quality and safety of medicinal plant products.

Research into plant-microbe interactions is opening new avenues for disease management in medicinal plants. Beneficial microorganisms, such as mycorrhizal fungi and plant-growth-promoting rhizobacteria, can enhance plant resistance to pathogens and improve overall plant health [173]. Inoculating growing media or applying these microorganisms as seed treatments shows promise in reducing disease incidence. The use of plant extracts and essential oils for pest and disease control is an area of growing interest, particularly for medicinal plants where chemical residues are a concern. Many medicinal plants produce compounds with antimicrobial or insecticidal properties. Extracts from plants like garlic, cinnamon, or thyme have shown efficacy against various pests and pathogens when applied as foliar sprays or soil drenches [174]. Precision agriculture techniques are increasingly being applied to medicinal plant cultivation. Drone-based imaging and sensors can detect early signs of pest infestation or disease outbreaks, allowing for targeted interventions. This technology can also optimize resource use, reducing plant stress and improving overall crop health [175].

Environmental manipulation can create conditions unfavorable to pests and diseases. For example, some pathogens are sensitive to soil pH. Adjusting soil acidity through liming or the application of organic amendments can suppress certain soilborne diseases. Similarly, altering light spectra in greenhouse environments can influence pest behavior and plant defense responses [176]. Quarantine measures are essential in preventing the introduction of new pests and diseases, especially when sourcing plant material from different regions. Implementing strict protocols for incoming plants, including isolation periods and thorough inspections, can safeguard existing crops from exotic threats. Education and training of growers and workers are crucial for the successful implementation of integrated pest and disease management strategies [173]. This includes recognizing early symptoms of pest damage or disease, understanding pest life cycles, and proper application of control measures. Ongoing research and extension services play a vital role in disseminating new knowledge and techniques to medicinal plant growers. In conclusion, effective pest and disease management in medicinal plant cultivation requires a holistic approach that integrates various strategies tailored to specific crops and local conditions. By combining cultural practices, biological control, judicious use of pesticides, and emerging technologies, growers can maintain plant health and product quality, while minimizing environmental impact [175]. Continuous monitoring, adaptation of strategies, and ongoing research are key to addressing the evolving challenges of pest and disease management in this important agricultural sector.

3.14 Conclusion and future perspective

Growing MAPs presents a multitude of challenges that span various aspects of cultivation, processing, and market dynamics [114]. These challenges require careful consideration and innovative solutions to ensure sustainable production and high-quality yields [172]. One of the primary challenges in cultivating MAPs is the maintenance of optimal growing conditions. These plants often have specific requirements for soil composition, pH levels, temperature, and humidity. Achieving and maintaining these conditions can be particularly difficult in outdoor settings where environmental factors are less controllable [176]. Climate change further complicates this issue, as shifting weather patterns and extreme events can disrupt established growing practices and necessitate adaptive strategies. Pest and disease management pose significant challenges in MAP cultivation. Many of these plants are susceptible to a range of pathogens and insect infestations that can severely impact crop yield and quality [170].

The use of chemical pesticides is often restricted due to concerns about residues in the final product, especially for plants intended for medicinal use [162]. This necessitates the development and implementation of integrated pest management strategies that rely on biological controls and cultural practices, which can be more labor-intensive and require specialized knowledge [168]. Water management is another critical challenge, particularly in regions facing water scarcity or unpredictable rainfall patterns. Many MAPs require precise irrigation schedules to optimize growth and the production of active compounds. Overwatering can lead to root rot and fungal diseases, while underwatering can stress plants and reduce yields [152]. Implementing efficient irrigation systems and water conservation techniques are essential but can be costly and technically demanding. Soil fertility management presents its own set of challenges. MAPs often have specific nutrient requirements that influence the production of desired compounds. Balancing soil nutrients without overfertilization is crucial, as excessive use of chemical fertilizers can lead to environmental pollution and may affect the quality of the plant material [171].

Organic and sustainable fertilization practices are increasingly important but require careful planning and execution [175]. Harvesting and postharvest handling are critical stages that significantly impact the final quality of MAPs. Timing the harvest to coincide with peak levels of active compounds requires precision and often specialized equipment. Postharvest processing, including drying and storage, must be carried out under controlled conditions to preserve the plant's medicinal properties and prevent contamination. Inadequate facilities or improper handling can lead to significant losses in both the quantity and quality of the harvested material [172]. Genetic diversity and varietal improvement pose ongoing challenges in the sector. Many MAPs have not undergone extensive breeding programs compared to major food crops. This limits the availability of high-yielding, disease-resistant varieties adapted to different growing conditions [174]. Developing new cultivars that maintain or enhance the desired medicinal properties, while improving agronomic traits, is a long-term process that requires significant investment in research and development. Regulatory compliance and quality control are increasingly important challenges in the MAP industry. Growers must navigate complex regulations regarding the cultivation, processing, and marketing of these plants, especially those used for medicinal purposes [176].

Meeting quality standards and obtaining necessary certifications can be costly and time-consuming, particularly for small-scale producers. Ensuring consistency in the levels of active compounds across different batches is crucial but challenging due to the influence of environmental factors on plant metabolism. Market volatility and price fluctuations present economic challenges for growers of MAPs. Demand for these products can be influenced by changing consumer preferences, scientific research findings, and regulatory decisions. This uncertainty makes it difficult for producers to plan long-term investments and can lead to boom-and-bust cycles in certain crops. Developing stable market channels and diversifying product offerings are strategies to mitigate these risks but require additional resources and expertise [142]. The shortage of skilled labor is a growing concern in the cultivation of MAPs. Many aspects of production, from planting to harvesting, require specialized knowledge and techniques. As rural populations decline in many regions and younger generations show less interest in agricultural careers, finding and retaining qualified workers becomes increasingly difficult.

This challenge is compounded by the seasonal nature of much of the work in this sector. Intellectual property rights and benefit-sharing present ethical and legal challenges, particularly when working with traditional medicinal plants [132]. Ensuring fair compensation for indigenous knowledge and genetic resources, while promoting innovation and research, can be complex. Balancing the interests of local communities, researchers, and commercial entities requires careful negotiation and the development of equitable frameworks [103]. Climate change adaptation is an overarching challenge that affects all aspects of MAP production [92]. Rising temperatures, changing precipitation patterns, and increased frequency of extreme weather events can disrupt established growing practices and alter plant physiology. Developing resilient cultivation systems and identifying or breeding climate-adapted varieties are crucial, but long-term endeavors.

The challenge of sustainable resource management is particularly acute for wild-harvested MAPs [54]. Overharvesting of popular species can lead to habitat destruction and biodiversity loss. Implementing sustainable wild-crafting practices and transitioning to cultivated sources, where possible, are necessary but complex processes that require cooperation between harvesters, conservationists, and regulatory bodies. Technology adoption and digital integration present both opportunities and challenges for the sector. While precision agriculture techniques and data-driven decision-making can improve efficiency and quality, the initial investment and learning curve can be significant barriers, especially for small-scale producers. Ensuring that technological advancements benefit all stakeholders in the value chain is an ongoing challenge. In conclusion, the cultivation of MAPs faces a diverse array of challenges that span agronomic, economic, regulatory, and environmental domains [162]. Addressing these challenges requires a multidisciplinary approach, combining traditional knowledge with modern scientific and technological advancements. Sustainable solutions must balance the needs of producers, consumers, and the environment, while ensuring the long-term viability of this important agricultural sector [110]. As global demand for natural products continues to grow, overcoming these challenges will be crucial in meeting market needs, while preserving biodiversity and supporting rural livelihoods.

References

[1]   Abbasi Khalaki, M., Moameri, M., Asgari Lajayer, B. and Astatkie, T. (2021). Influence of nano-priming on seed germination and plant growth of forage and medicinal plants. Plant Growth Regulation, 93, 13–28.

[2]   Abd_Allah, E. F., Hashem, A., Alqarawi, A. A., Bahkali, A. H. and Alwhibi, M. S. (2015). Enhancing growth performance and systemic acquired resistance of medicinal plant *Sesbania sesban* (L.) Merr using arbuscular mycorrhizal fungi under salt stress. Saudi Journal of Biological Sciences., 22, 274–283.

[3]   Adamczyk-Szabela, D., Markiewicz, J. and Wolf, W. M. (2015). Heavy metal uptake by herbs. IV. Influence of soil pH on the content of heavy metals in Valeriana officinalis L. Water, Air, and Soil Pollution, 226, 1–8.

[4]   Adhikari, B. (2021). Roles of alkaloids from medicinal plants in the management of diabetes mellitus. Journal of Chemistry 2021, 1(3), 2691525.

[5]   Aggarwal, S. (2021). Indian dye yielding plants: Efforts and opportunities. Natural Resources Forum, 45, 63–86.

[6]   Ahmed, A. A. and AL-Hamzi, E. H. (2012). Evaluation the bioactivity of some medicinal plants and arabic perfumes extracts on selected pathogenic fungi. Thamar University Journal of Natural & Applied Sciences, 5, 67–82.

[7]   Ahmed, K., Furusawa, Y., Tabuchi, Y., Emam, H. F., Piao, J. L., Hassan, M. A. and Kadowaki, M. (2012). Chemical inducers of heat shock proteins derived from medicinal plants and cytoprotective genes response. International Journal of Hyperthermia the Official Journal of European Society for Hyperthermic Oncology North American Hyperthermia Group, 28, 1–8.

[8]   Ahvazi, M., Khalighi-Sigaroodi, F., Charkhchiyan, M. M., Mojab, F., Mozaffarian, V. A. and Zakeri, H. (2012). Introduction of medicinal plants species with the most traditional usage in Alamut region. Iranian Journal of Pharmaceutical Research, 11, 185.

[9]   Li, H., Huang, C., Li, Y., Wang, P., Sun, J., Bi, Z. and Huang, X. (2024). Ethnobotanical study of medicinal plants used by the Yi people in Mile, Yunnan, China. Journal of Ethnobiology and Ethnomedicine, 20(1), 22.

[10]   Akthar, M. S., Degaga, B. and Azam, T. (2014). Antimicrobial activity of essential oils extracted from medicinal plants against the pathogenic microorganisms: A review. Journal of Issues ISSN, 2350, 1–7.

[11]   Al-aghabary, K., Zhu, Z. and Shi, Q. (2005). Influence of silicon supply on chlorophyll content, chlorophyll fluorescence, and antioxidative enzyme activities in tomato plants under salt stress. Journal of Plant Nutrition, 27, 2101–2115.

[12]   Alamgir, A. N. M. and Alamgir, A. N. M. (2017). Medicinal, non-medicinal, biopesticides, color-and dye-yielding plants; secondary metabolites and drug principles; significance of medicinal plants; use of medicinal plants in the systems of traditional and complementary and alternative medicines (CAMs). Therapeutic Use of Medicinal Plants and Their Extracts, 1, 61–104.

[13]   Albergaria, E. T., Oliveira, A. F. M. and Albuquerque, U. P. (2020). The effect of water deficit stress on the composition of phenolic compounds in medicinal plants. South African Journal of Botany, 131, 12–17.

[14] A, J. E., Maidarjav, A., Byambasuren, B. and Identification, N. D. (2024). Antimicrobial and plant growth promoting activities of endophytic fungi associated with *Cynomorium songaricum Rupr.*, a traditional medicinal plant in Mongolia. Diversity, 16(2), 122.

[15] Annan, K., Dickson, R. A., Amponsah, I. K. and Nooni, I. K. (2013). The heavy metal contents of some selected medicinal plants sampled from different geographical locations. Pharmacognosy Research, 5, 103.

[16] Ashraf, M., Mukhtar, N., Rehman, S. and Rha, E. S. (2004). Salt-induced changes in photosynthetic activity and growth in a potential medicinal plant Bishop's weed (*Ammi majus* L.). Photosynthetica, 42, 543–550.

[17] Asiminicesei, D. M., Vasilachi, I. C. and Gavrilescu, M. A. (2020). Heavy metal contamination of medicinal plants and potential implications on human health. Revue Roumaine Du Chimie, 71, 16–36.

[18] Atmani, D., Chaher, N., Berboucha, M., Ayouni, K., Lounis, H., Boudaoud, H. and Atmani, D. (2009). Antioxidant capacity and phenol content of selected Algerian medicinal plants. Food Chemistry, 112, 303–309.

[19] Banerjee, A. and Roychoudhury, A. (2017). Effect of salinity stress on growth and physiology of medicinal plants. In: Medicinal Plants and Environmental Challenges, 2, 177–188.

[20] Baričevič, D. and Zupančič, A. (2002). The impact of drought stress and/or nitrogen fertilization in some medicinal plants. Journal of Herbs, Spices & Medicinal Plants, 9, 53–64.

[21] Bariotakis, M., Georgescu, L., Laina, D., Oikonomou, I., Ntagounakis, G., Koufaki, M. I. and Pirintsos, S. A. (2019). From wild harvest towards precision agriculture: Use of Ecological Niche Modelling to direct potential cultivation of wild medicinal plants in Crete. Science of the Total Environment, 694, 133681.

[22] Barthwal, J., Smitha, N. A. I. and Kakkar, P. (2008). Heavy metal accumulation in medicinal plants collected from environmentally different sites. Biomedical and Environmental Sciences, 21, 319–324.

[23] Bazzazi, N., Khodambashi, M. and Mohammadi, S. H. (2013). The effect of drought stress on morphological characteristics and yield components of medicinal plant fenugreek. Isfahan University of Technology - Journal of Crop Production and Processing, 3, 11–23.

[24] Behera, B. and Bhattacharya, S. (2016). The importance of assessing heavy metals in medicinal herbs: A quantitative study. Cellmed, 6, 3–1.

[25] Bejarano, J. S. R., Rodrigues, T. S., Sánchez, C. M., Al-Ghanim, K. and Al-Saidi, M. (2020). Promoting sustainable businesses for strong local communities: Qatar's wild herbal plants industry. Energy Reporters, 6, 80–86.

[26] Beyk-Khormizi, A., Sarafraz-Ardakani, M. R., Hosseini Sarghein, S., Moshtaghioun, S. M., Mousavi-Kouhi, S. M. and Taghavizadeh Yazdi, M. E. (2023). Effect of organic fertilizer on the growth and physiological parameters of a traditional medicinal plant under salinity stress conditions. Hortic, 9, 701.

[27] Birhanu, Z. (2013). Traditional use of medicinal plants by the ethnic groups of Gondar Zuria District, North-Western Ethiopia. Journal of Natural Remedies, 1, 13(1), 46–53.

[28] Borges, C. V., Minatel, I. O., Gomez-Gomez, H. A. and Lima, G. P. P. (2017). Medicinal plants: Influence of environmental factors on the content of secondary metabolites. In: Medicinal Plants and Environmental Challenges, 2, 259–277.

[29] Borhannuddin Bhuyan, M. H. M., Hasanuzzaman, M., Nahar, K., Mahmud, J. A., Parvin, K., Bhuiyan, T. F. and Fujita, M. (2019). Plants behavior under soil acidity stress: Insight into morphophysiological, biochemical, and molecular responses. In: Plant Abiotic Stress Tolerance, 3, 35–82.

[30] Boyom, F. F., Ngouana, V., Zollo, P. H. A., Menut, C., Bessiere, J. M., Gut, J. and Rosenthal, P. J. (2003). Composition and anti-plasmodial activities of essential oils from some Cameroonian medicinal plants. Phytochemistry, 64, 1269–1275.

[31] Brechner, M. L., Albright, L. D. and Weston, L. A. (2007). Impact of a variable light intensity at a constant light integral: Effects on biomass and production of secondary metabolites by *Hypericum perforatum*. International Symposium on Medicinal and Nutraceutical Plants, 756, 221–228.

[32] Butnariu, M. (2021). Plants as source of essential oils and perfumery applications. In: Upadhyay, S. K. and Singh, S. P. (Eds.) Bioprospecting of Plant Biodiversity for Industrial Molecules, 2, 261–292. doi:10.1002/9781119718017.

[33] Cai, Y. Z., Sun, M., Xing, J., Luo, Q. and Corke, H. (2006). Structure–radical scavenging activity relationships of phenolic compounds from traditional Chinese medicinal plants. Life Science, 78, 2872–2888.

[34] Chandra, K. K., Kumar, N. and Chand, G. (2010). Studies on mycorrhizal inoculation on dry matter yield and root colonization of some medicinal plants grown in stress and forest soils. Journal of Environmental Biology, 31, 975.

[35] Chaplygin, V. A., Burachevskay, M. V., Minkina, T. M., Mandzhieva, S. S., Siromlya, T. I., Chernikova, N. P. and Dudnikova, T. S. (2024). Accumulation and distribution of heavy metals in soils and medicinal plants in the impact zone of Novocherkassk power station. Eurasian Soil Science, 57, 1746–1758.

[36] Chen, Y. M., Huang, J. Z., Hou, T. W. and Pan, I. C. (2019). Effects of light intensity and plant growth regulators on callus proliferation and shoot regeneration in the ornamental succulent Haworthia. Botanical Studies, 60, 1–8.

[37] Chengaiah, B., Rao, K. M., Kumar, K. M., Alagusundaram, M. and Chetty, C. M. (2010). Medicinal importance of natural dyes-a review. International Journal of Pharmtech Research, 2, 144–154.

[38] Cox-Georgian, D., Ramadoss, N., Dona, C. and Basu, C. (2019). Therapeutic and medicinal uses of terpenes. In: Joshee, N., Dhekney, S. A. and Prahlad, P. (Eds.) Medicinal Plants from Farm to Pharmacy, Springer, Cham 2, 333–359. https://doi.org/10.1007/978-3-030-31269-5

[39] Dar, R. A., Shahnawaz, M. and Qazi, P. H. (2017). General overview of medicinal plants: A review. The Journal of Phytopharmacology, 6, 349–351.

[40] Deshmukh, Y. and Khare, P. (2017). Effect of salinity stress on growth parameters and metabolites of medicinal plants: A review. In: Gupta, S. K. and M. R. Goyal (Eds.) Soil Salinity Management in Agriculture, Apple Academic Press, New York, 1st edition, 3, 197–234. doi: https://doi.org/10.1201/9781315365992

[41] Diaconu, D., Diaconu, R. and Navrotescu, T. (2012). Estimation of heavy metals in medicinal plants and their infusions. Ovidius University Annals of Chemistry, 23, 115–120.

[42] Dinu, C., Gheorghe, S., Tenea, A. G., Stoica, C., Vasile, G. G., Popescu, R. L. and Pascu, L. F. (2021). Toxic metals (As, Cd, Ni, Pb) impact in the most common medicinal plant (*Mentha piperita*). International Journal of Environmental Research and Public Health, 18, 3904.

[43] Dreshaj, A., Hidajete, N., Muzlijaj, H., Fekaj, F. and Beqiraj, I. (2013). Negative effects of heavy metals in medicinal plants. International Journal of Thermal Technologies, 3, 60–62.

[44] Du, L., Zhao, J., Abbas, F. and Liu, W. (2013). Higher nitrates, P and lower pH in soils under medicinal plants versus crop plants. Environmental Chemistry Letters, 11, 385–390.

[45] Duke, J. A. (1985). Medicinal plants. Science, 229, 1036–1036.

[46] Duke, J. A. (1993). Medicinal plants and the pharmaceutical industry. In: Janick, J. and Simon, J. E. (eds) Proceedings of the Second National Symposium: New crops, exploration, research and commercialization 1993, Proceedings of the Second National Symposium: New crops, exploration, research and commercialization 1991, Indianapolis, Indiana, 664–669. (USDA/ARS, Plant Science Institute, Bldg 001, Room 133 BARC-W, Beltsville, Maryland 20705-2350, USA).

[47] Dunford, N. T. and Vazquez, R. S. (2005). Effect of water stress on plant growth and thymol and carvacrol concentrations in Mexican oregano grown under controlled conditions. Journal of Applied Horticulture, 7, 20–22.

[48] El-Darier, S. M. and Youssef, R. S. (2000). Effect of soil type, salinity, and allelochemicals on germination and seedling growth of a medicinal plant *Lepidium sativum* L. The Annals of Applied Biology, 136, 273–279.

[49] Lama, Y. C., Ghimire, S. K. and Aumeeruddy-Thomas, Y. (2001). Medicinal Plants of Dolpo. Amchis' Knowledge and Conservation, WWF Nepal Program, Kathmandu.

[50] Emami Bistgani, Z., Barker, A. V. and Hashemi, M. (2023). Review on physiological and phytochemical responses of medicinal plants to salinity stress. Communications in Soil Science and Plant Analysis, 54, 2475–2490.

[51] Farnsworth, N. R., Akerele, O., Bingel, A. S., Soejarto, D. D. and Guo, Z. (1985). Medicinal plants in therapy. Bulletin of the World Health Organisation, 63, 965.

[52] Farooqi, A. H. A., Fatima, S., Khan, A. and Sharma, S. (2005). Ameliorative effect of chlormequat chloride and IAA on drought stressed plants of *Cymbopogon martinii* and *C. winterianus*. Plant Growth Regulation, 46, 277–284.

[53] Weyers, J. D. and Paterson, N. W. (2001). Plant hormones and the control of physiological processes. New Phytologist, 152, 375–407.

[54] Fennell, C. W., Light, M. E., Sparg, S. G., Stafford, G. I. and Van Staden, J. (2004). Assessing African medicinal plants for efficacy and safety: Agricultural and storage practices. Journal of Ethnopharmacology, 95, 113–121.

[55] Fonseca, J. M., Rushing, J. W., Rajapakse, N. C., Thomas, R. L. and Riley, M. B. (2006). Potential implications of medicinal plant production in controlled environments: The case of feverfew (*Tanacetum parthenium*). HortScience, 41, 531–535.

[56] Forouzandeh, M., Fanoudi, M., Arazmjou, E. and Tabiei, H. (2012). Effect of drought stress and types of fertilizers on the quantity and quality of medicinal plant Basil (*Ocimum basilicum* L.). Indian Journal of Innovations and Developments, 1, 696–699.

[57] Gedif, T. and Hahn, H. J. (2003). The use of medicinal plants in self-care in rural central Ethiopia. Journal of Ethnopharmacology, 87, 155–161.

[58] Getasetegn, M. and Tefera, Y. (2016). Biological activities and valuable compounds from five medicinal plants. Natural Products Chemistry and Research, 4, 220.

[59] Gharaghani, H., Shariatmadari, F. and Torshizi, M. A. (2015). Effect of fennel (*Foeniculum vulgare Mill.*) used as a feed additive on the egg quality of laying hens under heat stress. Brazilian Journal of Poultry Science, 17, 199–207.

[60] Gomez-Flores, R. and Tamez-Guerra, P. (2011). Sustainable agriculture for medicinal plants. Medicinal Plants and Sustainable Development, 25, 53–67.

[61] Gray, D. E., Pallardy, S. G., Garrett, H. E. and Rottinghaus, G. E. (2003). Acute drought stress and plant age effects on alkamide and phenolic acid content in purple coneflower roots. Planta Medica, 69, 50–55.

[62] Jahan, S., Anjali, K., Panwar, M., Mishra, R., Shankhdhar, S. C. and Shankhdhar, D. (2024). Integrative impacts of salicylic acid and water deficit stress on physiological processes of medicinal herb *Bacopa monnieri*. (L.). Plant Physiology Reports, 29(1), 65–75.

[63] Groom, N. (2012). The Perfume Handbook, Springer Science & Business Media, Hong kong.

[64] Guo, C., Lv, L., Liu, Y., Ji, M., Zang, E., Liu, Q. and Li, M. (2023). Applied analytical methods for detecting heavy metals in medicinal plants. Critical Reviews in Analytical Chemistry, 53, 339–359.

[65] Hadi, S. and Bremner, J. B. (2001). Initial studies on alkaloids from Lombok medicinal plants. Molecules, 6, 117–129.

[66] Halberstein, R. A. (2005). Medicinal plants: Historical and cross-cultural usage patterns. Annals of Epidemiology, 15, 686–699.

[67] Hamilton, A. C. (2004). Medicinal plants, conservation and livelihoods. Biodiversity Conservation, 13, 1477–1517.

[68] Haq, I. (2004). Safety of medicinal plants. Pakistan Journal of Medical Research, 43, 203–210.

[69]   Hashem, A. D. and Kaviani, B. (2010). In vitro proliferation of an important medicinal plant Aloe-A method for rapid production. Australian Journal of Crop Science, 4, 216–222.

[70]   Hassan, I. A. (2004). Interactive effects of salinity and ozone pollution on photosynthesis, stomatal conductance, growth, and assimilate partitioning of wheat (*Triticum aestivum* L.). Photosynthetica, 42, 111–116.

[71]   Hayati, A., Pramudya, M., Soepriandono, H., Maullani, A., Puspitasari, Y., Maulidah, S. and Dewi, F. R. P. (2023). Effects of medicinal plants rhizome on growth performance of tilapia (*Oreochromis niloticus*) exposed to micro plastics. AIP Conference Proceedings, 2554, 1.

[72]   Hayta, S., Polat, R. and Selvi, S. (2014). Traditional uses of medicinal plants in Elazığ (Turkey). Journal of Ethnopharmacology, 154, 613–623.

[73]   He, Y., Yu, J., Song, Z., Tang, Z., Duan, J. A., Zhu, H. and Cao, Z. (2024). Anti-oxidant effects of herbal residue from Shengxuebao mixture on heat-stressed New Zealand rabbits. Journal of Thermal Biology, 119, 103752.

[74]   Higashiuchi, K., Uno, Y., Kuroki, S., Hisano, M., Mori, T., Wong, C. W. and Itoh, H. (2016). Effect of light intensity and light/dark period on iridoids in *Hedyotis diffusa*. Environmental Control in Biology, 54, 109–116.

[75]   Hjouji, K., Haldhar, R., Alobaid, A. A., Taleb, M. and Rais, Z. (2024). Maximizing resource recovery: Anaerobic digestion of residual biomass from essential oil extraction in four aromatic and medicinal plants. Industrial Crops and Products, 216, 118820.

[76]   Hostettmann, K. and Marston, A. (2002). Twenty years of research into medicinal plants: Results and perspectives. Phytochemistry Reviews, 1, 275–285.

[77]   Hou, J. L., Li, W. D., Zheng, Q. Y., Wang, W. Q., Xiao, B. and Xing, D. (2010). Effect of low light intensity on growth and accumulation of secondary metabolites in roots of *Glycyrrhiza uralensis Fisch*. Biochemical Systematics and Ecology, 38, 160–168.

[78]   Ibtisham, F., Nawab, A., Niu, Y., Wang, Z., Wu, J., Xiao, M. and An, L. (2019). The effect of ginger powder and Chinese herbal medicine on production performance, serum metabolites and antioxidant status of laying hens under heat-stress condition. Journal of Thermal Biology, 81, 20–24.

[79]   Ijeabuonwu, A. M., Bernatoniene, J. and Pranskuniene, Z. (2024). Medicinal plants used to treat skin diseases and for cosmetic purposes in Norway. Plants, 13, 2821.

[80]   Jung, K., Kim, I. H. and Han, D. (2004). Effect of medicinal plant extracts on forced swimming capacity in mice. Journal of Ethnopharmacology, 93, 75–81.

[81]   Kant, R. and Kumar, A. (2022). Review on essential oil extraction from aromatic and medicinal plants: Techniques, performance and economic analysis. Sustainable Chemistry and Pharmacy, 30, 100829.

[82]   Kaya, C., Ak, B. E. and Higgs, D. (2003). Response of salt-stressed strawberry plants to supplementary calcium nitrate and/or potassium nitrate. Journal of Plant Nutrition, 26, 543–560.

[83]   Kinghorn, A. D. (1987). Biologically active compounds from plants with reputed medicinal and sweetening properties. Journal of Natural Products, 50, 1009–1024.

[84]   Kirakosyan, A., Seymour, E., Kaufman, P. B., Warber, S., Bolling, S. and Chang, S. C. (2003). Antioxidant capacity of polyphenolic extracts from leaves of *Crataegus laevigata* and *Crataegus monogyna* (Hawthorn) subjected to drought and cold stress. Journal of Agricultural and Food Chemistry, 51, 3973–3976.

[85]   Gaxiola, R. A., Li, J., Undurraga, S., Dang, L. M., Allen, G. J., Alper, S. L. and Fink, G. R. (2001). Drought-and salt-tolerant plants result from overexpression of the AVP1 H^+-pump. Proceedings of the National Academy of Sciences of the United States of America, 98, 11444–11449.

[86]   Kleinwächter, M. and Selmar, D. (2015). New insights explain that drought stress enhances the quality of spice and medicinal plants: Potential applications. Agronomy for Sustainable Development, 35, 121–131.

[87] Koocheki, A., Nassiri-Mahallati, M. and Azizi, G. (2008). Effect of drought, salinity, and defoliation on growth characteristics of some medicinal plants of Iran. Journal of Herbs, Spices & Medicinal Plants, 14, 37–53.

[88] Kose, M., Melts, I. and Heinsoo, K. (2022). Medicinal plants in semi-natural grasslands: Impact of Management. Plants, 11, 353.

[89] Kumar, S., Narula, A., Sharma, M. P. and Srivastava, P. S. (2004). In vitro propagation of *Pluchea lanceolata*, a medicinal plant, and effect of heavy metals and different aminopurines on quercetin content. In Vitro Cellular and Developmental Biology-Plant, 40, 171–176.

[90] Li, A., Li, S., Wu, X., Zhang, J., He, A., Zhao, G. and Yang, X. (2016). Effect of light intensity on leaf photosynthetic characteristics and accumulation of flavonoids in *Lithocarpus litseifolius* (Hance) *Chun*. (Fagaceae). Open Journal of Forestry, 6, 445–459.

[91] YAN, L., Craker, L. E. and Potter, T. (1995). Effect of light level on essential oil production of sage (*Salvia officinalis*) and thyme (*Thymus vulgaris*). International Symposium of Medicinal and Aromatic Plants, 426, 419–426.

[92] Liang, Z. L., Chen, F., Park, S., Balasubramanian, B. and Liu, W. C. (2022). Impacts of heat stress on rabbit immune function, endocrine, blood biochemical changes, antioxidant capacity and production performance, and the potential mitigation strategies of nutritional intervention. Frontiers in Veterinary Science, 9, 906084.

[93] Lubbe, A. and Verpoorte, R. (2011). Cultivation of medicinal and aromatic plants for specialty industrial materials. Industrial Crops and Products, 34, 785–801.

[94] Ma, Z., Li, S., Zhang, M., Jiang, S. and Xiao, Y. (2010). Light intensity affects growth, photosynthetic capability, and total flavonoid accumulation of Anoectochilus plants. HortScience, 45, 863–867.

[95] Mahasneh, Z. M., Abuajamieh, M., Abedal-Majed, M. A., Al-Qaisi, M., Abdelqader, A. and Al-Fataftah, A. R. A. (2024). Effects of medical plants on alleviating the effects of heat stress on chickens. International Journal of Poultry Science, 103(3), 103391.

[96] Maj, G., Najda, A., Klimek, K. and Balant, S. (2019). Estimation of energy and emissions properties of waste from various species of mint in the herbal products industry. Energies, 13, 55.

[97] Oladeji, O. M., Kopaopa, B. G., Mugivhisa, L. L. and Olowoyo, J. O. (2024). Investigation of heavy metal analysis on medicinal plants used for the treatment of skin cancer by traditional practitioners in Pretoria. Biological Trace Element Research, 202(2), 778–786.

[98] Maroyi, A. (2013). Traditional use of medicinal plants in south-central Zimbabwe: Review and perspectives. Journal of Ethnobiology and Ethnomedicine, 9, 1–18.

[99] Maryo, M., Nemomissa, S. and Bekele, T. (2015). An ethnobotanical study of medicinal plants of the Kembatta ethnic group in Enset-based agricultural landscape of *Kembatta Tembaro* (KT) Zone, Southern Ethiopia. Asian Journal of Plant Sciences, 5, 42–61.

[100] McGaw, L., Jäger, A., Grace, O., Fennel, C. and van Staden, J. (2005). Medicinal plants. In: Niekerk, A. (Eds.) Ethics in Agriculture – an African Perspective, Springer, Dordrecht, 67–83. https://doi.org/10.1007/1-4020-2989-6.

[101] Menezes-Benavente, L., Teixeira, F. K., Kamei, C. L. A. and Margis-Pinheiro, M. (2004). Salt stress induces altered expression of genes encoding antioxidant enzymes in seedlings of a Brazilian indica rice (*Oryza sativa* L.). Plant Science, 166, 323–331.

[102] Meng, X., Wen, Z., Qian, Y. and Yu, H. (2017). Evaluation of cleaner production technology integration for the Chinese herbal medicine industry using carbon flow analysis. Journal of Cleaner Production, 163, 49–57.

[103] Milenković, L., Ilić, Z. S., Šunić, L., Tmušić, N., Stanojević, L., Stanojević, J. and Cvetković, D. (2021). Modification of light intensity influence essential oils content, composition and antioxidant activity of *thyme, marjoram* and *oregano*. Saudi Journal of Biological Sciences., 28, 6532–6543.

[104] Mohamed, A. A. and Alotaibi, B. M. (2023). Essential oils of some medicinal plants and their biological activities: A mini review. Journal of Umm Al-Qura University for Applied Sciences, 9, 40–49.

[105] Mohamed, M. H., Harris, P. J. C., Henderson, J. and Senatore, F. (2002). Effect of drought stress on the yield and composition of volatile oils of drought-tolerant and non-drought-tolerant clones of Tagetes minuta. Planta Medica, 68, 472–474.

[106] Mondal, S., Sukul, S. and Sukul, N. C. (2012). Transfer of effect of heat shock and drug treatment from one plant to another through water. Journal of Alternative Medicine Research, 4, 179.

[107] Moussa, H. R. and Khodary, S. E. (2003). Effect of salicylic acid on the growth photosynthesis and carbohydrate metabolism in salt stressed maize plants. Applied Radiation and Isotopes, 35.

[108] Mukherjee, S., Chatterjee, N., Sircar, A., Maikap, S., Singh, A., Acharyya, S. and Paul, S. (2023). A comparative analysis of heavy metal effects on medicinal plants. Applied Biochemistry and Biotechnology, 195(4), 2483–2518.

[109] Muñoz-Acevedo, A., Torres, E. A., Gutiérrez, R. G., Cotes, S. B., Cervantes-Díaz, M. and Tafurt-García, G. (2015). Some Latin American plants promising for the cosmetic, perfume and flavor industries. Therapeutic Medicinal Plants: From Lab to the Market, 279–330.

[110] Nazari, M., Ghasemi-Soloklui, A. A., Kordrostami, M. and Aaha, L. (2023). Deciphering the response of medicinal plants to abiotic stressors: A focus on drought and salinity. Plant Stress,10,100255.

[111] Ndhlala, A. R., Van Staden, J. and Ncube, B. (2012). Ensuring quality in herbal medicines: Toxic phthalates in plastic-packaged commercial herbal products. South African Journal of Botany, 82, 60–66.

[112] Nishimura, T., Zobayed, S. M., T, K. and Goto, E. (2007). Medicinally important secondary metabolites and growth of *Hypericum perforatum* L. plants as affected by light quality and intensity. Environmental Control in Biology, 45, 113–120.

[113] Nkansah, M. A., Hayford, S. T., Borquaye, L. S. and Ephraim, J. H. (2016). Heavy metal contents of some medicinal herbs from Kumasi, Ghana. Cogent Environmental Science, 2, 1234660.

[114] Nurzyńska-Wierdak, R. (2013). Does mineral fertilization modify essential oil content and chemical composition in medicinal plants?. Acta Scientiarum Polonorum Hortorum Cultus, 12, 3–16.

[115] Olowoyo, J. O., Okedeyi, O. O., Mkolo, N. M., Lion, G. N. and Mdakane, S. T. R. (2012). Uptake and translocation of heavy metals by medicinal plants growing around a waste dump site in Pretoria, South Africa. South African Journal of Botany, 78, 116–121.

[116] Ozturk, M., Uysal, I., Gucel, S., Altundag, E., Dogan, Y. and Baslar, S. (2013). Medicinal uses of natural dye-yielding plants in Turkey. Research Journal of Textile and Apparel, 17, 69–80.

[117] Paee, F., Nasim, N. A. I., Sabran, S. F. and Zairi, M. N. M. (2019). Effect of different light intensities on growth rate in Mentha arvensis. IOP Conference Series: Earth and Environmental Science, 269, 012016.

[118] Pan, J. and Guo, B. (2016). Effects of light intensity on the growth, photosynthetic characteristics, and flavonoid content of *Epimedium pseudowushanense* BL Guo. Molecules, 21, 1475.

[119] Pandey, V., Tiwari, D. C., Dhyani, V., Bhatt, I. D., Rawal, R. S. and Nandi, S. K. (2021). Physiological and metabolic changes in two *Himalayan* medicinal herbs under drought, heat and combined stresses. Physiology and Molecular Biology of Plants, 27, 1523–1538.

[120] Pant, P., Pandey, S. and Dall'Acqua, S. (2021). The influence of environmental conditions on secondary metabolites in medicinal plants: A literature review. Chemistry & Biodiversity, 18, e2100345.

[121] Parzhanova, A. B., Petkova, N. T., Ivanov, I. G. and Ivanova, S. D. (2018). Evaluation of biologically active substance and antioxidant potential of medicinal plants extracts for food and cosmetic purposes. Journal of Pharmaceutical Sciences Research, 10, 1804–1809.

[122] Pennacchio, M., Jefferson, L. and Havens, K. (2010). Uses and Abuses of Plant-derived Smoke: Its Ethnobotany as Hallucinogen, Perfume, Incense, and Medicine, Oxford University Press, Newyork.

[123] Pereira, A. M. S., Bertoni, B. W., Menezes, J. A., Pereira, P. S. and Franca, S. C. (1998). Soil pH and production of biomass and wedelolactone in field grown *Eclipta alba*. Journal of Herbs, Spices & Medicinal Plants, 6, 43–48.

[124] Popović, V., Šarčević-Todosijević, L., Petrović, B., Ignjatov, M., Popović, D., Vukomanović, P. and Filipović, V. (2021). Economic justification application of medicinal plants in cosmetic and pharmacy for the drugs discovery. Introduction Of Medicinal Plants And Herbs, Nova Science Publishers, 63–105.

[125] Radanović, D., Antić-Mladenović, S. and Nastovski, T. (2006). Influence of soil characteristics and nutrient supply on medicinal and aromatic plants. 5, 43–56.

[126] Radulescu, C., Stihi, C., Popescu, I. V., Ionita, I., Dulama, I. D., Chilian, A. and Let, D. (2013). Assessment of heavy metals level in some perennial medicinal plants by flame atomic absorption spectrometry. Romanian Reports in Physics, 65, 246–260.

[127] Rafat Khafar, K., Mojtahedin, A., Rastegar, N., Kalvani Neytali, M. and Olfati, A. (2019). Dietary inclusion of *thyme* essential oil alleviative effects of heat stress on growth performance and immune system of broiler chicks. Iranian Journal of Applied Animal Science, 9, 509–517.

[128] Rajesh Arora, R. A., Archana Mathur, A. M. and Mathur, A. K. (2010). Emerging trends in medicinal plant biotechnology. In: Medicinal and Aromatic Plant Science and Biotechnology, 1–12.

[129] Ramawat, K. G., Dass, S. and Mathur, M. (2009). The chemical diversity of bioactive molecules and therapeutic potential of medicinal plants. In: Ramawat, K. G. (Eds.) Herbal Drugs: Ethnomedicine to Modern Medicine, Springer Berlin, Heidelberg, 7–32. https://doi.org/10.1007/978-3-540-79116-4.

[130] Sabee, M. M. S. M., Uyen, N. T. T., Ahmad, N. and Hamid, Z. A. A. (2021). Plastics packaging for pharmaceutical products. Reference Module in Materials Science and Materials Engineering, 316–329.

[131] Ahl HAH, S.-A. and Omer, E. A. (2011). Medicinal and aromatic plants production under salt stress. A review. Herba Polonica, 57, 72–87.

[132] Saikia, A. P., Ryakala, V. K., Sharma, P., Goswami, P. and Bora, U. (2006). Ethnobotany of medicinal plants used by Assamese people for various skin ailments and cosmetics. Journal of Ethnopharmacology, 106, 149–157.

[133] Sang, M., Liu, Q., Li, D., Dang, J., Lu, C., Liu, C. and Wu, Q. (2024). Heat Stress and Microbial Stress Induced Defensive Phenol Accumulation in Medicinal Plant *Sparganium stoloniferum*. International Journal of Molecular Sciences, 25, 6379.

[134] Sarma, H., Deka, S., Deka, H. and Saikia, R. R. (2011). Accumulation of heavy metals in selected medicinal plants. Reviews of Environmental Contamination and Toxicology, 214, 63–86.

[135] Schmelzer, G. H. and Gurib-Fakim, A., (eds). (2008). Medicinal Plants. Prota, Wageningen, Netherlands, 11.

[136] Schmidt, B. M. (2012). Responsible use of medicinal plants for cosmetics. HortScience, 47, 985–991.

[137] Sedghi, M., Nemati, A. and Esmaielpour, B. (2010). Effect of seed priming on germination and seedling growth of two medicinal plants under salinity. Emirates Journal of Food and Agriculture, 22, 130–139

[138] Séquin, M. (2021). The Chemistry of Plants: Perfumes, Pigments and Poisons, Royal Society of Chemistry, The United Kingdom by CPI Group (UK) Ltd, Croydon, UK.

[139] Seyedan, A., Alshawsh, M. A., Alshagga, M. A., Koosha, S. and Mohamed, Z. (2015). Medicinal plants and their inhibitory activities against pancreatic lipase: A review. Evid Based Complement Alternat Med, 2015(1), 973143.

[140] Shah, A., Niaz, A., Ullah, N., Rehman, A., Akhlaq, M., Zakir, M. and Suleman Khan, M. (2013). Comparative study of heavy metals in soil and selected medicinal plants. Journal of Chemistry, 2013 (1), 621265.

[141] Shahrajabian, M. H., Kuang, Y., Cui, H., Fu, L. and Sun, W. (2023). Metabolic changes of active components of important medicinal plants on the basis of traditional Chinese medicine under different environmental stresses. Current Organic Chemistry, 27, 782–806.

[142] Shi, J. Y., Yuan, X. F., Lin, H. R., Yang, Y. Q. and Li, Z. Y. (2011). Differences in soil properties and bacterial communities between the rhizosphere and bulk soil and among different production areas of the medicinal plant *Fritillaria thunbergii*. International Journal of Molecular Sciences, 12, 3770–3785.

[143] Sholikhah, E. N. (2016). Indonesian medicinal plants as sources of secondary metabolites for pharmaceutical industry. The Journal of Medical Sciences, 48, 226–239.

[144] Singh, K., Kumar, P. and Singh, N. V. (2020). Natural dyes: An emerging ecofriendly solution for textile industries. Pollution Research, 39, S87–S94.

[145] Singh, P. A., Bajwa, N., Chinnam, S., Chandan, A. and Baldi, A. (2022). An overview of some important deliberations to promote medicinal plants cultivation. Journal of Applied Research on Medicinal and Aromatic Plants, 31, 100400.

[146] Singh, R. (2015). Medicinal plants: A review. Journal of Plant Science, 8, 50–55.

[147] Song, F. L., Gan, R. Y., Zhang, Y., Xiao, Q., Kuang, L. and Li, H. B. (2010). Total phenolic contents and antioxidant capacities of selected Chinese medicinal plants. International Journal of Molecular Sciences, 11, 2362–2372.

[148] Song, X., Luo, J., Fu, D., Zhao, X., Bunlue, K., Xu, Z. and Qu, M. (2014). Traditional Chinese medicine prescriptions enhance growth performance of heat stressed beef cattle by relieving heat stress responses and increasing apparent nutrient digestibility. Asian-Australasian Journal of Animal Sciences, 27, 1513–1520.

[149] Stanojkovic-Sebic, A., Pivic, R., Josic, D., Dinic, Z. and Stanojkovic, A. (2015). Heavy metals content in selected medicinal plants commonly used as components for herbal formulations. The Indian Journal of Agricultural Sciences, 21, 317–325.

[150] Stević, T., Berić, T., Šavikin, K., Soković, M., Gođevac, D., Dimkić, I. and Stanković, S. (2014). Antifungal activity of selected essential oils against fungi isolated from medicinal plant. Industrial Crops and Products, 55, 116–122.

[151] Stevović, S., Ćalić, D., Surčinski-Mikovilović, V., Zdravković-Korać, S., Milojević, J. and Cingel, A. (2010). Correlation between environment and essential oil production in medicinal plants. The 2nd International Symposium on Medical Plants, Their Cultivation and Aspects of Uses, 5, 465–468.

[152] Svistova, I. D., Stekoľnikov, K. E., Paramonov, A. Y. and Kuvshinova, N. M. (2016). Effect of Medicinal Plants Cultivation on the Physicochemical Properties of Leached Chernozem. Eurasian Soil Science, 49, 194–197.

[153] Tschinkel, P. F., Melo, E. S., Pereira, H. S., Silva, K. R., Arakaki, D. G., Lima, N. V. and Nascimento, V. A. (2020). The hazardous level of heavy metals in different medicinal plants and their decoctions in water: A public health problem in Brazil. BioMed Research International, 2020(1), 1465051.

[154] Tungmunnithum, D., Thongboonyou, A., Pholboon, A. and Yangsabai, A. (2018). Flavonoids and other phenolic compounds from medicinal plants for pharmaceutical and medical aspects: An overview. Medicines, 5, 93.

[155] Turtola, S., Rousi, M., Pusenius, J., Yamaji, K., Heiska, S., Tirkkonen, V. and Julkunen-Tiitto, R. (2005). Clone-specific responses in leaf phenolics of willows exposed to enhanced UVB radiation and drought stress. Global Change Biology, 11, 1655–1663.

[156] Vardanega, R., Santos, D. T. and Meireles, M. A. A. (2014). Intensification of bioactive compounds extraction from medicinal plants using ultrasonic irradiation. Pharmacognosy Reviews, 8, 88.

[157] Volwiler, E. H. (1926). Medicinals and dyes. Industrial & Engineering Chemistry, 18, 1336–1337.

[158] Wang, C. L., Guo, Q. S., Zhu, Z. B. and Cheng, B. X. (2017). Physiological characteristics, dry matter, and active component accumulation patterns of Changium smyrnioides in response to a light intensity gradient. Pharmaceutical Biology, 55, 581–589.

[159] Wani, S. H., Kapoor, N. and Mahajan, R. (2017). Metabolic responses of medicinal plants to global warming, temperature and heat stress. In: Ghorbanpour, M. and Varma, A. (Eds.) Medicinal Plants and Environmental Challenges, Springer, Cham, 69–80. https://doi.org/10.1007/978-3-319-68717-9.

[160] Wannissorn, B., Jarikasem, S., Siriwangchai, T. and Thubthimthed, S. (2005). Antibacterial properties of essential oils from Thai medicinal plants. Fitoterapia, 76, 233–236.

[161] Wijesekera, R. O. B. (2017). The Medicinal Plant Industry, Routledge Boca raton, Florida.

[162] Xu, M. Y., Wu, K. X., Liu, Y., Liu, J. and Tang, Z. H. (2020). Effects of light intensity on the growth, photosynthetic characteristics, and secondary metabolites of Eleutherococcus senticosus Harms. Photosynthetica, 58, 3.

[163] Yaniv, Z. and Bachrach, U., (eds.) (2005). Handbook of Medicinal Plants, CRC Press, Binghamton, NY.

[164] Yoneda, Y., Shimizu, H., Nakashima, H., Miyasaka, J. and Ohdoi, K. (2017). Effects of light intensity and photoperiod on improving steviol glycosides content in *Stevia rebaudiana* (Bertoni) Bertoni while conserving light energy consumption. Journal of Applied Research on Medicinal and Aromatic Plants, 7, 64–73.

[165] Zhang, S., Zhang, L., Zou, H., Qiu, L., Zheng, Y., Yang, D. and Wang, Y. (2021). Effects of light on secondary metabolite biosynthesis in medicinal plants. Frontiers Plant Science, 12, 781236.

[166] Zhu, Z., Bao, Y., Yang, Y., Zhao, Q. and Li, R. (2024). Research progress on heat stress response mechanism and control measures in medicinal plants. International Journal of Molecular Sciences, 25, 8600.

[167] Marimuthu, T., Suganthy, M. and Nakkeeran, S. (2018). Common pests and diseases of medicinal plants and strategies to manage them. In: Singh, B. and Peter, K. V. (Eds.) New Age Herbals: Resource, Quality and Pharmacognosy, Springer, Singapore, 289–312. https://doi.org/10.1007/978-981-10-8291-7

[168] Carrubba, A., Lo Verde, G. and Salamone, A. (2015). Sustainable weed, disease, and pest management in medicinal and aromatic plants. Medicinal and Aromatic Plants of the World: Scientific, Production, Commercial and Utilization Aspects, 205–235.

[169] Sergeeva, V. (2015). Medicinal plants to control diseases and pests. In: Máthé, Á. (Eds.) Medicinal and Aromatic Plants of the World: Scientific, Production, Commercial and Utilization Aspects, Springer, Dordrecht, 257–271.https://doi.org/10.1007/978-94-017-9810-5

[170] Pandey, R. and Hembra, S. Pests and diseases of some commercial medicinal and aromatic plants and their management. In: Gopal, S., Bhat, J. A., Das, A. P. and Sumit C. (Eds.) Bioprospecting of Ethnomedicinal Plant Resources, Apple Academic Press, New York, 1st edition 2025, 375–397. https://doi.org/10.1201/9781003451488.

[171] Rajan, V. P. (2003). Insect Pests of Selected Medicinal Plants: Bionomics and Management (Doctoral Dissertation, Department of Agricultural Entomology, College of Horticulture, Vellanikkara.

[172] Pandey, V., Bhatt, I. D. and Nandi, S. K. (2019). Environmental stresses in Himalayan medicinal plants: Research needs and future priorities. Biodiversity and Conservation, 2019(28), 2431–2455.

[173] Kala, C. P. (2009). Medicinal plants conservation and enterprise development. Medicinal. Plants-International Journal of Phytomedicines and Related Industries, 1(2), 79–95.

[174] Ahmed, N., Alam, M., Saeed, M., Ullah, H., Junaid, M., Kanwal, M. and Ahmed, S. (2024). Role of plants in managing diseases. In: Öztürk, M., Ramaiah Sridhar, K., Sarwat, M., Altay, V., and Huerta-Martínez, F. M. (Eds.) Ethnic Knowledge and Perspectives of Medicinal Plants. Apple Academic Press, New York, 1 st edition. 579–604. https://doi.org/10.1201/9781003353089.

[175] Dwiastuti, M. E., Aji, T. G., Devy, N. F. and Hardiyanto, H. (2024). Prospects of medicinal plants to control citrus and other subtropical fruits postharvest diseases in Indonesia. In: Proceedings of the 1st International Conference On Food and Agricultural Sciences (ICFAS) 2022: Advanced Agricultural Technology to Deal with Climate Change Issues for Achieving Food Security, 24–25 November 2022, Bogor, Indonesia AIP Conference Proceedings , 2957, No. 1 AIP Publishing.

[176] Yetgin, A. (2024). Investigating medicinal plants for antimicrobial benefits in a changing climate. International Journal of Secondary Metabolite, 11(2), 364–377.

Nuraniye Eruygur* and Sanem Hoşbaş Coşkun

Chapter 4
Medicinal and aromatic plants that are toxic

Abstract: Toxic plants and the compounds they produce pose a significant health threat to both humans and animals. Many plants contain naturally occurring toxins that, while offering survival advantages to the plants, can lead to a range of harmful effects in humans when ingested, touched, or inhaled. These toxins can vary widely in their potency and the severity of their effects, ranging from mild symptoms such as nausea, vomiting, and skin irritation, to life-threatening conditions such as organ failure, respiratory distress, and even death. This chapter provides an in-depth exploration of some of the most notorious toxic plants, including oleander, castor bean, aconite, deadly nightshade, and tobacco, detailing the specific toxic compounds they contain and their effects on human health. For example, oleander contains cardiac glycosides that can cause fatal arrhythmias, while castor bean produces ricin, one of the most potent toxins known, capable of causing organ failure even with minimal exposure.

In addition to examining these plants, the chapter highlights the importance of recognizing toxic species in the environment and understanding the risks they pose. Effective prevention begins with education on plant identification and the implementation of safe practices for handling these plants. It discusses practical precautions such as wearing protective clothing, including gloves and long sleeves, when handling or gardening near potentially toxic species, and ensuring that toxic plants are stored and disposed of safely, out of reach of children and pets. Additionally, the chapter addresses the need for prompt medical intervention in cases of poisoning, emphasizing the importance of quick response, such as administering activated charcoal or using specific antidotes when available. The chapter further explores the role of awareness in reducing risks, stressing the need for community education on the identification of toxic plants in local environments and gardens. Public awareness campaigns can help prevent accidental exposure, which is especially important for households with children or pets. Finally, the chapter underscores the necessity of first aid knowledge, including recognizing the symptoms of poisoning and the immediate steps to take until professional medical help arrives.

In conclusion, the chapter emphasizes that while toxic plants are a natural part of the environment, with the right precautions, education, and preparedness, the risks they present can be significantly minimized. By fostering a deeper understanding of these plants and their harmful effects, we can create safer spaces for people and animals to coexist with the natural world. The balance between enjoying the

*Corresponding author: Nuraniye Eruygur,** Department of Pharmacognosy, Faculty of Pharmacy, Selcuk University, 42250 Konya, Turkey, e-mail: nuraniye.eruygur@selcuk.edu.tr, https://orcid.org/0000-0002-4674-7009
Sanem Hoşbaş Coşkun, Kelly Government Solutions, Bethesda, ABD

https://doi.org/10.1515/9783111469713-004

beauty and benefits of plants and protecting oneself from their toxic effects is critical for maintaining public health and safety.

Keywords: Secondary metabolites, natural toxins, alkaloids, phytotoxins

4.1 Introduction

Plants have been used for both medicinal and aromatic purposes throughout human history. However, some plants, in addition to their healing effects, can be dangerous if not handled with care due to the toxic compounds they contain. In this section, the properties, uses and potential risks of some medicinal and aromatic plants known to be toxic will be examined.

4.2 Toxic compounds and their effects

Poisonous plants are characterized by substances such as alkaloids, glycosides, and essential oils. These substances can cause various biochemical reactions in the body, resulting in both beneficial and harmful effects (Table 4.1). For example, while digitalis (foxglove) is used in the manufacture of medicines for heart disease, it can be fatal in the wrong dosage. While medicinal and aromatic plants offer a range of therapeutic benefits, their toxic compounds must be handled with care. These bioactive chemicals, such as alkaloids, glycosides, essential oils, saponins, and coumarins, can provide healing effects in controlled doses but become dangerous when misused. Therefore, proper dosing, professional guidance, and awareness of potential side effects are essential in the safe use of these plants.

4.2.1 Alkaloids

Alkaloids are nitrogen-containing compounds that can have strong physiological effects on humans. Many medicinal plants contain alkaloids that, in controlled doses, can have therapeutic effects, but in higher concentrations, they may become toxic. Atropine (from *Atropa belladonna*): used in medicine to dilate pupils and treat bradycardia (slow heart rate). However, excessive intake can cause hallucinations, seizures, coma and even death [1]. Nicotine (from *Nicotiana tabacum*): a stimulant found in tobacco that can cause addiction. In large doses, nicotine poisoning can lead to nausea, increased heart rate, paralysis or respiratory failure, and death [2]. Morphine (from *Papaver somniferum*): apowerful pain reliever derived from the opium poppy, but addiction and overdose can lead to respiratory depression and death [3].

Table 4.1: Toxic phytochemicals according to their secondary metabolite groups, showing their plant origin and toxic effects.

Secondary metabolite group	Phytochemical compound	Plant origin	Toxic effects
Alkaloids	Atropine	*Atropa belladonna* (deadly nightshade)	Dry mouth, blurred vision, hallucinations, and death in high doses
	Strychnine	*Strychnos nux-vomica* (nux vomica)	Muscle spasms, convulsions, and death from respiratory failure
	Coniine	*Conium maculatum* (hemlock)	Respiratory paralysis and death
Glycosides	Digoxin	*Digitalis purpurea* (foxglove)	Cardiac arrhythmias, heart failure, and death
	Amandin	*Prunus* spp. (bitter almonds)	Cyanide poisoning causing respiratory distress and coma
Proteins (toxins)	Ricin	*Ricinus communis* (castor bean)	Inhibits protein synthesis, organ failure, and death
	Abrin	*Abrus precatorius* (rosary pea)	Similar to ricin, inhibits protein synthesis, and fatal in small doses
Terpenoids	Aconitine	*Aconitum* spp. (monkshood, wolfsbane)	Nausea, vomiting, cardiac arrest, and death
Phenolic compounds	Cicutoxin	*Cicuta spp.* (water hemlock)	Seizures, vomiting, respiratory failure, death
Saponins	Ginsenosides	*Panax* spp. (ginseng)	High doses can cause vomiting, diarrhea, and toxicity
Cyanogenic glycosides	Amygdalin	*Prunus* spp. (bitter almond and cherry)	Cyanide release, respiratory failure, and death

4.2.2 Glycosides

Glycosides are molecules that consist of a sugar and another functional group, often a toxic compound. They are often found in medicinal plants, particularly those used for heart conditions. Digitalis glycosides (from *Digitalis purpurea*): used to treat heart conditions like atrial fibrillation. While therapeutic in low doses, an overdose can lead to life-threatening cardiac toxicity, arrhythmias, and cardiac arrest. Digitalis glycosides strengthen the heart muscle but can be lethal in overdose [4]. Cyanogenic glycosides (from *Prunus* species like cherry and apricot): release cyanide when metabolized, which can inhibit cellular respiration and lead to poisoning [5]. The toxic effects of cyanogenic glycosides are rapid breathing, seizures, and death. Fresh, cyanophoric plant materials have higher amounts of glycosides than processed foods that are often fed to pigs and poultry. Certain plants have the ability to collect high amounts of cyanogenic glycosides, which when consumed, turn into prussic acid. When stressed, stunted plants start to develop after drought breaks the danger of prussic acid toxicity in animals is significantly elevated. A powerful toxin, prussic acid enters an animal's circulation and travels throughout its body after consumption. The animal then dies from hypoxia as a result of the disruption of the mitochondria's electron transport chain (ETC), which prevents the use of oxygen [6].

4.2.3 Essential oils

Essential oils of some plants are used in aromatherapy but can be toxic in large quantities. Essential oils are highly concentrated plant extracts that contain volatile compounds. While they are used for therapeutic purposes such as in aromatherapy, some can be toxic if ingested or used improperly. Thujone (from *Artemisia absinthium* – wormwood): Known for its presence in absinthe. In large amounts, thujone can cause convulsions, seizures, neurological damage, hallucinations, and even brain damage [7]. Eugenol (from *Syzygium aromaticum* – cloves): commonly used for dental pain relief. However, eugenol in high doses can cause liver toxicity [8]. The essential oils like cinnamon oil or clove oil have effects of skin irritation and allergic reactions.

4.2.4 Saponins

Saponins are naturally occurring glycosides found in many plants. They have a soap-like property when mixed with water, and some can be toxic when ingested. Solanine (from *Solanum* species like potatoes and tomatoes): Solanine is a toxic glycoalkaloid found in green potatoes and can cause gastrointestinal irritation, diarrhea nausea, vomiting, and even death if consumed in large amounts. It can also lead to hemolysis (destruction of red blood cells) in extreme cases [9].

4.2.5 Coumarins

Coumarins are plant-derived chemicals that have anticoagulant properties, while some are used in medicine (like warfarin). Toilet soap and detergents, toothpaste, tobacco products, and some alcoholic drinks all contain coumarin, which is utilized as a fixative and enhancing factor in fragrances [10]. They can be toxic in large quantities. Dicoumarol (from *Melilotus* species): Found in spoiled sweet clover, it can lead to excessive bleeding and internal hemorrhage when ingested in large amounts. Primary lymphedema and lymphedema brought on by radiation therapy or surgery for breast cancer can both be effectively treated with coumarin. However, because of the potential for hepatotoxicity, which primarily manifests as mild to moderate transaminase increase, its clinical usage is restricted in a number of countries [11].

4.3 Poisonous medicinal plants

Throughout history, many poisonous plants have also been utilized for their medicinal properties. While they can provide significant health benefits when used correctly and in controlled doses, their toxic nature makes them potentially dangerous when misused. Poisonous medicinal plants such as *Digitalis, Atropa belladonna, Aconitum, Conium,* and *Nerium oleander* serve as a reminder that many plants hold both healing and dangerous properties (Table 4.2). While their toxic components can be beneficial in controlled medical use, their potential to cause harm, even in small amounts, necessitates careful handling, professional oversight, and clear public awareness. This section explores some of the most well-known poisonous medicinal plants, their uses, and their potential risks.

Table 4.2: Some common medicinal and aromatic plants that are known to be toxic, along with their potential toxic compounds and symptoms of toxicity.

Plant name	Toxic compounds	Symptoms of toxicity	Notes
Atropa belladonna	Tropane alkaloids (e.g., atropine and scopolamine)	Dilated pupils, blurred vision, tachycardia, dry mouth, urinary retention, seizures, and death in extreme cases	Highly toxic, especially when consumed in large quantities; historically used as a poison and in medicine for certain conditions
Aconitum (monkshood)	Aconitine alkaloids	Nausea, vomiting, abdominal pain, dizziness, arrhythmia, respiratory failure, and death	Used in traditional medicine, but highly toxic; even small doses can be fatal
Digitalis purpurea	Cardiac glycosides (e.g., digoxin)	Vomiting, diarrhea, heart arrhythmias, confusion, dizziness, and death	Widely used in heart treatment, but overdose can cause severe cardiac toxicity
Nerium oleander	Oleandrin (cardiac glycoside)	Vomiting, diarrhea, bradycardia, arrhythmia, heart block, and death	All parts of the plant are highly toxic, especially when ingested

Table 4.2 (continued)

Plant name	Toxic compounds	Symptoms of toxicity	Notes
Ricinus communis	Ricin, a highly toxic protein	Abdominal pain, vomiting, diarrhea, organ failure, and death	Castor bean plant, highly toxic when seeds are ingested; ricin is one of the strongest known poisons
Conium maculatum	Coniine, an alkaloid	Drooping eyelids, dilated pupils, weakness, respiratory failure, and death	Also known as poison hemlock; historically used as a poison in executions
Taxus baccata (yew)	Taxine alkaloids	Nausea, vomiting, dizziness, difficulty breathing, and death	Extremely toxic; every part of the plant is poisonous, with the exception of the red arils, which comprise the fruit's fleshy portion
Hyoscyamus niger	Tropane alkaloids (e.g., hyoscyamine)	Dilated pupils, tachycardia, dry mouth, confusion, hallucinations, and seizures	Known as henbane; used in folk medicine, but toxic at higher doses
Cicuta virosa (water hemlock)	Cicutoxin	Seizures, respiratory distress, vomiting, convulsions, and death	Ingestion of any part of the plant can be fatal
Silybum marianum (milk thistle)	Silymarin (in large doses)	Gastrointestinal distress, nausea, vomiting, and allergic reactions	Typically used for liver health; large doses can cause adverse effects
Lavandula stoechas	Camphor and thujone	Dizziness, seizures, confusion, respiratory issues, dermatitis, and neurotoxicity in excessive use	Camphor is toxic in high amounts, especially when ingested or applied topically in large doses
Ruta graveolens (rue)	Furanocoumarins (e.g., rutin)	Skin irritation, photosensitivity, nausea, vomiting, and liver damage	Toxic when ingested in large quantities or applied directly to the skin
Toxicodendron radicans (poison ivy)	Urushiol (a resin)	Skin irritation, rashes, itching, and blisters	Causes contact dermatitis upon exposure to the resin found in the plant

4.3.1 *Digitalis purpurea* (foxglove)

Foxglove contains cardiac glycosides, primarily digitoxin and digoxin. These compounds affect the sodium-potassium balance in heart cells, leading to stronger heart contractions. Foxglove has been used for centuries as a heart tonic in folk medicine. Digoxin, derived from foxglove, is used to treat certain heart conditions, such as heart failure and atrial fibrillation. It increases the force of heart contractions and slows down the heart rate [12].

Symptoms of poisoning include nausea, vomiting, diarrhea, confusion, visual disturbances (seeing halos or blurred vision), and potentially fatal heart arrhythmias. Ingesting a small amount of the plant can be lethal if not treated immediately [4]. Cardiac glycosides have been a key component in the therapy of congestive heart failure since William Withering's dissertation on the efficacy of the leaves of the common foxglove plant (*Digitalis purpurea*) in the late eighteenth century codified its use. The effectiveness and safety of this class of medications are still up for the question, despite their broad adoption into medical practice during the next 200 years. Furthermore, although the molecular target for the cardiac glycosides – the α-subunit of sarcolemmal Na^+-K^+-ATPase (or sodium pump) found on most eukaryotic cell membranes – has been known for several decades, it is still unclear whether the sympatholytic or positive inotropic effects of these agents are the mechanisms most relevant to alleviating heart failure symptoms [13].

4.3.2 *Atropa belladonna* (deadly nightshade)

Atropa belladonna contains tropane alkaloids, mainly atropine, scopolamine, and hyoscyamine. These alkaloids act as anticholinergics, blocking acetylcholine receptors in the nervous system. Historically, *belladonna* was used in Italy to dilate women's pupils (hence the name "beautiful lady" or "*belladonna*"). Atropine, extracted from *belladonna*, is used in medicine as a muscle relaxant, to treat bradycardia (slow heart rate), and as an antidote for certain types of poisonings (such as organophosphates). Scopolamine is used for motion sickness and postoperative nausea. Toxic effect symptoms include dry mouth, blurred vision, difficulty swallowing, confusion, hallucinations, seizures, and respiratory failure. Death can occur if large quantities are ingested. As little as 10–20 berries can be fatal for adults, while even fewer are deadly for children [14].

4.3.3 *Aconitum napellus* (monkshood, aconite)

Aconitine and related alkaloids are found in the *Aconitum* species, a potent neurotoxin and cardiotoxin. Aconitine interferes with sodium channels in nerves and muscles, leading to abnormal electrical activity. The wild plant is highly hazardous, particularly the roots and root tubers. Accidental eating of the wild plant or ingestion of a herbal decoction prepared from aconite roots can result in severe aconite poisoning. Aconite roots are only utilized in traditional Chinese medicine after being processed to lessen their harmful alkaloid content. Aconite alkaloids will be hydrolyzed into less toxic and nontoxic derivatives by soaking and boiling during processing or decoction production. However, the danger of poisoning is increased by using a dose that is higher than is advised and by improper processing [15]. Historically, aconite was used in traditional Chinese medicine for its analgesic and anti-inflammatory properties. It was also applied topically to treat joint pain and rheumatism. Some components of aconite are still used in modern Chinese herbal medicine in highly regulated, detoxified forms for pain relief and inflammation [16]. The toxic effect symptoms of aconitine poisoning include tingling or numbness of the face and limbs, vomiting, diarrhea, difficulty breathing, irregular heartbeats, paralysis, and death due to respiratory or cardiac arrest. Even a small amount of the raw plant (as little as 2 mg of aconitine) can be fatal.

4.3.4 *Conium maculatum* (hemlock)

Hemlock contains coniine, a neurotoxin that disrupts communication between the central nervous system and muscles by blocking nicotinic acetylcholine receptors. In ancient Greece, hemlock was infamously used as a method of execution, most notably in the case of the philosopher Socrates [17]. Despite its toxic nature, small amounts were used in the past to treat muscle spasms and arthritis. Due to its extreme toxicity, hemlock is no longer used in modern medicine. Toxic effect symptoms include muscle weakness, paralysis, respiratory failure, and death. The paralysis caused by hemlock moves from the extremities toward the core, ultimately leading to respiratory failure as the diaphragm is paralyzed. A very small dose, around 0.5 g of the plant, can be fatal [18].

4.3.5 *Nerium oleander* (oleander)

Oleander contains cardiac glycosides, primarily oleandrin and neriine, which have effects similar to those of digitalis. These compounds interfere with the electrolyte balance of heart cells, potentially leading to fatal arrhythmias. Oleander was used in folk medicine as a treatment for a variety of ailments, including heart conditions and skin diseases. Some research has explored the use of oleandrin in cancer treatment, though its extreme toxicity makes this highly experimental. Toxic effects of oleander poisoning include nausea, vomiting, abdominal pain, dizziness, arrhythmias, and potentially cardiac arrest [19]. Both ingestion and inhalation of oleander smoke can be fatal. Ingesting one leaf is potentially lethal for an adult, and even small amounts can kill a child [20].

4.3.6 *Datura stramonium* (jimsonweed)

Common names for *Datura stramonium* include prickly burr, fake castor oil plant, devil's cucumber, tolguacha, Jamestown weed, stinkweed, locoweed, thorn apple, moon flower, hell's bells, devil's trumpet, devil's weed, devil's snare, and jimsonweed. The Sanskrit language is the source of the genus name *Dhattūra*. Although the Neo-Latin species name *stramonium* has no recorded origin, Carolus Linnaeus used the word *stramonia* for a variety of *Datura* species in the seventeenth century. The tropane alkaloids atropine, hyoscyamine, and scopolamine – all of which are categorized as deliriants or anticholinergics – are present in hazardous concentrations in every section of *Datura* plants [21]. In traditional medicine, *D. stramonium* has long been used to treat a wide range of conditions such as asthma, muscle spasms, and motion sickness. Scopolamine is used in modern medicine as a sedative and anti-nausea medication. On the other hand, it is used recreationally to produce hallucinations and euphoria, a feeling of well-being [22]. As a result, it has also been used as an entheogenetic hallucinogen (of the anticholinergic/antimuscarinic, deliriant variety) to produce vivid images. Because of its effects on the body and mind, which are often subjectively seen as extremely unpleasant and can result in a profound and protracted state of disorientation that could be lethal, it is unlikely to ever become a major drug of abuse. It includes tropane alkaloids, which can be quite toxic and are what cause the deliriant effects. It is utilized to create medications that treat influenza, cough, asthma, nerve diseases, and swine flu.

Because *D. stramonium* seeds are analgesic, anthelmintic, and anti-inflammatory, they are used to cure toothaches, fever from inflammation, and stomach and intestinal discomfort brought on by worm infestation. Its fruit juice is administered to the scalp to cure hair loss and dandruff [23]. In addition to respiratory failure, flushing, mydriasis, sinus tachycardia, hyperpyrexia, decreased bowel activity, urinary retention, and neurological disorders with ataxia, impaired short-term memory, disorientation, confusion, hallucinations (visual and auditory), psychosis, agitated delirium, seizures, and coma, typical symptoms of *D. stramonium* poisoning include dry skin and mouth, hyperthermia, hallucinations, delirium, seizures, flushing, and coma. These signs are similar to those of atropine intoxication. Dry mouth, excessive thirst, convulsions, nausea, vomiting, elevated heart rate, unconsciousness, and breathing problems are some of the symptoms of the toxin, which can be lethal if consumed in high quantities.

4.3.7 *Ricinus communis* (castor bean)

One of the most hazardous naturally occurring compounds, ricin, is present in the seeds of the castor bean plant (*Ricinus communis*). Although reports on the amount of ricin in castor beans vary, it is most likely between 1% and 5%. Ricin is being investigated for therapeutic use in cell-based research, bone marrow transplantation, and cancer treatment. According to experimental data, malignant cells express more carbohy-drate-containing surface-lectin binding sites than nonmalignant cells, which may make them more vulnerable to ricin toxicity. Antibody-conjugated ricin has been studied as an immunotherapeutic treatment that targets cancer cells [24]. Castor oil, derived from the plant, is used as a laxative and in the treatment of skin conditions [25, 26]. The symptoms of ricin poisoning include severe abdominal pain, vomiting, diarrhea, and organ failure. Ingestion of even one or two castor beans can be fatal [27].

4.3.8 *Taxus baccata* (English yew)

Taxus baccata is a native evergreen non-resinous gymnosperm tree up to 20–28 m, often with multiple trunks and spreading, rounded or pyramidal anopy [28]. It con-tains taxine alkaloids, which are cardiotoxic and neurotoxic. Some species of yew (e.g., *Taxus brevifolia*) are used to produce the chemotherapy drug paclitaxel (Taxol), used in cancer treatment. The U.S. Department of Agriculture (USDA) and the National Cancer Institute (NCI) worked together on a plant screening program from 1960 to 1981 that gathered and examined 115,000 extracts from 15,000 plant species in order to find naturally occurring substances having anticancer properties. On the final day of his 1962 journey, USDA botanist Arthur Barclay collected samples from a solitary Pacific yew tree, *Taxus brevifolia*. Following his return, tests were conducted on crude extracts from fruit, twigs, bark, and needles; the results showed that the bark extract was cytotoxic. In 1964, samples of *T. brevifolia* were sent to Mansukh Wani and Mon-roe Wall, who were employed by the NCI at the Research Triangle Institute (Research Triangle Park, NC) under contract. By 1967, the active component of *T. brevifolia* bark had been separated, identified, and given the name Taxol based on the species from which it originated and the presence of hydroxyl groups [29]. Symptoms of yew poi-soning include vomiting, seizures, respiratory distress, and heart failure. Death can occur due to heart arrhythmias [30].

4.3.9 *Hyoscyamus niger* (black henbane)

Commonly referred to as henbane, *Hyoscyamus niger* L., a member of the Solanaceae family, is found throughout Asia and Europe. It may be found in India between 8,000 feet and 11,000 feet above sea level, from Kashmir to the Garhwal Himalayas. Tropane alkaloids such as scopolamine and hyoscyamine are found in *H. niger*. It contains nor-tropane alkaloids, such as calystegins, which have strong-to-moderate glycosidase in-hibitory effects, and is an excellent source of anticholinergic tropane alkaloids. There have also been reports of the existence of several non-alkaloidal components in addi-tion to alkaloids, such as lignan amides, lignans, withanolides, and tyramine deriva-tives [31]. The black henbane was used historically as a sedative, pain reliever, and to treat motion sickness. Scopolamine is still used to prevent nausea. Symptoms of henbane poisoning include hallucinations, agitation, dry mouth, increased heart rate, and coma. High doses can be lethal [32].

4.3.10 *Cicuta virosa* (water hemlock)

In North America and the UK, water hemlock is considered to be one of the most toxic plants. It is made up of a variety of species that are separated into two genera, *Cicuta* and *Oenanthe*. The plants often referred to as water hemlock are members of the Apiaceae family (the historical name Umbelliferae is an alternate family name allowed by the International Code of Botanical Nomenclature). This family's species are separated into the genera *Oenanthe* and *Cicuta*. Four species make up the genus *Cicuta*: *C. bulbifera* L., *C. douglasii* (DC.), *C. maculata* L., and *C. virosa* L. All save *C. bulbifera* have significant concentrations of cicutoxin, which is enough to be harmful to hosts. The primary toxin responsible for the unique neurological signs and symptoms is found in *O. crocata*, out of all the Oenanthe species. Cicutoxin and oenanthotoxin, the two main toxins, are members of a class of C17 conjugated polyacetylenes. In the central nervous system (CNS), they function as (noncompetitive) gamma-aminobutyric acid antagonists, causing unchecked neuronal depolarization that may result in seizures. Even a little amount of plant stuff can cause extreme intoxication if consumed [33]. It has been used in traditional medicine for pain relief, but its extreme toxicity limits its modern use. Symptoms of water hemlock ingestion include nausea, seizures, muscle twitching, respiratory failure, and death. Even a small amount can be fatal.

4.3.11 *Veratrum viride* (false hellebore)

Numerous *Veratrum* species are linked to toxicity in both humans and animals. They contain *Veratrum* alkaloids, such as veratridine, which have an impact on muscle and nerve sodium channels. Steroid alkaloids are the main poisons; some differ in their esterified acid moiety, while others have a modified steroid template. These alkaloids work by making nerve cells' sodium channels more permeable, which makes them fire constantly. Increased vagal nerve stimulation triggers the Bezold-Jarisch reaction, which is characterized by three responses: bradycardia, apnea, and hypotension. Traditionally used for treating high blood pressure and fever, but not commonly used in modern medicine due to its toxicity. Several *veratrum* extracts were promoted as antihypertensive medications in the clinical setting but were later taken off the market due to their limited therapeutic index. Vomiting and gastrointestinal discomfort are typical symptoms after ingesting *Veratrum* alkaloids. Cardiovascular consequences such bradycardia, hypotension, and aberrant cardiac conduction, as well as death from cardiovascular collapse, are therefore anticipated [34].

4.3.12 *Helleborus niger* (Christmas rose)

Since 2,500 BC, the ancient Greeks utilized *Helleborus niger* L. (*Helleboros melas*), a plant with potent diuretic, emetic, and narcotic properties that helped them treat psychiatric problems, leprosy, scabies, and deafness [35]. It contains cardenolides and saponins, both toxic to the heart and gastrointestinal system. It has a broad spectrum of pharmacological actions, including diuretic, immunostimulant, cardiotonic, antibacterial, anticancer, and emetic. Buffadenolides, flavonoids, and phenolic heterosides are among the several secondary metabolites found in *H. odorus* Waldst. et Kit. [36]. Traditionally used as a treatment for menstrual issues and as a diuretic, symptoms of *H. niger* poisoning include vomiting, diarrhea, dizziness, arrhythmias, and death [37].

4.3.13 *Mandragora officinarum* (mandrake)

Tropane alkaloids, such as scopolamine, hyoscyamine, and atropine, are found in the roots and rhizomes of the European mandrake (*Mandragora officinarum* L.), and they have long been used in medicine. The histological characteristics of the two species' roots and rhizomes are the same and comparable to those of *Atropa belladonna* roots, according to investigations. It was used in ancient times as an anesthetic and sedative. In small doses, it was also used to treat muscle spasms and arthritis. Symptoms of toxic effects include hallucinations, delirium, vomiting, and heart failure. Overdose can lead to death [38].

4.3.14 *Ageratina altissima* (white snakeroot)

The perennial plant *Ageratina altissima* (L.) King & H. Rob, often called *Eupatorium rugosum*, is a member of the Asteraceae family. Native to North America, it is also known by several indigenous names, such as "white snakeroot", "snakeroot," or "richweed." *A. altissima* contains several toxic components, among which are mainly pyrrolizidine alkaloids [123]. These alkaloids are a group of chemical compounds that are also found in other plants [124]. They are known for their hepatotoxic effect and their ability to cause toxic effects in humans, animals, and even insects that feed on these plants. Pyrrolizidine alkaloids harm liver function. The toxic effect is mainly caused by their biotransformation in the body into reactive metabolites that damage liver cells. These metabolites can react with DNA, proteins, and other cellular components, leading to loss of liver cell integrity, which disrupts normal liver function and can lead to chronic health problems. Consumption or contact with pyrrolizidine alkaloids can cause adverse effects on the digestive system, including nausea, vomiting, and diarrhea. It contains tremetol, a fat-soluble toxin. This plant was not used medicinally but caused historical mass poisoning known as "milk sickness" when cattle fed on the plant and passed the toxin through their milk. Symptoms of poisoning include vomiting, tremors, liver damage, and death [39].

4.3.15 *Bryonia alba* (white bryony)

Because *Bryonia spp.* roots and fruits are known to be toxic, the plant is included on the list of dangerous plants. The short-term toxicity of *Bryonia* species is well-known, and it seems to be caused by triterpenic cucurbitacins and their glycosides, which are found throughout the plant. Cucurbitacins have been suggested as possible anticancer medicines because of their powerful cytotoxic activity. Accidental envenomation may also be partially caused by bryodiofine, a poisonous protein found in *B. Dioica* fruits. More common in minors, bryony intoxication is typically reported to occur after consuming the fruits rather than the difficult-to-reach roots. Rarely are fatal problems noted; patients often arrive with pallor, sweating, convulsions, respiratory and cardiac problems, and symptoms related to the digestive tract, including nausea, vomiting, diarrhea, and abdominal discomfort. Activated carbon and diazepam should be used as part of a symptomatic therapy for acute poisoning in situations of convulsions. However, the long-term consequences of bryony use are yet unclear, particularly with regard to potential harmful kidney activities [40].

4.3.16 *Colchicum autumnale* (autumn crocus)

Colchicine is a neutral, lipophilic alkaloid with mild anti-inflammatory properties, derived from the plants *Colchicum autumnale* (autumn crocus, meadow saffron) and *Gloriosa superba* (glory lily). Historically, it has been used to treat acute gout and is FDA-approved for both gout prophylaxis and certain types of arthritis and the treatment of familial Mediterranean fever (FMF), where it helps reduce the risk of systemic amyloidosis. Colchicine may also be beneficial for other conditions like recurrent pericarditis, scleroderma, Behcet's syndrome, and Sweet's syndrome, though evidence for these uses is often limited and inconclusive. Despite its benefits, colchicine's use is restricted by its toxicity. It is generally safe when used according to established guidelines for FMF, but gastrointestinal side effects can occur even at recommended doses before acute gout pain relief is achieved. In higher doses, colchicine can cause severe systemic toxicity. Symptoms of poisoning include abdominal pain, nausea, vomiting, multi-organ failure, and death. Although acute colchicine poisoning is rare, it has a high mortality rate, making it crucial for clinicians to recognize and understand colchicine poisoning [41]. Colchicine intoxication involves multiple organs and is associated with a poor prognosis when large amounts of the drug are administered. Treatment is primarily supportive and symptomatic due to the rapid distribution and binding of colchicine to affected tissues. A novel approach using anti-colchicine antibodies has shown promise in experimental models. Key research areas include the impact of liver and kidney disease on colchicine metabolism, the use of colchicine levels for diagnosing intoxication and predicting outcomes, and the application of immunotoxicotherapy for colchicine poisoning in humans [42].

4.3.17 *Chelidonium majus* Linn. – Papaveraceae

Chelidonium majus is commonly known as greater celandine, nipplewort, tetterwort, or simply celandine. It belongs to the Papaveraceae family of perennial herbaceous plants, which includes poppies. In the genus *Chelidonium*, it is one of two species. Originally from Europe and western Asia, the species has spread significantly over North America. Because it contains a variety of isoquinoline alkaloids, the entire plant is hazardous in excessive amounts. The right dosage is necessary when using it in herbal medicine. Coptisine is the primary alkaloid found in the plant and root. Allocryptopine, stylopine, protopine, norchelidonine, berberine, chelidonine, sanguinarine, chelerythrine, methyl 2′-(7,8-dihydrosanguinarine-8-yl) acetate, and 8-hydroxydihydrosanguinarine are among the other alkaloids found. In rats, sanguinarine is especially harmful. Proteolytic enzymes and the phytocystatin chelidostatin, an inhibitor of cysteine protease, are also present in the distinctive latex [43].

In France, it is a popular folk cure for warts. Chickens are poisoned by the plant. Officially, the fresh herb is no longer utilized. There are no dose-finding trials available, and the clinical studies that have been published exhibit significant heterogeneity. Traditionally used to help with vision, greater celandine is now utilized as an antispasmodic, light sedative, and therapy for bronchitis, whooping cough, asthma, jaundice, gallstones, and gallbladder discomfort. Ringworm, corns, and warts are all treated topically using the latex. On tumor cells, Ukrain, a semisynthetic thiophosphate derivative of alkaloids from *C. majus*, has cytotoxic and cytostatic properties. *Chelidonium* causes strong exacerbation from motion, liver affections, cough with right-sided chest symptoms, and mental affections that are typical of these. In addition to relieving stomach discomfort, it also helps with other problems. Eating reduces mental symptoms [44].

4.4 Aromatic plants and poisons

Aromatic plants are known for their fragrant essential oils, which are often used in perfumes, cosmetics, and traditional medicine. While many aromatic plants are valued for their therapeutic benefits, some contain toxic compounds that can be harmful or even fatal when misused (Table 4.3). In this section, we will explore some common aromatic plants that possess both medicinal qualities and poisonous properties (Table 4.4).

Table 4.3: Toxic essential oils and their effects.

Essential oil	Toxicity/effects
Wintergreen	Contains methyl salicylate, toxic if ingested
Eucalyptus	Can cause respiratory distress if inhaled or ingested
Pennyroyal	Extremely toxic, especially to pets
Cinnamon	Skin irritant, toxic when ingested
Clove	Causes skin irritation and toxicity in large amounts

4.4.1 *Artemisia absinthium* (wormwood)

Artemisia absinthium, commonly known as wormwood and belonging to the Asteraceae family, holds a significant place in the history of medicine. Revered in medieval Europe as "the most important master against all exhaustions," this plant is widely recognized as medicinal across Europe, West Asia, and North America. The plant's raw materials, Absinthii herba and *Artemisiae absinthii* aetheroleum, are rich in biologically active compounds. These include essential oils, bitter sesquiterpenoid lactones, flavonoids, azulenes, phenolic acids, tannins, lignans, and other compounds contributing to its characteristic bitterness. In official European medicine, wormwood is employed in both allopathy and homeopathy. Traditional European and Asian practices utilize this species for various health conditions, such as gastrointestinal issues, helminthiasis (parasitic worm infections), anemia, insomnia, bladder diseases, fever, and difficult-to-heal wounds [45].

Wormwood's characteristic aroma comes from thujone, which is a major component of its essential oil. It has a bitter, strong smell. The essential oil is used in aromatherapy to treat digestive disorders and as a stimulant. It has been traditionally used to make absinthe, a well-known alcoholic beverage. Thujone is the primary toxic compound. It is a neurotoxin that affects the central nervous system by blocking GABA receptors, leading to overstimulation of the nervous system. Overconsumption of thujone can cause seizures, tremors, hallucinations, and in high doses, it can lead to death. Absinthe, which contains thujone, was banned in many countries for a long time due to its toxic effects, though it has been reintroduced with regulated thujone levels [46].

4.4.2 *Sassafras albidum* (sassafras)

Sassafras (*Sassafras albidum*), also known as white sassafras, is a medium-sized, aromatic tree that grows at a moderate rate and is easily recognized by its three distinct leaf shapes: entire, mitten-shaped, and three-lobed. In northern regions, it remains shrub-like, but it reaches its largest size in the Great Smoky Mountains, thriving in moist, well-drained sandy loam soils within open woodlands. Sassafras often serves as a pioneer species in abandoned fields, where it plays a significant role in supporting wildlife by providing browse material, often forming dense thickets through underground runners from the parent tree. While its soft, lightweight, and brittle wood has limited commercial use, the root bark is a source of sassafras oil, widely used in the perfume industry [47]. Essential oil: The bark and roots of sassafras are known for their distinctive fragrance, primarily due to safrole, an aromatic compound. Sassafras oil was historically used as a flavoring agent in root beer and other beverages. The oil is used in traditional medicine to treat wounds, skin problems, and colds. It has been used for its pleasant scent in perfumes and soap. Safrole is the main toxic compound found in sassafras. It is classified as a carcinogen and hepatotoxin (liver toxin). Safrole is toxic when ingested in large amounts. Long-term exposure or high doses can lead to liver damage and an increased risk of cancer. Due to its carcinogenic properties, the FDA banned the use of safrole-containing sassafras oil in foods and beverages in the 1960s [48].

4.4.3 *Lavandula angustifolia* (lavender)

Lavander is commonly known as English lavender, garden lavender, *Lavandula burnamii, L. dentate, L. dhofarensis, L. latifolia, L. officinalis L., or L. stoechas*. Lavender, originally native to the Mediterranean, Arabian Peninsula, Russia, and Africa, has been valued for its cosmetic and medicinal properties throughout history. Today, it is cultivated globally, with its fragrant flower oils widely utilized in aromatherapy and various products, including baked goods, candles, cosmetics, detergents, jellies, massage oils, perfumes, powders, shampoos, soaps, and teas. The most commonly used species is English lavender (*Lavandula angustifolia*), although other varieties, such as *L. burnamii, L. dentata, L. dhofarensis, L. latifolia,* and *L. stoechas*, are also utilized [49].

Lavender's distinctive, soothing scent comes from linalool and linalyl acetate. It is widely used in aromatherapy for its calming and stress-relieving effects. Lavender oil is commonly used to treat insomnia, anxiety, and skin ailments, and is a popular ingredient in perfumes and cosmetics. Although lavender is generally considered safe, linalool can be toxic when ingested in large amounts or applied excessively to the skin. Ingestion of large doses of lavender oil can cause nausea, vomiting, and central nervous system depression. Topically, it can cause allergic reactions in sensitive individuals. Lavender oil poisoning is rare but can occur, especially in children who accidentally ingest essential oils [50].

4.4.4 *Rosmarinus officinalis* (rosemary)

Rosmarinus officinalis L., commonly known as rosemary, is a medicinal plant from the Lamiaceae family. While it is well-known for its aromatic properties and culinary applications, rosemary also holds significant value among indigenous communities in regions where it grows naturally. Natural antioxidants found in rosemary extracts can prolong the shelf life of perishable goods. Notably, rosemary extract (E392) has been authorized by the European Union as a safe and efficient antioxidant for use in food preservation [51]. The essential oil of rosemary contains camphor, cineole, and alpha-pinene, which give it a strong, woody fragrance. Rosemary oil is used in aromatherapy to improve memory and concentration. It is also applied topically to treat muscle pain and improve circulation. Camphor is a potentially toxic compound found in high concentrations in rosemary oil. Ingesting or inhaling large amounts of camphor can be harmful. Camphor poisoning can lead to nausea, vomiting, seizures, and respiratory distress. In severe cases, it can cause death. While small doses in culinary use are generally safe, excessive exposure to rosemary oil or camphor-containing products can be toxic, especially for children [52].

4.4.5 *Mentha pulegium* (pennyroyal)

Mentha pulegium, commonly known as European pennyroyal, is also referred to as squaw mint, mosquito plant, or pudding grass. This flowering plant, part of the Lamiaceae family, is native to Europe, North Africa, and the Middle East. Its crushed leaves emit a strong fragrance resembling spearmint. Traditionally, pennyroyal has been used as a culinary herb, a folk remedy, and an abortifacient, while its essential oil is utilized in aromatherapy [53].

Pennyroyal is known for its strong, minty aroma due to pulegone, its primary component. The oil is used in traditional medicine for colds, fever, and menstrual problems. Pennyroyal oil has been used in folk medicine to induce menstruation and as an abortifacient, as well as a repellent for insects. Pulegone is a hepatotoxin and neurotoxin. It can cause severe liver and kidney damage, as well as nervous system effects when consumed in toxic amounts. Ingestion of even small amounts of pennyroyal oil can cause nausea, vomiting, abdominal pain, and liver failure. Larger doses can result in convulsions, respiratory failure, and death. Pennyroyal oil is highly toxic and has been linked to multiple cases of fatal poisoning [54, 55].

4.4.6 *Eucalyptus globulus* (eucalyptus)

Eucalyptus globulus is a species of shrub or flowering tree within the Myrtaceae family. The genus *Eucalyptus* encompasses over 700 species and has been utilized for various purposes throughout human history. Native primarily to Tunisia and Australia, eucalyptus is also found in Africa and regions ranging from tropical to southern temperate areas of the Americas. The genus includes four subspecies: *E. bicostata, E.pseudoglobulus, E. globulus*, and *E. maidenii*. Among these, *E. globulus* is a medium-to-large evergreen tree with broad leaves, capable of reaching heights up to 70 meters and a trunk diameter of 4–7 feet. This species is highly valued both nutritionally and therapeutically due to its distinct chemical composition. Esters, ethers, carboxylic acids, ketones, aldehydes, alcohols, hydrocarbons, monoterpenes, and sesquiterpenes are among the many different types of chemicals found in its essential oil. According to phytochemical analyses, oils isolated from the buds, branches, and fruits mostly include α-thujene, 1,8-cineole, and aromadendrene, but the leaf oil is rich in 1,8-cineole, α-pinene, p-cymene, cryptone, and spathulenol. Thanks to these bioactive compounds, Eucalyptus globulus exhibits significant antimicrobial, antifungal, antiviral, anti-inflammatory, analgesic, anti-nociceptive, and antioxidant properties [56].

Eucalyptus oil has a sharp, menthol-like scent, primarily due to cineole (eucalyptol), which has decongestant and anti-inflammatory properties. It is commonly used in steam inhalations to relieve cold symptoms, as well as in massage oils and topical balms for pain relief. Cineole can be toxic if consumed in large quantities or used improperly. Ingesting large amounts of eucalyptus oil can lead to nausea, vomiting, dizziness, muscle weakness, and respiratory problems. In severe cases, it can lead to coma or death. Eucalyptus oil should never be ingested in large amounts, particularly by children, as it can be extremely toxic [57].

4.4.7 *Myristica fragrans* (nutmeg)

Myristica fragrans, commonly known as nutmeg, is a plant that yields two spices: nutmeg and mace. Nutmeg refers to the seed kernel found within the fruit, while mace is the red, lacy aril that encases the kernel. This species belongs to the family Myristicaceae, under the order Magnoliales, which encompasses approximately 150 genera and over 3,000 species. Native to the Moluccas and indigenous to regions such as India, Indonesia, and Sri Lanka, *Myristica* species are now widely cultivated in tropical regions across both hemispheres, including South Africa [58]. Nutmeg is widely appreciated not only as a spice but also for its therapeutic properties. Known for its distinctive pleasant aroma and mildly warm flavor, it is a versatile ingredient used to enhance the taste of baked goods, confections, puddings, meats, sausages, sauces, vegetables, and beverages. Additionally, nutmeg is a key component in curry powders, teas, and soft drinks, and is often blended into milk or alcoholic beverages for added flavor.

Nutmeg contains myristicin, an aromatic compound responsible for its warm, spicy fragrance. In small quantities, nutmeg is used as a spice in cooking and baking, and its essential oil is used in perfumes and aromatherapy. Myristicin and safrole in nutmeg are both toxic when consumed in large doses. Large doses of nutmeg can cause hallucinations, nausea, dizziness, dry mouth, and in severe cases, convulsions and heart palpitations. Nutmeg toxicity can lead to coma or death in extreme cases. Nutmeg should be used in moderation due to the psychoactive and toxic effects of its essential oils when consumed in large amounts [59].

4.4.8 *Thuja occidentalis* (white cedar)

Thuja occidentalis, commonly known as arbor vitae or white cedar, is native to eastern North America and cultivated in Europe as an ornamental tree. It was first recognized as a medicinal plant by native Canadians during a sixteenth-century expedition, where it proved effective in treating scurvy-related weakness. In traditional medicine, *T. occidentalis* has been used to address conditions such as bronchial catarrh, enuresis, cystitis, psoriasis, uterine carcinomas, amenorrhea, and rheumatism. Today, it is primarily utilized in homeopathy, either as a mother tincture or in diluted form. Additionally, this medicinal plant is employed in evidence-based phytotherapy, often combined with other immunomodulating plants like *Echinacea purpurea*, *Echinacea pallida*, and *Baptisia tinctoria*. It is particularly used for treating acute and chronic upper respiratory tract infections and serves as an adjunct to antibiotics in managing severe bacterial infections, including bronchitis, angina, pharyngitis, otitis media, and sinusitis [60].

The essential oil of white cedar (*Thuja occidentalis*), widely used in traditional homeopathy and aromatherapy, owes its strong camphor-like scent to the compound thujone. This oil has been historically employed to alleviate respiratory conditions, treat skin infections, and support immune health. Thujone, a monoterpene ketone, is responsible for many of the therapeutic properties attributed to white cedar. However, its potent bioactivity also makes it a neurotoxin when consumed in large quantities, necessitating careful handling and controlled use. Thujone poisoning can lead to a range of toxic effects, with symptoms including convulsions, dizziness, hallucinations, and in severe cases, loss of consciousness or death. Research has shown that thujone acts on gamma-aminobutyric acid (GABA) receptors in the brain, disrupting inhibitory neurotransmission and causing excitotoxicity, which can result in seizures and neurological damage [61]. Cases of poisoning have been reported due to the ingestion of thujone-rich essential oils or herbal preparations, underscoring the importance of adhering to recommended dosages.

In regulated therapeutic contexts, such as aromatherapy, the use of white cedar oil is generally limited to external applications or diffusions at low concentrations to mitigate risks. Studies emphasize that internal use should only occur under the guidance of a qualified healthcare professional due to the narrow margin between its therapeutic and toxic doses. Furthermore, products containing thujone, including white cedar oil, are restricted in certain regions to prevent accidental poisoning [62].

To safely utilize the benefits of white cedar essential oil, consumers are advised to source products from reputable suppliers who provide detailed information about thujone content. Additionally, pregnant women, children, and individuals with neurological disorders should avoid using thujone-containing products, as they may be more susceptible to its adverse effects [63]. Adhering to safety guidelines and recognizing the symptoms of toxicity are crucial steps in harnessing the potential benefits of white cedar while avoiding its dangers.

4.4.9 *Illicium verum* (star anise)

Illicium verum Hook. f. (Illiciaceae) is an aromatic evergreen tree with purple-red flowers and star-shaped fruit that has a distinct anise scent. Native to southern China and Vietnam, its fruit, known as star anise, is a significant part of traditional Chinese medicine and is widely used as a spice. The uniquely shaped fruit has been traditionally used in Chinese medicine to treat conditions like vomiting, stomach aches, insomnia, skin inflammation, and rheumatic pain. Additionally, star anise essential oil has been used topically as an antibacterial and to cure rheumatism. Star anise is a popular spice that was brought to Europe in the seventeenth century. Anethole, a chemical component, gives it its unique licorice flavor.

Star anise is a rich source of lignans and seco-prezizaane-type sesquiterpenes. These compounds are chemically unique and occur exclusively in *Illicium* species, making them distinctive chemical markers for the genus. These constituents are known to exhibit a range of biological activities, including neurotoxic and neurotrophic effects [64]. Star anise has a sweet, licorice-like aroma due to anethole. It is widely used in cooking, perfumes, and traditional medicine. Star anise oil is used to relieve digestive problems, colds, and coughs. Star anise contains anisatin, which is toxic when ingested in large amounts. Japanese star anise (*Illicium anisatum*), which looks similar, is particularly toxic and should not be confused with the safe edible variety. Poisoning can cause nausea, vomiting, seizures, and neurotoxicity. The consumption of large quantities or contamination with the toxic species can lead to serious effects [65].

4.4.10 *Syzygium aromaticum* (clove)

Clove, also known as *Syzygium aromaticum,* is an evergreen tree of the Myrtaceae family. Indigenous to Indonesia's Maluku Islands, sometimes referred to as the Spice Islands, it is widely cultivated in tropical regions, including India, Sri Lanka, Madagascar, and Tanzania. The tree grows to a height of 8–12 meters and has large, oblong, and glossy leaves with fragrant flower buds that form in clusters [66]. The dried flower buds are the commercial "cloves," prized for their intense aroma and flavor.

These buds are harvested when they transition from green to pink and are then sun-dried. Clove is valued not only as a spice but also for its applications in traditional and modern medicine, cosmetics, and perfumery [67].

Cloves are a rich source of phytochemicals, including essential oils, tannins, flavonoids, and triterpenes. The primary bioactive compound is eugenol, which constitutes 70–85% of the essential oil and is responsible for clove's characteristic aroma and medicinal properties [68]. Other notable compounds include acetyl eugenol, β-caryophyllene, vanillin, and gallotannic acid. These phytochemicals contribute to clove's diverse biological activities, such as antioxidant, antimicrobial, antifungal, anti-inflammatory, analgesic, and anticancer effects [69, 70]. For instance, eugenol exhibits potent free radical scavenging and lipid peroxidation inhibition, making it a valuable natural antioxidant. Additionally, clove oil is widely used in dentistry for its anesthetic and antiseptic properties, particularly in treating toothaches and oral infections. The broad spectrum of activities attributed to cloves underscores its importance in both traditional medicine and modern pharmacology.

Clove oil is used in dentistry for its analgesic properties and in traditional medicine to treat infections and digestive issues. Eugenol in high concentrations can be toxic, causing liver damage and respiratory distress when ingested in large quantities. Overconsumption of clove oil can cause nausea, vomiting, abdominal pain, and liver failure. It can also lead to seizures in extreme cases [71].

4.4.11 *Juniperus sabina* (savin juniper)

Juniperus sabina, commonly known as savin juniper, is a low-growing, evergreen shrub belonging to the Cupressaceae family. It is native to mountainous regions of Central and Southern Europe, extending to parts of Central and Western Asia. The plant typically grows to a height of 1–2 meters, forming dense, spreading mats. Its leaves are scale-like, dark green, and emit a strong, pungent odor when crushed. The plant produces small, berry-like cones that mature from green to a bluish-black color. Due to its adaptability, *J. sabina* is often found on rocky slopes and poor soils, making it a vital component of erosion control in its native habitat. However, it is also noted for its toxicity, especially its essential oils, which limits its use in conventional landscaping and grazing regions [72].

J. sabina contains a diverse range of bioactive compounds, including essential oils rich in sabinene, sabinol, and thujone, as well as diterpenes, lignans, and flavonoids [73]. These phytochemicals confer the plant with notable biological properties, such as antimicrobial, antifungal, anti-inflammatory, and cytotoxic and abortifacient activities [74–75]. The essential oil of *J. sabina* is known for its strong cytotoxic and proapoptotic effects, which have shown potential in cancer studies. However, the presence of thujone, a toxic compound, poses safety concerns, particularly in medicinal applications. Traditionally, *J. sabina* has been used in folk medicine for the treatment of skin conditions, warts, and respiratory ailments, but its use is strictly regulated due to its high toxicity. The plant's potent bioactivity and phytochemical diversity make it a valuable subject for pharmacological research, particularly for developing antimicrobial agents and studying its cytotoxic mechanisms.

The strong-smelling oil from savin juniper contains sabinene, which gives it a sharp, pine-like fragrance. Traditionally, the oil has been used in medicine as an abortifacient and to treat skin ailments, though its use is highly discouraged due to its toxicity. Sabinyl acetate and sabinene are toxic compounds that can cause poisoning when ingested or applied in large amounts. Symptoms of toxicity include gastrointestinal irritation, kidney damage, and miscarriage. In large doses, it can lead to fatal poisoning [77].

4.4.12 *Pimpinella anisum* (anise)

An annual herbaceous plant in the Apiaceae family, *Pimpinella anisum* is sometimes referred to as anise or aniseed. Originating in Southwest Asia and the Eastern Mediterranean, it is cultivated globally for its aromatic seeds. The plant typically grows to a height of 50–80 cm and features feathery leaves and umbels of small, white flowers. The seeds, which are crescent-shaped with fine stripes, are the primary source of its economic and medicinal value. Anise seeds are rich in essential oil, which constitutes approximately 2–6% of their weight. The main volatile compound in the essential oil is trans-anethole, accounting for up to 90%, along with minor components like estragole, limonene, and γ-himachalene. The seeds also contain fatty acids, proteins, flavonoids, and phenolic acids, contributing to their therapeutic properties [78, 79].

Anise exhibits various pharmacological activities, including antioxidant, antimicrobial, antifungal, and spasmolytic effects. Traditionally, it has been used to treat digestive issues, respiratory conditions, and menstrual discomfort. Recent studies also highlight its potential as an anti-inflammatory, hypoglycemic, and estrogenic agent. Despite these benefits, *P. anisum* contains potentially toxic compounds, such as estragole, which has been implicated as a possible carcinogen when consumed in large amounts. High doses of anise oil can cause nausea, vomiting, pulmonary edema, and neurotoxic symptoms such as seizures. Allergic reactions and photosensitivity are also reported in sensitive individuals. Therefore, while anise remains a valuable plant in traditional and modern medicine, its usage should be moderated to avoid toxic effects [80].

Anise's characteristic licorice-like scent comes from anethole. It is used widely in cooking, aromatherapy, and traditional medicine for digestive relief. Anise oil is used to treat indigestion, cough, and colic. While generally considered safe, anise oil in large doses can be harmful, especially due to its high anethole content, which can be neurotoxic in large quantities. Overconsumption can cause nausea, vomiting, seizures, and respiratory problems, particularly in sensitive individuals [81].

4.4.13 *Lavandula stoechas* (French lavender)

Lavandula stoechas, commonly known as Spanish or French lavender, is a perennial aromatic shrub in the Lamiaceae family. Native to the Mediterranean region, it is widely cultivated for its ornamental and medicinal uses. The plant typically grows to a height of 30–100 cm and is characterized by narrow, grayish-green leaves and purple flowers arranged in spikes, topped with showy sterile bracts resembling wings. The essential oil of *L. stoechas* is its most studied component, consisting primarily of camphor, fenchone, and 1,8-cineole. These compounds contribute to its distinct camphoraceous aroma and therapeutic potential. Additionally, the plant contains flavonoids, phenolic acids (such as rosmarinic and caffeic acid), and tannins, which are responsible for its antioxidant and anti-inflammatory activities [82, 83].

Lavandula stoechas has been traditionally used for its antimicrobial, antispasmodic, and sedative effects, as well as for managing respiratory and digestive disorders. Its essential oil, rich in camphor and linalool, is characterized by a strong, fragrant aroma and exhibits potent antiseptic, anti-inflammatory, and neuroprotective properties. Widely used in aromatherapy, it is valued for alleviating stress and anxiety. Scientific studies have confirmed its biological activities, including significant antimicrobial, antioxidant, and neuroprotective effects, making it a valuable medicinal plant.

However, the high camphor content of *L. stoechas* poses toxicity risks if consumed, inhaled, or applied in large amounts. Camphor toxicity can result in neurotoxic effects such as dizziness, nausea, vomiting, confusion, and respiratory distress. Severe cases may lead to seizures or even death. Additionally, allergic reactions, including dermatitis, have been reported in sensitive individuals. While *L. stoechas* holds considerable therapeutic potential, careful dosage and administration are critical to ensure safety and mitigate its toxic effects [52, 84].

4.4.14 *Artemisia vulgaris* (mugwort)

Mugwort, or *Artemisia vulgaris,* is a perennial herbaceous plant that belongs to the Asteraceae family. Although it is indigenous to North Africa, Asia, and Europe, it has spread to many other regions of the world, including North America. The plant typically grows up to 1.5 meters in height, with deeply lobed leaves that are green on the upper surface and silvery-white underneath. Its small, yellowish to reddish flowers form in terminal panicles, and it is commonly found in wastelands, roadsides, and meadows. *A. vulgaris* has a long history of use in traditional medicine, culinary applications, and spiritual practices, often valued for its aromatic and medicinal properties [85].

The phytochemical profile of *A. vulgaris* is rich and diverse, contributing to its broad range of pharmacological activities. The plant contains essential oils dominated by monoterpenes and sesquiterpenes, including camphor, eucalyptol, and borneol. It also contains flavonoids (quercetin and luteolin), coumarins, phenolic acids (chlorogenic acid), tannins, and volatile compounds. These bioactive constituents are responsible for its antioxidant, antimicrobial, and anti-inflammatory properties. Additionally, the presence of artemisinin-related compounds links *A. vulgaris* to the genus's reputation for therapeutic applications [86, 87].

The medicinal properties of *A. vulgaris* are well-documented, particularly in traditional medicine systems. It has been used to treat gastrointestinal ailments, menstrual irregularities, and nervous disorders, as well as for wound healing and as an antiparasitic agent. Modern research has supported its potential as an antimicrobial, antimalarial, anti-inflammatory, and hepatoprotective agent. Furthermore, *A. vulgaris* has been studied for its neuroprotective effects, which may be attributed to its antioxidant and anti-inflammatory properties [88]. Its essential oil is also widely used in aromatherapy and as a natural pesticide, enhancing its relevance in both health and agricultural contexts.

Despite its medicinal benefits, *A. vulgaris* contains toxic compounds, such as thujone, that can pose health risks when consumed in excessive amounts. Thujone is a neurotoxin that, in high doses, can cause nausea, vomiting, dizziness, and seizures. Allergic reactions, including dermatitis and respiratory issues, are also common among sensitive individuals exposed to mugwort pollen, a significant allergen in many regions. Chronic or high-dose use of *A. vulgaris* preparations may result in hepatotoxicity and neurotoxicity, emphasizing the need for careful regulation of its use. Pregnant women are advised to avoid *A. vulgaris* due to its potential abortifacient properties [89]. *A. vulgaris* is a versatile plant with significant therapeutic potential due to its rich phytochemical composition and diverse pharmacological activities. However, its use should be approached cautiously, given the risks associated with its toxic constituents. Further studies are needed to better understand its mechanisms of action, therapeutic applications, and safety profiles to maximize its medicinal value while minimizing potential risks [90].

4.4.15 *Melaleuca alternifolia* (tea tree)

Melaleuca alternifolia, commonly known as tea tree, is a small tree or shrub belonging to the Myrtaceae family. Native to the subtropical coastal regions of Australia, particularly New South Wales and Queensland, this plant thrives in swamps and riverbanks. It typically grows to a height of 4–7 meters, with narrow, soft leaves and white or cream-colored bottlebrush-like flowers. The essential oil derived from its leaves is the primary source of its medicinal and commercial value. Traditionally, indigenous Australians have used tea tree leaves in herbal infusions to treat coughs and colds and as a poultice for wounds [91].

The phytochemical composition of *M. alternifolia* is dominated by terpenes and their derivatives, which account for its potent biological properties. The primary active component of tea tree oil (TTO) is terpinen-4-ol, which constitutes 30–40% of the oil. Other significant components include α-terpineol, γ-terpinene, and 1,8-cineole.

These constituents contribute to the oil's antimicrobial, anti-inflammatory, and anti-oxidant activities. The composition of TTO is standardized by the International Organization for Standardization (ISO 4730:2017) to ensure consistent therapeutic quality [92]. Tea tree oil exhibits a wide range of biological activities, making it a staple in modern natural medicine. Its antimicrobial properties are particularly well-documented, showing efficacy against a broad spectrum of bacteria, fungi, and viruses. Studies indicate that TTO disrupts microbial membranes, leading to cell lysis. It is commonly used in dermatology for treating acne, fungal infections, and wounds. Additionally, TTO demonstrates anti-inflammatory effects by modulating pro-inflammatory cytokines, further supporting its use in skin conditions like eczema and psoriasis [93].

Beyond dermatological applications, TTO has shown promise as an antiparasitic and antiviral agent. Recent studies highlight its potential in managing biofilm-associated infections, a challenging area in antimicrobial therapy. Its insecticidal properties also make it a valuable natural pesticide. Despite its medicinal uses, TTO should be used cautiously due to its potential toxicity and allergenicity [94].

While tea tree oil is generally considered safe for topical use in appropriate concentrations, it can be toxic when ingested or used in excessive amounts. Ingestion of TTO can cause central nervous system depression, leading to symptoms such as drowsiness, confusion, and ataxia. Severe cases of ingestion may result in coma. Dermal exposure to undiluted TTO can cause irritation, erythema, or allergic contact dermatitis in sensitive individuals. Additionally, certain components, such as 1,8-cineole, may pose respiratory risks, especially in children. Proper dilution and patch testing are recommended before use to minimize these risks [95].

Melaleuca alternifolia is a plant with significant therapeutic potential, particularly as an antimicrobial and anti-inflammatory agent. Its phytochemical richness underpins its diverse applications in medicine and cosmetics. However, its toxicity, especially in concentrated or ingested forms, warrants careful use. Ongoing research is needed to further elucidate its mechanisms of action and optimize its safe application in both traditional and modern medical practices.

4.4.16 *Pelargonium graveolens* (rose geranium)

Pelargonium graveolens, commonly known as rose geranium or geranium, is a perennial herb native to South Africa. This aromatic plant belongs to the Geraniaceae family and is widely cultivated for its fragrant leaves and essential oils. The plant typically grows up to 1 meter in height and has rounded, aromatic, and deeply lobed leaves. It produces small pink to red flowers with a distinctive sweet floral scent, which is why it is often used in perfumes and aromatherapy. In traditional medicine, rose geranium has been used to treat a variety of ailments, including skin conditions, anxiety, and menstrual disorders [96]. The plant is commonly grown in temperate climates and is commercially cultivated for its essential oil, which is rich in both monoterpenes and aldehydes, contributing to its therapeutic properties.

Phytochemically, rose geranium is notable for its essential oils, which contain a complex mixture of volatile compounds. The major constituents of rose geranium oil include citronellol, geraniol, and linalool, along with smaller amounts of phenyl ethanol and eugenol. These compounds are responsible for their distinctive rose-like fragrance and are associated with various biological activities. Studies have shown that geranium oil's chemical composition can vary depending on the growing conditions, such as geographical location, climate, and soil composition [97]. This variability in the oil's composition can affect its medicinal and commercial applications, highlighting the importance of quality control in its production.

Rose geranium exhibits a wide range of biological activities, most notably its antimicrobial, anti-inflammatory, and antioxidant properties. Numerous studies have demonstrated the antimicrobial efficacy of its essential oil against both gram-positive and gram-negative bacteria, fungi, and viruses [98, 99]. These antimicrobial effects make rose geranium oil a valuable component in topical treatments for wound healing, skin infections, and acne. In addition to its antimicrobial properties, *P. graveolens* essential oil has demonstrated strong anti-inflammatory properties by lowering cyclooxygenase enzyme activity and preventing the synthesis of pro-inflammatory cytokines [100]. This makes it useful for treating conditions like arthritis, inflammatory skin diseases, and general inflammation.

Moreover, rose geranium is often used in aromatherapy to alleviate symptoms of anxiety, stress, and depression. Several studies have found that the inhalation of its essential oil can produce a calming effect, potentially aiding in the reduction of anxiety levels and improving mood [101]. This makes it popular in the treatment of emotional disorders and in promoting relaxation. Additionally, the antioxidant activity of geranium oil has been confirmed through various assays, demonstrating its potential as a natural agent in preventing oxidative stress and related diseases such as cardiovascular conditions and aging [102].

Although rose geranium and its essential oil are generally considered safe when used appropriately, the plant does contain certain compounds that may pose health risks when misused. One of the primary toxic compounds found in rose geranium oil is citronellol, which can cause skin irritation or allergic reactions in sensitive individuals. Prolonged or excessive topical use of undiluted essential oil can lead to dermatitis or other allergic reactions, especially in individuals with sensitive skin [103, 104]. Ingesting large amounts of rose geranium essential oil can also lead to toxicity, manifesting as nausea, vomiting, and gastrointestinal distress. Due to its high concentration of active compounds, caution is advised when using the oil, especially during pregnancy or in children, as it may cause adverse effects such as uterine contractions or toxicity to the developing fetus [98].

In addition, rose geranium contains trace amounts of eugenol, a compound known for its toxicity in high doses. Eugenol can cause liver damage, respiratory problems, and central nervous system depression when consumed in large quantities. However, these toxic effects are generally not observed when the plant or its oil is used in moderation and in accordance with recommended dosage guidelines. It is important to note that the safety of essential oils, including rose geranium oil, depends on the method of use, dosage, and the individual's sensitivity. Therefore, it is essential to follow proper guidelines to avoid any adverse reactions [105].

Table 4.4: The common toxic essential oils, their plant origins, approximate toxic doses, and associated toxic effects.

Essential oil	Plant origin	Toxic dose	Toxicity/effects
Wintergreen	*Gaultheria procumbens*	Toxic if ingested in large amounts (~4–5 mL)	Contains methyl salicylate, which is similar to aspirin. Can cause seizures, nausea, vomiting, liver damage, and even death if ingested.
Eucalyptus	*Eucalyptus globulus*	Toxic if ingested > 2 mL (adults)	Can cause respiratory distress, nausea, vomiting, abdominal pain, dizziness, and in severe cases, coma or death.
Pennyroyal	*Mentha pulegium*	As little as 1–2 mL (can be fatal)	Highly toxic, especially to pets. Causes liver damage, kidney failure, seizures, and can be fatal if ingested.
Cinnamon	*Cinnamomum verum* or *Cinnamomum cassia*	Toxic in doses > 5 mL for adults	Skin irritant, especially when used undiluted. Ingesting large amounts can cause nausea, vomiting, diarrhea, and even liver toxicity.
Clove	*Syzygium aromaticum*	Toxic in doses > 3 mL for adults	Contains eugenol, which can cause skin irritation, liver damage, and respiratory issues in large amounts.
Birch	*Betula lenta*	Toxic if ingested in large amounts (>2–4 mL)	Contains methyl salicylate, which can cause salicylate toxicity (similar to aspirin). Symptoms include vomiting, tinnitus, and dizziness.
Sweet birch	*Betula alleghaniensis*	Toxic if ingested in large amounts	Similar to Birch, causes salicylate poisoning with symptoms like dizziness, vomiting, and organ damage.
Tansy	*Tanacetum vulgare*	Toxic in doses >5 mL	Contains thujone, which can cause convulsions, vomiting, and liver damage. Highly toxic if ingested.
Sage (common)	*Salvia officinalis*	Toxic in large doses (10 mL+)	High doses can cause dizziness, nausea, vomiting, and seizures due to thujone content.
Pine	*Pinus sylvestris*	Toxic if ingested in large amounts	When taken in excess, it might result in respiratory discomfort, nausea, vomiting, and diarrhea.

4.5 Safe use and precautions

The use of poisonous and aromatic plants in traditional medicine and modern phytotherapy has long been a cornerstone of therapeutic practices worldwide. While these plants offer numerous health benefits due to their rich chemical compositions, their improper or excessive use can pose significant health risks. To ensure the safe use of these plants, several precautions need to be considered. First and foremost, dosage control is critical. Since many poisonous plants contain bioactive compounds that can be toxic at higher concentrations, it is essential that these plants are used only under the guidance of a trained healthcare professional, particularly those with expertise in herbal medicine. Incorrect dosage or prolonged use can lead to adverse effects such as toxicity, organ damage, and even death [106]. Therefore, it is recommended that any therapeutic use of medicinal plants, especially those with toxic potential, be done under professional supervision, and always within the recommended dosage guidelines [107].

In addition to dosage control, monitoring for potential side effects is crucial. Some poisonous plants may have delayed or cumulative toxic effects that may not be immediately apparent. Side effects such as nausea, vomiting, dizziness, gastrointestinal discomfort, or more severe reactions like seizures or organ failure can occur, often as a result of prolonged exposure or misuse. Therefore, individuals using medicinal plants should be well-informed about the potential risks and signs of toxicity. Regular monitoring of their health is recommended, especially when using plants known to have potent pharmacological activities [108]. For instance, plants like *Atropa belladonna*, *Digitalis purpurea*, and *Aconitum* contain compounds that can cause toxicity at certain doses and lead to symptoms such as arrhythmias, visual disturbances, and paralysis if not carefully monitored [109].

Furthermore, increasing awareness and providing proper education on the safe use of poisonous and aromatic plants is essential. This includes training those who work with medicinal plants, such as herbalists, pharmacists, and practitioners of alternative medicine, to properly identify, dose, and apply these plants. Education should also be extended to the public to prevent unregulated use of potentially harmful plants. A major concern is the unregulated sale and use of herbal products, which may contain either incorrect species or harmful adulterants. Without adequate training, individuals may misuse these plants, unaware of their dangerous effects [110]. Awareness programs and training initiatives can help ensure that these plants are used safely, preventing poisoning and other adverse health outcomes.

Lastly, responsible use practices are necessary not only to prevent harm but also to promote the sustainability of these plants in the wild. Overharvesting and improper collection can lead to ecological imbalance, reducing the availability of these plants for future generations. Sustainable harvesting practices, combined with safe usage guidelines, are essential to preserve the benefits of poisonous and aromatic plants while minimizing risks to human health and the environment [111].

4.5.1 Safety guidelines and precautions

1. Proper dosage and safe use of aromatic plants

Aromatic plants and their derivatives, including essential oils and herbal supplements, can offer significant therapeutic and practical benefits. However, misuse or excessive consumption can lead to adverse effects, even with widely recognized safe plants. Essential oils, for instance, are highly concentrated extracts that should never be ingested or applied undiluted directly to the skin. Recommended dilution guidelines typically range between 1% and 3% in a carrier oil (e.g., coconut or jojoba oil) for topical applications to prevent skin irritation or sensitization derived from aromatic plants also require cautious usage. Even though they are "natural," their bioactive compounds can be harmful in high doses [61]. It is crucial to follow dosage instructions provided by reputable manufacturers or healthcare professionals. Excessive intake of supplements like lavender or rosemary can overwhelm the body and lead to toxicity. For example, common culinary herbs contain camphor and cineole, which are safe in culinary doses but can induce nausea, vomiting, or seizures if consumed excessively [112].

Additionally, some exhibit low toxicity at standard or small doses but become hazardous when consumed in large quantities. Nutmeg (*Myristica fragrans*), for example, is a popular spice in small culinary amounts but can cause severe toxic effects, including hallucinations, nausea, and central nervous system disturbances, when taken in large quantities due to its myristicin content [113]. To avoid adverse effects, aromatic plants should always be used in moderation and under professional guidance. Misuse of seemingly benign aromatic plants underscores the need for informed usage practices based on scientific understanding.

2. Avoiding prolonged use

While aromatic plants and their extracts are often used for therapeutic and practical purposes, prolonged or excessive use can lead to significant health risks. Certain plants, such as sassafras (*Sassafras albidum*) and pennyroyal (*Mentha pulegium*), contain potent compounds like safrole and pulegone, respectively, which are known to be toxic when used over extended periods. Safrole, a major constituent in sassafras oil, has been associated with liver damage and an increased risk of cancer, leading to restrictions on its use in many countries. Similarly, pulegone, found in pennyroyal oil, can cause severe liver and kidney damage, and even small amounts have been linked to toxic effects [114]. As such, the use of these plants and their derivatives should be limited to short durations and only under the guidance of a qualified healthcare provider.

In the realm of aromatherapy, prolonged exposure to essential oils such as eucalyptus (*Eucalyptus globulus*) and camphor (*Cinnamomum camphora*) can also pose risks [61]. While these oils are often used for their respiratory benefits, excessive or continuous inhalation can irritate the respiratory system, potentially leading to coughing, headaches, or exacerbation of preexisting respiratory conditions. To mitigate these risks, it is essential to incorporate regular breaks when using diffusers or inhalers, ensuring adequate ventilation in the environment. Proper education on the safe use of aromatic plants, especially concerning duration and intensity, is key to maximizing their benefits while minimizing potential harm [63].

3. Dilution and application methods

Proper dilution and application methods are crucial for the safe use of essential oils to prevent adverse reaction such as skin irritation, burns, or mucous membrane damage. When applying essential oils topically, they should always be diluted with a carrier oil like coconut, jojoba, or almond oil. A common and safe dilution ratio is 3–5 drops of essential oil per teaspoon (approximately 5 mL) of carrier oil. Undiluted application of potent oils, such as clove (*Syzygium aromaticum*), which is high in eugenol, can lead to skin burns, irritation, or allergic reactions, especially in individuals with sensitive skin [115].

For inhalation purposes, proper ventilation is essential to prevent irritation to the respiratory system. Direct inhalation of concentrated vapors should be avoided, as it can irritate mucous membranes and even cause headaches or nausea with prolonged exposure. Essential oils such as peppermint (*Mentha × piperita*) and tea tree (*Melaleuca alternifolia*) are particularly potent and can cause respiratory discomfort if inhaled excessively or without dilution in water or a diffuser. Using a diffuser in a well-ventilated space and limiting sessions to 30–60 min can help mitigate these risks. By adhering to these guidelines, essential oils can be used safely and effectively, whether for therapeutic, aromatic, or cosmetic purposes [61].

4. Age and vulnerable groups

Children are particularly susceptible to the toxic effects of essential oils and herbal remedies due to their smaller body size and developing systems. Even a small amount of oils such as eucalyptus (*Eucalyptus globulus*) and camphor (*Cinnamomum camphora*) can cause severe poisoning if ingested, leading to symptoms like seizures, respiratory distress, or even death in extreme cases. Topical use of these oils can also irritate sensitive skin or trigger allergic reactions. As a precaution, essential oils and herbal products should always be stored securely out of children's reach. Furthermore, the use of these remedies in children should only occur under the supervision of a pediatrician, with diluted formulations specifically designed for pediatric use to ensure safety [61].

Pregnant women should exercise extreme caution when using aromatic plants or essential oils, as some compounds can adversely affect pregnancy. For instance, pennyroyal (*Mentha pulegium*) contains pulegone, a known abortifacient, and sage (*Salvia officinalis*) contains thujone, which can induce uterine contractions. These compounds pose risks of miscarriage or premature labor, particularly during the early stages of pregnancy [116]. Essential oils with strong emmenagogue properties, such as fennel and anise, should also be avoided. Pregnant individuals are advised to consult healthcare providers before using any herbal or aromatic products, including over-the-counter preparations labeled as "natural," to prevent inadvertent harm to the mother or fetus [61].

Older adults and individuals with chronic health conditions, such as liver or kidney disease, are another vulnerable group requiring special consideration. Many aromatic plants and their extracts contain hepatotoxic or nephrotoxic compounds. For example, sassafras (*Sassafras albidum*) and pennyroyal are known to cause liver or kidney damage with prolonged use [63]. Aging reduces the body's ability to metabolize toxins effectively, increasing the risk of adverse effects from these substances. People with conditions like diabetes or cardiovascular disease should also be cautious, as some essential oils can interact with medications or exacerbate preexisting conditions [61]. Consulting a healthcare provider is imperative for elderly or chronically ill individuals before using aromatic plants, even in small amounts.

5. Skin sensitivity and allergies
Essential oils and plant extracts are powerful natural substances, but their improper use can cause skin sensitivity or allergic reactions. To minimize risks, it is essential to conduct a patch test before applying these products to larger areas of the skin. This involves applying a small, diluted amount of the oil or extract (e.g., 1–2 drops mixed with a teaspoon of carrier oil) to the inner forearm and waiting 24 h. If no adverse reactions – such as redness, itching, or swelling – occur during this time, the product is generally safe for topical use. This practice is recommended by dermatological guidelines and aromatherapy safety experts, including Tisserand and Young, as an effective first step to avoid allergic reactions [61].

Phototoxicity is another significant concern, especially with certain citrus oils like bergamot (*Citrus bergamia*), lemon (*Citrus limon*), and lime (*Citrus aurantiifolia*). These oils contain furanocoumarins, compounds that react with ultraviolet (UV) light to cause severe skin reactions such as redness, blistering, or hyperpigmentation. Research has shown that applying phototoxic oils to the skin and then exposing it to sunlight within 12–24 h can result in burns or long-lasting discoloration [117]. To prevent such reactions, it is advised to avoid direct sun exposure or use sunscreen on treated areas for at least 12 h after application. Additionally, phototoxic oils should be used in very low concentrations – typically below 0.5% for safe topical use [61]. By incorporating these safety measures, individuals can enjoy the benefits of essential oils and plant extracts while minimizing the risks associated with skin sensitivity and phototoxicity. These precautions are especially critical for those with sensitive skin or preexisting conditions like eczema or dermatitis.

6. Ingestion precautions
Ingesting essential oils or potent herbal preparations should be approached with extreme caution, as improper use can lead to severe health risks, including organ failure or death. Consulting with a qualified herbalist, naturopath, or medical professional is essential before consuming these substances. For instance, oils like wintergreen (*Gaultheria procumbens*), which contain high levels of methyl salicylate, are highly toxic in even small amounts. Studies have shown that ingesting as little as one teaspoon of wintergreen oil can be fatal, particularly in children, due to its potent blood-

thinning and organ-damaging properties [61]. Professional guidance ensures appropriate dosage and identifies potential contraindications, such as interactions with medications or preexisting conditions.

Self-medication with toxic medicinal plants is equally dangerous, as some herbs have a very narrow therapeutic margin, meaning the difference between a therapeutic dose and a toxic dose is minimal. Foxglove (*Digitalis purpurea*), for example, contains digitalis compounds used in heart medications, but improper dosage can cause severe cardiac arrhythmias or death. Similarly, aconite (*Aconitum spp.*), which contains aconitine, is used in traditional medicine for pain and fever relief but is highly poisonous and can lead to respiratory paralysis if misused [63]. These plants and their derivatives are typically reserved for controlled medical settings where dosages are carefully monitored to prevent toxicity. Individuals are strongly advised never to self-medicate with such potent plants without proper supervision and prescription. Strict adherence to expert guidance and avoidance of unsupervised use are crucial for safely benefiting from these powerful yet potentially dangerous substances. Such precautions protect against unintended poisoning and ensure that any therapeutic use is both safe and effective.

7. Interaction with medications

The use of aromatic plants, herbal supplements, and their extracts is widespread due to their perceived natural benefits, but their interaction with prescription medications can lead to significant health risks. These interactions occur because many plants contain bioactive compounds that can alter the way drugs are metabolized, absorbed, or excreted in the body. This highlights the critical importance of understanding herb-drug interactions before integrating these natural remedies into a health regimen.

One of the most well-documented examples is *St. John's Wort* (*Hypericum perforatum*), commonly used for mild depression or anxiety. This plant induces the enzyme CYP3A4 in the liver, which can accelerate the metabolism of certain drugs, reducing their effectiveness. Medications impacted include oral contraceptives, antidepressants, and immunosuppressants like cyclosporine [118]. Studies have shown that women taking oral contraceptives alongside *St. John's Wort* experienced breakthrough bleeding and unplanned pregnancies, underscoring its significant interaction risk [119].

Similarly, plants like ginger (*Zingiber officinale*) and garlic (*Allium sativum*) have anticoagulant properties that can increase the risk of bleeding when taken with blood-thinning medications such as warfarin or aspirin. Garlic, in particular, inhibits platelet aggregation, which can compound the effects of anticoagulants and lead to severe bleeding episodes. Clinical trials have demonstrated that high doses of garlic supplements significantly prolong bleeding time in patients taking warfarin, making it imperative to monitor these interactions closely [63]. Another notable herb is ginseng (*Panax ginseng*), which is often used to boost energy and immune function. However, it can interfere with medications like insulin and oral hypoglycemic drugs, potentially causing dangerous fluctuations in blood sugar levels. It may also reduce the effectiveness of anticoagulants by promoting blood clotting [119].

Essential oils can also present risks. For example, peppermint oil (*Mentha × piperita*), often used for digestive issues, can inhibit the activity of enzymes like CYP2C19, potentially altering the metabolism of drugs such as omeprazole. Additionally, grapefruit oil contains compounds that inhibit CYP3A4, much like *St. John's Wort*, leading to elevated levels of certain medications in the bloodstream and increasing the risk of side effects [61]. To mitigate these risks, it is essential to inform healthcare providers about all herbal supplements or essential oils being used. A detailed discussion can help identify potential interactions and adjust medication dosages accordingly. For instance, patients on anticoagulants should avoid high doses of garlic or ginger, while those on oral contraceptives may need to explore alternatives if taking *St. John's Wort*.

Healthcare providers can also utilize databases and updated resources on herb-drug interactions to provide accurate guidance. Tools like the Natural Medicines Comprehensive Database offer valuable information about known and potential interactions, ensuring safer integration of natural remedies into a patient's treatment plan [120]. In conclusion, while aromatic plants and herbal supplements can offer therapeutic benefits, their interactions with medications necessitate caution and professional guidance. By consulting healthcare professionals and using reliable information sources, individuals can safely incorporate these natural remedies without compromising their health.

8. Safe storage

Proper storage of essential oils and herbs is critical to ensuring their safety and efficacy while minimizing the risk of accidental poisoning or misuse. Many aromatic substances are highly concentrated and can be toxic if ingested, particularly by vulnerable groups like children and pets. For example, oils like pennyroyal (*Mentha pulegium*), tea tree (*Melaleuca alternifolia*), and lavender (*Lavandula angustifolia*) are known to be hazardous when consumed, especially for cats and dogs. Pennyroyal contains pulegone, a compound toxic to the liver, while tea tree oil can cause neurological symptoms like tremors and seizures in pets, even in small amounts [61]. To prevent such incidents, essential oils and herbs should always be stored in a cool, dark place, securely out of reach of children and animals.

Clear and consistent labeling is another crucial aspect of safe storage. All bottles and containers should be properly labeled with the name of the oil or herb, its concentration, and any safety warnings. This practice helps prevent accidental ingestion or inappropriate application. For instance, some essential oils are phototoxic, while others should never be ingested, and clear labeling ensures that users are reminded of these precautions. According to the National Poison Control Center, one of the most common causes of accidental poisoning with essential oils is confusion due to improperly labeled containers [120].

In addition to labeling, consider using child-resistant caps for essential oil bottles. These are particularly important in households with young children, who may be drawn to the pleasant smells of the oils and attempt to ingest them. Moreover, always return oils to their designated storage area immediately after use to minimize exposure risks. Maintaining an organized storage system not only improves safety but also helps preserve the oils' therapeutic properties by protecting them from heat, light, and air, which can degrade their quality over time [63]. Finally, educating household members about the potential risks associated with essential oils and herbs can further enhance safety. Make sure everyone understands the importance of proper handling, storage, and usage to prevent accidents. By combining secure storage, clear labeling, and education, you can enjoy the benefits of these natural remedies while ensuring the safety of your family and pets.

9. Identifying toxic plants

One of the most critical aspects of using plants for medicinal or culinary purposes is the accurate identification of species. Many toxic plants closely resemble nontoxic or medicinal ones, posing a significant risk to those who are untrained in botany. For instance, star anise (*Illicium verum*), commonly used in cooking and traditional medicine, can be confused with Japanese star anise (*Illicium anisatum*), which contains potent neurotoxins and is not safe for consumption. Such mix-ups can result in severe health consequences, including seizures or poisoning. Reputable suppliers who clearly label and verify their products are essential to avoid these potentially life-threatening errors [61].

The danger of misidentification increases significantly when wild-harvested plants are involved. Collecting plants from the wild is particularly risky because many toxic species mimic the appearance of edible or medicinal herbs. For example, water hemlock (*Cicuta maculata*), one of the most toxic plants in North America, is often mistaken for wild parsley or other edible herbs due to its similar appearance. Hemlock ingestion can cause respiratory failure and death within hours. Likewise, wild garlic (*Allium ursinum*) has been confused with lily of the valley (*Convallaria majalis*), a highly toxic plant, leading to accidental poisonings [63]. Foraging should only be attempted by those with expertise in plant identification, and when in doubt, it is better to err on the side of caution.

To further ensure safety, individuals can rely on resources like regional plant identification guides, workshops, or mobile applications designed to differentiate between safe and toxic species. In addition, consulting with experienced herbalists or botanists before using wild plants is advisable. Clear labeling, reputable sourcing, and education on plant identification are critical measures to prevent dangerous mix-ups and ensure safe usage of herbal remedies and culinary ingredients [121].

10. Emergency preparedness

When using aromatic plants, herbs, or essential oils, it is vital to recognize the signs of poisoning to ensure prompt and effective treatment in emergencies. Common symptoms of poisoning include nausea, vomiting, dizziness, seizures, respiratory distress, and confusion. For instance, ingestion of toxic plants like water hemlock (*Cicuta maculata*) can result in severe convulsions and respiratory failure, while exposure to highly concentrated essential oils such as wintergreen (*Gaultheria procumbens*) can cause symptoms of salicylate poisoning, including hyperventilation and lethargy [61]. Early recognition of these signs can be lifesaving, making familiarity with the risks associated with the specific plants or oils you use essential for preparedness.

In the event of suspected poisoning, contacting your local poison control center or seeking immediate medical attention is crucial. Poison control centers are equipped to provide specific advice tailored to the substance involved, whether it is an essential oil, herbal extract, or plant ingestion. Having the product's packaging or a sample of the plant available can help healthcare providers quickly identify the toxin and administer appropriate treatment. For example, if a child ingests tea tree oil (*Melaleuca alternifolia*), which is toxic even in small amounts, providing the packaging to medical professionals can help guide interventions, such as gastric lavage or supportive care [122].

Preparedness also includes preventive measures. Store potentially hazardous plants and oils securely, away from children and pets, and ensure all containers are clearly labeled. Additionally, keeping emergency contact numbers, such as those for poison control centers, readily accessible can save valuable time during a crisis. For households that frequently use aromatic plants or oils, creating a first-aid kit with activated charcoal and other recommended supplies can provide an extra layer of readiness [63]. By combining awareness of poisoning symptoms, access to emergency resources, and proactive safety measures, individuals can minimize the risks associated with the use of aromatic plants and essential oils while being prepared to respond effectively in emergencies.

4.6 Conclusions

To sum up, poisonous plants and the substances they contain pose a serious and frequently disregarded threat to human health and welfare. Even though many of these plants are beautiful, beneficial to the environment, and even have therapeutic qualities, it is impossible to overlook the risk of injury. We can better grasp the risks these toxins provide and the significance of being vigilant by knowing their nature, whether they are cardiac glycosides found in oleander, ricin found in castor beans, or alkaloids found in aconite and deadly nightshade.

The importance of recognizing hazardous plants and the vital role that education plays in averting unintentional poisoning have been discussed in this chapter. People can identify these dangerous species and take proactive measures to prevent exposure if they are properly informed. The risk of intoxication can be significantly decreased by taking sensible precautions, such as wearing protective clothes when handling potentially toxic plants, keeping plants safely out of children's and pets' reach, and properly disposing of hazardous plant components. In the event of poisoning, prompt and knowledgeable medical interventions are also essential. Understanding first aid protocols and being aware of the signs of plant poisoning might mean the difference between life and death. When combined with appropriate medical care, early intervention can lessen the consequences of poisoning and avoid long-term damage.

In the end, reducing these hazards requires promoting a society that is aware of the dangers posed by hazardous plants and their chemicals. We can make both human and animal habitats safer by striking a balance between appreciating the beauty of nature and being mindful of its possible hazards. While hazardous plants will always exist in the natural environment, we may prevent harmful consequences and cohabit with them in a safe and knowledgeable way by managing them carefully, being informed, and being prepared.

References

[1] Das, J. and Sarkar, B. R. (2023). Ethnomedicinal and pharmacological importance of *Atropa belladonna*: A review. International Journal of Pharmacognosy, 10(8), 448–453.

[2] Yildiz, D. (2004). Nicotine, its metabolism and an overview of its biological effects. Toxicon, 43, 619–632.

[3] Karch, S. B. (2019). Drug Abuse Handbook, CRC press, Boca Raton, Florida.

[4] Withering, W. (2014). An Account of the Foxglove, and Some of Its Medical Uses, Cambridge University Press, Cambridge.

[5] Poulton, J. E. (1983). Cyanogenic compounds in plants and their toxic effects. Handbook of Natural Toxins, 1, 117–157.

[6] Deen, A. U., Kumari, V., Sharma, A. N., Mondal, G. and Singh, G. P. (2018). Understanding cyanogenic glycoside toxicity in livestock: A review. International Journal of Chemical Studies, 6(5), 1559–1561.

[7] Pelkonen, O., Abass, K. and Wiesner, J. (2013). Thujone and thujone-containing herbal medicinal and botanical products: Toxicological assessment. Regulatory Toxicology and Pharmacology, 65(1), 100–107.

[8] Nejad, S. M., Özgüneş, H. and Başaran, N. (2017). Pharmacological and toxicological properties of eugenol. Turkish Journal of Pharmaceutical Sciences, 14(2), 201–206.

[9] Friedman, M. (2006). Potato glycoalkaloids and metabolites: Roles in the plant and in the diet. Journal of Agricultural and Food Chemistry, 54(23), 8655–8681.

[10] Lake, B. (1999). Coumarin metabolism, toxicity and carcinogenicity: Relevance for human risk assessment. Food and Chemical Toxicology, 37(4), 423–453.

[11] Pitaro, M., Croce, N., Gallo, V., Arienzo, A., Salvatore, G. and Antonini, G. (2022). Coumarin-induced hepatotoxicity: A narrative review. Molecules, 27(24), 9063.

[12] Smith, T. W. (1975). Digitalis toxicity: Epidemiology and clinical use of serum concentration measurements. The American Journal of Medicine, 58(4), 470–476.

[13] Hauptman, P. J. and Kelly, R. A. (1999). Digitalis. Circulation, 99(9), 1265–1270.

[14] Lee, M. R. (2007). Solanaceae IV: *Atropa belladonna*, deadly nightshade. Journal of the Royal College of Physicians of Edinburgh, 37(2), 77–84.

[15] Chan Thomas, Y. K. (2009). Aconite poisoning. Human & Experimental Toxicology, 28(12), 795–797.

[16] Yeung, H. C. (1995). Handbook of Chinese Herbs and Formulas, Institute of Chinese Medicine, Los Angeles, CA.

[17] Bloch, E. (2001). Hemlock poisoning and the death of Socrates: Did Plato tell the truth? Journal of the International Plato Society, 1, 2–7.

[18] Watt, J. M. and Breyer-Brandwijk, M. G. (1962). The Medicinal and Poisonous Plants of Southern and Eastern Africa Being an Account of Their Medicinal and Other Uses, Chemical Composition,

Pharmacological Effects and Toxicology in Man and Animal, E. & S. Livingstone, Teviot Place, Edinburgh: E. & S. Livingstone Ltd.

[19] Langford, S. D. and Boor, P. J. (1996). Oleander toxicity: An examination of human and animal toxic exposures. Toxicology, 109(1), 1–13.

[20] Bandara, V., Weinstein, S. A., White, J. and Eddleston, M. (2010). A review of the natural history, toxinology, diagnosis and clinical management of *Nerium oleander* (common oleander) and *Thevetia peruviana* (yellow oleander) poisoning. Toxicon, 56(3), 273–281.

[21] Glatstein, M., Alabdulrazzaq, F. and Scolnik, D. (2016). Belladonna alkaloid intoxication: The 10-year experience of a large tertiary care pediatric hospital. American Journal of Therapeutics, 23(1), e74–e7.

[22] Sayyed, A. (2014). Phytochemistry, pharmacological and traditional uses of *Datura stramonium* L. review. Journal of Pharmacognosy and Phytochemistry, 2(5), 123–125.

[23] Soni, P., Siddiqui, A. A., Dwivedi, J. and Soni, V. (2012). Pharmacological properties of *Datura stramonium* L. as a potential medicinal tree: An overview. Asian Pacific Journal of Tropical Biomedicine, 2(12), 1002–1008.

[24] Audi, J., Belson, M., Patel, M., Schier, J. and Osterloh, J. (2005). Ricin poisoning: A comprehensive review. Jama, 294(18), 2342–2351.

[25] Tunaru, S., Althoff, T. F., Nüsing, R. M., Diener, M. and Offermanns, S. (2012). Castor oil induces laxation and uterus contraction via ricinoleic acid activating prostaglandin EP3 receptors. Proceedings of the National Academy of Sciences, 109(23), 9179–9184.

[26] Parvizi, M. M., Saki, N., Samimi, S., Radanfer, R., Shahrizi, M. M. and Zarshenas, M. M. (2024). Efficacy of castor oil cream in treating infraorbital hyperpigmentation: An exploratory single-arm clinical trial. Journal of Cosmetic Dermatology, 23(3), 911–917.

[27] Bradberry, S. M., Dickers, K. J., Rice, P., Griffiths, G. D. and Vale, J. A. (2003). Ricin poisoning. Toxicological Reviews, 22, 65–70.

[28] Thomas, P. and Polwart, A. (2003). *Taxus baccata* L. Journal of Ecology, 91(3), 489–524.

[29] Weaver, B. A. (2014). How Taxol/paclitaxel kills cancer cells. Molecular Biology of the Cell, 25(18), 2677–2681.

[30] Labossiere, A. W. and Thompson, D. F. (2018). Clinical toxicology of yew poisoning. Annals of Pharmacotherapy, 52(6), 591–599.

[31] Begum, S., Saxena, B., Goyal, M., Ranjan, R., Joshi, V. B. and Rao, C. V. (2010). Study of anti-inflammatory, analgesic and antipyretic activities of seeds of *Hyoscyamus niger* and isolation of a new coumarinolignan. Fitoterapia, 81(3), 178–184.

[32] Adamse, P., Van Egmond, H., Noordam, M., Mulder, P. and De Nijs, M. (2014). Tropane alkaloids in food: Poisoning incidents. Quality Assurance and Safety of Crops & Foods, 6(1), 15–24.

[33] Schep, L. J., Slaughter, R. J., Becket, G. and Beasley, D. M. G. (2009). Poisoning due to water hemlock. Clinical Toxicology, 47(4), 270–278.

[34] Schep, L. J., Schmierer, D. M. and Fountain, J. S. (2006). Veratrum poisoning. Toxicological Reviews, 25, 73–78.

[35] Sara Vitalini, A. B. and Fico, G. (2011). Study on secondary metabolite content of *Helleborus niger* L. leaves. Fitoterapia, 82(2), 152–154.

[36] Andzela Brajanovska, B. B. (2018). *Helleborus sp.* an ethnopharmacological and toxicological review. Macedonian Pharmaceutical Bulletin, 64(1), 3–9.

[37] Jean, B. (1999). Toxic Plants Dangerous to Humans and Animals, Intercept Limited, Andover-UK.

[38] Benítez, G., Leonti, M., Böck, B., Vulfsons, S. and Dafni, A. (2023). The rise and fall of mandrake in medicine. Journal of Ethnopharmacology, 303, 115874.

[39] Patočka MCO, J. and Navrátilová, Z. An expert view on the poisonous plant *Ageratina altissima*. 06.02. 2024.

[40] Yamani, A., Bunel, V., Antoine, M. H., Husson, C., Stévigny, C., Duez, P. . . . and Nortier, J. (2015). Substitution between *Aristolochia* and *Bryonia* genus in North-Eastern Morocco: Toxicological implications. Journal of Ethnopharmacology, 166, 250–260.

[41] Finkelstein, Y., Aks, S. E., Hutson, J. R., Juurlink, D. N., Nguyen, P., Dubnov-Raz, G. . . . and Bentur, Y. (2010). Colchicine poisoning: The dark side of an ancient drug. Clinical Toxicology, 48(5), 407–414.

[42] Putterman, C., Ben-Chetrit, E., Caraco, Y., & Levy, M. (1991). Colchicine intoxication: clinical pharmacology, risk factors, features, and management. Seminars in Arthritis and Rheumatism, 21 (3), 143–145.

[43] Rogelj, B., Popovič, T., Ritonja, A., Štrukelj, B. and Brzin, J. (1998). Chelidocystatin, a novel phytocystatin from *Chelidonium majus*. Phytochemistry, 49(6), 1645–1649.

[44] Boddupalli, R. S. (2021). A review on most important poisonous plants and their medicinal properties. Journal of Medicinal Botany, 5, 1–13.

[45] Szopa, A., Pajor, J., Klin, P., Rzepiela, A., Elansary, H. O., Al-Mana, F. A. . . . and Ekiert, H. (2020). *Artemisia absinthium* L. – Importance in the history of medicine, the latest advances in phytochemistry and therapeutical, cosmetological and culinary uses. Plants, 9(9), 1063.

[46] Olsen RW. (2000). GABA and the thujone content of absinthe: An explanation for its effects on the human brain. Trends in Neurosciences, 23(3), 156–161.

[47] Griggs, M. M., Burns, R. and Honkala, B. (1990). Sassafras albidum (Nutt.) Nees. Burns, RM; Honkala, BH, Technical coordinators. Silvics of North America, 2, 773–777.

[48] Miller, J. A. and Miller, E. C. (1983). The metabolic activation and nucleic acid adducts of naturally-occurring carcinogens: Recent results with ethyl carbamate and the spice flavors safrole and estragole. British Journal of Cancer, 48(1), 1–15.

[49] Basch, E., Foppa, I., Liebowitz, R., Nelson, J., Smith, M., Sollars, D. and Ulbricht, C. (2004). Lavender (Lavandula angustifolia miller). Journal of Herbal Pharmacotherapy, 4(2), 63–78.

[50] Prashar, A., Locke, I. C. and Evans, C. S. (2004). Cytotoxicity of lavender oil and its major components to human skin cells. Cell Proliferation, 37(3), 221–229.

[51] Andrade, J. M., Faustino, C., Garcia, C., Ladeiras, D., Reis, C. P. and Rijo, P. (2018). Rosmarinus officinalis L.: An update review of its phytochemistry and biological activity. Future Science OA, 4(4), FSO283.

[52] Mack, R. B. (1994). Camphor: New perspectives on an old remedy. Emergency Medicine News, 16, 22–24.

[53] Miraj, S. and Kiani, S. (2016). Study of pharmacological effect of Mentha pulegium: A review. Der Pharmacia Lettre, 8(9), 242–24.

[54] Woolf, A. (1999). Essential oil poisoning. Journal of Toxicology: Clinical Toxicology, 37(6), 721–727.

[55] Anderson, I. B., Mullen, W. H., Meeker, J. E., Khojasteh-Bakht, S. C., Oishi, S., Nelson, S. D. and Blanc, P. D. (1996). Pennyroyal toxicity: Measurement of toxic metabolite levels in two cases and review of the literature. Annals of Internal Medicine, 124(8), 726–734.

[56] Hayat, U., Jilani, M. I., Rehman, R. and Nadeem, F. (2015). A Review on Eucalyptus globulus: A new perspective in therapeutics. International Journal of Chemical and Biochemical Sciences, 8, 85–91.

[57] Bhowal, M. and Gopal, M. (2015). Eucalyptol: Safety and pharmacological profile. Journal of Pharmaceutical Sciences, 5, 125–131.

[58] Asgarpanah, J. and Kazemivash, N. (2012). Phytochemistry and pharmacologic properties of *Myristica fragrans* Hoyutt.: A review. African Journal of Biotechnology, 11(65), 12787–12793.

[59] Abernethy, M. K. and Becker, L. B. (1992). Acute nutmeg intoxication. The American Journal of Emergency Medicine, 10(5), 429–430.

[60] Naser, B., Bodinet, C., Tegtmeier, M. and Lindequist, U. (2005). Thuja occidentalis (Arbor vitae): A review of its pharmaceutical, pharmacological and clinical properties. Evidence-Based Complementary and Alternative Medicine, 2(1), 69–78.

[61] Tisserand, R. and Young, R. (2013). Essential Oil Safety: A Guide for Health Care Professionals, Elsevier Health Sciences, Churchill Livingstone, London.

[62] European Medicines Agency (EMA). (2015). Assessment report on *Thuja occidentalis* L. folium aetheroleum.

[63] Duke, J. (2002). Handbook of Medicinal Herbs, CRC press, Boca Raton.

[64] Wang, G. W., Hu, W. T., Huang, B. K. and Qin, L. P. (2011). *Illicium verum*: A review on its botany, traditional use, chemistry and pharmacology. Journal of Ethnopharmacology, 136(1), 10–20.

[65] Zou, Q., Huang, Y., Zhang, W., Lu, C. and Yuan, J. (2023). A comprehensive review of the pharmacology, chemistry, traditional uses and quality control of star anise (Illicium verum Hook. F.): An aromatic medicinal plant. Molecules, 28(21), 7378.

[66] Cortés-Rojas, D. F., De Souza, C. R. F. and Oliveira, W. P. (2014). Clove (Syzygium aromaticum): A precious spice. Asian Pacific Journal of Tropical Biomedicine, 4(2), 90–96.

[67] Milind, P. and Deepa, K. (2011). Clove: A champion spice. International Journal of Research in Ayurveda and Pharmacy, 2(1), 47–54.

[68] Chaieb, K., Hajlaoui, H., Zmantar, T., Kahla-Nakbi, A. B., Rouabhia, M., Mahdouani, K. and Bakhrouf, A. (2007). The chemical composition and biological activity of clove essential oil, *Eugenia caryophyllata* (*Syzygium aromaticum* L. Myrtaceae): A short review. Phytotherapy Research: An International Journal Devoted to Pharmacological and Toxicological Evaluation of Natural Product Derivatives, 21(6), 501–506.

[69] Prashar, A., Locke, I. C. and Evans, C. S. (2006). Cytotoxicity of clove (*Syzygium aromaticum*) oil and its major components to human skin cells. Cell Proliferation, 39(4), 241–248.

[70] Gülçin, İ., Elmastaş, M. and Aboul-Enein, H. Y. (2012). Antioxidant activity of clove oil–A powerful antioxidant source. Arabian Journal of Chemistry, 5(4), 489–499.

[71] Özbek, Z. A. and Ergönül, P. G. (2022). Clove (Syzygium aromaticum) and eugenol toxity. In: Chemistry, Functionality and Applications, Academic Press, London, 267–314.

[72] Adams, R. P. (2014). Junipers of the World: The Genus Juniperus, Trafford Publishing, Trafford Publishing, Vancouver.

[73] Asili, J., Emami, S. A., Rahimizadeh, M., Fazly-Bazzaz, B. S. and Hassanzadeh, M. K. (2010). Chemical and antimicrobial studies of *Juniperus sabina* L. and Juniperus foetidissima Willd. essential oils. Journal of Essential Oil Bearing Plants, 13(1), 25–36.

[74] Akarca, G., Şevik, R., Kilinç, M., Denizkara, A. J. and Aşçıoğlu, Ç. (2023). Chemical composition and biological activity of juniper (*Juniperus sabina* L.) essential oil growing in the Aegean region of Türkiye. Journal of Food Safety & Food Quality/Archiv Für Lebensmittelhygiene, 74(6), 165–170.

[75] Kavaz, D. and Faraj, R. E. K. E. (2023). Investigation of composition, antioxidant, antimicrobial and cytotoxic characteristics from Juniperus sabina and Ferula communis extracts. Scientific Reports, 13(1), 7193.

[76] Soltanian, S., Sheikhbahaei, M. and Mohamadi, N. (2017). Cytotoxicity evaluation of methanol extracts of some medicinal plants on P19 embryonal carcinoma cells. Journal of Applied Pharmaceutical Science, 7(7), 142–149.

[77] Nejatbakhsh, F., Aghababaei, Z., Shirazi, M., Mazaheri, M. and Ghaemi, M. (2022). Medicinal plants with abortifacient or emmenagogue activity: A narrative review based on traditional Persian medicine. Jundishapur Journal of Natural Pharmaceutical Products, 17(2), e119559.

[78] Singh, G., Kapoor, I. P. S., De Heluani, C. S. and Catalan, C. A. N. (2008). Chemical composition and antioxidant potential of essential oil and oleoresins from anise seeds (*Pimpinella anisum* L.). International Journal of Essential Oil Therapeutics, 2, 122–130.

[79] Shojaii, A. and Abdollahi Fard, M. (2012). Review of pharmacological properties and chemical constituents of Pimpinella anisum. International Scholarly Research Notices, 2012(1), 510795.

[80] Sun, W., Shahrajabian, M. H. and Cheng, Q. (2019). Anise (*Pimpinella anisum* L.), a dominant spice and traditional medicinal herb for both food and medicinal purposes. Cogent Biology, 5(1), 1673688.

[81] Wichtl, M. (Ed.) (2004). Herbal Drugs and Phytopharmaceuticals: A Handbook for Practice on a Scientific Basis, CRC press, Medpharm GmbH Scientific Publishers, Stuttgart.

[82] Aziz, Z. A., Ahmad, A., Setapar, S. H. M., Karakucuk, A., Azim, M. M., Lokhat, D. . . . and Ashraf, G. M. (2018). Essential oils: Extraction techniques, pharmaceutical and therapeutic potential-a review. Current Drug Metabolism, 19(13), 1100–1110.

[83] Kırmızıbekmez, H., Demirci, B., Yeşilada, E., Başer, K. H. C. and Demirci, F. (2009). Chemical composition and antimicrobial activity of the essential oils of *Lavandula stoechas* L. *ssp. stoechas* growing wild in Turkey. Natural Product Communications, 4(7), 1934578X0900400727.

[84] Insawang, S., Pripdeevech, P., Tanapichatsakul, C., Khruengsai, S., Monggoot, S., Nakham, T. . . . and Panuwet, P. (2019). Essential oil compositions and antibacterial and antioxidant activities of five *Lavandula stoechas* cultivars grown in Thailand. Chemistry & Biodiversity, 16(10), e1900371.

[85] Xiao, J., Liu, P., Hu, Y., Liu, T., Guo, Y., Sun, P. . . . and Wang, Y. (2023). Antiviral activities of *Artemisia vulgaris* L. extract against herpes simplex virus. Chinese Medicine, 18(1), 21.

[86] Trinh, P. T. N., Tien, L. X., Danh, T. T., Le Hang, D. T., Hoa, N. V., Yen, T. T. B. and Dung, L. T. (2024). Antioxidant, Anti-Inflammatory, and Anti-Bacterial Activities of *Artemisia vulgaris* L. Essential Oil in Vietnam. Natural Product Communications, 19(8), 1934578X241275782.

[87] Sharma, K. R. and Adhikari, S. (2023). Phytochemical analysis and biological activities of *Artemisia vulgaris* grown in different altitudes of Nepal. International Journal of Food Properties, 26(1), 414–427.

[88] Siwan, D., Nandave, D. and Nandave, M. (2022). *Artemisia vulgaris* Linn: An updated review on its multiple biological activities. Future Journal of Pharmaceutical Sciences, 8(1), 47.

[89] Sharifi-Rad, J., Herrera-Bravo, J., Semwal, P., Painuli, S., Badoni, H., Ezzat, S. M. . . . and Cho, W. C. (2022). *Artemisia spp.*: An update on its chemical composition, pharmacological and toxicological profiles. Oxidative Medicine and Cellular Longevity, 2022(1), 5628601.

[90] Anwar, F., Ahmad, N. and Alkharfy, K. M. (2016). Mugwort (*Artemisia vulgaris*) oils. In: Essential Oils in Food Preservation, Flavor and Safety, London, Academic Press, 573–579.

[91] Carson, C. F., Hammer, K. A. and Riley, T. V. (2006). *Melaleuca alternifolia* (tea tree) oil: A review of antimicrobial and other medicinal properties. Clinical Microbiology Reviews, 19(1), 50–62.

[92] Southwell, I. A., Hayes, A. J., Markham, J. and Leach, D. N. (1993, March). The search for optimally bioactive Australian tea tree oil. In: International Symposium on Medicinal and Aromatic Plants, Vol. 344, Acta Horticulture, 256–265.

[93] Hammer, K. A., Carson, C. F. and Riley, T. V. (2003). Antifungal activity of the components of Melaleuca alternifolia (tea tree) oil. Journal of Applied Microbiology, 95(4), 853–860.

[94] Carson, C. and Riley, T. V. (2001). Safety, efficacy and provenance of tea tree (Melaleuca alternifolia) oil. Contact Dermatitis (01051873), 45(2), 65–70.

[95] Hammer, K. A., Carson, C. F., Riley, T. V. and Nielsen, J. B. (2006). A review of the toxicity of Melaleuca alternifolia (tea tree) oil. Food and Chemical Toxicology, 44(5), 616–625.

[96] Hsouna, A. B. and Hamdi, N. (2012). Phytochemical composition and antimicrobial activities of the essential oils and organic extracts from *Pelargonium graveolens* growing in Tunisia. Lipids in Health and Disease, 11, 1–7.

[97] Dzamic, A. M., Sokovic, M. D., Ristic, M. S., Grujic, S. M., Mileski, K. S. and Marin, P. D. (2014). Chemical composition, antifungal and antioxidant activity of Pelargonium graveolens essential oil. Journal of Applied Pharmaceutical Science, 4(3), 001–005.

[98] Boukhris, M., Simmonds, M. S., Sayadi, S. and Bouaziz, M. (2013). Chemical composition and biological activities of polar extracts and essential oil of rose-scented geranium, *Pelargonium graveolens*. Phytotherapy Research, 27(8), 1206–1213.

[99] Asgarpanah, J. and Ramezanloo, F. (2015). An overview on phytopharmacology of *Pelargonium graveolens* L. Indian Journal of Traditional Knowledge, 14(4), 558–563.

[100] Jaradat, N., Hawash, M., Qadi, M., Abualhasan, M., Odetallah, A., Qasim, G. . . . and Al-Maharik, N. (2022). Chemical markers and pharmacological characters of *Pelargonium graveolens* essential oil from Palestine. Molecules, 27(17), 5721.

[101] Karimi, N., Hasanvand, S., Beiranvand, A., Gholami, M. and Birjandi, M. (2024). The effect of Aromatherapy with Pelargonium graveolens (*P. graveolens*) on the fatigue and sleep quality of critical care nurses during the COVID-19 pandemic: A randomized controlled trial. Explore, 20(1), 82–88.

[102] Medjdoub, H., Bouali, W., Semaoui, M., Benaissa, A., Chaib, F. and Azzi, A. (2025). Chemical composition, antioxidant and antibacterial activities of the essential oil of Pelargonium graveolens L'Hér. Chemical Papers, 79, 1367–1374.

[103] Roman, S., Voaides, C. and Babeanu, N. (2023). Exploring the sustainable exploitation of bioactive compounds in *Pelargonium sp.*: Beyond a fragrant plant. Plants, 12(24), 4123.

[104] Sarkic, A. and Stappen, I. (2018). Essential oils and their single compounds in cosmetics – A critical review. Cosmetics, 5(1), 11.

[105] Amel, H. A., Kamel, H., Meriem, F. and Abdelkader, K. (2022). Traditional uses, botany, phytochemistry, and pharmacology of *Pelargonium graveolens*: A comprehensive review. Tropical Journal of Natural Product Research, 6(10), 1547–1569.

[106] Fong, H. H. S., Pauli, G. F., Bolton, J. L., Van Breemen, R. B., Banuvar, S., Shulman, L. . . . and Farnsworth, N. R. (2006). Evidence-Based Herbal Medicine: Challenges in Efficacy and Safety Assessments. In: Current Review of Chinese Medicine, 11–26.

[107] Bone, K. and Mills, S. (2012). Principles and Practice of Phytotherapy: Modern Herbal Medicine, Elsevier Health Sciences, Churchill Livingstone.

[108] Jităreanu, A., Trifan, A., Vieriu, M., Caba, I. C., Mârțu, I. and Agoroaei, L. (2022). Current trends in toxicity assessment of herbal medicines: A narrative review. Processes, 11(1), 83.

[109] Bogusz, M. J. and Al-Tufail, M. (2008). Toxicological aspects of herbal remedies. In: Handbook of Analytical Separations, Vol. 6, Elsevier Science BV, New York, 589–610.

[110] Boullata, J. I. and Nace, A. M. (2000). Safety issues with herbal medicine. *Pharmacotherapy*. The Journal of Human Pharmacology and Drug Therapy, 20(3), 257–269.

[111] Zhou, X., Li, C. G., Chang, D. and Bensoussan, A. (2019). Current status and major challenges to the safety and efficacy presented by Chinese herbal medicine. Medicines, 6(1), 14.

[112] McKenna, D. J., Jones, K., Hughes, K. and Tyler, V. M. (2012). Botanical Medicines: The Desk Reference for Major Herbal Supplements, Routledge Routledge Taylor & Francis Group, New York and London.

[113] Hallström, H. and Thuvander, A. (1997). Toxicological evaluation of myristicin. Natural Toxins, 5(5), 186–192.

[114] Voigt, V., Franke, H. and Lachenmeier, D. W. (2024). Risk assessment of pulegone in foods based on benchmark dose–response modeling. Foods, 13(18), 2906.

[115] Preedy, V. R. (Ed.) (2015). Essential Oils in Food Preservation, Flavor and Safety, Academic press, London.

[116] Kumar, S., Kavitha, T. K. and Angurana, S. K. (2019). Kerosene, camphor, and naphthalene poisoning in children. Indian Journal of Critical Care Medicine: Peer-reviewed, Official Publication of Indian Society of Critical Care Medicine, 23(Suppl 4), S278.

[117] Navarra, M., Miroddi, M. and Calapai, G. (2015). Chapter 9: Phototoxiciy of Essential oils, In: Bagetta, G., Cosentino, M., and Sakurada, T. (eds) Aromatherapy: Basic Mechanisms and Evidence Based Clinical Use, CRC Press, Boca Raton, 191–213.

[118] Henderson, L., Yue, Q. Y., Bergquist, C., Gerden, B. and Arlett, P. (2002). St John's wort (*Hypericum perforatum*): Drug interactions and clinical outcomes. British Journal of Clinical Pharmacology, 54(4), 349–356.

[119] Izzo, A. A. and Ernst, E. (2001). Interactions between herbal medicines and prescribed drugs: A systematic review. Drugs, 61, 2163–2175.

[120] Brazier, N. C. and Levine, M. A. (2003). Drug-herb interaction among commonly used conventional medicines: A compendium for health care professionals. American Journal of Therapeutics, 10(3), 163–169.

[121] European Medicines Agency (EMA). (2003). Guidelines on safety and identification of medicinal plants 201 World Health Organization. In: WHO Guidelines on Good Agricultural and Collection Practices [GACP] for Medicinal Plants, World Health Organization, Geneva, 7–12.

[122] Nelson, L. S., Shih, R. D., Balick, M. J. and Lampe, K. F. (2007). Handbook of Poisonous and Injurious Plants, New York Botanical Garden, New York, 55–306.

[123] Bischoff, K., and Smith, M. C. (2011). Toxic plants of the northeastern United States. Veterinary Clinics: Food Animal Practice, 27(2), 459–480.

[124] Smith, L. W., and Culvenor, C. C. J. (1981). Plant sources of hepatotoxic pyrrolizidine alkaloids. Journal of Natural Products, 44(2), 129–152.

Part II: **Effect of stress factors on medicinal and aromatic plants**

Gülen Özyazıcı* and Negar Valizadeh

Chapter 5
Impact of drought stress on the medicinal and aromatic plants' biochemistry

Abstract: Today, medicinal plants have a wide range of uses not only in the field of health but also in different sectors such as perfumery, cosmetics, food industry and phytotherapy. In addition to their potential to treat diseases, these plants are also valued for their contributions to different industries. Medicinal plants with various phytochemical compounds such as secondary metabolites, for example, alkaloids, terpenoids, phenols, steroids, flavonoids, tannins, glycosides, volatile oils and aromatic compounds are exposed to abiotic stress such as drought. Drought, one of the abiotic factors, causes a decrease in plant height, plant leaf area, number, and such other decreases, thus not only changing the plant structurally and anatomically, but also leading to fluctuations in its chemical components. The quality and quantity of the components of secondary metabolites synthesized by the plant help to cope with the harmful effects of stress for adaptation and defense. Numerous studies have shown that drought affects the accumulation of secondary metabolites in different organs of the plant and causes an increase or decrease in different plant species and even in different species of the same genus. Since the main aim in medicinal plants is not only to increase the yield of seeds, leaves, and flowers but also to increase the production of active ingredients such as essential oils, the cultivation and management of medicinal plants under stress conditions is different from other crops. This study provides a summary of recent literature covering the studies on the morphology, physiology, and biochemistry of medicinal and aromatic plants under drought stress.

Keywords: water deficit, drought stress, plant secondary metabolites, essential oil, morphology, plant physiology

*Corresponding author: Gülen Özyazıcı, Department of Field Crops, Faculty of Agriculture, Siirt University, Siirt, Türkiye, e-mail: gulenozyazici@siirt.edu.tr, https://orcid.org/0000-0003-2187-6733
Negar Valizadeh, Research Division of Natural Resources, East Azerbaijan Agricultural and Natural Resources Research and Education Center, Agricultural Research, Education and Extension Organization (AREEO), Tabriz, Iran

https://doi.org/10.1515/9783111469713-005

5.1 Introduction

Drought, which occurs as a natural result of climate change and is one of the abiotic stress factors, is one of the most important stress factors affecting agricultural production all over the world. When usable areas on earth are classified, drought, which is a natural stress factor, comes first with a share of 26% [1], followed by mineral stress with 20%, cold and frost stress with 15%, while 10% of the area is not exposed to any stress factor [2].

Drought is an important factor affecting the growth and metabolic activities of plant species. Drought conditions inhibit plant growth, disrupt plant-water balance, reduce water use efficiency, and negatively affect the physiological processes in the plant. Drought stress in plants results from disruption of water flow in the xylem. Obstruction of water flow leads to a decrease in cell turgor pressure and, therefore, affects cell elongation and expansion. Important physiological developments such as cell growth and division, enzyme synthesis, and protein synthesis slow down [3]. As a result, growth and development are also significantly affected, and yield is reduced. Drought stress causes a decrease in plant development parameters such as plant height, leaf number, leaf size, root length, leaf area, etc. and these are the first visible symptoms of drought [4–6]. Research has shown that the responses of medicinal and aromatic plants to salinity and drought stress have common points [1]. In order to access water resources in drought stress situations, plants accelerate root development in the early stages of stress and, conversely, slow down shoot development [7]. Under drought conditions, reduction of plant leaf surface and reduction of transpiration increase tolerance to drought stress. In many plants, dry conditions accelerate the aging process and abscission of old leaves, while roots adapt by developing the root system to reach water in the deep layers of the soil [8, 9]. Moreover, responses such as shortening of height, less fruit/seed/biomass yield, and changes in the reproductive process occur [10]. Examples of water stress reducing plant growth in medicinal and aromatic plants include *Hypericum brasiliense* Choisy [11]. *Catharanthus roseus* [12] and *Bupleurum chinense* DC plant species [13].

While drought generally has a negative effect on plant growth and development, there are studies on the positive effect of water deficit when it comes to the biosynthesis of secondary metabolites, enzyme activities and solute accumulation [14]. Some of these responses may be related to the plant's ability to survive under restrictive conditions. Although the metabolic responses of most cultivated plants to water deficit or drought have been studied, studies on this subject in medicinal and aromatic plants are few or relatively new. Water is an important factor in the yield and development of medicinal and aromatic plants, as well as other plants. Unlike traditionally cultivated crop plants, water and plant relationships in medicinal and aromatic plants are not fully understood. This section focuses on the effects of drought stress or water deficiency on the biochemistry of plants.

5.2 Effect of drought or water deficiency on the morphology of medicinal plants

Drought stress is one of the most important abiotic stresses that show dramatic changes in plant growth and yield. Drought occurs when the available water in the soil decreases. It also occurs due to inadequate rainfall and continuous water loss through the process of transpiration and evaporation [15]. Water availability is regarded as a critical determinant influencing plant growth. The commercial medicinal value of an aromatic plant depends on the presence of secondary metabolites, which are impacted by water scarcity [16, 17]. Drought first manifests itself with reduced cell development, especially in the stem and leaves. Stem growth stops and leaf growth is also restricted. One of the first signs of water deficiency is a decrease in turgor pressure, which leads to inhibition of cell growth and development, especially in stems and leaves. In addition, nutrient uptake decreases under drought stress conditions due to decreased soil moisture, which causes a decrease in the diffusion rate of nutrients from the soil matrix to the absorbent root surface, limiting leaf growth and development. As a result, leaf area, light interception, and overall photosynthetic capacity are reduced [18]. The degree of stress tolerance varies from one plant species to another. Water stress is an important factor that limits the growth of the plant in the initial stages. It negatively affects the plant morphology of medicinal and aromatic plants. Dry matter accumulation decreases in all plant organs under drought stress, but different organs are affected to varying degrees. Therefore, plants growing in drought conditions are morphologically smaller than plants growing under normal conditions.

Drought stress leads to plant dehydration, stomatal closure [19], and restricted gas exchange, subsequently resulting in metabolic inhibition, reduced photosynthetic rates, and ultimately plant death [12]. However, a plant's ability to survive under such stressful conditions depends on its species, growth stage, and the duration and intensity of water deficit [12]. Moisture deficiency triggers various structural changes in plants that are crucial for responding to drought stress. These include morphological adaptations (such as reduced growth rates, development of deeper root systems, and alterations in the root-to-shoot ratio to avoid desiccation [20] as well as physiological and metabolic responses (such as stomatal closure, accumulation of antioxidants, and the expression of stress-specific genes).

Drought stress reduces plant height, leaf length, leaf weight, leaf area, and fresh and dry weight in lemongrass species [21]. In a study investigating the effect of water deficit on the morphology of *Salvia sclarea* populations, it was reported that yield and yield components decreased under stress conditions [22]. In their study on various levels of drought stress on *Plantago psyllium* L., *Achillea millefolium* L., *Salvia officinalis* L., *Calendula officinalis* L., and *Matricaria chamomilla* L. plants, it was reported that increasing drought stress led to a decrease in shoot weight and plant height compared to nonstress conditions [23]. They reported that growth and yield decreased in thyme (*Thymus vulgaris* L.), Japanese mint (*Mentha cordifolia* Opiz.), and Mexican marigold (*Tagetes minuta* L.) under drought stress [24–26]. When water stress was applied to *Ocimum basilicum* L. (sweet basil) at 100%, 85%, 70%, and 55% of field capacity, shoot branching was limited under severe drought stress (55% of field capacity) [27].

Water limitation reduces the yield of medicinal and aromatic plants in three ways: First, drought-induced leaf area expansion can be limited by temporary leaf wilting or premature leaf senescence. Secondly, it can limit the grain yield of medicinal and aromatic plants by reducing radiation utilization efficiency, and thirdly, by reducing the harvest index. This can occur even without a strong reduction in total medicinal and aromatic plant dry matter accumulation if a short stress period coincides with the critical developmental stage around flowering [28]. Numerous studies have shown that grain yield can be significantly reduced as a result of water deficit during the reproductive period in coriander [29] and Mexican marigold [30]. Similarly, Petropoulos et al. [31] noted that leaf and root weight, and leaf number (35% of field capacity) were significantly reduced due to water stress. In the studies of Miao et al. [32], irrigation treatments in *Tulipa edulis* L., an important medicinal plant with anticancer properties, were control (80% of field capacity), continuous drought (50% of field capacity) and alternating wetting and drying (50% and 80% of field capacity, AWD). They found that persistent drought significantly inhibited plant growth and yield compared to the control. Khorasaninejad et al. [33] reported that mint plant is moderately tolerant to water stress (85% field capacity) because it does not have adverse effects on some growth parameters of this plant.

5.3 Effect of drought or water deficiency on the physiology of medicinal plants

The response of plants to drought depends significantly on the drought severity, plant developmental stage, genotype, and physiological status [34, 35]. Under water stress, plants attempt to maintain the water potential of their tissues by closing their stomata and reducing water losses through transpiration. Stress-induced growth decline may result from a decrease in meiosis and photosynthesis as a result of stomata closure, or from a decrease in cell development brought on by a drop in turgor pressure [36]. Furthermore, plant growth and photosynthesis are reduced due to decreased CO_2 levels and increased formation and accumulation of ABA, proline, mannitol and radical scavenging compounds (ascorbate, glutathione, α-tocopherol, etc.), stress proteins, and mRNAs [37]. However, these are at the cellular level and cannot be seen with the naked eye. The changes caused by drought in plants above and below the ground are presented in Figure 5.1.

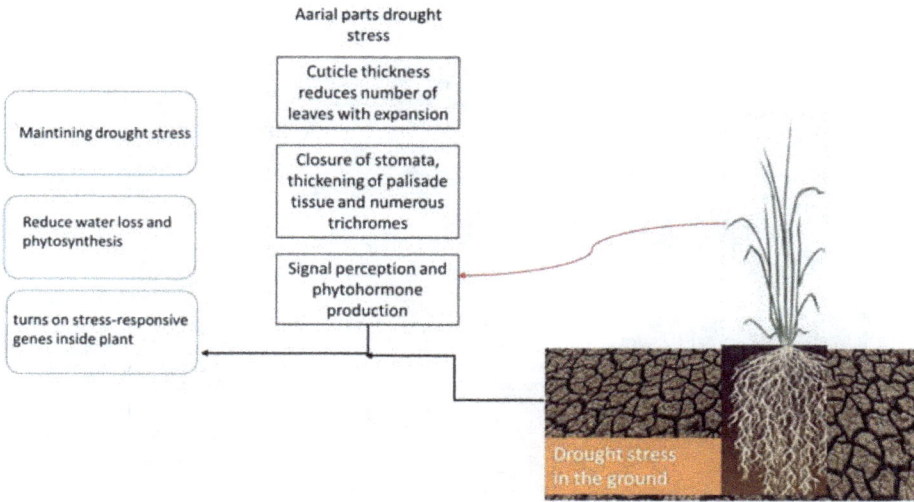

Figure 5.1: Changes caused by drought in plants above and below ground.

Studies have shown that drought affects the physiological, biochemical, and morphological properties of medicinal plants both qualitatively and quantitatively, but the effect varies, depending on the genotype, and irrigation regime and characteristics. Considering global climate change, it becomes clear how important it is to include drought-resistant medicinal plant genotypes in future selection and breeding programs.

Drought stress causes oxidative stress in plants. During periods of insufficient water, light-chlorophyll interactions in the chloroplast cause oxidative stress in the vegetative tissues of the plant. Plants have a complex defense mechanism consisting of lipid-soluble and membrane-bound antioxidants, water-soluble antioxidants, and enzymatic antioxidants against the harmful effects of oxidative stress. Plants exposed to drought stress can combat oxidative stress as a result of the activation of some or all of their antioxidant defense systems [38-41]. Increasing the synthesis and accumulation of osmolytes such as proline, glycine betaine, and polyamines are other defense mechanisms that reduce osmotic stress in plant cells [42]. Protein content decreases in plants under drought stress, which is associated with increased activity of protein degrading enzymes and accumulation of free amino acids such as proline [43]. If the balance between the production of free radicals and the plant antioxidant defense system is disrupted, oxidative stress destroys cell membranes and other organelles [44]. However, long-term and sometimes even short-term stress can cause visible damage to plants and even death if the capacities of defense mechanisms are [45, 46]. The effects of drought on plants and the defense mechanisms formed by plants are summarized in Figure 5.2.

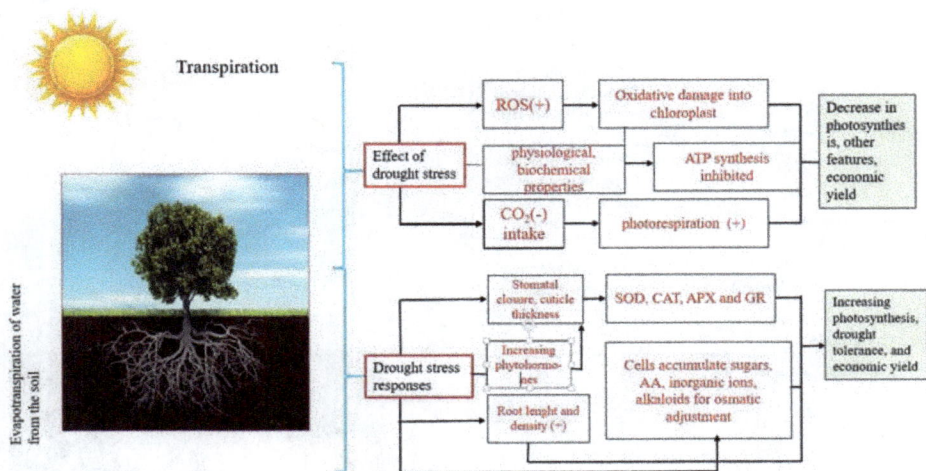

Figure 5.2: Negative effects and adaptations of plants to drought stress, modified from Seleiman et al. [47]; (–) means decrease and (+) means increase.

There are more than 200 species of the *Thymus* genus, which is an important medicinal and spice plant, and these species respond differently to water deficiency. Moradi et al. [48] conducted a study to determine and physiologically evaluate the response of eleven populations of various species of thyme (*Thymus daenensis, T. kotchyanous, T. vulgaris, T. serpyllum, T. capitata,* and *T. zygis*) to water deficit stress. The findings showed that populations had significantly different root/shoot ratios under drought conditions, with leaf water potential decreasing from –3.4 bar in irrigated plants to –10.5 bar in droughted plants. Moradi et al. [48] found a significant negative relationship between water content and water potential and determined that *T. serpyllum* was more resistant than other thyme species and that the Spanish population of *T. vulgaris* was susceptible.

5.4 Effect of drought or water deficiency on secondary metabolites of medicinal plants

Droughts occur annually in many parts of the world and often cause significant damage to crop production. Drought, one of the important abiotic stresses, is known to increase the amount of secondary metabolites in plants. On the other hand, drought stress can cause oxidative stress due to its formation. In order to protect against the harmful effects of active oxygen species, plants have developed a complex antioxidant system that includes enzymatic antioxidants and nonenzymatic antioxidants. Accumulation of secondary metabolites is known as a defense mechanism of plants, and plants can respond and adapt to water stress by changing their cellular metabolism under stress conditions [49]. The chemical structures of some common secondary metabolites produced in medicinal and aromatic plants are presented (Figure 5.3).

In studies conducted with medicinal and aromatic plants, it is thought that the genetic characteristics of the plant, and its anatomical and morphological development stages, as well as stress factors play a role in the formation of its phytochemical composition, unlike traditional plant products [50]. Since stress-related metabolism largely affects all other metabolic events, it is known to also affect the synthesis and accumulation of secondary metabolites [51].

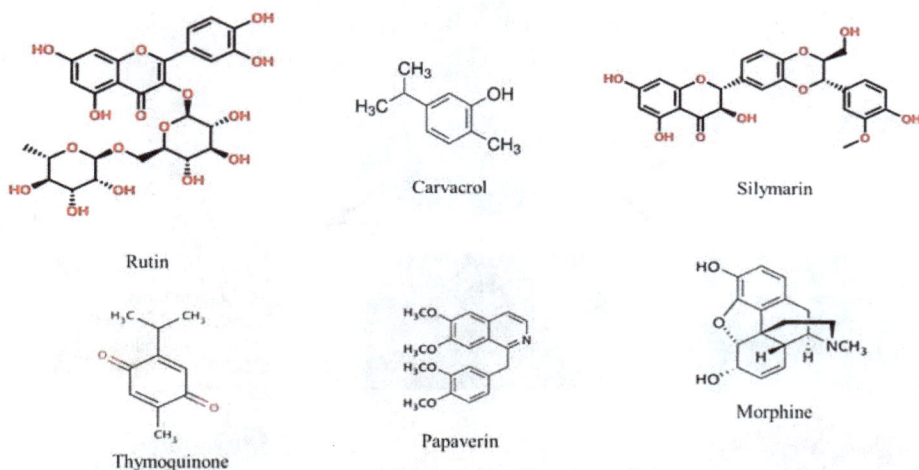

Figure 5.3: Chemical structures of some common secondary metabolites produced in medicinal and aromatic plants.

Many studies have shown that drought increases the amounts of secondary metabolites in a wide range of plant species, including hesperidin in *Rehmannia glutinosa* [52], indole alkaloid in leaves and roots of *Catharanthus roseus* [15], rutin, quercetin, betulinic acid in *H. brasiliense* Choisy [11], and saikosaponin a and c in *Bupleurum chinense* DC [13]. In *S. miltiorrhiza*, whose roots are widely used in traditional Chinese medicine, it was determined that drought stress reduced both shoot and root dry weight and water stress reduced the yield of tanshinone IIA; on the other hand, the contents of other active components, except rosmarinic acid, and the yield of salvianolic acid B increased under water stress [53].

Drought stress in the early vegetative stages of *Spigelia anthelmia*, which is used locally as an anthelmintic, reduced its growth but did not affect the alkaloid content [54]. Superoxide dismutase and peroxidase antioxidant enzyme activities of *Hyoscyamus niger* increased in root and leaf under water deficit, while hyoscyamine and scopolamine decreased under moderate and severe water deficit. The use of plant-growth-promoting rhizobacteria (PGPR) reduced the negative effect of drought severity on alkaloid abundance [55]. Similarly, studies revealed that water stress increased the tannin, saponin, and flavonoid content of *Bryophyllum pinnatum* but decreased the alkaloid content [56].

In order to increase the quantity and quality of secondary metabolites that determine the economic value of medicinal plants, the water content available to plants should be kept under control. Studies have reported that appropriate levels of water stress increase secondary metabolite content in medicinal plants [43, 57, 58] (Figure 5.4).

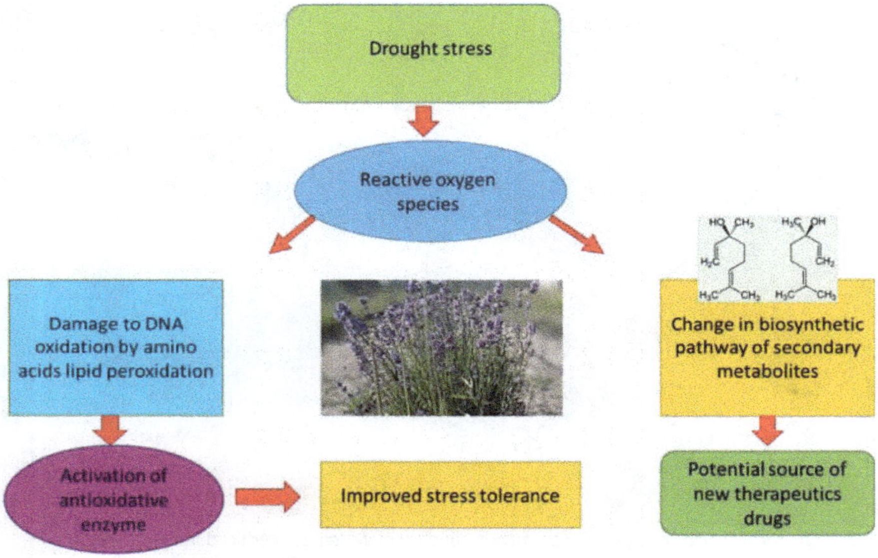

Figure 5.4: Diagram showing the effect of drought stress on DNA, proteins and lipid modified from Bistgani et al. [59].

The amount and composition of essential oils are affected by various abiotic stress factors. In *Dracocephalum moldavica* L. plants exposed to different drought treatments, the highest essential oil content of 0.58% was detected in the moderate drought treatment [60]. When three different irrigation regimes were applied to two different *Salvia* species, it was found that the highest essential oil content (2.20%) was in the moderate drought application [61].

Some phytochemicals can only be seen as a product of the response mechanism under stress conditions, depending on their synthesis. For example, Kılıç and Kaya [62], although α-pinene, sabinene, limonene were not detected in basil essential oil under normal conditions, they were determined at different rates under drought stress. However, it has been reported that some plants respond differently to drought stress; the oil yield of *Lavandula latifolia* and *Salvia sclarea* decreases and it has no effect on *Mentha piperita*, *Salvia lavandulifolia*, *Thymus capitatus*, and *Thymus masti-china* [63]. In general, drought stress or water deficiency can affect plant growth, vola-tile oil content, and components, depending on the species (Table 5.1). The appearance of *Melissa officinalis* and *Rosmarinus officinalis* plants is given in Figures 5.5 and 5.6.

Table 5.1: The effect of non-drought stress and severe stress on essential oil percentage (%).

Scientific name	Control	Mild drought stress	Moderate drought stress	Severe drought stress	Reference
Cymbopogon nardus	0.32	0.39	0.36	–	[21]
Melissa officinalis	0.10	0.14	0.28	0.30	[64]
Lavandula latifolia	2.85	2.67	–	–	[63]
Mentha piperita	2.22	2.13	–	–	
Salvia lavandulifolia	2.41	2.47	–	–	
Salvia sclarea	2.77	1.31	–	–	
Thymus capitatus	4.63	4.46	–	–	
Thymus daenensis	1.91	–	2.47	2.89	[65]
Origanum vulgare	2.2	–	–	3.2	[66]
Tagetes minuta	0.90	1.10	1.11	1.12	[26]
Ocimum basilicum	–	1.15	0.61	0.70	[67]
Ocimum × africanum	–	2.79	2.72	2.76	
Ocimum americanum		0.70	0.76	0.54	
Rosmarinus officinalis	0.73	–	0.84	0.87	[68]
Dracocephalum moldavica		0.23	0.22	0.17	[69]

Sage (*Salvia officinalis* L.) is a species sensitive to drought, and severe drought can cause a decrease in the activity of enzymes involved in the biosynthesis of phenolic compounds [70]. The main components of the essential oil of peppermint, one of the most important and widely used medicinal and aromatic plants worldwide, show different responses to drought stress at different growth stages (Figure 5.7). The essential oil content of plants exposed to mild water stress (60 ± 5% field capacity) increases, while moderate water stress (40 ± 5% field capacity) significantly reduces the essential oil content [71]. Drought stress decreased the growth, seed yield and yield components, total fatty acid content, and especially petroselinic acid content of cumin (*Carum carvi* L.), while it increased the essential oil components [72].

The effects of drought stress on alkaloid, glaucoside, and glucosinolate components of some medically and economically important plants vary, depending on the plant spe-cies and the type of component. When plants are exposed to various stress conditions, alkaloid concentrations often increase. This is well known, probably due to the passively

Figure 5.5: *Melissa officinalis.*

Figure 5.6: *Rosmarinus officinalis.*

increased biosynthesis rate caused by greatly elevated NADPH concentrations in stressed plants [73]. In *Papaver somniferum* plant, drought stress caused an increase in the concentration of alkaloids (morphine, codeine, and papaverine) [74]. In response to

Figure 5.7: Peppermint (*Mentha piperita*) and sage (*Salvia officinalis*).

water scarcity, plants usually change their secondary metabolism, which is leading to an increase in nitrogen-containing compounds such as alkaloids. Morphine biosynthesis depends on nitrogen availability and is therefore affected by drought conditions through changes in enzymatic activity and metabolic pathways. Moderate drought stress can increase morphine accumulation by upregulating essential biosynthetic genes, while severe stress can inhibit growth and reduce overall alkaloid yield. In addition to this, nitrogen metabolism plays an important role in alkaloid production because nitrogen-containing precursors such as tyrosine and ornithine are essential for morphine biosynthesis. The interaction between drought stress and nitrogen availability suggests that optimizing water and nutrient management may be a strategy to increase alkaloid production in poppy (*Papaver somniferum*) under drought-prone conditions.

In the plant *Withania somnifera*, different effects were observed on different secondary metabolites. Withanolide compounds increased, while withanolide and 12-deoxywithastramonolide levels decreased. For all that, the concentration of Withaferin A compound increased in the roots and leaves. An increase in asiaticoside and madecassoside components was observed in the leaves of *Centella asiatica* under low temperature and drought conditions. In cassava (*Manihot esculenta*) plant, drought stress appears to cause a strong increase in the concentration of cyanogenic glycosides, especially in tuberous roots and leaves. In *Lupinus angustifolius* (narrow-leaved lupin), a significant increase in quinolizidine alkaloids in seeds occurred. In *Eucalyptus cladocalyx*, drought conditions increased the levels of cyanogenic glycosides in dried leaves and oil content. A strong increase in the concentration of indole alkaloids was recorded in *Catharanthus roseus*. In general, it appears that drought stress significantly increases the production of certain secondary metabolites in these plants, and this may be related to the defense mechanisms of the plants [75]. In *Brassica napus*, drought stress caused a large increase in the concentration of glucosinolates [76]. In rapeseed, the biochemical properties of the seed were greatly altered in plants exposed to drought during flowering. Drought during the early vegetative and flowering stages caused a slight increase in seed protein concentration. Depending on its timing, significant effects of drought stress were observed on the accumulation of secondary metabolites (i.e., phenolics and glucosinolates) that are of great importance for rapeseed meal quality. *Achnatherum inebrians* is a grass species that produces alkaloids such as ergonovine and ergine (lysergic acid amide). These alkaloids, which are classified as nitrogen-containing compounds, play an important role in plant defense by deterring herbivores and insects. Ergonovine and ergine biosynthesis is affected by environmental factors such as drought stress and nitrogen availability. In response to drought stress, the plant can increase alkaloid production, resulting in increased accumulation of these nitrogenous metabolites [77]. However, the extent of alkaloid production depends on the balance between nitrogen uptake and stress adaptation. Understanding the relationship between drought, nitrogen metabolism, and alkaloid production in *Achnatherum inebrians* is important to manage its ecological impact and potential applications in biotechnology.

The biosynthesis and accumulation of active substances such as silymarin, found in milkthistle (*Silybum marianum* (L.) Gaertn) seeds in plant tissues, are highly affected by environmental conditions [78]. Moderate and severe drought stress increases silymarin content, which is attributed to more silymarin, silybin, isosilybin, and silychristin content in stressed plants, while silydianin content decreases.

Safflower (*Carthamus tinctorius* L.) is an important plant both as a medicinal and oil plant (Figure 5.8). The pharmacological properties of the safflower plant are mainly due to its ability to accumulate some active secondary metabolites, mainly phenolic and flavonoid compounds. Drought stress decreases the seed yield and oil content of the safflower plant; on the contrary, the amounts of vanillic and caffeic acids, and rutin and quercetin in flower and seed extracts increase. The presence of these compounds causes an increase in antioxidant capacity [79]. *Carthamus tinctorius* is a multipurpose plant that can grow in arid and semiarid environments due to its tolerance to drought stress, salinity, and low and high temperatures. Although saf-

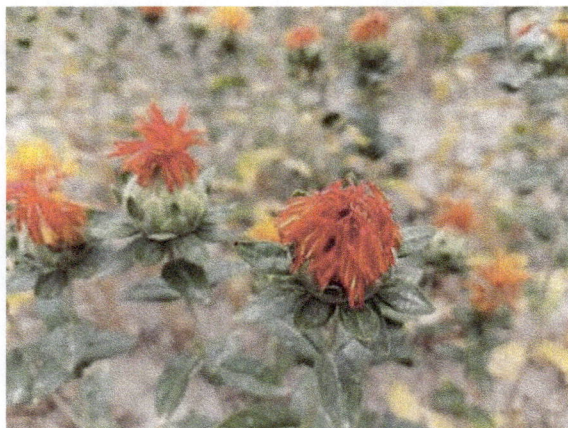

Figure 5.8: Safflower (*Carthamus tinctorius*) flower.

flower can grow in arid and semiarid climates, drought stress reduces plant height and yield, leaf chlorophyll content and leaf area, photosynthetic rate, yield components, oil content and yield, and fatty acid composition. Increased root/shoot ratio and root growth are some of the drought adaptation mechanisms of safflower.

5.5 Different approaches to mitigate the negative effects of drought stress on plants

Drought stress is an inevitable factor that exists without obvious warning in various environments that inhibits plant biomass production, quality, and energy. Its cumulative and subtle effect seriously affects plant morphological, physiological, biochemical, and molecular characteristics with negative impact on photosynthetic capacity. Plants that are coping with water limitation develop a variety of complex resistance and adaptation mechanisms, including physiological and biochemical responses that differ, depending on the species level. The strategies adopted by water deficient plants are reduction in transpiration loss by changing stomatal conductance and distribution, leaf curling, change in root-shoot ratio, increase in root length, accumulation of solutes, and osmotic and hormonal regulation. Planting time, plant genotype, and soil and nutrient management practices can help reduce yield losses in plants exposed to drought stress. Nevertheless, the use of drought-tolerant transgenic plants is the most popular approach to reducing drought stress.

Various strategies can be used for drought stress resistance. Among these, pre-planting or post-planting chemical applications, plant growth regulators (PGR), and bacterial inoculations are of great importance to increase drought resistance in different growth stages of the plant. Another application is the use of nanoparticles (NPs). NPs with sizes ranging from 1 to 100 nanometers have high surface energy and surface-to-volume ratio, which makes them highly efficient for many purposes by enhancing their other biological activities [80, 81]. Nanoparticles have emerged as a promising tool to reduce the negative effects of drought stress in plants. Due to their physicochemical properties, nanoparticles can enhance plant growth, improve water use efficiency, and regulate stress-related biochemical pathways. Nanoparticles can enhance water exchange, increase root growth, and improve water absorption efficiency. Silicon (SiO_2) and carbon-based nanoparticles increase the water retention capacity of soil and reduce water loss due to drought. It regulates stress-related hormones; silver (Ag) and zinc oxide (ZnO) nanoparticles balance the drought response by affecting abscisic acid (ABA) and cytokinin levels. Silicon (Si) nanoparticles prevent premature aging by reducing stress-induced ethylene production. Copper (Cu) and selenium (Se) nanoparticles increase antioxidant enzyme activity (SOD, CAT, and POD). Iron oxide (Fe_3O_4) nanoparticles improve proline, glycine, and betaine synthesis, which helps maintain cell hydration. Nanoparticles are applied as seed priming, controlled nutrient release nanofertilizers in drought stress management. Ghavam [82] reported that silver NP applications in *Thymus daenensis* and *Thymus vulgaris L.* increased the ability to withstand drought stress and increased germination and root length in saline conditions (200 mM). Titanium dioxide (TiO_2) NPs applied to *Verbascum sinuatum* plant alleviated the negative effects of drought [83]. Application of titanium dioxide (10 ppm) to the leaves of *Dracocephalum moldavica* L., growing under drought conditions, resulted in increased shoot dry biomass and essential oil content [84].

5.6 Case studies

Due to a worldwide water shortage and the rising use of herbal medicines, studies on drought stress affecting the composition of secondary metabolites in medicinal plants is crucial. Even though drought stress is generally considered as the main factor responsible for serious yield losses in agricultural production, this is different for medicinal and aromatic plants [85].

Total flavonoid and rosmarinic acid contents were not affected in *Melissa officinalis* L. and *Thymus vulgaris* L., grown under different water stress conditions. On the other hand, *Melissa officinalis* L. plants gave lower biomass weight under low water stress [86].

Tatarai et al. [16] investigated the effects of drought stress on two-year-old *Thymus citriodorus* plants by treating them with different concentrations of polyethylene glycol (PEG-6000) (0%, 2%, and 4%) for 15 days under greenhouse conditions. *Thymus citriodorus* exhibited a morphological drought avoidance mechanism by reducing shoot fresh weight to protect root system development, which enhanced root absorptive capacity and sustained plant growth. Additionally, thyme plants minimized tissue dehydration through stomatal closure and improved root water uptake. Regarding essential oil composition, the levels of geraniol and diisobutyl phthalate increased under drought stress, while pseudophytol content decreased. Although thymol was not the main component under control or mild stress conditions, its content increased under severe drought stress. Furthermore, carvacrol levels rose by 31.7% under severe drought stress compared to control plants.

Lotfi et al. [87] investigated the effects of drought stress on morphological traits, proline accumulation, soluble carbohydrates, and yield to determine the drought tolerance threshold of tarragon (*Artemisia dracunculus* L.). Stress treatments were applied at four levels: T1 (100% field capacity), T2 (80% field capacity), T3 (60% field capacity), and T4 (40% field capacity). The study revealed that drought stress significantly impacted morphological traits, flowering shoot yield, proline accumulation, and soluble carbohydrate content. As drought stress increased, plant height, crown diameter, leaf length, leaf width, leaf surface area, stem diameter, longest lateral shoot length, root length and development, shoot yield, and dry leaf yield decreased. The highest values for plant height, crown diameter, leaf length, leaf width, leaf surface area, stem diameter, longest lateral shoot length, root length, shoot yield, and dry leaf yield were observed under T1 (unstressed conditions). In contrast, the highest levels of proline, soluble carbohydrates, and root development were recorded under T4 (40% field capacity). While drought stress negatively affected most morphological traits and flowering shoot yield, it led to an increase in root length, proline accumulation, and soluble carbohydrates in flowering shoots.

The effect of drought stress on the quality and quantity yield of *Thymus vulgaris* was evaluated under field and laboratory conditions [88]. The study that included five different stress treatments showed that drought stress significantly affected plant height, flowering shoot yield, oil percentage, oil yield, thymol percentage, carvacrol percentage, chlorophyll a amount, chlorophyll b, proline, soluble sugars, sodium, magnesium, iron, and relative water content. The highest (2.22%) and lowest (0.74%) essential oil percentages were observed at 20% and 100% of the field capacity, respectively. The maximum thymol percentage was at 80% (42.37%), 60% (42.52%), and 40% (41.4%) of the field capacity.

Irrigation causes significant changes in the morphological and biochemical properties of *Ocimum* species [89]. The same researchers reported that while fresh and dry biomass yields increased in irrigated plants, there was no significant change in essential oil yield and composition of main compounds among the treatments. However, a slight increase was observed in camphor, nerol, and *trans*-β-caryophyllene ratios. On the other hand, drought stress increased EO content, polyphenol content, and antioxidant capacity. Moreover, drought stress had a positive effect on 1,8-cineole and eugenol ratios. Morphological and biochemical variations were also detected among basil species. Accordingly, higher biomass and essential oil yield among the species were obtained from *O. basilicum* and *O. × africanum*, respectively.

Plants respond to various abiotic and biotic signals that affect their growth and development. Although the responses vary from plant to plant, the growth and development of medicinal plants, in short, their responses to environmental stresses depend on the genotype. In their study with 10 different fennel genotypes, Poudineh et al. [90] stated that water stress has different effects on different varieties and causes various physiological and biological changes in fennel plants, one of which is the accumulation of reactive oxygen species (ROS) in the cell.

Torun et al. [91] investigated the physiological and biochemical responses of ninety-day-old *Hypericum perforatum* seedlings by exposing them to three weeks of drought. The results revealed that it decreased leaf length, relative water content, osmotic potential, chlorophyll fluorescence, increased lipid peroxidation, hydrogen peroxide, proline content, superoxide dismutase, catalase and glutathione reductase, and decreased peroxidase and ascorbate peroxidase activities.

Basil plants changed the number and size of stomata, depending on the severity of drought they were exposed to for 3 months, and accordingly, partial changes occurred in their phytochemical contents [62]. When drought levels were compared with the control subject, obvious phytochemical changes were observed.

Baudoin et al. [92] applied five different irrigation regimes (severe over-irrigation, moderate over-irrigation, standard irrigation, moderate under-irrigation, and severe under-irrigation) to Oregano (thyme) and rosemary plants, and while Oregano phytochemical ratios increased significantly under moderate under-irrigation, there was no change in rosemary phytochemical ratios. Researchers reported that the fact that the phytochemical ratios of rosemary did not change under drought stress may be due to some morphological characteristics of rosemary, and the plant's ability to keep stomatal opening under tight control throughout the day and its ability to develop tolerance to stress by activating some mechanisms in its leaves.

In a study evaluating three irrigation regimes in ten black cumin (*Nigella sativa* L.) genotypes, water stress increased the activities of carotenoids, proline, total soluble carbohydrates, malondialdehyde, hydrogen peroxide, and catalase and ascorbate peroxidase, but decreased the relative water content and chlorophyll content. These physiological changes varied according to the genotypes [93].

In a separate study, under lysimeter conditions at Shahid Sadoughi's combating desertification research station, Rad et al. [94] investigated the effect of three different water constraints (100%, 70%, and 40% of field capacity) on *Eucalyptus camaldulensis* Dehnh. Results revealed that mild drought stress resulted in increased essential oil yield, water use efficiency, and 1,8-cineole production, but reduced or stopped the production of many other compounds.

In a water stress study of Mexican marigold (*Tagetes minuta* L.) at 100%, 75%, 50%, and 25% of field capacity, growth responses, oxidative stress indicators, and phytochemical variations were recorded in stressed and unstressed plants. Photosynthetic pigments and relative water content decreased in stressed plants, but malondialdehyde, osmolyte compounds, and total phenol contents increased with increasing water limitation. Catalase, guaiacol peroxidase, ascorbate peroxidase, and polyphenol oxidase activities were also increased in stressed *T. minuta* plants, in response to drought stress. Drought stress did not have a significant effect on the essential oil content of *T. minuta*, but the essential oil composition was significantly affected. Drought stress changed the proportions of essential oil components and induced the synthesis of new components, including 1,8-cineole and germacrene D. *T. minuta* can resist water stress up to 75% of its field capacity [26]. Thakur and Thakur [95] tested *Chlorophytum borivilianum*, *Stevia rebaudiana*, *Withania somnifera*, and *Andrographis paniculata* plants under 50% water deficit and different stress periods and showed different potentials in terms of growth, yield, and physiological characteristics. The negative effect of stress on growth, yield, photosynthetic rate, canopy temperature decrease, and chlorophyll fluorescence (Fv/Fm) ratio was higher in *Stevia rebaudiana* and *Andrographis paniculata* compared to *Chlorophytum borivilianum* and *Withania somnifera*.

Zhang et al. [96] investigated the effects of different water stress levels on root biomass, secondary metabolites and endogenous hormones in roots, relative water content, and tissue density in leaves of *Stellaria dichotoma* L. var. *lanceolata* Bge. The findings showed that in root biomass, total saponin content first increased and then decreased with increasing drought intensity. The researchers reported that moderate water stress (60–70% or 80–90% field capacity) was suitable for root biomass formation and secondary metabolite accumulation, which were influenced by endogenous hormones and water status.

In anise plant from the Umbelliferae family, drought stress decreased yield and yield components (seed yield, number of branches per plant, number of seeds, number of umbels, and thousand seed weight) and physiological traits such as chlorophyll content, relative water content, quantum efficiency of photosystem II, and cell membrane stability, while increasing leaf temperature. Otherwise, moderate drought severity increased anise essential oil content, while severe drought decreased it [97].

Drought stress is a major environmental constraint that severely limits crop productivity. Water scarcity caused significant decreases in umbels per plant, umbels per umbel, fruits per umbel, 1,000 fruit weight, biological yield, and finally fruit yield, despite increases in fruit essential oil content under moderate drought stress conditions [98]. In their greenhouse study, Soltanbeigi et al. [99] investigated the effects of different irrigation regimes and nutrient sources on the growth parameters and essential oil components of *Salvia officinalis*. Yield decreased significantly as drought stress increased, and essential oil content was observed in moderate and then severe drought stress.

In another study, Mirniyam et al. [100] investigated the seed yield, essential oil constituents, polyphenolic composition, and antioxidant capacity of ajowan (*Trachyspermum ammi* L.) populations under three (normal, moderate, and heavy) irrigation regimes. The results revealed that both essential oil and seed yield showed significant decreases as a result of water stress, while total phenolic and flavonoid contents increased under drought stress treatment.

An effective method to solve the water deficit problems in arid regions, which have increased in recent years as a result of global climate change, is the development of drought-resistant species. Shams et al. [101] conducted a study to develop drought-tolerant ecotypes in *Lallemantia royleana* (Benth.) plants collected from Kalat in Khorasan Razavi province, Zakheh in Kurdistan province, Kondor in Alborz province, and Jupar in Kerman province. Their studies revealed that drought-tolerant ecotypes produced greater dry matter and seed yields under drought conditions. Relative water content, photosynthetic pigment content, seed yield, seed oil amount, and omega-6 fatty acid contents decreased under drought conditions in all ecotypes, while ascorbate peroxidase, catalase, superoxide dismutase and peroxidase activities, and phenol and proline amounts increased. Tavosi et al. [102] they examined the effect of plant characteristics of coneflower (*Echinaceae purpurea*) in drought conditions. Their findings showed that drought stress caused a significant decrease in the growth characteristics of different coneflower, chlorophyll a, carotenoid, and chlorophyll b content. In addition, severe drought stress (40% field capacity) caused a significant decrease in the phytochemical compounds of coneflower; secondary metabolites were affected not only by genetics but also by changing environmental factors.

Antioxidant activity of *Cuminum cyminum* L. seeds, one of the most common aromatic plants of Mediterranean cuisine, increased under dry conditions. While the essential oil content increased at moderate drought severity, it decreased as the drought severity increased, and total phenol content also increased under drought conditions. Moderate drought improved the number of umbels per plant and the number of umbels per umbel and seed yield of cumin seeds compared to normal conditions, but severe drought reduced it. This showed that cumin plant is moderately resistant to drought [103].

Leaf area, and dry and fresh leaf weight were significantly decreased in *Hibiscus esculentus* L under different irrigation regimes. On the other hand, protein content decreased as a result of drought-affecting protein biosynthesis and degradation. Application of salicylic acid and ascorbic acid to plants under drought stress alleviated the effects of stress [104]. Protein and sugar content of *Satureja hortensis* grown under three different irrigation regimes were negatively affected. Drought stress affected protein biosynthesis, decreasing the amount of protein and sugar content due to the photosynthetic process [105]. Similar to *Hibiscus esculentus*, *Satureja hortensis* also alleviated the negative effects of drought. Antioxidant enzyme activities, essential oil yield, and abscisic acid content of hyssop (*Agastache foeniculum* [Pursh] Kuntze) were found to be high under drought conditions [106].

In *Aloe vera* (L.) Burm.f., under severe drought stress, the impairment of the ability of leaves to synthesize assimilates caused growth suppression, while mild drought stress increased total phenolic and flavonoid content. Increasing leaf thickness, leaf biomass, and gel production of the plant associated with mild drought severity increased. It also increased the photochemical activity in the leaves and changed the amount of all secondary metabolites of vanillic acid produced. Mild water restriction can be applied for secondary metabolite productivity and for better growth of aloe plant [107].

Increased water stress in *Chrysanthemum morifolium* caused an increase in phenolic compounds such as chlorogenic acid, rutin, ferulic acid, quercetin, apigenin, and luteolin. Investigating the expression of genes that play a role in the formation of these metabolites under drought conditions and understanding the accumulation mechanism of polyphenols against water stress may create new perspectives [108].

On the other hand, Mustafavi et al. [109] reported that many of the biochemical properties of the valerian plant were significantly affected by water stress. The potassium, zinc, and iron contents of the leaves increased as the amount of available water decreased to 70%, and the amount of these elements decreased as the level of drought increased further. Interestingly, while the aboveground biomass and root biomass of the valerian plant decreased with drought, its essential oil content increased. The fact that belowground organ development and essential oil production are affected differently by drought levels requires caution in irrigation in production.

The less studied *S. dolomitic* species of sage plant has gained importance due to its antiplasmodial and anti-inflammatory properties. Moderate and severe drought increases the production of sesquiterpenes, an important class of terpenoids for their intended use, which has led to the application of controlled drought in the production of secondary metabolites of this plant [110].

Drought and/or heat stress induced the accumulation of proline, sugars, glycine betaine, and sugar alcohols (osmolites), including inositol and mannitol, in *M. piperita* and *C. roseus* plants, while total phenol, flavonoid, and saponin contents decreased in response to drought and/or heat stress, but the levels of other secondary metabolites (including tannins, terpenoids, and alkaloids) increased under stress in both plants. Researchers have emphasized that the application of abiotic stress (drought and/or heat stress) could be a strategy to increase the content of therapeutic secondary metabolites of these plants [111].

Nanoparticles (NP) and growth regulators are increasingly being used to reduce the negative effects of drought stress. In coriander (*Coriandrum sativum* L.) plants exposed to drought stress, chlorophyll content decreases, whereas total soluble sugar, superoxide dismutase, and peroxidase activities increase [112]. When coriander plants were sprayed with salicylic acid and silicon-NPs to reduce the negative effects of drought, increases in chlorophyll content, total soluble sugar, and activity of antioxidant enzymes occurred. Moderate drought significantly increased total phenolic content and total flavonoid content, essential oil content, and essential oil yield with Si-NPs. Foliar application of silicon nanoparticles was determined to be more effective than salicylic acid for improving the antioxidant potential and EO efficiency of coriander plant. Similarly, Mahmoud et al. [113] reported that the application of silicon, zinc, and zeolite nanoparticles only positively affected the morphological, physiological, and biochemical properties of coriander plant under drought stress. In a study investigating the combined effects of drought stress and nanosilicon application on the morphological traits and essential oil content and composition of hemp (*Cannabis sativa* L.), maximum plant height, number of nodes, and number of flowering branches were recorded in 1.5 mM nanosilicon and 100% field capacity application, while the lowest fresh and dry above-ground biomass was recorded in severe drought stress (40% field capacity [114]. Mild water stress (80% field capacity) and foliar application of 1.5 mM nanosilicon provided the maximum essential oil content, while the highest cannabidiol content in essential oil was detected in severe water stress (40% field capacity) and 0.5 mM nanosilicon application. The findings showed that nanosilicon application improved the morphological characteristics of the cannabis plant and changed its biochemical content and components under dry conditions.

In cichory (*Cichorium intybus*), root growth and cumulative inulin yield decreases as drought duration increases [115]. The percentage of total inulin in roots increased under mild drought stress and decreased under severe drought stress. Bat et al. [116] stated that drought stress decreased the leaf area, relative water content in leaf tissues, and membrane durability index of echinacea (*Echinacea purpurea* L.) plant, and increased malondialdehyde level and ion leakage in leaf tissues, while it did not affect the leaf chlorophyll ratio.

It was determined that water stress and temperature increase negatively affected seed production in *Fagopyrum tataricum* [117]. The use of mycorrhiza and vermicompost is recommended under stress conditions. Mycorrhiza, applied to buckwheat under different stress conditions, was effective on phytochemicals, while worm compost increased aboveground biomass and seed yield.

In chamomile (*Matricaria recutita* L.), drought conditions caused a decrease in plant height, flower yield, shoot weight, and apigenin content, but had no significant effect on oil content or oil composition, maintaining the potential for biomass production. Despite the decrease in the agronomic properties of chamomile, the phytochemical properties of the plant did not change, indicating that chamomile is a moderately drought-resistant medicinal plant [118]. Shoot fresh and dry weight, root fresh weight, and shoot length of rosemary (*Rosmarinus officinalis*) plant did not show any difference under drought stress conditions (75% field capacity) compared to normal conditions (100% field capacity) [119]. On the other hand, under drought stress conditions at 75% of field capacity, root length increased, and root dry weight, leaf area, and leaf number decreased significantly under drought conditions. In contrast to these changes in roots and leaves, quercetin, trans-ferulic acid, hesperidin, eugenol, hesperetin, and rosmarinic acid amounts increased under drought stress at 25% field capacity. Kharazi and Asgharzadeh [120] reported that in *Nigella sativa* L., plant growth traits decreased with increasing drought severity, but foliar salicylic acid application alleviated the adverse effects of drought stress.

In *Rosa damascena* Herrm., grown in drought conditions, flower yield decreased, and irrigation regime significantly affected essential oil yield and some components in essential oil. Drought stress increased the amount of citronellol and geraniol in essential oil, and decreased the amount of nonadecane, eicosane, and heneicosan [121]. This situation proves that the components in the essential oil of *R. damascena* can be changed and managed by water stress. The response of Damascus rose to drought and the mechanisms that mediate this response are unknown. In a study conducted by water-restricted *R. damascena*, it was determined that water stress significantly reduced the fresh and dry weights of the plant and all photosynthetic parameters, except leaf temperature [122]. Apoplastic water fraction did not change significantly in response to water stress. *R. damascena* underwent an osmotic adjustment in response to water stress, resulting from active accumulation of soluble carbohydrates and, to a lesser extent, proline under mild stress and tissue dehydration (passive osmotic adjustment) under severe stress. Farahani et al. [123] showed that the quality and quantity of *Rosa damascena* could be increased by foliar application of potassium silicate under water deficit stress equal to 50% and 25% of plant water requirement. Plant biomass of *Stevia rebaudiana*, an economically important medicinal plant, decreased after drought treatments [124]. The photosynthetic properties decreased by drought included intercellular CO_2, net photosynthesis, chlorophylls, carotenoids, and water use efficiency, followed by the decrease in carbohydrates. Under water stress, reactive oxygen species accumulated and hydrogen peroxide production increased in plants.

Drought stress also caused the accumulation of proline and glycine betaine. The results showed that carbohydrates and plant growth were reduced. These results indicate that water deficiency is an important criterion for *Stevia* plants used in vegetative parts.

In *Lycoris aurea*, grown under drought conditions, increased plant growth is restricted, causing an increase in the fresh weight of the bulb, a decrease in the chlorophyll content, and a decrease in the maximum net photosynthesis rate of the leaves [125]. In contrast, galanthamine and lycorine alkaloids in the bulb increased. Mild water stress increased the galanthamine and lycorine contents to the maximum level. These results indicate that *L. aurea* has a water requirement during vegetative growth periods, and plants should be subjected to mild water restriction in order to increase their alkaloid content in advanced growth stages.

Interestingly, in the saffron plant (*Crocus sativus* L.), one of the most expensive spice plants in the world, the increase in the severity of drought stress caused an increase in secondary metabolites (crocin, picrocrocin, and safranal) [126–128]. In contrast, the dry weight of corm decreased due to drought stress. Methyl jasmonate and auxin applications were reported to have the potential to reduce the negative effects of drought stress. However, more research is needed on this subject to understand these effects comprehensively.

5.7 Conclusions

Drought affects the morphological, physiological, and biochemical characteristics of medicinal plants. Medicinal plants respond to drought not only by decreasing yields but also by changing the amount and content of secondary metabolites. This situation causes medicinal plants to lose their economic importance. In drought or water deficiency, signals are transmitted from roots to leaves via xylem vascular bundles and stomata partially close. This causes gas exchange in the cell to slow down, free radicals such as hydrogen peroxide and superoxide, and increase reactive oxygen species. As a result, the biosynthesis of secondary compounds (such as volatile oil content, phenols, terpenoids, alkaloids, glycosides, and flavonoids) decreases, cells and tissues are damaged and, depending on the severity of drought, the death of the plant occurs. Therefore, the development of drought-resistant varieties should be the primary goal in breeding programs for medicinal, aromatic, and spice plants, whose secondary metabolites are of economic importance.

References

[1]   Tiryaki, İ. (2018). Adaptation mechanisms of some field plants against to salt stress. KSU Journal of Natural Sciences, 21(5), 800–808.

[2]   Blum, A. and Jordan, W. R. (1985). Breeding crop varieties for stress environments. Critical Reviews in Plant Sciences, 2(3), 199–238.

[3]   Farooq, M., Hussain, M., Wahid, A. and Siddique, K. H. M. (2012). Drought stress in plants: An overview. Plant Responses to Drought Stress: From Morphological to Molecular Features, 1–33.

[4]   Deblonde, P. M. K. and Ledent, J.-F. (2001). Effects of moderate drought conditions on green leaf number, stem height, leaf length and tuber yield of potato cultivars. European Journal of Agronomy, 14, 31–41.

[5]   Bettaieb, I., Zakhama, N., Wannes, W. A., Kchouk, M. and Marzouk, B. (2009). Water deficit effects on Salvia officinalis fatty acids and essential oils composition. Scientia Horticulturae, 120(2), 271–275.

[6]   Nasir, M. W. and Toth, Z. (2022). Effect of drought stress on potato production: A review. Agronomy, 12, 635.

[7]   Kim, Y., Chung, Y. S., Lee, E., Tripathi, P., Heo, S. and Kim, K. H. (2020). Root response to drought stress in rice (Oryza sativa L.). International Journal of Molecular Sciences, 21(4), 1513.

[8]   Mahajan, S. and Tuteja, N. (2005). Cold, salinity and drought stresses: An overview. Archives of Biochemistry & Biophysics, 444(2), 139–158.

[9]   Elena, M., Katarína, K., Ivana, V. and Zuzana, K. (2019). Responses of medicinal plants to abiotic stresses. In: Handbook of Plant Crop Stress, 4th Edition, CRC Press, Boca Raton, Florida, USA.

[10] Hossain, A., Pamanick, B., Venugopalan, V. K., Ibrahimova, U., Rahman, M. A., Siyal, A. L., Maitra, S., Chatteriee, S. and Aftab, T. (2022). Emerging roles of plant growth regulators for plants adaptation to abiotic stress–induced oxidative stress. Emerging Plant Growth Regulators in Agriculture Academic Press, 1, 1–72.

[11] De Abreu, I. N. and Mazzafera, P. (2005). Effect of water and temperature stress on the content of active constituents of Hypericum brasiliense Choisy. Plant Physiology and Biochemistry, 43(3), 241–248.

[12] Jaleel, C. A., Gopi, R., Sankar, B., Gomathinayagam, M. and Panneerselvam, R. (2008). Differential responses in water use efficiency in two varieties of Catharanthus roseus under drought stress. Comptes Rendus Biologies, 331(1), 42–47.

[13] Zhu, Z., Liang, Z., Han, R. and Wang, X. (2009). Impact of fertilization on drought response in the medicinal herb Bupleurum chinense D.C.: Growth and saikosaponin production. Industrial Crops and Products, 29(2–3), 629–633.

[14] Singh-Sangwan, N., Farooqi, A. H. A., Shabih, F. and Sangwan, R. S. (2001). Regulation of essential oil production in plants. Plant Growth Regulators, 34, 3–2.

[15] Jaleel, C. A., Manivannan, P., Kishorekumar, A., Sankar, B., Gopi, R., Somasundaram, R. and Panneerselvam, R. (2007). Alterations in osmoregulation, antioxidant enzymes and indole alkaloid levels in Catharanthus roseus exposed to water deficit. Colloids and Surfaces B: Biointerfaces, 59(2), 150–157.

[16] Tátrai, Z. A., Sanoubar, R., Pluhár, Z., Mancarella, S., Orsini, F. and Gianquinto, G. (2016). Morphological and physiological plant responses to drought stress in Thymus citriodorus. International Journal of Agronomy, 2016(1), 4165750.

[17] Yadav, B., Jogawat, A., Rahman, M. S. and Narayan, O. P. (2021). Secondary metabolites in the drought stress tolerance of crop plants: A review. Gene Reports, 23, 101040.

[18] Rouphael, Y., Cardarelli, M., Schwarz, D., Franken, P. and Colla, G. (2012). Effects of drought on nutrient uptake and assimilation in vegetable crops. In: Aroca, R. (editor) Plant Responses to Drought Stress, Springer, Berlin, Heidelberg, 171–195.

[19] Pirasteh-Anosheh, H., Saed-Moucheshi, A., Pakniyat, H. and Pessarakli, M. (2016). Stomatal responses to drought stress. Water Stress and Crop Plants: A Sustainable Approach, 1, 2440.

[20] Hund, A., Ruta, N. and Liedgens, M. (2009). Rooting depth and water use efficiency of tropical maize inbred lines, differing in drought tolerance. Plant & Soil, 318, 311–325.

[21] Singh-Sangwan, N., Farooqi, A. H. A. and Singh-Sangwan, R. (1994). Effect of drought stress on growth and essential oil metabolism in lemon grasses. New Phytologist, 128(1), 173–179.

[22] Asadi, S., Lebaschy, M. H., Khourgami, A. and Rad, A. H. S. (2012). Effect of drought stress on the morphology of three *Salvia sclarea* populations. Annals of Biological Research, 3(9), 4503–4507.

[23] Lebaschi, M. H. and Sharifi Ashurabadi, A. (2004). Growth indices of some medicinal plants under different water stresses. Iranian Journal of Medicinal and Aromatic Plants Research, 20, 249–261.

[24] Letchamo, W., Marquard, R., Holzl, J. and Gosselin, A. (1994). Effects of water supply and light intensity on growth and essential oil of two *Thymus vulgaris* selections. Angewandte Botanik, 68, 83–88.

[25] Misra, A. and Srivastava, N. K. (2000). Influence of water stress on Japanese mint. Journal of Herbs, Spices, and Medicinal Plants, 7, 51–58.

[26] Babaei, K., Moghaddam, M., Farhadi, N. and Pirbalouti, A. G. (2021). Morphological, physiological and phytochemical responses of Mexican marigold (*Tagetes minuta* L.) to drought stress. Scientia Horticulturae, 284, 110116.

[27] Hassani, A. and Omidbeigi, R. (2002). The effect of water stress on some morphological, physiological and metabolic characteristics of basil. Journal of Agricultural Science, 12, 47–59.

[28] Farahani, H. A., Valadabadi, S. A., Daneshian, J., Shiranirad, A. H. and Khalvati, M. A. (2009). Medicinal and aromatic plants farming under drought conditions. Journal of Horticulture and Forestry, 1(6), 086–092.

[29] Aliabadi, F. H., Lebaschi, M. H., Shiranirad, A. H., Valadabadi, A. R. and Daneshian, J. (2008). Effects of arbuscular mycorrhizal fungi, different levels of phosphorus and drought stress on water use efficiency, relative water content and proline accumulation rate of coriander (*Coriandrum sativum* L.). Journal of Medicinal Plants Research, 2(6), 125–131.

[30] Mohamed, M. A. H., Harris, P. J. C., Henderson, J. and Senatore, F. (2002). Effect of drought stress on the yield and composition of volatile oils of drought tolerant and non-drought-tolerant clones of *Tagetes minuta*. Planta Medica, 68(5), 472–474.

[31] Petropoulos, S. A., Daferera, D., Polissiou, M. G. and Passam, H. C. (2008). The effect of water deficit stress on the growth, yield and composition of essential oils of parsley. Scientia Horticulturae, 115(4), 393–397.

[32] Miao, Y., Zhu, Z., Guo, Q., Ma, H. and Zhu, L. (2015). Alternate wetting and drying irrigation-mediated changes in the growth, photosynthesis and yield of the medicinal plant *Tulipa edulis*. Industrial Crops and Products, 66, 81–88.

[33] Khorasaninejad, S., Mousavi, A., Soltanloo, H., Hemmati, K. and Khalighi, A. (2011). The effect of drought stress on growth parameters, essential oil yield and constituent of peppermint (*Mentha piperita* L.). Journal of Medicinal Plants Research, 5(22), 5360–5365.

[34] Pinheiro, C. and Chaves, M. M. (2011). Photosynthesis and drought: Can we make metabolic connections from available data?. J Experimental Botany, 62, 869–882.

[35] Moursi, Y. S., Thabet, S. G., Amro, A., Dawood, M. F., Baenziger, P. S. and Sallam, A. (2020). Detailed genetic analysis for identifying QTLs associated with drought tolerance at seed germination and seedling stages in barley. Plants, 9(11), 1425.

[36] Akbari, S., Kafi, M. and Rezvan Beidokhti, S. (2017). Effect of drought stress on growth and morphological characteristics of two garlic (*Allium sativum* L.) ecotypes in different planting densities. Journal of Agroecology, 9(2), 559–574.

[37] Abobatta, W. F. (2020). Plant responses and tolerance to combined salt and drought stress. In: Hasanuzzaman, M. & Tanveer, M. Salt and Drought Stress Tolerance in Plants: Signaling Networks and Adaptive Mechanisms, Springer Nature, Switzerland AG, 17–52.

[38] Lima, A. L. S., DaMatta, F. M., Pinheiro, H. A., Totola, M. R. and Loureiro, M. E. (2002). Photochemical responses and oxidative stress in two clones of *Coffea canephora* under water deficit conditions. Environmental and Experimental Botany, 47, 239–247.

[39] Pinheiro, H. A., DaMatta, F. M., Chaves, A. R. M., Fontes, E. P. B. and Loureiro, M. E. (2004). Drought tolerance in relation to protection against oxidative stress in clones of *Coffea canephora* subjected to long-term drought. Plant Science, 167, 1307–1314.

[40] Ramachandra Reddy, A., Chaitanya, K. V., Jutur, P. P. and Sumithra, K. (2004). Differential antioxidative responses to water stress among five mulberry (*Morus alba* L.) cultivars. Environmental and Experimental Botany, 52, 33–42.

[41] Safaei Chaeikara, S., Marzvan, S., Jahangirzadeh Khiavi, S. and Rahimi, M. (2020). Changes in growth, biochemical, and chemical characteristics and alteration of the antioxidant defense system in the leaves of tea clones (*Camellia sinensis* L.) under drought stress. Scientia Horticulturae, 265, 109257.

[42] Pirzad, A., Shakiba, M. R., Zehtab-Salmasi, S., Mohammadi, S. A., Darvishzadeh, R. and Samadi, A. (2011). Effect of water stress on leaf relative water content, chlorophyll, proline and soluble carbohydrates in *Matricaria chamomilla* L. Journal of Medicinal Plants Research, 5, 2483–2488.

[43] Hosseini, M. S., Samsampour, D., Ebrahimi, M., Abadía, J. and Khanahmadi, M. (2018). Effect of drought stress on growth parameters, osmolyte contents, antioxidant enzymes and glycyrrhizin synthesis in licorice (*Glycyrrhiza glabra* L.) grown in the field. Phytochemistry, 156, 124–134.

[44] Valentovic, P., Luxova, M., Kolarovic, L. and Gasparikova, O. (2006). Effect of osmotic stress on compatible solutes content, membrane stability and water relations in two maize cultivars. Plant Soil Environment, 52, 184.

[45] Kalefetoğlu, T. and Ekmekçi, Y. (2005). The effects of drought on plants and tolerance mechanisms. Gazi University Journal of Science, 18(4), 723–740.

[46] Atkinson, N. J. and Urwin, P. E. (2012). The interaction of plant biotic and abiotic stresses: From genes to the field. Journal of Experimental Botany, 63(10), 3523–3543.

[47] Seleiman, M. F., Al-Suhaibani, N., Ali, N., Akmal, M., Alotaibi, M., Refay, Y., Dindaroglu, T., Abdul-Wajid, H. H. and Battaglia, M. L. (2021). Drought stress impacts on plants and different approaches to alleviate its adverse effects. Plants, 10, 259.

[48] Moradi, P., Ford-Lloyd, B. and Pritchard, J. (2014). Plant-water responses of different medicinal plant thyme (*Thymus* spp.) species to drought stress condition. Australian. Journal of Crop Science, 8(5), 666–673.

[49] Gulen, H. and Eris, A. (2004). Effect of heat stress on peroxidase activity and total protein content in strawberry plants. Plant Science, 166(3), 739–744.

[50] Lakušić, B., Ristić, M., Slavkovska, V., Stojanović, D. and Lakušić, D. (2013). Variations in essential oil yields and compositions of *Salvia officinalis* (Lamiaceae) at different developmental stages. Botanica Serbica, 37(2), 127–139.

[51] Elmas, S. (2021). Responses of *Salvia officinalis* (common sage) to some abiotic stress factors. Journal of the Institute of Science and Technology, 11(2), 943–959.

[52] Chung, I. M., Kim, J. J., Lim, J. D., Yu, C. Y., Kim, S. H. and Hahn, S. J. (2006). Comparison of resveratrol, SOD activity, phenolic compounds and free amino acids in *Rehmannia glutinosa* under temperature and water stress. Environmental and Experimental Botany, 56(1), 44–53.

[53] Liu, H., Wang, X., Wang, D., Zou, Z. and Liang, Z. (2011). Effect of drought stress on growth and accumulation of active constituents in *Salvia miltiorrhiza* Bunge. Industrial Crops and Products, 33(1), 84–88.

[54] Umebese, C. E., Okunade, K. I. and Orotope, O. M. (2012). Impact of water deficit stress on growth and alkaloid content of organs of *Spigelia anthelmia* (L.). Ife Journal of Science, 14(2), 357–362.

[55] Ghorbanpour, M., Hatami, M. and Khavazi, K. (2013). Role of plant growth promoting rhizobacteria on antioxidant enzyme activities and tropane alkaloid production of *Hyoscyamus niger* under water deficit stress. TurkishJournal of Biology, 37(3), 350–360.

[56] Umebese, C. E. and Falana, F. D. (2013). Growth, phytochemicals and antifungal activity of Bryophyllum pinnatum L. subjected to water deficit stress. African Journal of Biotechnology, 12(47), 6599–6604.

[57] Cheng, L., Han, M., Yang, L., Yang, L., Sun, Z. and Zhang, T. (2018). Changes in the physiological characteristics and baicalin biosynthesis metabolism of *Scutellaria baicalensis* Georgi under drought stress. Industrial Crops and Products, 122, 473–482.

[58] Abd Elbar, O. H., Farag, R. E. and Shehata, S. A. (2019). Effect of putrescine application on some growth, biochemical and anatomical characteristics of *Thymus vulgaris* L. under drought stress. Annals of Agricultural Research, 64, 129–137.

[59] Bistgani, Z. E., Barker, A. V. and Hashemi, M. (2024). Physiology of medicinal and aromatic plants under drought stress. The Crop Journal, 12, 330–339.

[60] Khaleghnezhad, V., Yousefi, A. R., Tavakoli, A., Farajmand, B. and Mastinu, A. (2021). Concentrations-dependent effect of exogenous abscisic acid on photosynthesis, growth and phenolic content of *Dracocephalum moldavica* L. under drought stress. Planta, 253(6), 1–18.

[61] Khodadadi, F., Ahmadi, F. S., Talebi, M., Moshtaghi, N., Matkowski, A., Szumny, A. and Rahimmalek, M. (2022). Essential oil composition, physiological and morphological variation in *Salvia abrotanoides* and *S. yangii* under drought stress and chitosan treatments. Industrial Crops and Products, 187, 115429.

[62] Kılıç, S. and Kaya, H. (2023). The effect of drought on micro-morphological structures and secondary metabolite content of basil (*Ocimum basilicum* L.). Turkish Journal of Forestry, 24(1), 18–24.

[63] García-Caparrós, P., Romero, M. J., Llanderal, A., Cermeño, P., Lao, M. T. and Segura, M. L. (2019). Effects of drought stress on biomass, essential oil content, nutritional parameters, and costs of production in six Lamiaceae species. Water, 11(3), 573.

[64] Abbaszadeh, B., Farahani, H. A. and Morteza, E. (2009). Effects of irrigation levels on essential oil of balm (*Melissa officinalis* L.). American-Eurasian Journal of Sustainable Agriculture, 3(1), 53–56.

[65] Bahreininejad, B., Razmjou, J. and Mirza, M. (2013). Influence of water stress on morpho-physiological and phytochemical traits in *Thymus daenensis*. International Journal of Plant Production, 7(1), 151–166.

[66] Azizi, A., Yan, F. and Honermeier, B. (2009). Herbage yield, essential oil content and composition of three oregano (*Origanum vulgare* L.) populations as affected by soil moisture regimes and nitrogen supply. Industrial Crops and Products, 29(2–3), 554–561.

[67] Mulugeta, S. M. and Radácsi, P. (2022). Influence of drought stress on growth and essential oil yield of *Ocimum* species. Horticulturae, 8, 175.

[68] Farhoudi, R. (2013). Effect of drought stress on chemical constituents, photosynthesis and antioxidant properties of *Rosmarinus officinalis* essential oil. Journal of Medicinal Plants and By-products, 2(1), 17–22.

[69] Karimzadeh Asl, K., Ghorbanpour, M., Marefatzadeh Khameneh, M. and Hatami, M. (2018). Influence of drought stress, biofertilizers and zeolite on morphological traits and essential oil constituents in *Dracocephalum moldavica* L. Journal of Medicinal Plants, 17(67), 91–112.

[70] Bettaieb, I., Hamrouni-Sellami, I., Bourgou, S., Limam, F. and Marzouk, B. (2011). Drought effects on polyphenol composition and antioxidant activities in aerial parts of *Salvia officinalis* L. Acta Physiologiae Plantarum – Acta Physiol Plant, 33(4), 1103–1111.

[71]   Abdi, G., Shokrpour, M. and Salami, S. A. (2019). Essential oil composition at different plant growth development of peppermint (*Mentha x piperita* L.) under water deficit stress. Journal of Essential Oil Bearing Plants, 22(2), 431–440.

[72]   Laribi, B., Bettaieb, I., Kouki, K., Sahli, A., Mougou, A. and Marzouk, B. (2009). Water deficit effects on caraway (*Carum carvi* L.) growth, essential oil and fatty acid composition. Industrial Crops and Products, 30(3), 372–379.

[73]   Yahyazadeh, M., Meinen, R., Hänsch, R., Abouzeid, S. and Selmar, D. (2018). Impact of drought and salt stress on the biosynthesis of alkaloids in *Chelidonium majus* L. Phytochemistry, 152, 204–212.

[74]   Szabo, B., Tyihak, E., Szabo, L. G. and Botz, L. (2003). Mycotoxin and drought stress induced change of alkaloid content of *Papaver somniferum* plantlets. Acta Botanica Hungarica, 45(3/4), 409–417.

[75]   Shil, S. and Dewanjee, S. (2022). Impact of drought stress signals on growth and secondary metabolites (SMs) in medicinal plants. The Journal of Phytopharmacology, 11(5), 371–376.

[76]   Bouchereau, A., Clossais-Besnard, N., Bensaoud, A., Leport, L. and Renard, M. (1996). Water stress effects on rapeseed quality. European Journal of Agronomy, 5, 19–30.

[77]   Zhang, X. X., Li, C. J. and Nan, Z. B. (2011). Effect of salt and drought stress on alkaloid production in endophyte- infected drunken horse grass (*Achnatherum inebrians*). Biochemistry and Systematics Ecology, 39(4), 476.

[78]   Keshavarz Afshar, R., Chaichi, M. R., Ansari Jovini, M., Jahanzad, E. and Hashemi, M. (2015). Accumulation of silymarin in milk thistle seeds under drought stress. Planta, 242, 539–543.

[79]   Yeloojeh, K. A., Saeidi, G. and Sabzalian, M. R. (2020). Drought stress improves the composition of secondary metabolites in safflower flower at the expense of reduction in seed yield and oil content. Industrial Crops and Products, 154, 112496.

[80]   Seleiman, M. F., Almutairi, K. F., Alotaibi, M., Shami, A., Alhammad, B. A. and Battaglia, M. L. (2020). Nano fertilization as an emerging fertilization technique: Why modern agriculture can benefit from its use?. Plants, 10, 2.

[81]   Punetha, A., Kumar, D., Suryavanshi, P., Padalıa, R. and Kt, V. (2022). Environmental abiotic stress and secondary metabolites production in medicinal plants: A review. Journal of Agricultural Sciences, 28(3), 351–362.

[82]   Ghavam, M. (2019). Effect of silver nanoparticles on tolerance to drought stress in *Thymus daenensis* Celak and *Thymus vulgaris* L. in germination and early growth stages. Environmental Stresses in Crop Sciences, 12(2), 555–566.

[83]   Karamian, R., Ghasemlou, F. and Amiri, H. (2020). Physiological evaluation of drought stress tolerance and recovery in *Verbascum sinuatum* plants treated with methyl jasmonate, salicylic acid and titanium dioxide nanoparticles. Plant Biosystems-An International Journal Dealing with All Aspects of Plant Biology, 154(3), 277–287.

[84]   Mohammadi, H., Esmailpour, M. and Gheranpaye, A. (2016). Effects of TiO_2 nanoparticles and water-deficit stress on morpho-physiological characteristics of dragonhead (*Dracocephalum moldavica* L.) plants. Acta Agriculturae Slovenicais, 107, 385396.

[85]   Kleinwächter, M. and Selmar, D. (2015). New insights explain that drought stress enhances the quality of spice and medicinal plants: Potential applications. Agronomy for Sustainable Development, 35, 121–131.

[86]   Nemeth-Zambori, E., Pluhar, Z., Szabo, K., Malekzadeh, M., Radácsi, P., Inotai, K., Komáromi, B. and Seidler-Lozykowska, K. (2016). Effect of water supply on growth and polyphenols of lemon balm (*Melissa officinalis* L.) and thyme (*Thymus vulgaris* L.). Acta Biologica Hungarica, 67, 64–74.

[87]   Lotfi, M., Abbaszadeh, B. and Mirza, M. (2014). The effect of drought stress on morphology, proline content and soluble carbohydrates of tarragon (*Artemisia dracunculus* L.). Iranian Journal of Medicinal and Aromatic Plants Research, 30(1), 19–29.

[88]   Sarajuoghi, M., Abbaszadeh, B. and Ardakani, M. R. (2014). Investigation morphological and physiological response of *Thymus vulgaris* L. to drought stress. Journal of Biodiversity and Environmental Sciences, 5(2), 486–492.

[89]   Mulugeta, S. M., Gosztola, B. and Radácsi, P. (2022). Morphological and biochemical responses of selected *Ocimum* species under drought. Herba Polonica, 68(4), 1–10.

[90]   Poudineh, Z., Fakheri, B. A., Sirosmehr, A. R. and Shojaei, S. (2018). Effect of drought stress on the morphology and antioxidant enzymes activity of *Foeniculum vulgare* cultivars in Sistan. Indian Journal of Plant Physiology, 23, 283–292.

[91]   Torun, H., Eroğlu, E., Yalçın, V. and Usta, E. (2021). Physicochemical and antioxidant responses of st. john's wort (*Hypericum perforatum* L.) under drought stress. Düzce University Journal of Science & Technology, 9(1), 40–50.

[92]   Baudoin, D. C., Bush, E., Gauthier, T., Hernandez, A. B. and KirkBallard, H. (2022). Effects of irrigation and drought on growth and essential oil production in *O. vulgare* and *R. officinalis*. American Journal of Plant Sciences, 13(5), 659–667.

[93]   Bayati, P., Karimmojeni, H., Razmjoo, J., Pucci, M., Abate, G., Baldwin, T. C. and Mastinu, A. (2022). Physiological, biochemical, and agronomic trait responses of *Nigella sativa* genotypes to water stress. Horticulturae, 8, 193.

[94]   Rad, M. H., Jaimand, K., Assareh, M. H. and Soltani, M. (2014). Effects of drought stress on the quantity and quality of essential oil and water use efficiency in eucalyptus (*Eucalyptus camaldulensis* Dehnh.). Iranian Journal of Medicinal and Aromatic Plants Research, 29(4), 772–782.

[95]   Thakur, A. and Thakur, C. L. (2018). Evaluation of four medicinal herb species under conditions of water-deficit stress. Indian Journal of Plant Physiology, 23(3), 459–466.

[96]   Zhang, W., Cao, Z., Xie, Z., Lang, D., Zhou, L., Chu, Y. and Zhao, Y. (2017). Effect of water stress on roots biomass and secondary metabolites in the medicinal plant *Stellaria dichotoma* L. var. *lanceolata* Bge. Scientia Horticulturae, 224, 280–285.

[97]   Mehravi, S., Hanifei, M., Gholizadeh, A. and Khodadadi, M. (2023). Water deficit stress changes in physiological, biochemical and antioxidant characteristics of anise (*Pimpinella anisum* L.). Plant Physiology and Biochemistry, 201, 107806.

[98]   Peymaei, M., Sarabi, V. and Hashempour, H. (2024). Improvement of the yield and essential oil of fennel (*Foeniculum vulgare* Mill.) using external proline, uniconazole and methyl jasmonate under drought stress conditions. Scientia Horticulturae, 323, 112488.

[99]   Soltanbeigi, A., Yıldız, M., Dıraman, H., Terzi, H., Sakartepe, E. and Yıldız, E. (2021). Growth responses and essential oil profile of *Salvia officinalis* L. influenced by water deficit and various nutrient sources in the greenhouse. Saudi Journal of Biological Sciences, 28(12), 7327–7335.

[100] Mirniyam, G., Rahimmalek, M., Arzani, A., Matkowski, A., Gharibi, S. and Szumny, A. (2022). Changes in essential oil composition, polyphenolic compounds and antioxidant capacity of ajowan (*Trachyspermum ammi* L.) populations in response to water deficit. Foods, 11(19), 3084.

[101] Shams, H., Omidi, H. and Sahandi, M. S. (2022). The impact of phytochemical, morpho-physiological, and biochemical changes of *Lallemantia royleana* (Benth.) on drought tolerance. Plant Production Science, 25(4), 440–457.

[102] Tavosi, R., Sayyari, M. and Azizi, A. (2024). Impact of drought stress on some growth and phytochemical characteristics of the coneflower (*Echinacea purpurea* L.). Environmental Stresses in Crop Sciences, 17(3), 619–637.

[103] Rebey, I. B., Jabri-Karoui, I., Hamrouni-Sellami, I., Bourgou, S., Limam, F. and Marzouk, B. (2012). Effect of drought on the biochemical composition and antioxidant activities of cumin (*Cuminum cyminum* L.)seeds. Industrial Crops and Products, 36(1), 238–245.

[104] Amin, B., Mahleghah, G., Mahmood, H. M. R. and Hossein, M. (2009). Evaluation of interaction effect of drought stress with ascorbate and salicylic acid on some of physiological and biochemical parameters in okra (*Hibiscus esculentus* L.). Research Journal of Biological Sciences, 4(4), 380–387.

[105] Yazdanpanah, S., Baghizadeh, A. and Abbassi, F. (2011). The interaction between drought stress and salicylic and ascorbic acids on some biochemical characteristics of *Satureja hortensis*. African Journal of Agricultural Research, 6(4), 798–807.

[106] Saeedfar, S., Negahban, M., Soore, S. and M, M. (2015). The effect of drought stress on the essential oil content and some of the biochemical characteristics of anise hyssop (*Agastache foeniculum* [Pursh] Kuntze). European Journal of Molecular Biotechnology, 8(2), 103–114.

[107] Habibi, G. (2018). Effects of mild and severe drought stress on the biomass, phenolic compounds production and photochemical activity of *Aloe vera* (L.) Burm. F. Acta Agriculturae Slovenica, 111(2), 463–476.

[108] Hodaei, M., Rahimmalek, M., Arzani, A. and Talebi, M. (2018). The effect of water stress on phytochemical accumulation, bioactive compounds and expression of key genes involved in flavonoid biosynthesis in *Chrysanthemum morifolium* L. Industrial Crops and Products, 120, 295–304.

[109] Mustafavi, S. H., Shekari, F. and Maleki, H. H. (2016). Influence of exogenous polyamines on antioxidant defence and essential oil production in valerian (*Valeriana officinalis* L.) plants under drought stress. Acta Agriculturae Slovenica, 107(1), 81–91.

[110] Caser, M., Chitarra, W., D'Angiolillo, F., Perrone, I., Demasi, S., Lovisolo, C. and Scariot, V. (2019). Drought stress adaptation modulates plant secondary metabolite production in *Salvia dolomitica* Codd. Industrial Crops and Products, 129, 85–96.

[111] Alhaithloul, H. A., Soliman, M. H., Ameta, K. L., El-Esawi, M. A. and Elkelish, A. (2019). Changes in ecophysiology, osmolytes, and secondary metabolites of the medicinal plants of *Mentha piperita* and *Catharanthus roseus* subjected to drought and heat stress. Biomolecules, 10(1), 43.

[112] Afshari, M., Pazoki, A. and Sadeghipour, O. (2023). Biochemical changes of coriander (*Coriandrum sativum* L.) plants under drought stress and foliar application of salicylic acid and silicon nanoparticles. Journal of Medicinal Plants and By-products, 12(3), 197–207.

[113] Mahmoud, A. W. M., Rashad, H. M., Esmail, S. E., Alsamadany, H. and Abdeldaym, E. A. (2023). Application of silicon, zinc, and zeolite nanoparticles – A tool to enhance drought stress tolerance in coriander plants for better growth performance and productivity. Plants, 12(15), 2838.

[114] Rezghiyan, A., Esmaeili, H. and Farzaneh, M. (2025). Nanosilicon application changes the morphological attributes and essential oil compositions of hemp (*Cannabis sativa* L.) under water deficit stress. Scientific Reports, 15(1), 3400.

[115] Afzal, S. F., Yar, A. K., Ullah, R. H., Ali, B. G., Ali, J. S., Ahmad, J. S. and Fu, S. (2017). Impact of drought stress on active secondary metabolite production in *Cichorium intybus* roots. Journal of Applied Environmental and Biological Sciences, 7(7), 39–43.

[116] Bat, M., Tunçtürk, R. and Tunçtürk, M. (2019). Effect of drought stress and seaweed applications on some physiological parameters in echinacea (*Echinacea purpurea* L.). KSU Journal of Natural Sciences, 23(1), 99–107.

[117] Aubert, L. and Quinet, M. (2022). Comparison of heat and drought stress responses among twelve Tartary buckwheat (*Fagopyrum tataricum*) varieties. Plants, 11(11), 1517.

[118] Baghalian, K., Abdoshah, S., Khalighi-Sigaroodi, F. and Paknejad, F. (2011). Physiological and phytochemical response to drought stress of German chamomile (*Matricaria recutita* L.). Plant Physiology and Biochemistry, 49(2), 201–207.

[119] Tamadon, L. and Riasat, M. (2021). Effect of drought stress on some morphophysiological characteristics and phenolic compounds of rosemary plant (*Rosmarinus officinalis* L.). Environmental Stresses in Crop Sciences, 14(2), 439–448.

[120] Kharazi, M. E. and Asgharzadeh, A. (2023). Effects of drought stress, salicylic acid, and polyamines on plant growth yield and oil and essential oil content of *Nigella sativa* L. Iranian Journal of Medicinal and Aromatic Plants, 39(2), 237–253.

[121] Sharifi Ashourabadi, E., Tabaei-Aghdaei, S. R., Mirza, M., Nadery, M. and Nadery, B. (2024). Effect of water deficit stress on yield and essential oil components of rosa damascena herrm. Journal of Medicinal Plants and By-Products, 13(2), 285–291.

[122] Al-Yasi, H., Attia, H., Alamer, K., Hassan, F., Ali, E., Elshazly, S. and Hessini, K. (2020). Impact of drought on growth, photosynthesis, osmotic adjustment, and cell wall elasticity in Damask rose. Plant Physiology and Biochemistry, 150, 133–139.

[123] Farahani, H., Sajedi, N., Madani, H., Changizi, M. and Naeini, M. R. (2021). Effect of potassium silicate on water use efficiency, quantitative traits and essential oil yield of damask rose (*Rosa damascena* Miller) under water deficit stress. Iranian Journal of Horticultural Science, 52(1), 171–182.

[124] Hajihashemi, S. and Sofo, A. (2018). The effect of polyethylene glycol-induced drought stress on photosynthesis, carbohydrates and cell membrane in *Stevia rebaudiana* grown in greenhouse. Acta Physiologiae Plantarum, 40(8), 142.

[125] Liang, J., Quan, M., She, C., He, A., Xiang, X. and Cao, F. (2020). Effects of drought stress on growth, photosynthesis and alkaloid accumulation of *Lycoris aurea*. Pakistan Journal of Botany, 52(4), 1137–1142.

[126] Aboueshaghi, R. S., Omidi, H. and Bostani, A. (2023). Assessment of changes in secondary metabolites and growth of saffron under organic fertilizers and drought. Journal of Plant Nutrition, 46(3), 386–400.

[127] Ahmadian, A., Esmaeilian, Y., Tavassoli, A., Fernández-Gálvez, J. and Caballero-Calvo, A. (2024). Application of a superabsorbent hydrogel for improving water productivity and quality of saffron (*Crocus sativus* L.) under water deficit conditions. Scientia Horticulturae, 336, 113411.

[128] Ziaei, S. M., Khashei Siuki, A., Ahmadian, A. and Salarian, A. (2025). Reduction of drought stress effects on saffron (*Crocus sativus* L.) using phytohormones. Journal of Medicinal Plants and By-Products, 14(1), 54–62.

Fatemeh Ahmadi*

Chapter 6
Impact of salinity stress on medicinal and aromatic plant biotechnology

Abstract: Salinity stress is a critical environmental challenge affecting the growth, development, and metabolic processes of medicinal and aromatic plants (MAPs). These plants are highly valued for their secondary metabolites, which have extensive applications in medicine, cosmetics, and the food industry. Salinity disrupts key physiological functions, including photosynthesis, water relations, and nutrient balance, while inducing oxidative stress and altering metabolite production. Despite these challenges, MAPs exhibit remarkable physiological, biochemical, and molecular adaptive mechanisms, such as osmotic adjustment, ion homeostasis, antioxidant defenses, and regulation of stress-responsive genes. This chapter provides a comprehensive analysis of the effects of salinity on MAPs, highlighting changes in growth metrics, photosynthesis, and secondary metabolite production. It delves into molecular mechanisms that enable salt tolerance, such as the SOS pathway, ion transporters, and transcription factors. Furthermore, the chapter explores strategies to enhance salt resilience in MAPs, including the use of plant growth regulators, beneficial microorganisms, genetic engineering, and agronomic practices like mulching and silicon supplementation. The integration of advanced biotechnological tools, such as CRISPR/Cas9 and omics approaches, is also discussed to optimize salt tolerance and metabolite production. By leveraging these insights, sustainable solutions for MAP cultivation in saline environments can be achieved, ensuring the continued economic and therapeutic significance of these plants.

Keywords: antioxidant defenses, CRISPR/Cas9, genetic engineering, plant growth regulators, salt tolerance mechanisms

6.1 Introduction

High soil salt content presents a formidable ecological hurdle, severely impacting crop development and output [1]. This phenomenon results from excessive salt concentrations in the plant's environment, interfering with vital physiological processes. Originally confined to dry regions, this problem has now expanded globally, driven by both natural events and human activities [2, 3]. Researchers gauge soil salinity by

---

*Corresponding author: Fatemeh Ahmadi, School of Agriculture and Environment, University of Western Australia, Crawley, WA 6009, Australia, e-mail: fatemeh.ahmadi@uwa.edu.au

https://doi.org/10.1515/9783111469713-006

measuring the soil solution's electrical conductivity (EC) [4]. The USDA Salinity Laboratory considers soil to be saline when its saturated paste extract (EC_e) exceeds 4 dS/m, though many plants react negatively to lower levels, with field conditions potentially exacerbating these effects [5]. Salt-induced stress manifests through osmotic and ionic mechanisms. The former restricts water uptake, creating drought-like conditions even in wet soils [6]. The latter involves cellular damage from ion overload, particularly Na^+ and Cl^-, disrupting plant metabolic functions. Salinity types include dryland and irrigation-induced, based on salt accumulation methods. Dryland salinity occurs in nonirrigated areas due to natural phenomena like groundwater movements [7]. Irrigation salinity results from repeated use of salt-rich water, leading to salt buildup [8]. Furthermore, salinity is classified as primary or secondary based on its source. Primary salinity evolves naturally through geological processes, while secondary salinity results from human interventions such as land clearing and poor irrigation management [9, 10].

Salt-affected soils are classified based on their electrical conductivity, sodium content, and pH levels, as three primary types: saline, sodic, and saline-sodic soils [11, 12]. These soil conditions significantly challenge plant survival by creating complex physiological stress mechanisms. Plants respond to salinity through sophisticated adaptive strategies at molecular and cellular levels [13]. They develop multiple defense mechanisms to counteract salt stress, including accumulating protective osmolytes, regulating ion homeostasis, enhancing antioxidant defenses, modifying gene expression, and producing stress-responsive hormones like abscisic acid (ABA) [14, 15]. The primary survival strategies involve maintaining water balance, preventing toxic ion accumulation, and protecting cellular structures from oxidative damage. These adaptive responses enable plants to survive and potentially thrive in challenging saline environments [16]. Understanding these intricate plant responses provides crucial insights for developing salt-tolerant crop varieties, ultimately supporting agricultural productivity in regions with challenging soil conditions [17, 18].

6.2 Importance of medicinal and aromatic plants

For centuries, medicinal and aromatic plants (MAPs) have been closely linked to human health and cultural heritage. These plants, celebrated for their diverse bioactive compounds, have served as the cornerstone of traditional medicine and continue to hold significant importance in modern healthcare, pharmaceutical advancements, and a wide range of industrial applications [19]. The significance of MAPs goes beyond their medicinal properties, encompassing economic, ecological, and cultural importance. The use of plants for healing predates written history, with early civilizations, including those in Egypt, China, India, and Greece, developing detailed herbal medicine systems. For instance, the Egyptian Ebers Papyrus, a document from 1550 BCE, describes over 850 plant-based remedies. Similarly, traditional Chinese medicine and Ayurveda have harnessed MAPs for thousands of years, creating complex pharmacopeia and therapeutic approaches [20].

MAPs form a vital component of global biodiversity, with more than 50,000 medicinal species among the estimated 422,000 flowering plants. This biodiversity is essential for ecological stability and represents a vast resource for discovering new medicines. However, growing demand for MAPs has resulted in overharvesting, necessitating robust conservation measures [21]. Conservation strategies include in situ preservation of natural habitats, ex situ conservation in botanical gardens and gene banks, adopting sustainable harvesting techniques, and promoting cultivation to alleviate pressure on wild populations [22].

The therapeutic value of MAPs arises from their phytochemical composition, which includes alkaloids (e.g., morphine from *Papaver somniferum*), glycosides (e.g., digoxin from *Digitalis lanata*), terpenoids (e.g., artemisinin from *Artemisia annua*), phenolics (e.g., curcumin from *Curcuma longa*), and flavonoids (e.g., quercetin from various sources) [23]. These compounds often exhibit synergistic effects, enhancing their medicinal efficacy. For example, garlic (*Allium sativum*) derives its therapeutic properties from organosulfur compounds like allicin, known for its antibacterial, antifungal, and antiviral properties [24].

MAPs are extensively used in managing diverse health conditions. Examples include *Ginkgo biloba* and garlic for cardiovascular health, paclitaxel from Pacific yew and vincristine from Madagascar periwinkle for cancer treatment, bitter melon and fenugreek for diabetes, Echinacea species for respiratory ailments, and St. John's wort for mental health concerns. Additionally, MAPs are a cornerstone of drug discovery, contributing directly or indirectly to about 25% of modern medicines. The process involves studying traditional applications, isolating bioactive components, determining their chemical structures, and conducting preclinical and clinical trials. Notable drugs derived from MAPs include aspirin (from *Salix* species), morphine, quinine (from *Cinchona* species), and artemisinin [25].

Aromatic plants, valued for their ability to produce essential oils, find extensive applications across various fields, including medicine, perfumery, cosmetics, and aromatherapy [26]. These essential oils are renowned for their wide-ranging biological activities, such as the potent antimicrobial properties of tea tree oil (*Melaleuca alternifolia*), the anti-inflammatory effects associated with lavender oil (*Lavandula angustifolia*), and the calming, anxiolytic benefits evidenced in research on lavender oil inhalation [27]. The global essential oil market, which was valued at USD 7.03 billion in 2020, is projected to experience consistent growth due to increasing demand across industries. Similarly, the herbal medicine market, initially valued at USD 83 billion in 2019, is expected to witness exponential growth, with predictions estimating its value at USD 550 billion by 2030 [28]. These markets underscore the economic importance of MAPs, which support agriculture, industrial processing, international trade, and employment, especially in developing nations [29, 30].

Despite their immense potential, MAPs face challenges such as maintaining consistent quality, standardizing cultivation and processing, navigating diverse regulatory frameworks, addressing conservation needs alongside growing demand, and safeguarding intellectual property rights related to traditional knowledge [31]. Future directions for research include refining phytochemical analysis with advanced technologies, employing omics approaches like genomics and metabolomics, utilizing biotechnological methods to produce plant-derived compounds, validating traditional uses through rigorous clinical trials, and developing sustainable cultivation and production practices [32].

The contributions of MAPs to healthcare, the economy, and cultural heritage remain indispensable. Their role in modern drug discovery, the growing demand for natural products, and their potential to address global health issues reinforce their significance [33]. As research progresses, integrating traditional knowledge with modern science will be crucial to ensuring their sustainable use and conservation for future generations [34]. The multidisciplinary importance of MAPs – from traditional medicine and global trade to biodiversity and cultural preservation – requires a holistic approach to their study and sustainable utilization. Advancing our understanding through collaborative efforts promises to unlock even greater potential from these extraordinary plants [35].

6.3 Salinity effect on medicinal plants

Salinity stress severely impacts MAPs, affecting their growth, photosynthesis, and metabolite production [36, 37]. These plants develop complex adaptive mechanisms to survive in saline environments. Studying these responses offers valuable insights for enhancing plant resilience and developing salt-tolerant cultivation methods, showcasing the remarkable adaptability of plants under environmental stress [38].

6.3.1 Effects on growth and development

Salinity stress significantly impairs plant growth, particularly in medicinal species, by disrupting essential physiological processes [39]. High salt levels interfere with chlorophyll production and photosynthetic efficiency, leading to stunted development that is summarized in Table 6.1. Research on plants like moringa demonstrates how salt exposure can dramatically reduce growth metrics and biomass allocation, underscoring the profound impact of salt stress on plant health [37].

Salinity stress also affects plants at the germination stage. Higher salt concentrations often delay or entirely inhibit seed germination, which can significantly reduce the establishment of crops and overall yield [40]. Additionally, salt stress interferes with the flowering process in several MAPs. As flowering is often a key determinant

Table 6.1: Impact of salinity on the growth and development of medicinal plants.

Parameter	Observation	Example species
Seed germination	Reduced germination rate under high salinity levels	*Catharanthus roseus*
Chlorophyll Content and biomass	Decreased chlorophyll concentration and photosynthetic efficiency	*St. John's wort*
Allocation	Shift toward root biomass; reduction in shoot biomass	*Moringa oleifera*
Flowering	Delayed or inhibited flowering due to ionic and osmotic stress	*Trachyspermum ammi*

of reproductive success and medicinal compound production, reduced flowering under salinity stress can severely affect yield, particularly in species that derive medicinal properties from their flowers [41].

6.3.2 Impact on photosynthesis and water relations

Salinity stress disrupts the photosynthetic machinery of MAPs, reducing both the chlorophyll content and photosynthetic efficiency. For example, salinity-stressed *St. John's wort* plants displayed an 18.9% reduction in chlorophyll levels compared to unstressed controls [42]. Furthermore, salinity affects critical parameters related to photosynthetic performance. Reductions in F_v/F_m (maximum quantum yield of PSII), F_v/F_o (maximum primary yield of PSII photochemistry), and PI (performance index) under salinity stress highlight significant losses in photosynthetic capacity, leading to reduced biomass and growth [43].

Salt stress disrupts plant water uptake by altering soil solution dynamics and creating artificial drought conditions, despite water availability [44, 45]. To counteract this osmotic challenge, plants synthesize osmoprotectants like amino acids and carbohydrates, which help maintain cellular equilibrium in saline environments [46]. This adaptive response enables plants to mitigate some of the harmful effects of salt exposure, showcasing their resilience to environmental stressors.

6.3.3 Ionic stress and nutrient imbalance

Salt stress in plants leads to an overaccumulation of sodium and chloride ions, disrupting key cellular processes in Figure 6.1. Studies on moringa have shown increased Na^+ and Cl^- levels under saline conditions, accompanied by reduced potassium uptake [47, 48]. This ionic imbalance interferes with the absorption of essential nutrients like potassium, calcium, and magnesium [49]. Consequently, plants experience impaired growth, metabolic disturbances, and decreased productivity due to the combined effects of ion toxicity and nutrient deficiencies [50].

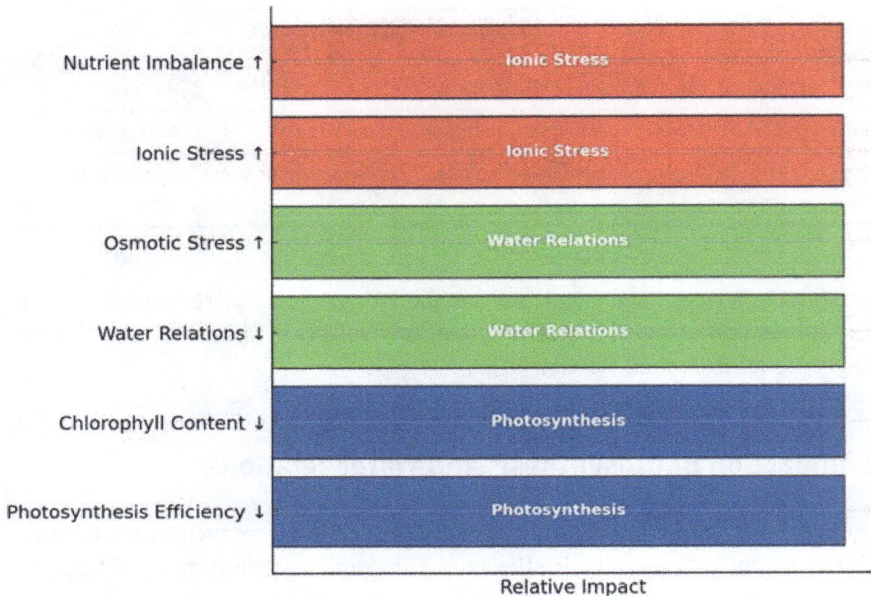

Figure 6.1: Effect of soil salinity on photosynthesis and water retention.

6.3.4 Oxidative stress and antioxidant response

Salt stress triggers excessive production of reactive oxygen species (ROS) in plants, threatening cellular components. Plants respond by activating antioxidant defenses, including enzymes like superoxide dismutase and catalase, as well as nonenzymatic compounds such as ascorbic acid and flavonoids [51, 52]. These mechanisms help neutralize ROS, protecting cellular structures and enhancing plant resilience under saline conditions. Studies on medicinal plants have shown increased antioxidant activity correlating with salt exposure levels, demonstrating plants' adaptive strategies against oxidative stress [53, 54].

6.3.5 Impact on secondary metabolite production

Salt stress paradoxically affects medicinal plants, often stunting growth, while boosting the production of valuable secondary metabolites [55]. Under saline conditions, these plants frequently increase their synthesis of compounds like tannins, saponins, and phenols [56]. This response involves changes in key biosynthetic pathways that are sensitive to environmental stress [57]. Research has shown that salt exposure can enhance essential oil production in various aromatic plants, potentially serving as a survival mechanism [58]. This phenomenon has significant implications for the medicinal plant industry, suggesting that controlled salinity could be used to stimulate the production of high-value bioactive compounds [59]. By integrating stress management into cultivation practices, growers may enhance both the therapeutic efficacy and economic value of medicinal plants. This approach represents a promising strategy for optimizing medicinal plant production in challenging environments [60].

6.4 Molecular responses to salinity stress

Medicinal plants adapt to salt stress through complex molecular mechanisms, involving changes in gene expression, protein activity, and metabolite production. Proteomic studies have revealed stress-induced alterations in proteins crucial for tolerance, photosynthesis, and secondary metabolite synthesis [61]. Salt exposure triggers the upregulation of genes linked to osmolyte production, ion transport, and antioxidant defenses. This molecular plasticity not only enhances salt tolerance but also promotes the accumulation of valuable secondary metabolites. Understanding these intricate molecular responses provides insights into how medicinal plants adapt to saline conditions, while potentially increasing their therapeutic and economic value [62].

6.5 Salt stress and primary metabolites in medicinal plants

6.5.1 Amino acids

Salt stress triggers significant shifts in plant amino acid metabolism. Many amino acids, including alanine, arginine, and glycine, increase under saline conditions, with proline showing the most dramatic rise. Nonprotein amino acids and amides also accumulate. Proline's buildup, observed in various medicinal plants, results from reduced proline oxidase activity and serves as a key osmoprotectant [60]. This helps maintain cellular balance and structure under salt stress. The overall increase in free amino acids, seen in plants like *Catharanthus roseus*, aids in osmotic adjustment and provides resources for energy and biosynthesis during stress. These metabolic changes highlight plants' adaptive strategies for surviving in saline environments, demonstrating the crucial role of amino acids in stress resilience [57].

6.5.2 Proteins

The elevation of free amino acids in plants experiencing salt stress is partially explained by the breakdown of proteins. For example, protein degradation has been observed in *Catharanthus roseus* exposed to salinity [63]. In chamomile and sweet marjoram, salt stress reduced the levels of soluble proteins, likely due to protein aggregation within the cells [64]. In *Achillea fragratissima*, a decline in crude protein synthesis was noted at 4,000 ppm salinity [65]. However, some studies indicate an increase in protein synthesis under higher salinity levels, suggesting that plants may store nitrogen in proteins for use during recovery from stress [66].

6.5.3 Carbohydrates

Salinity disrupts carbohydrate metabolism in plants, typically leading to imbalances due to reduced photosynthesis and nutrient availability. For instance, fennel plants exhibited a decrease in carbohydrate content under saline conditions [67]. In contrast, plants such as *Salvia officinalis* and *Satureja hortensis* showed an increase in carbohydrates as salinity levels rose, highlighting species-specific responses to salt stress [68, 69].

6.5.4 Lipids

Salt stress significantly impacts the lipid profile of plants, influencing fatty acid and oil synthesis. For example, in *Ricinus communis*, salinity reduced oil yield in roots but increased oil content in shoots [70]. In *Coriandrum sativum*, salinity stress led to a notable decrease in total fatty acid content, with reductions in key fatty acids such as α-linolenic and linoleic acids as NaCl concentrations increased [71–75].

6.6 Study of alkaloids through proteomic and other approaches

Proteomic studies have enhanced our knowledge of secondary metabolite production in medicinal plants, revealing complex mechanisms behind the synthesis of therapeutically valuable compounds [76, 77]. Researchers are using advanced techniques like cell cultures and metabolic engineering to increase the yield of these naturally limited substances. Studies on plants such as *Catharanthus roseus* have shown how various factors, including phytohormones and salt stress, affect protein expression and alkaloid production [78, 79]. This research has identified key enzymes and proteins involved in metabolite synthesis, providing insights into how plants respond to stress by altering their biochemical pathways. These findings open new possibilities for enhancing the production of medicinal compounds for pharmaceutical and commercial use [80]. In *Chelidonium majus*, two-dimensional gel electrophoresis revealed 21 proteins associated with stress signaling, nucleic acid binding, and defense responses, shedding light on the molecular mechanisms underpinning salinity tolerance [81]. Studies on *Papaver somniferum* identified codeinone reductase as a central enzyme in morphine biosynthesis, emphasizing its critical role in stress-induced secondary metabolite production [82].

These advancements in proteomics provide invaluable insights into the complex molecular pathways that regulate secondary metabolite production, paving the way for optimized cultivation practices and biotechnological innovations to maximize the medicinal and economic potential of these compounds [83, 84].

6.7 Phenolic compounds during stress

Phenolic compounds have emerged as important indicators of salt stress in plants. These diverse molecules, numbering around 9,000, play crucial roles in plant defense, particularly in neutralizing ROS generated during stress conditions [85]. Salt stress disrupts photosynthetic processes, leading to increased ROS production and oxidative damage. In response, plants synthesize various phenolic compounds, including phenolic acids, flavonoids, and proanthocyanidins, which act as antioxidants [85]. Research on crop plants consistently shows elevated phenolic content under saline conditions. For example, spearmint and *Achillea fragratissima* exhibit higher levels of phenolic acids and tannins when exposed to salt stress. *Matricaria chamomilla* demonstrates increased production of specific phenolic acids like protocatechuic, chlorogenic, and caffeic acids. Similarly, *Nigella sativa* and *Mentha pulegium* show a positive correlation between phenolic accumulation and salinity levels [86]. In Nigella, cultivated in saline soils, compounds such as quercetin, apigenin, and trans-cinnamic acid are found in higher concentrations, illustrating the plant's adaptive response to salt stress through enhanced phenolic synthesis.

6.8 Strategies for improving salt tolerance in MAPs

The growing issue of soil salinity poses a major threat to the cultivation of MAPs, which are valued for their therapeutic and economic significance. To address this, various strategies have been developed, ranging from traditional agronomic methods to advanced biotechnological approaches, aimed at mitigating the impact of salinity on MAPs [86]

Table 6.2: Strategies to improve salt tolerance in medicinal plants.

Strategy	Mechanism	Example plants
Use of plant growth regulators	Enhances stress tolerance through hormonal regulation and antioxidant activity	*Basil* and *chamomile*
Application of Mycorrhizae	Improves nutrient uptake and water relations under saline conditions	*Marjoram*
Genetic engineering	Overexpression of salt-tolerance genes like SOS1 and P5CS	*Artemisia annua*
Exogenous osmoprotectants	Protects cellular structures and maintains osmotic balance	*Chamomile* and *sage*

(Table 6.2). This section explores these strategies, focusing on their mechanisms and effectiveness in enhancing salt tolerance [87].

6.8.1 Exogenous application of plant growth regulators

Plant growth regulators (PGRs) have shown promise in enhancing the salt tolerance of MAPs by influencing key physiological processes [88]. Among these, salicylic acid (SA) has demonstrated particular efficacy in alleviating salt stress effects. Research on *St. John's wort*, exposed to saline conditions, revealed that SA treatment significantly boosted growth parameters and photosynthetic performance [89]. SA application led to marked improvements in chlorophyll levels, photosystem efficiency, and electron transport. Additionally, SA treatment reduced stress hormone levels, while increasing antioxidant enzyme activity, thereby improving the plant's ability to manage oxidative stress under saline conditions. These findings underscore the potential of PGRs, especially SA, as a practical approach to enhancing salt tolerance in medicinal plants, offering a promising strategy for maintaining crop productivity in salt-affected areas [90].

Gibberellic acid (GA3): Gibberellic acid has shown considerable potential in alleviating the detrimental effects of salinity on MAPs. For instance, in basil (*Ocimum basilicum*), GA3 application under saline conditions significantly improved growth metrics such as plant height, leaf area, and biomass (both fresh and dry weight). Moreover, GA3 enhanced the production of essential oils and their primary constituents, including linalool and methyl chavicol. These benefits are attributed to GA3's role in preserving membrane stability, enhancing antioxidant enzyme activity, and regulating osmolyte accumulation, which collectively contribute to improved stress resilience [91, 92].

Brassinosteroids (BRs): Brassinosteroids have also demonstrated promising effects on salt tolerance in MAPs. For example, foliar application of 24-epibrassinolide in chamomile (*Matricaria chamomilla*) under saline conditions enhanced growth, photosynthetic efficiency, and essential oil content. BRs improve salt tolerance by activating the plant's antioxidant defense mechanisms, enhancing water relations, and promoting the accumulation of compatible solutes. These combined effects enable plants to maintain physiological stability and adapt to saline environments more effectively [93].

The application of PGRs, including SA, GA3, and BRs, underscores their potential as valuable tools for mitigating salt stress in MAPs, ultimately supporting improved growth, productivity, and secondary metabolite synthesis in challenging environments.

6.8.2 Use of beneficial microorganisms

Utilizing beneficial microorganisms, such as arbuscular mycorrhizal fungi (AMF) and plant growth-promoting rhizobacteria (PGPR), offers an eco-friendly approach to enhance salt tolerance in MAPs [94]. These microorganisms improve plant resilience to salinity stress by enhancing nutrient uptake, increasing water absorption, balancing hormones, and activating stress-related biochemical pathways. AMF enhances root–soil interactions, facilitating phosphorus uptake, while PGPR promotes root growth, produces beneficial phytohormones, and bolsters antioxidant defenses 95]. For instance, inoculating Origanum majorana with Glomus mosseae helps mitigate salinity effects by improving growth and essential oil production. Similarly, in basil (*Ocimum basilicum*), PGPR strains like Pseudomonas putida and *Bacillus lentus* enhance growth and essential oil yield under salt stress by improving antioxidant enzyme activity and nutrient absorption [96]. Overall, these beneficial microbes play a crucial role in supporting plant health and productivity in saline environments, aligning with sustainable agricultural practices.

6.8.3 Genetic approaches

Genetic strategies hold great potential for improving salt tolerance in MAPs by combining traditional breeding with advanced biotechnological methods. Traditional breeding allows for the selection and crossing of salt-tolerant varieties, while modern techniques such as genetic engineering and marker-assisted selection enable precise targeting of genes linked to salinity resistance [97]. These approaches can lead to the development of MAPs that are more resilient to saline conditions, enhancing their medicinal and economic value. Research has identified specific genes that contribute to salt tolerance, such as the Na^+/H^+ antiporter gene (NHX1), which has been successfully introduced into *Artemisia annua*, resulting in improved growth and artemisinin production under salt stress [98]. Overall, these genetic advancements provide promising pathways for enhancing the resilience of MAPs against salinity, supporting sustainable agricultural practices.

RNA Interference (RNAi) technology: RNAi has been used to suppress genes that hinder salt tolerance [99]. For example, silencing the *SmMYB39* gene in *Salvia miltiorrhiza* enhanced the plant's tolerance to salt stress. This was achieved by increasing the production of phenolic acids and tanshinones, which are important medicinal compounds [100].

6.8.4 CRISPR/Cas9 gene editing

The CRISPR/Cas9 genome editing system offers precise and efficient modifications, making it a promising approach for enhancing salinity tolerance in MAPs. While its application in MAPs is still developing, this tool shows great promise for creating salt-tolerant plants with improved medicinal properties [101].

6.8.5 Agronomic practices

Agronomic interventions are essential for managing the effects of salinity on MAPs by improving soil conditions and ensuring better plant growth in saline environments [102].

Irrigation management: Efficient irrigation techniques can control soil salinity levels. Drip irrigation, for example, has been shown to maintain lower salinity in the root zone compared to traditional methods like furrow irrigation. Studies on *Rosmarinus officinalis* (rosemary) revealed that drip irrigation not only improved plant growth but also increased the yield and quality of essential oils under saline conditions [103].

6.8.6 Use of mulches

Mulching is an effective agronomic strategy to mitigate salinity stress. Organic mulches help reduce water loss through evaporation and prevent salt buildup in the root zone. For example, straw mulch applied to *Salvia officinalis* (sage) significantly enhanced growth and essential oil yield under saline conditions. This practice also improved soil moisture retention, minimized temperature fluctuations, and boosted soil biological activity [104, 105].

6.8.7 Application of organic amendments

Organic amendments, such as compost and biochar, have proven effective in improving soil health and mitigating salinity-induced stress in MAPs. These amendments enhance soil structure, promote water retention, and alleviate the negative effects of salt stress [106]. For instance, in *Foeniculum vulgare* (fennel), vermicompost application significantly enhanced plant-growth-boosted essential oil yield, and improved antioxidant activity under saline conditions. These improvements were attributed to enhanced soil properties, increased nutrient availability, and stimulation of beneficial microbial activity [107, 108].

6.8.8 Silicon supplementation

Silicon (Si) supplementation has emerged as an effective method for enhancing plant tolerance to salinity [109]. While not essential for growth, Si plays a vital role in mitigating abiotic stresses, including salt stress. In basil, Si treatment has been shown to alleviate the negative impacts of salinity, resulting in improved growth, enhanced photosynthesis, and increased essential oil production [110]. These benefits are linked to heightened antioxidant enzyme activity, better water absorption, reduced sodium accumulation, and improved potassium balance. Similarly, in peppermint (*Mentha piperita*), Si application leads to better growth and essential oil yield, while reinforcing antioxidant defenses and maintaining membrane integrity under saline conditions. Overall, Si supplementation proves beneficial for promoting resilience in plants facing salinity stress [111].

6.8.9 Application of polyamines

Polyamines, including putrescine, spermidine, and spermine, are crucial for regulating plant growth and stress responses, especially under saline conditions [112]. Their application has been recognized for enhancing salt tolerance in MAPs by stabilizing cell membranes, reducing oxidative damage from ROS, and balancing ion levels. Additionally, polyamines promote the expression of stress-related genes, improve nutrient uptake, and assist in osmotic adjustment, thereby strengthening plants' ability to cope with salinity [113]. For example, in chamomile, putrescine application improved growth, flower yield, and essential oil content under salt stress by increasing antioxidant enzyme activity and enhancing photosynthetic efficiency [114]. In sage, spermidine application helped mitigate salinity effects by improving water retention and promoting compatible solute accumulation, which supported growth and essential oil production. Overall, polyamines present a promising strategy for sustainable MAP cultivation in saline environments due to their multifunctional roles in enhancing plant resilience [115].

6.8.10 Nanofertilizers and nanoparticles

Nanofertilizers and nanoparticles are innovative tools for enhancing nutrient uptake efficiency and improving plant resilience to salinity. In *Mentha piperita* (peppermint), zinc oxide nanoparticles (ZnO NPs) significantly enhanced growth, essential oil yield, and antioxidant activity under saline conditions. These improvements were linked to increased accumulation of compatible solutes, better photosynthesis, and more stable nutrient levels [116]. Similarly, in basil, iron oxide nanoparticles (Fe_3O_4 NPs) alleviated salt stress by improving growth, increasing essential oil production, and boosting antioxidant enzyme activity. These nanoparticles also supported secondary metabolite synthesis and helped maintain membrane stability in plants subjected to salinity [117].

6.8.11 Application of melatonin

Melatonin is recognized as an effective enhancer of salt tolerance in various plants, including medicinal and aromatic species. Acting both as a growth regulator and antioxidant, melatonin helps mitigate salinity stress, which has garnered significant research interest. In lemon balm (*Melissa officinalis*), applying melatonin has been shown to improve growth, essential oil yield, and antioxidant activity under saline conditions [118]. This enhancement is linked to melatonin's ability to promote the accumulation of compatible solutes like proline, which aids in osmotic regulation, stabilizes chloroplast structures, and maintains ionic balance by reducing sodium uptake and increasing potassium retention. These mechanisms collectively improve the plant's resilience to salinity [84].

Melatonin's potential to enhance salt tolerance, combined with its ability to boost secondary metabolite production, makes it a promising solution for the sustainable cultivation of MAPs in saline environments, offering both therapeutic and economic benefits [119]. Similarly, in *Lavandula angustifolia* (lavender), melatonin treatment alleviated the adverse effects of salt stress on growth and essential oil production. The protective mechanism involved heightened antioxidant enzyme activity, improved water relations, and stabilization of membrane integrity [120, 121].

The strategies for improving salt tolerance in MAPs are diverse, ranging from basic agronomic practices to advanced biotechnological techniques. Exogenous applications of PGRs, beneficial microorganisms, silicon, polyamines, and melatonin have shown great potential for enhancing salt tolerance in MAPs by modulating physiological and biochemical processes, boosting antioxidant defenses, improving water relations, and maintaining ionic equilibrium under saline conditions [123].

Genetic approaches, such as identifying and overexpressing salt-tolerance genes and utilizing advanced gene editing techniques, provide long-term solutions for developing MAP varieties with enhanced salt tolerance. These methods not only improve plant resilience but also increase the production of valuable secondary metabolites [124]. Practical agronomic methods, including efficient irrigation management, mulching, and organic amendments, offer easily implementable solutions to manage soil salinity in MAP cultivation [125]. These practices contribute to improved growth and sustainable soil management.

The emerging field of nano-fertilizers and nanoparticles offers promising outcomes in enhancing nutrient efficiency and plant stress tolerance. However, further studies are required to fully understand their long-term implications and optimize their use across different MAP species [126, 127].

As soil salinity continues to pose challenges, integrating these diverse strategies will be vital for the sustainable cultivation of MAPs [128]. Future research should aim to explore the combined effects of different approaches, tailoring holistic solutions to specific plant species and growing conditions. Moreover, it is essential to examine the impact of these salt tolerance strategies on the quality and quantity of medicinal compounds, ensuring no compromise on their therapeutic value [129, 130]. By adopting these strategies, it is possible to enhance the growth, yield, and medicinal properties of MAPs under saline conditions, addressing environmental challenges sustainably [131, 132].

6.9 Physiological and biochemical responses to salinity stress

Medicinal plants utilize various physiological and biochemical mechanisms to adapt to and endure salinity stress. This section focuses on the primary adaptive strategies these plants employ to thrive in saline conditions [133].

6.9.1 Water relations and osmotic adjustment

Salinity stress profoundly affects plant water relations by lowering the osmotic potential of the soil, thereby restricting water uptake and creating conditions akin to drought, even when soil water is available. To overcome this challenge, medicinal plants deploy adaptive mechanisms aimed at maintaining water balance [134]. One critical strategy involves the accumulation of compatible solutes, such as proline, glycine betaine, and soluble sugars. These molecules act as osmoprotectants, helping to regulate osmotic potential within cells and sustain cellular functions under stress. For instance, in *Moringa oleifera*, exposure to salinity stress resulted in a significant increase in proline and soluble sugar levels. This accumulation was closely associated with a reduction in osmotic potential in both roots and leaves, highlighting their role in osmoregulation and the plant's ability to adapt to saline conditions [135]. These mechanisms not only enable plants to maintain water uptake but also support overall stress resilience, ensuring survival and productivity in saline environments. Relative water content (RWC) typically declines under salinity stress. In a study involving five *Coleus* species, RWC was observed to decrease significantly with rising salinity levels. For instance, in *Coleus aromaticus*, RWC dropped from 103% to 77.6% under high salinity. Similarly, leaf water potential (LWP) also declined. The same study reported a reduction in LWP from −0.02 MPa to −1.2 MPa in *Coleus aromaticus* and from −0.05 MPa to −2.14 MPa in *Coleus forskohlii* under high salinity conditions [136].

6.9.2 Ion homeostasis and nutrient balance

Salinity stress disrupts ion balance in medicinal plants, leading to ionic toxicity and nutrient deficiencies that hinder growth and development [137]. This stress typically results in excessive accumulation of sodium (Na^+) and chloride (Cl^-) ions, which can interfere with cellular functions and exacerbate nutrient imbalances. For example, in Moringa oleifera, salinity causes a significant rise in Na^+ and Cl^- levels while decreasing potassium (K^+) concentrations [138]. A balanced K^+/Na^+ ratio is crucial for salt tolerance, as it supports essential enzymatic functions; however, salinity often lowers this ratio due to increased Na^+ uptake and limited K^+ absorption. The high levels of Na^+ disrupt the uptake of other vital nutrients like K^+, calcium (Ca^{2+}), and magnesium (Mg^{2+}) through competitive interactions, leading to further nutrient shortages and metabolic disturbances that impair growth [139]. To counter these effects, maintaining ion homeostasis is essential. Strategies such as selective ion transport, compartmentalizing Na^+ into vacuoles, and enhancing the uptake of K^+ are vital for improving salt tolerance in medicinal plants. These mechanisms highlight the importance of managing salinity stress for sustainable cultivation of medicinal plants in saline environments.

6.10 Molecular mechanisms of salt tolerance

Salinity represents a major environmental challenge, severely impacting plant growth, physiological processes, and overall productivity [140]. To cope with salinity stress, plants have evolved intricate molecular mechanisms that enable them to maintain ion homeostasis, regulate osmotic balance, and mitigate oxidative damage. These adaptations in-

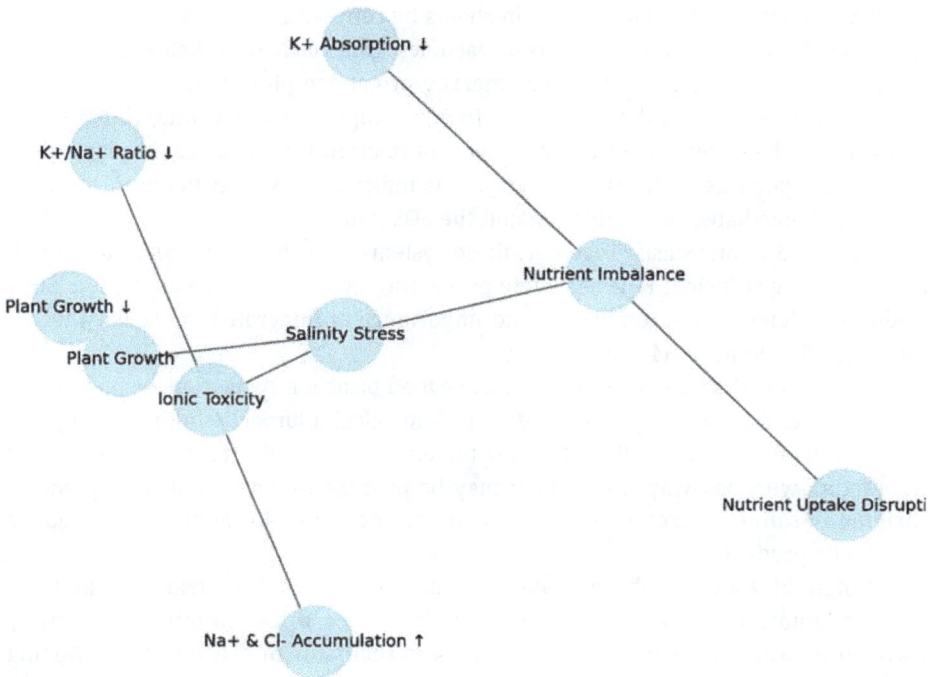

Figure 6.2: Effect of soil salinity on the ion homeostasis and nutrient balance.

clude the activation of ion transport systems to control Na^+ and K^+ distribution, the accumulation of osmoprotectants such as proline and glycine betaine to adjust cellular osmotic potential, and the enhancement of antioxidant defenses to counteract ROS. Together, these mechanisms allow plants to survive and adapt to high-salt environments, ensuring their functionality under adverse conditions [141].

A key mechanism for salt tolerance is the Salt Overly Sensitive (SOS) pathway, which is activated when plants detect high salinity. Calcium sensors such as SOS3 and SCaBP8 recognize salt stress and activate SOS2, a protein kinase. This, in turn, phosphorylates SOS1, a Na^+/H^+ antiporter, which facilitates the removal of excess Na^+ from cells [142]. SOS1, composed of an N-terminal transmembrane region and a cytosolic regulatory C-terminal domain, is activated through phosphorylation by SOS2. This prevents the interaction of SOS1 with inhibitory proteins like 14-3-3, thereby improving the plant's salt tolerance. Beyond cellular ion regulation, SOS1 also plays a role in controlling Na^+ transport between roots and shoots, reducing Na^+ buildup in leaves and improving overall salt tolerance [143, 144].

The ability to sustain a proper K^+/Na^+ balance is a critical determinant of salt tolerance, ensuring the stability of cellular processes and overall plant health. A key player in this balance is the HIGH-AFFINITY K^+ TRANSPORTER 1 (*HKT1*) family, which actively minimizes Na^+ accumulation in shoots by retrieving Na^+ ions from the xylem sap and redirecting them to the roots or vacuoles. This mechanism helps reduce toxic levels of Na^+ in above-ground tissues, thereby protecting photosynthetic machinery and other essential cellular functions. In Arabidopsis thaliana, mutations in the AtHKT1 gene have demonstrated enhanced salt tolerance, particularly in plants deficient in components of the SOS pathway. This indicates a synergistic interaction between HKT1-mediated Na^+ regulation and the SOS pathway, which is involved in Na^+ extrusion and homeostasis. Together, these systems coordinate to maintain ion balance, ensuring sufficient K^+ levels while preventing excessive Na^+ accumulation. Such findings underscore the complexity and importance of integrated molecular mechanisms in salt tolerance [145, 146].

The role of HKT1 transporters extends beyond basic ion regulation, as they represent a potential target for genetic and biotechnological approaches aimed at improving salt tolerance in MAPs. By enhancing the efficiency of HKT1 transporters or their interaction with pathways like SOS, it may be possible to develop MAPs capable of thriving in saline environments, while maintaining optimal growth and secondary metabolite production.

Protein kinases and phosphatases are vital regulators of ion transport and salt stress adaptation. For example, the GSK3-like kinase BIN2 inhibits SOS2 activity through phosphorylation, acting as a negative regulator of salt tolerance. During post-stress recovery, SOS3 and SCaBP8 facilitate BIN2's movement to the membrane, where it represses SOS2 activity by phosphorylating it at a specific residue (T172). This process enables growth-related signaling proteins like BES1 and BZR1 to promote recovery, showcasing BIN2's role in balancing stress tolerance and growth recovery [147, 148].

Calcium signaling is central to plant responses to salinity, involving various calcium-dependent proteins and transporters that mediate stress signaling. For example, AtANN1, a calcium-binding membrane protein, regulates Na^+ influx and K^+ efflux in root cells. Plants lacking AtANN1 exhibit excessive Na^+ uptake and K^+ loss, leading to compromised root growth under salinity stress. Similarly, AtANN4, another calcium-binding protein and putative transporter, plays a crucial role in maintaining cytosolic calcium levels and activating the SOS pathway during salt stress [149, 150].

Osmotic adjustment represents one of the most critical strategies that plants employ to adapt to salinity stress. This adaptation involves the accumulation of compatible solutes, such as proline, glycine betaine, and soluble sugars, which function as osmoprotectants. These solutes play multiple roles: they reduce the osmotic potential within cells, maintain water uptake, stabilize cellular structures, and protect enzymes and proteins from denaturation caused by ionic and osmotic stress [151]. The biosynthesis of these solutes is tightly regulated by specific genes and enzymes, such as P5CS (Δ1-pyrroline-5-carboxylate synthetase) in proline synthesis, which shows heightened expression under salinity stress. Similarly, BADH (betaine aldehyde dehydrogenase), crucial for glycine betaine synthesis, is upregulated in response to saline environments, enabling plants to better manage osmotic and ionic imbalances [152].

Salinity stress also intensifies the generation of ROS, such as superoxide anions, hydrogen peroxide, and hydroxyl radicals. These ROS can severely damage cellular components, including membranes, proteins, and nucleic acids, leading to impaired growth and metabolic dysfunction [153]. To mitigate this, plants activate a dual-layered antioxidant defense system comprising enzymatic and nonenzymatic components.

The enzymatic defense includes superoxide dismutase (SOD), which converts superoxide radicals into hydrogen peroxide, catalase (CAT), which decomposes hydrogen peroxide into water and oxygen, and ascorbate peroxidase (APX), which uses ascorbic acid as a substrate to neutralize hydrogen peroxide. Glutathione peroxidase (GPX) also contributes to detoxifying peroxides. Under salinity stress, the activity and gene expression of these enzymes are significantly upregulated, enhancing the plant's ability to neutralize ROS and prevent oxidative damage.

Complementing enzymatic antioxidants, nonenzymatic antioxidants such as ascorbic acid, glutathione, flavonoids, and phenolic compounds provide additional protection. These molecules scavenge free radicals and reinforce cellular defenses. The synthesis and regulation of these antioxidants involve complex transcriptional, posttranscriptional, and signaling pathways, ensuring a rapid and effective response to oxidative stress [154].

By integrating osmotic adjustment and robust antioxidant defenses, plants demonstrate a multifaceted response to salinity stress. These mechanisms not only ensure cellular homeostasis but also contribute to overall plant resilience, offering valuable insights for breeding and biotechnological strategies aimed at enhancing the salt tolerance of crops and medicinal plants. These approaches hold promise for sustainable agriculture and the cultivation of salt-tolerant plant species in challenging environments.

Plant hormones coordinate various salt stress responses. ABA is a key hormone that is upregulated under salinity, triggering adaptive responses such as stomatal closure, the induction of stress-responsive genes, and the accumulation of osmolytes [155]. The ABA signaling pathway involves receptors, protein phosphatases, and transcription factors (TFs) working in a complex network. Other hormones, including jasmonic acid (JA), SA, and ethylene, also contribute to salt tolerance, often interacting with ABA signaling to fine-tune responses.

TFs are essential for regulating salt-responsive gene expression. TF families such as AP2/ERF, bZIP, MYB, NAC, and WRKY are particularly important. For example, DREB (Dehydration-Responsive Element Binding) TFs from the AP2/ERF family regulate genes associated with stress responses. Overexpression of certain *DREB* genes has improved salt tolerance in various species. Similarly, NAC family members, such as *ANAC092*, are involved in processes like senescence and salt stress adaptation [156, 157]. Posttranscriptional and posttranslational modifications fine-tune plant responses to salinity. MicroRNAs (miRNAs) regulate gene expression by targeting transcripts related to ion transport, osmolyte production, and antioxidant defenses [158]. At the protein level, modifications like phosphorylation, ubiquitination, and sumoylation adjust the activity and stability of critical proteins involved in salt stress responses.

Epigenetic mechanisms, including DNA methylation and histone modifications, play significant roles in salt tolerance. Salt stress often alters methylation patterns, leading to changes in gene expression. Histone acetylation and methylation also regulate the activity of salt-responsive genes. For instance, histone deacetylase HDA6 has been shown to modulate stress-related gene expression in *Arabidopsis* [159, 160].

Membrane lipid composition is dynamically adjusted under salinity to maintain fluidity and support the function of membrane-bound proteins. Similarly, modifications in the cell wall, such as changes in pectin content and composition, contribute to maintaining integrity and regulating ion transport under salt stress [161]. Omics technologies have provided new insights into salt tolerance mechanisms. Transcriptomics has identified salt-responsive genes, proteomics has uncovered protein abundance and modifications, and metabolomics has highlighted shifts in metabolic pathways under salinity, including changes in primary and secondary metabolite profiles [162]. Salt tolerance in plants involves a coordinated interplay of physiological, biochemical, and genetic processes [163]. These include ion homeostasis, osmotic adjustment, antioxidant defenses, hormonal signaling, transcriptional regulation, and cellular adaptations. Understanding these mechanisms is crucial for improving crop resilience and productivity under salinity stress. Future research should focus on integrating these mechanisms into cohesive strategies to enhance salt tolerance in plants, ensuring global food security [164].

6.11 Biotechnological approaches for improving salt tolerance

Biotechnology offers innovative solutions to address the challenges posed by soil salinization, enabling the development of salt-tolerant crop varieties [165]. These approaches range from advanced genetic engineering techniques to improved plant breeding methods supported by genomics (Figure 6.3). The goal is to create crops that maintain productivity under saline conditions, thereby addressing global food security concerns [166]. Plant cell cultures provide a valuable tool for identifying salt-tolerant traits. Screening large populations of cells under controlled conditions, using pure salts, salt mixtures, or seawater, has been effective in selecting salt-tolerant lines [167, 168]. Traits identified in cell cultures are often expressed in regenerated plants and inherited through subsequent generations, demonstrating the potential of this technique for developing salt-tolerant crops [169]. Cell culture-based selection offers the advantage of consistent selection pressure and rapid evaluation of large populations, which can be challenging in field-based methods [170]. Utilizing wild germplasm is a valuable strategy for introducing salt tolerance traits into economically important crop species. Wild relatives often possess resilience traits that have been lost during domestication and breeding for higher yields [171]. Advanced hybridization techniques, such as embryo rescue and protoplast fusion, enable breeders to overcome crossing barriers between distantly related species. These techniques allow for the transfer of genetic diversity from wild species to cultivated crops, introducing novel salt-tolerance genes. For instance, salt tolerance traits from wild relatives have been successfully incorporated into wheat and rice, improving their ability to grow in saline soils [172, 173].

Recombinant DNA technology has revolutionized the development of salt-tolerant crops. This approach involves identifying, cloning, and transferring salt-tolerance genes into target species. Genomic tools like forward and reverse genetics in model plants such as *Arabidopsis thaliana* are often used to identify candidate genes [174]. Once isolated, these genes can be introduced into crops through transformation techniques like *Agrobacterium*-mediated transformation or particle bombardment. Despite its success in creating transgenic crops with enhanced salt tolerance, commercialization often faces challenges such as regulatory hurdles and public skepticism [175].

Improving salt tolerance in plants heavily relies on effective ion transport mechanisms. The SOS pathway, initially studied in Arabidopsis, has been a focal point for genetic modifications aimed at maintaining ion balance by promoting the exclusion of sodium (Na^+) from cells [176]. Overexpressing components like SOS1, which encodes a Na^+/H^+ antiporter, have been shown to enhance salt tolerance in various species. Additionally, manipulating genes from the HIGH-AFFINITY K^+ TRANSPORTER 1 (HKT1) family has helped improve Na^+ exclusion and maintain K^+/Na^+ ratios under saline conditions [177]. Osmolytes such as proline, glycine betaine, and soluble sugars are crucial for osmotic adjustment and cellular protection during salinity stress. Genetic engineering efforts have focused on increasing the production of these compounds [114]. For example, overexpressing the enzyme P5CS, which is involved in proline synthesis, has significantly enhanced salt tolerance in multiple plant species. Similarly, modifying non-glycine betaine-producing plants to synthesize this osmolyte has proven effective in improving resilience to salinity. Salt stress also triggers the production of ROS, which can damage plant cells. Strengthening antioxidant defenses is vital for mitigating this damage, and genetic engineering has targeted both enzymatic antioxidants like superoxide dismutase (SOD) and catalase (CAT), as well as nonenzymatic antioxidants such as flavonoids [121, 132]. Overexpressing genes responsible for these antioxidants has led to notable improvements in salt tolerance among plants.

High-throughput genomics technologies have accelerated the discovery of salt-tolerance genes. Tools like genome-wide association studies (GWAS) and quantitative trait loci (QTL) mapping have identified key genes and alleles associated with salinity

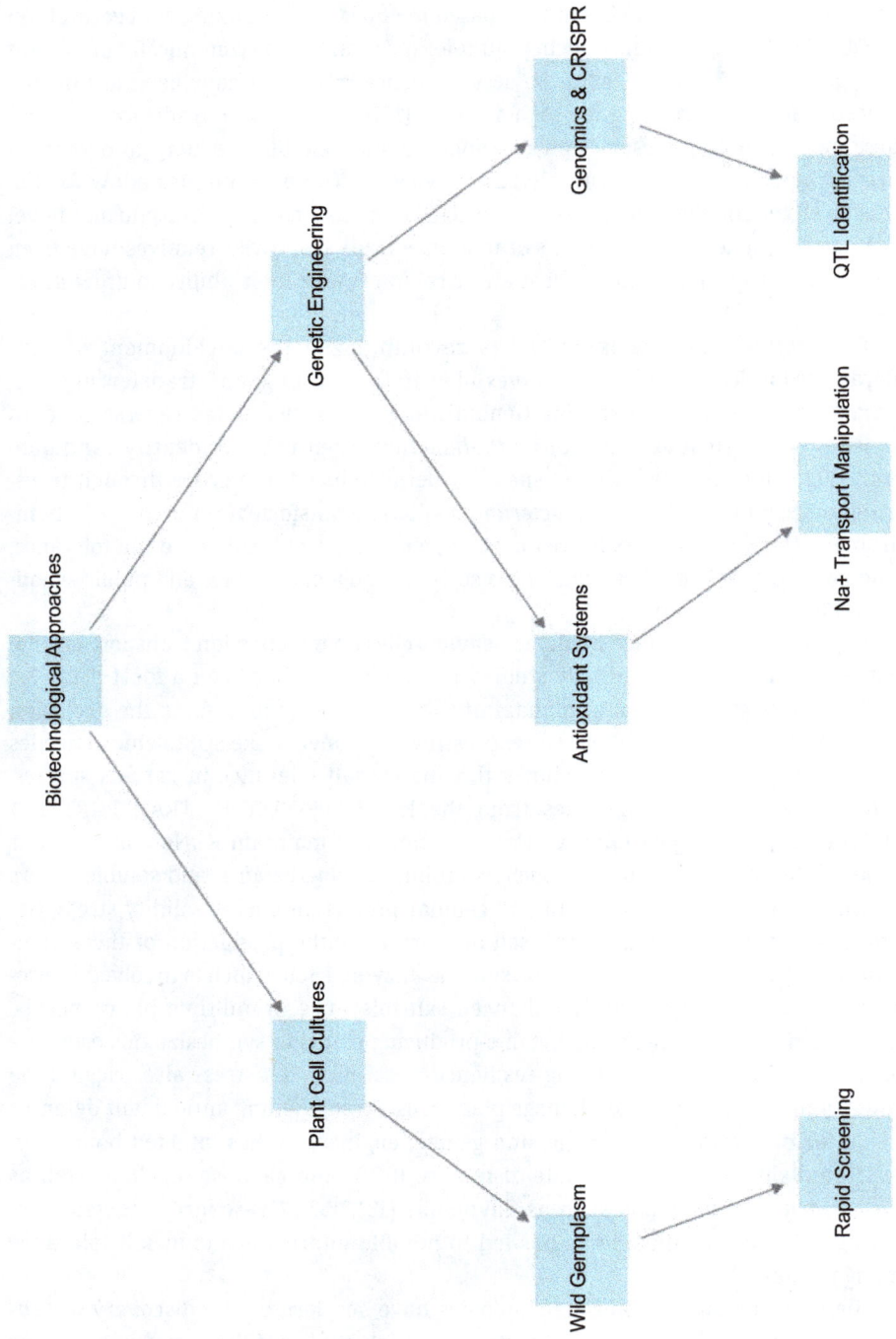

Figure 6.3: Biotechnological approaches for improving salt tolerance.

tolerance in crops like rice, providing valuable markers for breeding programs [84, 105]. Transcriptomics has further revealed complex gene expression networks, offering new targets for genetic improvement [16].

The advent of CRISPR/Cas9 has enabled precise gene editing to improve salt tolerance. This technique allows for targeted modifications of genes involved in salinity responses, such as those regulating Na^+ transport. For instance, CRISPR/Cas9-mediated modifications in rice improved salt tolerance by enhancing ion homeostasis. As it avoids the introduction of foreign DNA, this method may address regulatory and public acceptance concerns surrounding genetically modified crops [114, 122]. Despite these advances, challenges remain. Salt tolerance is a complex trait involving multiple genes and pathways, making single-gene modifications often insufficient for achieving robust and stable improvements. Future efforts will likely focus on multigene engineering and the manipulation of regulatory networks for comprehensive tolerance. Additionally, it is crucial to ensure that increased salt tolerance does not compromise yield or quality under nonstress conditions [75, 85]. Biotechnological approaches offer transformative solutions for developing salt-tolerant crops. From the utilization of cell cultures for screening to advanced genetic engineering techniques, these strategies have yielded promising results in enhancing salinity resilience [23]. As research continues to uncover the molecular basis of salt tolerance and refine genomic tools, the potential for creating more resilient crops will only grow. However, biotechnology should complement traditional breeding and agronomic practices, forming part of a holistic approach to address soil salinization and ensure global food security [117].

6.11.1 Genetic engineering strategies

Genetic engineering has become a transformative tool in improving salt tolerance in plants by enabling precise modifications to enhance their resilience to saline environments [152]. This approach focuses on isolating and transferring key salt-tolerant genes, particularly those involved in pathways critical for managing salinity stress. One such pathway is the SOS pathway, which regulates ion homeostasis and prevents sodium toxicity. Enhancing the expression of SOS1, a gene encoding a plasma membrane Na^+/H^+ antiporter, has shown significant results, with transgenic plants exhibiting better growth and yield under saline conditions due to improved Na^+ efflux and ionic balance. Similarly, the *HKT1* family of transporters, which facilitates the removal of excess sodium from xylem tissues, has been a key target, as modifications to *HKT1* expression have effectively reduced sodium buildup in shoots and enhanced overall salt tolerance [164]. Another critical strategy involves targeting genes associated with osmotic adjustment, such as P5CS (Δ1-pyrroline-5-carboxylate synthetase), which is vital for proline synthesis. Overexpression of P5CS increases proline levels, allowing plants to better manage osmotic stress. Similarly, enhancing glycine betaine production through genetic modifications has proven effective in enabling plants to maintain osmotic balance in saline conditions [127].

Efforts to enhance salt tolerance in plants have increasingly focused on strengthening antioxidant defense systems to combat oxidative damage caused by ROS generated during salinity stress [105]. Overexpressing genes for enzymes such as superoxide dismutase (SOD), catalase (CAT), and ascorbate peroxidase (APX) have been shown to improve a plant's capacity to detoxify ROS, thereby reducing oxidative stress. Recent advancements in genome editing technologies, particularly CRISPR/Cas9, have provided new opportunities for enhancing salt tolerance [136]. This technique allows for precise modifications of specific genes or regulatory regions without introducing foreign DNA. CRISPR/Cas9 has been successfully used to edit genes associated with salinity tolerance, leading to crops that can better withstand salt stress, while addressing concerns related to traditional transgenic methods [36]. These developments highlight the potential of genetic engineering in tackling the challenges posed by salinity, offering innovative solutions for cultivating salt-tolerant crops and medicinal plants, which is essential for sustainable agricultural productivity in saline environments [74].

6.11.2 Identification of salt-tolerant genes

The discovery of genes associated with salt tolerance is essential for developing crops resilient to salinity. Researchers have employed various methods to identify and understand the genes involved in these mechanisms. GWAS have become a powerful approach to identifying QTLs linked to salinity tolerance [110]. For example, studies in rice have revealed several QTLs that influence salt tolerance traits, providing a valuable resource for breeding programs. Transcriptomics, such as RNA sequencing, has highlighted the complexity of gene networks activated during salinity stress, offering new candidates for genetic improvement. Research in *Melia azedarach* identified pivotal genes, including *CBL7*, *SAPK10*, *EDL3*, and *AKT1*, through Gene Ontology (GO) analysis, which are crucial for salinity responses. Similarly, in rice, genes like *OsSTL1* and *OsSTL2* have been validated for their roles in salt tolerance, with *OsSTL1* found to be homologous to the *SRP1* gene in *Arabidopsis* [162].

TFs are also critical regulators of salinity tolerance. In rice, TFs such as *OsWRKY53* and *OsMKK10.2* have been identified as key contributors to salt resilience. Furthermore, zinc finger proteins have been implicated in regulating salinity responses, with several genes from this family associated with salt tolerance traits [42]. Additionally, wild relatives of cultivated crops offer a valuable reservoir of salt-tolerant genes that have often been lost during domestication. Advanced hybridization techniques now enable the transfer of these traits from wild species to domesticated crops, improving their ability to withstand salinity [118].

6.11.3 Use of plant growth regulators

PGRs have shown great potential for enhancing plant tolerance to salt stress when applied externally. These compounds influence various physiological and biochemical processes to improve plant performance under adverse conditions. SA, for instance, has demonstrated efficacy in mitigating salt stress by boosting photosynthesis, increasing chlorophyll levels, and enhancing antioxidant defenses [12]. In *Hypericum perforatum* (St. John's wort), SA treatment increased photochemical efficiency by 133% and improved the activity of PSII by 294%. It also reduced ABA levels by 32% and boosted SOD activity by 15.4%.

Gibberellic acid (GA3) is another effective PGR that alleviates the impact of salinity. In *Ocimum basilicum* (basil), GA3 application improved plant growth metrics, including height, leaf area, and biomass. It also increased essential oil content, particularly components like linalool and methyl chavicol, by maintaining membrane integrity and enhancing antioxidant activity [56]. Brassinosteroids (BRs), such as 24-epibrassinolide, have also shown potential in mitigating salinity stress. In *Matricaria chamomilla* (chamomile), BR application enhanced growth, photosynthetic efficiency, and essential oil production by improving water relations and boosting antioxidant defenses.

Osmoprotectants play a key role in helping plants maintain water balance under saline conditions. Compounds such as proline, glycine betaine, and trehalose assist in osmotic adjustment, while protecting cellular components. Proline accumulation, both natural and externally applied, helps plants maintain osmotic equilibrium and stabilize proteins and membranes under salt stress [5]. Similarly, glycine betaine has proven effective in improving salinity tolerance, particularly in genetically engineered crops that do not naturally produce it [93]. Trehalose, some sugar with protective properties, has been widely recognized for stabilizing cellular structures during stress.

Antioxidants mitigate the damaging effects of ROS produced under salt stress [63]. Exogenous application of antioxidants such as ascorbic acid (vitamin C) and α-tocopherol (vitamin E) has been shown to enhance salt tolerance. Ascorbic acid scavenges ROS and supports enzymatic activity, while α-tocopherol protects membranes from oxidative damage, improving plant resilience under stress conditions [95].

Genetic engineering has played a pivotal role in enhancing plant responses to salinity. One major focus has been modifying ion transport systems, particularly the SOS pathway, which regulates ion balance and Na^+ exclusion. Overexpressing *SOS1*, encoding a plasma membrane Na^+/H^+ antiporter, has led to improved salinity tolerance in multiple species by reducing sodium toxicity [73]. Similarly, enhancing the expression of *HKT1* transporters, such as *OsHKT1;5* in rice, has demonstrated success in limiting sodium accumulation in shoots and maintaining ion homeostasis [62].

Another important target is osmolyte biosynthesis. Increasing the production of proline, glycine betaine, and trehalose through genetic engineering has significantly enhanced plant resilience to salinity. For instance, overexpression of *P5CS*, a key enzyme in proline synthesis, has resulted in greater proline accumulation and improved plant performance under salt stress [139]. Similarly, introducing glycine betaine biosynthesis pathways into nonproducing crops has been effective in enhancing their stress tolerance.

Improving antioxidant defense mechanisms through genetic modifications has also yielded positive results. Overexpression of genes-encoding enzymes such as superoxide dismutase (SOD), catalase (CAT), and ascorbate peroxidase (APX) has bolstered the ability of plants to neutralize ROS, reducing oxidative damage and enhancing overall tolerance [21, 48].

Identifying genes associated with salt tolerance is a key step in advancing crop resilience to salinity. Various strategies have been employed to uncover and study these genes [93]. Among them, GWAS have been highly effective in pinpointing QTLs linked to salinity adaptation. For instance, multiple QTLs related to salt tolerance have been identified in rice. One study [37] identified eight QTLs associated with alkalinity tolerance during the seedling stage and highlighted *OsIRO3*, a bHLH TF, as a key regulator of the Fe-deficiency response in rice. Another investigation [19] involving 179 Vietnamese rice landraces uncovered 26 QTLs associated with nine traits related to salinity, with 10 QTLs linked to hormonal pathways and transcriptional regulation under stress.

Transcriptomics, including RNA sequencing, provides deeper insights into how plants respond to salt stress at the genetic level [62]. For example, four critical genes (*CBL7*, *SAPK10*, *EDL3*, and *AKT1*) were identified in *Melia azedarach* as central to salt stress responses through Gene Ontology (GO) analysis. In rice, *OsSTL1* and *OsSTL2* have been confirmed as crucial salt-tolerant genes, with *OsSTL1* sharing homology with the *SRP1* gene in *Arabidopsis* [32]. Additionally, TFs like *OsWRKY53* and *OsMKK10.2* have been recognized as important regulators of salinity responses in rice, while zinc finger proteins have also been implicated in stress adaptation [42].

PGRs offer significant potential for improving plant responses to salt stress by modulating physiological and biochemical pathways. SA, for instance, has been shown to enhance plant photosynthesis, chlorophyll production, and antioxidant activity under saline conditions [33]. In *Hypericum perforatum* (St. John's wort), SA application improved the photochemical efficiency of PSII by 294% and reduced ABA levels by 32%. SA also increased superoxide dismutase (SOD) activity by 15.4%, further contributing to stress mitigation [125].

Gibberellic acid (GA3) has demonstrated efficacy in countering salt stress. In *Ocimum basilicum* (basil), GA3 treatment under saline conditions boosted plant growth parameters, including height and leaf area, while increasing essential oil yield. Components like linalool and methyl chavicol were particularly enhanced due to GA3's role in maintaining membrane integrity and regulating osmolyte levels [43, 74]. Similarly, brassinosteroids (BRs), such as 24-epibrassinolide, have been effective in improving salt tolerance in plants like *Matricaria chamomilla* (chamomile), enhancing growth, photosynthesis, and essential oil content by supporting antioxidant systems and water relations [112].

Using osmoprotectants and antioxidants is another reliable approach to enhancing plant tolerance to salinity [51]. Osmoprotectants, including proline, glycine betaine, and trehalose, help plants manage osmotic stress and protect cellular structures. Proline is widely recognized for its dual role as an osmolyte and a stabilizer of cellular proteins and membranes. Many plants naturally accumulate proline under stress, and its external application has been shown to further improve salt tolerance [63]. Similarly, glycine betaine enhances osmotic adjustment and has been introduced into nonproducing plants through genetic engineering. Trehalose, a disaccharide, is also effective in stabilizing cellular functions and has gained attention for its role in improving salinity adaptation [51].

Antioxidants play a vital role in combating oxidative stress caused by salinity-induced ROS [47]. Exogenous application of antioxidants such as ascorbic acid (vitamin C) and α-tocopherol (vitamin E) has been shown to protect plants from oxidative damage. Ascorbic acid acts as a direct scavenger of ROS and supports enzymatic activities, while α-tocopherol safeguards cellular membranes from peroxidation, enhancing plant resilience under saline conditions [36, 85].

6.12 Conclusion and key points

Efforts to develop salt-tolerant crops require integrating diverse strategies, from physiological and biochemical assessments to advanced molecular techniques. Physiological markers, such as growth rates, water relations, and ion content analysis, offer a broad understanding of plant responses to salinity. Biochemical markers, including the accumulation of proline and antioxidant activities, provide insights into cellular-level adaptations. Molecular markers, such as SSRs and SNPs, enable direct analysis of genetic traits linked to salt tolerance. For example, combining GWAS with expression QTL analysis in rice has revealed regulators like *STG5*, which influence Na^+/K^+ balance by modulating *OsHKT* gene expression. Similarly, in wheat, integrating morphological traits with SSR markers has been effective for evaluating salt tolerance in recombinant inbred lines.

These approaches demonstrate the multifaceted nature of salt tolerance and highlight the importance of combining multiple methodologies to effectively identify and develop salt-tolerant crops. By leveraging such integrative strategies, researchers can address the complex challenges posed by soil salinity, ensuring sustainable agricultural productivity in the face of environmental changes.

References

[1]   Abd_Allah, E. F., Hashem, A., Alqarawi, A. A., Bahkali, A. H. and Alwhibi, M. S. (2015). Enhancing growth performance and systemic acquired resistance of medicinal plant *Sesbania sesban* (L.) Merr using arbuscular mycorrhizal fungi under salt stress. Saudi Journal of Biological Sciences, 22, 274–283.

[2]   Adıgüzel, P., Nyirahabimana, F., Shimira, F., Solmaz, İ. and Taşkın, H. (2023). Applied biotechnological approaches for reducing yield gap in melon grown under saline and drought stresses: An overview. Journal of Soil Science and Plant Nutrition, 23, 139–151.

[3]   Afolayan, A. J., Grierson, D. S., Kambizi, L., Madamombe, I., Masika, P. J. and Jäger, A. K. (2002). In vitro antifungal activity of some South African medicinal plants. South African Journal of Botany, 68, 72–76.

[4]   Aghaei, K. and Komatsu, S. (2013). Crop and medicinal plants proteomics in response to salt stress. Frontiers Plant Science, 4, 8–15.

[5]   Ahmadi, F., Kariman, K., Mousavi, M. and Rengel, Z. (2024). Echinacea: Bioactive compounds and agronomy. Plants, 13(9), 1235.

[6]   Ali, H. M., Siddiqui, M. H., Basalah, M. O., Al-Whaibi, M. H., Sakran, A. M. and Al-Amri, A. (2012). Effects of gibberellic acid on growth and photosynthetic pigments of *Hibiscus sabdariffa* L. under salt stress. African Journal of Biotechnology, 11(4), 800–804.

[7]   Al-Rekaby, L. S. and Mohsin Atiyah, K. (2020). Effect of Nano and bio fertilizer on production of bioactive compounds of *Solidago canadensis* L. Journal of Physics: Conference Series, 1664(1), 012120, IOP Publishing.

[8]   Amiri, H., Banakar, M. H. and Hemmati Hassan Gavyar, P. (2024). Polyamines: New plant growth regulators promoting salt stress tolerance in plants. Journal of Plant Growth Regulation, 3, 1–18.

[9]   Amirifar, A., Hemati, A., Asgari Lajayer, B., Pandey, J. and Astatkie, T. (2022). Impact of various environmental factors on the biosynthesis of alkaloids in medicinal plants. Environmental Challenges and Medicinal Plants: Sustainable Production Solutions under Adverse Conditions, 4, 229–248.

[10]   Arzani, A. (2008). Improving salinity tolerance in crop plants: A biotechnological view. In Vitro Cellular & Developmental Biology-Plant, 44, 373–383.

[11]   Arzani, A. and Ashraf, M. (2016). Smart engineering of genetic resources for enhanced salinity tolerance in crop plants. Critical Reviews in Plant Sciences, 35(3), 146–189.

[12]   Ashraf, M. A., Iqbal, M., Rasheed, R., Hussain, I., Riaz, M. and Arif, M. S. (2018). Environmental stress and secondary metabolites in plants: An overview. Plant Metabolites Regul Environ Stress, 4, 153–167.

[13]   Ashraf, M. J. B. A. (2009). Biotechnological approach of improving plant salt tolerance using antioxidants as markers. Biotechnology Advance, 27(1), 84–93.

[14]   Ashraf, M. and Akram, N. A. (2009). Improving salinity tolerance of plants through conventional breeding and genetic engineering: An analytical comparison. Biotechnology Advance, 27(6), 744–752.

[15]   Ashraf, M., Iqbal, M., Hussain, I. and Rasheed, R. (2015). Physiological and biochemical approaches for salinity tolerance. Managing Salt Tolerance in Plants: Molecular and Genomic Perspectives, 3, 79–85.

[16]   Ashraf, M., Mukhtar, N., Rehman, S. and Rha, E. S. (2004). Salt-induced changes in photosynthetic activity and growth in a potential medicinal plant Bishop's weed (*Ammi majus* L.). Photosynthetica, 42, 543–550.

[17]   Aslam, M. S. and Ahmad, M. S. (2016). Worldwide importance of medicinal plants: Current and historical perspectives. Recent Advances in Biology and Medicine, 2(2016), 909.

[18] Assaf, M., Korkmaz, A., Ş, K. and Kulak, M. (2022). Effect of plant growth regulators and salt stress on secondary metabolite composition in *Lamiaceae* species. South African Journal of Botany, 144, 480–493.

[19] Ates, M. T., Yildirim, A. B. and Turker, A. U. (2021). Enhancement of alkaloid content (galanthamine and lycorine) and antioxidant activities (enzymatic and non-enzymatic) under salt stress in summer snowflake (*Leucojum aestivum* L.). South African Journal of Botany, 140, 182–188.

[20] Atta, K., Mondal, S., Gorai, S., Singh, A. P., Kumari, A., Ghosh, T. and Jespersen, D. (2023). Impacts of salinity stress on crop plants: Improving salt tolerance through genetic and molecular dissection. Frontiers Plant Science, 14, 1241736.

[21] Badria, F. A. (2002). Melatonin, serotonin, and tryptamine in some Egyptian food and medicinal plants. Journal of Medicinal Food, 5(3), 153–157.

[22] Bajguz, A., Bajguz, A. J. and Tryniszewska, E. A. (2013). Recent advances in medicinal applications of brassinosteroids – A group of plant hormones. Studies in Natural Products Chemistry, 40, 33–49.

[23] Banerjee, A. and Roychoudhury, A. (2017). Effect of salinity stress on growth and physiology of medicinal plants. Medicinal Plants and Environmental Challenges, 1, 177–188.

[24] Batista, V. C. V. et al. (2019). Salicylic acid modulates primary and volatile metabolites to alleviate salt stress-induced photosynthesis impairment on medicinal plant Egletes viscosa. Environmental and Experimental Botany, 167, 103870.

[25] Bennett, B. C. et al. (2000). Introduced plants in the *indigenous pharmacopoeia* of Northern South America. Economic Botany, 5, 90–102.

[26] Beyk-Khormizi, A. et al. (2023). Effect of organic fertilizer on the growth and physiological parameters of a traditional medicinal plant under salinity stress conditions. Horticulturae, 9(6), 701.

[27] Bhardwaj, S. et al. (2021). Nitric oxide: A ubiquitous signal molecule for enhancing plant tolerance to salinity stress and their molecular mechanisms. Journal of Plant Growth Regulation, 40(6), 2329–2341.

[28] Bhat, M. A. et al. (2008). Salinity stress enhances production of solasodine in *Solanum nigrum* L. Chemical and Pharmaceutical Bulletin, 56(1), 17–21.

[29] Bistgani, Z. E. et al. (2019). Effect of salinity stress on the physiological characteristics phenolic compounds and antioxidant activity of *Thymus vulgaris* L. and Thymus daenensis Celak. Industrial Crops and Products, 135, 311–320.

[30] Borges, C. V., Minatel, I. O., Gomez-Gomez, H. A. and Lima, G. P. P. 2017. Medicinal plants: Influence of environmental factors on the content of secondary metabolites. Medicinal Plants and Environmental Challenges, 3, 259–277.

[31] Briskin, D. P. (2000). Medicinal plants and phytomedicines. Linking plant biochemistry and physiology to human health. Plant Physiology, 124(2), 507–514.

[32] Caliskan, O., Radusiene, J., Temizel, K. E., Staunis, Z., Cirak, C., Kurt, D. and Odabas, M. S. (2017). The effects of salt and drought stress on phenolic accumulation in greenhouse-grown *Hypericum pruinatum*. Italian Journal of Agronomy, 12(3), 132–141.

[33] Caniato, R., Filippini, R., Piovan, A., Puricelli, L., Borsarini, A. and Cappelletti, E. M. (2003). Melatonin in plants. Developments in Tryptophan and Serotonin Metabolism, 1, 593–597.

[34] Carillo, P., Annunziata, M. G., Pontecorvo, G., Fuggi, A. and Woodrow, P. (2011). Salinity stress and salt tolerance. Abiotic Stress in Plants – Mechanisms and Adaptations, 1, 21–38.

[35] Chanchal Malhotra, C. H., Kapoor, R. and Ganjewala, D. (2016). Alleviation of abiotic and biotic stresses in plants by silicon supplementation. Scientia, 13(2), 59–73.

[36] Chen, G., Huo, Y., Tan, D. X., Liang, Z., Zhang, W. and Zhang, Y. (2003). Melatonin in Chinese medicinal herbs. Life Science, 73(1), 19–26.

[37] Chen, J. T., Aroca, R. and Romano, D. (2021). Molecular aspects of plant salinity stress and tolerance. International Journal of Molecular Sciences, 22(9), 4918.

[38] Chen, T., Shabala, S., Niu, Y., Chen, Z. H., Shabala, L., Meinke, H. and Zhou, M. (2021). Molecular mechanisms of salinity tolerance in rice. The Crop Journal, 9(3), 506–520.

[39] Chowdhury, J. B., Jain, S. and Jain, R. K. (1993). Biotechnological approaches for developing salt tolerant field crops. Journal of Plant Biochemistry and Biotechnology, 2, 1–7.

[40] Dagar, J. C., Minhas, P. S. and Mukesh Kumar, M. K. (2011). Cultivation of medicinal and aromatic plants in saline environments. CABI Reviews, 2011, 1–11.

[41] De Almeida, C. P. H. et al. (2022). Salinity and medicinal plants: Challenges and strategies for production. Scientific Electronic Archives, 15(8), 61–73.

[42] Deshmukh, Y. and Khare, P. Effect of salinity stress on growth parameters and metabolites of medicinal plants: A review. In: S. K. Gupta and Megh R. Goyal (Eds.) Soil Salinity Management in Agriculture, Apple Academic Press, Incorporated, 2017, 197–234.

[43] Egamberdieva, D., Shrivastava, S. and Varma, A., (Eds). (2015).Plant-Growth-Promoting Rhizobacteria (PGPR) and Medicinal Plants, Springer International Publishing, Cham, 42.

[44] Emami Bistgani, Z. E. et al. (2023). Review on physiological and phytochemical responses of medicinal plants to salinity stress. Communications in Soil Science and Plant Analysis, 54(18), 2475–2490.

[45] Etesami, H., Fatemi, H. and Rizwan, M. (2021). Interactions of nanoparticles and salinity stress at physiological, biochemical and molecular levels in plants: A review. Ecotoxicology and Environmental Safety, 225, 112769.

[46] Ezawa, S. and Tada, Y. (2009). Identification of salt tolerance genes from the mangrove plant *Bruguiera gymnorhiza* using Agrobacterium functional screening. Plant Science, 176(2), 272–278.

[47] Falleh, H. et al. (2012). Effect of salt treatment on phenolic compounds and antioxidant activity of two *Mesembryanthemum edule provenances*. Plant Physiology and Biochemistry, 52, 1–8.

[48] Farnsworth, N. R. and Soejarto, D. D. (1991). Global importance of medicinal plants. The Conservation of Medicinal Plants, 4, 25–51.

[49] Farouk, S. et al. (2020). Silicon supplementation mitigates salinity stress on Ocimum basilicum L. via improving water balance, ion homeostasis, and antioxidant defense system. Ecotoxicology & Environmental Safety, 206, 111396.

[50] Fathi, A. et al. (2023). Effect of plant hormones on antioxidant response and essential oil production of peppermint (*Mentha Piperita*) at different levels of salinity stress. Gesunde Pflanzen, 75(6), 2611–2622.

[51] Garg, D. et al. (2023). Nano-biofertilizer formulations for agriculture: A systematic review on recent advances and prospective applications. Bioengineering, 10(9), 1010.

[52] Gengmao, Z. et al. (2014). The physiological and biochemical responses of a medicinal plant (*Salvia miltiorrhiza* L.) to stress caused by various concentrations of NaCl. PLoS One, 9(2), e89624.

[53] Ghassemi-Golezani, K. and Abdoli, S. (2022). Physiological and biochemical responses of medicinal plants to salt stress. In: Tariq Aftab (Ed.), Environmental Challenges and Medicinal Plants: Sustainable Production Solutions under Adverse Conditions, Springer International Publishing, Cham, 153–181.

[54] Gohari, G. et al. (2023). Mitigation of salinity impact in spearmint plants through the application of engineered chitosan-melatonin nanoparticles. International Journal of Biological Macromolecules, 224, 893–907.

[55] Golkar, P. and Taghizadeh, M. (2018). In vitro evaluation of phenolic and osmolite compounds, ionic content, and antioxidant activity in safflower (*Carthamus tinctorius* L.) under salinity stress. Plant Cell Tissue Organ Cult (PCTOC), 134, 357–368.

[56] Gosal, S. S. et al. (2009). Biotechnology and drought tolerance. Journal of Crop Improvement, 23(1), 19–54.

[57] Guo, M. et al. (2022). CRISPR-Cas gene editing technology and its application prospect in medicinal plants. Chinese Medicine, 17(1), 33.

[58] Gupta, B. and Huang, B. (2014). Mechanism of salinity tolerance in plants: Physiological, biochemical, and molecular characterization. International Journal of Genomics, 2014, 701596.

[59] Gupta, S. et al. (2023). Biotechnological intervention for sugarcane improvement under salinity. Sugar Technology, 25(1), 15–31.

[60] Haider, M. Z., Ashraf, M. A., Rasheed, R., Hussain, I., Riaz, M., Qureshi, F. F. and Hafeez, A. (2023). Impact of salinity stress on medicinal plants. In: Azamal Husen, Muhammad Iqbal (Eds.), Medicinal Plants: Their Response to Abiotic Stress, Springer Nature Singapore, Singapore, 199–239.

[61] Hamilton, A. C. (2004). Medicinal plants, conservation and livelihoods. Biodiversity Conservation, 13, 1477–1517.

[62] Hao, D. C. and Xiao, P. G. (2015). Genomics and evolution in traditional medicinal plants: Road to a healthier life. Evolution Bioinform, 11, EBO–S31326.

[63] Hao, S., Wang, Y., Yan, Y., Liu, Y., Wang, J. and Chen, S. (2021). A review on plant responses to salt stress and their mechanisms of salt resistance. Horticulturae, 7(6), 132.

[64] Hao, X., Cao, H., Wang, Z., Jia, X., Jin, Z. and Pei, Y. (2024). Hydrogen sulfide improves plant drought tolerance by regulating the homeostasis of reactive oxygen species. Plant Growth Regulation, 3, 1–19.

[65] Hassan, K. M. et al. (2024). Silicon: A powerful aid for medicinal and aromatic plants against abiotic and biotic stresses for sustainable agriculture. Horticulturae, 10(8), 806.

[66] Hassan, W. et al. (2017). Oxidative stress and antioxidant potential of one hundred medicinal plants. Curr Topics Med Chem, 17(12), 1336–1370.

[67] Hernández, J. A. (2019). Salinity tolerance in plants: Trends and perspectives. International Journal of Molecular Sciences, 20(10), 2408.

[68] Hosseini, S. J. et al. (2023). Functional quality, antioxidant capacity and essential oil percentage in different mint species affected by salinity stress. Chemistry & Biodiversity, 20(4), e202200247.

[69] Huang, J. et al. (2000). Genetic engineering of glycinebetaine production toward enhancing stress tolerance in plants: Metabolic limitations. Plant Physiology, 122(3), 747–756.

[70] Huie, C. W. (2002). A review of modern sample-preparation techniques for the extraction and analysis of medicinal plants. Analytical and Bioanalytical Chemistry, 373, 23–30.

[71] Hussain, K. et al. (2010). What molecular mechanism is adapted by plants during salt stress tolerance?. African Journal of Biotechnology, 9(4), 54–63.

[72] Iqbal, R. and Khan, T. (2022). Application of exogenous melatonin in vitro and in planta: A review of its effects and mechanisms of action. Biotechnol Letters, 44(8), 933–950.

[73] Islam, S. and Mohammad, F. (2020). Triacontanol as a dynamic growth regulator for plants under diverse environmental conditions. Physiology and Molecular Biology of Plants, 26(5), 871–883.

[74] Jaleel, C. A. et al. (2008). Soil salinity alters the morphology in *Catharanthus roseus* and its effects on endogenous mineral constituents. EurAsian Journal of BioSciences, 2(1), 18–25.

[75] Jaleel, C. A. et al. (2008). Soil salinity alters growth, chlorophyll content, and secondary metabolite accumulation in Catharanthus roseus. Turkish Journal of Biology, 32(2), 79–83.

[76] Jamalian, S. et al. (2013). Abscisic acid-mediated leaf phenolic compounds, plant growth and yield in strawberry under different salt stress regimes. Theoretical and Experimental Plant Physiology, 25, 291–299.

[77] Jamwal, K. et al. (2018). Plant growth regulator mediated consequences of secondary metabolites in medicinal plants. Journal of Applied Research on Medicinal and Aromatic Plants, 9, 26–38.

[78] Jeyasri, R. et al. (2023). Methyl jasmonate and salicylic acid as powerful elicitors for enhancing the production of secondary metabolites in medicinal plants: An updated review. Plant Cell Tissue Organ Cult (PCTOC), 153(3), 447–458.

[79] Jha, U. C. et al. (2019). Salinity stress response and 'omics' approaches for improving salinity stress tolerance in major grain legumes. Plant Cell Reports, 38, 255–277.

[80] Joseph, B. and Jini, D. (2010). Proteomic analysis of salinity stress-responsive proteins in plants. Asian Journal of Plant Sciences, 9(6), 307.

[81] Kaur-Sawhney, R. et al. (2003). Polyamines in plants: An overview. Journal of Molecular Cell Biology, 2, 1–12.

[82]   Khan, Z. et al. (2024). Exogenous melatonin induces salt and drought stress tolerance in rice by promoting plant growth and defense system. Scientific Reports, 14(1), 1214.

[83]   Khanam, D. and Mohammad, F. (2018). Plant growth regulators ameliorate the ill effect of salt stress through improved growth, photosynthesis, antioxidant system, yield and quality attributes in *Mentha piperita* L. Acta Physiologiae Plantarum, 40(11), 188.

[84]   Kolář, J. and Macháčková, I. (2005). Melatonin in higher plants: Occurrence and possible functions. Journal of Pneal Research, 39(4), 333–341.

[85]   Koocheki, A., Nassiri-Mahallati, M. and Azizi, G. (2008). Effect of drought, salinity, and defoliation on growth characteristics of some medicinal plants of Iran. Journal of Herbs, Spices & Medicinal Plants, 14(1–2), 37–53.

[86]   Krishna, R. et al. (2022). Biotechnological interventions in tomato (*Solanum lycopersicum*) for drought stress tolerance: Achievements and future prospects. Biotechnology, 11(4), 48.

[87]   Kumar, A., et al. 2018. Biotechnological tools for enhancing abiotic stress tolerance in plants. Eco-friendly Agro-biological Techniques for Enhancing Crop Productivity, 147–172.

[88]   Lamsaadi, N. et al. (2024). Different approaches to improve the tolerance of aromatic and medicinal plants to salt-stressed conditions. Journal of Applied Research on Medicinal and Aromatic Plants, 2, 100532.

[89]   Lang, L. et al. (2017). Quantitative trait locus mapping of salt tolerance and identification of salt-tolerant genes in *Brassica napus* L. Frontiers Plant Science, 8, 1000.

[90]   Lemma, D. T. and Abewoy, D. (2021). Role of organic and inorganic fertilizers on the performance of some medicinal plants. International Journal of Plant Breeding and Crop Science Research, 8(1), 1016–1024.

[91]   Lemoine, P., Bablon, J. C. and Da Silva, C. (2019). A combination of melatonin, vitamin B6 and medicinal plants in the treatment of mild-to-moderate insomnia: A prospective pilot study. Complementary Therapies in Medicine, 45, 104–108.

[92]   Liang, W., Ma, X., Wan, P. and Liu, L. (2018). Plant salt-tolerance mechanism: A review. Biochem Biophys Res Commun Plants, 495(1), 286–291.

[93]   Liao, R. and Zhu, J. (2022). Amino acid promotes selenium uptake in medicinal plant Plantago asiatica. Physiology and Molecular Biology of Plants, 28(5), 1005–1012.

[94]   Linić, I. et al. (2019). Involvement of phenolic acids in short-term adaptation to salinity stress is species-specific among Brassicaceae. Plants, 8(6), 155.

[95]   Mahajan, M. and Pal, P. K. (2023). Drought and salinity stress in medicinal and aromatic plants: Physiological response, adaptive mechanism, management/amelioration strategies, and an opportunity for production of bioactive compounds. Advances in Agronomy, 182, 221–273.

[96]   Maryum, Z. et al. (2022). An overview of salinity stress, mechanism of salinity tolerance and strategies for its management in cotton. Frontiers Plant Science, 13, 907937.

[97]   Mehrafarin, A. et al. (2012). Trigonelline alkaloid, a valuable medicinal metabolite plant. Journal of Medicinal Plants, 11(41), 12–29.

[98]   Mir, T. A. et al. (2022). Influence of salinity on the growth, development, and primary metabolism of medicinal plants. In: Tariq Aftab (Ed.), Environmental Challenges and Medicinal Plants: Sustainable Production Solutions under Adverse Conditions, Springer International Publishing, Cham, 339–353.

[99]   Miransari, M. et al. (2022). The physicochemical approaches of altering growth and biochemical properties of medicinal plants in saline soils. Applied Microbiology and Biotechnology, 106(5), 1895–1904.

[100] Miransari, M. et al. (2021). The biological approaches of altering the growth and biochemical properties of medicinal plants under salinity stress. Applied Microbiology and Biotechnology, 3, 1–13.

[101] Moghaddam, M. et al. (2020). Seed germination, antioxidant enzymes activity and proline content in medicinal plant *Tagetes minuta* under salinity stress. Plant Biosystems, 154(6), 835–842.

[102] Mohammadi, H. et al. (2019). Morphophysiological and biochemical response of savory medicinal plant using silicon under salt stress. Annales Universitatis Mariae Curie-Sklodowska, Sectio C – Biologia, 72(2), 29–40.

[103] Mohammed, A. H. (2019). Importance of medicinal plants. Research in Pharmacy and Health Sciences, 5(2), 124–125.

[104] Mondal, H. K. and Kaur, H. (2017). Effect of salt stress on medicinal plants and its amelioration by plant growth promoting microbes. International Journal of Bio-resource and Stress Management, 8(3), 477–487.

[105] Muhammad, Z. and Hussain, F. (2010). Effect of NaCl salinity on the germination and seedling growth of some medicinal plants. Pakistan Journal of Botany, 42(2), 889–897.

[106] Munir, N. et al. (2022). Strategies in improving plant salinity resistance and use of salinity resistant plants for economic sustainability. Critical Reviews in Environmental Science and Technology, 52(12), 2150–2196.

[107] Munns, R. and Tester, M. (2008). Mechanisms of salinity tolerance. Annual Review of Plant Biology, 59(1), 651–681.

[108] Murch, S. J. et al. (1997). Melatonin in feverfew and other medicinal plants. Lancet, 350(9091), 1598–1599.

[109] Mustafavi, S. H. et al. (2018). Polyamines and their possible mechanisms involved in plant physiological processes and elicitation of secondary metabolites. Acta Physiologiae Plantarum, 40, 1–19.

[110] Nandy, S. et al. (2023). Role of polyamines in molecular regulation and cross-talks against drought tolerance in plants. Journal of Plant Growth Regulation, 42(8), 4901–4917.

[111] Nawaz, M. et al. (2022). Trehalose: A promising osmoprotectant against salinity stress – Physiological and molecular mechanisms and future prospective. Molecular Biology Reports, 49(12), 11255–11271.

[112] Nazari, M. et al. (2023). Deciphering the response of medicinal plants to abiotic stressors: A focus on drought and salinity. Plant Stress, 2, 100255.

[113] Negrão, S. et al. (2011). Recent updates on salinity stress in rice: From physiological to molecular responses. Critical Reviews in Plant Sciences, 30(4), 329–377.

[114] Niazian, M. (2019). Application of genetics and biotechnology for improving medicinal plants. Planta, 249, 953–973.

[115] Ahmet, O. et al. (2004). Effects of salt stress and water deficit on plant growth and essential oil content of lemon balm (*Melissa officinalis* L.). Pakistan Journal of Botany, 36(4), 787–792.

[116] Palevitch, D. (1998). Agronomics and medicinal plants. In: Medicinal Plants: Their Role in Health & Biodiversity, University of Pennsylvania Press, Philadelphia, 103–119.

[117] Pandey, D. K. et al. (2018). Arbuscular mycorrhizal fungi: Effects on secondary metabolite production in medicinal plants. Fungi and Their Role in Sustainable Development: Current Perspectives, 3, 507–538.

[118] Pardo, J. M. (2010). Biotechnology of water and salinity stress tolerance. Current Opinion in Biotechnology, 21(2), 185–196.

[119] Parekh, J., Jadeja, D. and Chanda, S. (2005). Efficacy of aqueous and methanol extracts of some medicinal plants for potential antibacterial activity. Turkish Journal of Biology, 29(4), 203–210.

[120] Pasandi Pour, A., Farahbakhsh, H. and Saffari, M. (2014). Response of Fenugreek plants to short-term salinity stress in relation to lipid peroxidation, antioxidant activity and protein content. Journal of Ethno-Pharmaceutical Products, 1(1), 45–52.

[121] Peleg, Z., Apse, M. P. and Blumwald, E. (2011). Engineering salinity and water-stress tolerance in crop plants: Getting closer to the field. In: Advances in Botanical Research, Elsevier Science & Technology, Academic Press, 57, 405–443.

[122] Prasad, A. et al. (1998). Response of *Artemisia annua* L. to soil salinity. Journal of Herbs, Spices & Medicinal Plants, 5(2), 49–55.

[123] Prinsloo, G. and Nogemane, N. (2018). The effects of season and water availability on chemical composition, secondary metabolites and biological activity in plants. Phytochemistry Reviews, 17(4), 889–902.

[124] Qin, Y. Y. et al. (2024). Aerial signaling by plant-associated Streptomyces setonii WY228 regulates plant growth and enhances salt stress tolerance. Microbiological Research, 286, 127823.

[125] Rabie, G. H. and Almadini, A. M. (2005). Role of bioinoculants in development of salt-tolerance of Vicia faba plants under salinity stress. African Journal of Biotechnology, 4(3), 210.

[126] Raei, Y. and Alami-Milani, M. (2014). Organic cultivation of medicinal plants: A review. Journal of Biodiversity and Environmental Sciences, 4, 6–18.

[127] Ram, D., Ram, M. and Singh, R. (2006). Optimization of water and nitrogen application to menthol mint (*Mentha arvensis* L.) through sugarcane trash mulch in a sandy loam soil of semi-arid subtropical climate. Bioresource Technology, 97(7), 886–893.

[128] Rani, S. et al. (2021). Biotechnological interventions for inducing abiotic stress tolerance in crops. Plant Genetic, 27, 100315.

[129] Rao, A. et al. (2013). Antioxidant activity and lipid peroxidation of selected wheat cultivars under salt stress. Journal Meidicinal Plants Reserarch, 7(4), 155–164.

[130] Rasool Hassan, B. A. (2012). Medicinal plants (importance and uses). Pharmaceutica Analytica Acta, 3(10), 2153–2435.

[131] Reiter, R. J. et al. (2011). Melatonin: New applications in clinical and veterinary medicine, plant physiology and industry. Neuroendocrinology Letters, 32(5), 575–587.

[132] Reiter, R. J. et al. (2007). Melatonin in edible plants (phytomelatonin): Identification, concentrations, bioavailability and proposed functions. More on Mediterranean Diets, 2, 211–230.

[133] Sabagh, A. E. et al. (2021). Prospective role of plant growth regulators for tolerance to abiotic stresses. Plant Growth Regulators: Signalling Under Stress Conditions, 1, 1–38.

[134] Saha, K. et al. (2004). Evaluation of antioxidant and nitric oxide inhibitory activities of selected Malaysian medicinal plants. Journal of Ethnopharmacology, 92(2–3), 16–26.

[135] Ahl HAH, S.-A. and Omer, E. A. (2011). Medicinal and aromatic plants production under salt stress: A review. Herba Polonica, 57(2), 130–141.

[136] Sciuchetti, L. A. (1961). Influence of gibberellic acid on medicinal plants. Journal Pharmaceutical Sciences, 50(12), 981–998.

[137] Seifikalhor, M. et al. (2019). Calcium signaling and salt tolerance are diversely entwined in plants. Plant Signaling and Behavior, 14(11), 1665455.

[138] Sengar, K. et al. (2013). Biotechnological and genomic analysis for salinity tolerance in sugarcane. International Journal of Biotechnology and Bioengineering Research, 4(5), 407–414.

[139] Shabala, S. and Munns, R. 2017. Salinity stress: Physiological constraints and adaptive mechanisms. In: Plant Stress Physiology, Cabi, Wallingford UK, 24–63.

[140] Shahzad, B. et al. (2022). Salt stress in brassica: Effects, tolerance mechanisms, and management. Journal of Plant Growth Regulation, 5, 1–15.

[141] Shahzad, S. M. et al. (2015). Alleviation of abiotic stress in medicinal plants by PGPR. In: Plant-Growth-Promoting Rhizobacteria (PGPR) and Medicinal Plants, Springer International Publishing Switzerland, 135–166.

[142] Shaukat, K. et al. Role of salicylic acid–induced abiotic stress tolerance and underlying mechanisms in plants. In: Tariq Aftab (Ed.), Emerging Plant Growth Regulators in Agriculture, Academic Press, 2022, 73–98.

[143] Singh, A. and Dwivedi, P. (2018). Methyl-jasmonate and salicylic acid as potent elicitors for secondary metabolite production in medicinal plants: A review. Journal of Pharmacognosy and Phytochemistry, 7(1), 750–757.

[144] Singh, M. et al. (2021). Salinity tolerance mechanisms and their breeding implications. Journal of Genetic Engineering and Biotechnology, 19(1), 173.

[145] Singh, R. K. et al. (2010). Varietal improvement for abiotic stress tolerance in crop plants: Special reference to salinity in rice. Abiotic Stress Adaptation in Plants: Physiological, Molecular and Genomic Foundation, 3, 387–415.

[146] Singla-Pareek, S. L. et al. (2003). Genetic engineering of the glyoxalase pathway in tobacco leads to enhanced salinity tolerance. Proceedings of the National Academy of Sciences of the United States of America, 100(25), 14672–14677.

[147] Sohrabi, O. et al. (2024). Exploring the effects of medicinal plant extracts on tomato (*Solanum lycopersicum* L.) morphology, biochemistry, and plant growth regulators under greenhouse conditions. International Journal of Horticultural Science and Technology, 11(3), 285–298.

[148] Stafford, G. I. et al. (2005). Effect of storage on the chemical composition and biological activity of several popular South African medicinal plants. Journal of Ethnopharmacology, 97(1), 107–115.

[149] Sunil, L. et al. (2023). Proteomics response of medicinal plants to salt stress. In: Stress-responsive Factors and Molecular Farming in Medicinal Plants, Springer Nature Singapore, Singapore, 227–241.

[150] Süntar, I. (2020). Importance of ethnopharmacological studies in drug discovery: Role of medicinal plants. Phytochemistry Reviews, 19(5), 1199–1209.

[151] Tabatabaie, S. J. and Nazari, J. (2007). Influence of nutrient concentrations and NaCl salinity on the growth, photosynthesis, and essential oil content of peppermint and lemon verbena. Turkish Journal of Agriculture and Forestry, 31(4), 245–253.

[152] Taylor, J. L. S. et al. (2001). Towards the scientific validation of traditional medicinal plants. Plant Growth Regulation, 34, 23–37.

[153] Tejesvi, M. V. et al. (2007). Genetic diversity and antifungal activity of species of Pestalotiopsis isolated as endophytes from medicinal plants. Fungal Divers, 24(3), 1–18.

[154] Thakur, M. and Kumar, R. (2021). Mulching: Boosting crop productivity and improving soil environment in herbal plants. Journal of Applied Research on Medicinal and Aromatic Plants, 20, 100287.

[155] Thakur, M. et al. (2019). Improving production of plant secondary metabolites through biotic and abiotic elicitation. Journal of Applied Research on Medicinal and Aromatic Plants, 12, 1–12.

[156] Turan, S., Cornish, K. and Kumar, S. (2012). Salinity tolerance in plants: Breeding and genetic engineering. Australian Journal of Crop Science, 6(9), 1337–1348.

[157] Tuteja, N. (2007). Mechanisms of high salinity tolerance in plants. Methods Enzymology, 428, 419–438.

[158] Van Zelm, E., Zhang, Y. and Testerink, C. (2020). Salt tolerance mechanisms of plants. Annual Review of Plant Biology, 71(1), 403–433.

[159] Wahid, A. and Rasul, E. (1997). Identification of salt tolerance traits in sugarcane lines. Field Crops Research, 54(1), 9–17.

[160] Wan, Y. X. et al. (2024). Elucidating the mechanism regarding enhanced tolerance in plants to abiotic stress by Serendipita indica. Plant Growth Regulation, 103(2), 271–281.

[161] Wang, W. X. et al. (2000). Biotechnology of plant osmotic stress tolerance physiological and molecular considerations. In: S. Sorvari and others (Eds.), IV International Symposium on in Vitro Culture and Horticultural Breeding, 560, Acta Horticulturae, 285–292.

[162] Waśkiewicz, A., Muzolf-Panek, M. and Goliński, P. (2013). Phenolic content changes in plants under salt stress. In: Parvaiz Ahmad (Ed.), Ecophysiology and Responses of Plants under Salt Stress, Springer, New York, NY, 283–314.

[163] Wu, X. et al. (2021). Melatonin: Biosynthesis, content, and function in horticultural plants and potential application. Scientia Horticulturae, 288, 110392.

[164] Yang, L. et al. (2018). Response of plant secondary metabolites to environmental factors. Molecules, 23(4), 762.

[165] Yang, X. et al. (2008). Genetic engineering of the biosynthesis of glycinebetaine leads to increased tolerance of photosynthesis to salt stress in transgenic tobacco plants. Plant Molecular Biology, 66, 73–86.

[166] Yao, W. et al. (2020). Identification of salt tolerance-related genes of *Lactobacillus plantarum* D31 and T9 strains by genomic analysis. Annals Microbiol, 70, 1–14.

[167] Yasin, N. A. et al. (2018). Halotolerant plant-growth promoting rhizobacteria modulate gene expression and osmolyte production to improve salinity tolerance and growth in Capsicum annum L. Environmental Science and Pollution Research, 25, 23236–23250.

[168] Yavas, I. (2021). Effect of salinity on morphological, physiological and biochemical properties of medicinal plants. Medicinal Aromatic Plants, 2, 45.

[169] Yeo, A. (1998). Molecular biology of salt tolerance in the context of whole-plant physiology. Journal of Experimental Botany, 49(323), 915–929.

[170] Yoshida, K. (2002). Plant biotechnology – Genetic engineering to enhance plant salt tolerance. Journal of Bioscience & Bioengineering, 94(6), 585–590.

[171] Yousefi, F. et al. (2021). Foliar application of polyamines improve some morphological and physiological characteristics of rose. Folia Horticulturae, 33(1), 147–156.

[172] Yu, Z. et al. (2020). How plant hormones mediate salt stress responses. Trends in Plant Science, 25(11), 1117–1130.

[173] Zafar, S. et al. (2024). Nano-biofertilizer an eco-friendly and sustainable approach for the improvement of crops under abiotic stresses. Environmental and Sustainability Indicators, 1, 100470.

[174] Zahra, S. T. et al. (2024). Salt-Tolerant Plant Growth-Promoting Bacteria (ST-PGPB): An effective strategy for sustainable food production. Current Microbiology, 81(10), 304.

[175] Zhao, Y. et al. (2022). Arbuscular mycorrhizal fungi and production of secondary metabolites in medicinal plants. Mycorrhiza, 32(3), 221–256.

[176] Zhou, Y. et al. (2018). Effects of salt stress on plant growth, antioxidant capacity, glandular trichome density, and volatile exudates of Schizonepeta tenuifolia Briq. International Journal of Molecular Sciences, 19(1), 252.

[177] Zubek, S. et al. (2012). Arbuscular mycorrhizal fungi and soil microbial communities under contrasting fertilization of three medicinal plants. Applied Soil Ecology, 59, 106–115.

Negar Valizadeh* and Gülen Özyazıcı

Chapter 7
Impact of heavy metal on the medicinal and aromatic plants' biochemistry

Abstract: Medicinal plants have long been vital to human health and well-being, serving as critical resources for food, medicine, and environmental sustenance. These plants are characterized by their active biochemical compounds, including alkaloids, flavonoids, and essential oils, which possess therapeutic and aromatic properties. Their importance spans traditional and modern medicine, as well as the pharmaceutical, food, and cosmetic industries. However, environmental factors, particularly abiotic stresses such as heavy metal contamination, significantly influence the production of these bioactive compounds. Heavy metals such as lead, cadmium, mercury, and arsenic are among the most harmful environmental pollutants, originating from industrial activities, mining, and agricultural practices. These metals accumulate in soil and water, disrupting plant growth, metabolic functions, and secondary metabolite production. While some metals, like copper and zinc, are essential micronutrients, their excessive concentrations, alongside toxic metals, can lead to oxidative stress, stunted growth, and bioaccumulation in plants and the food chain, posing serious risks to human and ecological health. Medicinal plants are particularly susceptible to heavy metal contamination due to their bioaccumulation tendencies. Studies have revealed significant heavy metal levels in commonly used species like mint, lavender, and rosemary, potentially jeopardizing their medicinal quality. However, this absorption ability also positions medicinal plants as key agents in phytoremediation, a biotechnological approach to mitigating environmental pollution. By tolerating and sequestering heavy metals, these plants help remediate contaminated soils while maintaining some level of secondary metabolite production. Despite their phytoremediation potential, heavy metals can alter the composition and yield of essential oils, impacting their medicinal and aromatic properties. These changes depend on factors such as metal concentration, plant species, and metabolic pathways. While low metal concentrations may stimulate essential oil production, higher levels typically inhibit it, underscoring the complex relationship between heavy metals and plant biochemistry. To address these challenges, advanced detection technologies, improved agricultural practices, and stringent regulations are essential for minimizing heavy metal

*Corresponding author: Negar Valizadeh, Research Division of Natural Resources, East Azerbaijan, Agricultural and Natural Resources Research and Education Center, Agricultural Research, Education and Extension Organization (AREEO), Tabriz, Iran, e-mail: n.valizadeh@areeo.ac.ir, https://orcid.org/0000-0003-3066-2534
Gülen Özyazıcı, Department of Field Crops, Faculty of Agriculture, Siirt University, Siirt, Türkiye

https://doi.org/10.1515/9783111469713-007

contamination. Furthermore, future research must focus on understanding the genetic and biochemical mechanisms underlying these effects, broadening the study scope to include diverse medicinal plant species. Such efforts will optimize the use of medicinal plants in polluted environments and ensure their sustainability for human and industrial applications.

Keywords: contamination, herbal medicine, heavy metals stress, plant secondary metabolites

7.1 Introduction

Plants have long been fundamental to human survival, providing food, oxygen, and medicinal benefits even before human civilization emerged. Medicinal plants, in particular, serve as valuable genetic resources and play a crucial role in shaping plant ecosystems. Their ability to adapt to diverse climates makes them a highly significant natural asset for any country. These plants are indispensable in the pharmaceutical industry, traditional medicine, and food production, as they contain bioactive compounds with therapeutic properties that contribute to health and wellbeing. The active components in medicinal plants can exert healing effects directly or indirectly and have been used for centuries to treat ailments and support overall health [1].

In modern times, medicinal plants continue to be utilized in both traditional and contemporary medicine. They serve as key ingredients in pharmaceuticals, herbal extracts, and essential oil production. These plants are particularly rich in bioactive compounds such as alkaloids, flavonoids, terpenoids, and essential oils, all of which contribute to disease treatment and support various bodily functions [2].

Aromatic medicinal plants are especially valued for their distinctive scents and therapeutic benefits. These plants are known as primary sources of essential oils and perfumes due to the presence of volatile compounds in their leaves, flowers, and other parts. Well-known aromatic plants such as lavender, mint, rosemary, and thyme are widely used as flavoring agents and fragrances in industries such as food, cosmetics, and personal care. Beyond their aromatic qualities, the essential oils derived from these plants are recognized for their antibacterial and antioxidant properties, making them vital in both cosmetic formulations and medicinal applications [3].

Plants subjected to abiotic stresses display a range of physiological, biochemical, and molecular responses. One of their primary defense mechanisms is the biosynthesis of secondary metabolites, including essential oils. Given the widespread application of these bioactive compounds in the pharmaceutical, cosmetic, and food industries, it is crucial to understand how environmental factors influence their production. While the effects of specific abiotic stresses, such as salt stress, have been extensively researched, there is limited information on how toxic pollutants affect the synthesis of active compounds in medicinal plants. Meanwhile, pollution of soil and water by toxic and heavy metals has become an escalating environmental concern. These metals accumulate in the food chain and within living organisms, making them some of the most hazardous environmental pollutants. Their presence in the environment is rising due to industrial activities, agricultural practices, and urban expansion, potentially posing serious health risks to humans and other organisms [4].

Heavy metals are characterized as elements with an atomic number above 20 and a density exceeding five grams per cubic centimeter. Some of these, including copper, zinc, nickel, molybdenum, manganese, and iron, are essential micronutrients that contribute to plant growth, redox reactions, electron transfer, and various metabolic functions. However, when present in excessive amounts, these metals can disrupt metabolic activities and inhibit plant development. In contrast, nonessential metals such as lead, cadmium, chromium, and mercury can be highly toxic to plants, even in small concentrations [5].

While heavy metals naturally exist in soil due to the weathering of parent rocks, human activities have significantly increased their concentration in the environment. Industrialization, metal smelting, mining, improper waste disposal, and the extensive use of chemical fertilizers containing heavy metals have all contributed to their accumulation in ecosystems [6].

Lead from sources such as paints, batteries, and fossil fuels; cadmium, which is a byproduct of zinc mining and phosphate fertilizer usage; and mercury, mainly emitted by chemical industries and power plants, are frequently introduced into the environment. Furthermore, mining operations and the irrigation of crops with contaminated water intensify arsenic pollution [7]. These persistent pollutants can disperse over vast distances through air and water currents, leading to extensive ecological and human health risks. Consequently, heavy metal emissions resulting from human activities present a major environmental concern.

The toxicity of heavy metals varies depending on factors such as their chemical state, concentration, and bioavailability. Once released into soil, water, and air, these metals can easily enter the food chain and be absorbed by living organisms. Their interactions with cellular structures can severely disrupt essential biological processes, including enzyme activity, oxidative balance, and DNA integrity. This disruption can result in oxidative stress, impaired growth, and in severe cases, plant death. In humans and animals, exposure to these metals can cause neurological disorders and kidney damage.

Due to their low degradability, heavy metals persist in the environment for extended periods, increasing the likelihood of long-term exposure and bioaccumulation [8]. High concentrations of these metals in soil pose a significant threat by degrading soil structure, reducing biological activity and fertility, decreasing agricultural productivity, lowering crop quality, and raising metal concentrations in food products. Ultimately, this contamination can have severe implications for human health by introducing toxic elements into the food supply [9].

As a result, ongoing research aims to further explore the toxic effects of heavy metals, improve detection techniques, and develop effective methods for remediating contaminated environments [10]. Heavy metal pollution is a pressing environmental issue with significant implications for medicinal plants and, consequently, human health. Many medicinal plants have a natural capacity to absorb heavy metals from the soil, leading to the accumulation of these elements in their tissues. This is particularly concerning in areas where soil and water contamination levels are high. Metals such as lead, cadmium, mercury, and arsenic are among the most prevalent pollutants that accumulate in plant structures, including roots, stems, and leaves. Research indicates that certain plants, such as mint, lavender, and rosemary, are especially prone to absorbing these metals [11, 12].

Interestingly, this absorption ability has an advantage – medicinal plants can be used in phytoremediation, a biotechnological strategy for reducing environmental pollution. Certain plant species are capable of withstanding environmental stressors, including heavy metals, making them valuable tools for mitigating contamination. Their ability to synthesize secondary metabolites and antioxidant compounds helps counteract the damaging effects of metal toxicity [4]. Research has shown that medicinal plants in polluted environments can absorb and store heavy metals, such as cadmium and lead, while still maintaining the production of bioactive compounds [13].

A study conducted by Luo et al. [14], titled "Heavy Metal Contaminations in Herbal Medicines," revealed that 30.51% of the analyzed herbal medicine samples contained at least one heavy metal exceeding permissible concentration limits. While 70.93% of these samples were within an acceptable risk threshold, arsenic was identified as the most hazardous element. The research indicated that plants such as *Tetradium ruticarpum*, *Plantago asiatica*, and *Desmodium styracifolium* exhibited the highest contamination risks.

Similarly, another study by Abou-Arab et al. [15], titled "Accumulation of Heavy Metals in Selected Medicinal Plants," investigated heavy metal accumulation in 88 medicinal plant species. Findings from this study demonstrated that certain species contained metal concentrations surpassing the permissible limits set by international regulatory bodies. These results emphasize the need for strict monitoring of contamination levels in medicinal plants, as prolonged consumption could lead to serious health hazards.

Given the essential role of medicinal plants in healthcare, industry, and medicine, controlling heavy metal contamination is of paramount importance. Recommended strategies to mitigate contamination include adopting improved agricultural techniques, using uncontaminated soils, and enforcing strict regulatory measures to monitor plant-based products. Additionally, advancements in rapid contamination detection technologies and the implementation of innovative phytoremediation techniques can significantly help minimize the adverse effects of heavy metal exposure.

7.2 Heavy metals and their effects on the environment

Heavy metals are a class of elements characterized by their high density compared to water. While some are essential in trace amounts, many become toxic at elevated concentrations, posing risks to both ecosystems and human health. Common heavy metals include lead, cadmium, mercury, arsenic, zinc, copper, nickel, chromium, iron, aluminum, beryllium, cobalt, and manganese. These metals are widely recognized as environmental pollutants due to their persistence and potential for bioaccumulation.

Although heavy metals naturally occur in the environment, their levels have significantly risen due to human activities. Industrial processes such as mining, metal refining, battery manufacturing, and paint production contribute heavily to their accumulation. Agricultural practices, including pesticide application, and the disposal of municipal and industrial wastewater further exacerbate contamination. The combustion of fossil fuels also releases heavy metals into the air, contributing to widespread pollution. In particular, gold mining operations and emissions from burning fossil fuels are major contributors to mercury pollution, allowing it to enter and persist in the environment [16].

Arsenic primarily contaminates the environment through groundwater pollution, which can result from both industrial processes and natural geological activities. It is widely regarded as a major environmental pollutant in numerous regions across the globe [17]. Due to their distinct physical and chemical characteristics, heavy metals have a tendency to persist in ecosystems, leading to long-term adverse effects on living organisms. These metals dissolve easily in water, allowing plants to absorb them through their root systems, ultimately introducing them into the food chain. By interacting with plant proteins and enzymes, heavy metals can interfere with essential metabolic and physiological functions. Furthermore, their accumulation in plant and animal tissues can contribute to growth abnormalities, decreased agricultural yields, and the formation of harmful biotoxins in the food supply [18].

Among heavy metals, lead is considered particularly hazardous due to its widespread use in products such as batteries, paints, and fuel additives. This metal enters the soil and water, where it can be absorbed by plants, disrupting their physiological processes. Lead exposure hinders plant growth, alters metabolic activities, and increases oxidative stress by affecting antioxidant enzymes [19]. A study conducted by Usman et al. [20] examined how different lead concentrations influence the antioxidant response of *Tetraena qataranse*. The findings revealed that while low lead levels stimulated plant growth, higher concentrations (100 mg/L or 1,600 mg Pb per kg of soil) significantly inhibited root and shoot development. Additionally, lead accumulation was found to be much greater in the roots (2,784 mg/kg) than in the shoots (1,141.6 mg/kg).

The activity of antioxidant enzymes, including superoxide dismutase (SOD), catalase (CAT), ascorbate peroxidase (APX), guaiacol peroxidase (GPX), and glutathione reductase (GR), showed an upward trend with increasing lead concentrations. This indicates that *Tetraena qataranse* possesses the ability to absorb significant amounts of lead while mitigating its toxic effects through enhanced antioxidant defense mechanisms, thereby improving its chances of survival.

Cadmium is another highly toxic heavy metal known for its detrimental impact on plant growth and physiological functions. It primarily enters biological systems through soil and water contamination. This metal disrupts essential plant processes by interfering with nutrient absorption and damaging cellular structures. One of the most significant consequences of cadmium exposure in plants is the decline in bioactive compounds, which directly influences the medicinal value of these plants [18].

Several studies have explored the impact of cadmium on secondary metabolite production in medicinal plants. For example, research examining *Gynura procumbens* under cadmium and copper stress reported inhibited growth and modifications in secondary metabolite synthesis, potentially affecting the plant's medicinal properties [21]. Similarly, a study focusing on *Salvia miltiorrhiza* found that higher cadmium concentrations in the soil led to increased cadmium accumulation in the plant's roots and leaves, which in turn influenced the biosynthesis of key active compounds [22].

Cadmium is highly mobile in soil, with its movement influenced by several factors, including pH levels, organic matter content, cation exchange capacity, and the mineral composition of the soil. In acidic soils with low organic matter, cadmium availability tends to increase. For example, studies on *Hypericum* species collected from acidic soils in eastern Austria revealed higher cadmium concentrations compared to those in calcareous soils [23]. Additionally, research suggests that while organic matter can retain cadmium within the soil solution, it may also facilitate the transfer of cadmium from the plant's roots to its aerial parts [24].

Exposure to cadmium triggers intense oxidative stress in plants by suppressing key enzymatic activities and promoting excessive production of reactive oxygen species (ROS). This oxidative stress can result in DNA damage, protein degradation, and lipid peroxidation in cell membranes, ultimately compromising both the quality and productivity of medicinal plants. Studies have further demonstrated that cadmium accumulation in the flowering parts of *Matricaria chamomilla* is influenced by climatic variations [25].

Certain medicinal plants possess a natural tendency to accumulate cadmium and are classified as hyperaccumulators. Research on *Hypericum perforatum* has shown considerable variation in cadmium uptake, with concentrations among 56 examined accessions ranging from 0.04 to 7.8 ppm [26] in Figure 7.1. Figure 7.1 shows a view of the flowering period of the *Hypericum perforatum* plant under field conditions.

Certain plant species, including yarrow (*Achillea millefolium*), German chamomile (*Matricaria chamomilla*), and tobacco (*Nicotiana* sp.), are recognized as hyperaccumulators of cadmium [27]. Among them, the cadmium content in tobacco leaves varies between 0.3 and 2.2 ppm, depending on geographical location [28]. Mistletoe (*Viscum*

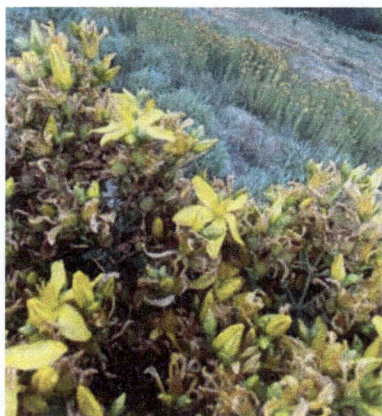

Figure 7.1: *Hypericum perforatum* L.

album) also exhibits variable cadmium accumulation, influenced by the host tree it grows on. This plant tends to store lower cadmium levels when attached to fruit trees like apple and hawthorn but accumulates higher amounts on hosts such as pine and willow [29].

To minimize the detrimental effects of cadmium toxicity, proper soil management and the selection of plant accessions with lower cadmium uptake play a crucial role. For instance, poppy species cultivated in contaminated soils demonstrated significant cadmium accumulation; however, when grown in uncontaminated conditions, their cadmium levels decreased considerably [30]. Additionally, applying cadmium in a soluble form at the initial stages of an experiment leads to increased plant uptake, as the metal has not yet had sufficient time to stabilize in the soil matrix [31].

Given the detrimental impact of cadmium on the quality of medicinal plants, it is crucial to monitor pollution sources and adopt effective soil and plant management strategies. Selecting suitable cultivation sites and growing heavy metal-resistant plant species can help minimize cadmium entry into the food chain and maintain the medicinal properties of plants. Research indicates that introducing copper into soil at elevated levels (up to 300 ppm) can lead to its accumulation in the aerial parts of chamomile, reaching concentrations as high as 271 ppm, without causing visible toxicity symptoms in the plant [32].

Moreover, utilizing composts rich in copper – containing up to 760 ppm of this metal – has been found to have no adverse effects on the growth of medicinal species such as peppermint (*Mentha × piperita*) and dill (*Anethum graveolens*). These plants only absorb minimal amounts of copper (around 12 ppm) into their tissues [33]. Studies conducted on milk thistle (*Silybum marianum*) cultivated in highly polluted soils demonstrated a decline in seed yield (Figure 7.2). Despite the significant presence of heavy metals, essential compounds such as fats and silymarin remained unaffected, ensuring that the final product was free from contamination [34].

Figure 7.2: *Silybum marianum* seeds.

Likewise, clary sage (*Salvia sclarea*) cultivated in soils with elevated levels of cadmium, lead, and zinc did not exhibit any trace of contamination in its extracted essential oil [35]. Similarly, essential oils obtained from different peppermint species (*Mentha × piperita* and *Mentha arvensis*) grown in highly polluted environments were also devoid of heavy metal contamination. This trait suggests their potential use in the gradual rehabilitation of contaminated soils [36].

Mercury, a highly toxic heavy metal, primarily enters ecosystems through industrial emissions and pollution from fossil fuel combustion. It readily dissolves in water, making it easily absorbable by plants and subsequently entering the food chain. Mercury exposure can disrupt enzymatic functions, impair photosynthetic efficiency, and ultimately hinder plant growth. Additionally, its accumulation in plant tissues poses significant risks to both human and animal health upon consumption [37].

Mercury, as a heavy and toxic metal, plays a crucial role in environmental pollution. Research indicates that plant absorption of mercury is minimal, leading to generally negligible concentrations in medicinal plants. For instance, an analysis of wild chamomile flowers collected across Slovakia between 1995 and 2003 found an average mercury concentration of 0.04 ppm [38]. Likewise, plant samples from *Vaccinium* species, birch, and willow, gathered in Finland and northeastern Russia, contained mercury levels below 0.04 ppm [39]. A comprehensive study by Gasser et al. [29] in Germany also reported that mercury concentrations in nearly 120 medicinal herb samples remained under 0.1 mg/kg.

Mercury accumulation is more prevalent in algae. For example, research along Portugal's northern coast revealed that *Fucus* species exhibited higher mercury levels than the surrounding sediments [40].

Arsenic, another hazardous heavy metal, is commonly found in groundwater and surface water. It is easily absorbed by plants, primarily accumulating in their leaves and roots, where it can degrade plant quality and overall performance. Arsenic exposure can trigger oxidative stress and cause severe cellular damage, ultimately leading to stunted growth and lower yields [41]. Studies from Poland indicate that arsenic levels in *Mentha × piperita* range between 0.05 and 0.13 mg/kg, while in *Urtica dioica*, concentrations vary from 0.09 to 0.24 mg/kg (Figure 7.3) [42].

Figure 7.3: Leaves of *Mentha piperita* and *Urtica dioica*.

In Austria, a study analyzing plant samples grown in chernozem soil – a fertile, humus-rich black soil – found that arsenic concentrations in the leaves and branches ranged from 1.2 to 2.0 mg/kg, while the soil itself contained 18 mg/kg of arsenic [43]. Similarly, for *Mentha spicata* (spearmint), the reported average arsenic content is approximately 0.2 mg/kg (Figure 7.4) [44].

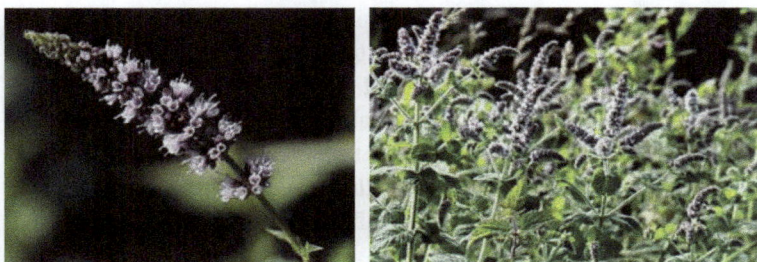

Figure 7.4: Leaves and flowers of *Mentha spicata*.

Furthermore, analyses of fennel seed samples from India have detected arsenic concentrations ranging from 0.51 to 0.59 mg/kg [45]. Research conducted in Bulgaria has revealed that commonly consumed tea bags contain arsenic levels between 0.02 and 0.25 mg/kg [46]. Additionally, significant arsenic accumulation has been reported in macroalgae and sea moss, with concentrations varying from 20 to 100 mg/kg [47]. These results underscore the need for stringent monitoring of arsenic levels in food and medicinal products, particularly in edible and therapeutic plants.

Although nickel is an essential micronutrient for certain plants, excessive amounts can be harmful. At high concentrations, it interferes with protein and enzyme structures, disrupting vital metabolic activities such as respiration and photosynthesis. Moreover, nickel exposure can induce genetic and structural alterations in plants, potentially leading to long-term damage. Such effects are particularly concerning for medicinal and aromatic plants, as their bioactive compound composition must remain stable and of high quality [18]. Consequently, identifying sources of contamination and mitigating heavy metal levels in the environment is crucial.

7.3 Mechanisms of heavy metal uptake and accumulation in plants

The soil solution, which serves as the medium for plant mineral absorption, consists of a complex blend of elements. Consequently, interactions between these elements frequently occur in plants. Due to the chemical resemblance between cadmium and zinc, their interactions have been widely observed in various plant species. Research on plants cultivated in different soil types has shown that higher zinc levels in flax seeds (*Linum usitatissimum*) lead to a significant decrease in cadmium accumulation [48]. A pot experiment conducted in North Dakota, USA, further confirmed that soil enrichment with zinc can lower cadmium content in flax seeds. However, when cadmium and zinc were applied together, cadmium concentration in the seeds increased compared to when cadmium was added alone [49]. The application of monoammonium phosphate in flax seeds has been found to elevate cadmium concentrations while decreasing zinc levels [48]. Chizzola and Mitteregger [50] explored cadmium-zinc interactions in chamomile plants, revealing that soil enrichment with zinc significantly lowered cadmium accumulation in the aerial parts. However, this reduction was insufficient for plants in contaminated areas to match the cadmium levels of those in uncontaminated regions. In the case of *Achillea millefolium* (yarrow), the concurrent addition of cadmium and zinc did not significantly influence cadmium accumulation [51].

Cadmium-manganese interactions have also been documented in lettuce, where an increase in cadmium concentration within the nutrient solution led to greater manganese absorption and translocation to the aerial parts [52]. Similarly, nutrient solution studies indicated that introducing a small amount of cadmium enhanced iron uptake in sorghum [24]. In *Picea abies* (Norway spruce), elevated calcium levels reduced cadmium and zinc accumulation, whereas the presence of cadmium or copper hindered calcium uptake [53]. Additionally, interactions between phosphorus and zinc at both the plant and soil levels have been reported [54].

7.4 Processes of heavy metal uptake by roots

Roots serve as the primary interface between plants and soil, playing a crucial role in the absorption of heavy metals. This process occurs through two main mechanisms: active and passive uptake. Active uptake involves the transport of heavy metal ions into root cells via specialized membrane proteins, such as H^+ ions and H^+-ATPase. This energy-driven mechanism enables plants to absorb heavy metals efficiently from the soil. In contrast, passive uptake depends on diffusion or nonspecific ion channels, allowing heavy metals to enter the plant without energy expenditure [55].

Several factors influence the efficiency of heavy metal absorption by roots. Soil pH is a major determinant, as it affects the availability of heavy metals for uptake. In acidic conditions, heavy metals are more likely to exist in their ionic forms, making them more accessible to plants. Additionally, soil characteristics such as organic matter content and cation exchange capacity also play a role in regulating heavy metal absorption [7].

Dinu et al. [12] examined how peppermint (*Mentha piperita*) responds to soil contamination by heavy metals such as cadmium, lead, nickel, and arsenic. The study assessed the movement of these metals from the soil into different plant parts, including roots, stems, and leaves, and compared the findings with a control group grown in uncontaminated soil. Peppermint seedlings were exposed for 3 months to two metal mixtures with similar concentrations. The first mixture contained arsenic and cadmium, while the second included arsenic, cadmium, nickel, and lead. The metal concentrations were 23.7 mg/kg for arsenic, 5 mg/kg for cadmium, 136 mg/kg for nickel, and 95 mg/kg for lead.

The results showed that cadmium, nickel, and lead accumulated in different plant tissues, with arsenic remaining undetectable. The accumulation pattern followed the order: roots > stems > leaves. During the first month, no significant differences in growth, development, or chlorophyll content were observed when compared to the control plants. However, after 3 months, signs of phytotoxicity began to appear. Analysis of metal translocation and transfer factors yielded values below 1, indicating that *M. piperita* primarily retained metals within its root system. These findings suggest that peppermint can effectively stabilize heavy metals in its roots during short-term exposure, demonstrating its tolerance to contaminated conditions when grown in nutrient-rich garden soil. This research highlights peppermint's potential as a metal-resistant plant, particularly for applications where root-based metal stabilization is needed.

7.5 Transport and accumulation in various plant tissues

Once absorbed by the roots, heavy metals are transported to different plant tissues via the vascular system. In flowering plants, this movement primarily occurs through the xylem, which carries metals from roots to stems and leaves, and the phloem, which facilitates transport between stems and leaves. During this process, heavy metals accumulate in various plant parts. While roots, stems, and leaves serve as the primary sites of accumulation, their specific distribution varies based on plant species, metal type, and environmental conditions. For instance, cadmium predominantly accumulates in roots, whereas lead and zinc are more commonly found in stems and leaves [6].

A study conducted in 2021, titled *Hysteresis of Heavy Metal Uptake in Dandelions*, examined how heavy metals are absorbed and distributed within dandelion plants. The results indicated that these metals could accumulate in different plant tissues, potentially affecting the plant's medicinal properties [56].

7.6 The effect of soil type and environmental conditions on heavy metal uptake

Soil composition and environmental conditions are key factors influencing the uptake and accumulation of heavy metals in plants. Various physical and chemical characteristics of the soil, such as pH, moisture content, organic matter, mobile cation concentrations, and microbial activity, significantly affect the extent to which plants absorb these metals. For example, in acidic soils, heavy metal ions are more readily available, facilitating their uptake by plants. Conversely, in alkaline soils, heavy metals tend to form mineral complexes or insoluble precipitates, reducing their bioavailability.

In addition to soil properties, environmental factors like temperature, light, and moisture play a crucial role in heavy metal absorption and transport within plants. Higher temperatures can stimulate enzymatic activity and metabolic processes, thereby enhancing metal uptake. Similarly, moisture levels influence metal bioavailability, with heavy metals being more accessible to plants in wetter soil conditions [13, 57].

A study investigated the effects of different exposure durations to thiram on the uptake of manganese, cobalt, nickel, copper, zinc, cadmium, and lead in basil (*Ocimum basilicum* L.). In addition, the research assessed various physiological parameters of the plants. The experiment was conducted using two common agricultural soil types found in rural areas of Poland. The methodology included soil analysis to determine bioavailable and total metal forms, measurements of chlorophyll content, and evaluations of gas exchange in plants. Metal concentrations were quantified using atomic absorption spectroscopy.

Analysis of variance indicated that thiram treatment significantly affected metal transfer from the soil to basil plants, influencing metal concentrations in both the roots and aerial parts. These effects were most pronounced on the 14th day after fungicide application. Thiram altered metal uptake and distribution within basil by modifying the microbial composition of the rhizosphere. The impact was more noticeable in plants grown in mineral soils than in organic soils, which have a higher buffering capacity [58].

7.7 The effects of heavy metals on physiology, metabolism, and secondary metabolites in medicinal and aromatic plants

Heavy metals have a significant impact on various plant physiological functions, including growth, photosynthesis, respiration, and metabolic activities. One of the primary effects is their disruption of the photosynthetic process. Metals such as cadmium, lead, and zinc can inhibit key photosynthetic enzymes, including RuBisCO (ribulose-1,5-bisphosphate carboxylase/oxygenase), and interfere with chlorophyll production. This disruption reduces photosynthetic efficiency, limiting the energy available for plant growth [59].

Additionally, heavy metals influence plant respiration by altering mitochondrial structure and function, thereby reducing energy efficiency during the respiratory process [55]. These changes lead to decreased ATP production and an accumulation of free radicals, which can damage cellular membranes, proteins, and DNA. Heavy metal exposure also negatively affects plant growth, particularly in the early stages of development, leading to lower germination rates, reduced root and shoot elongation, and decreased dry weight [60, 61].

The adverse effects of heavy metals are particularly concerning for medicinal and aromatic plants, as their optimal growth is essential for synthesizing bioactive compounds used in pharmaceutical and aromatic industries. These plants are known for producing secondary metabolites with therapeutic and aromatic properties. Heavy metal exposure can interfere with the production of these metabolites by affecting enzyme activity and disrupting metabolic pathways. Key affected compounds include alkaloids (e.g., morphine, atropine, and papaverine), flavonoids (e.g., quercetin, apigenin, and luteolin), terpenoids (e.g., menthol, thymol, and linalool), and anthocyanins (Figures 7.5–7.7).

For instance, in aromatic plants such as mint and lemon, heavy metals can impair the biosynthesis of terpenoid compounds, which contribute to their characteristic aromas. Metals like cadmium and lead inhibit the activity of enzymes involved in terpenoid production, leading to a reduction in these vital compounds [62, 63].

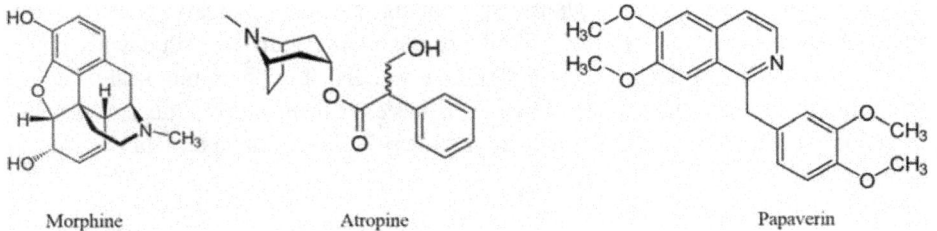

Morphine   Atropine   Papaverin

Figure 7.5: Alkaloids' chemical structures.

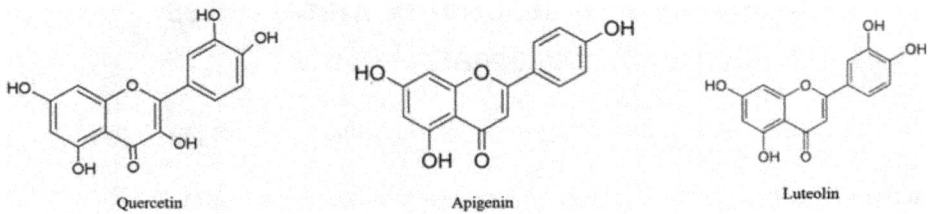

Quercetin   Apigenin   Luteolin

Figure 7.6: Flavonoids' chemical structures.

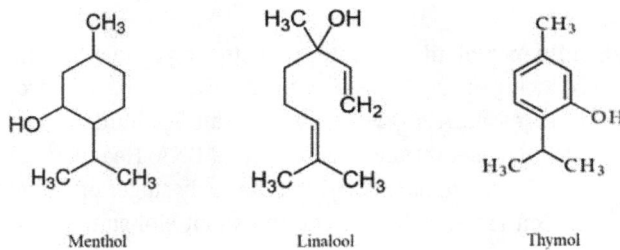

Menthol   Linalool   Thymol

Figure 7.7: Terpenoids' chemical structures.

Heavy metals can significantly influence the biochemical composition of medicinal and aromatic plants, leading to notable reductions in anthocyanins, flavonoids, and terpenoids. Anthocyanins and flavonoids, which belong to the phenolic compound group, play a crucial role in protecting plants from environmental stress while also providing antioxidant benefits to counteract free radical damage. Research suggests that heavy metal exposure leads to decreased anthocyanin and flavonoid concentrations in plants such as thyme, basil, and sage. This reduction is likely attributed to disruptions in their biosynthetic pathways, triggered by oxidative stress resulting from heavy metal accumulation [64].

Similarly, terpenoid production in aromatic plants, including basil, mint, and pennyroyal, can be significantly impaired by heavy metals. These compounds are essential not only for the characteristic scent of these plants but also for their medicinal properties, such as anti-inflammatory and anticancer activities. A decline in terpenoid biosynthesis negatively affects both the therapeutic efficacy and aromatic qualities of these plants, diminishing their overall value in medicinal and aromatic applications [65].

Antioxidant activity is a fundamental medicinal property of plants, playing a vital role in shielding them from damage caused by free radicals. When exposed to heavy metal contamination, plants experience elevated levels of free radicals, placing considerable stress on their antioxidant defense systems. While some plants may temporarily enhance their antioxidant responses to counteract the damage, excessive heavy metal accumulation can overwhelm these systems, ultimately reducing their effectiveness and making plants more vulnerable to oxidative stress [55].

In addition to affecting antioxidant properties, heavy metals also influence the antibacterial activity of medicinal and aromatic plants. These plants naturally produce antibacterial compounds that help them resist harmful bacteria and microorganisms. However, exposure to heavy metals can alter the composition of these bioactive substances, leading to a decline in antibacterial potency and a subsequent reduction in the plants' therapeutic value. Studies have demonstrated that heavy metals such as cadmium and lead can impair the antibacterial properties of plants, thereby diminishing their medicinal efficacy [7].

Furthermore, bioactive compounds – including alkaloids, glycosides, and essential oils – are critical components of medicinal and aromatic plants, contributing to their therapeutic benefits. These compounds not only offer health-promoting effects but also define the unique characteristics of each plant. Heavy metal contamination disrupts the biosynthetic pathways responsible for producing these bioactive substances. For instance, alkaloids present in plants like *Datura* and *Peganum harmala* decrease under heavy metal stress due to the inhibited activity of key enzymes, such as alkaloid synthase, which are essential for their synthesis [8].

Likewise, glycosides, which are found in plants such as ginseng and thyme, are adversely affected by heavy metal stress. Alterations in the plant's internal environment due to metal exposure result in decreased glycoside production [7]. Essential oils, primarily composed of terpenoid compounds, are particularly vulnerable to heavy metal contamination. Plants like mint and pennyroyal experience a significant reduction in essential oil synthesis when exposed to metals such as cadmium and lead. This decline negatively affects both the aromatic properties and medicinal efficacy of these plants [65].

7.8 Plant defense mechanisms against heavy metals

Plants utilize various defense strategies to withstand stress caused by heavy metal exposure. These adaptive mechanisms involve alterations in their antioxidant systems and the production of signaling molecules to mitigate heavy metal toxicity. The plant's response to heavy metals is highly complex and depends on factors such as contamination levels and the specific type of metal involved. One of the most notable responses to heavy metal stress is the increased activity of antioxidant defense systems.

Heavy metal exposure leads to the generation of free radicals within plant tissues, which can harm cellular membranes, proteins, and DNA. To counteract this damage, plants enhance their antioxidant defenses. Key antioxidant enzymes, including SOD, CAT, and peroxidase (POD), play a crucial role in neutralizing free radicals and minimizing oxidative harm. Research indicates that plants such as wheat and maize exhibit a significant rise in antioxidant enzyme activity, particularly SOD and CAT, when subjected to heavy metal stress as a means of reducing oxidative damage [66, 67].

Under extreme heavy metal contamination, the efficiency of antioxidant defense systems may decline significantly, compromising the plant's ability to neutralize free radicals and mitigate oxidative stress. In such conditions, cellular membranes, proteins, and DNA become increasingly vulnerable to damage, leading to the breakdown of cellular structures and disruption of essential physiological processes. The weakened defense mechanisms can manifest as cellular stress symptoms, inhibited growth, decreased synthesis of vital biochemical compounds, and, in severe cases, cell death.

Beyond enzymatic antioxidants, nonenzymatic compounds also play a vital role in plant defense. These include flavonoids, phenols, and vitamin C, which help counteract heavy metal-induced oxidative stress through their antioxidant properties. Research suggests that exposure to heavy metals such as cadmium and lead can trigger an increase in the levels of these antioxidant compounds in certain plants, including beans and sainfoin, as part of their adaptive response to stress [67].

Plants synthesize specific signaling compounds to adapt to heavy metal stress. These molecules function as internal messengers, enabling plants to regulate their responses to adverse conditions. Among these compounds are plant hormones such as auxins, cytokinins, salicylic acid (SA), and ethylene. SA is a key signaling molecule involved in plant responses to environmental stress, including heavy metal exposure. This hormone plays a vital role in strengthening plant defense mechanisms against metal toxicity. Research indicates that elevated levels of SA can enhance the activity of antioxidant enzymes, thereby alleviating the harmful effects of heavy metals on plant health [68, 69]. Ethylene is another essential hormone produced in response to heavy metal stress. It contributes to stress adaptation by regulating critical physiological processes, such as root development and tissue growth, which help plants better withstand unfavorable conditions [70].

7.9 Molecular and genetic responses to heavy metal contamination

Plants adapt to heavy metal stress through various molecular and genetic modifications. These adaptations involve changes in gene expression, activation of genetic resistance pathways, and, in some cases, genetic engineering techniques aimed at enhancing tolerance. Heavy metals such as cadmium, lead, zinc, and copper influence multiple signaling pathways, leading to alterations in the expression of specific genes that regulate plant defense responses.

One of the primary responses to heavy metal exposure is the increased expression of genes involved in the production of antioxidants and protective enzymes. For example, exposure to heavy metals can stimulate the upregulation of genes responsible for synthesizing SOD, CAT, and POD. These enzymes are critical in neutralizing free radicals, reducing oxidative stress, and safeguarding plant cells from heavy metal toxicity [71, 72].

Additionally, heavy metals can activate genes associated with hormonal signaling pathways, including those governing ethylene, SA, and auxins. These hormones play an essential role in regulating plant defense responses to heavy metal stress [68]. To survive in polluted environments, plants employ complex genetic mechanisms that regulate the uptake, transport, storage, and detoxification of heavy metals within their cells and tissues.

A crucial component of the plant defense system against heavy metal stress is the role of transporter proteins, which regulate the movement of heavy metals within the plant. These proteins help transfer metals from the roots to other tissues or confine them within specific compartments for safe storage. For instance, metallothionein (MT) proteins and heavy metal ATPases (HMA) play a significant role in absorbing heavy metals and isolating them within designated plant structures.

Additionally, plants can minimize heavy metal uptake by stabilizing these contaminants in the soil. This process involves the secretion of chemical compounds, such as organic acids and amino acids, which bind with heavy metals to reduce their mobility. One such mechanism, known as phytostabilization, allows plants to release these substances into the surrounding soil, thereby preventing further absorption of heavy metals [73].

Another way plants resist heavy metal stress is by making structural and functional adjustments to their cell membranes. Some plants reinforce their membranes to block heavy metal entry, while others modify their structure to reduce metal absorption. Given the significant role of medicinal and aromatic plants, genetic engineering has emerged as a promising strategy to enhance their ability to tolerate heavy metals. This approach involves introducing genetic modifications that strengthen plant defense mechanisms and improve their capacity for metal uptake and detoxification.

One genetic engineering technique involves incorporating genes that encode MT proteins and heavy metal transporters into medicinal plants. These genes help plants effectively sequester heavy metals in specific tissues or expel them altogether. For example, transferring HMA or MT genes into medicinal plants has been found to enhance their resistance to metals such as cadmium and lead. Additionally, genetic modifications can target hormonal signaling pathways to improve stress resilience. Enhancing the activity of SA and ethylene signaling pathways, for instance, enables plants to better manage the adverse effects of heavy metals. Research has shown that increasing the production of these hormones through genetic engineering significantly improves the tolerance of medicinal and aromatic plants to heavy metal contamination.

Furthermore, cutting-edge technologies like CRISPR/Cas9 offer precise gene-editing tools to enhance plant resistance. This method allows for the targeted modification of genes responsible for heavy metal detoxification, thereby strengthening the plant's ability to tolerate and manage these contaminants more effectively [74].

7.10 Management and control of heavy metal contamination in medicinal and aromatic plants

To mitigate the effects of heavy metals on plants, various agricultural and management strategies can be employed. One crucial approach is reducing soil contamination by addressing pollution sources. Effective management of industrial and agricultural waste plays a vital role in preventing heavy metals from entering the soil, thereby minimizing long-term environmental damage.

Additionally, techniques such as cover cropping and crop rotation help limit heavy metal exposure. Cover crops assist in immobilizing heavy metals within the soil, reducing their uptake by medicinal and aromatic plants. Likewise, crop rotation and diversifying plant cultivation in a given area can prevent excessive accumulation of heavy metals in the soil. For instance, planting crops that absorb and retain heavy metals in their root systems can act as a protective barrier, shielding medicinal plants from contamination [75, 76]. Effective soil and water remediation techniques play a vital role in managing heavy metal contamination in affected regions. One commonly applied approach is phytoremediation, which utilizes plants to cleanse environments polluted with heavy metals. This method encompasses various processes, including phytostabilization, hyperaccumulation, and phytodetoxification. Among these, phytostabilization is particularly significant, as it involves the use of plants to restrict heavy metals within the soil, thereby preventing their dispersion and minimizing environmental hazards (Figure 7.8).

PHYTOROMEDIATION TYPES

Phytovolatilization

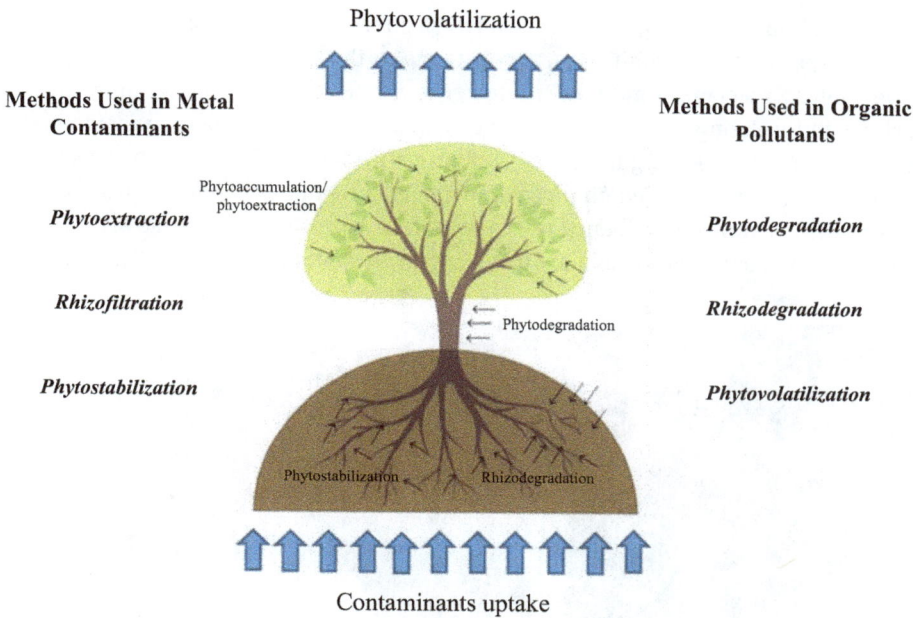

Figure 7.8: Uptake mechanisms by plants with phytoremediation technology [77].

Hyperaccumulation is a process in which certain plants absorb significant amounts of heavy metals and store them within their tissues [77]. This mechanism is particularly effective for plant species thriving in metal-contaminated soils. Studies have identified plants such as *Brassica juncea*, *Helianthus annuus*, and *Thlaspi caerulescens* as highly efficient at extracting heavy metals from the soil and accumulating them within their tissues [78]. These species play a crucial role in remediating soils polluted with metals like cadmium, lead, zinc, and copper.

Beyond phytoremediation, another effective method for detoxifying contaminated soils involves the application of soil amendments. These substances help reduce the mobility and bioavailability of heavy metals in the soil. For example, adding organic materials such as compost and biofertilizers assists in stabilizing heavy metals and limiting their movement. Likewise, using mineral amendments like lime and gypsum can alter soil pH levels, thereby decreasing heavy metal availability to plants [79].

7.11 Selection of heavy metal-resistant plants

An effective approach to mitigating the adverse effects of heavy metals is the selection of plant species that exhibit natural resistance to these contaminants. Such plants have evolved specialized mechanisms to cope with heavy metal stress, either by absorbing and isolating these metals within their tissues or by preventing their uptake at the cellular level. Among medicinal and aromatic plants, species such as *Lavandula angustifolia* (lavender), *Mentha piperita* (peppermint), and *Ocimum basilicum* (basil) have shown significant resilience to heavy metal exposure. These plants are capable of growing and adapting even in environments with contaminated soils (Figure 7.9).

Figure 7.9: Flower spike of *Lavandula angustifolia*.

Plants employ various mechanisms to resist heavy metal stress, including the production of antioxidant compounds, alterations in metal transport pathways, and the synthesis of chemical substances that confine heavy metals within specific tissues. For example, certain medicinal plants can limit heavy metal absorption or stabilize these contaminants in the soil by secreting compounds such as organic acids and phosphates [80, 81].

A study by Pandey et al. [81] investigated the adaptation mechanisms of clary sage (*Salvia sclarea* L.) when exposed to high zinc concentrations (900 micromoles of Zn) over an eight-day period in a hydroponic system. The research aimed to understand the plant's zinc tolerance strategies by assessing factors such as nutrient absorption, leaf pigmentation, phenolic compound content, photosynthetic efficiency, and structural changes in the leaves. To analyze zinc distribution and essential element levels – including calcium, magnesium, iron, manganese, and copper – the study utilized inductively coupled plasma mass spectrometry (ICP-MS). Findings revealed that *S. sclarea*, as a zinc-accumulating species, counteracts toxic zinc levels by increasing the concentrations of iron, calcium, and manganese ions in its leaves. This adjustment helps sustain photosynthetic efficiency and supports the functionality of photosystems I (PSI) and II (PSII).

Furthermore, increased zinc levels notably boosted the synthesis of phenolic compounds and anthocyanins in the leaves, which play a vital role in detoxifying zinc and reducing oxidative stress. Although the plants were exposed to high zinc concentrations, indicators of damage, including lipid peroxidation and electrolyte leakage, exhibited only slight elevations. These results suggest that *S. sclarea* could be a practical and cost-efficient option for phytoextraction or phytostabilization of zinc-polluted soils. Moreover, genetic engineering holds great promise for enhancing plant tolerance to heavy metals. The incorporation of specific genes that aid in detoxifying heavy metals can strengthen a plant's capacity to absorb and neutralize these contaminants. For instance, introducing MT and HMA genes into medicinal plants has been found to improve their resistance to heavy metals such as cadmium and lead [74, 82].

7.12 Case studies

Extensive research has explored the impact of heavy metals like cadmium, lead, and copper, which are major contaminants in agricultural soils. For example, Es-Sabihi et al. [83] studied the role of SA in alleviating copper toxicity in *Salvia officinalis* L. Their findings indicated that copper stress significantly hindered both stem and root growth while also depleting calcium, phosphorus, and potassium levels in leaves and roots. Interestingly, exposure to copper stress increased essential oil yield by 16.66% compared to control plants. However, the application of SA enhanced plant growth and replenished calcium, phosphorus, and potassium content in leaves and roots. Notably, treatment with 0.5 mM SA led to a 116.66% increase in essential oil content relative to untreated plants under copper stress. In terms of essential oil composition, copper stress resulted in a 19% decline in oxygenated monoterpenes, particularly α-thujone, camphor, and 1,8-cineole. However, SA application effectively restored these compounds to levels observed in nonstressed plants. These findings indicate that SA, especially at 0.5 mM concentration, can effectively mitigate copper-induced stress while enhancing growth, yield, and essential oil quality in *S. officinalis* L.

A study by Pirooz et al. [84] examined the effects of nitric oxide and silicon, both separately and together, on rosmarinic acid and essential oil production in *S. officinalis* leaves under normal and copper-stressed conditions. The findings revealed that high copper levels led to a decline in biomass and polyphenol content. However, moderate copper concentrations, particularly at 200 μM, were associated with an increase in polyphenol levels, essential oil production, and antioxidant activity in the leaves.Similarly, research by Elzaawely et al. [85] found that treating shell ginger (*Alpinia zerumbet* (Pers.) B.L. Burtt & R.M. Sm.) with 500 mM copper sulfate resulted in a reduction in total essential oil yield. Interestingly, despite the overall decrease, certain essential oil components, including 1,8-cineole, linalool, camphor, borneol, and cumin aldehyde, showed increased concentrations.

Lajayer et al. [86] investigated the impact of different copper and zinc concentrations on the growth, nutrient composition, and essential oil production of *Mentha pulegium* L. Their study revealed that the best growth outcomes, including plant height, shoot dry weight, essential oil content, and yield, were observed when 5 mg/kg copper and 10 mg/kg zinc were applied. The combined use of these metals at these concentrations enhanced the uptake of essential nutrients such as potassium, manganese, iron, copper, and zinc in the aerial parts of the plants. Moreover, notable increases were recorded in essential oil constituents, including pulegone, cis-isopulegone, α-pinene, sabinene, 1,8-cineole, and thymol. These findings suggest that appropriate levels of copper and zinc not only promote plant growth and nutrient absorption but also boost essential oil yield and quality in *M. pulegium* L.

Babashpour-Asl et al. [87] examined the impact of selenium nanoparticles on *Coriandrum sativum* L. under cadmium-induced stress. Cadmium was introduced at concentrations of 0, 4, and 8 mg/L, while selenium nanoparticles were applied as a foliar spray at 0, 20, 40, and 60 mg/L. The findings revealed that cadmium stress led to increased cadmium accumulation in both roots and shoots of coriander; however, the application of selenium nanoparticles significantly reduced this uptake. Exposure to the highest cadmium concentration resulted in decreased root and shoot biomass, chlorophyll content, and relative water content (RWC), whereas selenium nanoparticles improved these parameters. Additionally, cadmium stress elevated proline and malondialdehyde (MDA) levels, while selenium nanoparticle treatment lowered MDA levels, thereby reducing lipid peroxidation. Changes in essential oil composition were also observed, particularly in compounds such as *n*-decanal, 2*E*-dodecanal, 2*E*-decanal, and *n*-nonane, in response to cadmium stress and selenium supplementation. The study concluded that selenium nanoparticles alleviated cadmium-induced stress by enhancing growth, biochemical properties, and essential oil quality in coriander plants.

Farajzadeh Memari-Tabrizi et al. [88] investigated the effects of silicon nanoparticles on *Satureja hortensis* L. cultivated in cadmium-contaminated soil. Cadmium stress significantly reduced root and shoot biomass, along with RWC, while increasing cadmium accumulation in plant tissues and proline levels. Interestingly, moderate cadmium stress led to an increase in total phenolic content (TPC), total flavonoid content (TFC), and essential oil production. Foliar application of silicon nanoparticles (1.5 and 2.25 mM) under cadmium stress improved plant growth and boosted essential oil yield. Key essential oil components, including carvacrol, γ-terpinene, *p*-cymene, and thymol, were influenced by both cadmium stress and silicon nanoparticle treatments. Overall, the study suggested that applying silicon nanoparticles at concentrations of 1.5–2.25 mM mitigated cadmium-induced stress by enhancing physiological and biochemical characteristics in *S. hortensis* L.

A study examined the effects of cadmium and lead on seed germination, growth characteristics, and essential oil composition in *Ocimum basilicum* L [89]. Soil was treated with cadmium (0, 5, 10, and 20 mg/kg) and lead (0, 100, 200, and 400 mg/kg) over a 2-month period. The findings indicated that exposure to these heavy metals adversely impacted seed germination, flowering, stem growth, leaf area, and dry biomass. Gas chromatography-mass spectrometry (GC-MS) analysis identified 38 compounds in the essential oil, with major constituents including estragole, 2,6-octadienal, caryophyllene oxide, caryophyllene, phthalic acid, and geranial. Although cadmium and lead stress negatively affected plant growth and morphology, they also enhanced essential oil yield and modified its composition. These results suggest that *O. basilicum* may hold promise for phytoremediation in contaminated soils.

Poursaeid et al. [90] also reported that cadmium exposure stimulated essential oil production in basil. Their study found that different cadmium concentrations increased the synthesis of key compounds such as geranial, linalool, and estragole in a dose-dependent manner. These findings indicate that cadmium stress, at specific levels, can influence the biosynthesis of active metabolites in basil plants.

Similarly, Mohammed et al. [91] investigated the effects of cadmium-contaminated irrigation on two mint species: spearmint (*Mentha × piperita* L.) and mint (*M. spicata var. crispa* L.). Their research revealed significant phytotoxic effects, including abnormal growth patterns and a decline in total chlorophyll levels due to cadmium exposure.

Youssef [92] also observed that applying cadmium (5, 10, 15, 20, and 25 ppm) and lead (100, 350, 750, 1,000, and 1,500 ppm) to *Ocimum basilicum* L. (basil) increased essential oil yield. However, a separate study on *Melissa officinalis* L. (lemon balm) found that soil contamination with cadmium at concentrations of 10, 20, and 30 mg/kg over 3 months led to a significant decrease in essential oil production [93]. This decline was likely attributed to structural and functional damage that interfered with essential oil biosynthesis.

Another study assessed the essential oil yield of vetiver grass cultivated in Botswana's mine tailings and explored the effects of chelating agents such as ethylenediaminetetraacetic acid (EDTA) and arbuscular mycorrhizal fungi (AMF) on oil production. The findings revealed that vetiver grass grown in mine tailings produced a higher quantity of essential oil compared to those in uncontaminated soils. In sterilized soil, the oil yield was only 0.26%, whereas it increased to 0.86% in mine tailings. Further enhancements were observed with the addition of EDTA or AMF, raising the yield to 0.95% and 0.89%, respectively. These improvements were attributed to greater heavy metal uptake, which induced stress in the plants and stimulated the production of secondary metabolites, including essential oils. Notably, despite the elevated heavy metal concentrations in mine tailings, the extracted essential oils contained negligible amounts of these metals, highlighting the suitability of vetiver grass for thriving in polluted environments while producing high-quality essential oils [94].

Amirmoradi et al. [95] examined how different concentrations of cadmium (10, 20, 40, 60, 80, and 100 ppm) and lead (100, 300, 600, 900, 1,200, and 1,500 ppm) in irrigation water affected peppermint (*M. piperita*). Their study demonstrated that as cadmium and lead levels increased, essential oil content decreased significantly, accompanied by visible symptoms of phytotoxicity. Likewise, Azimychetabi et al. [96] investigated the impact of cadmium on peppermint and found that it altered the composition of essential oils. Specifically, pulegone and menthofuran concentrations rose, whereas menthol levels declined.

Kunwar et al. [65] investigated the effects of lead (500, 600, 750, and 900 ppm), copper (270, 300, 500, and 700 ppm), and cadmium (6, 10, 20, and 30 ppm) on *M. spicata* and *O. basilicum*. Their findings indicated that in *O. basilicum*, total essential oil yield, particularly its main component linalool, increased, whereas methyl chavicol levels declined. In contrast, *M. spicata* exhibited no significant alterations in either essential oil content or composition.

Additionally, Sulastri and Tampubolon [97] examined cadmium's influence on various plant species, including *Vetiveria zizanioides, Cymbopogon citratus, C. nardus, Curcuma xanthorrhiza, Pogostemon cablin*, and *Alpinia galanga*. The study highlighted species-specific variations, with essential oil production in *V. zizanioides* doubling, while other species showed minimal or no significant changes.

Sa et al. [98] investigated how different lead concentrations in soil influenced the growth of *Mentha crispa*. Their findings showed that increased lead contamination significantly enhanced essential oil production and modified its chemical profile. Notably, the proportion of carvone, the dominant component of mint essential oil, increased to 90% in lead-contaminated soils. Prasad et al. [99] carried out a pot culture experiment to assess how increasing chromium and lead levels (30.0 and 60.0 mg/kg of soil) influenced yield, essential oil composition, and heavy metal accumulation in three mint species: *M. piperita, M. arvensis*, and *M. citrata*. The study found that while *M. arvensis* exhibited no significant changes in fresh weight yield under chromium and lead exposure, its essential oil yield declined considerably compared to the control. Conversely, *M. piperita* demonstrated increased fresh weight yield, root biomass, and essential oil production under higher heavy metal concentrations, whereas *M. citrata* experienced reductions in these parameters.

The application of chromium and lead also led to significant alterations in essential oil composition. In *M. arvensis* and *M. piperita*, levels of α-pinene, β-pinene, sabinene, β-myrcene, limonene, menthone, and isomenthone changed notably, while in *M. citrata*, sabinene, pinene, and linalyl acetate concentrations were affected. Additionally, heavy metal accumulation in both the aerial parts and roots increased across all three mint species, with the highest levels observed in the roots. Based on these results, *M. piperita* was identified as the most suitable species for cultivation in chromium- and lead-contaminated soils, followed by *M. arvensis* and *M. citrata*.

Zheljazkov et al. [100] examined how cadmium, lead, copper, and their combinations affected *Anethum graveolens* L. (dill), *Mentha × piperita* L. (mint), and *Ocimum basilicum* L. (basil). Their findings indicated that exposure to these heavy metals led to a reduction in menthol content within mint essential oil and a decline in total oil yield in basil. Likewise, at the highest tested copper concentration (150 mg/L), a significant decrease in dill oil content was observed.

In a separate study, Nabi et al. [101] explored the impact of nickel on menthol mint (*Mentha arvensis* L.) by growing seedlings in soils treated with nickel at concentrations of 20, 40, 60, 80, and 100 mg/kg. The results revealed a biphasic response, where lower nickel levels (20 mg/kg) stimulated essential oil production, while higher concentrations led to a decline. Interestingly, at 20 mg/kg, menthol content decreased, whereas menthone and menthyl acetate levels increased.

Biswas et al. [102] investigated the effects of arsenic exposure using disodium hydrogen arsenate [$Na_2HAsO_4 \cdot 7\,H_2O$] at soil concentrations of 10, 50, and 150 ppm. Their study on basil (*Ocimum basilicum*) demonstrated a dose-dependent effect on essential oil production. While lower arsenic concentrations (10 and 50 ppm) increased oil yield, a higher level (150 ppm) resulted in a decline. Additionally, arsenic exposure influenced oil composition, with linalool levels rising and 1,8-cineole and methyl eugenol concentrations decreasing.

Several studies have investigated the impact of soils contaminated with complex heavy metal mixtures on essential oil content. For example, Scora and Chang [103] found that the composition of peppermint (*Mentha piperita*) essential oil remained unchanged when the plant was cultivated in soils containing cadmium, chromium, copper, nickel, lead, and zinc. Similarly, Pandey et al. [104] studied *Cymbopogon martinii* (palmarosa) grown in soil contaminated with tannery effluent, which contained substantial amounts of chromium, nickel, lead, and cadmium. Their results indicated that despite the presence of these heavy metals, palmarosa's essential oil yield remained unaffected.

In another study, Gautam and Agrawal [105] examined *Cymbopogon citratus* (lemongrass) cultivated in soil mixed with sludge containing heavy metals and supplemented with wastewater effluent at different concentrations (5%, 10%, and 15% by weight). The findings suggested that lower concentrations (5% and 10%) enhanced total essential oil production, while the effects of higher concentrations were not specified.

Additionally, Gharib et al. [106] conducted a comparative analysis of wild mint (*Mentha longifolia*) collected from both polluted and non-polluted areas along the Nile River in Egypt. Their study revealed that plants from polluted regions produced higher essential oil yields. Furthermore, antioxidant activity, assessed using the DPPH (2,2-diphenyl-1-picrylhydrazyl) free radical scavenging assay, was significantly greater in oils from polluted sites. However, the oil composition varied, with menthone levels increasing and pulegone concentrations decreasing in plants exposed to pollution.

Givianrad and Hashemi [107] analyzed the chemical makeup of *Tanacetum polycephalum* Sch.Bip., a species belonging to the Asteraceae family. The plant samples were collected from different distances around the Veshnavah mine in Qom, Iran. Their study found that the concentration of chemical compounds increased as the distance from the mine increased. The primary heavy metals detected in the soil samples from these regions were copper and silver. The research suggests that heavy metals significantly impact essential oil production and composition, depending on factors such as the type and concentration of the metal, the plant species, and surrounding environmental conditions. At lower concentrations, certain metals, including copper, cadmium, and lead have been shown to enhance both the total yield of essential oils and the presence of specific components within them. For instance, increased essential oil production has been noted in plants such as sage, coriander, and wild mint when exposed to minimal levels of heavy metals.

On the other hand, when present in higher concentrations, these metals exhibit phytotoxic effects, leading to inhibited growth, disruptions in essential oil biosynthesis, and undesirable changes in the oils' chemical composition. Additional negative effects include reduced chlorophyll content and indications of cellular toxicity. For example, in plants like shell ginger (*Alpinia zerumbet*) and basil (*Ocimum basilicum*), excessive heavy metal exposure resulted in lower essential oil yields, although certain compounds showed an increase under these conditions. Studies also indicate that plant responses to heavy metals vary considerably between species. Some, such as *Mentha piperita* and *Vetiveria zizanioides*, display greater tolerance or even utilize heavy metals to enhance essential oil production. In contrast, species like *Melissa officinalis* experience a significant decline in essential oil yield when exposed to the same conditions. In summary, while low levels of heavy metals can enhance essential oil production, excessive amounts generally lead to reduced yields and alterations in chemical composition. These findings emphasize the importance of evaluating environmental conditions, metal concentrations, and plant species when dealing with contaminated soils. Furthermore, although low levels of heavy metals may offer benefits, their potential risks to both the environment and human health must be carefully considered.

7.13 Conclusions

Plants exhibit various responses to heavy metal-induced stress, ranging from inhibited growth to biochemical adjustments, such as changes in antioxidant enzyme activity. Secondary metabolites, including essential oils, play a vital role in shielding plants from harmful substances and are notably influenced by these stressors. Research suggests that exposure to heavy metals, known for their toxicity, can impact both the quantity and composition of essential oils. However, the results remain inconsistent, with no definitive patterns identified. The fluctuations in essential oil yield and composition in medicinal plants under heavy metal stress are influenced by multiple factors, such as the plant species, the type of essential oil, and the concentration of heavy metals. Studies indicate that while low levels of heavy metals may promote essential oil production, higher concentrations generally have an inhibitory effect.

This decline is linked to disruptions in the metabolic pathways responsible for essential oil synthesis, which may arise from factors such as altered enzymatic activity, increased free radical production, or direct effects on genes regulating these processes. Due to the limited scope of research and the small number of species studied, two important aspects require further investigation. Firstly, heavy metals impact gene expression, particularly those involved in essential oil biosynthesis, leading to either upregulation or downregulation and modifications in associated biochemical pathways. Secondly, recent studies have explored the feasibility of growing medicinal plants in heavy metal-contaminated soils as an alternative to food crops. Research on heavy metal bioaccumulation indicates that many medicinal plants can absorb substantial amounts of these metals; however, this contamination typically does not extend to their essential oils. Despite this, cultivating or harvesting plants in polluted environments poses challenges, as it may lead to unpredictable variations in essential oil yield and composition, potentially affecting their quality and therapeutic properties. The limited research available and the narrow focus on specific plant species hinder the broader applicability of findings, highlighting significant knowledge gaps.

To overcome these limitations, future studies should explore the effects of heavy metals on a wider variety of medicinal plant species and further examine the underlying mechanisms driving these impacts. A deeper understanding of this subject could aid in developing strategies to sustainably cultivate medicinal plants in contaminated environments while ensuring the preservation of their beneficial properties.

References

[1]   Vaou, N., Stavropoulou, E., Voidarou, C., Tsigalou, C. and Bezirtzoglou, E. (2021). Towards advances in medicinal plant antimicrobial activity: A review study on challenges and future perspectives. Microorganisms, 9(10), 2041.

[2]   Aware, C. B., Patil, D. N., Suryawanshi, S. S., Mali, P. R., Rane, M. R., Gurav, R. G. and Jadhav, J. P. (2022). Natural bioactive products as promising therapeutics: A review of natural product-based drug development. South African Journal of Botany, 151(Part B), 512–528.

[3]   Boukhatem, M. N. and Setzer, W. N. (2020). Aromatic herbs, medicinal plant-derived essential oils, and phytochemical extracts as potential therapies for coronaviruses: Future perspectives. Plants, 9(6), 800.

[4]   Hubai, K. and Kováts, N. (2024). Interaction between heavy metals posed chemical stress and essential oil production of medicinal plants. Plants, 13(20), 2938.

[5]   Rubio, C., Lucas, J. R. D., Gutiérrez, A. J., Glez-Weller, D., Pérez Marrero, B., Caballero, J. M., Revert, C. and Hardisson, A. (2012). Evaluation of metal concentrations in *Mentha* herbal teas (*Mentha piperita*, *Mentha pulegium*, and *Mentha* species) by inductively coupled plasma spectrometry. Journal of Pharmaceutical and Biomedical Analysis, 71, 11–17.

[6]   Ali, B. and Gill, R. A. (2022). Editorial: Heavy metal toxicity in plants: Recent insights on physiological and molecular aspects, volume II. Frontiers in Plant Science, 13, 1016257.

[7]   Zhao, K., Fu, W., Ye, Z. and Zhang, C. (2015). Contamination and spatial variation of heavy metals in the soil-rice system in Nanxun County, Southeastern China. International Journal of Environmental Research and Public Health, 12(2), 1577–1594.

[8]   Sharma, P. and Dubey, R. S. (2005). Lead toxicity in plants. Brazilian Journal of Plant Physiology, 17(1), 1–19.

[9]   Nagajyoti, P. C., Lee, K. D. and Sreekanth, T. V. M. (2010). Heavy metals, occurrence and toxicity for plants: A review. Environmental Chemistry Letters, 8(3), 199–216.

[10]   Kaur, R., Sharma, R., Thakur, S., Chandel, S. and Chauhan, S. K. (2024). Exploring the combined effect of heavy metals on accumulation efficiency of *Salix alba* raised on lead and cadmium contaminated soils. International Journal of Phytoremediation, 26(9), 1486–1499.

[11] Chen, Y.-G., He, X.-L.-S., Huang, J.-H., Luo, R., Ge, H.-Z., Wołowicz, A., Wawrzkiewicz, M., Gładysz-Płaska, A., Li, B., Yu, Q.-X., Kołodyńska, D., Lv, G.-Y. and Chen, S.-H. (2021). Impacts of heavy metals and medicinal crops on ecological systems, environmental pollution, cultivation, and production processes in China. Ecotoxicology and Environmental Safety, 219, 112336.

[12] Dinu, C., Gheorghe, S., Tenea, A. G., Stoica, C., Vasile, G. G., Popescu, R. L., Serban, E. A. and Pascu, L. F. (2021). Toxic metals (As, Cadmium, Ni, Pb) impact in the most common medicinal plant (*Mentha piperita*). International Journal of Environmental Research and Public Health, 18(8), 3904.

[13] Asiminicesei, D.-M., Fertu, D. I. and Gavrilescu, M. (2024). Impact of heavy metal pollution in the environment on the metabolic profile of medicinal plants and their therapeutic potential. Plants, 13(6), 913.

[14] Luo, L., Wang, B., Jiang, J., Fitzgerald, M., Huang, Q., Yu, Z. et al. (2021). Heavy metal contaminations in herbal medicines: Determination, comprehensive risk assessments, and solutions. Frontiers in Pharmacology, 11, 595335.

[15] Abou-Arab, A. A. K. and Abou Donia, M. A. (2000). Heavy metals in Egyptian spices and medicinal plants and the effect of processing on their levels. Journal of Agricultural and Food Chemistry, 48(6), 2300–2304.

[16] Tchounwou, P. B., Yedjou, C. G., Patlolla, A. K. and Sutton, D. J. (2012). Heavy metal toxicity and the environment. Experientia Supplementum, 101, 133–164.

[17] Singh, N. K., Raghubanshi, A. S. and Upadhyay, A. K. (2016). Arsenic and other heavy metal accumulation in plants and algae growing naturally in contaminated area of West Bengal, India. Ecotoxicology and Environmental Safety, 130, 224–233.

[18] Zhao, H., Guan, J., Liang, Q., Zhang, X., Hu, H. and Zhang, J. (2021). Effects of cadmium stress on growth and physiological characteristics of sassafras seedlings. Scientific Reports, 11, 9913.

[19] Xu, J., Zhang, J., Lv, Y., Xu, K., Lu, S., Liu, X. and Yang, Y. (2020). Effect of soil mercury pollution on ginger (*Zingiber officinale Roscoe*): Growth, product quality, health risks and silicon mitigation. Ecotoxicology and Environmental Safety, 195, 110472.

[20] Usman, K., Abu-Dieyeh, M. H., Zouari, N. and Al-Ghouti, M. A. (2020). Lead (Pb) bioaccumulation and antioxidative responses in Tetraena qataranse. Scientific Reports, 10, 17070.

[21] Ibrahim, M. H., Chee Kong, Y. and Mohd Zain, N. A. (2017). Effect of cadmium and copper exposure on growth, secondary metabolites and antioxidant activity in the medicinal plant sambung nyawa (*Gynura procumbens* (Lour.) Merr). Molecules, 22(10), 1623.

[22] Fu, H., Yuan, J., Liu, R. and Wang, X. (2023). Effects of cadmium on the synthesis of active ingredients in *Salvia miltiorrhiza*. Open Life Sciences, 18(1), 20220603.

[23] Chizzola, R. and Lukas, B. (2005). Variability of the cadmium content in *Hypericum* species collected in eastern Austria. Water, Air, & Soil Pollution, 170(1–4), 331–343.

[24] Pinto, A. P., Mota, A. M., De Varennes, A. and Pinto, F. C. (2004). Influence of organic matter on the uptake of cadmium, zinc, copper and iron by sorghum plants. The Science of the Total Environment, 326(1–3), 239–247.

[25] Salamon, I., Labun, P., Skoula, M. and Fabian, M. (2007). Cadmium, lead and nickel accumulation in chamomile plants grown on heavy metal-enriched soil. Acta Horticulturae, 749, 231–237.

[26] Schneider, M. and Marquard, R. (1996). Aufnahme und Akkumulation von Cadmium und weiterer Schwermetalle bei *Hypericum perforatum* L. und *Linum usitatissimum* L. Zeitschrift Für Arznei- & Gewürzpflanzen, 1(2), 111–116.

[27] Doroszewska, T. and Berbeć, A. (2004). Variation for cadmium uptake among *Nicotiana* species. Genetic Resources and Crop Evolution, 51(4), 323–333.

[28] Lugon-Moulin, N., Martin, F., Krauss, M. R., Ramey, P. B. and Rossi, L. (2006). Cadmium concentration in tobacco (*Nicotiana tabacum* L.) from different countries and its relationship with other elements. Chemosphere, 63(7), 1074–1086.

[29] Gasser, U., Klier, B., Kuhn, A. V. and Steinhoff, B. (2009). Current findings on the heavy metal content in herbal drugs. Pharmeuropa Scientific Notes, 1, 37–50.

[30] Chizzola, R. (2001). Micronutrient composition of *Paliuver somnilemni L.*, grown under low cadmium stress conditions. Journal of Plant Nutrition, 24(11), 1663–1677.

[31] Pluquet, E., Filipinski, M. and Gruppe, M. (1990). Zur Cadmiumaufnahme von Kulturpflanzen aus geogen und anthropogen mit Cadmium angereicherten Böden. Zeitschrift Für Kulturtechnik Und Landentwicklung, 31(1), 105–111.

[32] Grejtovsky, A., Markusova, K. and Eliasova, A. (2006). The response of chamomile (*Matricaria chamomilla* L.) plants to soil zinc supply. Plant, Soil and Environment, 52(1), 1–7.

[33] Zheljazkov, V. D. and Warman, P. R. (2004). Application of high-Cu compost to dill and peppermint. Journal of Agricultural and Food Chemistry, 52, 2615–2622.

[34] Zheljazkov, V. D. and Nikolov, S. (1996). Accumulation of Cadmium, Pb, Cu, Mn, and Zn by *Sylibum marianum* L. grown on polluted soils. Acta Horticulturae, 426, 297–308.

[35] Zheljazkov, V. D. and Nielsen, N. E. (1996). Growing clary sage (*Salvia sclarea* L.) in heavy metal-polluted areas. Acta Horticulturae, 426, 309–328.

[36] Zheljazkov, V. D., Jeliazkova, E. A. and Craker, L. E. (1999). Heavy metal uptake by mint. Acta Horticulturae, 500, 111–117.

[37] Al-Khayri, J. M., Banadka, A., Rashmi, R., Nagella, P., Alessa, F. M. and Almaghasla, M. I. (2023). Cadmium toxicity in medicinal plants: An overview of the tolerance strategies, biotechnological and omics approaches to alleviate metal stress. Frontiers in Plant Science, 13, 1047410.

[38] Salamon, I. and Plackova, A. (2007). Environmental risks associated with the production and collection of chamomile flowers. Acta Horticulturae, 749, 211–215.

[39] Reimann, C., Roller, F., Frengstad, B., Kashulina, G., Niskavaara, H. and Englmaier, P. (2001). Comparison of the element composition in several plant species and their substrate from a 1,500,000-km^2 area in Northern Europe. The Science of the Total Environment, 278, 87–112.

[40] Cairrao, E., Pereira, M. J., Pastorinho, M. R., Morgado, F., Scares, A. M. V. M. and Guimarães, L. (2007). *Fucus* spp. as a mercury contamination bioindicator in coastal areas (Northwestern Portugal). Bulletin of Environmental Contamination and Toxicology, 79, 388–395.

[41] Emamverdian, A., Ding, Y., Hasanuzzaman, M., Barker, J., Liu, G., Li, Y. and Mokhberdoran, F. (2023). Insight into the biochemical and physiological mechanisms of nanoparticles-induced arsenic tolerance in bamboo. Frontiers in Plant Science, 14, 1121886.

[42] Fijalek, Z., Soltyk, K., Lozak, A., Kominek, A. and Ostapczuk, P. (2003). Determination of some micro- and macroelements in preparations made from peppermint and nettle leaves. Pharmazie, 58, 480–482.

[43] Tlustoš, P., Szakova, J., Vysloužilová, M., Pavlíková, D., Weger, J. and Jarovská, H. (2007). Variation in the uptake of arsenic, cadmium, lead, and zinc by different species of willows (*Salix* spp.) grown in contaminated soils. Central European Journal of Biology, 2, 254–275.

[44] Choudhury, R. P., Kumar, A. and Garg, A. N. (2006). Analysis of Indian mint (*Mentha spicata*) for essential, trace, and toxic elements and its antioxidant behavior. Journal of Pharmaceutical and Biomedical Analysis, 41, 825–832.

[45] Garg, C., Khan, S. A., Ansari, S. H. and Garg, M. (2010). Efficacy and safety studies of *Foeniculum vulgare* through evaluation of toxicological and standardization parameters. International Journal of Pharmacy and Pharmaceutical Sciences, 2, 43–45.

[46] Arpadjan, S., Celik, G., Taskesen, S. and Gücer, S. (2008). Arsenic, cadmium, and lead in medicinal herbs and their fractionation. Food and Chemical Toxicology, 46(8), 2871–2875.

[47] Guedon, D., Brum, M., Bizet, D., Bizo, S., Bourny, E., Compagnon, P. A. and Urizzi, P. (2007). Impurities in herbal substances, herbal preparations, and herbal medicinal products. IV. Heavy (toxic) metals. Pharma Pratiques, 18, 231–268.

[48] Grant, C. A. and Bailey, L. D. (1997). Effects of phosphorus and zinc fertilizer management on cadmium accumulation in flaxseed. Journal of the Science of Food and Agriculture, 73(3), 307–314.

[49] Moraghan, J. T. (1993). Accumulation of cadmium and selected elements in flax seed grown on a calcareous soil. Plant and Soil, 150(1), 61–68.

[50] Chizzola, R. and Mitteregger, U. S. (2005). Cadmium and zinc interactions in trace element accumulation in chamomile. Journal of Plant Nutrition, 28, 1383–1396.

[51] Chizzola, R. (2005). Cadmium and micronutrient accumulation in yarrow. Phyton, 45, 159–171.

[52] Ramos, I., Esteban, E., Lucena, J. J. and Garate, A. (2002). Cadmium uptake and sub-cellular distribution in plants of Lactuca sp. Cadmium-Mn Interaction. Plant Science, 162, 761–767.

[53] Österhas, A. H. and Greger, M. (2003). Accumulation of, and interactions between, calcium and heavy metals in wood and bark of Picea abies. Journal of Plant Nutrition and Soil Science, 166, 246–253.

[54] Marschner, H. (1995). Mineral Nutrition of Higher Plants, 2nd ed., Academic Press, Harcourt Press and Co, London.

[55] Ali, H., Khan, E. and Sajad, M. A. (2013). Phytoremediation of heavy metals-concepts and applications. Chemosphere, 91(7), 869–881.

[56] Adamczyk-Szabela, D., Lisowska, K. and Wolf, W. M. (2021). Hysteresis of heavy metals uptake induced in Taraxacum officinale by thiuram. Scientific Reports, 11, 20151.

[57] Xu, D., Shen, Z., Dou, C., Dou, Z., Li, Y., Gao, Y. and Sun, Q. (2022). Effects of soil properties on heavy metal bioavailability and accumulation in crop grains under different farmland use patterns. Scientific Reports, 12, 9211.

[58] Adamczyk-Szabela, D., Romanowska-Duda, Z., Lisowska, K. and Wolf, W. M. (2017). Heavy metal uptake by herbs: V. Metal accumulation and physiological effects induced by thiuram in Ocimum basilicum L. Water, Air, & Soil Pollution, 228, 334.

[59] Singh, S., Parihar, P., Singh, R., Singh, V. P. and Prasad, S. M. (2016). Heavy metal tolerance in plants: Role of transcriptomics, proteomics, metabolomics, and ionomics. Frontiers in Plant Science, 6, 1143.

[60] Mansoor, S., Ali, A., Kour, N., Bornhorst, J., AlHarbi, K., Rinklebe, J., Moneim, D. A. E., Ahmad, P. and Chung, Y. S. (2023). Heavy metal induced oxidative stress mitigation and ROS scavenging in plants. Plants, 12(16), 3003.

[61] Emamverdian, A., Ding, Y., Mokhberdoran, F. and Xie, Y. (2015). Heavy metal stress and some mechanisms of plant defense response. The Scientific World Journal, 756120. https://doi.org/10.1155/2015/756120.

[62] Asgari Lajayer, B., Ghorbanpour, M. and Nikabadi, S. (2017). Heavy metals in contaminated environment: Destiny of secondary metabolite biosynthesis, oxidative status, and phytoextraction in medicinal plants. Ecotoxicology and Environmental Safety, 145, 377–390.

[63] Asare, M. O., Száková, J. and Tlustoš, P. (2023). The fate of secondary metabolites in plants growing on Cadmium-, As-, and Pb-contaminated soils-A comprehensive review. Environmental Science and Pollution Research, 30, 11378–11398.

[64] Goncharuk, E. A. and Zagoskina, N. V. (2023). Heavy metals, their phytotoxicity, and the role of phenolic antioxidants in plant stress responses with focus on cadmium: Review. Molecules, 28(9), 3921.

[65] Kunwar, G., Pande, C., Tewari, G., Singh, C. and Kharkwal, G. C. (2015). Effect of heavy metals on terpenoid composition of Ocimum basilicum L. and Mentha spicata L. Journal of Essential Oil Bearing Plants, 18(4), 818–825.

[66] Yuan, Z., Cai, S., Yan, C., Rao, S., Cheng, S., Xu, F. and Liu, X. (2024). Research progress on the physiological mechanism by which selenium alleviates heavy metal stress in plants: A review. Agronomy, 14(8), 1787.

[67]   Feng, D., Wang, R., Sun, X., Liu, L., Liu, P., Tang, J., Zhang, C. and Liu, H. (2023). Heavy metal stress in plants: Ways to alleviate with exogenous substances. Science of the Total Environment, 897, 165397.

[68]   Chen, S., Zhao, C.-B., Ren, R.-M. and Jiang, J.-H. (2023). Salicylic acid had the potential to enhance tolerance in horticultural crops against abiotic stress. Frontiers in Plant Science, 14, 1141918.

[69]   Sharma, A., Sidhu, G. P. S., Araniti, F., Bali, A. S., Shahzad, B., Tripathi, D. K., Brestic, M., Skalicky, M. and Landi, M. (2020). The role of salicylic acid in plants exposed to heavy metals. Molecules, 25(3), 540.

[70]   Bücker-Neto, L., Paiva, A. L. S., Machado, R. D., Arenhart, R. A. and Margis-Pinheiro, M. (2017). Interactions between plant hormones and heavy metals responses. Genetics and Molecular Biology, 40(1 suppl 1), 373–386.

[71]   Cruz, Y., Villar, S., Gutiérrez, K., Montoya Ruiz, C., Gallego, J. L., Del Pilar Delgado, M. and Saldarriaga, J. F. (2021). Gene expression and morphological responses of Lolium perenne L. exposed to cadmium (Cadmium^{2+}) and mercury (Hg^{2+}). Scientific Reports, 11, 11257.

[72]   Cong, W., Miao, Y., Xu, L., Zhang, Y., Yuan, C., Wang, J., Zhuang, T., Lin, X., Jiang, L., Wang, N., Ma, J., Sanguinet, K. A., Liu, B., Rustgi, S. and Ou, X. (2019). Transgenerational memory of gene expression changes induced by heavy metal stress in rice (Oryza sativa L.). BMC Plant Biology, 19, 282.

[73]   Chaudhary, K., Agarwal, S. and Khan, S. (2018). Role of phytochelatins (PCs), metallothioneins (MTs), and heavy metal ATPase (HMA) genes in heavy metal tolerance. In: Prasad, R. (ed.) Mycoremediation and Environmental Sustainability, Fungal Biology, Springer, Cham, 45–65. https://doi.org/10.1007/978-3-319-77386-5_2.

[74]   El-Sappah, A. H., Zhu, Y., Huang, Q., Chen, B., Soaud, S. A., Abd Elhamid, M. A., Yan, K., Li, J. and El-Tarabily, K. A. (2024). Plants' molecular behavior to heavy metals: From criticality to toxicity. Frontiers in Plant Science, 15, 1423625.

[75]   Hailu, T. A., Devkota, P., Osoko, T. O., Singh, R. K., Zak, J. C. and Van Gestel, N. (2024). No-till and crop rotation are promising practices to enhance soil health in cotton-producing semiarid regions: Insights from citizen science. Soil Systems, 8(4), 108.

[76]   Osman, H. E., Quronfulah, A. S., El-Morsy, M. H., Alamoudi, W. M. and El-Hamid, H. T. A. (2024). Bioenergy crop rotation for phytoremediation of heavy metal contaminated soils at Mahd AD'Dahab mine, Kingdom of Saudi Arabia. Journal of Taibah University for Science, 18(1), 2357257.

[77]   Bhat, S. A., Bashir, O., Ul Haq, S. A., Amin, T., Rafiq, A., Ali, M., Américo-Pinheiro, J. H. P. and Sher, F. (2022). Phytoremediation of heavy metals in soil and water: An eco-friendly, sustainable, and multidisciplinary approach. Chemosphere, 303(Pt 1), 134788.

[78]   Vera Tomé, F., Blanco Rodríguez, P. and Lozano, J. C. (2009). The ability of Helianthus annuus L. and Brassica juncea to uptake and translocate natural uranium and 226Ra under different milieu conditions. Chemosphere, 74(2), 293–300.

[79]   Lwin, C. S., Seo, B. H., Kim, H. U., Owens, G. and Kim, K. R. (2018). Application of soil amendments to contaminated soils for heavy metal immobilization and improved soil quality-a critical review. Soil Science and Plant Nutrition, 64(2), 156–167.

[80]   Dobrikova, A., Apostolova, E., Hanć, A., Yotsova, E., Borisova, P., Sperdouli, I., Adamakis, I. S. and Moustakas, M. (2021). Tolerance mechanisms of the aromatic and medicinal plant Salvia sclarea L. to excess zinc. Plants, 10(2), 194.

[81]   Pandey, J., Verma, R. K. and Singh, S. (2019). Suitability of aromatic plants for phytoremediation of heavy metal contaminated areas: A review. International Journal of Phytoremediation, 21(5), 405–418.

[82]   Koźmińska, A., Wiszniewska, A., Hanus-Fajerska, E. and Muszyńska, E. (2018). Recent strategies of increasing metal tolerance and phytoremediation potential using genetic transformation of plants. Plant Biotechnology Reports, 12(1), 1–14.

[83]   Es-sbihi, F. Z., Hazzoumi, Z., Benhima, R. and Amrani Joutei, K. (2020). Effects of salicylic acid on growth, mineral nutrition, glandular hairs distribution, and essential oil composition in *Salvia officinalis* L. grown under copper stress. Environmental Sustainability, 3, 199–208.

[84]   Pirooz, P., Amooaghaie, R., Ahadi, A., Sharififar, F. and Torkzadeh-Mahani, M. (2022). Silicon and nitric oxide synergistically modulate the production of essential oil and rosmarinic acid in *Salvia officinalis* under Cu stress. Protoplasma, 259(4), 905–916.

[85]   Elzaawely, A. A., Xuan, T. D. and Tawata, S. (2005). Changes in essential oil, kava pyrones and total phenolics of Alpinia zerumbet (Pers.) BL Burtt. & RM Sm. leaves exposed to copper sulphate. Environmental and Experimental Botany, 59(3), 347–353.

[86]   Lajayer, H. A., Savaghebi, G. and Hadian, J. Others. (2017). Comparison of copper and zinc effects on growth, micro- and macronutrients status and essential oil constituents in pennyroyal (*Mentha pulegium* L.). Brazilian Journal of Botany, 40, 379–388.

[87]   Babashpour-Asl, M., Farajzadeh-Memari-Tabrizi, E. and Yousefpour-Dokhanieh, A. (2022). Foliar-applied selenium nanoparticles alleviate cadmium stress through changes in physio-biochemical status and essential oil profile of coriander (*Coriandrum sativum* L.) leaves. Environmental Science and Pollution Research, 29, 80021–80031.

[88]   Farajzadeh Memari-Tabrizi, E., Yousefpour-Dokhanieh, A. and Babashpour-Asl, M. (2021). Foliar-applied silicon nanoparticles mitigate cadmium stress through physio-chemical changes to improve growth, antioxidant capacity, and essential oil profile of summer savory (*Satureja hortensis L.*). Plant Physiology and Biochemistry, 165, 71–79.

[89]   Fattahi, B., Arzani, K., Souri, M. K. and Barzegar, M. (2019). Effects of cadmium and lead on seed germination, morphological traits, and essential oil composition of sweet basil (*Ocimum basilicum* L.). Industrial Crops and Products, 138, 111584.

[90]   Poursaeid, M., Iranbakhsh, A., Ebadi, M. and Fotokian, M. H. (2021). Morpho-physiological and phytochemical responses of basil (*Ocimum basilicum* L.) to toxic heavy metal cadmium. Notulae Botanicae Horti Agrobotanici Cluj-Napoca, 49, 11902.

[91]   Mohammed, N. A., Ali, W. N., Younis, Z. M., Zeebaree, P. J. and Qasim, M. J. (2024). Response of two mint cultivars peppermint (*Mentha piperita* L.) and curly mint (*Mentha spicata* var. *crispa*) to different levels of cadmium contamination. Pakistan Journal of Life and Social Sciences, 22, 99–116.

[92]   Youssef, N. A. (2020). Changes in the morphological traits and the essential oil content of sweet basil (*Ocimum basilicum* L.) as induced by cadmium and lead treatments. International Journal of Phytoremediation, 23(3), 291–299.

[93]   Kilic, S. and Kilic, M. (2017). Effects of cadmium-induced stress on essential oil production, morphology and physiology of lemon balm (*Melissa officinalis* L., Lamiaceae). Applied Ecology and Environmental Research, 15, 1653–1669.

[94]   Kereeditse, T. T., Pheko-Ofitlhile, T., Ultra, V. U. and Dinake, P. (2023). Effects of heavy metals on the yield of essential oil from vetiver grass cultivated in mine tailings amended with EDTA and arbuscular mycorrhizal fungi. Natural Product Communications, 18(3), 1–12.

[95]   Amirmoradi, S., Moghaddam, P. R., Koocheki, A., Danesh, S. and Fotovat, A. (2012). Effect of cadmium and lead on quantitative and essential oil traits of peppermint (*Mentha piperita* L.). Notulae Scientia Biologicae, 4(4), 101–109.

[96]   Azimychetabi, Z., Nodehi, M. S., Moghadam, T. K. and Motesharezadeh, B. (2021). Cadmium stress alters the essential oil composition and the expression of genes involved in their synthesis in peppermint (*Mentha piperita* L.). Industrial Crops and Products, 168, 113602.

[97]   Sulastri, Y. S. and Tampubolon, K. (2019). Aromatic plants: Phytoremediation of cadmium heavy metal and the relationship to essential oil production. International Journal of Science and Technology Research, 8, 1064–1069.

[98] Sá, R. A., Alberton, O., Gazim, Z. C., Laverde, A., Caetano, J., Amorin, A. C. and Dragunski, D. C. (2015). Phytoaccumulation and effect of lead on yield and chemical composition of *Mentha crispa* essential oil. Desalination and Water Treatment, 53, 3007–3017.

[99] Prasad, A., Singh, A. K., Chand, S., Chanotiya, C. S. and Patra, D. D. (2010). Effect of chromium and lead on yield, chemical composition of essential oil, and accumulation of heavy metals of mint species. Communications in Soil Science and Plant Analysis, 41(18), 2170–2186.

[100] Zheljazkov, V. D., Craker, L. E. and Xing, B. (2006). Effects of Cadmium, Pb, and Cu on growth and essential oil contents in dill, peppermint, and basil. Environmental and Experimental Botany, 58(1), 9–16.

[101] Nabi, A., Naeem, M., Aftab, T. and Khan, M. M. A. (2020). Alterations in photosynthetic pigments, antioxidant machinery, essential oil constituents and growth of menthol mint (*Mentha arvensis* L.) upon nickel exposure. Brazilian Journal of Botany, 43, 721–731.

[102] Biswas, S., Koul, M. and Bhatnagar, A. K. (2015). Effect of arsenic on trichome ultrastructure, essential oil yield, and quality of *Ocimum basilicum* L. Medicinal Plant Research, 5(1), 1–9.

[103] Scora, R. W. and Chang, A. C. (1997). Essential oil quality and heavy metal concentrations of peppermint grown on a municipal sludge-amended soil. Journal of Environmental Quality, 26, 975–979.

[104] Pandey, J., Chand, S., Pandey, S., Raj, K. and Patra, D. D. (2015). Palmarosa [*Cymbopogon martinii* (Roxb.) Wats.] as a putative crop for phytoremediation, in tannery sludge polluted soil. Ecotoxicology and Environmental Safety, 122, 296–302.

[105] Gautam, M. and Agrawal, M. (2017). Influence of metals on essential oil content and composition of lemongrass (*Cymbopogon citratus* (D.C.) Stapf.) grown under different levels of red mud in sewage sludge amended soil. Chemosphere, 175, 315–322.

[106] Gharib, F. A., Mansour, K. H., Ahmed, E. Z. and Galal, T. M. (2021). Heavy metals concentration, and antioxidant activity of the essential oil of the wild mint (*Mentha longifolia* L.) in the Egyptian watercourses. International Journal of Phytoremediation, 23(6), 641–651.

[107] Givianrad, M. H. and Hashemi, A. (2014). A survey of the effect of some heavy metals in plant on the composition of the essential oils close to Veshnaveh-Qom mining area. Oriental Journal of Chemistry, 30(2), 737–743.

Esra Uçar, Gamze Tüzün, Burak Tüzün*, and Elyor Berdimurodov

Chapter 8
Metabolic and hormonal responses of medicinal and aromatic plants to abiotic stress

Abstract: Throughout their life cycle, plants are exposed to various biotic and abiotic stresses, such as drought, salinity, low temperatures, and pathogen attacks. In order to survive and adapt, they produce a variety of hormonal and metabolic responses. These responses are shaped by the interaction of genetic and environmental factors, and regulate their fundamental life processes such as growth, development, and reproduction. Plant hormones are chemical signaling molecules that regulate processes such as plant growth, development, environmental adaptation, and stress management. The major plant hormones include auxins, cytokinins, gibberellins, abscisic acid (ABA), ethylene, salicylic acid, and jasmonates. Auxins regulate cell elongation and tropic movements, while gibberellins promote seed germination and flowering. Cytokinins stimulate cell division and delay senescence. ABA plays a crucial role in stomatal closure and water balance under certain stress conditions. Ethylene is involved in processes such as ripening and leaf abscission, whereas jasmonates and salicylic acid activate defense mechanisms. The interactions among these hormones are critical for enabling plants to adapt to environmental conditions and develop optimal growth strategies. Molecular docking calculations have been evaluated to understand plant responses to environmental stress factors such as drought, salinity, and pathogens, as well as to analyze biomolecular interactions. The binding affinities of natural or synthetic compounds with defense proteins (e.g., 1HJO) have been examined, elucidating stress mechanisms, identifying biologically active compounds, and developing innovative strategies to enhance plant stress tolerance.

Keywords: abiotic stress, biotic stress, hormone, medicinal and aromatic plants

---

*Corresponding author: Burak Tüzün**, Plant and Animal Production Department, Technical Sciences Vocational School of Sivas, Sivas Cumhuriyet University, 58140 Sivas, Turkey,
e-mail: theburaktuzun@yahoo.com, https://orcid.org/0000-0002-0420-2043
Esra Uçar, Plant and Animal Production Department, Technical Sciences Vocational School of Sivas, Sivas Cumhuriyet University, 58140 Sivas, Turkey
Gamze Tüzün, Department of Chemistry, Faculty of Science, Cumhuriyet University, 58140 Sivas, Turkey
Elyor Berdimurodov, Chemical and Materials Engineering, New Uzbekistan University, 54 Mustaqillik Ave, Tashkent 100007, Uzbekistan; Faculty of Chemistry, National University of Uzbekistan, Tashkent 100034, Uzbekistan

https://doi.org/10.1515/9783111469713-008

8.1 Introduction

Plants may experience a variety of unfavorable environmental circumstances during their life cycle. These factors may have a detrimental impact on their growth and development and can also restrict their output. Stress factors are environmental elements that have a detrimental effect on the proper growth and development of plants. The literature often refers to these elements that cause such adversities as "biotic and abiotic stresses" [1]. Plants may suffer physiological and biochemical damage as a result of biotic and abiotic stress causes. Plants have molecular defense mechanisms and initially adjust to lessen the effects of these damages. There are three categories that this reaction mechanism might be placed into. The first one is about the homeostasis of macromolecules and ions. It makes sure that there is a balance of big molecules like proteins, lipids, carbohydrates, and nucleic acids, as well as ions like sodium, potassium, calcium, and chloride. The second comprises the creation of protective molecules, which neutralize reactive oxygen species (ROS) via antioxidants such as ascorbic acid, tocopherols, and flavonoids. Osmolytes like proline, betaine, and sorbitol also help by keeping water levels balanced, preserving the structure of proteins and membranes. Heat shock proteins (HSPs) and pathogenesis-related (PR) proteins increase the ability of cells to withstand thermal or biotic stimuli. LEA proteins, which contain hydrophilic amino acids, help preserve cellular proteins and membrane structures from losing water. Phytohormones, including abscisic acid (ABA), salicylic acid (SA), jasmonic acid, and ethylene, are able to detect stress signals and begin the process of activating the genes that are associated to them. Secondary metabolites, such as phenolics, alkaloids, and terpenoids, have antioxidant, antibacterial, and insect-repellent activities, which help protect plants against pests and diseases [2–7]. The third part is the production and detoxification of ROS [7]. These are molecules that are chemically active and are produced during different metabolic processes that take place in plant cells. The primary forms are superoxide anion (O_2^-), hydrogen peroxide (H_2O_2), hydroxyl radical (OH·), and singlet oxygen ($^1O_2$) [8, 9]. Plants often produce them in little quantities under normal circumstances. However, when plants are under stress, they tend to produce more ROS. During the light processes of photosynthesis in chloroplasts, excess light energy may transform oxygen molecules into ROS. Enzymes, including oxidases and lipoxygenases, may help produce ROS [10–14].

Plants respond to stress in both metabolic and hormonal ways. They produce hormones such as ABA, ethylene, jasmonate, SA, cytokinin, gibberellin, and auxin. ABA causes stomata to close, which helps to reduce water loss. It also causes seeds to become dormant, which stops them from germinating under difficult circumstances. Additionally, ABA increases the production of proteins that help seeds tolerate stress [9, 15–18]. Ethylene is produced when there is both biotic and abiotic stress. It alters the cell wall during times of stress, which helps to protect against mechanical damage. In the case of a pathogen assault, it encourages programmed cell death in the regions that are affected in order to stop the disease from spreading. It encourages the creation of proteins that help the body defend itself against stress. Ethylene stimulates leaf abscission to save energy and supports deeper root development under water stress conditions, such as drought and salt. Ethylene also enhances the synthesis of PR (pathogenesis-related) proteins, which help to boost the immune system of plants [19–25]. Jasmonate protects against injury and pathogen assaults, whereas SA is useful in plant immunity [26]. Plant survival and production are significantly affected by environmental stress factors such as drought, salt, heavy metals, and diseases. These stressors disturb the balance of the cell, which leads to the creation of ROS. These ROS may cause oxidative damage to cellular components, including membranes, DNA, and proteins. Plants use a variety of defensive mechanisms, including enzymes, proteins, and phytochemicals, to reduce the impact of these threats and keep their cells stable. Molecular docking simulations have become a valuable computational technique for studying biomolecular interactions and discovering useful chemicals that improve the ability of plants to withstand stress. Molecular docking helps to find bioactive chemicals and explains their functions in plant metabolism by measuring the binding affinities of natural or synthesized ligands with important defense proteins, including superoxide dismutase, catalase, and HSPs. For example, the 1HJO protein is renowned for its antioxidant activity and plays an important role in detoxifying ROS, which helps plants withstand biotic and abiotic stressors. This method gives important information on plant stress biology and presents new ideas for sustainable agriculture and biotechnological progress.

8.2 Metabolic and hormonal responses to abiotic stress

Plants respond to stress by increasing or regulating the synthesis of various phytohormones. Stress factors such as drought, extreme cold, salinity, and waterlogging induce stress in plants, prompting them to engage in a struggle for survival. In response to stress conditions, plants exhibit: morphological and physiological changes, cytological changes, biomolecular responses, hormonal responses, and genetic responses [27].

8.3 Water stress

8.3.1 Drought stress

Drought causes plants to be unable to absorb enough water from the soil, which results in water loss from plant cells and a drop in turgor pressure. This reduces the rate of cell division and restricts growth. Plants also seal their stomata to stop losing more water, in addition to the reduction in turgor pressure. Although this helps to save water, it also restricts the amount of carbon dioxide that can be absorbed, which leads to a decrease in the rate of photosynthesis. Chlorophyll degradation occurs as a result of drought circumstances, which in turn decreases the ability to perform photosynthesis. At the same time, it causes oxidative stress and destroys the organelles of cells. Roots that cannot absorb water are unable to access the soil solution, which results in a lack of nutrients for the plant. When the amount of water in the soil decreases, the concentration of dissolved salts rises, which leads to an increase in osmotic stress. When there is a drought, the levels of ABA rise, which causes the stomata to shut and decreases the amount of water that is lost [16, 28]. Because the enzymes that are involved in ABA production are mostly found in leaf tissues, the buildup of ABA takes place in the vascular tissues of the leaves [29]. Plants do not have a central nervous system, yet they are nonetheless able to communicate stress signals between their roots and shoots via their vascular system. Hydraulic signals, electrical currents, calcium waves, ROS, and hormone-like peptides are all involved in long-distance communication in response to drought stress [30, 31]. In order to maintain a stable osmotic equilibrium, plants create osmoprotective substances such as proline, trehalose, polysaccharides, and betaine. These metabolites build up throughout the plant due to stress, which reduces the cell water potential and helps the plant retain water. This mechanism, known as osmotic adjustment, serves to maintain cell turgor [32–35]. The plant's roots usually grow longer in order to access groundwater, which is a structural characteristic of the plant. In addition, it decreases the area of the leaf, thickens the waxy cuticle layer in the epidermis, and generally decreases the number of stomata in order to limit water loss [16]. Figure 8.1 provides a schematic representation of drought stress.

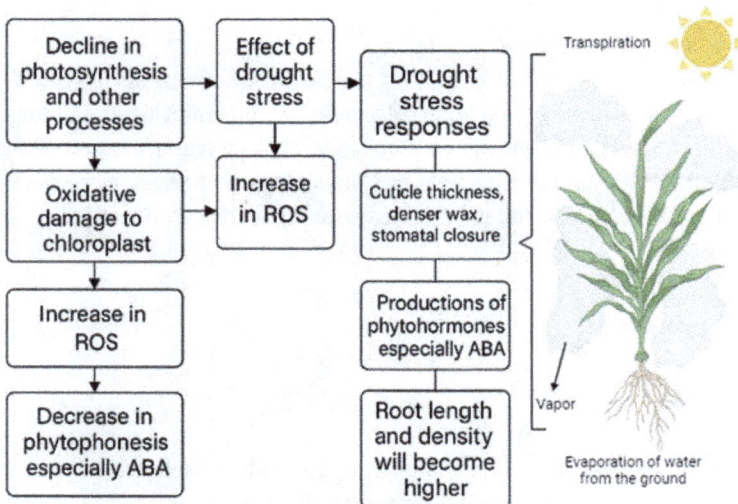

Figure 8.1: Drought stress in plants.

8.3.2 Waterlogging stress

In excessive waterlogging, soil pores are filled with water, and oxygen levels decrease. As a result, it becomes difficult for the roots to absorb oxygen. Some plants develop resistance to waterlogging through aerenchyma (air tissue) and adventitious roots. Aerenchyma tissue enables the diffusion of oxygen from the plant's aerial parts to the submerged areas, allowing the roots to maintain aerobic respiration [36]. Insufficient oxygen intake prevents plants from performing aerobic respiration in the mitochondria, and the inhibition of photosynthesis leads to energy deficiency, which results in ATP production limited to glycolysis. While the end products of aerobic respiration are CO_2 and H_2O, anaerobic respiration produces lactic acid and ethanol [37–41]. As a result, energy efficiency decreases, and metabolism is disrupted. The lactic acid and ethanol produced from anaerobic respiration damage the roots, hindering growth. Roots submerged in water are unable to absorb nutrients properly, slowing down growth. Due to reduced water uptake by the roots in waterlogging, anaerobic respiration leads to decreased energy production, preventing the ion pumps necessary for stomatal opening from functioning, causing stomata to close. Additionally, stressed roots produce high amounts of ABA and ethylene, and the hormones reaching the leaves play a role in stomatal closure [42, 43]. Stomata can also close for different reasons, such as high CO_2 levels in the soil, which may cause the plant to close its stomata to maintain gas balance. Stomata typically facilitate water vaporization from the plant through transpiration. However, during waterlogging, due to excessive air humidity and water in plant tissues, transpiration is not necessary. As a result, stomata close. Additionally, during waterlogging, stomata close to prevent pathogens, such as fungi and bacteria, from entering the plant when they find reproductive opportunities. Stomata tend to remain closed to reduce photosynthesis rate as well. When roots cannot absorb sufficient nutrients, the raw materials required for photosynthesis cannot be transported to the leaves, and the plant reduces its photosynthesis rate to lessen metabolic load. As a result, ROS production increases, damaging the cell membrane. Following root decay, the transport of water and nutrients is hindered, leading the plant to death [44–47].

8.4 Temperature stress

8.4.1 High temperature (heat shock)

As a result of high temperature, the physical structure of lipids in cell membranes is disrupted, and membrane permeability increases [48]. This leads to an imbalance of ions inside and outside the cell. Cellular functions are impaired, and cell death occurs due to damage to the cell membrane. Additionally, high temperature causes protein denaturation, resulting in the loss of protein function. The disruption of protein structure affects enzymes involved in photosynthesis and respiration. There is an increase in the number of HSPs, which work to prevent improper protein aggregation and support the normal folding of cellular proteins under stress situations [49–54]. High temperature also affects chloroplasts, which are sensitive to heat. Chlorophylls degrade, stomata close, and carbon dioxide uptake decreases [55]. As a result of increased respiration due to high temperature, carbon dioxide reserves are depleted, and photosynthesis decreases. Consequently, energy production drops, and metabolism slows down. The rise in temperature also increases transpiration. Stomata close to counteract this, and photosynthesis slows down. Otherwise, it leads to wilting in the plant, and in extreme cases, desiccation [56–58]. Decreased water content within the cell leads to an increase in ROS, resulting in lipid peroxidation, protein, and DNA damage [59]. As a result of these effects, even if the plant does not die, pollen viability decreases, and fertilization does not occur [60]. An increase in temperature causes variations in plant hormone levels. The production of ABA, which induces stomatal closure, increases, leading to slower growth, while ethylene production induces leaf abscission [18]. If the stress level is mild, metabolic adaptation and recovery are observed in the plant. If it is moderate, growth and reproduction are hindered, and if severe stress occurs, it leads to cellular death and desiccation of the plant.

8.4.2 Low-temperature stress

Plants' physiological and biochemical processes are significantly impacted by low temperatures, which hinders their growth and development. Cold stress occurs when temperatures are between around 0 °C and 15 °C, while freezing happens when temperatures are below 0 °C. Lipids that melt at high temperatures become solid at low temperatures, which makes them less fluid. The membrane gets stiffer and less permeable. The reduction in membrane permeability affects the movement of water and ions, which decreases cellular functioning [61, 62]. Ions like Na, K, and Ca are essential for plants to deal with biotic and abiotic stress situations. If there are not enough of these ions, it may cause harm to the plants [58, 60, 63]. Stomatal closure and inadequate gas exchange occur when energy-dependent systems, such as ion pumps (e.g., ATPase), are unable to operate [64, 65]. High protoplasmic viscosity is present [66]. Chloroplasts are sensitive to both high and low temperatures. In most plants, the temperature needed for photosynthesis is typically lower than the temperature necessary for respiration [54]. When temperatures are low, photosynthetic enzymes become less active, which causes stomata to close and reduces the amount of carbon dioxide that is absorbed. At the same time, when the amount of water in the soil drops [66] and the water in the roots gets immobilized under freezing temperatures, the stomata shut to reduce water loss via restricted transpiration. The formation of ROS is increasing. If freezing happens, the water outside the cell turn into ice crystals. These sharp-edged crystals rip the membrane, which cause the cell to die. Because of the presence of dissolved chemicals, the freezing point of intracellular water is lower than that of pure water. As a result, the freezing process starts mostly in the areas outside of the cells. When ice crystals develop outside of the cell, the concentration of dissolved chemicals in the surrounding medium rises, which reduces the osmotic potential. As a consequence, the water within the cell flows outward in order to balance the difference in pressure, and an increase in osmotic pressure may be seen. When freezing occurs, it disrupts the action of enzymes, which has a detrimental impact on activities, including photosynthesis, respiration, and protein synthesis [61]. Leaf and root development halt, and growth slows down. Low temperatures cause changes in hormone levels, increasing the ABA level, which leads to stomatal closure, while the increase in ethylene promotes leaf abscission. A plant's resistance to low temperatures varies, depending on its developmental stage, meaning the metabolic changes occurring within it. If this occurs during a period of increased sugar and protein concentrations in the cells, it reduces ice formation within the cells, thus enhancing frost tolerance [66]. In response to low temperatures, plants typically increase the amount of soluble proteins in their tissues to cope with stress. Some of these accumulating proteins exhibit antifreeze properties and are referred to as antifreeze proteins (AFPs); they can alter the shape of ice crystals [67, 68]. Figure 8.2 shows cold stress schematically.

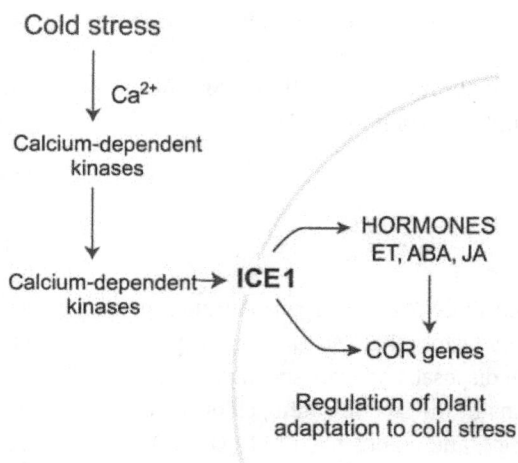

Cold stress

Ca^{2+}

Calcium-dependent
kinases

Calcium-dependent → **ICE1**
kinases

→ HORMONES
ET, ABA, JA

→ COR genes

Regulation of plant
adaptation to cold stress

Figure 8.2: Cold stress in plants.

8.5 Light stress

Light is an important component for the germination of seeds, the growth of leaves, the elongation of plant height, the timing of blooming, and other activities that occur throughout the development and life cycle of plants [69]. Plants have photoreceptors that are able to detect and react to the intensity, direction, and quality of light. Chromophores, which are photopigments, are found in these light-sensitive proteins. They help the body perceive and respond to light [70]. Plants may get stressed if they are exposed to too much or too little light intensity. Photosynthetic activity is greatly affected by the quality and intensity of light. At first, a rise in light intensity causes the rate of photosynthesis to increase and the plant's need for CO_2 to grow, but only up to a certain point. In order to satisfy this demand, plants expand their stomata to take in more carbon dioxide (CO_2). However, as the stomata open, transpiration increases, which causes the plant to lose water. Plants maintain water balance by transporting the ABA hormone to the leaves when the water loss in the leaves reaches a threshold level. This is done by signals sent from the roots to the leaves, which causes the stomata to shut. This system protects against drought stress. Certain plants are more vulnerable to strong light stress when their cytokinin levels drop [71]. When plants are stressed, it might alter the allocation of energy, which can lead to the creation of singlet oxygen from triplet chlorophyll molecules. This may expose the plant to oxidative stress [72, 73]. The plant type and ambient circumstances might cause this impact to be different. Plants can carry out photosynthesis most effectively at lower light levels when the temperature and nutritional conditions are right. On the other hand, plants may experience stress as a result of changes in environmental circumstances. Plants may experience light stress as a consequence of fluctuations in light intensity caused by climate change, which may expose them to either low or high light intensity. This has a deleterious effect on the plant's metabolism of antioxidants and photosynthesis. As a result, the plant undergoes changes at the biochemical and molecular levels [70, 74].

8.6 Salt stress

Excessive salinity in the soil occurs when there is a higher concentration of salts in the soil solution. It is one of the abiotic stress factors that causes stress in plants. Improper fertilization, pesticide applications, waste disposal, and excessive irrigation are all factors that can lead to the accumulation of ions in the soil, including chloride and sulfate, as well as cations like Na^+, K^+, Ca^{2+}, and Mg^{2+}, and anions like Cl^-, HCO_3^-, CO_3^{2-}, SO_4^{2-}, and NO_3^-, all of which can be found in high concentrations. When plant cells take in too many of these ions, it leads to ionic imbalances [75, 76]. Because of the excessive salinity, sodium and chloride ions start to build up in the tissues of the plant. This disrupts the ionic equilibrium and negatively impacts the plant's metabolic processes. As a result, the enzymes are not able to work properly. The roots of the plant have a harder time taking in water, which causes the turgor pressure to drop and the plant to start withering. Growth stops as turgor pressure decreases. When a plant wilts, its stomata shut, which restricts the amount of carbon dioxide that can be absorbed, and leads to a decrease in photosynthesis [77–79]. When salt builds up, it causes oxidative stress on cellular components, which may lead to lipid peroxidation, protein degradation, and even DNA damage [80]. Plants produce enzymes like superoxide dismutase and catalase in order to reduce the stress that ROS cause [81]. Salt stress causes cells to lose water and destabilizes the plasma membrane, which leads to damage to the cell membrane and the subsequent release of ions. Calcium ions (Ca^{2+}), protein kinases, and phospholipids are all involved in the signaling pathway for salt tolerance. Plants often create ABA when they are exposed to high saline levels in order to cover their stomata and prevent water loss [82–84]. Plants try to survive by increasing the amount of water that is retained in their cells. They do this by accumulating osmolytes, which include proline, sugars, and betaines [85]. Proline is important for avoiding protein dehydration because it binds its hydrophobic ends to proteins and its hydrophilic ends to water molecules [86].

8.7 Nutrient stress

Hydrogen, carbon, and oxygen, which together make up around 95% of plant biomass and are mostly derived from air and water, are among the minimum of 17 basic elements that plants need to maintain normal growth. The other 14 elements, which are nitrogen, potassium, calcium, magnesium, phosphorus, sulfur, chlorine, boron, iron, manganese, zinc, copper, nickel, and molybdenum, are taken in directly from the earth. While sodium, cobalt, and silicon are not regarded essential for plant growth, some experts suggest that a total of 20 elements are required for optimum plant development [87–90]. When plants do not get nutrients like nitrogen, phosphorus, and potassium, they become stressed, which causes them to respond with changes in their metabolism and hormones. Nitrogen is not present in the parent rock and mostly comes from the environment and organic matter. It is essential for physiological and biochemical activities, such as the synthesis of proteins and chlorophyll, root respiration, and fruit production [88, 91, 92]. Plants may experience a variety of challenges when they have too much or too little of certain nutritional components. When plants do not have enough nutrients, their main reaction is to try to adapt. At first, they encourage the main root to grow longer and for more lateral roots to develop so that they may take up more nutrients. Additionally, the number of root hairs increases. Excess nitrogen, responsible for the formation of green tissue, leads to delayed flowering and fruit formation. As the plant height increases significantly, lodging or breakage may occur. Excess nitrogen also contributes to the development of fungal diseases [89, 92, 93]. In nitrogen deficiency, vegetative growth is retarded. Since nitrogen is an element in the chlorophyll structure, its deficiency leads to chlorophyll degradation, resulting in a decline in photosynthesis.

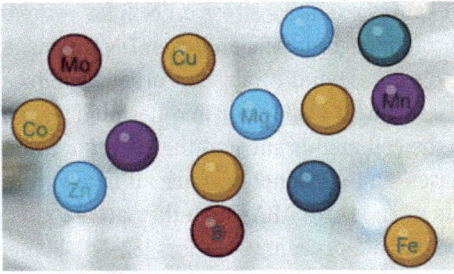

Figure 8.3: Nutrient stress in plants.

The plant's color shifts from dark green to light green, and in severe stages of insufficiency, chlorosis develops. The rates of flowering and fruit set decrease [90, 94]. Chlorophyll generation is dependent on potassium; therefore, a lack of potassium may result in chlorosis. It affects the opening and shutting of stomata in leaves, which causes water to be lost via transpiration. Additionally, it has an effect on how much water root cells absorb. As a result of these factors, a deficit causes a drop in turgor pressure, which leads to water stress in the plant [89, 92, 94, 95]. Phosphorus is another essential element for plant life. Since phosphorus plays a role in flower and fruit formation, its deficiency damages generative organs. As phosphorus increases plants' resistance to diseases and pests, its deficiency reduces the plant's resistance to diseases [90, 92, 94, 95]. Figures 8.3 and 8.4 show nutrient stress schematically.

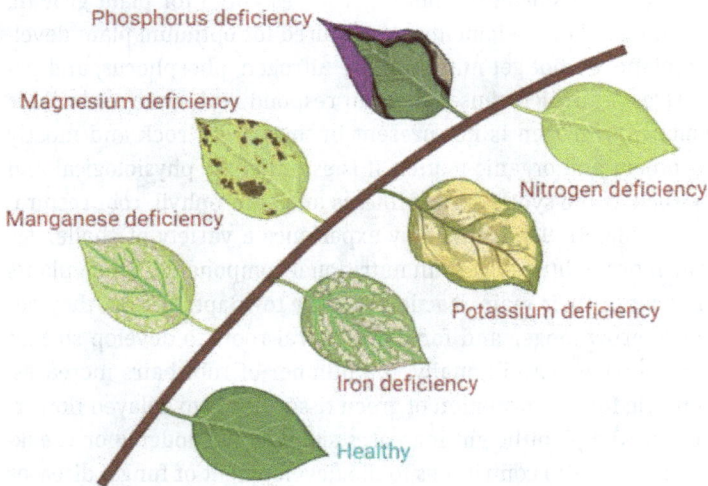

Figure 8.4: Nutrient stress in plants.

8.8 Heavy metal stress

Environmental pollution has become a problem due to the development of technology, industrialization, high traffic, and other things. Heavy metals are one of the contaminants that are of great concern since they pose a hazard to the health of living beings. Heavy metals have a specific gravity greater than 5 g/cm^3 and an atomic number greater than 20, including more than 60 metals such as Fe, Mn, Cu, Zn, Hg, Ni, Cr, Cd, Co, Mo, Pb, Hg, and Al. Plants need small quantities of certain of these heavy metals (Fe, Mn, Co, Zn, Cu, Ni, and Mo) in order to flourish. They are involved in a number of physiological and biochemical activities, including photosynthesis, respiration, carbon and nitrogen metabolism, cell division, and nitrogen fixation [96]. On the other hand, when plants absorb a large amount of heavy metals, it interferes with their physiological functioning, which may lead to harmful consequences and slow their growth [97, 98]. The level of toxicity of these metals is different for each plant species and depends on the chemical structure of the metal. The increase in heavy metals in the soil primarily disrupts root respiration, mineral absorption, and enzyme activities, leading to damage in the root structure. As root development is negatively affected, the uptake of essential nutrients and water for the plant is also impaired, resulting in disruptions in the plant's growth [98, 99]. Metals have the ability to replace each other, and some metals can bind to the magnesium (Mg) element, which is essential for the chlorophyll molecule. As a result, this causes the chlorophyll molecules to break down [96, 100, 101] and the breakdown of photosynthesis enzymes and chlorophyll structure due to stress negatively affects photosynthesis. The disruption of enzyme activity slows down metabolic processes. Heavy metals impair the function of stomata, reducing transpiration. This affects the cell's water balance, leading to osmotic stress. Protein synthesis and hormonal balance are disrupted. Eventually, membrane stability begins to deteriorate, leading to cellular damage [102, 103]. The intensity and duration of stress in plants play a critical role in their resilience. As the stress level increases, the plant's capacity for adaptation decreases. In stress conditions that exceed the threshold of resilience, initially unnoticed damages gradually become visible, and irreversible structural or functional damage may occur. Additionally, these stresses can pass to humans through plant-based food sources, leading to chronic and harmful health issues [104, 105]. Some plant species are capable of accumulating heavy metals in their tissues without causing any harm (hyperaccumulators), and it has been reported that they can contribute to the reduction of pollution (phytoremediation). Hyperaccumulators can accumulate 50–500 times more heavy metals compared to the soil [98, 106, 107]. Some plants avoid heavy metal damage by trapping metals within the cell walls of root hairs, effectively blocking their movement to aerial parts [108].

8.9 Molecular docking calculation for stress

For the purpose of comparing the biological activities of molecules to those of biological materials, molecular docking calculations are carried out. Molecular docking calculations were performed using the Maestro Molecular Modeling Platform (version 13.4) created by Schrödinger [109]. Calculations are comprised of a number of different processes. There is a distinct approach to each phase. The first stage is the usage of the protein preparation module [110] to prepare the proteins. This module is responsible for determining the active sites that are present in the proteins. In the next stage, the molecules that have been investigated are prepared. The molecules are first optimized using the Gaussian software tool. After that, the LigPrep module [111] is prepared for calculations using the optimized structures. After preparation, the Glide ligand docking module [112, 113] is utilized to study the interactions that occur between the compounds and the cancer protein. Throughout all of the computations, the OPLS4 technique is used to do the calculations. In conclusion, an ADME/T study, which stands for absorption, distribution, metabolism, excretion, and toxicity, is carried out in order to investigate the possible pharmacological effects of the compounds that are investigated. The Qik-prop module [114] of the Schrödinger program is used to forecast the effects and responses of chemicals in human metabolism.

Molecular docking calculations are a powerful computational approach to understand plant responses to environmental stress factors and to identify biomolecular interactions. When plants are exposed to various stress conditions such as drought, salinity, heavy metals, and pathogens, enzymes, proteins, and phytochemicals that regulate resistance mechanisms against these stresses come into play [115]. The molecular docking method analyzes the interactions of these biomolecules at the atomic level and allows the identification of effective compounds against plant stress.

These calculations are usually performed to evaluate the binding affinity of natural or synthetic ligands with proteins that play a critical role in the plant defense system (e.g., superoxide dismutase, catalase, and HSPs). Thus, by determining which compounds interact more strongly with target proteins under a certain stress condition, potential compounds that can increase plant stress tolerance can be identified [116].

In addition, molecular docking studies enable the screening of plant bioactive compounds and the elucidation of the roles these compounds play in plant metabolism. For example, by examining the interactions of compounds such as proline and ABA with plant target proteins against drought stress, mechanisms that reduce water loss and maintain cellular homeostasis can be revealed [117].

As a result, molecular docking calculations are an important tool in understanding plant stress biology and developing new biotechnological strategies to increase plant resistance. This approach contributes to the design of plant stress-resistant genotypes or biotechnological interventions, providing innovative solutions in the fields of sustainable agriculture and plant biotechnology [118].

1HJO protein is a protein that plays an important role in cellular processes and is often associated with antioxidant defense mechanisms [119]. In studies on protein structures, 1HJO has been associated with enzymes that play a role especially in oxidative stress conditions. This is of critical importance in terms of providing protection against biotic and abiotic stress factors in plants.

Throughout their life cycle, plants are exposed to a variety of stressors. These pressures are often classified into two primary categories: abiotic stress and biotic stress. Environmental conditions such as drought, salt, excessive heat, extreme cold, and ultraviolet radiation may all produce abiotic stress [120]. These kinds of stressors disturb the cellular equilibrium of plants and lead to an increase in the creation of free radicals. Living elements such as harmful organisms (pathogens and insects) and competitive plants are responsible for biotic stress.

When plants are under stress, they produce more molecules known as ROS. ROS may cause damage to cellular membranes, mutations in DNA, and damage to protein structures [121]. The 1HJO protein is important for detoxifying ROS since it has antioxidant action. These characteristics improve the capacity of plants to adapt and endure under challenging environmental circumstances.

The functions of the 1HJO protein include antioxidant protection, protecting the structure of proteins that are denatured during stress, and increasing the adaptive capacity of plants by regulating genetic expression during the stress response [122] (Table 8.1).

In the field of agriculture, increasing the stress tolerance of plants is a critical strategy for maintaining productivity [123]. Promoting the production of proteins such as 1HJO through biotechnological methods or ensuring that they are expressed more genetically can increase the resistance of plants to stress conditions such as drought and salinity. For example, the increase in 1HJO protein in plants exposed to drought stress can help plants limit water loss and protect cell membranes. In salinity stress, this protein, which regulates ion balance and plays a role in ROS detoxification, can maintain plant health and growth [124] (Figures 8.5–8.7).

Molecular docking calculations have been evaluated to understand plant responses to environmental stress factors such as drought, salinity, and pathogens, and to analyze their biomolecular interactions. By examining the binding affinities of natural or synthetic compounds with defense proteins (e.g., 1HJO), stress mechanisms have been elucidated, biologically active compounds have been identified, and innovative strategies have been developed to increase plant stress tolerance [125].

The 1YET protein is a molecule that is involved in the defensive systems that plants generate in response to environmental and biological challenges [126]. This protein has a regulatory impact, particularly in plant metabolism, and is essential for maintaining cellular homeostasis and guaranteeing survival under stressful situations.

Throughout their life cycle, plants are subjected to abiotic challenges, including drought, salt, temperature variations, and ultraviolet radiation, as well as biotic stresses, such as diseases and insects [127]. These stressors generate abnormalities in the meta-

Table 8.1: Numerical values of the docking parameters of the molecule against protein.

1HJO	Docking score	Glide ligand efficiency	Glide hbond	Glide evdw	Glide ecoul	Glide emodel	Glide energy	Glide einternal	Glide posenum
Abscisic acid	−4.42	−0.23	0.00	−21.23	−3.41	−0.07	−30.29	−24.64	6.52
Aminocyclopropane carboxylic acid	−5.36	−0.77	−0.11	−11.17	−6.34	−0.10	−32.15	−17.51	0.11
Ascorbic acid	−5.62	−0.47	−0.06	−19.21	−15.66	−0.09	−47.53	−34.86	10.06
Benzyladenine	−6.04	−0.36	−0.26	−34.28	−2.38	0.00	−50.68	−36.67	1.07
Citric acid	−6.95	−0.53	−0.60	−4.59	−12.64	−0.19	−43.46	−17.22	13.46
EDTA	−2.79	−0.14	−0.54	−22.57	−0.85	−0.14	−32.00	−23.43	8.84
Epigallocatechin gallate	−6.40	−0.19	−0.26	−27.29	−19.03	−0.04	−65.09	−46.33	5.14
Geldanamycin	−4.78	−0.12	−0.32	−27.88	−10.69	0.00	−48.00	−38.58	3.46
Glutathione	−7.95	−0.40	−0.67	−25.43	−16.21	−0.26	−81.48	−41.64	10.08
Jasmonic acid	−6.18	−0.41	−0.09	−23.21	−6.00	−0.09	−50.72	−29.20	3.68
Kinetin	−5.14	−0.32	−0.25	−32.30	−5.31	−0.07	−48.02	−37.61	3.95
Phytochelatins	−7.95	−0.40	−0.67	−25.43	−16.21	−0.26	−81.48	−41.64	10.08
Proline	−5.56	−0.69	−0.12	−13.44	−8.20	−0.11	−37.75	−21.64	0.07
Quercetin	−6.94	−0.32	0.00	−27.92	−15.19	0.00	−61.77	−43.11	1.84
Radicicol	−3.81	−0.15	−0.63	−21.72	−8.49	−0.06	−38.74	−30.22	0.30
Rutin	−8.51	−0.20	−0.56	−42.01	−28.19	−0.11	−108.78	−70.20	13.96
Salicylic acid	−8.47	−0.85	−0.29	−10.75	−13.88	−0.10	−52.55	−24.64	2.62
trans-Zeatin	−6.28	−0.39	−0.96	−29.93	−11.26	−0.18	−57.37	−41.20	1.46
Trehalose	−7.47	−0.32	0.00	−26.93	−22.04	0.00	−63.61	−48.97	9.20
Trolox	−5.68	−0.32	−0.13	−18.62	−0.40	−0.09	−31.59	−19.02	0.95

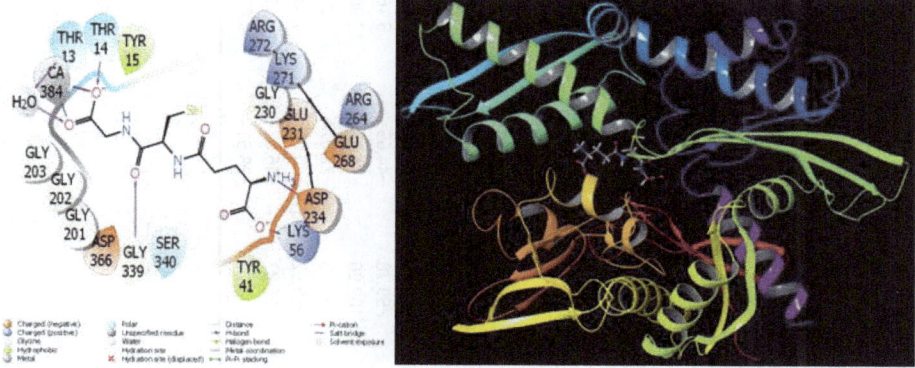

Figure 8.5: Illustration of the interaction of the phytochelatin molecules with protein 1HJO.

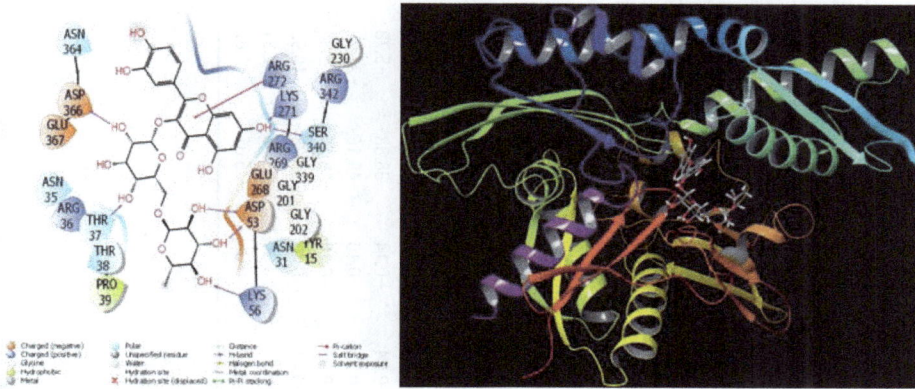

Figure 8.6: Illustration of the interaction of the rutin molecule with protein 1HJO.

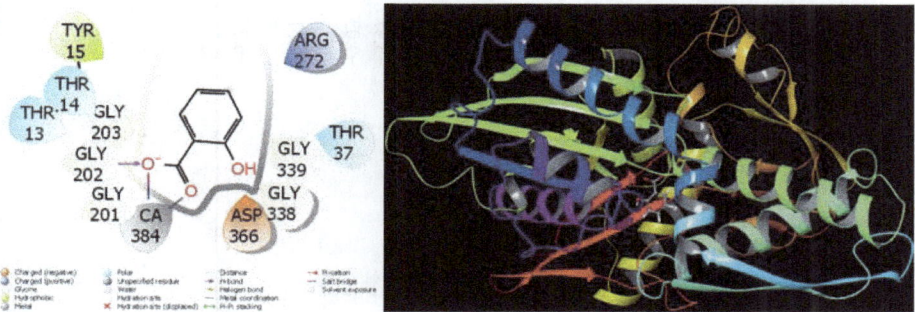

Figure 8.7: Illustration of the interaction of the salicylic acid molecule with protein 1HJO.

Table 8.2: Numerical values of the docking parameters of the molecule against protein.

1YET	Docking score	Glide ligand efficiency	Glide hbond	Glide evdw	Glide ecoul	Glide emodel	Glide energy	Glide einternal	Glide posenum
Abscisic acid	-5.50	-0.29	-0.06	-24.69	-2.46	0.00	-36.38	-27.15	4.27
Aminocyclopropane carboxylic acid	-3.90	-0.56	-0.40	-6.18	-5.44	-0.12	-18.36	-11.62	0.34
Ascorbic acid	-4.80	-0.40	-0.47	-14.50	-9.93	0.00	-32.87	-24.43	0.76
Benzyladenine	-6.22	-0.37	-0.16	-32.43	-7.45	-0.03	-55.41	-39.87	0.96
Citric acid	-4.39	-0.34	-0.47	-7.65	-11.64	-0.06	-25.67	-19.30	6.99
EDTA	-2.38	-0.12	-0.48	-17.52	-10.17	-0.05	-34.42	-27.69	4.27
Epigallocatechin gallate	-7.44	-0.23	-0.16	-38.85	-20.03	-0.01	-82.81	-58.88	5.91
Geldanamycin	-8.55	-0.21	-0.80	-47.88	-15.84	-0.03	-95.54	-63.72	2.70
Glutathione	-5.49	-0.27	-0.46	-27.20	-9.70	0.00	-52.07	-36.90	6.18
Jasmonic acid	-5.08	-0.34	-0.29	-20.02	-5.22	0.00	-33.34	-25.24	3.94
Kinetin	-5.87	-0.37	-0.16	-31.03	-7.12	-0.03	-51.95	-38.15	1.50
Phytochelatins	-5.49	-0.27	-0.46	-27.20	-9.70	0.00	-52.07	-36.90	6.18
Proline	-5.19	-0.65	-0.16	-13.86	-3.96	0.00	-26.87	-17.83	0.48
Quercetin	-7.02	-0.32	-0.35	-26.51	-17.99	0.00	-60.74	-44.50	10.06
Radicicol	-5.97	-0.24	-0.14	-38.88	-5.78	0.00	-59.78	-44.65	3.16
Rutin	-9.03	-0.21	-0.10	-49.16	-27.22	-0.04	-109.90	-76.38	13.90
Salicylic acid	-5.74	-0.57	-0.16	-16.74	-5.34	0.00	-31.06	-22.07	1.39
trans-Zeatin	-6.61	-0.41	-0.46	-29.08	-10.92	-0.03	-56.92	-40.00	1.41
Trehalose	-7.15	-0.31	0.00	-23.76	-22.91	0.00	-66.87	-46.67	6.01
Trolox	-6.27	-0.35	0.00	-26.65	-1.87	0.00	-40.27	-28.51	0.02

bolic processes of plants and lead to the buildup of damaging chemicals termed ROS. If ROS accumulates, it may harm cellular membranes, proteins, and DNA.

The 1YET protein enhances the ability of plants to defend themselves against the buildup of ROS by improving their antioxidant defense systems. This protein is critical for intracellular signaling pathways, detects stress signals, and controls the production of genes that help plants adapt [128]. Furthermore, the 1YET protein helps to minimize the consequences of oxidative stress by maintaining the redox equilibrium in cells (Table 8.2).

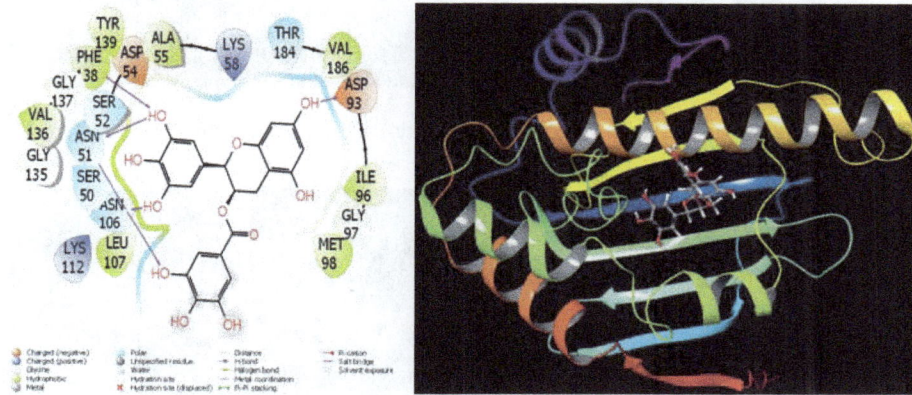

Figure 8.8: Illustration of the interaction of the epigallocatechin gallate molecule with protein 1YET.

Under abiotic stress conditions, such as drought or salinity, the 1YET protein activates mechanisms that support the retention of water within the cell. In addition, the protein regulates ion balance and alleviates the negative effects of salinity stress. Against biotic stresses, the 1YET protein increases the production of phenolic compounds and phytochemicals as part of the plant's defense response.

This protein also regulates plant growth processes and energy metabolism. It increases the plant's chances of survival by ensuring the optimal use of energy resources under stress conditions [129]. Increasing the 1YET protein through genetic engineering approaches is a promising strategy for developing plant species with high stress tolerance in agricultural production (Figures 8.8–8.10).

The 3KDJ protein is a protein that helps plants defend themselves against environmental stress situations [130]. This protein is vital for the plant stress response and is useful in helping plants adapt to both biotic and abiotic stressors. Abiotic stressors are generated by environmental variables such drought, salt, high heat, cold, and ultraviolet radiation. On the other hand, biotic stresses are induced by harmful organisms, diseases, and insects [131]. The 3KDJ protein helps to correct cellular abnormalities that are produced by various stress events.

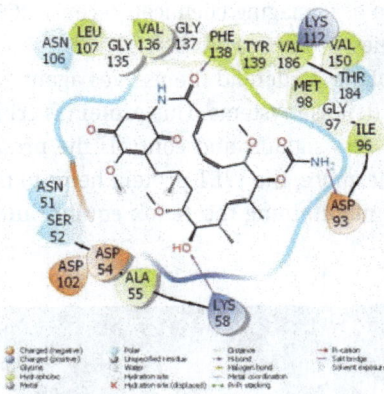

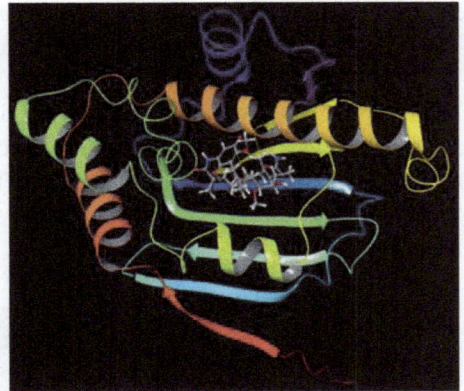

Figure 8.9: Illustration of the interaction of the geldanamycin molecule with protein 1YET.

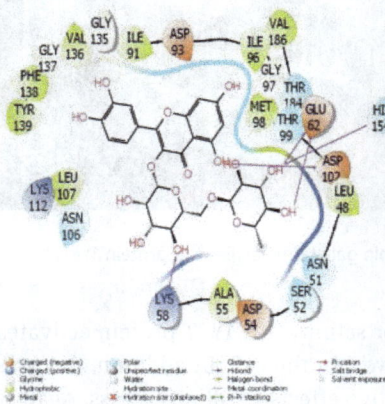

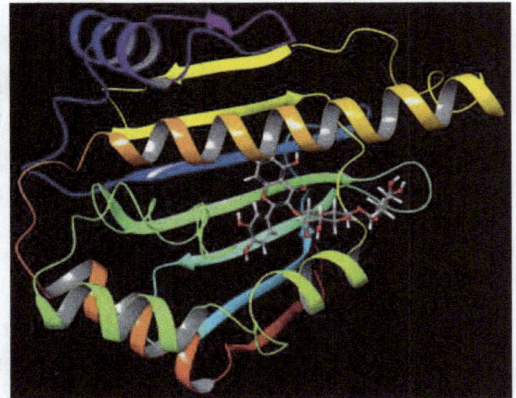

Figure 8.10: Illustration of the interaction of the geldanamycin molecule with protein 1YET.

ROS, which are damaging chemicals, build up in plants when they are under abiotic stress conditions. Cell membranes, DNA, and proteins may all be harmed by these chemicals. The 3KDJ protein is essential for detoxifying ROS and keeping cells in a state of homeostasis [132]. This enables the plant to reduce the harmful impacts of stress and continue to live. Furthermore, this protein modifies the plant's metabolic activities via controlling gene expression when the plant is under stress [133] (Table 8.3).

Under biotic stress conditions, the 3KDJ protein activates mechanisms that support the plant's immune system. It increases the resistance of plants to harmful organisms, especially by increasing defense responses against pathogens [134]. In this process, the plant's production of phenolic compounds and other defense metabolites increases. In addition, this protein coordinates defense mechanisms in the plant by regulating cellular signaling pathways [135].

Table 8.3: Numerical values of the docking parameters of the molecule against protein.

3KDJ	Docking score	Glide ligand efficiency	Glide hbond	Glide evdw	Glide ecoul	Glide emodel	Glide energy	Glide einternal	Glide posenum
Abscisic acid	−8.19	−0.43	−0.36	−25.89	−5.06	−0.07	−51.87	−30.94	3.97
Aminocyclopropane carboxylic acid	−4.91	−0.70	−0.28	−7.90	−7.11	−0.14	−24.95	−15.01	0.26
Ascorbic acid	−5.67	−0.47	−0.45	−12.69	−17.03	−0.03	−42.21	−29.72	5.30
Citric acid	−3.90	−0.30	−0.38	−11.27	−5.17	−0.10	−21.95	−16.44	11.31
EDTA	−2.34	−0.12	−0.44	−22.47	−1.34	−0.16	−28.48	−23.81	7.55
Jasmonic acid	−5.31	−0.35	−0.33	−22.98	−4.11	−0.04	−38.45	−27.09	3.45
Kinetin	−4.78	−0.30	−0.02	−30.95	−1.93	0.00	−41.70	−32.88	3.63
Radicicol	−7.34	−0.29	−0.50	−27.13	−6.86	0.00	−35.69	−33.99	0.37
Salicylic acid	−5.51	−0.55	−0.33	−10.17	−11.21	−0.03	−33.52	−21.37	0.44
Trehalose	−7.16	−0.31	0.00	−18.47	−17.12	0.00	−39.59	−35.59	8.88

Increasing the 3KDJ protein through genetic modification may contribute to the development of plants with high stress tolerance in agricultural production. Such pro-

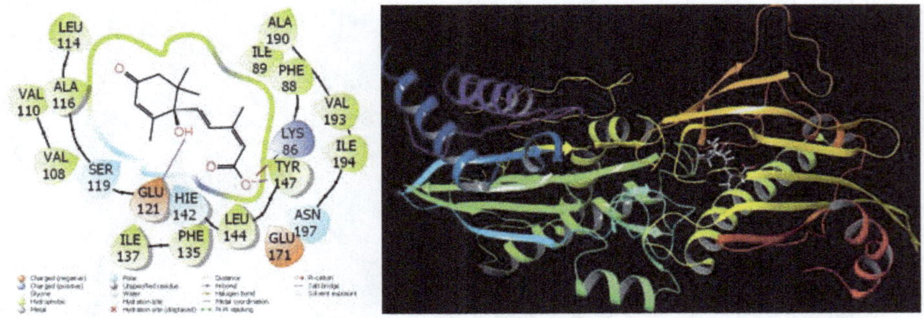

Figure 8.11: Illustration of the interaction of the molecule abscisic acid with protein 3KDJ.

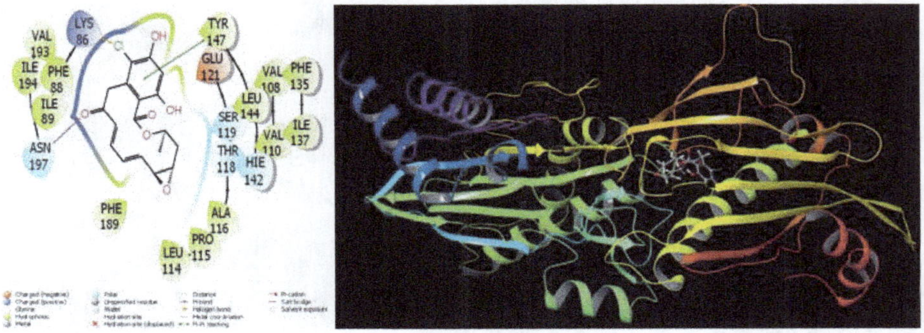

Figure 8.12: Illustration of the interaction of the molecule abscisic acid with protein 3KDJ.

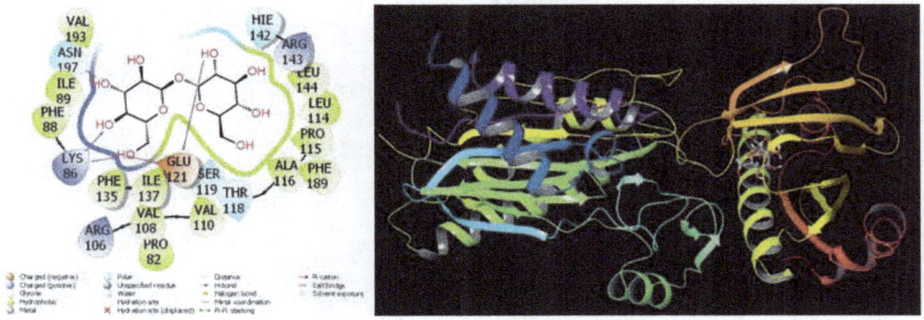

Figure 8.13: Illustration of the interaction of the molecule trehalose with protein 3KDJ.

teins are promising targets for maintaining plant productivity in adverse environmental conditions such as drought or salinity. In the future, a better understanding of the molecular mechanisms of the 3KDJ protein may offer innovative solutions in the field of plant biotechnology (Figures 8.11–8.13).

8.10 Conclusion

Plants develop various responses to stress factors at both hormonal and metabolic levels. Phytohormones such as ethylene, ABA, SA, and jasmonates function as key regulators of stress responses, while metabolic responses include the management of ROS, osmolyte accumulation, and the activation of antioxidant defense mechanisms. These adaptations are crucial for supporting plant survival and growth under stressful conditions.

Molecules with high binding energy to 3KDJ protein showed strong inhibitory properties against plant stress. Molecules binding to 1YET protein showed mitigating effects on plant stress responses by reducing ROS. Inhibition of 1HJO protein increased plant resistance under stress conditions by maintaining cellular homeostasis. The tested molecules regulated stress signaling pathways by selectively binding to the active sites of all three proteins. Molecular modeling of these proteins may increase the usability of target-specific inhibitors in agricultural biotechnology.

References

[1]   Büyük, İ., Soydam-Aydın, S. and Aras, S. (2012). Molecular responses of plants to stress conditions. Türk Hijyen Ve Deneysel Biyoloji Dergisi, 69(2), 97–110.

[2]   Naikoo, M. I., Dar, M. I., Raghib, F., Jaleel, H., Ahmad, B., Raina, A. . . . and Naushin, F. (2019). Role and regulation of plants phenolics in abiotic stress tolerance: An overview. In: Plant Signaling Molecules, 157–168.

[3]   Tuladhar, P., Sasidharan, S. and Saudagar, P. (2021). Role of phenols and polyphenols in plant defense response to biotic and abiotic stresses. In: Biocontrol Agents and Secondary Metabolites, Woodhead Publishing, Woodhead Publishing, 419–441.

[4]   Singh, P., Choudhary, K. K., Chaudhary, N., Gupta, S., Sahu, M., Tejaswini, B. and Sarkar, S. (2022). Salt stress resilience in plants mediated through osmolyte accumulation and its crosstalk mechanism with phytohormones. Frontiers in Plant Science, 13, 1006617.

[5]   Ahmad, P., Abd_Allah, E. F., Alyemeni, M. N., Wijaya, L., Alam, P., Bhardwaj, R. and Siddique, K. H. (2018). Exogenous application of calcium to 24-epibrassinosteroid pre-treated tomato seedlings mitigates NaCl toxicity by modifying ascorbate–glutathione cycle and secondary metabolites. Scientific Reports, 8(1), 13515.

[6]   Ahmad, P., Alyemeni, M. N., Wijaya, L., Alam, P., Ahanger, M. A. and Alamri, S. A. (2017). Jasmonic acid alleviates negative impacts of cadmium stress by modifying osmolytes and antioxidants in faba bean (Vicia faba L.). Archives of Agronomy and Soil Science, 63(13), 1889–1899.

[7]   Isah, T. (2019). Stress and defense responses in plant secondary metabolites production. Biological Research, 52(39), 1–25.

[8]   Karabulut, H. and Gülay, M. Ş. (2016). Serbest radikaller. Mehmet Akif Ersoy University Journal of Health Sciences Institute, 4(1), 50–59.

[9]   Li, S., Liu, S., Zhang, Q., Cui, M., Zhao, M., Li, N. . . . and Wang, L. (2022). The interaction of ABA and ROS in plant growth and stress resistances. Frontiers in Plant Science, 13, 1050132.

[10]   Fedoreyeva, L. I. (2024). ROS as signaling molecules to initiate the process of plant acclimatization to abiotic stress. International Journal of Molecular Sciences, 25(21), 11820.

[11]   Hasanuzzaman, M., Bhuyan, M. B., Parvin, K., Bhuiyan, T. F., Anee, T. I., Nahar, K. . . . and Fujita, M. (2020). Regulation of ROS metabolism in plants under environmental stress: A review of recent experimental evidence. International Journal of Molecular Sciences, 21(22), 8695.

[12]   Huang, H., Ullah, F., Zhou, D. X., Yi, M. and Zhao, Y. (2019). Mechanisms of ROS regulation of plant development and stress responses. Frontiers in Plant Science, 10, 800.

[13]   Krumova, K. and Cosa, G. (2016). Overview of reactive oxygen species. In: Nonell, S. & Flors, C. (eds.) Singlet Oxygen: Applications in Biosciences and Nanosciences, The Royal Society of Chemistry, Cambridge, 1–21.

[14]   Anonym (2024) chrome-extension://efaidnbmnnnibpcajpcglclefindmkaj/ https://www.mustafaaltini sik.org.uk/21-adsem-01.pdf

[15]   Finkelstein, R. R., Gampala, S. S. and Rock, C. D. (2002). Abscisic acid signaling in seeds and seedlings. The Plant Cell, 14(suppl_1), S15–S45.

[16]   Cutler, S. R., Rodriguez, P. L., Finkelstein, R. R. and Abrams, S. R. (2010). Abscisic acid: Emergence of a core signaling network. Annual Review of Plant Biology, 61(1), 651–679.

[17]   Verslues, P. E., Agarwal, M., Katiyar-Agarwal, S., Zhu, J. and Zhu, J. K. (2006). Methods and concepts in quantifying resistance to drought, salt, and freezing, abiotic stresses that affect plant water status. The Plant Journal, 45(4), 523–539.

[18]   Kumlay, A. M. and Eryiğit, T. (2011). Bitkilerde büyüme ve gelişmeyi düzenleyici maddeler: Bitki hormonları. Journal of the Institute of Science and Technology, 1(2), 47–56.

[19]   Abeles, F. B., Morgan, P. W. and Saltveit, M. E. Jr. (1992). Ethylene in Plant Biology, Academic Press, Harcourt Brace Jovanovich, Publishers, San Diego, California 92101–4311.

[20]   Steffens, B. (2014). The role of ethylene and ROS in salinity, heavy metal, and flooding responses in rice. Frontiers in Plant Science, 5, 685.

[21]   Vidhyasekaran, P. (2014). Ethylene Signaling System in Plant Innate Immunity. In: Plant Hormone Signaling Systems in Plant Innate Immunity. Signaling and Communication in Plants, Vol 2, Springer, Dordrecht, 195–244. https://doi.org/10.1007/978-94-017-9285-1_4.

[22]   Chen, H., Bullock, D. A., Alonso, J. M. and Stepanova, A. N. (2021). To fight or to grow: The balancing role of ethylene in plant abiotic stress responses. Plants (Basel), 11(1), 33.

[23]   Li, N., Han, X., Feng, D., Yuan, D. and Huang, L. J. (2019). Signaling crosstalk between salicylic acid and ethylene/jasmonate in plant defense: Do we understand what they are whispering?. International Journal of Molecular Sciences, 20(3), 671.

[24]   Ma, B., Chen, H., Chen, S. Y. and Zhang, J. S. (2014). Roles of ethylene in plant growth and responses to stresses. In: Phytohormones: A Window to Metabolism, Signaling and Biotechnological Applications, 81–118.

[25]   Kitajima, S. and Sato, F. (1999). Plant pathogenesis-related proteins: Molecular mechanisms of gene expression and protein function. The Journal of Biochemistry, 125(1), 1–8.

[26]   Wang, J., Wu, D., Wang, Y. and Xie, D. (2019). Jasmonate action in plant defense against insects. Journal of Experimental Botany, 70(13), 3391–3400.

[27]   Yüksel, B. and Aksoy, Ö. (2017). Su stresi koşullarında bitkilerde gözlenen değişimler. Türk Bilimsel Derlemeler Dergisi, 10(2), 1–5.

[28]   Takahashi, F., Kuromori, T., Urano, K., Yamaguchi-Shinozaki, K. and Shinozaki, K. (2020). Drought stress responses and resistance in plants: From cellular responses to long-distance intercellular communication. Frontiers in Plant Science, 11, 556972.

[29]   Kuromori, T., Seo, M. and Shinozaki, K. (2018). ABA transport and plant water stress responses. Trends in Plant Science, 23(6), 513–522.

[30]   Takahashi, F. and Shinozaki, K. (2019). Long-distance signaling in plant stress response. Current Opinion in Plant Biology, 47, 106–111.

[31]   Takahashi, F., Hanada, K., Kondo, T. and Shinozaki, K. (2019). Hormone-like peptides and small coding genes in plant stress signaling and development. Current Opinion in Plant Biology, 51, 88–95.

[32]   Bartels, D. and Sunkar, R. (2005). Drought and salt tolerance in plants. Critical Reviews in Plant Sciences, 24(1), 23–58.

[33]   Verslues, P. E. and Sharma, S. (2010). Proline metabolism and its implications for plant-environment interaction. The Arabidopsis Book/American Society of Plant Biologists, 8, 1–23.

[34]   Krasensky, J. and Jonak, C. (2012). Drought, salt, and temperature stress-induced metabolic rearrangements and regulatory networks. Journal of Experimental Botany, 63(4), 1593–1608.

[35]   Hasanuzzaman, M., Bhuyan, M. B., Nahar, K., Hossain, M. S., Mahmud, J. A., Hossen, M. S. . . . and Fujita, M. (2018). Potassium: A vital regulator of plant responses and tolerance to abiotic stresses. Agronomy, 8(3), 31.

[36]   Armstrong, W. (1980). Aeration in higher plants. In: Advances in Botanical Research, Vol. 7, Academic Press, 225–332.

[37]   Anonym 2024b. http://www.biyolojiportali.com/konu-anlatimi/8/24/aerobik-solunum-oksijenli-solunum

[38]   Chirkova, T. and Yemelyanov, V. (2018). The study of plant adaptation to oxygen deficiency in Saint Petersburg University. Biological Communications, 63(1), 17–31.

[39]   Drew, M. C. (1997). Oxygen deficiency and root metabolism: Injury and acclimation under hypoxia and anoxia. Annual Review of Plant Biology, 48(1), 223–250.

[40]   Vartapetian, B. B. and Jackson, M. B. (1997). Plant adaptations to anaerobic stress. Annals of Botany, 79(suppl_1), 3–20.

[41]   Bailey-Serres, J. and Voesenek, L. A. C. J. (2008). Flooding stress: Acclimations and genetic diversity. Annual Review of Plant Biology, 59(1), 313–339.

[42]   Liu, W., Zhang, Y., Fang, X., Tran, S., Zhai, N., Yang, Z. . . . and Xu, L. (2022). Transcriptional landscapes of de novo root regeneration from detached Arabidopsis leaves revealed by time-lapse and single-cell RNA sequencing analyses. Plant Communications, 3(4), 100306.

[43]   Yemelyanov, V. V., Puzanskiy, R. K. and Shishova, M. F. (2023). Plant life with and without oxygen: A metabolomics approach. International Journal of Molecular Sciences, 24(22), 16222.

[44]   Cowan, I. R. and GD, F. (1977). Stomatal function in relation to leaf metabolism and environment. Symposia of the Society for Experimental Biology, 31, 471–505. (U.S.A.; DA, BIBL. 1 P. 1/2).

[45]   Cowan, I. R. (1978). Stomatal behaviour and environment. In: Advances in Botanical Research, Vol. 4, Academic Press, 117–228.

[46]   Aasamaa, K., Söber, A. and Rahi, M. (2001). Leaf anatomical characteristics associated with shoot hydraulic conductance, stomatal conductance and stomatal sensitivity to changes of leaf water status in temperate deciduous trees. Functional Plant Biology, 28(8), 765–774.

[47]   Elliott-Kingston, C., Haworth, M., Yearsley, J. M., Batke, S. P., Lawson, T. and McElwain, J. C. (2016). Does size matter? Atmospheric CO_2 may be a stronger driver of stomatal closing rate than stomatal size in taxa that diversified under low CO_2. Frontiers in Plant Science, 7, 1253.

[48]   Niu, Y. and Xiang, Y. (2018). An overview of biomembrane functions in plant responses to high-temperature stress. Frontiers in Plant Science, 9, 915.

[49] Gusta, L. V. and Chen, T. H. H. (1987). The physiology of water and temperature stress. Wheat and Wheat Improvement, 13, 115–150.

[50] Türkan, I., Bor, M., Özdemir, F. and Koca, H. (2005). Differential responses of lipid peroxidation and antioxidants in the leaves of drought-tolerant P. acutifolius Gray and drought-sensitive P. vulgaris L. subjected to polyethylene glycol mediated water stress. Plant Science, 168(1), 223–231.

[51] Escribá, P. V., González-Ros, J. M., Goñi, F. M., Kinnunen, P. K., Vigh, L., Sánchez-Magraner, L. . . . and Barceló-Coblijn, G. (2008). Membranes: A meeting point for lipids, proteins and therapies. Journal of Cellular and Molecular Medicine, 12(3), 829–875.

[52] Scharf, K. D., Berberich, T., Ebersberger, I. and Nover, L. (2012). The plant heat stress transcription factor (Hsf) family: Structure, function and evolution. Biochimica Et Biophysica Acta (Bba)-gene Regulatory Mechanisms, 1819(2), 104–119.

[53] Bernfur, K., Rutsdottir, G. and Emanuelsson, C. (2017). The chloroplast-localized small heat shock protein Hsp21 associates with the thylakoid membranes in heat-stressed plants. Protein Science, 26(9), 1773–1784.

[54] Başaran, F. and Akçin, Z. T. A. (2022). Sıcaklık Faktörünün Bitkiler Üzerindeki Etkileri ve Yüksek Sıcaklık Stresi. Bahçe, 51(2), 139–147.

[55] Yavaş, İ. and İlker, E. (2020). Çevresel stres koşullarına maruz kalan bitkilerde fotosentez ve fitohormon seviyelerindeki değişiklikler. Bahri Dağdaş Bitkisel Araştırma Dergisi, 9(2), 295–311.

[56] Xiong, F. S., Mueller, E. C. and Day, T. A. (2000). Photosynthetic and respiratory acclimation and growth response of Antarctic vascular plants to contrasting temperature regimes. American Journal of Botany, 87, 700–710.

[57] Georgieva, K., Szigeti, Z., Sarvari, E., Gaspar, L., Maslenkova, L., Peeva, V. . . . and Tuba, Z. (2007). Photosynthetic activity of homoiochlorophyllous desiccation tolerant plant Haberlea rhodopensis during dehydration and rehydration. Planta, 225, 955–964.

[58] Kabay, T. and Şensoy, S. (2017). Yüksek Sıcaklığın Fasulyede Enzim, Klorofil ve İyon Değişimine Etkisi. Ege Üniversitesi Ziraat Fakültesi Dergisi, 54(4), 429–437.

[59] Asada, K. (2006). Production and scavenging of reactive oxygen species in chloroplasts and their functions. Plant Physiol, 141, 391–396.

[60] Zushi, K., Kajiwara, S. and Matsuzoe, N. (2012). Chlorophyll a fluorescence OJIP transient as a tool to characterize and evaluate response to heat and chilling stress in tomato leaf and fruit. Scientia Horticulturae, 148, 39–46.

[61] Anonym (2024). chrome-extension://efaidnbmnnnibpcajpcglclefindmkaj/ https://flora.com.tr/dosya lar/bitkilerde-stres.pdf.

[62] Aslantaş, R., Karakurt, H. and Karakurt, Y. (2010). Bitkilerin Düşük Sıcaklıklara Dayanımında Hücresel ve Moleküler Mekanizmalar. Journal of Agricultural Faculty of Atatürk University, 41(2), 157–167.

[63] Kacar, B., Katkat, B. and Öztürk, Ş. (2006). Bitki Fizyolojisi. Nobel Yayım Dağıtım, 2, 493–533.

[64] Astolfi, S. and Zuchi, S. (2013). Adequate S supply protects barley plants from adverse effects of salinity stress by increasing thiol contents. Acta Physiologiae Plantarum, 35, 175–181.

[65] Doğru, A. (2006). Kolza (Brassica Napus L. Ssp. Oleifera)'nın Bazı Kışlık Çeşitlerinde Düşük Sıcaklık Toleransı Ile Ilgili Fizyolojik Ve Biyokimyasal Parametrelerin Araştırılması, Unpublished doctoral thesis, Hacettepe University, Institute of Science, Ankara.

[66] Küden, A. B., Küden, A., Paydaş, S., Kaşka, N. and İmrak, B. (1998). Bazı ılıman iklim meyve tür ve çeşitlerinin soğuğa dayanıklılığı üzerinde çalışmalar. Tr. J. Of Agriculture Ve Forestry, 22, 101–109.

[67] Ertürk, Y. and Güleryüz, M. (2007). Erzincan koşullarında bazı yerli ve yabancı kayısı çeşitlerinin düşük sıcaklıklara dayanım derecelerinin belirlenmesi. Tarım Bilimleri Derg, 13(2), 128–136.

[68] Pearce, R. S. (2001). Plant freezing and damage. Annals of Botany, 87, 417–424.

[69] Yang, C., Xie, F. and Li, L. (2021). Phenotypic Study of Photomorphogenesis in Arabidopsis Seedlings. In: Plant Photomorphogenesis: Methods and Protocols, 41–47.

[70] Kaya, C. (2022). Işık Stresinin Bitkiler Üzerindeki Etkisi. Bitkilerde Abiyotik Ve Biyotik Stres Yönetimi, Chapter 7, Iksad Publications, Ankara, ISBN: 978–625-6380-66–0.

[71] Cortleven, A., Nitschke, S., Klaumunzer, M., AbdElgawad, H., Asard, H., Grimm, B., Riefler, M. and Schmulling, T. (2014). A novel protective function for cytokinin in the light stress response is mediated by the Arabidopsis Histidine Kinase2 and Arabidopsis Histidine Kinase3 receptors. Plant Physiology, 164, 1470–1483.

[72] Krieger-Liszkay, A. (2005). Singlet oksijen production in photosynthesis. Journal of Experimental Botany, 56, 33346.

[73] Doğru, A. (2020). Bitkilerde Aktif Oksijen Türleri ve Oksidatif Stres. International Journal of Life Sciences and Biotechnology, 3(2), 205–226.

[74] Yang, B., Tang, J., Yu, Z., Khare, T., Srivastav, A., Datir, S. and Kumar, V. (2019). Light stress responses and prospects for engineering light stress. Journal of Plant Growth Regulation, 38, 1489–1506.

[75] Mudgal, V., Madaan, N., Mudgal, A., Singh, A. and Kumar, P. (2010). Comparative study of the effects of salinity on plant growth, nodulation, and legheamoglobin content in kabuli and desi cultivars of Cicer arietinum (L.). Atlas Journal of Biology, 1(1), 1–4.

[76] Corwin, D. L. (2021). Climate change impacts on soil salinity in agricultural areas. European Journal of Soil Science, 72(2), 842–862.

[77] Sheldon, A. R., Dalal, R. C., Kirchhof, G., Kopittke, P. M. and Menzies, N. W. (2017). The effect of salinity on plant-available water. Plant and Soil, 418(1), 477–491.

[78] Mishra, P., Mishra, J. and Arora, N. K. (2021). Plant growth promoting bacteria for combating salinity stress in plants–recent developments and prospects: A review. Microbiological Research, 252, 126861.

[79] Koç, F. N., Çetinkaya, H. and Seçkin Dinler, B. (2022). Bitkilerde Tuz Stresinin Etkileri, Savunma Cevapları Ve Inyal Iletim Yolu. Chapter 4, Bitkilerde Abiyotik Ve Biyotik Stress Yönetimi, İKSAD Publishing House, Ankara.

[80] Zhao, C., Zhang, H., Song, C., Zhu, J. K. and Shabala, S. (2020). Mechanisms of plant responses and adaptation to soil salinity. The Innovation, 1(1), 100017.

[81] Ahmad, R., Hussain, S., Anjum, M. A., Khalid, M. F., Saqib, M., Zakir, I. . . . and Ahmad, S. (2019). Oxidative stress and antioxidant defense mechanisms in plants under salt stress. In: Plant Abiotic Stress Tolerance: Agronomic, Molecular and Biotechnological Approaches, 191–205.

[82] Zhu, J. K. (2002). Salt and drought stress signal transduction in plants. Annual Reviews of Plant Biology, 53, 247–273.

[83] Xiong, L., Schumaker, K. S. and Zhu, J. K. (2002). Cell signaling during cold, drought, and salt stress. The Plant Cell, 14(suppl_1), S165–S183.

[84] Cao, Y. R., Chen, S. Y. and Zhang, J. S. (2008). Ethylene signaling regulates salt stress response: An overview. Plant Signaling and Behavior, 3, 761–763.

[85] Abid, M., Zhang, Y. J., Li, Z., Bai, D. F., Zhong, Y. P. and Fang, J. B. (2020). Effect of salt stress on growth, physiological and biochemical characters of four kiwifruit genotypes. Scientia Horticulturae, 271, 109473.

[86] Hao, S., Wang, Y., Yan, Y., Liu, Y., Wang, J. and Chen, S. (2021). A review on plant responses to salt stress and their mechanisms of salt resistance. Horticulturae, 7(6), 132.

[87] White, R. E. (2006). Principles and Practice of Soil Science: The Soil as a Natural Resource, 4th Edition, Wiley-Blackwell Scientific Publication, London, United Kingdom.

[88] Fageria, N. K. (2009). The Use of Nutrients in Crop Plants, CRC Pres, Boca Raton, Florida, New York.

[89] Kacar, B. and Katkat, V. (2010). Bitki Besleme. 5. Baskı, Nobel Yayın Dağıtım Tic, Ltd. Şti, Kızılay-Ankara.

[90] Bolat, İ. and Kara, Ö. (2017). Bitki Besin Elementleri: Kaynakları, İşlevleri, Eksik ve Fazlalıkları. Journal of Bartin Faculty of Forestry, 19(1), 218–228.

[91] Kantarcı, M. D. (2000). Toprak Ilmi. Iü Toprak Ilmi Ve Ekoloji Anabilim Dalı, I Ü Yayın No. 4261, Orman Fakültesi Yayın No. 462, İstanbul, 420s, Istanbul University Faculty of Forestry Publications.

[92] Boşgelmez, A., Boşgelmez, İ. İ., Savaşçı, S. and Paslı, N. (2001). Ekoloji – II (Toprak). Başkent Klişe Matbaacılık, Kızılay-Ankara.

[93] Küçükyumuk, Z. and Bayındır, Ü. (2023). Bitki Beslenmesi ve Bitki Hastalıkları İlişkisi. Turkish Journal of Science and Engineering, 5(1), 42–49.

[94] Aktaş, M. and Ateş, A. (1998). Bitkilerde Beslenme Bozuklukları Nedenleri Tanınmaları. Nurol Matbaacılık A.Ş, Ostim-Ankara.

[95] McCauley, A., Jones, C. and Jacobsen, J. (2009). Nutrient Management. Nutrient management module 9 Montana State University Extension Service. Publication, 4449–9, 1–16.

[96] Yang, L., Ji, J., Harris-Shultz, K. R., Wang, H., Wang, H., Abd-Allah, E. F., Luo, Y. and Hu, X. (2016). The dynamic changes of the plasma membrane proteins and the protective roles of nitric oxide in rice subjected to heavy metal cadmium stress. Frontiers in Plant Sciences, 7, 190.

[97] Bayçu, G., Özden, H., Gören-Sağlam, N. and Rognes, S. (2008). Effect of Cd, Pb, chilling and drought treatments on activity of five antioxidant enzymes and free proline level in Albizzia leaves, FESPB congress, 17–24 August, Tampere- Finland. Physiologia Plantarum, 133.

[98] Dalcorso, G., Fasani, E. and Furini, A. (2013). Recent advances in the analysis of metal hyperaccumulation and hypertolerance in plants using proteomics. Frontiers in Plant Science, 4, 280.

[99] Liu, H., Wang, C., Xie, Y., Luo, Y., Sheng, M. and Xu, F. (2020). Ecological responses of soil microbial abundance and diversity to cadmium and soil properties in farmland around an enterprise-intensive region. Journal of Hazardous Materials, 392, 122478.

[100] Lin, A. J., Zhang, X. H., Chen, M. M. and Cao, Q. (2007). Oxidative stress and DNA damages induced by cadmium accumulation. Journal of Environmental Sciences, 19, 596–602.

[101] Turfan, N. (2022). Bitkilerde Ağir Metal Stresi. Bitkilerde Abiyotik Ve Biyotik Stres Yönetimi, Chapter 6, ISBN: 978-625-6380-66-0, İKSAD Publishing House, 164–192.

[102] Haktanır, K. and Arcak, S. (1998). Çevre Kirliliği. Ankara Üniversitesi Ziraat Fakültesi, Ders Kitabı, Ankara. Yayın No: 1503, 457s.

[103] Sieprawska, A., Kornaś, A. and Filek, M. (2015). Involvement of selenium in protective mechanisms of plants under environmental stress conditions–review. Acta Biologica Cracoviensia Series Botanica, 57, 1–12.

[104] Turhan, Ş., Turfan, N. and Kurnaz, A. (2022). Heavy metal contamination and health risk evaluation of chestnut (Castaneasativa Miller) consumed in Turkey. International Journal of Environmental Health Research, 33(11), 1091–1101.

[105] Sultana, M. S., Rana, S., Yamazaki, S., Aono, T. and Yoshida, S. (2017). Health risk assessment for carcinogenic and noncarcinogenic heavy metal exposures from vegetables and fruits of Bangladesh. Cogent Environmental Science, 3(1291107), 1–17.

[106] Singh, S., Parihar, P., Singh, R., Singh, V. P. and Prasad, S. M. (2016). Heavy metal tolerance in plants: Role of Transcriptomics, Proteomics, Metabolomics, and Ionomics. Frontiers in Plant Sciences, 6, 1143.

[107] Kumar, V., Singh, J. and Kumar, P. (2019). Heavy metals accumulation in crop plants: Sources, response mechanisms, stress tolerance and their effects. In: Kumar, V., Kumar, R., Singh, J. & Kumar, P. ((eds)) Contaminants in Agriculture and Environment: Health Risks and Remediation, Vol. 1, Agro Environ Media, Haridwar, India, 38–57.

[108] Kumar, S. S., Kadier, A., Malyan, S. K., Ahmad, A. and Bishnoi, N. R. (2017). Phytoremediation and rhizoremediation: Uptake, mobilization and sequestration of heavy metals by plants. In: Singh, D.P. and Singh, H.B (eds) Plant-Microbe Interactions in Agro-Ecological Perspectives, Springer, Singapore, 367–394.

[109] Schrödinger Release. (2022–4). Maestro, Schrödinger, LLC, New York, NY, 2022.

[110] Schrödinger Release. (2022–4). Protein Preparation Wizard, Epik, Schrödinger, LLC, New York, NY, 2022. Impact, Schrödinger, LLC, New York, NY; Prime, Schrödinger, LLC, New York, NY, 2022.

[111] Schrödinger Release. (2022–4). LigPrep, Schrödinger, LLC, New York, NY, 2022.

[112] Shahzadi, I., Zahoor, A. F., Tüzün, B., Mansha, A., Anjum, M. N., Rasul, A. . . . and Mojzych, M. (2022). Repositioning of acefylline as anti-cancer drug: Synthesis, anticancer and computational studies of azomethines derived from acefylline tethered 4-amino-3-mercapto-1, 2, 4-triazole. Plos One, 17(12), e0278027.

[113] El Faydy, M., Lakhrissi, L., Dahaieh, N., Ounine, K., Tüzün, B., Chahboun, N. . . . and Zarrouk, A. (2024). Synthesis, biological properties, and molecular docking study of novel 1, 2, 3-triazole-8-quinolinol hybrids. ACS Omega, 9(23), 25395–25409.

[114] Schrödinger Release. (2022–4). QikProp, Schrödinger, LLC, New York, NY, 2022.

[115] Yadav, B., Dubey, R., Gnanasekaran, P. and Narayan, O. P. (2021). OMICS approaches towards understanding plant's responses to counterattack heavy metal stress: An insight into molecular mechanisms of plant defense. Plant Gene, 28, 100333.

[116] Arbona, V., Manzi, M., De Ollas, C. and Gómez-Cadenas, A. (2013). Metabolomics as a tool to investigate abiotic stress tolerance in plants. International Journal of Molecular Sciences, 14(3), 4885–4911.

[117] Muhammad Aslam, M., Waseem, M., Jakada, B. H., Okal, E. J., Lei, Z., Saqib, H. S. A. . . . and Zhang, Q. (2022). Mechanisms of abscisic acid-mediated drought stress responses in plants. International Journal of Molecular Sciences, 23(3), 1084.

[118] Şimşek, Ö., Isak, M. A., Dönmez, D., Dalda Şekerci, A., İzgü, T. and Kaçar, Y. A. (2024). Advanced biotechnological interventions in mitigating drought stress in plants. Plants, 13(5), 717.

[119] Elrobh, M., Alanazi, M., Abduljaleel, Z., El-Huneidi, W. and Khan, W. (2017). Isolation, sequencing and in silico characterization of promoter region of hspa6 gene from Arabian camel (Camelus dromedaries). Journal of Animal and Plant Sciences, 27(6), 1806–1815.

[120] Oshunsanya, S. O., Nwosu, N. J. and Li, Y. (2019). Abiotic stress in agricultural crops under climatic conditions. In: Sustainable Agriculture, Forest and Environmental Management, 71–100.

[121] Juan, C. A., Pérez de la Lastra, J. M., Plou, F. J. and Pérez-Lebeña, E. (2021). The chemistry of reactive oxygen species (ROS) revisited: Outlining their role in biological macromolecules (DNA, lipids and proteins) and induced pathologies. International Journal of Molecular Sciences, 22(9), 4642.

[122] Dutta, T., Singh, H., Gestwicki, J. E. and Blatch, G. L. (2021). Exported plasmodial J domain protein, PFE0055c, and PfHsp70-x form a specific co-chaperone-chaperone partnership. Cell Stress and Chaperones, 26(2), 355–366.

[123] Nguyen, H. C., Lin, K. H., Ho, S. L., Chiang, C. M. and Yang, C. M. (2018). Enhancing the abiotic stress tolerance of plants: From chemical treatment to biotechnological approaches. Physiologia Plantarum, 164(4), 452–466.

[124] Ahanger, M. A., Tomar, N. S., Tittal, M., Argal, S. and Agarwal, R. (2017). Plant growth under water/ salt stress: ROS production; antioxidants and significance of added potassium under such conditions. Physiology and Molecular Biology of Plants, 23, 731–744.

[125] Godoy, F., Olivos-Hernández, K., Stange, C. and Handford, M. (2021). Abiotic stress in crop species: Improving tolerance by applying plant metabolites. Plants, 10(2), 186.

[126] Miller, G., Shulaev, V. and Mittler, R. (2008). Reactive oxygen signaling and abiotic stress. Physiologia Plantarum, 133(3), 481–489.

[127] Atkinson, N. J. and Urwin, P. E. (2012). The interaction of plant biotic and abiotic stresses: From genes to the field. Journal of Experimental Botany, 63(10), 3523–3543.

[128] Conde, A., Chaves, M. M. and Gerós, H. (2011). Membrane transport, sensing and signaling in plant adaptation to environmental stress. Plant and Cell Physiology, 52(9), 1583–1602.

[129] Farooq, M., Wahid, A., Kobayashi, N. S. M. A., Fujita, D. B. S. M. A. and Basra, S. M. (2009). Plant drought stress: Effects, mechanisms and management. In: Sustainable Agriculture, 153–188.

[130] Janicki, M., Marczak, M., Cieśla, A. and Ludwików, A. (2020). Identification of novel inhibitors of a plant group A protein phosphatase type 2C using a combined in silico and biochemical approach. Frontiers in Plant Science, 11, 526460.

[131] Umar, O. B., Ranti, L. A., Abdulbaki, A. S., Bola, A. L., Abdulhamid, A. K., Biola, M. R. and Victor, K. O. (2021). Stresses in plants: Biotic and abiotic. In: Current Trends in Wheat Research, 1–8.

[132] Hewage, K. A. H., Yang, J. F., Wang, D., Hao, G. F., Yang, G. F. and Zhu, J. K. (2020). Chemical manipulation of abscisic acid signaling: A new approach to abiotic and biotic stress management in agriculture. Advanced Science, 7(18), 2001265.

[133] MacRae, T. H. (2010). Gene expression, metabolic regulation and stress tolerance during diapause. Cellular and Molecular Life Sciences, 67, 2405–2424.

[134] Prasannath, K. (2017). Plant defense-related enzymes against pathogens: A review. AGRIEAST: Journal of Agricultural Sciences, 11(1), 38–48.

[135] DeFalco, T. A. and Zipfel, C. (2021). Molecular mechanisms of early plant pattern-triggered immune signaling. Molecular Cell, 81(17), 3449–3467.

Part III: **Pharmaceutical use of medicinal plants**

İlayda Bersu Kul, Gamze Tüzün, Burak Tüzün*, and
Elyor Berdimurodov

Chapter 9
Medicinal and aromatic plants used in burn treatment

Abstract: Burn injuries remain a significant public health concern worldwide, particularly in low- and middle-income countries. This chapter explores the therapeutic potential of medicinal and aromatic plants in burn treatment, emphasizing their anti-inflammatory, antimicrobial, analgesic, and wound-healing properties. Traditional remedies such as Aloe vera, Calendula officinalis, Centella asiatica, Hypericum perforatum, and Syzygium aromaticum are evaluated based on their phytochemical composition, mechanisms of action, and clinical applications. The integration of traditional knowledge with modern pharmacological science is highlighted, supported by molecular, omic, and in silico research approaches. Furthermore, innovative formulations—ranging from herbal gels and ointments to nanotechnology-based wound dressings—are discussed in terms of efficacy, stability, and safety. The chapter also addresses the pharmacogenetic implications of plant-based therapies and the challenges of standardization, toxicity, and regulatory frameworks. Overall, medicinal plants represent a promising and sustainable alternative or complement to conventional burn treatments, offering culturally rooted and scientifically supported solutions for wound management.

Keywords: burn treatment, medicinal plants, aromatic plants, wound healing, anti-inflammatory, phytochemicals, antimicrobial, traditional medicine

*Corresponding author: Burak Tüzün, Plant and Animal Production Department, Technical Sciences Vocational School of Sivas, Sivas Cumhuriyet University, Sivas 58140, Turkey,
e-mail: theburaktuzun@yahoo.com, https://orcid.org/0000-0002-0420-2043
İlayda Bersu Kul, Department of Molecular Biology and Genetics, Science Faculty, Sivas Cumhuriyet University, Sivas 58140, Turkey
Gamze Tüzün, Department of Chemistry, Faculty of Science, Sivas Cumhuriyet University, Sivas 58140, Turkey
Elyor Berdimurodov, Chemical and Materials Engineering, New Uzbekistan University, 54 Mustaqillik Ave, Tashkent 100007, Uzbekistan; Faculty of Chemistry, National University of Uzbekistan, Tashkent 100034, Uzbekistan

https://doi.org/10.1515/9783111469713-009

9.1 Introduction

Burns are serious pathological conditions that occur as a result of traumatic effects of thermal, chemical, electrical activity or radiation to the skin tissue, triggering inflammatory responses and increasing the risk of infection. Millions of people worldwide experience burn injuries every year, and this situation poses a serious public health problem, especially in low- and middle-income countries due to limited healthcare services. The main goals in burn treatment are to accelerate wound healing, prevent infection, control inflammation and manage pain. However, the limitations of modern treatment methods related to cost, accessibility, and side effects have increased interest in herbal treatments. Medicinal and aromatic plants exhibit anti-inflammatory, antibacterial, analgesic, and wound-healing properties through bioactive compounds such as flavonoids, phenolic acids, alkaloids, tannins, essential oils, and saponins. These properties provide multifaceted effects that support the healing of both acute and chronic burn wounds. Plants such as *Aloe vera, Calendula officinalis, Hypericum perforatum, Mentha piperita*, and *Syzygium aromaticum*, which have been used in traditional medicine for centuries are prominent examples in this field. The effects of these plants, such as reducing burn-related inflammation, preventing infection and accelerating the regeneration of wound tissue, have also been supported by modern scientific research. For example, *Aloe vera* gel offers moisturizing, anti-inflammatory, and cell-regenerating effects thanks to its active ingredients such as polyphenols and anthraquinone derivatives, while *Hypericum perforatum* has antimicrobial and antioxidant effects with its hypericin content. Historically, plants used in the treatment of burns represent one of the oldest medical practices of humanity. The therapeutic effects of these plants have been passed down from generation to generation in different cultural traditions such as traditional Chinese medicine, Ayurveda, Middle Eastern folk medicine, and Anatolian medicine and have survived to the present day. Modern medicine integrates this traditional knowledge into pharmaceutical practices, supported by biochemical analyses and clinical studies. For example, *Hypericum perforatum* extract has found its place in modern pharmacy in the form of commercial ointments and creams. Similarly, clinical studies on the wound-healing effects of *Calendula officinalis* have popularized the use of this plant in medical preparations.

9.2 General properties of medicinal and aromatic plants used in burn treatment

Medicinal and aromatic plants in burn treatment attract attention as an area where traditional knowledge and modern science meet. The main factor affecting the use of these plants is the wide pharmacological effects provided by their rich phytochemical content and easy accessibility. Multifaceted benefits such as wound healing, infection control, reduction of inflammation, and pain management have enabled medicinal and aromatic plants to gain an important place in burn treatment [1]. General Properties of medicinal and aromatic plants used in burn treatment are given in Figure 9.1.

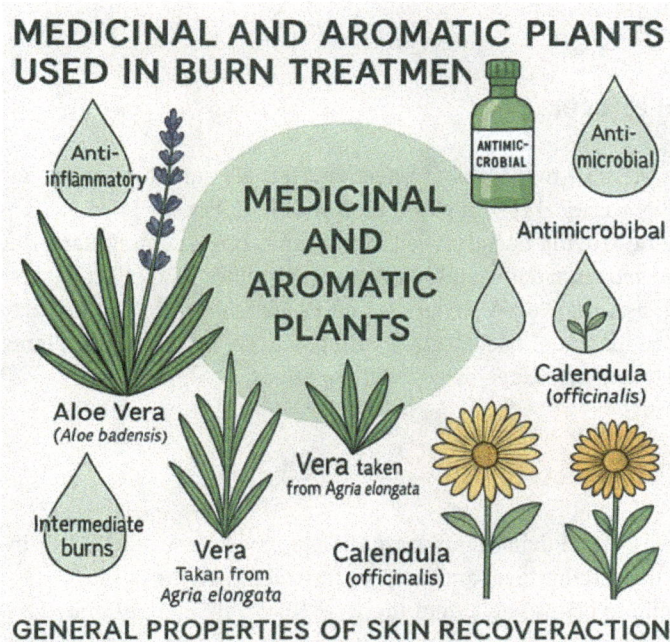

Figure 9.1: General properties of medicinal and aromatic plants used in burn treatment.

9.2.1 Phytochemical content and mechanisms of action

Medicinal and aromatic plants perform their various biological effects through the bioactive compounds they contain. These phytochemicals provide a series of therapeutic benefits in burn treatment by showing both systemic and local effects. Especially anti-inflammatory, antimicrobial, wound healing, and analgesic properties reveal the pharmacological potential of these plants [2]:

1. Anti-inflammatory effects: Inflammation seen in burn wounds is one of the most important processes that delays the healing process and leads to complications. Medicinal and aromatic plants control inflammation and create an environment suitable for tissue regeneration.
2. Flavonoids: They reduce inflammation by suppressing the secretion of inflammatory cytokines (e.g., TNF-α and IL-6). For example, *Aloe vera* and *Calendula offici-*

nalis (*Calendula*) exhibit strong anti-inflammatory properties thanks to their high flavonoid content.

3. Phenolic acids: They suppress inflammation caused by oxidative stress. *Hypericum perforatum* (St. John's wort) plays an effective role in controlling inflammation with its phenolic components [3].

9.2.2 Antimicrobial effects

Burn areas are highly vulnerable to the risk of infection. Herbal components with antimicrobial properties play a critical role in preventing wound infections [4].

Essential oils act by destroying bacterial cell membranes or disrupting bacterial metabolism. *Syzygium aromaticum* (clove) oil has a strong antimicrobial effect against common pathogens such as *Staphylococcus aureus* and *Escherichia coli*. Tannins, prevent the growth and reproduction of bacteria by binding with cellular proteins. Plants such as *Mentha piperita* (mint) are widely used for these effects.

9.2.3 Wound-healing effects

The wound-healing effects of medicinal and aromatic plants are associated with increased proliferation at the cellular level and support for collagen synthesis [5].

Polysaccharides promote cell renewal and increase fibroblast activity. *Aloe vera* accelerates the epithelialization process with the polysaccharides it contains. Essential fatty acids contribute to tissue regeneration by preserving the integrity of cell membranes. *Hypericum perforatum* is rich in fatty acids that support the restructuring of wound tissue.

9.2.4 Analgesic effects

Management of intense pain in burn wounds is a critical part of treatment. Some compounds found in medicinal plants have pain-reducing properties [6].

Menthol reduces pain signals in nerve endings by having a local anesthetic effect. *Mentha piperita* (mint) is therefore frequently used in pain management in burn treatment. Phenol derivatives have an inhibitory effect on pain transmission. *Syzygium aromaticum* exhibits analgesic effects thanks to its phenol derivatives.

9.2.5 Advantages and disadvantages of herbal treatments

The use of medicinal and aromatic plants should be evaluated with both their advantages and limitations. Although there are many positive aspects that have made these approaches widespread in burn treatment, some disadvantages should also be taken into consideration.

9.2.5.1 Advantages

Natural and safe use: Herbal treatments generally carry less risk of toxicity due to their natural origin. Allergic reactions or side effects are quite rare when used in the right dosages [7].

Accessibility and affordability: Medicinal and aromatic plants can be found naturally in wide geographical areas and are low-cost. This makes them widely preferred, especially in low-income communities.

Multifaceted mechanisms of action: Many medicinal plants act on more than one biological process at the same time. For example, *Aloe vera* offers anti-inflammatory, antibacterial, and moisturizing effects together.

Traditional knowledge and cultural origins: Herbal treatments have been accepted in society because they are based on traditional knowledge passed down through generations. Modern science has supported this information and created a more reliable basis.

9.2.5.2 Disadvantages

Lack of standardization: The effectiveness and reliability of medicinal plants may vary depending on the type of plant used, the region where it grows, and the processing methods. This leads to uncertainties in terms of dosage and effectiveness [8].

Lack of clinical data: In cases where information based on traditional use is not sufficiently supported by scientific research, the effectiveness of these treatments may be questioned.

Risk of side effects and interactions: Although herbal treatments are natural, they may have toxic effects when used in the wrong doses. In addition, they may interact with pharmacological drugs and cause negative results.

Need for long-term treatment: Herbal treatments generally show effects over a longer period of time than pharmacological treatments. This may be a disadvantage, especially in acute burn cases.

9.3 Medicinal and aromatic plants used in burn treatment

Medicinal and aromatic plants used in burn treatment contain many bioactive compounds that accelerate wound-healing processes, reduce inflammation, prevent infection, and are effective in pain management. These plants have been at the center of traditional treatment practices throughout history and are now the subject of modern pharmacological research [9]. Medicinal and aromatic plants used in burn treatment are shown in Figure 9.2.

Figure 9.2: Medicinal and aromatic plants used in burn treatment.

9.3.1 *Aloe vera*

Aloe vera is a plant belonging to the Liliaceae family, widely grown in tropical and subtropical regions. The therapeutic properties of the plant are attributed to the gel-form content found especially in the inner part of its leaves.

Phytochemical content and mechanisms: *Aloe vera* gel is rich in polysaccharides, glycoproteins, enzymes, amino acids, vitamins (A, C, and E), and minerals. In particular, a polysaccharide called acemannan in its content draws attention with its immunomodulatory and wound-healing effects. In addition, salicylic acid and sterols, which have anti-inflammatory effects, suppress inflammation in burn wounds and accelerate tissue repair [10].

9.3.1.1 Clinical effects

Moisturizing and epithelialization effect: *Aloe vera* accelerates the epithelialization process in burn wounds by increasing cell proliferation and supporting fibroblast activity. Its moisturizing effect preserves the elasticity of damaged tissues and prevents the wound area from drying out [11].

Anti-inflammatory and antiseptic properties: The anti-inflammatory effect of *Aloe vera* occurs through the suppression of inflammatory cytokines (TNF-α and IL-1β). In addition, the natural pH level of the gel prevents the proliferation of microorganisms, reducing the risk of infection [12].

Aloe vera is applied topically, especially in first- and second-degree burns. Studies show that burn wounds treated with *Aloe vera* gel heal faster and have significantly reduced pain levels compared to conventional treatments [13].

9.3.2 *Calendula officinalis* (*Calendula*)

Calendula officinalis is a plant belonging to the Asteraceae family and has long been used medicinally for its wound healing and antiseptic properties. Also known as "*Calendula*," this plant is widely used in traditional medicine, especially for skin problems [14].

Phytochemical content and mechanisms: *Calendula* flowers contain bioactive components such as flavonoids, saponins, carotenoids, triterpenes and volatile oils. The anti-inflammatory effect of flavonoids occurs through the suppression of prostaglandin synthesis, while carotenoids support tissue repair and regeneration [15].

Cell regenerative properties: *Calendula* accelerates wound healing by increasing fibroblast proliferation and collagen synthesis. This cell regenerative property is especially effective in the formation of granulation tissue [16].

Antiseptic effect: *Calendula*'s essential oils have antimicrobial effects against both gram-positive and gram-negative bacteria. In this way, it reduces the risk of infection and ensures safe wound closure.

Usage areas: *Calendula*'s tincture, ointment, and cream preparations are widely used in the treatment of superficial burns. Clinical studies have shown that *Calendula* ointments shorten the wound closure time and reduce scarring.

9.3.3 *Centella asiatica* (gotu kola)

Centella asiatica is a tropical plant belonging to the Apiaceae family and has a long history in Asian traditional medicine, especially in the treatment of skin diseases. The wound-healing effects of the plant have also been supported by modern science [17].

Phytochemical content and mechanisms: *Centella asiatica* is rich in triterpenoid saponins (especially asiaticoside, madecassoside), flavonoids and phenolic compounds. These compounds support collagen synthesis and angiogenesis by increasing fibroblast activity at the cellular level.

Collagen synthesis-enhancing effect: *Centella asiatica* improves wound strength by increasing Type I and Type III collagen synthesis in wound tissue. This feature ensures that the scar tissue formed after burns is more regular.

Anti-inflammatory and antioxidant properties: Asiaticoside and madecassoside reduce inflammation by suppressing the release of inflammatory mediators. It also reduces oxidative stress thanks to its free radical scavenging properties.

Centella asiatica is used in first- and second-degree burns in cream and gel forms. In clinical studies, faster epithelialization and granulation were observed in burn wounds where *Centella asiatica* was applied. In addition, it was reported that scar formation was less.

In addition to commonly used plants such as *Aloe vera, Calendula officinalis,* and *Centella asiatica,* other important plants such as *Hypericum perforatum, Syzygium aromaticum* (clove), and *Lavandula angustifolia* (lavender) also stand out with their therapeutic properties. St. John's wort has strong antimicrobial and anti-inflammatory effects thanks to the hypericin and hyperforin compounds it contains, making it a valuable option in preventing post-burn infections. In addition, due to its cell regenerative properties, it contributes to tissue healing, especially in deep burns. Clove, on the other hand, has a strong bioactive compound, eugenol, which relieves pain with its local anesthetic effect, while also reducing the risk of infection with its antiseptic properties. Lavender, on the other hand, has calming and anti-inflammatory effects due to its linalool and linalyl acetate content; these properties are used to both reduce pain and support the healing process in burn wounds [18].

Herbal treatments are not limited to the use of a single plant. In traditional medicine and modern practices, combination treatments that include more than one plant are seen to yield more effective results. For example, herbal formulations using *Aloe vera* and *Calendula officinalis* together accelerate the healing process thanks to the synergistic effects of both plants. Similarly, topical applications using lavender oil and clove oil together have been effective in both managing pain and reducing inflammation. Since herbal pastes and oils used in traditional mixtures are less processed than modern formulations, they can preserve natural phytochemical diversity, which can contribute to treatment [19]. Herbal treatments also vary in terms of formulation forms. Herbal creams, ointments, gels, and tinctures are frequently used, especially in burn treatment. While *Aloe vera*-based gels attract attention with their moisturizing and healing properties, ointments containing *Calendula officinalis* accelerate wound closure by supporting granulation tissue. In addition, essential oils such as lavender and clove play a complementary role in pain and stress management with aromatherapy applications. Sterile plant extracts combined with dressing materials offer a practical and effective option for healing burn wounds.

The integration of traditional and modern medical approaches is also important in the use of herbal treatments. The methods used in traditional medicine are based on thousands of years of knowledge and are generally supported by modern research. For example, the wound healing properties of *Centella asiatica* in Ayurvedic medicine and the skin rejuvenating effects of *Aloe vera* in Chinese medicine are frequently emphasized. In Anatolian medicine, *Hypericum perforatum* oil prepared with olive oil is still a popular option in burn treatment. Modern research has proven the effectiveness of these traditional practices in laboratory and clinical studies, enabling the development of more reliable treatment methods [20].

Herbal treatments attract attention not only with their effectiveness but also with their safety profiles and standardization problems. The amount and quality of active compounds contained in such products vary depending on many factors such as the plant's growing conditions, harvest time, and processing methods. For example, the hypericin content of *Hypericum perforatum* may vary in different geographical regions, which directly affects the therapeutic effect of the product. In addition, some plants have potential side effects; for example, *Hypericum perforatum* is known to increase sensitivity to the sun. Therefore, standardization studies are of great importance for the safe use of herbal products.

In the future, research on medicinal and aromatic plants used in burn treatment is expected to be combined with innovative approaches such as nanotechnology and biomaterials. Formulating herbal extracts into nanoemulsions can allow active ingredients to reach tissues more quickly and in a more targeted manner. In addition, biodegradable wound dressings enriched with plants such as *Aloe vera* have the potential to accelerate the healing process while reducing the risk of infection. With the advancement of pharmacogenetic research, it will also be possible to develop individualized herbal treatment approaches. Such innovations will further strengthen the integration of herbal treatments with modern medicine.

9.4 Molecular basis of plant action mechanisms

Medicinal and aromatic plants exhibit anti-inflammatory, antimicrobial, and wound-healing effects through their biologically active components. These effects occur through multifaceted mechanisms at the cellular and molecular levels, depending on the chemical composition of the plants. These active components, called phytochemicals, provide therapeutic effects such as suppressing inflammation, preventing microorganism growth and accelerating tissue regeneration by intervening in biological processes. The mechanisms of action of plants used in burn treatment are of great importance, especially in terms of regulating inflammation, preventing infections and supporting the wound healing process [21].

Anti-inflammatory mechanisms: Burns are pathological conditions in which an inflammatory response is triggered as a result of thermal, chemical or mechanical trauma to the tissue. Controlling inflammation is a critical process to facilitate tissue healing after burns. Medicinal plants exert their anti-inflammatory effects through various phytochemicals. Compounds such as flavonoids, tannins, alkaloids, and terpenoids stand out in this mechanism [22].

Suppression of pro-inflammatory cytokines: Flavonoids and phenolic compounds suppress the production of pro-inflammatory cytokines (e.g., TNF-α, IL-6, and IL-1β) that play a role in the inflammatory process. These compounds regulate macrophage activity and reduce the severity of inflammation by inhibiting inflammatory pathways such as nuclear factor kappa B (NF-κB). For example, aloemodin found in *Aloe vera* has an anti-inflammatory effect by suppressing the NF-κB signaling pathway [23].

Reducing inflammation with antioxidant effect: Another important trigger of inflammation in burns is oxidative stress. Antioxidant compounds such as flavonoids and carotenoids prevent lipid peroxidation and tissue damage by neutralizing reactive oxygen species (ROS). This mechanism helps alleviate the inflammatory response and protect tissues [24].

Enzyme inhibition: Plant compounds reduce the production of prostaglandins and leukotrienes by inhibiting cyclooxygenase (COX) and lipoxygenase (LOX) enzymes that play a role in inflammation. For example, asiaticoside in *Centella asiatica* provides an anti-inflammatory effect by inhibiting the COX-2 enzyme [25].

Antimicrobial mechanisms: Burn wounds require antimicrobial treatment due to their susceptibility to infection. Medicinal plants prevent the growth of microorganisms through antimicrobial compounds to reduce the risk of infection. These compounds work with different mechanisms of action on bacteria, fungi, and viruses.

Effect on cell wall and membrane: Essential oils and phenolic compounds increase permeability in bacterial cell membranes, causing loss of intracellular contents. *Lavandula angustifolia* (lavender) essential oil disrupts the lipid layer on the cell membrane, leading to bacterial lysis. This mechanism, which is especially effective against gram-negative bacteria, is important in preventing infections.

Inhibition of protein synthesis: Plant compounds bind to ribosomal RNA, inhibiting protein synthesis and stopping bacterial proliferation. Eugenol, found in *Syzygium aromaticum* (clove), binds to bacterial ribosomes and exerts this effect.

Interference with enzymatic functions: Plant alkaloids bind to the active sites of bacterial enzymes, stopping metabolic processes. Terpenoids found in tea tree oil inhibit enzymes that play critical roles in energy production, such as ATP synthase [26].

Preventing biofilm formation: Bacteria can become resistant to treatment by forming a biofilm in the wound. Herbal compounds break this resistance by preventing bacterial cells from adhering to the surface or by breaking down the biofilm layer. *Hypericum perforatum* (St. John's wort) has strong antibacterial properties that prevent biofilm formation.

9.4.1 Cellular mechanisms in wound healing

Processes such as fibroblast proliferation, collagen synthesis, and angiogenesis are of critical importance in tissue repair after burns. Medicinal plants accelerate wound healing as they contain bioactive compounds that regulate these processes [27].

Fibroblast proliferation and extracellular matrix production: In wound healing, fibroblasts play an important role in restoring tissue integrity by synthesizing extracellular matrix proteins such as collagen and elastin. Madecassoside found in *Centella asiatica* supports collagen production by increasing fibroblast activity and accelerates wound healing.

Collagen synthesis: Collagen is necessary to increase the strength of wound tissue. Triterpenoids found in medicinal plants contribute to the wound healing process by inducing collagen synthesis. For example, *Calendula officinalis* activates fibroblasts, increases collagen production, and accelerates the formation of granulation tissue.

Angiogenesis: New blood vessel formation is a critical process for the nutrition and oxygenation of tissues after burns. Herbal compounds stimulate angiogenesis by activating the vascular endothelial growth factor (VEGF) signaling pathway. Polysaccharides found in *Aloe vera* promote the proliferation of endothelial cells and support the formation of new blood vessels.

Support for epithelialization: Medicinal plants accelerate the epithelialization process by increasing the migration and proliferation of epidermal cells. For example, *Lavandula angustifolia* accelerates wound closure by stimulating epithelial cell renewal on the wound surface.

9.4.2 Innovative research methods in herbal treatments

The combination of traditional herbal treatment methods with modern science requires the use of innovative research approaches. These approaches facilitate the understanding of herbal treatments at the molecular and biochemical level, allowing for the development of more effective and safe therapeutic options. Advanced research methods, especially omics technologies and in silico modeling, play a critical role in understanding the mechanisms of action of herbal compounds and determining their pharmacological potential.

9.4.2.1 Omic technologies: genomic, proteomic, and metabolomic approaches

Omic technologies are scientific methods that systematically and comprehensively analyze the biological processes of living systems. These technologies allow us to better understand the biological effects of herbal compounds by examining the effects of herbal treatments at the cellular and molecular level:

1. Genomic approaches: Genomic analyses play an important role in understanding the genetic structures of medicinal and aromatic plants. These methods allow the identification of the genetic material responsible for the biological activities of plants. For example, genes responsible for the anti-inflammatory effects of a particular plant can be identified, and the expression profiles of these genes can be examined to determine under which conditions the plant is more effective. In addition, genomic technologies are also used to understand the resistance mechanisms of plants to environmental stress factors such as climate change [28].

2. Proteomic approaches: Proteomics analyzes the structure, quantity, and function of proteins affected by herbal compounds. This approach is valuable for understanding how plants used in burn treatment interact with specific proteins involved in collagen synthesis or inflammation processes, for example. Proteomic analyses can reveal which biological pathways herbal treatments modulate and the role of this modulation in therapeutic effects [29].

3. Metabolomic approaches: Metabolomics is a discipline that studies the effects of plant compounds at the intracellular and extracellular metabolite level. This method is important for understanding how the active compounds contained in plants contribute to changes in human metabolism. For example, metabolomic analyses can show how polyphenols found in plants such as *Aloe vera* suppress inflammatory processes. In addition, metabolomic data can be used to understand the synergistic effects of combinations of different plants.

9.4.2.2 In silico modeling: computer simulations and artificial intelligence applications

In silico modeling is an advanced research method that simulates the interaction of plant compounds with biological systems in a computer environment. This method offers a rapid and low-cost screening process before experiments are conducted in a laboratory environment and is an important tool in determining potential pharmacological targets of plant therapies:

1. Molecular docking studies: Molecular docking is an in silico modeling method that predicts how plant compounds bind to specific proteins. For example, the binding capacities of compounds responsible for the anti-inflammatory effects of a plant extract used in burn treatment with target proteins such as NF-κB or COX-2 that play

a role in inflammatory processes can be analyzed with this method. These analyses play an important role in prioritizing potential therapeutic candidates [30].

2. Pharmacokinetic and toxicity simulations: It is possible to predict information about the absorption, distribution, metabolism, and excretion processes (ADME) of plant compounds through pharmacokinetic simulations. Such simulations play a critical role in understanding the safety profiles of herbal products and in dosage optimization. Furthermore, potential toxic effects can be analyzed using in silico toxicity prediction models, and formulations can be adapted to reduce side effects [31].

3. Artificial intelligence and machine learning: Artificial intelligence and machine learning algorithms help predict the effects of herbal treatments by analyzing relationships in large data sets. For example, data sets created with patient data and literature reviews can be used to determine the most effective herbal extracts for a particular type of burn. These methods can also predict the effects of combinations of herbs used in burn treatment and provide recommendations to optimize the most effective combinations [32].

Omic technologies and in silico modeling strengthen the scientific basis of herbal treatments by providing a deeper understanding of their biological mechanisms. These innovative approaches are accelerating the integration of herbal treatments with modern medicine and providing new opportunities for personalized medicine

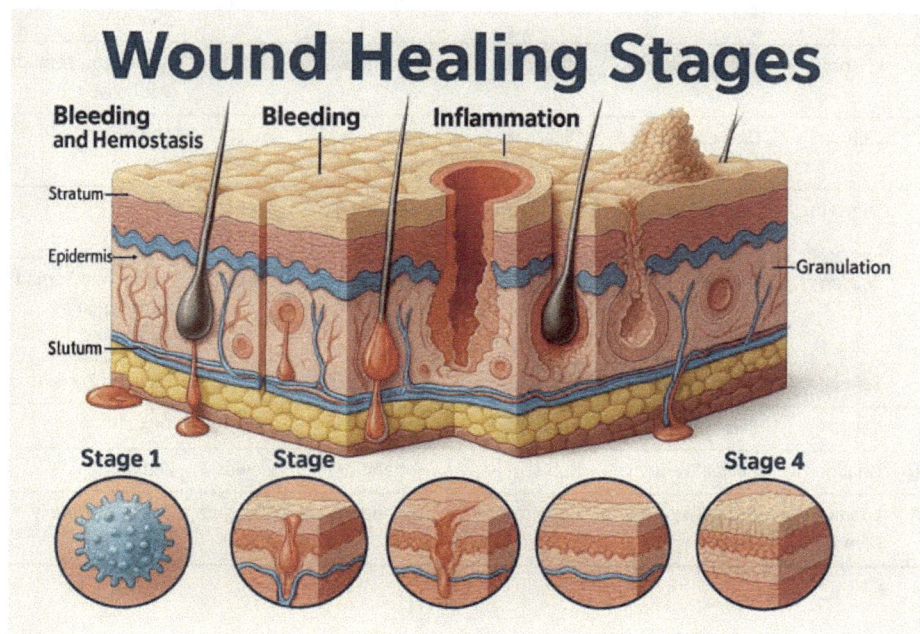

Figure 9.3: Burn treatment healing process.

applications. In the future, the wider use of these methods will contribute to the development of sustainable and innovative treatment options, as well as optimize the efficacy and safety profiles of herbal treatments [33].

Due to the involvement of blood vessels, the tissue has a granular texture (granulation tissue). Finally, within the granulation tissue, differentiated fibroblastic cells (myofibroblasts) begin to remodel the extracellular matrix approximately 1–2 weeks after injury. Extracellular matrix remodeling accompanied by resident cell apoptosis leads to the formation of an acellular scar. Medicinal plants and their metabolites used in the treatment of different types of wounds are shown in Figure 9.3 and Table 9.1 [34].

Table 9.1: Plants used in burn treatment.

No.	Traditional name	Scientific name	Family	Characteristics	Part used	Dosage form
1	Henna	*Lawsonia inermis* L.	Lythraceae	Cold and dry	Leaf	Natool
2	Hofariqan	*Hypericum perforatum* L.	Hypericaceae	Hot and dry	Leaf	Zemad
3	Ass	*Myrtus communis* L.	Myrtaceae	Cold and dry	Leaf/fruit	Natool, Duk, Marham, Qeiroot, and Zemad (with olive oil)
4	Khobazi	*Malva sylvestris* L.	Malvaceae	Cold and wet	Leaf	Natool and Zemad (with olive oil)
5	Zaitoon	*Olea europaea* L.	Oleaceae	Hot and dry (ripe fruits)	Fruit	Zemad
6	Semsem	*Sesamum indicum* L.	Pedaliaceae	Hot and wet	Seed/ seed oil	Zemad
7	Sousan	*Iris* spp.	Iridaceae	Hot and dry	Leaf/bulb	Natool and Zemad (with Dokar el ward)
8	Sanober	*Pinus pinea* L.	Pinaceae	Hot and dry	Bark/leaf	Tela and Zemad
9	Selgh	*Beta vulgaris* L.	Amaranthaceae	Hot and wet	Leaf	Tela
10	Ghalioon	*Gallium verum* L.	Rubiaceae	Hot and dry	Flower	Zemad
11	Loban (kondor)	*Boswellia carterii* Birdw.	Burseraceae	Hot and dry	Oleogum resin	Zemad (with oily base)

Table 9.1 (continued)

No.	Traditional name	Scientific name	Family	Characteristics	Part used	Dosage form
12	Abou khalsa	*Arnebia euchroma* (Royle) I.M. Johnst./ *Alkanna tinctoria* Tausch	Boraginaceae	Hot and dry	Root	Marham, Qeiroot (with olive oil), and Tela (with Dokar el ward)
13	Lebleb	*Convolvulus arvensis* L.	Convolvulaceae	Dry and hot/ cold*	Leaf	Natool, Qeiroot, and Zemad (with Dokar el ward)
14	Narjes	*Narcissus tazetta* L.	Amaryllidaceae	Hot and dry	Bulb	Zemad (with honey)
15	Nil	*Indigofera tinctoria* L.	Leguminosae	Hot and dry	–	Zemad (with honey)
16	Zairnera	*Ocimum basilicum* L.	Lamiaceae	Hot and dry	Leaf	Tela (with rosewater)
17	Quessus	*Hedera helix* L.	Araliaceae	Hot	Flower/ leaf	Qeiroot (with olive oil)
18	Quash'a zarirah	*Acorus calamus* L.	Acoraceae	Hot and dry	–	Zemad (with Dokar el ward)

9.5 Formulation and application methods of herbal products

In burn treatment, medicinal and aromatic plants are used in various pharmaceutical formulations suitable for topical application to accelerate wound healing, prevent infection, and reduce inflammation. While these formulations are designed to increase the bioavailability and effectiveness of herbal ingredients, the application method and dosage determinations are also critical for the success of the treatment. In addition, nanotechnological approaches offered by modern technologies further improve the effectiveness of these products and offer innovative treatment options [35].

9.5.1 Pharmaceutical formulations

Herbal ingredients are applied by being converted into different pharmaceutical forms for use in burn treatment. These forms provide advantages such as preserving the chemical stability of the ingredients, providing controlled release to target tissues and ease of application:

1. Creams and ointments: Creams and ointments are generally formulated on an oil-water or water-oil emulsion basis and are the most common pharmaceutical products in which herbal extracts are used as carrier systems. Creams based on *Aloe vera* gel, *Calendula* extract, and *Centella asiatica* are used to support cell renewal and form a protective layer on the skin surface. Ointments, in particular, prevent moisture loss due to their high oil content and accelerate the healing process by providing an occlusive effect on the wound [36].

2. Gels: Hydrophilic gels are applied directly to the wound and provide a moisturizing and cooling effect. *Aloe vera* in gel form is widely used in burn treatment and supports epithelialization. In addition, gels stand out as an effective carrier system because they provide a homogeneous distribution of essential oils or herbal extracts in high concentration [37].

3. Essential oils: Essential oils such as lavender oil, tea tree oil, and clove oil are used in burn treatment due to their antimicrobial and anti-inflammatory properties. These oils are usually diluted with carrier oils (such as coconut oil or jojoba oil) and applied directly to the wound or integrated into creams [38].

4. Tinctures and extracts: Herbal tinctures are concentrated plant extracts prepared using alcohol-based solvents. For example, *Hypericum perforatum* (St. John's wort) tincture is used in burn treatment due to its antimicrobial and wound-healing effects. Considering the drying effect of alcohol, these products are usually applied diluted or with other carrier products [39].

5. Dressing materials: Herbal extracts can be used in dressing materials by integrating them with biomaterials applied to the wound. Modern dressing products with increased antimicrobial and regenerative properties have been developed, especially by incorporating herbal components (e.g., *Centella asiatica* or *Calendula*) into alginate- or collagen-based wound dressings [40].

9.5.2 Dosage and application methods

Appropriate dosage and application method are of great importance for herbal products to provide effective treatment. The concentration, application frequency, and treatment duration of the products used in burn treatment are determined depending on the pharmacokinetic and pharmacodynamic properties of the herbal components used [41]:

1. Concentration and application dose: Products with high bioactive content such as essential oils are generally used in low concentrations (1–5%) to avoid potential toxic effects. For example, a cream containing lavender oil can be safely used on burns by formulating it at a concentration of 2–3%.

2. Application frequency: Topical products are usually applied 2–3 times a day to ensure a continuous effect of the active ingredients on the tissue. However, this

frequency can be increased or decreased depending on the degree of the burn, the risk of infection, and individual patient characteristics.

3. Treatment duration: Herbal treatments applied in the acute phase are generally used for 7–14 days. Chronic burn cases or treatments aimed at preventing scar formation may require longer-term applications.

9.5.3 Nanotechnological approaches

In recent years, innovative approaches based on nanotechnology have been developed to increase the effectiveness of herbal extracts and optimize tissue penetration. These methods provide specific delivery of active ingredients to target tissues while minimizing systemic side effects [42]. A visual about nanotechnological approaches is given in Figure 9.4:

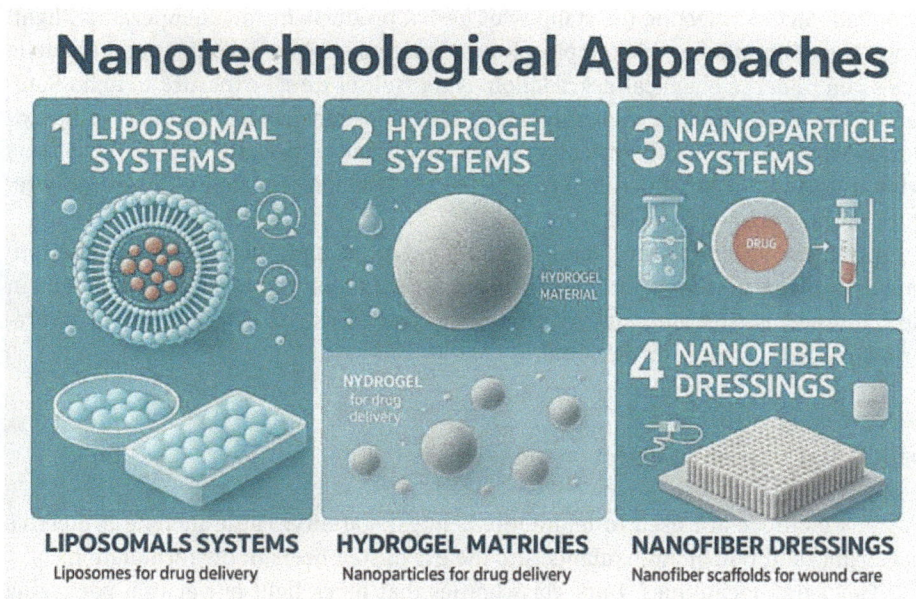

Figure 9.4: Nanotechnological approaches.

1. Liposomal systems: Liposomes formed by coating herbal extracts with phospholipids increase tissue penetration and provide controlled release of active ingredients. Liposomal formulations of *Aloe vera* extract allow anti-inflammatory and regenerative effects to last longer.

2. Nanoparticle systems: Silver nanoparticles or polymeric nanoparticles are combined with herbal ingredients to increase antibacterial effects. For example, silver nanoparticle formulations prepared with tea tree oil are used to reduce the risk of infection.

3. Hydrogel materials: Nanotechnological hydrogel structures support wound healing by providing a moisturizing environment and allow controlled release of herbal extracts. These materials are an effective option especially for chronic wounds and second-degree burns.

4. Nanofiber dressings: Integrating herbal extracts into nanofiber wound dressings prepared by electrospinning supports both infection control and tissue regeneration. For example, nanofiber dressings containing *Centella asiatica* extract accelerate wound healing by increasing collagen synthesis.

9.5.4 Factors affecting chemical stability

The main factors affecting the stability of herbal products include temperature, light, oxygen, and humidity. For example, photosensitizers such as flavonoids and phenolic acids can undergo chemical degradation as a result of direct exposure to light. Similarly, terpenoids found in essential oils can lose their effectiveness by undergoing oxidation when exposed to high temperatures or oxygen. Especially in herbal products with antioxidant or antimicrobial properties, such chemical changes can seriously reduce product effectiveness [43].

pH is also an important factor in stability. For example, acidic or basic environments can lead to isomerization, hydrolysis, or oxidation of bioactive compounds in herbal products. Therefore, it is of great importance to keep the pH within an appropriate range in product formulation.

9.5.4.1 Stability enhancement methods

Various methods are used to maintain chemical stability. These include protective packaging, addition of antioxidants, and the use of stabilizers in the formulation:

1. Packaging techniques: Dark glass bottles that block light or vacuum packaging that reduces oxygen permeability can increase the stability of essential oils and herbal extracts.

2. Use of stabilizers: The effect of free metal ions that can cause oxidation can be reduced by adding stabilizers (e.g., chelating agents such as EDTA) to the formulation of herbal products.

3. Addition of antioxidants: Natural antioxidants such as vitamin E (tocopherol) or ascorbic acid can help protect bioactive compounds in herbal products from oxidation.

9.5.4.2 Importance of storage conditions

Storage conditions of herbal products directly affect product stability and efficacy. Ideal storage conditions should include low temperature, a dark environment, and low humidity [44].

Essential oils should generally be stored in a cool, dry environment. A constant temperature between 15 and 25 °C reduces the risk of oxidation. Formulations containing water, such as creams and gels, should generally contain a preservative to protect them against the risk of microbial contamination and should be consumed shortly after opening. Powdered herbal extracts should be stored in packages that are completely isolated from air to prevent moisture from affecting them.

9.5.4.3 Stability tests and quality control

Stability tests should be applied to ensure chemical stability and determine the shelf life of the product. These tests are usually performed under accelerated conditions (high temperature, high humidity, exposure to light) and the physicochemical properties of the product are analyzed at regular intervals. For example:

Changes in the concentrations of bioactive compounds are monitored with methods such as HPLC (high-performance liquid chromatography) or GC-MS (gas chromatography-mass spectrometry). Physical parameters such as color, odor, and viscosity provide clues about the stability of the product.

9.6 Clinical research and evidence-based practices

The use of medicinal and aromatic plants in burn treatment forms the basis of evidence-based practices with a process extending from historical experiences to scientific research. Modern phytotherapy is based on clinical studies aimed at verifying traditional knowledge with scientific methods. These studies guide clinical practice by examining the effectiveness and safety of herbal treatments. In addition, the synthesis of existing research using systematic methods such as meta-analysis and literature reviews enriches the knowledge in this field and supports clinical decision-making processes [45].

9.6.1 Clinical studies

Clinical studies conducted with medicinal and aromatic plants used in burn treatment indicate that herbal products can be used effectively and safely. The results of these studies reveal that certain herbal components offer positive effects in reducing inflammation, preventing infection and accelerating wound healing:

1. *Aloe vera* (*Aloe barbadensis* Miller): A large number of clinical studies conducted with aloe vera show that this plant is an effective option in burn treatment. In a randomized controlled trial conducted especially on second-degree burns, it was found that aloe vera gel application provided faster epithelialization and reduced pain levels compared to conventional silver sulfadiazine cream. This effect is thought to be related to the moisturizing effect of the polysaccharides contained in aloe vera and the support of epithelial cell proliferation [46].

2. *Calendula officinalis* (*Calendula*): Clinical studies with *Calendula* extract have confirmed the antiseptic, anti-inflammatory, and wound-healing effects of this plant. One study observed that a calendula-based ointment shortened the time to burn wound closure and reduced scar formation. This effect was reported to be related to the flavonoid and triterpene compounds contained in calendula increasing collagen synthesis [47].

3. *Lavandula angustifolia* (lavender): Lavender oil has been studied in the treatment of burns for its antimicrobial and wound healing effects. A randomized clinical trial evaluated the effects of a topical product containing lavender oil on mild thermal burns. The study showed that lavender oil application reduced inflammation and accelerated wound closure. Active compounds such as linalool and linalyl acetate in lavender oil are thought to provide these effects [48].

4. *Melaleuca alternifolia* (tea tree oil): Tea tree oil has been studied as a potential agent for infection control in burns. A clinical study has shown that a topical formulation containing tea tree oil reduced the microbial load in infected burns and accelerated wound healing. This effect was reported to be due to the antimicrobial effects of tea tree oil [49].

5. *Centella asiatica* (gotu kola): *Centella asiatica*, known for its collagen synthesis-enhancing effects, has been found to be particularly effective in preventing hypertrophic scar formation. In a clinical study, application of a cream containing *Centella asiatica* extract increased the elasticity of post-burn scar tissue and improved scar appearance [50].

9.6.2 Effectiveness of phytotherapeutic products compared to conventional treatments

Herbal treatments have been shown to offer many advantages over conventional pharmacological agents. For example, products containing *Aloe vera* or *Calendula* have been shown to provide similar or better clinical results compared to commonly used topical burn creams such as silver sulfadiazine, while also having fewer side effects. The high cost of traditional treatments and the sometimes adverse effects such as toxicity have led to the emergence of phytotherapeutic products as complementary or alternative treatments. However, the effectiveness of phytotherapeutic products depends on the standardization of herbal ingredients, correct dosage, and method of application [51].

9.6.2.1 Meta-analyses and literature reviews

Meta-analyses and systematic literature reviews provide a wealth of information evaluating the effectiveness of herbal treatments in burn healing. Such studies combine the results of individual clinical trials to provide an overall effect size and provide strong evidence for clinical application [52]:

1. Effectiveness of *Aloe vera*: A meta-analysis on *Aloe vera* has shown that this plant accelerates wound healing and reduces infection rates in the treatment of second-degree burns. The studies included in the review highlighted *Aloe vera*'s moisturizing effects and cell regeneration-promoting properties.
2. *Calendula* and lavender: A literature review of the effects of *Calendula officinalis* and *Lavandula angustifolia* in burn treatment found that these plants were effective in reducing inflammation and accelerating wound healing. However, these reviews also point to the need for larger, randomized controlled trials.
3. Comparison of phytotherapeutic products: Comparative meta-analyses of traditional and herbal treatments show that phytotherapeutic products are a strong treatment option, especially in low and moderate burns. It has also been stated that herbal treatments are safer in terms of side effects, but the lack of standard dosage and formulation causes some difficulties.

The mechanisms of herbal compounds enable the customization and individualization of treatment processes. For example, the molecular pathways underlying the anti-inflammatory effects of plants may be related to the inhibition of pro-inflammatory cytokines and enzymes. This suggests that plants not only accelerate wound healing but also control inflammatory responses caused by traumas such as burns. Therefore, the identification of phytochemicals with anti-inflammatory properties may lead to the discovery of new agents to be included in treatment protocols [53].

However, the effectiveness of herbal treatments in wound healing depends on different biological processes occurring at the cellular level. For example, the presence of phytochemical compounds that increase fibroblast proliferation and migration may accelerate wound healing. Furthermore, knowing how processes such as angiogenesis and collagen synthesis can be modulated by herbal compounds will help us understand the effectiveness of these treatments in more depth. In this context, more molecular studies are needed on how plants such as *Centella asiatica* and *Aloe vera* accelerate the healing process by increasing collagen synthesis and supporting the formation of microvessels at the wound site. Such studies indicate that phytotherapeutic approaches are not only an alternative treatment method, but also an essential component of the treatment process [54].

In addition to the efficacy of herbal treatments supported by clinical data, the safety profiles of these treatments are also of great importance. One of the greatest challenges in the use of phytotherapeutic treatments is that potential toxicity and side effects of herbal compounds should be considered in addition to their efficacy. In particular, it has been reported that herbal products used in burn treatments may cause adverse effects such as allergic reactions, dermal irritations, or phototoxicity when applied to the skin. For this reason, herbal treatments should be compared with pharmaceutical products not only in terms of efficacy but also in terms of safety. Before integrating herbal treatments into pharmaceutical treatments, extensive toxicity tests and collection of clinical safety data can ensure the safe use of these treatments [55].

It is necessary to develop appropriate formulation techniques for herbal compounds to cross the skin barrier. In this context, nanotechnology and nanomedicine applications offer an important strategy to increase the bioavailability of herbal extracts and ensure tissue penetration. Formulating herbal compounds with nanocarriers allows these compounds to be directed to the areas where they will have an effect, making the treatment process more targeted. Such innovative approaches can increase the effectiveness of phytotherapeutic treatments and provide faster results compared to traditional pharmaceutical treatments [56]. Studies conducted in different ethnic groups and age groups can determine how herbal treatments interact with genetic, environmental, and biological differences and further improve treatment processes. For example, whether herbal treatments differ in terms of safety and effectiveness between elderly patients and pediatric patients should be analyzed in more depth. Such studies will allow herbal treatment options to be applied effectively and safely to a wider population.

9.7 Safety and side effects

Although herbal treatments are an important part of traditional medicine, they can pose some safety issues when used in medical applications. The safety of phytotherapeutic treatments depends not only on the effectiveness of the herbal compounds, but also on the possible toxic effects of these compounds. The safety of herbal products used in specific clinical situations such as burn treatment requires careful evaluation of both therapeutic efficacy and potential risks. In this context, incorrect dosage or inappropriate use of herbal treatments may cause toxic effects. Safety and side effects are shown in Figure 9.5.

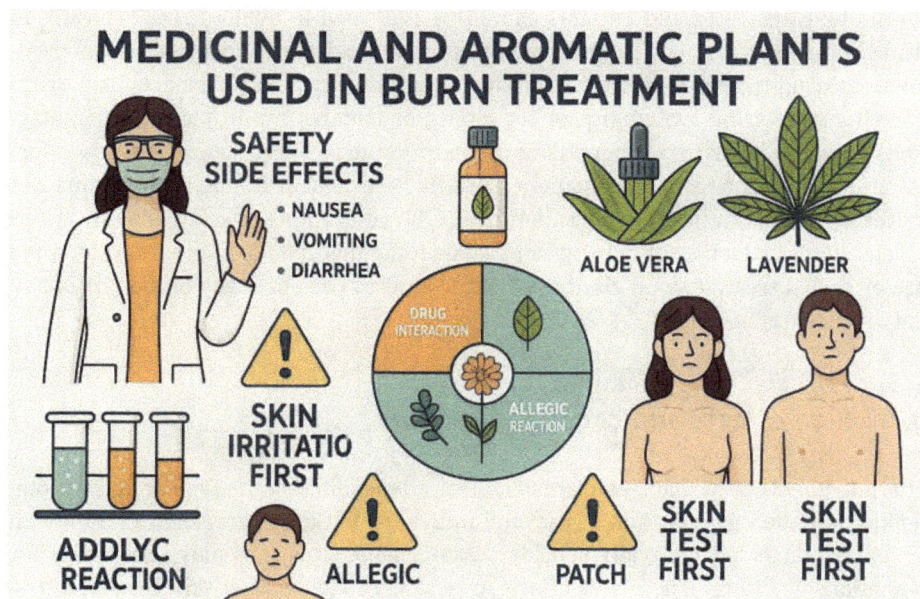

Figure 9.5: Safety and side effects.

9.7.1 Toxicological risks

Toxicological evaluation of herbal products is of great importance in determining both efficacy and potential harmful effects. Although herbal treatments are generally considered natural and safe, compounds contained in plants have the potential to cause unexpected or harmful effects on biological systems. Plants used in burn treatment may cause toxic effects when used in inappropriate dosages or with incorrect application methods. These toxic effects may be due to adverse reactions of herbal compounds on the skin and body. For example, excessive use of *Aloe vera* may cause irritation in the digestive system or allergic reactions on the skin. In addition, high-dose consumption or long-term use of some herbal compounds may adversely affect liver and kidney functions. Therefore, each herbal product should be subjected to toxicological tests before use in the pharmaceutical field [57].

In addition, phototoxicity can occur as a significant side effect of herbal treatments. Phototoxicity is the harmful reaction that some herbal compounds cause on the skin when exposed to ultraviolet rays. In particular, some plants such as St. John's wort (*Hypericum perforatum*) can show phototoxic reactions that can cause redness, irritation, and burns on the skin when exposed to the sun. The use of herbal treatments in the treatment of burns, when combined with sun exposure, can negatively affect the treatment process. Therefore, the phototoxic potential of herbal treatments should be considered and patients should be informed to avoid direct exposure to sunlight during treatment. Another important toxicological risk factor of herbal treatments is drug interactions. Herbal products can interact with pharmaceutical drugs and change their bioavailability, effectiveness, or metabolism. In particular, interactions between herbal treatments and prescription drugs used can negatively affect the treatment process and the patient's health status. For example, some plants can increase the effectiveness of drugs that have a blood-thinning effect, while others can accelerate the metabolism of drugs and weaken the therapeutic effect. Therefore, the use of herbal treatments in clinical practice requires careful consideration of potential drug interactions [58].

9.7.2 Side effects and contraindications

The side effects of herbal treatments may vary depending on the type of herbal compound used, dosage, duration of use, and individual patient characteristics. Although herbal products are generally considered safe, some situations may pose a risk for potential side effects. These side effects are usually mild, such as skin reactions, gastrointestinal disorders, headache, or allergic reactions, but in some cases they can have more serious consequences. For example, *Aloe vera* gel has moisturizing and soothing effects on the skin, but it can cause itching or rashes in some patients. Similarly, plants such as *Calendula officinalis* can sometimes cause allergic skin reactions, which can adversely affect the treatment [59].

Special health conditions, such as pregnancy, breastfeeding, or childhood, may further complicate the use of herbal products. During pregnancy, some herbal compounds have been reported to cause uterine contractions or adverse effects on the fetus. For example, some plants may increase the risk of premature birth by stimulating the uterus. In addition, it is thought that some herbal products may pass to babies through milk and cause toxic effects for breastfeeding mothers. For this reason, the use of herbal treatments for women who are pregnant and breastfeeding should be carefully evaluated [60]. Children also constitute a special group in the use of herbal treatments. Since children have different skin structure and immune systems than adults, the use of some herbal products may have adverse effects. The use of herbal treatment methods in children is especially important in terms of determining age-appropriate dosages and carefully monitoring potential allergic reactions. In children, some herbs, such as essential oils such as tea tree oil, may cause skin irritation or more serious allergic reactions [61].

The use of herbal treatments in individuals with chronic diseases should also be considered carefully. In particular, conditions such as diabetes, hypertension, or liver/kidney diseases may affect the metabolism and effectiveness of herbal products. Herbal products may interact with drugs used in the treatment of these diseases and affect the treatment process. For example, some herbal treatments may affect blood sugar levels and increase the risk of hypoglycemia in diabetic patients. The safety of herbal treatments is not limited to the toxicological profile of individual compounds, but a broader assessment of the interactions of these compounds with biological systems is also required. These interactions should be carefully considered, especially in terms of the way herbal products are metabolized in the body and their effects on various biochemical pathways. The metabolic functioning of herbal products after they are taken into the body can directly affect the bioavailability of the active compounds and the effects of these compounds. In particular, pharmacokinetic properties are an important factor determining the effectiveness of herbal treatment. In this context, it has been reported that some herbal compounds may have effects especially on liver enzymes and as a result, may accelerate the metabolism of pharmaceutical drugs and weaken their therapeutic effects. For example, St. John's wort (*Hypericum perforatum*) can reduce plasma levels of some drugs by acting on the cytochrome P450 enzyme system, which can lead to a decrease in treatment efficacy. Such pharmacokinetic interactions pose a significant risk for the clinical use of herbal treatments [62]. Another important factor in terms of the safety of herbal treatments is the growing conditions and environmental factors of the plants. The soil and environment in which the plants grow can directly affect the amount and quality of the active compounds they contain. For example, the presence of toxic substances such as heavy metals in the soil can cause the plant to absorb these substances and ultimately cause adverse effects on human health. In addition, microbiological contamination may occur during the use of some herbal treatments due to environmental factors. Therefore, the methods used in the production of medicinal and aromatic plants in particular must comply with the safety standards determined by international health authorities such as the Food and Drug Administration and the World Health Organization [63].

In addition, individual biological differences that affect the effectiveness of herbal treatments should also be taken into account. People's genetic structures, immune responses, and metabolic rates may vary, which may lead to herbal treatments producing different results for each individual. Genetic variations can affect the effectiveness and safety of a herbal product. For example, some individuals may be genetically more sensitive to herbal compounds, which may cause side effects to become more pronounced. Such individual differences emphasize the importance of personalized treatment approaches. Future research may use genetic and biomarker analyses to customize herbal treatments to individuals and further increase the effectiveness of these treatments. In addition, the formulations and methods of use of herbal products are also important factors in terms of safety. Some carriers or auxiliary ingredients used to increase the effectiveness of herbal compounds may cause allergic reactions or skin irritations in the user. In particular, intensive use of intense and volatile compounds such as essential oils may cause undesirable reactions on the skin. In addition, alcohol-based solvents or preservatives used in the formulations of herbal products may also create potential side effects. Therefore, careful selection of ingredients used in the formulation of herbal products is critical for user safety [64]. Clinical observations and literature reviews indicate that herbal treatments may cause some side effects after long-term use, especially in individuals with chronic diseases. For example, some herbal products used in the treatment of diabetes may cause sudden drops in blood sugar levels, which increase the risk of hypoglycemia. Similarly, antihypertensive herbal treatments have been reported to cause excessive hypotension, especially in individuals with low blood pressure. Such situations require that herbal treatments be used only under the supervision of a health professional and taking into account the general health status of the patient.

9.8 Integration of traditional knowledge and modern science

9.8.1 Ethnobotany and traditional knowledge

Ethnobotany is a discipline that studies the relationship of plants with people's cultures, societies, and health practices. This field investigates the use of plants not only as food or building materials but also as therapeutic agents. Traditional knowledge is a knowledge that is passed down through generations among the people and is based on observations and experiences of the healing properties of plants. This knowledge is a kind of "natural experiment" that societies develop as a result of their interaction with nature and is usually passed down from generation to generation by local people. This traditional knowledge about the healing properties of plants often combines with oral culture, rituals, and social beliefs to form various folk treatments [65].

However, although traditional knowledge is often thought to be shaped by observations and experiments without any scientific basis, modern science has begun to test the accuracy of this knowledge and has investigated the effectiveness of traditional treatment methods more systematically. Today, pharmacological and toxicological studies conducted to verify the efficacy and safety of traditional herbal treatments allow us to understand the biological effects of herbal compounds at the molecular level. This integration makes it possible to base traditional knowledge on scientific foundations and to include this knowledge in the treatment options of modern medicine.

In ethnobotanical studies, after traditional uses reveal the healing properties of a certain plant species, the active compounds obtained from these plants are examined with modern biotechnology, and the efficacy and safety profiles of these compounds are evaluated with pharmacological studies. For example, the traditional use of *Aloe vera* in burn treatment has been confirmed by modern scientific research and the anti-inflammatory, moisturizing and wound-healing properties of its gel have been proven. Such studies ensure that traditional treatment methods are supported by scientific evidence and the effectiveness of plants used among the public is accepted in the medical field [66].

In addition, the integration of traditional knowledge with modern science also contributes to the development of sustainable and environmentally friendly medical practices. Herbal treatments generally have a lower side effect profile than synthetic drugs and offer production processes that are sensitive to environmental effects. This situation increases the need to develop treatments from natural and renewable sources and encourages a pharmaceutical production approach that does not harm nature. In addition, thanks to biotechnological innovations, it has become possible to purify active compounds obtained from traditional plants and use them more efficiently in pharmaceutical preparations.

9.8.2 Cultural and regional diversity

Plants used in different cultures and geographies are a reflection of people's interactions with ecosystems and their health needs. When ethnobotanical studies examine the different uses and adaptations of a particular plant across cultures, they reveal how the biological and pharmacological properties of these plants vary and the environmental factors behind this diversity. Each culture sees plants as solutions to different health problems, and the effectiveness of the plants used often varies depending on geographical features, soil structure, and climatic conditions. For example, plants growing in tropical regions generally have antimicrobial, antifungal, and immune-boosting properties, while plants growing in cold climates offer more antioxidant, anti-inflammatory, and analgesic properties. This geographical diversity is based on the biodiversity of plants that develop in their local ecosystems and traditional practices with these plants [67]. Another important factor explaining regional diversity is the climatic and environmental conditions in which plants grow. The active ingredients of plants are directly related to the environment in which they grow. For example, the type of soil, climate, and water resources in which a plant grows can affect the type and concentration of chemical compounds in that plant. The same plant species can produce different biochemical compounds in different geographical regions and under different growing conditions, which can change its therapeutic effects. This situation highlights the importance of considering local environmental factors in determining the therapeutic effects of a plant. Modern science aims to standardize herbal treatment methods and increase the reliability of these treatments by taking these environmental variations into account [68]. In addition, cultural diversity indicates that the ways plants are used can also differ. For example, the same plant can be processed in different ways in different geographies, and the methods of application may vary. Plants used in Ayurvedic medicine in India are used more as herbal supplements or massage oils in the West, while similar plants can be processed as medicinal teas or extracts in Chinese medicine. However, the adaptation of traditional uses of plants to modern practices is not limited to determining the effects of these plants alone. It is also a process that should take into account cultural practices and local health needs. The balance between respect for traditional practices and modern scientific standards can ensure that these herbal treatments are widely accepted and their effectiveness is increased [69]. Consequently, the integration of traditional knowledge with modern science not only scientifically validates the effectiveness of herbal treatments, but also provides a broader societal benefit by preserving the sustainability and cultural diversity of these treatment methods. Ethnobotanical studies combine traditional knowledge with modern biotechnological tools to make herbal treatments safer, more effective, and more accessible. This integration strengthens the role of herbal treatments in global health, while preserving cultural diversity, paving the way for the development of a treatment approach strengthened by scientific knowledge.

9.9 Future research areas and innovation

9.9.1 Pharmacogenetics and personalized medicine

Pharmacogenetics is a field that personalizes drug treatment options based on the genetic structure of individuals and offers a revolutionary approach in medical treatment processes. The integration of herbal treatments with pharmacogenetics stands out as one of the most important and innovative areas of research to be conducted in this field. Since the genetic structure of each individual is different, the effectiveness and side effect profile of herbal treatments may also vary at the individual level. Genetic polymorphism is an important factor that shows differences in biological responses and affects treatment processes. In this context, the use of herbal treatment methods as a part of personalized medicine can provide higher success rates in the treatment process [70]. For example, it is understood that the active ingredients contained in plants such as *Aloe vera* or *Centella asiatica* may have different effects depending on the genetic structure of individuals. While the cell regenerative effect of *Aloe vera* is genetically stronger in some individuals, this effect may be weaker in others. Pharmacogenetic analyses can personalize treatment processes by determining which herbal treatments are more compatible with the genetic makeup of individuals. Similarly, research on genetic factors that modulate the biological activity of plants can help optimize personal treatment plans. In this direction, in the future, genetic tests and biomarkers can be used to determine the most appropriate herbal treatment approaches for each individual [71].

9.9.2 Biodegradable and smart materials

Biodegradable materials are materials that are environmentally friendly, biodegradable, and compatible with biological systems. The use of biodegradable wound dressings in burn treatment has emerged as an important innovation in recent years. When these materials are enriched with herbal ingredients, they become more effective in both accelerating wound healing and reducing the risk of infection. Biodegradable wound dressings are applied directly to the wound, supporting the healing process and at the same time providing a continuous release of natural treatment ingredients to the target area. The wound-healing properties of plants such as *Aloe vera, Calendula officinalis,* and *Centella asiatica* are ideal components to be used in these dressings. These herbal compounds reduce inflammation in the wound area, accelerate cellular regeneration, and help restructure skin tissue [72].

One of the most striking features of plant-based biodegradable materials is that they are naturally absorbed by the body and dissolve over time, eliminating the need for patient monitoring and intervention during the treatment process. Smart materials, on the other hand, offer an innovative approach that optimizes the treatment process by responding to environmental factors. Such materials are sensitive to external variables such as temperature, pH, or enzymatic activity, allowing the controlled release of therapeutic agents when necessary. For example, smart materials with plant-based ingredients applied to a wound site can be designed to release their contents depending on the stages of wound healing. The development of such innovative materials could accelerate wound healing and provide safer, more effective, and personalized solutions for treatment.

9.9.3 Combined use of herbal treatments

Combining phytotherapeutics with pharmacological agents is an important strategy that strengthens the effects of herbal products in pharmacological treatment processes. Traditional herbal treatments generally show stronger effects when supported by various treatment approaches rather than alone. In burn treatment, the combination of herbal compounds with pharmaceutical agents can accelerate the treatment process and provide better results. Especially in the treatment of burns, the combination of herbal compounds with antibacterial properties with pharmaceutical antibiotics can greatly reduce the risk of infection [73].

Herbal treatments can increase the effectiveness of pharmacological agents and reduce the side effect profile. For example, a treatment process supported by analgesic herbs can improve pain management while also preventing possible side effects by reducing the dosage of pharmacological analgesics. In addition, it is possible that active compounds in plants contribute to the treatment by interacting with pharmaceutical drugs such as accelerating or inhibiting their metabolism. An approach such as combining the anti-inflammatory effects of *Aloe vera* with nonsteroidal anti-inflammatory drugs can create an effective synergy in the treatment process. These combinations not only increase the therapeutic effect, but can also balance the possible toxic effects of the treatment process. Combining traditional herbal treatments with pharmaceutical drugs can combine the strengths of both treatment methods, resulting in fewer side effects and faster recovery times. However, careful examination of such combinations in clinical trials is important for the safe management of interactions of both herbal therapeutic agents and pharmaceutical drugs.

9.10 Conclusion

This study evaluated the use of medicinal and aromatic plants in burn treatment in light of traditional knowledge and modern science. While burn treatment requires a multidisciplinary approach in both acute and chronic processes, herbal treatments offer multifaceted benefits such as reducing inflammation, preventing infection, relieving pain and accelerating wound healing thanks to the bioactive compounds they contain. The effectiveness of plants such as *Aloe vera, Calendula officinalis, Hypericum perforatum, Mentha piperita,* and *Syzygium aro*maticum in these areas has been supported by both historical and modern scientific evidence.

However, the effectiveness of herbal treatments depends on factors such as bioavailability of phytochemicals, dosage standardization, and safety profiles. In addition, innovative approaches such as nanotechnology and biomaterial applications hold promise for increasing the therapeutic potential of herbal compounds. However, further increase in clinical research and detailed evaluation of toxicity risks are necessary.

As a result, herbal treatments in burn treatment can be considered as a complementary or alternative option to modern pharmacological treatments. This approach offers great potential, especially in low- and middle-income areas, in terms of low cost, accessibility, and safety. Future studies may increase the knowledge in this area and offer new approaches to herbal burn treatment.

References

[1] Jahromi, M. A. M., Zangabad, P. S., Basri, S. M. M., Zangabad, K. S., Ghamarypour, A., Aref, A. R. and Hamblin, M. R. (2018). Nanomedicine and advanced technologies for burns: Preventing infection and facilitating wound healing. Advanced Drug Delivery Reviews, 123, 33–64.

[2] Mssillou, I., Bakour, M., Slighoua, M., Laaroussi, H., Saghrouchni, H., Amrati, F. E. Z. and Derwich, E. (2022). Investigation on wound healing effect of Mediterranean medicinal plants and some related phenolic compounds: A review. Journal of Ethnopharmacology, 298, 115663.

[3] Bahadur, S. and Fatima, S. (2024). Essential oils of some potential medicinal plants and their wound healing activities. Current Pharmaceutical Biotechnology, 25(14), 1818–1834.

[4]   Bittner Fialová, S., Rendeková, K., Mučaji, P., Nagy, M. and Slobodníková, L. (2021). Antibacterial activity of medicinal plants and their constituents in the context of skin and wound infections, considering European legislation and folk medicine – A review. International Journal of Molecular Sciences, 22(19), 10746.

[5]   Budovsky, A., Yarmolinsky, L. and Ben-Shabat, S. (2015). Effect of medicinal plants on wound healing. Wound Repair and Regeneration, 23(2), 171–183.

[6]   Maver, T., Kurečič, M., Smrke, D. M., Kleinschek, K. S. and Maver, U. (2018). Plant-derived medicines with potential use in wound treatment. In: Philip F. Builders (ed) Herbal Medicine: Biomolecular and Clinical Aspects, intechopen, 10.

[7]   Bandaranayake, W. M. (2006). Quality control, screening, toxicity, and regulation of herbal drugs. In: Dr. Iqbal Ahmad, Farrukh Aqil, Dr. Mohammad Owais (eds) Modern Phytomedicine: Turning Medicinal Plants into Drugs, WILEY-VCH Verlag GmbH & Co. KGaA, Weinheim, 25–57.

[8]   Hamilton, A. C. (2004). Medicinal plants, conservation and livelihoods. Biodiversity & Conservation, 13, 1477–1517.

[9]   Schilrreff, P. and Alexiev, U. (2022). Chronic inflammation in non-healing skin wounds and promising natural bioactive compounds treatment. International Journal of Molecular Sciences, 23(9), 4928.

[10]   Riaz, S., Hussain, S., Syed, S. K. and Anwar, R. (2021). Chemical characteristics and therapeutic potentials of Aloe vera. RADS Journal of Biological Research & Applied Sciences, 12(2), 160–166.

[11]   Movaffagh, J., Bazzaz, B. S. F., Taherzadeh, Z., Hashemi, M., Moghaddam, A. S., Abbas Tabatabaee, S. . . . Jirofti, N. (2022). Evaluation of wound-healing efficiency of a functional Chitosan/Aloe vera hydrogel on the improvement of re-epithelialization in full thickness wound model of rat. Journal of Tissue Viability, 31(4), 649–656.

[12]   Park, M. Y., Kwon, H. J. and Sung, M. K. (2011). Dietary aloin, aloesin, or aloe-gel exerts anti-inflammatory activity in a rat colitis model. Life Science, 88(11–12), 486–492.

[13]   Atiba, A., Abdo, W., Ali, E., Abd-Elsalam, M., Amer, M., Abdel Monsef, A. and Mahmoud, A. (2022). Topical and oral applications of Aloe vera improve healing of deep second-degree burns in rats via modulation of growth factors. Biomarkers, 27(6), 608–617.

[14]   Sapkota, B. and Kunwar, P. (2024). A review on traditional uses, phytochemistry and pharmacological activities of Calendula officinalis Linn. Natural Products Communications, 19(6), 1934578X241259021.

[15]   Dhingra, G., Dhakad, P. and Tanwar, S. (2022). Review on phytochemical constituents and pharmacological activities of plant Calendula officinalis Linn. Biological Sciences, 2(2), 216–228.

[16]   Nicolaus, C., Junghanns, S., Hartmann, A., Murillo, R., Ganzera, M. and Merfort, I. (2017). In vitro studies to evaluate the wound healing properties of Calendula officinalis extracts. Journal of Ethnopharmacology, 196, 94–103.

[17]   Seevaratnam, V., Banumathi, P., Premalatha, M. R., Sundaram, S. P. and Arumugam, T. (2012). Functional properties of Centella asiatica (L.): A review. International Journal of Pharmaceutical Sciences Research, 4(5), 8–14.

[18]   Adepoju, A., Ogunkunle, T., Femi-Adepoju, A. and Ejigboye, E. (2024). Scientific common names (SCNS) for selected medicinal plants: An improved method of Botany: SCIENTIFIC COMMON NAMES FOR MEDICINAL PLANTS. Arabian Journal of Medicinal and Aromatic Plants, 10(1), 189–251.

[19]   Jain, A., Yadav, S. and Khan, J. (2024). Revolutionizing wound healing: Unleashing Nanostructured lipid carriers embodied with herbal medicinal plant. Current Pharmaceutical Biotechnology.

[20]   Singh, H., Kumar, S. and Arya, A. (2023). Ethno-dermatological relevance of medicinal plants from the Indian Himalayan region and its implications on cosmeceuticals: A review. Journal of Drug Research in Ayurvedic Sciences, 8(2), 97–112.

[21] Vitale, S., Colanero, S., Placidi, M., Di Emidio, G., Tatone, C., Amicarelli, F. and D'Alessandro, A. M. (2022). Phytochemistry and biological activity of medicinal plants in wound healing: An overview of current research. Molecules, 27(11), 3566.

[22] Nunes, C. D. R., Barreto Arantes, M., Menezes de Faria Pereira, S., Leandro da Cruz, L., De Souza Passos, M., Pereira de Moraes, L. and Barros de Oliveira, D. (2020). Plants as sources of anti-inflammatory agents. Molecules, 25(16), 3726.

[23] Leyva-López, N., Gutierrez-Grijalva, E. P., Ambriz-Perez, D. L. and Heredia, J. B. (2016). Flavonoids as cytokine modulators: A possible therapy for inflammation-related diseases. International Journal of Molecular Sciences, 17(6), 921.

[24] Ozougwu, J. C. (2016). The role of reactive oxygen species and antioxidants in oxidative stress. International Journal of Research, 1(8), 1–8.

[25] Dey, R., Dey, S., Samadder, A., Saxena, A. K. and Nandi, S. (2022). Natural inhibitors against potential targets of cyclooxygenase, lipoxygenase and leukotrienes. Combinatorial Chemistry & High Throughput Screening, 25(14), 2341–2357.

[26] Smitha Grace, S. R., Chandran, G. and Chauhan, J. B. (2019). Terpenoids: An activator of "fuel-sensing enzyme AMPK" with special emphasis on antidiabetic activity. Plant and Human Health, Volume 2: Phytochemistry and Molecular Aspects, Springer, Gewerbestrasse 11, 6330 Cham, Switzerland, 227–244.

[27] Shedoeva, A., Leavesley, D., Upton, Z. and Fan, C. (2019). Wound healing and the use of medicinal plants. Evidence-Based Complementary and Alternative Medicine, 2019(1), 2684108.

[28] Tasneem, S., Liu, B., Li, B., Choudhary, M. I. and Wang, W. (2019). Molecular pharmacology of inflammation: Medicinal plants as anti-inflammatory agents. Pharmacological Research, 139, 126–140.

[29] Cánovas, F. M., Dumas-Gaudot, E., Recorbet, G., Jorrin, J., Mock, H. P. and Rossignol, M. (2004). Plant proteome analysis. Proteomics, 4(2), 285–298.

[30] Chen, X., Ung, C. Y. and Chen, Y. (2003). Can an in silico drug-target search method be used to probe potential mechanisms of medicinal plant ingredients?. Natural Product Reports, 20(4), 432–444.

[31] Sucharitha, P., Reddy, K. R., Satyanarayana, S. V. and Garg, T. (2022). Absorption, distribution, metabolism, excretion, and toxicity assessment of drugs using computational tools. In: Parihar, A., Khan, R., Kumar, A., Kaushik, A. K. and Gohel, H. (eds.) Computational Approaches for Novel Therapeutic and Diagnostic Designing to Mitigate SARS-CoV-2 Infection, Academic Press, 335–355. doi: https://doi.org/10.1016/C2020-0-04145-9.

[32] Gupta, R., Srivastava, D., Sahu, M., Tiwari, S., Ambasta, R. K. and Kumar, P. (2021). Artificial intelligence to deep learning: Machine intelligence approach for drug discovery. Mol Divers, 25, 1315–1360.

[33] Buriani, A., Garcia-Bermejo, M. L., Bosisio, E., Xu, Q., Li, H., Dong, X. and Hylands, P. J. (2012). Omic techniques in systems biology approaches to traditional Chinese medicine research: Present and future. Journal of Ethnopharmacology, 140(3), 535–544.

[34] Albahri, G., Badran, A., Hijazi, A., Daou, A., Baydoun, E., Nasser, M. and Merah, O. (2023). The therapeutic wound healing bioactivities of various medicinal plants. Life, 13(2), 317.

[35] Roshni, P. T. and Rekha, P. D. (2024). Essential oils: A potential alternative with promising active ingredients for pharmaceutical formulations in chronic wound management. Inflammopharmacology, 32, 3611–3630, 1–20.

[36] Das, T. (2013). Formulation and evaluation of a herbal cream for wound healing activity. International Journal of Pharmacy and Pharmaceutical Sciences, 6(2), 693–697.

[37] He, J. J., McCarthy, C. and Camci-Unal, G. (2021). Development of hydrogel-based sprayable wound dressings for second-and third-degree burns. Adv Nanobiomed Res, 1(6), 2100004.

[38]   Low, W. L., Kenward, K., Britland, S. T., Amin, M. C. and Martin, C. (2017). Essential oils and metal
       ions as alternative antimicrobial agents: A focus on tea tree oil and silver. International Wound
       Journal, 14(2), 369–384.
[39]   Pageau, A. (2020). Tinctures for use in Aromatherapy. Aromatherapy Journal, 2020(4), 73–78.
[40]   Gokarneshan, N. (2019). Application of natural polymers and herbal extracts in wound
       management. In: Dr. Iqbal Ahmad, Farrukh Aqil, Dr. Mohammad Owais (eds) Advanced Textiles for
       Wound Care, Elsevier, Duxford, CB22 4QH, United Kingdom, 541–561.
[41]   Pferschy-Wenzig, E. M. and Bauer, R. (2015). The relevance of pharmacognosy in pharmacological
       research on herbal medicinal products. Epilepsy & Behavior, 52, 344–362.
[42]   Alexander, A., Patel, R. J., Saraf, S. and Saraf, S. (2016). Recent expansion of pharmaceutical
       nanotechnologies and targeting strategies in the field of phytopharmaceuticals for the delivery of
       herbal extracts and bioactives. Journal of Controlled Release, 241, 110–124.
[43]   Narayana, D. A. and Dobriyal, R. M. (2009). Shelf-life of herbal remedies: Challenges and
       approaches. In: Pulok K Mukherjee and Peter J Houghton (eds) Evaluation of Herbal Medicinal
       Products[Internet], Britain, An imprint of RPS Publishing, South Atkinson Road, Suite 200, Grayslake,
       IL 60030–7820, USA, 369–379.
[44]   Bansal, G., Suthar, N., Kaur, J. and Jain, A. (2016). Stability testing of herbal drugs: Challenges,
       regulatory compliance and perspectives. Phytotherapy Research, 30(7), 1046–1058.
[45]   Halberstein, R. A. (2005). Medicinal plants: Historical and cross-cultural usage patterns. Annals of
       Epidemiology, 15(9), 686–699.
[46]   Huang, Y. N., Chen, K. C., Wang, J. H. and Lin, Y. K. (2024). Effects of aloe vera on burn injuries: A
       systematic review and meta-analysis of randomized controlled trials. Journal of Burn Care &
       Research, Irae, 061.
[47]   Shafeie, N., Naini, A. T. and Jahromi, H. K. (2015). Comparison of different concentrations of
       Calendula officinalis gel on cutaneous wound healing. Biomedical & Pharmacology Journal, 8(2),
       979–992.
[48]   Hajiali, H., Summa, M., Russo, D., Armirotti, A., Brunetti, V., Bertorelli, R. . . . Mele, E. (2016).
       Alginate–lavender nanofibers with antibacterial and anti-inflammatory activity to effectively
       promote burn healing. Journal of Materials Chemical B, 4(9), 1686–1695.
[49]   Halcón, L. and Milkus, K. (2004). Staphylococcus aureus and wounds: A review of tea tree oil as a
       promising antimicrobial. American Journal of Infection Control, 32(7), 402–408.
[50]   Arribas-López, E., Zand, N., Ojo, O., Snowden, M. J. and Kochhar, T. (2022). A systematic review of the
       effect of Centella asiatica on wound healing. International Journal of Environmental Research and
       Public Health, 19(6), 3266.
[51]   Mensah, M. L., Komlaga, G., Forkuo, A. D., Firempong, C., Anning, A. K. and Dickson, R. A. (2019).
       Toxicity and safety implications of herbal medicines used in Africa. Herbal Medicine, 63(5),
       1992–0849.
[52]   George, B., Bhatia, N. and Suchithra, T. V. (2021). Burgeoning hydrogel technology in burn wound
       care: A comprehensive meta-analysis. European Polymer Journal, 157, 110640.
[53]   Lam, P., Cheung, F., Tan, H. Y., Wang, N., Yuen, M. F. and Feng, Y. (2016). Hepatoprotective effects of
       Chinese medicinal herbs: A focus on anti-inflammatory and anti-oxidative activities. International
       Journal of Molecular Sciences, 17(4), 465.
[54]   Liu, E., Gao, H., Zhao, Y., Pang, Y., Yao, Y., Yang, Z. . . . Guo, J. (2022). The potential application of
       natural products in cutaneous wound healing: A review of preclinical evidence. Frontiers in
       Pharmacology, 13, 900439.
[55]   Okaiyeto, K. and Oguntibeju, O. O. (2021). African herbal medicines: Adverse effects and cytotoxic
       potentials with different therapeutic applications. International Journal of Environmental Research
       and Public Health, 18(11), 5988.

[56] Bonifacio, B. V., Da Silva, P. B., Ramos, M. A. D. S., Negri, K. M. S., Bauab, T. M. and Chorilli, M. (2014). Nanotechnology-based drug delivery systems and herbal medicines: A review. International Journal of Nanomedicine, 9, 1–15. https://doi.org/10.2147/IJN.S52634.

[57] Van Wyk, A. S. and Prinsloo, G. (2020). Health, safety and quality concerns of plant-based traditional medicines and herbal remedies. South African Journal of Botany, 133, 54–62.

[58] Fu, P. P., Xia, Q., Zhao, Y., Wang, S., Yu, H. and Chiang, H. M. (2013). Phototoxicity of herbal plants and herbal products. Journal of Environmental Science and Health, Part C, 31(3), 213–255.

[59] Ekor, M. (2014). The growing use of herbal medicines: Issues relating to adverse reactions and challenges in monitoring safety. Frontiers in Pharmacology, 4, 177.

[60] Amer, M. R., Cipriano, G. C., Venci, J. V. and Gandhi, M. A. (2015). Safety of popular herbal supplements in lactating women. Journal of Human Lactation, 31(3), 348–353.

[61] Niggemann, B. and Grüber, C. (2003). Side-effects of complementary and alternative medicine. Allergy, 58(8), 707–716.

[62] Yang, C. S., Sang, S., Lambert, J. D. and Lee, M. J. (2008). Bioavailability issues in studying the health effects of plant polyphenolic compounds. Molecular Nutrition & Food Research, 52(S1), S139–S151.

[63] Siqueira, J. O., Nair, M. G., Hammerschmidt, R., Safir, G. R. and Putnam, A. R. (1991). Significance of phenolic compounds in plant-soil-microbial systems. Critical Reviews in Plant Sciences, 10(1), 63–121.

[64] Liao, Y., Li, Z., Zhou, Q., Sheng, M., Qu, Q., Shi, Y. . . . Shi, X. (2021). Saponin surfactants used in drug delivery systems: A new application for natural medicine components. International Journal of Pharmaceutics, 603, 120709.

[65] Nolan, J. M. and Turner, N. J. (2011). Ethnobotany: The study of people-plant relationships. Ethnobiology, 9, 133–147.

[66] Chelu, M., Musuc, A. M., Popa, M. and Calderon Moreno, J. (2023). Aloe vera-based hydrogels for wound healing: Properties and therapeutic effects. Gels, 9(7), 539.

[67] Smith-Hall, C., Larsen, H. O. and Pouliot, M. (2012). People, plants and health: A conceptual framework for assessing changes in medicinal plant consumption. Journal of Ethnobiology and Ethnomedicine, 8, 1–11.

[68] Briskin, D. P. (2000). Medicinal plants and phytomedicines. Linking plant biochemistry and physiology to human health. Plant Physiology, 124(2), 507–514.

[69] Khare, C. P., Ed. (2011). Indian Herbal Remedies: Rational Western Therapy, Ayurvedic and Other Traditional Usage, Botany, Springer science & business media, Springer Verlag Berlin Heidelberg New york.

[70] Singh, P. (2023). Pharmacogenomics advances: Customizing drug therapies for individual patients. Journal of Advanced Research in Pharmaceutical Sciences and Pharmacology Interventions, 6(1), 21–27.

[71] Malsagova, K. A., Butkova, T. V., Kopylov, A. T., Izotov, A. A., Potoldykova, N. V., Enikeev, D. V. and Kaysheva, A. L. (2020). Pharmacogenetic testing: A tool for personalized drug therapy optimization. Pharmaceutics, 12(12), 1240.

[72] Sanjarnia, P., Picchio, M. L., Solis, A. N. P., Schuhladen, K., Fliss, P. M., Politakos, N. and Osorio-Blanco, E. R. (2024). Bringing innovative wound care polymer materials to the market: Challenges, developments, and new trends. Advanced Drug Delivery Reviews, 115217.

[73] Yuan, H., Ma, Q., Ye, L. and Piao, G. (2016). The traditional medicine and modern medicine from natural products. Molecules, 21(5), 559.

Serkan Kapancik*, Atteneri López Arencibia, and Burak Tuzun

Chapter 10
Medicinal and aromatic plants used in respiratory diseases

Abstract: In ancient times, people sought remedies in plants found in nature to treat their illnesses. The use of medicinal plants by people is quite old. The information obtained about these plants is based on the experiences that emerged after their use, and as a result of medical developments, the therapeutic effectiveness of many plants on diseases has been proven by scientific studies today. A wide variety of medicinal and aromatic plants are frequently used in the treatment of many diseases such as cancer, immune system diseases, endocrine diseases, cardiovascular diseases, digestive disorders, kidney and liver diseases, central nervous system diseases, gastrointestinal diseases, inflammatory diseases, skin diseases, and respiratory diseases. One of the diseases for which medicinal and aromatic plants are used for treatment purposes is respiratory system diseases. Mortality rates due to respiratory diseases are quite high worldwide. In order to reduce the mortality rates caused by respiratory diseases, alternative treatment methods that will contribute to the treatment of respiratory diseases are needed. For this reason, studies on the role of medicinal and aromatic plants in the treatment of respiratory diseases are increasing day by day. In this part of the book, we will discuss the roles of medicinal and aromatic plants in the treatment of respiratory diseases such as chronic obstructive pulmonary disease (COPD), asthma, pneumonia, and lung cancer in the light of scientific research.

Keywords: medicinal and aromatic plants, chronic obstructive pulmonary disease, asthma, pneumonia, lung cancer

10.1 Introduction

Medicinal and aromatic plants have great therapeutic importance worldwide through the bioactive molecules they contain. In ancient times, people sought remedies in plants in nature in order to treat their diseases. Although medicinal plants' use by

---

*Corresponding author: Dr. Serkan Kapancik, Department of Biochemistry, School of Medicine, Sivas Cumhuriyet University, Sivas, Turkey, e-mail: serkankapancik@gmail.com, https://orcid.org/0000-0003-3019-4275
Atteneri López Arencibia, Universidad De La Laguna, La Laguna, Tenerife, Islas Canarias 38203, Spain
Burak Tuzun, Plant and Animal Production Department, Technical Sciences Vocational School of Sivas, Sivas Cumhuriyet University, Sivas, Turkey

https://doi.org/10.1515/9783111469713-010

people is quite old, the information obtained about these plants was also based on the experience that emerged after their use. Through experience, it began to be revealed which plants could be used in the treatment of which disease. With the development of pharmacology in recent years, the use of medicinal plants for therapeutic purposes has increased due to the decrease in the therapeutic capabilities of synthetic drugs synthesized in laboratories. The biggest reason for this is the low side effects of medicinal plants compared to synthetic drugs produced pharmacologically. Medicinal and aromatic plants, which have a wide variety, are frequently used in the treatment of many diseases such as cancer, immune diseases, endocrine diseases, cardiovascular diseases, digestive disorders, kidney and liver diseases, central nervous system diseases, gastrointestinal diseases, and inflammatory diseases [1, 2]. Respiratory system diseases, one of the diseases in which medicinal and aromatic plants are used for treatment, refer to the group of diseases that occur in the organs and tissues related to respiration, which prevent individuals from maintaining their respiratory functions in a healthy way. In order to be used for therapeutic purposes, medicinal and aromatic plants are subjected to the extraction method as shown in Figure 10.1, and their extracts are obtained in this way.

Figure 10.1: Obtaining plant extract through extraction of plants (created via BioRender.com).

Mortality rates from respiratory diseases continue to occur at high rates despite advances in medicine. The rate of contracting chronic obstructive pulmonary disease (COPD), which is in the group of respiratory diseases, is quite high. This number has reached 65 million. Mortality rates from COPD rank fourth when compared to mortality rates from other diseases worldwide. Asthma is a respiratory disease that affects approximately 14% of children in the world and is also frequently encountered in adults. Pneumonia is especially fatal in children under the age of 5 and causes the death of millions of people. Lung cancer is an important respiratory disease that causes the death of 1.4 million people every year and has higher mortality rates compared to other types of cancer. Tuberculosis is another respiratory disease that causes high mortality. Approximately 1.4 million people die each year due to tuberculosis. In general, approximately 4 million people die from chronic respiratory diseases each year [1, 3]. In this chapter, we will discuss the therapeutic effects of medicinal and aromatic plants for each respiratory disease.

10.2 COPD

COPD is a disease based on tissue damage caused by oxidative stress and oxidant accumulation caused by inflammation. Since the treatments associated with this disease are not fully sufficient, in a study investigating whether *Ocimum sanctum* leaf extract has a therapeutic role in COPD, it was observed that in a model of COPD formation in mice exposed to cigarette smoke, as in Figure 10.2, the administration of the plant extract increased antioxidant capacity and decreased oxidant capacity. This resulted in a decrease in inflammation, and in this respect, *Ocimum sanctum* leaf extract exhibited a protective effect against COPD formation. Due to these effects, it has been reported that *Ocimum sanctum* leaf extract can be used in COPD [4].

Again, the effects of the alcohol extracts of the medicinal plants *Mikania glomerata* Spreng, *Plantago major*, *Equisetum arvense*, and *Arctium lappa* on the disease were investigated in a COPD rat model created by exposure to cigarette smoke. When the animals were exposed to the process in Figure 10.2 for 2 months and the plant extracts obtained during this period were given to the rats, it was shown that the macrophages and mast cells in the lungs of the rats were suppressed compared to the group that was not given the plant extract. This situation indicates that the inflammation levels of the animals that were given the plant extracts decreased, and it has been reported that these plants may have a therapeutic and protective role for COPD because the plant extract application prevents inflammation through the decrease in inflammatory mediator levels [5]. In the COPD mouse model where *Epilobium pyrricholophum* extracts were applied, it was shown that clinical symptoms were suppressed as a result of the inhibition of cell aggregation with inflammatory properties. Although it is known that *Epilobium pyrricholophum* extracts mediate a decrease in the levels of radical, especially in immune cells, and reduce myeloperoxidase and cytokine release, it has also been reported that these plant extracts have an effect on reducing inflammation by suppressing NFκB expression in the COPD mouse model. In this respect, it has been suggested that the *Epilobium pyrricholophum* plant may contribute to the treatment of COPD by suppressing both the immune and the inflamma-

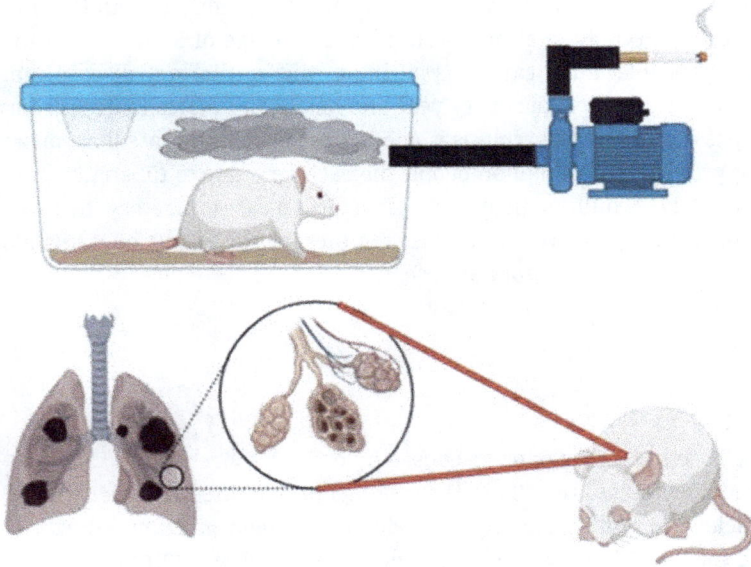

Figure 10.2: COPD mouse model with cigarette smoke exposure (created via BioRender.com).

tion genes [6]. The effect of ethanol extracts obtained from the *Alisma orientale* Juzep-zuk on the COPD mouse model was also investigated in another study. In this study, a mouse COPD model was created as in Figure 10.3 after the application of lipopolysac-charide as a spray, in addition to the intratracheal application of elastase to the mice.

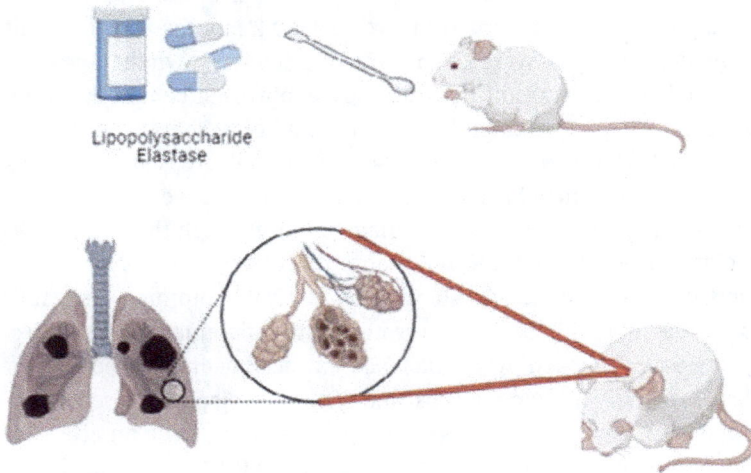

Lipopolysaccharide
Elastase

Figure 10.3: COPD mouse model induced with lipopolysaccharide or elastase (created via BioRender. com).

Then, by looking at the levels of genes and proteins related to inflammation, it was investigated whether ethanol extracts obtained from the tuber of *Alisma orientale* Juzepzuk had a therapeutic role on COPD disease. It was determined that inflammation in the lungs decreased, following the application of the plant extract in the COPD model. In addition, it was observed that the levels of proinflammatory cytokines such as TNF-α, TGF-β, and IL-6, which are related to inflammation, decreased. Based on these results, it was reported that the *Alisma orientale* Juzepzuk reduces the pathological aspects of the disease by suppressing lung inflammation in COPD [7]. *Lonicera japonica* flower has antioxidant properties. It is mostly used in East Asia for its therapeutic properties. Research has also been conducted on the effectiveness of *Lonicera japonica* flower in COPD. For this study, a COPD model was developed in mice. For this model, after applying lipopolysaccharide and smoking solution to mice, it was determined that the expression of IL-6 and TNF-α, which are genes related to inflammation, decreased in animals exposed to *Lonicera japonica* flower treatment. However, the application of *Lonicera japonica* flower microparticles also caused serious decreases in inflammatory-related cells. In addition, it was reported that the treatment of this medicinal plant mediated a decrease in the expression of Caspase3, which may play a role in apoptosis that may be induced as a result of the increase in oxidant amounts in the lungs of COPD model mice. As a result, it has been suggested that the inhalation of plant flower microparticles may be a promising treatment strategy for COPD treatment [8]. The stem, leaves and root of the *Celastrus orbiculatus* Thunb. plant are used to contribute to the treatment of diseases. Therefore, *Celastrus orbiculatus* Thunb. is also included in the class of medicinal plants. The therapeutic effects of *Celastrus orbiculatus* Thunb., its in vitro effects on COPD, have also been investigated. For this purpose, A549 lung cancer cell lines were exposed to cigarette smoke extract to create an inflammatory cell model, and then when the stem, leaves, and root extracts of the plant were applied to A549 cells, it was shown that the levels of proinflammatory factors decreased in a dose-dependent manner. However, it was emphasized that the stem part of the plant may be effective for COPD compared to the root and leaf parts [9]. It was aimed to scientifically reveal its role in COPD by examining the effect of *Pseudognaphalium affine* (D.Don) Anderb. which is used by humans for cough, COPD, and asthma, on an in vivo COPD mouse model. In this study, it was observed that *Pseudognaphalium affine* (D.Don) Anderb. extract application in a COPD mouse model suppressed damage in the lungs of animals. *Pseudognaphalium affine* (D.Don) Anderb. achieved this by reducing the levels of proinflammatory cytokines that cause the emergence of inflammatory effect. In particular, application of the plant's extract has been the most important mechanism in preventing inflammation through suppression of protein levels as a result of suppressing the expression of NF-κB. Based on this, the extract of *Pseudognaphalium affine* (D.Don) Anderb. has been reported to be effective against COPD due to the inhibition of this pathway by suppressing proinflammatory cytokines and causing a decrease in NF-κB levels [10]. *Isodon suzhouensis* is a plant that is both edible and used for medical treatment. Due to its inflammation-reducing properties, the effectiveness of plant extracts was investigated in mice with COPD models in order to determine whether it has therapeutic properties. The therapeutic activities of *Isodon suzhouensis* and its active component, glycocalycin A, were investigated in animals, with COPD models. It was determined that *Isodon suzhouensis* extracts improved lung functions in mice. However, it was observed that they suppressed inflammation by reducing IL-1β and TNF-α levels. It was reported that *Isodon suzhouensis* extracts prevented the development of COPD by suppressing inflammation, prevented apoptosis of cells in lung tissue, and reduced the expression levels of proteins in the JAKs/STATs pathway. Glycocalysin A has been shown to suppress proinflammatory factors in a COPD mouse model and to have an ameliorating effect on COPD by inhibiting the JAKs/STATs pathway [11]. It is known that *Azadirachta indica* A. Juss. leaf has antioxidant effects and antiinflammatory properties. Due to these properties of this medicinal plant, its effect on COPD has also been investigated in a study. It has been determined that the extract of plant leaf reduces the levels of reactive oxygen species and the number of inflammatory cells in bronchoalveolar lavage fluid against inflammation caused by cigarette smoke and lipopolysaccharide. In addition, it has been determined that the levels of proinflammatory IL-6 and TNF-α in bronchoalveolar lavage fluid are reduced by the plant extract. It has been reported that plant leaf extract reduces the inducible nitrite oxide synthase expression in lung tissue in a COPD mouse model, prevents ERK and JNK activation, and suppresses phosphorylation of TNF-α, and has the potential to be used in the treatment of COPD due to these findings [12]. *Thymus vulgaris* L. is known for its therapeutic properties for respiratory diseases. It has been suggested that the application of *Thymus vulgaris* L., known as a traditional medicinal plant, may have therapeutic effects for COPD by reducing NF-κB levels, IL-1beta, and IL-8 levels. In addition, it has been shown that the mucociliary-beating frequency, which is impaired in COPD, can be increased by mediating an induce, increasing the Ca2 + and cAMP levels with the application of *Thymus vulgaris* L. extract, and thus can be used as a support for COPD treatment [13]. *Myrciaria cauliflora* is an edible fruit and is also used as a treatment for asthma. It has been suggested that *Myrciaria cauliflora* can also be used as a treatment for COPD due to its antiinflammatory effects [14]. *Lilium longiflorum* Thunb plant is also one of the medicinal plants known to have anti-inflammatory properties. In a COPD mouse model created with porcine pancreas elastase and cigarette smoke extract, oral administration of fermented *lilium longiflorum* Thunb bulb extract has been shown to suppress inflammation by suppressing the infiltration of immune cells and the producing inflammatory mediators, thus preventing lung damage. In addition, fermented *Lilium longiflorum* Thunb bulb extract has been shown to mediate a decrease in the levels of IL-8 and IL-6, which are proinflammatory factors that are increased, in levels with cigarette smoke extract and lipopolysaccharide in the epithelial cell line H292 cells. It has been reported that fermented *Lilium longiflorum* Thunb bulb extract is effective in preventing and slowing down inflammation in an animal model of COPD [15]. Baru nut is a species that contains high antioxidants and phenols with therapeutic properties. It has been reported that the baru nut ethanol extract reduces reactive oxygen species in NCI-H441 and A549 lung epithelial cells and contributes to wound healing. Therefore, it has been mentioned that baru nut may be a beneficial species for diseases such as COPD, which are based on oxidative stress [16]. Scientific studies have shown that the *Zataria multiflora* plant has a high antioxidant content and can contribute to the treatment of respiratory diseases due to its inflammation-suppressing effects. It has been shown that COPD patients who were given the extract of the *Zataria multiflora* plant had inflammation-reducing effects and that the plant extract had an effect on the treatment of COPD [17]. Anti-inflammatory drugs are used in COPD patients. The effect of these drugs is based on the principle of inhibiting inflammation by inhibiting cyclooxygenase-2 (COX-2). Thus, the increase in inflammation in the lungs of COPD patients can be prevented. In the study investigating the relationship between COX-2 and the extracts obtained from kersen leaf (*Muntingia calabura*) and Legetan warak (*Adenostemma lavenia*), it was shown that these plant extracts inhibit COX-2 and thus suppress inflammation [18]. It has been reported that COPD patients who were given capsules containing the hydroalcoholic extract of rosemary had healing effects and showed an increase in cognitive functions [19]. In a study investigating the effects of *Angelicae dahuricae* Radix on a mouse COPD model created with lipopolysaccharide and cigarette smoke extract, it was reported that the plant extract caused suppression of COPD disease. In this study, it was suggested that *Angelicae dahuricae* Radix extract suppressed the level of inflammatory cells that increased in the bronchoalveolar lavage fluid after application to animals, with a COPD model, thus causing an inhibitory effect on COPD [20]. The therapeutic effects of the *Siraitia grosvenorii* plant were investigated in a scientific study in the COPD mouse model created with cigarette smoke extract and lipopolysaccharide and in BEAS-2B cells to which lipopolysaccharide was applied, as in Figure 10.4.

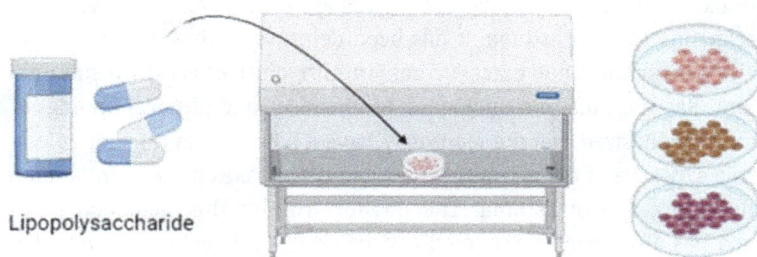

Lipopolysaccharide

Figure 10.4: COPD cell line model induced with lipopolysaccharide (created via BioRender.com).

In this study, it was shown that *Siraitia grosvenorii* extract mediates the preservation of viability in BEAS-2B cells treated with lipopolysaccharide and reduces the expression and levels of inflammation-related cytokines in these cells. An increase in the infiltration of immune cells into the respiratory tract of animal with a COPD model was observed, and it was shown that *Siraitia grosvenorii* extract suppressed this condition. In addition, plant extracts mediated a decrease in cytokine release in the bronchoalveolar fluid of mice, with a COPD model. Using these findings, it was reported that

Siraitia grosvenorii has an anti-inflammatory activity and that it could be a potential herbal medicine for the treatment of COPD due to these properties [21].

10.3 Asthma

The effectiveness of hexane, methanol, and ethyl acetate extracts of *Asystasia gangetica* T. Adams leaf in the treatment of asthma was investigated through a study. Since *Asystasia gangetica* T. Adams leaf is known to be used for asthma among the public, this study revealed the scientific role of the plant in the treatment of the disease. It was determined that the plant extracts prevented the contraction induced by using spasmogens. In addition, the extracts also mediated the relaxation of tracheal strips contracted using histamine. Among the extracts of *Asystasia gangetica* T. Adams leaf, methanol extract has the highest anti-inflammatory effects on mice. Based on the evidences obtained from study, it has been stated that *Asystasia gangetica* T. Adams leaf may be effective in asthma [22]. In the ovalbumin-induced asthma mouse model, it has been reported that *Ocimum basilicum* leaves have a therapeutic effect. In this study, the application of plant extracts in asthma model caused inhibition in PLA2, TP, IgE, and IL-4 levels, while there was an increase in the IFN-γ/IL-4 ratio. The fact that *Ocimum basilicum* leaves mediate the improvement in inflammatory and immunological factors is important in terms of the use of this plant in asthma and its therapeutic potential in asthma disease [23]. *Inula racemosa* Hook. F. is a plant, and its roots are often used for antiseptic purposes, as an anti-inflammatory, digestive, and antipyretic drug, and have important therapeutic effects. *Inula racemosa* Hook. F. root extracts have been investigated in vivo and in vitro for their possible therapeutic roles in asthma. It has been determined that the root of the plant, especially its petroleum ether extracts, has an antigonist effect on histamine-induced contractions. Based on the immunological, biochemical, and physical findings of the study, it has been emphasized that the plant may have a potential therapeutic role in asthma [24]. Chronic exposure of the airways, where breathing takes place to inflammation, mediates the emergence of asthma. The possible role of the medicinal plant *L. aspera* in asthma has also been investigated and it has been determined that dried whole plant extracts have therapeutic properties for asthma. Methanol extracts were extracted from dried whole plant parts of *L. aspera* and these methanol extracts were used in the study. It has been shown that *L. aspera* methanol extract may have an important therapeutic effect for asthma due to its bronchodilator, inflammation suppressor, antihistamine, mast cell stabilizing, and anticholinergic activity in asthma models [25]. It has been reported that the methanol extract of Moringa oleifera Lam. leaves prevents inflammation in the respiratory tract, causes bronchoconstriction and may be useful against asthma due to these medicinal effects. In order to demonstrate the therapeutic effects of *Moringa oleifera* Lam. leaf methanol extract against asthma, an ovalbumin-induced asthma model was used in guinea pigs. Application of plant leaf methanol extracts to asthma model animals caused improvement in the lung functions of these animals [26]. In the study investigating the effects of water, ethanol, and petroleum ether extracts of *Solanum xanthocarpum* flowers on asthma in vivo and in vitro, it was stated that the ethanol extract of *Solanum xanthocarpum* flowers may have a potential role in the treatment of asthma because it provides mast cell stabilization, produces antihistaminic effects, and produces effects in reducing capillary permeability [27]. It is known that the fruit of the *Solanum nigrum* Linn plant is used, especially for asthma, among the public. Therefore, in the study conducted to reveal the role of the fruit of the *Solanum nigrum* Linn plant in the treatment of asthma, petroleum ether, ethanol, and water extracts of the fruit of the plant were extracted. It was determined that the petroleum ether extracts of the fruits of the plant suppressed catalepsy, induced by clonidine. In addition, the petroleum ether extract of the fruit of the plant had a reducing effect on the increase in eosinophils and leukocyte cells caused by milk allergen, and also showed reducing effects on histamine-induced contractions. Considering the experimental results of this study, it was reported that the petroleum ether extract of the fruit of the plant could prevent the symptoms caused by asthma [28]. *Carica papaya* leaves are a medicinal plant used in traditional medicine to suppress inflammation. The therapeutic effect of *Carica papaya* leaves was investigated in an ovalbumin-induced asthma mouse model. The application of *Carica papaya* leaves extract to these asthma model mice reduced the infiltration of inflammatory cells in the lungs of the animals and also prevented alveolar thickening. In this study, it was determined that the extract of *Carica papaya* leaves reduced the number of leukocytes in the bronchoalveolar lavage fluid and in the blood of asthma model animals. It was also shown that the application of the plant extract suppressed the expression levels of IL-4, NF-κB, iNOS, TNF-α, IL-5, and eotaxin, thereby improving inflammation levels [29]. The antioxidant and anti-inflammatory effects of ethanol extracts of *Paeonia* and *Schisandra* plants on rats with asthma model were investigated in a scientific study. It was determined that after the application of ethanol extracts of *Paeonia* and *Schisandra* plants, eosinophils in the tracheal tissues of asthma model rats was reduced and that it mediated healing in the mucosal tissue. In vivo, it was shown that the application of ethanol extracts of the plants caused an increase in the antioxidant levels of rats, an increase in serum and erythrocyte SOD activity, and a decrease in MDA levels. It was reported that the application of ethanol extracts of *Paeonia* and *Schisandra* plants reduced NF-κB p65 protein expression in asthma model rats, and therefore, according to the findings of this study, *Paeonia* and *Schisandra* medicinal plants mediate antioxidant effects in vivo and can be used to treat asthma because they prevent the development of inflammation [30]. In the study investigating the therapeutic efficacy of the ethanol extract of the *Viola mandshurica* W. Becker plant in an asthma mouse model, induced by ovalbumin, it was determined that the application of the plant extract mediated a decrease in the levels of IgE, IL-13, and IL4 in the blood serum and bronchoalveolar lavage fluid of the animals and prevented eosinophilia and mucus secretion. It was predicted that the ethanol extract of this plant is a medicinal plant that can be used in the treatment of asthma and can be a useful lead material for the development of asthma drugs [31]. It has been shown that the ethanol extract of the *Polyscias fruticosa* plant may be useful in asthma due to its antihis-taminic and mast cell stabilizing effects in an ovalbumin-mediated asthma model in guinea pigs. It has been determined that the application of the ethanol extract of the *Polyscias fruticosa* plant to the animal model has a reducing effect against histamine-induced bronchospasm, reducing the recovery time and preventing mast cell degranulation [32]. *Perilla frutescens* (L.) Britton is used as a medicinal plant among the public for the treatment of asthma. It has been determined that the use of *Perilla* leaf extract is effective in reducing cells and cytokines related to inflammation in bronchoalveolar lavage fluid and plays a positive role in the healing of lung tissue, and that it has a role in the suppression of inflammation on the airway in the mouse asthma model, induced with ovalbumin. It has been shown that *Perilla* leaf extract mediates the inhibition of inflammation in RBL-2H3 cells, induced by antigen, and in human peripheral blood mononuclear cells, induced with ovalbumin. In addition, in the gene expression analyses performed in vivo and in vitro, it has been determined that *Perilla* leaf extract reduces the expressions of genes related to inflammation and phosphorylates the proteins synthesized from these genes, which is effective in eliminating inflammatory effects. Based on the data obtained from the findings, the study concluded that the use of *Perilla* leaf extract has a therapeutic effect in inhibiting allergic inflammation [33]. *Aster yomena* is a traditionally used medicinal plant. It is used especially in the treatment of asthma, cough, and insect bites. In a study investigating the therapeutic effect of the *Aster yomena* plant for asthma, it was shown that the application of the alcohol extract of the plant in a mouse asthma model, created with ovalbumin, suppressed the levels of enzymes involved in the production of inflammatory mediators and thus produced therapeutic effects on asthma. However, it was reported that there was a decrease in the levels of cytokines and eosinophil counts in the bronchoalveolar lavage fluid after the application of the plant extract to asthma model mice. It was determined that the *Aster yomena* plant mediated healing in the lungs by eliminating the sensitivity on the respiratory tract. Based on these findings, it was stated that the *Aster yomena* plant is a natural agent that can be used for the treatment of bronchial asthma [34]. *Duchesnea chrysantha* plant is a medicinal plant class with anti-oxidant properties and anti-inflammatory therapeutic effects. In the study investigating the disease-treating effects of *Duchesnea chrysantha* plant in an asthma mouse model, created through ovalbumin, first, the *Duchesnea chrysantha* plant was pulverized and ethanol extract was obtained. It was shown that *Duchesnea chrysantha* plant extract mediated the suppression of leukocytosis and eosinophilia in the bronchoalveolar lavage fluid of asthma model animal and inhibited mucus secretion. *Duchesnea chrysantha* plant extract suppressed the expression of IL-5, IL-13, IL-4, and eotaxin from inflammation-related factors. It was reported that results indicate the anti-asthmatic effect of *Duchesnea chrysantha* plant extract [35]. It is known that *Nigella sativa* has an antihistamine and a relaxant effect on tracheal chains. In addition, the effects of boiled *Nigella sativa* seed extract on asthma have also been investigated. In the study, asthma patients were divided into two groups and the first group was given the plant extract and the other group was given a placebo solution. At the beginning of the treatment and twice at 45-

day intervals after the treatment, parameters such as asthma severity, symptom frequency, and lung function were evaluated. In the two evaluations after the treatment, patients given *Nigella sativa* plant seed extract experienced a decrease in asthma severity, frequency of recurrence, and chest wheezing, while lung functions also improved. This situation caused a decrease in the need for medication in patients, while no change was reported in the placebo group. It has been suggested that these findings indicate that *Nigella sativa* plant seeds may contribute to the prevention of asthma [36]. The *Echinodorus scaber* Rataj plant is used as a medicinal plant among the public, especially in the treatment of respiratory tract diseases where inflammation is present. It has been reported that the hydroethanolic extract of the leaves of the *Echinodorus scaber* Rataj plant has pharmacological effects through its anti-inflammatory effects in asthma. In a study investigating the effects of the plant leaf extract on an ovalbumin-induced allergic asthma mouse model, it was determined that it reduced the number of immune-related cells such as eosinophils, neutrophils, leukocytes, and mononuclear cells. It has also been reported that the *Echinodorus scaber* Rataj plant leaf extract caused a decrease in the levels of IL-5, IL-13, and IL-4 cytokines in the bronchoalveolar lavage fluid and IgE levels in the blood plasma, which increased after the allergic asthma model in mice. The anti-inflammatory effect of the *Echinodorus scaber* Rataj plant leaf extract has been attributed to these biochemical changes [37]. *Zataria multiflora* Boiss is a plant traditionally used to treat cough and respiratory disorders. In addition to its antioxidant properties, it also has anti-inflammatory properties. In a study conducted to reveal the role of the *Zataria multiflora* Boiss plant's leaf and stem extract in asthma, it was determined that it improved respiratory function in asthmatic patients and increased lung function. It was reported that the *Zataria multiflora* Boiss plant's leaf and stem extract induced a decrease in patients' inflammation-related cells and had a therapeutic effect for asthmatic patients due to these properties [38]. In another study investigating the effects of hydroethanolic extract of *Curcuma longa* and curcumin on asthma model rats, it was shown that *Curcuma longa* and curcumin have anti-inflammatory and antioxidant properties. It has been suggested that *Curcuma longa* and curcumin may have a therapeutic effect in the treatment of asthma through their anti-inflammatory properties [39].

10.4 Pneumonia

Magnolia officinalis is a medicinal plant. *Magnolia officinalis* bark extract is used especially in the treatment of fever, cold, cough, and bronchitis. In a study investigating the therapeutic properties of *Magnolia officinalis* bark extract in a mouse pneumonia model induced by influenza virus A, it was determined that the expression levels of inflammation factors such as IL-6, nitric oxide, and TNF-α in the serums of animal given the plant extract, decreased. In addition, it has also been reported that *Magnolia officinalis* bark extract has a suppressive effect on pneumonia in mice. In particular, it has been determined that it is effective in reducing the proximity of cells in these tissues to apoptosis by mediating a decrease NF-κB and TLR3 expression in the tissues infected with influenza virus A of mice. In this respect, it has been suggested that *Magnolia officinalis* bark extract can be used effectively in the treatment of pneumonia through its anti-inflammatory roles and its effects on suppressing apoptosis [40]. *Moringa oleifera* is a tree species that plays a role in suppressing inflammation with the help of its rich phytochemicals. *Moringa oleifera* is a powerful antioxidant, in addition to its roles in preventing inflammation. The therapeutic roles of the *Moringa oleifera* leaves ethanolic extract in inflammation of lung cells induced by lipopolysaccharide in mice were examined in vitro in W138 cells. It was determined that the ethanolic extract of plant leaves suppressed IL-6 and IL-1β levels in lung cells. Since the suppression of these inflammation-related cytokines by the ethanolic extract of plant leaves may also play an inhibitory role in inflammation occurring in pneumonia, it has been reported that *Moringa oleifera* plant can be used in diseases related to inflammation and these study results support its traditional use [41]. Green chemistry synthesis was performed with the help of copper ions from the water extract of *Alhagi maurorum* plant, and the effectiveness of this synthesized green chemistry synthesis product in pneumonia was investigated. For this purpose, a pneumonia model was first used in BALB/c mice. For the pneumonia model, the disease was created in mice by injecting mycoplasma pneumonia. The effects on inflammatory factors such as, TNF-α, IL-8, IL-1, IL-6, and TGF were examined. It was determined that the product obtained as a result of green chemistry synthesis had suppressive properties on inflammation-related factors, and thus, it produced inflammatory inhibitory effects in pneumonia model mice. It also mediated decreases in the number of inflammation-related cells. As a result of this study, it was reported that green chemistry synthesis, with the help of copper ions from the water extract of *Alhagi maurorum* plant, exhibited healing effects in pneumonia mouse model [42]. Acute pneumonia is especially fatal in elderly individuals and children with weakened immune systems. Acute pneumonia is an inflammation-related pathological condition, resulting from inflammation of the lung tissue. The therapeutic efficacy of *Symplocos prunifolia* extract was investigated in A549 and RAW264.7 cells after inflammation induced by lipopolysaccharide. In this in vitro study, it was determined that *Symplocos prunifolia* extract reduced nitric oxide in RAW 264.7 cells stimulated with lipopolysaccharide. In addition, this plant extract reduced the cyclooxygenase-2 enzyme and nitric oxide synthase enzyme expression levels. However, it was determined that *Symplocos prunifolia* extract suppressed the expressions of cyclooxygenase-2, inducible nitric oxide synthase, and inflammation-related proteins in LPS-stimulated A549 cells. It has been suggested that *Symplocos prunifolia* extract suppresses the activation of NF-κB, MAPK, and PI3K/Akt signaling pathways in order to suppress inflammation, and therefore it is a candidate for therapeutic efficacy in acute pneumonia [43].

10.5 Lung cancer

Lung cancer is a disease that is frequently seen and has a high mortality rate. Despite technological developments, the frequency of deaths due to lung cancer has increased, while the mortality rate has remained high. The increase in life expectancy in lung cancer depends on the effectiveness of treatment. Therefore, many studies are being conducted on the therapeutic effects of medicinal plants in lung cancer. If we talk about these studies, we will first discuss the role of the extracts of *Erythrophleum succirubrum*, *Croton oblongifolius*, and *Bridelia ovata* plants, extracted with 50% ethanol and ethyl acetate, in the treatment of lung cancer. In the study conducted on A549, it was determined that ethyl acetate extracts of *Erythrophleum succirubrum*, *Croton oblongifolius*, and *Bridelia ovata* plants had cytotoxic effects on cell lines. In addition, it was determined that the ethanolic extract of *Erythrophleum succirubrum* plant had cytotoxic effects on lung cancer cells. In addition, when these plant extracts were combined with chemotherapy drugs currently used for cancer treatment, they also mediated an increase in the synergistic effectiveness of these drugs. It was determined that the cytotoxic effect of the extracts of *Bridelia ovata*, *Croton oblongifolius*, and *Erythrophleum succirubrum* plants was achieved by inducing apoptosis. It was also determined that these plant extracts had anticancer effects on samples taken from tumor tissues of lung cancer patients. Based on these data, it was suggested that the ethyl acetate extracts of *Erythrophleum succirubrum*, *Croton oblongifolius*, and *Bridelia ovata* plants and the ethanolic extract of *Erythrophleum succirubrum* plant have significant potential in the treatment of lung cancer [44]. Another plant whose therapeutic activity in lung cancer has been investigated is the *Teucrium polium* plant. *Teucrium polium* is a medicinal plant that has been used among the public for a long time for complaints such as diabetes and indigestion. In the study investigating the effectiveness of the *Teucrium polium* plant in non-small cell lung cancer, H322 and A549 lung cell lines were used. The effect of the plant extract on cell cycle and apoptosis in these lung cell lines was examined. It was determined that the extract of the *Teucrium polium* plant suppressed proliferation and de-regulated the progression of the cell cycle in H322 and A549 lung cell lines. Therefore, it was stated that the extract of the plant is a good therapeutic that it can be used for the treatment of lung cancer [45]. It is known that *Luffa acutangula* and *Lippia nodiflora* plants have different biological activities for many diseases. In particular, it is suggested that they have anticancer activity. Research on *Lippia nodiflora* and *Luffa acutangula* plants has shown that leaf extracts of these plants can contribute to the treatment of lung cancer. In this study, when *Lippia nodiflora* and *Luffa acutangula* plant leaf extracts were applied to NCI-H460 lung cancer cell lines, dramatic changes occurred in the proliferation of these cells. In addition, mitochondrial depolarization occurred in these cells. In addition, it was determined that *Luffa acutangula* and *Lippia nodiflora* plant leaf extracts induced apoptosis in NCI-H460 lung cancer cells [46]. *Scutellaria barbata* is a plant in the class of medicinal plants used among the public. It is especially used in suppressing inflammation and as a diuretic. *Scutellaria barbata* extracts have been shown to have anticancer activity in different types of cancers. The effects of *Scutellaria barbata* extracts on lung cancer was also examined in A549. It was determined that *Scutellaria barbata* ethanol extracts significantly suppressed the proliferation of A549. It did this by revealing cytotoxic effects and inducing apoptosis in lung cancer cells. In addition, it was determined that *Scutellaria barbata* ethanol extracts mediated changes in the expressions of many genes involved in DNA damage, cell cycle control mechanisms, regulation of nucleic acid binding, and protein phosphorylation. It has been reported that expression changes in these genes mediate the death of lung cancer cells. In this respect, the effectiveness of the *Scutellaria barbata* plant in lung cancer has been demonstrated by molecular mechanisms [47]. It has been shown that the extract of *Kalanchoe tubiflora* in n-butanol has an inhibitory effect on cell proliferation. It has been determined that the water extract of *Kalanchoe tubiflora* mediates cell cycle arrest in A549 lung cancer cells. The in vivo effects of the water extract of *Kalanchoe tubiflora* have also been investigated on nude mice. In this study, A549-xenografted nude mouse models were created by implating A549 cells into nude mice, as in Figure 10.5.

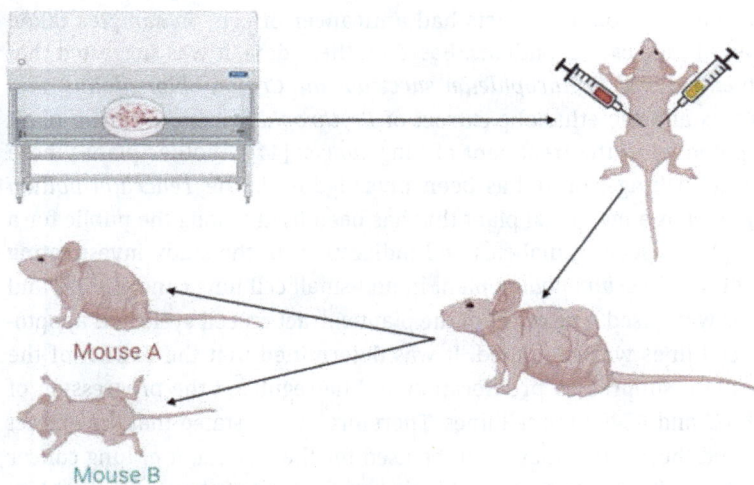

Figure 10.5: Lung cancer xenograft nude mouse model created by implantation of lung cancer cell lines (created via BioRender.com) (Mouse A: mouse dies because it does not receive treatment; Mouse B: mouse that can survive after treatment with medicinal and aromatic plant extracts).

It was found that the water extract of *Kalanchoe tubiflora* caused the shrinkage of lung cancer tumors generated by A549 non-small cell lines in these nude mice. Based on these results, it was reported that *Kalanchoe tubiflora* plant has antitumor activity for lung cancer [48]. *Salvia miltiorrhiza* Bunge is a medicinal plant that contains many components and is used especially for the treatment of cardiovascular patients, but also has anti-inflammatory, anticancer, and anti-allergic effects. The effects of methanol extract of plant roots on proliferation and apoptosis in A549 were investigated for its role in lung cancer. It was determined that methanol extract of plant roots inhibited the proliferation of non-small cell lung cancer cells in a dose-dependent manner, and also induced early and late apoptosis in these cells, as analyzed by flow cytometry. It caused interruption in the G2/M phase of the cell cycle. It was determined that methanol extract of *Salvia miltiorrhiza* Bunge roots mediated an increase in the expression of p53, PARP1, p21, and caspase-3/9. Bcl-2, one of the apoptosis-related proteins, has been shown to induce cell apoptosis by causing a decrease in Bcl-xl expression and an increase in Bax expression. In light of these results, it has been said that methanol extract of plant roots has the ability to induce apoptosis in lung cancer cells and may be a complementary treatment method to inhibit tumor growth [49]. Another study investigating the effectiveness of *Morinda citrifolia* vegetable leaves as a complementary treatment in the treatment of lung cancer compared the anticancer activity of the plant with erlotinib to reveal its anticancer activity in lung cancer. For this purpose, a BALB/c mouse model was used, in which lung cancer was induced in vivo. It was determined that the ethanol extract of *Morinda citrifolia* vegetable leaves inhibited tumor growth in a lung cancer animal model at a dose-dependent level that could be considered significant. It also caused an increase in the immune cells in lung cancer model mice and a decrease in the levels of EGFR, which has an important biological role in lung cancer, as shown in Figure 10.6. The extract of *Morinda citrifolia* vegetable leaves also showed an effect on inflammation and mediated a decrease in the levels of cyclooxygenase 2, an enzyme associated with inflammation. As a result, it was stated that 50% ethanol extract of *Morinda citrifolia* vegetable leaves strengthened immunity and suppressed cell proliferation, thus showing anticancer activity against lung cancer [50].

Therapeutic activity of *Punica granatum* leaf extract against non-small cell lung cancer was also investigated on cell lines. For this study, A549 and H1299 were used, while anticancer activity study of *Punica granatum* leaf extract was also performed on LL/2. *Punica granatum* leaf extract was shown to suppress cell proliferation in lung cancer cell lines in a dose-dependent manner. In cell cycle analyses performed by flow cytometry, it was determined that *Punica granatum* leaf extract inhibited the cell cycle in G2/M phase in H1299 cell lines in a dose-dependent manner. In addition, it was determined that *Punica granatum* leaf extract caused an increase in apoptosis, mediated by mitochondria, and could prevent cell migration and cell invasion of H1299 cells. Based on these anticancer effects of *Punica granatum* leaf extract application on lung cancer cell lines, it has been indicated that this plant can be used safely and effectively in the treatment of lung cancer [51]. *Nigella sativa* is also used by humans in traditional medicine for the treatment of many diseases. It is known that *Nigella sativa* extracts have high antioxidant properties and anti-inflammatory effects. The relationship between plant seed extract and seed oil and lung cancer was investigated in vitro in lung cancer cell lines. In this study conducted on lung cancer cells without A549 small cells, it was

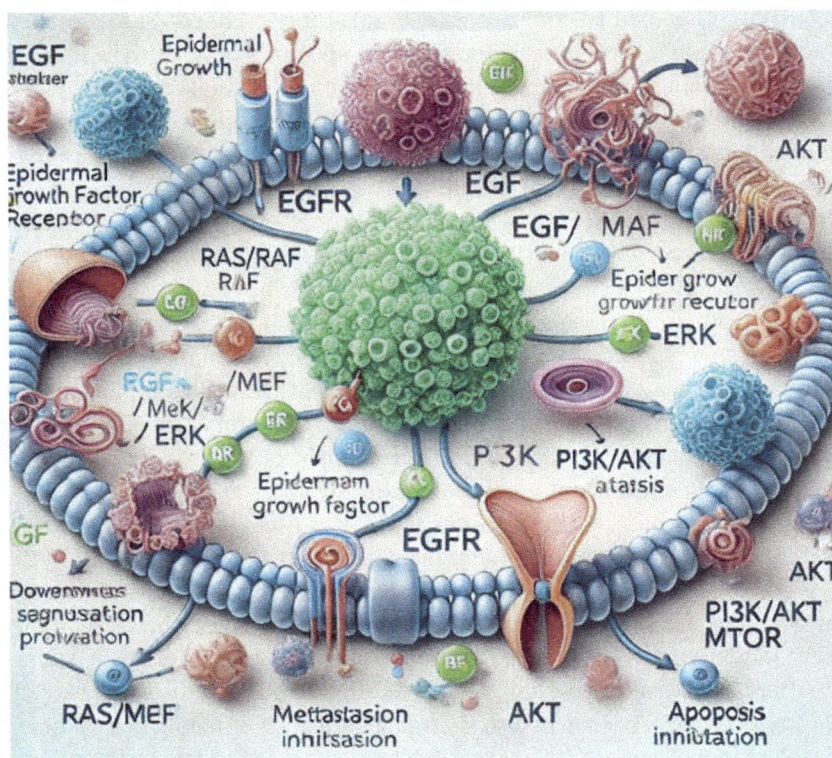

Figure 10.6: EGFR signaling pathway and biological effects in lung cancer (created via BioRender.com).

determined that after the application of *Nigella sativa* seed extract and seed oil, A549 cells lost their typical morphological appearance and appeared smaller than they should be. It was revealed with the help of viability tests that *Nigella sativa* seed extract and seed oil dramatically reduced the viability of lung cancer cells [52]. *Artemisia judaica* L. has a high antioxidant capacity and active effects in suppressing inflammation. In addition, it has roles related to apoptosis. Therapeutic effects of *Artemisia judaica* L. plant extract in lung cancer were evaluated in vitro and in vivo. It was determined that *Artemisia judaica* L. plant extract caused cytotoxic effects at a good IC50 dose compared to doxorubicin in A549 cells. In addition, in the analysis performed on cell cycle, it was determined that it caused cell cycle to arrest in G2/M phase. It caused a decrease in the expression of proteins with anti-apoptotic effect and an increase in the protein expression with apoptotic effect. It was determined that *Artemisia judaica* L. plant extract dramatically reduced tumor size in mice with xenograft lung cancer model. When this antitumor effect is compared with the antitumor effect caused by doxorubicin, it has been shown that it causes a tumor suppression of 54% compared to doxorubicin treatment in the xenograft model. In this study, in the docking study conducted on the

active components of the plant extract of *Artemisia judaica* L., it has also been shown
that the active components of the plant bind to the active site of the epidermal growth
factor receptor [53]. Curcumin, obtained from the *Curcuma longa* plant, has anticancer
activity in many different types of cancers. The effectiveness of the *Curcuma longa*
plant extract, from which curcumin is obtained, on A549 was also investigated. For this
purpose, three different n-hexane, dichloromethane, and methanol extracts of the *Curcuma longa* plant were prepared and applied to A549 cancer cells. It was determined
that the *n*-hexane extract of the plant had dose-related cytotoxicity in A549 cells. In addition, it was shown that the *n*-hexane extract of the *Curcuma longa* plant also inhibited
telomerase activity. It has been suggested that the *n*-hexane extract of the plant is a potential source for drug studies in lung cancer due to its cytotoxic effect and telomerase
inhibitory roles [54]. *Asparagus racemosus* plant has shown anticancer effects in many
types of cancer. The therapeutic roles of *Asparagus racemosus* root methanol and chloroform extracts in lung cancer were also investigated in A549. There was a change in
the morphology of A549 lung cancer cells, to which the root extracts of the plant were
applied. A549 cells changed from their standard shape to round and small. In addition,
it has been shown that *Asparagus racemosus* root methanol and chloroform extracts
have cytotoxic effects on A549 cells. In this respect, it has been suggested that *Asparagus
racemosus* root extract may be a candidate for drug development studies since it mediates a decrease in cell growth in lung cancer [55]. There are rapid developments in the
development of in vitro disease models in lung cancer studies. With the development of
organoid technology, organs are being re-developed on microfluidics and biomaterials,
as in Figure 10.7, and these models are used for the treatment of diseases [56].

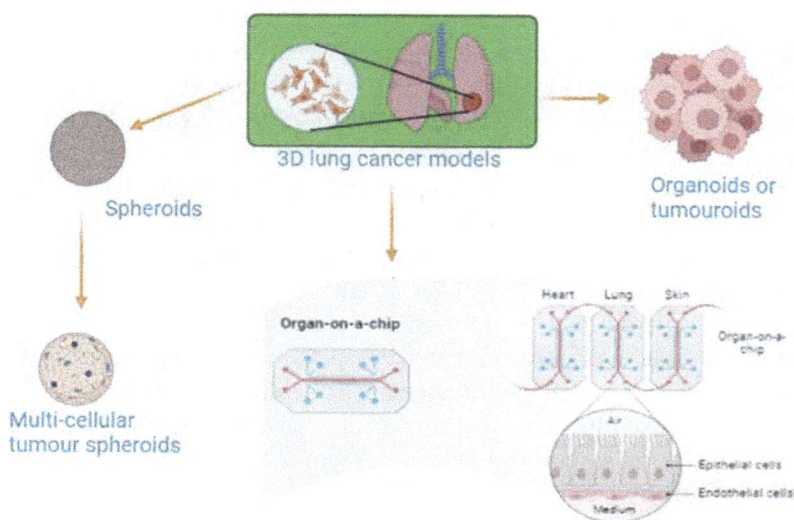

Figure 10.7: Lung model on a microfluidic chip with 3D lung cancer model (created via BioRender.com).

The efficacy of *Eleutherine bulbosa* bulbs against lung cancer was also investigated in a 3D in vitro cell line model created, as in Figure 10.5.3, in addition to 2D as a result of technological developments. First, ethyl acetate, chloroform, and n-hexane extracts of *Eleutherine bulbosa* bulbs were obtained. It was determined that chloroform extract created higher levels of cytotoxic effects compared to other extracts. In addition, it was reported that chloroform extract of *Eleutherine bulbosa* bulbs prevented the formation of colonies in A549 and caused an increase in apoptosis levels in A549 cells. Chloroform extract suppressed the formed spheroid size, mediated a decrease in the stem cell ratio as a result of inducing apoptosis and caused the cell cycle to pause in the S phase. As a result of the study, it was determined that the chloroform extract of *Eleutherine bulbosa* bulbs suppressed cell proliferation in a 2D and 3D in vitro lung cancer model, created with the help of A549 cells for lung cancer. Therefore, it was reported that the chloroform extract of *Eleutherine bulbosa* bulbs could be an agent that can be used in lung cancer [57]. Another technique used effectively for lung cancer is nanotechnological methods. In another study where copper nanoparticles were synthesized with the help of *Thymus fedtschenkoi* leaf extract and their effectiveness in lung cancer was investigated, it was determined that these nanoparticles were antiproliferative for lung cancer cell lines. It was also determined that these particles did not cause any cytotoxicity up to high doses in HUVECs, which are human umbilical cord endothelial cells and not cancer. For these reasons, it is anticipated that these nanoparticles, synthesized with the help of *Thymus fedtschenkoi* leaf extract, can be used as a drug for lung cancer after in vivo studies are carried out [58]. *Tagetes erecta* is a medicinal plant with anticancer effects. The effectiveness of *Tagetes erecta* plant in lung cancer was investigated with the help of hydroalcoholic extracts obtained from the flowers and leaves of the plant. It was determined that the flower extract of *Tagetes erecta* plant has cytotoxic effects on lung cancer cells. In addition, anticancer activities were examined in vivo in a lung cancer mouse model. For this purpose, first, LLC lung cancer cells were implanted into C57BL/6 mice to create a xenograft lung cancer model. The administration of *Tagetes erecta* plant flower extract to this lung cancer mouse model mediated a decrease in tumor growth. In the microscopic analysis of tumor tissues removed from the animals, a decrease in mitosis was detected, while an increase in necrotic areas was observed [59]. *Melissa officinalis* has antitumor properties. When aqueous, methanolic, ethanolic, hydromethanolic, and hydroethanolic extracts of this plant were applied to lung cancer cells, it was found that the proliferation of lung cancer cells was suppressed. When compared to other types of extracts, it was determined that the ethanolic extract of the plant had the strongest anticancer activity. It was shown that plant ethanolic extract affected the cell cycle in lung cancer cells, causing a decrease in pro-caspase3 levels and an increase in P53 levels in these cells [60].

References

[1] Petrovska, B. B. (2012). Historical review of medicinal plants' usage. Pharmacognosy Reviews, 6(11), 1.

[2] Maleš, I., Pedisić, S., Zorić, Z., Elez-Garofulić, I., Repajić, M., You, L. . . . Dragović-Uzelac, V. (2022). The medicinal and aromatic plants as ingredients in functional beverage production. Journal of Functional Foods, 96, 105210.

[3] Marciniuk, D., Ferkol, T., Nana, A., de Oca, M. M., Rabe, K., Billo, N. and Zar, H. (2014). Respiratory diseases in the world. Realities of today–opportunities for tomorrow. African Journal of Respiratory Medicine, 9(1), 4–13.

[4] Srivastava, A., Subhashini, Pandey, V., Yadav, V., Singh, S. and Srivastava, R. (2023). Potential of hydroethanolic leaf extract of Ocimum sanctum in ameliorating redox status and lung injury in COPD: An in vivo and in silico study. Scientific Reports, 13(1), 1131.

[5] Possebon, L., Lebron, I. D. S. L., Da Silva, L. F., Paletta, J. T., Glad, B. G., Sant'Ana, M. . . . Girol, A. P. (2018). Anti-inflammatory actions of herbal medicines in a model of chronic obstructive pulmonary disease induced by cigarette smoke. Biomedicine & Pharmacotherapy, 99, 591–597.

[6] Jung, S. Y., Kim, G. D., Choi, D. W., Shin, D. U., Eom, J. E., Kim, S. Y. . . . Shin, H. S. (2021). Epilobium pyrricholophum extract suppresses porcine pancreatic elastase and cigarette smoke extract-induced inflammatory response in a chronic obstructive pulmonary disease model. Foods, 10(12), 2929.

[7] Kim, K. H., Song, H. H., Ahn, K. S., Oh, S. R., Sadikot, R. T. and Joo, M. (2016). Ethanol extract of the tuber of Alisma orientale reduces the pathologic features in a chronic obstructive pulmonary disease mouse model. Journal of Ethnopharmacology, 188, 21–30.

[8] Park, Y. C., Jin, M., Kim, S. H., Kim, M. H., Namgung, U. and Yeo, Y. (2014). Effects of inhalable microparticle of flower of Lonicera japonica in a mouse model of COPD. Journal of Ethnopharmacology, 151(1), 123–130.

[9] Yang, N., Wang, H., Lin, H., Liu, J., Zhou, B., Chen, X. . . . Li, P. (2020). Comprehensive metabolomics analysis based on UPLC-Q/TOF-MS E and the anti-COPD effect of different parts of Celastrus orbiculatus Thunb. RSC Advances, 10(14), 8396–8420.

[10] Ye, X., Luo, S., Chang, X., Fang, Y., Liu, Y., Zhang, Y. and Li, H. (2022). Pseudognaphalium affine extract alleviates COPD by inhibiting the inflammatory response via downregulation of NF-κB. Molecules, 27(23), 8243.

[11] Zhai, K., Wang, W., Zheng, M., Khan, G. J., Wang, Q., Chang, J. . . . Cao, H. (2023). Protective effects of Isodon suzhouensis extract and glaucocalyxin A on chronic obstructive pulmonary disease through SOCS3–JAKs/STATs pathway. Food Frontiers, 4(1), 511–523.

[12] Lee, J. W., Ryu, H. W., Park, S. Y., Park, H. A., Kwon, O. K., Yuk, H. J. . . . Ahn, K. S. (2017). Protective effects of neem (Azadirachta indica A. Juss.) leaf extract against cigarette smoke-and lipopolysaccharide-induced pulmonary inflammation. International Journal of Molecular Medicine, 40(6), 1932–1940.

[13] Nabissi, M., Marinelli, O., Morelli, M. B., Nicotra, G., Iannarelli, R., Amantini, C. . . . Maggi, F. (2018). Thyme extract increases mucociliary-beating frequency in primary cell lines from chronic obstructive pulmonary disease patients. Biomedicine & Pharmacotherapy, 105, 1248–1253.

[14] Zhao, D. K., Shi, Y. N., Petrova, V., Yue, G. G., Negrin, A., Wu, S. B. . . . Kennelly, E. J. (2019). Jaboticabin and related polyphenols from jaboticaba (*Myrciaria cauliflora*) with anti-inflammatory activity for chronic obstructive pulmonary disease. Journal of Agricultural and Food Chemistry, 67(5), 1513–1520.

[15] Eom, J. E., Kim, G. D., Kim, Y. I., Min Lim, K., Song, J. H., Kim, Y. . . . Shin, H. S. (2023). Bulb of Lilium longiflorum Thunb extract fermented with Lactobacillus acidophilus reduces inflammation in a

chronic obstructive pulmonary disease model. Journal of Microbiology and Biotechnology, 33(5), 634.

[16] Coco, J. C., Ataide, J. A., Sake, J. A., Tambourgi, E. B., Ehrhardt, C. and Mazzola, P. G. (2022). In vitro antioxidant and wound healing properties of baru nut extract (Dipteryx alata Vog.) in pulmonary epithelial cells for therapeutic application in chronic pulmonary obstructive disease (COPD). Natural Product Research, 36(17), 4469–4475.

[17] Ghorani, V., Khazdair, M. R., Mirsadraee, M., Rajabi, O. and Boskabady, M. H. (2022). The effect of two-month treatment with *Zataria multiflora* on inflammatory cytokines, pulmonary function testes and respiratory symptoms in patients with chronic obstructive pulmonary disease (COPD). Journal of Ethnopharmacology, 293, 115265.

[18] Iswantini, D. and Tuwalaid, B. (2021, September). The Potency of Legetan warak (Adenostemma lavenia) and Kersen Leaf (Muntingia calabura) Extract as a Candidate for Chronic Obstructive Pulmonary Disease (COPD) Herbal Medicine. In 2nd International Conference on Science, Technology, and Modern Society (ICSTMS 2020) (pp. 447–452). Atlantis Press.

[19] Safarabadi, A. M., Gholami, M., Kordestani-Moghadam, P., Ghaderi, R. and Birjandi, M. (2024). The effect of rosemary hydroalcoholic extract on cognitive function and activities of daily living of patients with chronic obstructive pulmonary disease (COPD): A clinical trial. Explore, 20(3), 362–370.

[20] Kwak, H. G. and Lim, H. B. (2011). Inhibitory effects of Angelicae Dahuricae Radix extract on COPD induced by cigarette smoke condensate and lipopolysaccharide in mice. Korean Journal of Medicinal Crop Science, 19(5), 380–387.

[21] Kim, M. S., Kim, D. S., Yuk, H. J., Kim, S. H., Yang, W. K., Park, G. D. . . . Sung, Y. Y. (2023). Siraitia grosvenorii extract attenuates airway inflammation in a murine model of chronic obstructive pulmonary disease induced by cigarette smoke and lipopolysaccharide. Nutrients, 15(2), 468.

[22] Akah, P. A., Ezike, A. C., Nwafor, S. V., Okoli, C. O. and Enwerem, N. M. (2003). Evaluation of the anti-asthmatic property of Asystasia gangetica leaf extracts. Journal of Ethnopharmacology, 89(1), 25–36.

[23] Eftekhar, N., Moghimi, A., Mohammadian Roshan, N., Saadat, S. and Boskabady, M. H. (2019). Immunomodulatory and anti-inflammatory effects of hydro-ethanolic extract of Ocimum basilicum leaves and its effect on lung pathological changes in an ovalbumin-induced rat model of asthma. BMC Complementary and Alternative Medicine, 19, 1–11.

[24] Vadnere, G. P., Gaud, R. S., Singhai, A. K. and Somani, R. S. (2009). Effect of Inula racemosa root extract on various aspects of asthma. Pharmacologyonline, 2, 84–94.

[25] Limbasiya, K. K., Modi, V. R., Tirgar, P. R., Desai, T. R. and Bhalodia, P. N. (2012). Evaluation of Anti asthmatic activity of dried whole plant extract of Leucas aspera using various experimental animal models. International Journal of Phytopharmacology, 3(3), 291–298.

[26] Suresh, S., Chhipa, A. S., Gupta, M., Lalotra, S., Sisodia, S. S., Baksi, R. and Nivsarkar, M. (2020). Phytochemical analysis and pharmacological evaluation of methanolic leaf extract of Moringa oleifera Lam. in ovalbumin induced allergic asthma. South African Journal of Botany, 130, 484–493.

[27] Vadnere, G. P., Gaud, R. S. and Singhai, A. K. (2008). Evaluation of anti-asthmatic property of Solanum xanthocarpum flower extracts. Pharmacologyonline, 1, 513–522.

[28] Nirmal, S. A., Patel, A. P., Bhawar, S. B. and Pattan, S. R. (2012). Antihistaminic and antiallergic actions of extracts of Solanum nigrum berries: Possible role in the treatment of asthma. Journal of Ethnopharmacology, 142(1), 91–97.

[29] Inam, A., Shahzad, M., Shabbir, A., Shahid, H., Shahid, K. and Javeed, A. (2017). Carica papaya ameliorates allergic asthma via down regulation of IL-4, IL-5, eotaxin, TNF-α, NF-κB, and iNOS levels. Phytomedicine, 32, 1–7.

[30] Chen, X., Huang, Y., Feng, J., Jiang, X. F., Xiao, W. F. and Chen, X. X. (2014). Antioxidant and anti-inflammatory effects of Schisandra and Paeonia extracts in the treatment of asthma. Experimental and Therapeutic Medicine, 8(5), 1479–1483.

[31]   Lee, M. Y., Yuk, J. E., Kwon, O. K., Kim, H. S., Oh, S. R., Lee, H. K. and Ahn, K. S. (2010). Anti-inflammatory and anti-asthmatic effects of Viola mandshurica W. Becker (VM) ethanolic (EtOH) extract on airway inflammation in a mouse model of allergic asthma. Journal of Ethnopharmacology, 127(1), 159–164.

[32]   Asumeng Koffuor, G., Boye, A., Kyei, S., Ofori-Amoah, J., Akomanin Asiamah, E., Barku, A. . . . Kumi Awuku, A. (2016). Anti-asthmatic property and possible mode of activity of an ethanol leaf extract of Polyscias fruticosa. Pharmaceutical Biology, 54(8), 1354–1363.

[33]   Yang, H., Sun, W., Fan, Y. N., Li, S. Y., Yuan, J. Q., Zhang, Z. Q. . . . Hou, Q. (2021). Perilla leaf extract attenuates asthma airway inflammation by blocking the syk pathway. Mediators of Inflammation, 2021(1), 6611219.

[34]   Sim, J. H., Lee, H. S., Lee, S., Park, D. E., Oh, K., Hwang, K. A. . . . Kim, H. R. (2014). Anti-asthmatic activities of an ethanol extract of Aster yomena in an ovalbumin-induced murine asthma model. Journal of Medicinal Food, 17(5), 606–611.

[35]   Yang, E. J., Lee, J. S., Yun, C. Y., Kim, J. H., Kim, J. S., Kim, D. H. and Kim, I. S. (2008). Inhibitory effects of Duchesnea chrysantha extract on ovalbumin-induced lung inflammation in a mouse model of asthma. Journal of Ethnopharmacology, 118(1), 102–107.

[36]   Boskabady, M. H., Javan, H., Sajady, M. and Rakhshandeh, H. (2007). The possible prophylactic effect of Nigella sativa seed extract in asthmatic patients. Fundamental & Clinical Pharmacology, 21(5), 559–566.

[37]   Rosa, S. I. G., Rios-Santos, F., Balogun, S. O., De Almeida, D. A. T., Damazo, A. S., Da Cruz, T. C. D. . . . De Oliveira Martins, D. T. (2017). Hydroethanolic extract from Echinodorus scaber Rataj leaves inhibits inflammation in ovalbumin-induced allergic asthma. Journal of Ethnopharmacology, 203, 191–199.

[38]   Alavinezhad, A., Ghorani, V., Rajabi, O. and Boskabady, M. H. (2022). Zataria multiflora extract influenced asthmatic patients by improving respiratory symptoms, pulmonary function tests and lung inflammation. Journal of Ethnopharmacology, 285, 114888.

[39]   Shakeri, F., Soukhtanloo, M. and Boskabady, M. H. (2017). The effect of hydro-ethanolic extract of Curcuma longa rhizome and curcumin on total and differential WBC and serum oxidant, antioxidant biomarkers in rat model of asthma. Iranian Journal of Basic Medical Sciences, 20(2), 155.

[40]   Wu, X. N., Yu, C. H., Cai, W., Hua, J., Li, S. Q. and Wang, W. (2011). Protective effect of a polyphenolic rich extract from Magnolia officinalis bark on influenza virus-induced pneumonia in mice. Journal of Ethnopharmacology, 134(1), 191–194.

[41]   Hamdy, N. M. (2023). Effect of Moringa Oleifera Lam. Leaf extract in treating pneumonia. Egyptian Journal of Desert Research, 73(2), 423–442.

[42]   Liyuan, T., Lijun, Z., Wei, H., Meixuan, J., Man, Z., Zhihui, Y. . . . Yakun, W. (2022). Green synthesised CuNPs using Alhagi maurorum extract and its ability to amelioration of Mycoplasma pneumoniae infected pneumonia mice model. Journal of Experimental Nanoscience, 17(1), 585–598.

[43]   Kim, S. W., Jee, W., Park, S. M., Park, Y. R., Bae, H., Na, Y. C. . . . Jang, H. J. (2024). Anti-inflammatory Effect of Symplocos prunifolia extract in an in vitro model of acute Pneumonia. Plant Foods for Human Nutrition, 79(4), 893–900.

[44]   Poofery, J., Khaw-On, P., Subhawa, S., Sripanidkulchai, B., Tantraworasin, A., Saeteng, S. . . . Banjerdpongchai, R. (2020). Potential of Thai herbal extracts on lung cancer treatment by inducing apoptosis and synergizing chemotherapy. Molecules, 25(1), 231.

[45]   Alachkar, A. and Al Moustafa, A. E. (2011). Teucrium polium plant extract provokes significant cell death in human lung cancer cells. Health, 3(06), 366.

[46]   Vanajothi, R., Sudha, A., Manikandan, R., Rameshthangam, P. and Srinivasan, P. (2012). Luffa acutangula and Lippia nodiflora leaf extract induces growth inhibitory effect through induction of apoptosis on human lung cancer cell line. Biomedicine & Preventive Nutrition, 2(4), 287–293.

[47]   Yin, X., Zhou, J., Jie, C., Xing, D. and Zhang, Y. (2004). Anticancer activity and mechanism of Scutellaria barbata extract on human lung cancer cell line A549. Life Sciences, 75(18), 2233–2244.

[48]   Hsieh, Y. J., Huang, H. S., Leu, Y. L., Peng, K. C., Chang, C. J. and Chang, M. Y. (2016). Anticancer activity of Kalanchoe tubiflora extract against human lung cancer cells in vitro and in vivo. Environmental Toxicology, 31(11), 1663–1673.

[49]   Ye, Y. T., Zhong, W., Sun, P., Wang, D., Wang, C., Hu, L. M. and Qian, J. Q. (2017). Apoptosis induced by the methanol extract of Salvia miltiorrhiza Bunge in non-small cell lung cancer through PTEN-mediated inhibition of PI3K/Akt pathway. Journal of Ethnopharmacology, 200, 107–116.

[50]   Lim, S. L., Goh, Y. M., Noordin, M. M., Rahman, H. S., Othman, H. H., Bakar, N. A. A. and Mohamed, S. (2016). Morinda citrifolia edible leaf extract enhanced immune response against lung cancer. Food & Function, 7(2), 741–751.

[51]   Li, Y., Yang, F., Zheng, W., Hu, M., Wang, J., Ma, S. . . . Yin, W. (2016). Punica granatum (pomegranate) leaves extract induces apoptosis through mitochondrial intrinsic pathway and inhibits migration and invasion in non-small cell lung cancer in vitro. Biomedicine & Pharmacotherapy, 80, 227–235.

[52]   Al-Sheddi, E. S., Farshori, N. N., Al-Oqail, M. M., Musarrat, J., Al-Khedhairy, A. A. and Siddiqui, M. A. (2014). Cytotoxicity of Nigella sativa seed oil and extract against human lung cancer cell line. Asian Pacific Journal of Cancer Prevention, 15(2), 983–987.

[53]   Goda, M. S., Nafie, M. S., Awad, B. M., Abdel-Kader, M. S., Ibrahim, A. K., Badr, J. M. and Eltamany, E. E. (2021). In vitro and in vivo studies of anti-lung cancer activity of Artemesia judaica L. crude extract combined with LC-MS/MS metabolic profiling, docking simulation and HPLC-DAD quantification. Antioxidants, 11(1), 17.

[54]   Mohammad, P., Nosratollah, Z., Mohammad, R., Abbas, A. and Javad, R. (2010). The inhibitory effect of Curcuma longa extract on telomerase activity in A549 lung cancer cell line. African Journal of Biotechnology, 9(6), 912–919.

[55]   Biswas, D., Mathur, M., Bhargava, S., Malhotra, H. and Malhotra, B. (2018). Anticancer activity of root extracts in nonsmall cell lung cancer Asparagus racemosus A549 cells. Asian J Pharm Pharmacol, 4, 764–770.

[56]   Kim, S. K., Kim, Y. H., Park, S. and Cho, S. W. (2021). Organoid engineering with microfluidics and biomaterials for liver, lung disease, and cancer modeling. Acta Biomaterialia, 132, 37–51.

[57]   Zakaria, N. H., Saad, N., Che Abdullah, C. A. and Mohd. Esa, N. (2023). The Antiproliferative Effect of Chloroform Fraction of Eleutherine bulbosa (Mill.) Urb. on 2D-and 3D-Human Lung Cancer Cells (A549) Model. Pharmaceuticals, 16(7), 936.

[58]   Dehnoee, A., Javad Kalbasi, R., Zangeneh, M. M., Delnavazi, M. R. and Zangeneh, A. (2024). Characterization, anti-lung cancer activity, and cytotoxicity of bio-synthesized copper nanoparticles by Thymus fedtschenkoi leaf extract. Journal of Cluster Science, 35(3), 863–874.

[59]   González, A. S. C., Valencia, M. G., Cervantes-Villagrana, R. D., Zapata, A. B. and Cervantes-Villagrana, A. R. (2023). Cytotoxic and antitumor effects of the hydroalcoholic extract of tagetes erecta in lung cancer cells. Molecules, 28(20), 7055.

[60]   Magalhães, D. B., Castro, I., Lopes-Rodrigues, V., Pereira, J. M., Barros, L., Ferreira, I. C. . . . Vasconcelos, M. H. (2018). Melissa officinalis L. ethanolic extract inhibits the growth of a lung cancer cell line by interfering with the cell cycle and inducing apoptosis. Food & Function, 9(6), 3134–3142.

Amra Alispahić*, Emina Boškailo, Alema Dedić,
and Hurija Džudžević-Čančar

Chapter 11
Medicinal and aromatic plants with antioxidant properties

Abstract: Reactive oxygen and nitrogen species (ROS/RNS) are two substances that are both naturally produced in the human body. They are required for the delivery of an oxidative burst to immune cells to kill microorganisms. However, their overproduction leads to several detrimental processes, including aging and cancer. Substances that eliminate the effects of free radicals are called antioxidants. Increased plant intake can be beneficial here because plants contain numerous natural antioxidants, mostly polyphenolics and flavonoids. These biologically active components in herbal essential oils have been used as therapeutic agents, as they are natural sources of antioxidants. They inactivate free radicals, reduce oxidative stress, and have been used in the pharmaceutical, cosmetic, and food research fields. In fact, studies have shown that there is a positive relationship between the total phenol content of medicinal plants and aromatic plants and their antioxidant capacity. Today, reliability concerns on synthetic antioxidants are increasing. Therefore, the interest of the health and food industry in aromatic plants and the natural antioxidants obtained from these plants has also increased. Aromatic and medicinal plants have been used in many fields, such as food, medicine, cosmetics, and spices since the beginning of human history. This chapter focuses on the antioxidant properties of medicinal and aromatic plants, as aromatic plants are widely considered to be rich sources of antioxidants.

Keywords: aromatic plants, medicinal plants, free radicals, antioxidants

11.1 Introduction

The human body converts at least 5% of the oxygen intake by breathing into reactive oxygen species (ROS). It has become clear in recent decades that ROS can have detri-

---

*Corresponding author: Amra Alispahić**, Department of Chemistry in Pharmacy, University of Sarajevo-Faculty of Pharmacy, Sarajevo, Bosnia and Herzegovina, e-mail: amra.alispahic@ffsa.unsa.ba
Emina Boškailo, Department of Ecology and Environmental Protection, Faculty of Social Sciences Dr. Milenko Brkić, Herzegovina University, Mostar, Bosnia and Herzegovina; International Society of Engineering Science and Technology, Nottingham, UK
Alema Dedić, Hurija Džudžević-Čančar, Department of Chemistry in Pharmacy, University of Sarajevo-Faculty of Pharmacy, Sarajevo, Bosnia and Herzegovina

https://doi.org/10.1515/9783111469713-011

mental impacts on human health under some circumstances. By neutralizing the impacts of free radicals, antioxidants can help the body protect itself from a variety of harmful factors and illnesses. Recent research on antioxidants and free radicals has opened a new era of health management against a number of illnesses [1]. Recent advancements in pharmaceuticals and functional foods derived from medicinal and nutritional plants (fruits and vegetables) have improved all facets of life, including extending human lifespan, reducing the need for synthetic antibiotics, and alleviating physical disorders [2]. To prevent oxidative deterioration and prolong food storage, the food industry mostly uses synthetic antioxidants such as propyl gallates, butyl hydroxytoluene, tertiary butyl hydroxyquinone, and butyl hydroxy anisole [3]. Despite their excellent efficacy, stability, and affordability, these synthetic antioxidants may have mutagenic, carcinogenic, and teratogenic adverse effects [3].

Free radicals are produced by a variety of external or environmental factors during the course of regular metabolic processes in both humans and animals. ROS are unstable and have a short lifetime. They may therefore readily interact with a wide range of biological components found in plant, animal, and human organism, including proteins, lipids, carbohydrates, and nucleic acids. They consequently contribute to the emergence of numerous illnesses and conditions (cancer, liver, deficiency in immune system, etc.), particularly aging in people [4].

The industry has been searching for natural sources of antioxidants due to consumer preferences and the understanding that antioxidants eliminate free radicals. As a result, medicinal and aromatic plants have gained increasing importance. Since plants are the primary natural source of antioxidants, plants, essential oils, and plant extracts are regarded as significant antioxidizing agents [5]. Flavonoids, coumarins, tocopherols, phenolic acids, cinnamic acid, and other phenolic and polyphenolic compounds are the most prevalent types of natural antioxidants. Furthermore, flavonoids and polyphenolic antioxidants derived from natural plants are efficient in defending against the free radicals that are produced within the human body, which is of great importance [6].

Under normal circumstances, the body's powerful antioxidant defense system prevents the harm that oxygen radicals inflict. According to studies, certain phenolic antioxidants can stop or prevent oxidative stress-induced cell death. Numerous aromatic and therapeutic plants from various plant groups have been shown to contain high levels of phenolic and flavonoid compounds [7].

Some medicinal and aromatic plants that have economic and industrial value in some countries of the world are produced using tissue or cell techniques under controlled conditions, as an alternative method, because of limited production quantities, challenges in obtaining products of standard quality, distance of production areas from industrial areas, and high extraction and purification costs [8].

In artificial nutritional settings and aseptic conditions, micropropagation is the process of creating new plants from plant parts (seed, leaf, root, stem, sprout, embryo, callus, etc.) that have the capacity to produce a whole plant [9]. Numerous plant sources have been evaluated for antioxidants due to growing interest in finding natural alternatives to synthetic antioxidants. The assessment of antioxidant activity is made more difficult by nutraceuticals that double the effect of natural antioxidants that stabilize foods and optimize health benefits. As a result, phytomedicine is gaining popularity again, and numerous species of medicinal plants are currently being evaluated for their pharmacological potential [9]. These plants, particularly those that contain high levels of phenolic components such as phenolic acids, flavonoids, tannins, stilbenes, and anthocyanins, have long been employed as sustainable, safe, and effective natural antioxidants or free radical scavengers [10]. These phenols are primarily thought to support the antioxidant activity of food and medicinal plants, which helps organism to fight a variety of pathological ailments like diabetes, cancer, aging, cardiovascular disease, and other degenerative diseases [10]. The food industry has been using aromatic and medicinal herbs for a long time, which serve a variety of functions, to enhance the flavor and taste of food products. Their aromatic properties also play a particularly important role. There are several plants that are rich in compounds, known to have antioxidant properties. Furthermore, plants also possess various antimicrobial and antiviral properties that vary, depending on the plant species, the type of microorganisms, and the concentration of essential oil in the plant [11, 12].

11.2 Oxidative stress

Oxidative stress is a global concept in biology, medicine, biochemistry, and nutritional science. Firstly, the term "oxidative stress" was introduced and explained in 1985, and since then, it has received significant attention through various research, especially in recent years [13]. The occurrence of oxidative stress is directly related to oxygen, one of the most abundant elements on the Earth, without which there is no life, but which can still be toxic to living beings, in certain states.

Oxidation and reduction reactions in living systems form the basis for numerous biochemical metabolic processes. One of the most important oxidation processes is the respiratory chain, which occurs in the mitochondria of eukaryotic cells. During this process, carbohydrate molecules, such as glucose, are oxidized to CO_2 and water, and the energy released during oxidation is used to create adenosine triphosphate (ATP), which is then use as an energy source in cells. This is where oxygen plays a key role in mammals, since it is the final electron acceptor in mitochondrial electron transport. However, during this process, toxic metabolites of ROS are also produced, which if they leave the mitochondria, cause cellular damage through the oxidation of biological molecules in the cytoplasm [13].

Oxidation process is the loss of one or more electrons from an atom, while reduction process is the acceptance of one or more electrons in an atom. A reductant (antioxidant) is a substance that donates electrons, while an oxidant is a substance that accepts electrons [14]. An imbalance in cellular redox processes toward oxidation or an excessive production of free radicals is known as oxidative stress [15]. Individual characteristics, including genetics, gender, age, lifestyle, habits, and most importantly, diet, affect the body's defense mechanisms' capacity to reduce oxidative stress and enhance oxidative state. Many cells can withstand oxidative stress, and in certain cases, it is essential for their function (e.g., endothelium, lung, and blood cells) [16].

A class of extremely reactive chemical entities containing one or more unpaired electrons in the outer shell is known as free radicals [14]. In addition to interacting with other radicals, free radicals can also interact with non-radical molecules by removing or accepting electrons. While a chain reaction of generated radicals happens with non-radicals, resulting in oxidative stress, the first scenario involves a radical-radical reaction that ends without producing oxidative stress or cell damage. Although many chemical species include unpaired electrons, molecules, and chemical species that contain carbon, nitrogen, and oxygen perform the most significant roles in the human body [15]. ROS are among the most significant subgroups of extremely reactive chemical species. More than a hundred human diseases, including atherosclerosis, arthritis, ischemia, disorders of the central nervous system, gastritis, cancer, and AIDS, are proved to be caused by free radicals [17]. Environmental pollution, radiation, chemicals, poisons, deep-fried and very spicy foods, and physical stress all produce free radicals, which weaken the immune system, alter gene expression, and produce aberrant proteins. Natural antioxidants may be required as free radical scavengers because of the immune system's fatigue in a number of diseases [18].

11.2.1 Reactive oxygen species

ROS are normally generated in essential physiological processes in biological organisms. Although ROS have an adverse effect on the organism, they also play an important role in the mechanisms of cellular repair and regeneration (e.g., in apoptosis), and can be secondary messengers, signaling molecules, and catalysts for the modulation of protein structures. They also participate in wound healing and immune response, and in the mobilization of cellular transport systems. Whether ROS will serve as a beneficial biological agent or as an initiator of oxidative damage depends on the balance of ROS generation and scavenging reactions [19]. ROS include both radicals and non-radical oxygen derivatives. The most important radicals of physiological and pathophysiological processes in humans include superoxide radical ($O_2^{\bullet-}$), hydroxyl radical ($OH^{\bullet}$), and hydroperoxyl radical ($HOO^{\bullet}$), while non-radicals include hydrogen peroxide (H_2O_2), singlet oxygen ($^1O_2$), and ozone (O_3) (Figure 11.1) [19].

In addition to ROS, other free radicals are lipid radical ($L^{\bullet}$), lipid peroxyl radical ($LOO^{\bullet}$), lipid alkyl radical ($LO^{\bullet}$), and protein radical ($P^{\bullet}$). Reactive nitrogen species (RNS) are nitrogen dioxide ($NO_2^{\bullet}$), nitrogen oxide ($NO^{\bullet}$), and peroxynitrite ($ONOO^-$). Non-radicals include lipid hydroperoxide ($LOOH$), iron-oxygen complex ($Fe=O$), and hypochlorite ($HOCl$) [15].

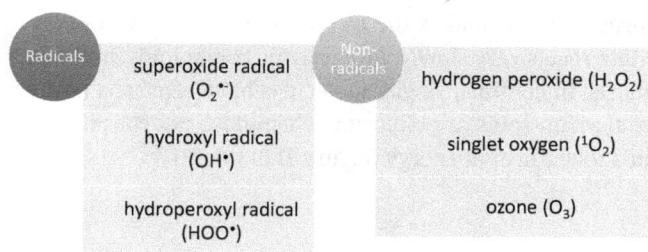

Figure 11.1: Reactive oxygen species (ROS).

Overproduction of ROS causes oxidative damage to biomolecules like DNA, protein modification, and lipid peroxidation, which can contribute to the development of diseases because of increased apoptosis; ischemia damage to muscle tissue, necrosis, inflammation; and also lead to insulin resistance [19]. Diseases associated with oxidative stress include cardiovascular diseases, atherosclerosis, diabetes and similar endocrine diseases, neurodegenerative diseases such as Parkinson's and Alzheimer's diseases, cancer, gastrointestinal diseases, etc. ROS significantly affect human ontogenesis, as well as the aging process itself [19, 20]. $HO^{\bullet}$ is considered the most reactive ROS and is responsible for many pathological processes. However, its half-life is very short (10 ns), which means that it reacts only with molecules that are in the immediate vicinity of the site where it was generated. On the other hand, the half-life of $^{1}O_2$ in aqueous solution is approximately 4 μs, which allows it to diffuse over a distance of 150–220 nm. Therefore, $^{1}O_2$ can react at various sites outside of its site of origin, allowing it to affect surrounding molecules and organelles more widely than $HO^{\bullet}$. However, this distance is not sufficient for extracellularly generated $^{1}O_2$ to penetrate the cell interior. Therefore, $^{1}O_2$ generated inside the cell can damage various cellular components, including DNA and organelles [21].

11.2.2 Sources and generation of free radicals

Free radicals can originate from both external (exogenous) and internal (endogenous) sources. Internal sources include phagocytes, xanthine oxidase, mitochondria, arachidonic acid pathways, ischemia/reperfusion, exercise, inflammation, xanthine oxidase, and processes involving iron and other transition metals. However, external sources include things like cigarette smoke, toxins in the environment, radiation, UV light, some medications, pesticides, anesthetics, industrial solvents, and ozone [22].

Oxygen is an element, whose molecule at temperatures compatible for life is in the lowest energy state, the triplet state. In order to more easily enter into chemical reactions, it must first pass into the singlet state by absorbing energy [19]. In the outer π nonbonding orbital, oxygen has two unpaired electrons (biradical) with equal spins, which is a paramagnetic property that makes the oxygen molecule more reactive. This is why oxygen in this state reacts very slowly with organic compounds, but reacts extremely quickly with radicals. In contrast, singlet molecules have electrons in their orbitals in pairs, where the electron spins are opposite, which does not cause a magnetic moment and results in a lower orbital energy (Figure 11.2) [19, 23].

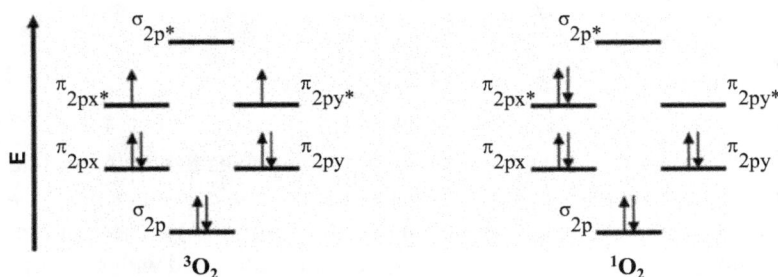

Figure 11.2: Triplet and singlet state of oxygen (electronic configuration).

Singlet oxygen therefore readily reacts with and oxidizes most organic compounds. The state in which the electrons are paired is more energetically rich in the case of singlet oxygen. If one electron is received, triplet oxygen can change to superoxide radical [23].

In mammalian cells, about 95% of oxygen is metabolized to water by a tetravalent reaction:

$$O_2 + 4H^+ + 4e^- \rightarrow 2H_2O$$

However, at least 5% of oxygen undergoes a gradual one-electron transfer reduction, whereby free radicals are formed as intermediates:

Reaction 1: $O_2 + e \rightarrow O_2^{\bullet-}$

Reaction 2: $O_2^{\bullet-} + e \rightarrow H_2O_2$

Reaction 3: $H_2O_2 + e \rightarrow {}^{\bullet}OH$

Reaction 4: ${}^{\bullet}OH + e \rightarrow H_2O$

The given reactions explain the formation of ROS during cellular respiration, which takes place in the mitochondria. Under normal conditions, reactive intermediates do not leave this process until the end of the reaction, but in certain pathophysiological conditions, ROS may leave this complex and initiate oxidative damage [14].

11.3 Consequences of oxidative stress on human organism

11.3.1 Lipid peroxidation

The two most prevalent ROS that have the ability to profoundly impact lipids are hydroperoxyl and hydroxyl radicals. The smallest, most mobile, water-soluble, and chemically most reactive kind of ROS is the hydroxyl radical. This short-lived molecule can be created from O_2 during cellular metabolism and under a variety of stressors. About 50 hydroxyl radicals are produced by a cell every second. Each cell produces over 4 million hydroxyl radicals in a day, which can either attack or neutralize biomolecules [24]. The development of atherogenesis in blood arteries, as a result of oxidative stress caused by lipid peroxidation, raises the risk of heart attack. At the systemic level, ROS play an active role in cardiovascular contractility, hemostasis, angiogenesis, immunological and cognitive control, blood pressure regulation, and platelet activation, in response to injury [23]. Intrinsic or dietary phospholipids reside in the circulation as stable antioxidant-conjugate lipoproteins, which are selectively oxidized through a complex series of enzymatic and nonenzymatic pathways, resulting in unstable lipoids in serum and cell plasma. Although these oxidized metabolites play their part in signaling and cellular metabolism, their large amounts can cause atherogenic problems. Cells also include opposing processes to preserve redox balance, such as antioxidants and stabilizers. But in conditions like atherosclerotic cardiovascular disease, the redox balance changes in favor of pro-inflammatory mechanisms, which either directly or indirectly inhibits the antioxidant activity and causes the system to produce more free radicals [25]. Three steps make up the entire lipid peroxidation process: start, propagation, and termination. Prooxidants such hydroxyl radical abstract allylic hydrogen at the first stage of lipid peroxidation, creating a carbon-centered lipid radical (L'). Lipid hydroperoxide (LOOH) and a freshly formed L' (which restarts the chain reaction) are produced when the lipid radical (L') quickly combines with oxygen to form a lipid peroxyl radical (LOO'), which extracts hydrogen from another lipid molecule. Antioxidants like vitamin E provide the LOO' species a hydrogen atom in the termination reaction, creating a matching vitamin E radical that then combines with another LOO' to create non-radical products. Following the initiation of lipid peroxidation, chain reactions will continue to spread until the termination products are produced [24].

11.3.2 Protein oxidation

The two main types of oxidative protein changes are irreversible oxidation and reversible oxidation, which can be specifically brought on by RNS and ROS. Protein carbonyls and 3-nitrotyrosine are involved in irreversible oxidation, whereas products of cysteine modification, including sulfonic acid, nitroso thiols, and *S*-glutathione, are involved in reversible oxidation. Arginine, histidine, lysine, proline, threonine, and cysteine are among the amino acid residues that generate protein carbonyls, which are frequently employed as biomarkers to quantify protein oxidation and oxidative stress in aging and disease states [26].

RNS and a protein's tyrosine residue combine to generate nitrotyrosine, typically 3-nitrotyrosine. Glutathione (GSH), which, although a primary cellular antioxidant, can also affect proteins through the creation of mixed disulfides, leading to functional changes in target proteins. This can shield the target protein from permanent and irreversible harm, but it is also frequently linked to negative consequences on target protein function. This implies that some protein oxidation processes can still be advantageous and reduce cellular oxidative damage [26].

11.3.3 DNA oxidation

Nucleic acids and 2-deoxyribose are also targets of ROS and other reactive chemical species, resulting in DNA damage manifested through mutagenesis and carcinogenesis. The most frequent ROS that causes oxidation at the DNA level is the hydroxyl radical [27]. The most prevalent and well-characterized indicators for oxidative DNA and RNA lesions among the several nucleoside oxidation products are the guanosine oxidation products 8-hydroxyguanosine (8-OHG) and 8-oxo-7,8-dihydroguanosine (8-oxoG). Despite being less deadly to cells than genetic changes, oxidative RNA damage is linked to a number of age-related illnesses, including cancer, type 2 diabetes, and neuropsychiatric disorders. Certain studies suggest that patients suffering from schizophrenia or depression have increased concentrations of oxidatively damaged RNA in their urine. Although cells possess DNA and RNA repair mechanisms, if oxidative stress exceeds the repair capacity, damage accumulation occurs in the form of base mismatches, abnormal cell signaling, and the synthesis of irregular protein structures [27].

11.4 Defense of the organism against ROS

The human body has several defenses against ROS, including free radicals. Because they target distinct oxidants or act in separate areas of the cells, these systems work in concert. A system of enzymes that lower the concentration of the most harmful oxidants, including glutathione peroxidases, superoxide dismutase, and catalase, is one of the main processes. Since catalase breaks down hydrogen peroxide, superoxide dismutase is especially crucial because it catalyzes the transformation of superoxide radicals into hydrogen peroxide and oxygen. Selenium-containing glutathione peroxidases are crucial for reducing hydroperoxides, particularly those produced by lipid oxidation [28].

Maintaining enzymatic defense against free radicals is mostly dependent on nutrition. The structure and function of the aforementioned enzymes depend on minerals like zinc, copper, manganese, and selenium. Enzymatic defenses may be less efficient if certain minerals are deficient [28]. Small molecules that function as antioxidants, reacting with oxidizing chemicals to reduce their harmfulness, are the second line of defense. Normal metabolism produces some of these antioxidants, including ubiquinol, glutathione, and uric acid. The only known fat-soluble antioxidant made by animal cells is ubiquinol, which is crucial for shielding cells from oxidative damage. Vitamins E and C are examples of antioxidants that are present in diet. Plant pigments also include certain antioxidants. One example of a carotenoids is vitamin A, which is a vital component of the receptors for eyesight and crucial for healthy embryonic development, making it an essential nutrient in the human body. Antioxidants have no caloric value in the diet [29].

11.4.1 Free radicals and antioxidants

ROS and RNS are the two main components of free radicals, which are byproducts of many biological activities. ROS are produced in the human body as a result of a variety of environmental factors, xenobiotics, and human-caused causes that change an organism's biological activity [30, 31]. ROS (hydroxyl, superoxide, peroxyl, and hydrogen peroxide) can cause damage to biomolecules (lipids, proteins, enzymes, and nucleic acids) and can cause a variety of disorders if they are produced in excess of normal physiological levels [32]. Naturally occurring in people, animals, and plants, antioxidants shield cells from the damaging effects of free radicals [33]. Antioxidants scavenge free radicals and prevent organisms from producing them in excess [34].

There are two main classes of antioxidants [35]:
– Natural antioxidants
– Synthetic antioxidants

Fruits and vegetables typically include natural antioxidants such as vitamin C (ascorbic acid), vitamin E (tocopherols and tocotrienols), carotenoids, and polyphenols. Among the tocopherols and tocotrienols found in vitamin E, α-tocopherol has been the subject of the most research. Functionally, α-tocopherol is more active, acting against peroxyl radicals and quenching singlet oxygen (Figure 11.3) [36]. Because it contains the enediol group [37], vitamin C is a well-known natural antioxidant that has the ability to scavenge ROS. There are around 700 known naturally occurring carotenoids from plants that have antioxidant properties [36]. The most prevalent polyphenolics in plants include flavonoids, stilbenes, phenolic acids, and lignans [38]. On the other hand, flavonoids are strong metal chelators and scavengers of free radicals [39]. Numerous artificial antioxidants have been incorporated into a broad range of dietary items and cosmetics. However, overuse of synthetic antioxidants may result in mutagenicities and

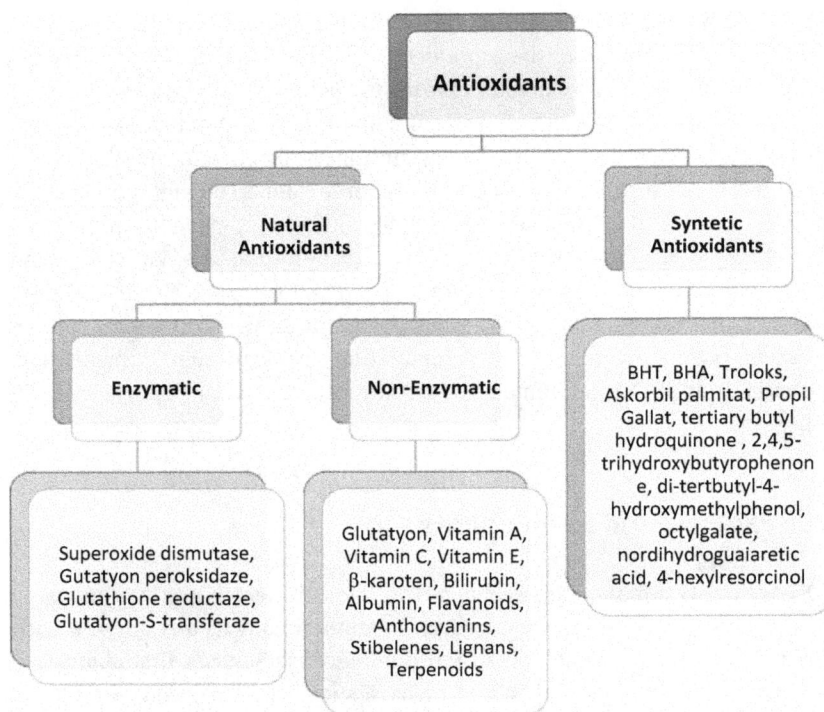

Figure 11.3: Classification of antioxidants.

toxicities, which could be detrimental to health [40]. Nonetheless, a large range of natural antioxidants have distinct characteristics, such as their components, modes of action, and target locations [41]. The ability of plants and animals to naturally produce proteins, enzymes, and secondary metabolites is one of the primary characteristics of natural antioxidant enzymes. Antioxidative enzymes, such as catalase (CAT), glutathione peroxidase (GPx), superoxide dismutase (SOD), and others, biocatalysis metabolic pathways by converting ROS and RNS into stable compounds [42]. High-molecular-weight substances that prevent metal production, catalyzed by free radicals, include albumin, transferrin, and ceruloplasmin [43]. Water-soluble antioxidants and lipid-soluble antioxidants are the two subcategories of low-molecular-weight molecules. Ascorbic acid, uric acid, and certain polyphenols are water-soluble antioxidants, while tocopherol, quinines, carotenoids, bilirubin, and certain polyphenols are lipid-soluble antioxidants [44]. The antioxidative qualities of minerals and micronutrients such as manganese, copper, zinc, and selenium, among others, have been extensively established [45]. Well-known stable antioxidants, vitamins A, C, and E are crucial in reducing the possibility of peroxidation-induced damage to the biological system. Vitamin C removes various types of radicals, such as $OH^{\bullet-}$, H_2O_2 and $O_2^{\bullet-}$. Vitamin C donates one electron to $O_2^{\bullet-}$ to produce $ASC^{\bullet-}$ (ascorbyl radical anion) or loses an electron to form its oxidized form, DHA (dehydroascorbic acid) [46, 47]. Natural antioxidants are mostly found in fruits, vegetables, and medicinal plants. Consumers have recently become quite interested in new spices and herbs as natural antioxidant sources, and some of these have been discussed here [48].

11.4.2 Antioxidants action mechanism

Antioxidants have varying mechanisms of action. The reactivity of substances with free radicals produced during the lipid oxidation process, which results in the formation of inert molecules, is the most significant mechanism of antioxidant activity. As real antioxidants, these substances typically react with alkoxyl or peroxyl free radicals that are produced during the breakdown of lipid hydroperoxides [49]. Lipid peroxides are stabilized by other antioxidants, which stop them from breaking down into free radicals. For antioxidants, two primary mechanisms of action have been proposed. The major antioxidant neutralizes free radicals by donating an electron through the first method, known as chain breaking. The second preventative mechanism eliminates ROS/RNS initiators by quenching the chain initiation step through the action of secondary antioxidants (Table 11.1) [49].

Table 11.1: Mechanism of antioxidant activity [49, 50].

Antioxidant class	Mechanism of action	Examples of antioxidants
Real antioxidants	Inactivation of free lipid radicals	Phenolic compounds
Hydroperoxide stabilizers	Preventing the decomposition of hydroperoxide into free radicals	Phenolic compounds
Synergists	Strengthening the activity of real antioxidants	Citric acid and ascorbic acid
Metal chelators	Binding of heavy metals into inactive components	Phosphoric, ascorbic and citric acid
"Quenchers" or singlet oxygen extinguishers	Transfer of singlet oxygen to triplet oxygen	Carotenoids (beta-carotene, lycopene, and lutein)
Substances that reduce hydroperoxides	Reduction of hydroperoxide in a non-radical way	Proteins and amino acids

Antioxidants, capable of neutralizing free radicals, act at different levels of defense such as prevention, radical scavenging, repair and adaptation. The first line of defense consists of superoxide dismutase, catalase, glutathione reductase, glutathione peroxidase, selenoprotein, transferrin, lactoferrin, ferritin, some minerals Mn, Zn, Cu, and Se, and non-enzymatic proteins, which are considered preventive antioxidants and limit the formation of free radicals. Superoxide dismutase converts superoxide radical (O_2^-)

into hydrogen peroxide (H_2O_2). Catalase catalyzes the decomposition of hydrogen peroxide (H_2O_2) into water (H_2O) and molecular oxygen (O_2). Glutathione peroxidase is a selenium-dependent enzyme that detoxifies lipid hydroperoxides to alcohols. Cytosolic superoxide dismutase is a Cu-containing enzyme that removes superoxide radicals from the cytosol. Selenium is an essential element for the removal of peroxides from the cytosol and cell membranes. Zinc is a component of several enzymes such as alcohol dehydrogenase, carbonic anhydrase, alkaline phosphatase, and cytosolic superoxide dismutase, and also plays an important role in growth and reproduction [38].

The second line of defense includes glutathione (GSH), vitamin E, vitamin C, uric acid, bilirubin, albumin, carotenoids, and flavonoids, which have radical scavenging activity. Glutathione scavenges ROS such as lipid peroxyl radical, peroxynitrite, and hydrogen peroxide. It also helps in detoxification of inhaled oxidizing air pollutants. Vitamin E protects polyunsaturated fatty acid and low-density lipoproteins by scavenging peroxyl radical intermediates generated in lipid peroxidation reactions. It prevents coronary heart disease and atherosclerosis. Vitamin C quenches radicals such as singlet oxygen, superoxide radical, and hydroxyl radical. β-carotene helps in scavenging singlet oxygen. Flavonoids inhibit lipoxygenase and lipid peroxidation. The third line of defense includes a group of enzymes required for the repair mechanism of damaged DNA, proteins, and lipids. These enzymes are capable of stopping the chain propagation of the lipid peroxyl radical, for example, DNA repair enzymes, proteases, lipases, transferases, and methionine sulfoxide reductase. The fourth line of defense is an adaptation in which immunology plays an important role in the production and reaction of free radicals with appropriate antioxidants [38, 50]. Flavonoids and polyphenols achieve their antioxidant effect in several ways, including direct scavenging and scavenging of free radicals, reduction of leukocyte immobilization, and regulation of nitric oxide and xanthine oxidase activities. Several flavonoids, including quercetin, reduce ischemia-reperfusion injury by interfering with the inducible activity of nitric oxide synthase. Nitric oxide itself can be considered as a radical that is directly scavenged by flavonoids. Therefore, it is assumed that the scavenging of nitric oxide plays a role in the therapeutic effects of flavonoids. The significant effects of polyphenols are the result of radical scavenging, but another possible mechanism of action is the interaction with various enzyme systems such as superoxide dismutase, catalase, and glutathione peroxidase [38]. Furthermore, *in vitro* studies have shown antiproliferative activity of polyphenols through inhibition of polyamine biosynthesis and signal transduction enzymes such as protein tyrosine kinase, protein kinase C and phosphoinositide 3-kinase, induction of apoptosis and cell cycle arrest in the G1/G2 phase, differentiation of transformed cells, and rehabilitation of cellular homeostasis [51].

11.5 Methods for determination of antioxidative activity

Methods for measuring antioxidant activity in plant extracts, food, and biological systems can be divided into several ways according to:
1. test system (*in vivo* and *in vitro*),
2. detection method (spectrophotometric, fluorimetric, and chemiluminescent),
3. directness of determination (direct and indirect),
4. presence of lipids in the system (the degree of inhibition of lipid substrate oxidation and measurement of the antioxidant capacity of free radicals in systems that do not contain lipids), and
5. reaction mechanism (methods based on hydrogen atom transfer (HAT) reactions and methods based on electron transfer reactions that take place between antioxidant compounds and free radicals) [51].

11.5.1 Methods based on hydrogen atom transfer

Methods based on HAT are based on a reaction in which the antioxidant and the substrate compete for peroxyl radicals created by the terminal decomposition of the azo component, and the result is obtained on the basis of a kinetic curve. The methods are composed of synthetically produced free radicals, antioxidants, and oxidants. In these methods, the hydrogen atom donating capacity of the antioxidant is measured [52].

This group of methods includes:
- IOU method (English inhibited oxygen uptake method)
- Inhibition of induced lipid autooxidation
- TRAP method (total radical trapping antioxidant parameter)
- ORAC method (oxygen radical absorbance capacity assay)
- CBA method (crocin bleaching assay)
- Fluo-lip
- HORAC method (hydroxyl radical antioxidant capacity assay)

11.5.2 Methods based on electron transfer

Methods based on the transfer of one electron are based on a redox reaction with an antioxidant as an indicator of the end point of the reaction. This method includes two components in its reaction mixture: oxidant and antioxidant [52]. They are based on the following electron transfer reaction:

oxidant + e^- (from antioxidant) → reduced oxidant + oxidized antioxidant

The oxidant receives an electron from the antioxidant, which results in a color change of the oxidant. The intensity of the color change is proportional to the concentration of the antioxidant. The end point of the reaction is reached when the color change ceases. After that, the direction of the change in absorbance as a function of the antioxidant concentration is plotted.

This group of methods includes:
- FCR method (total phenols assay by Folin-Ciocalteu reagent)
- FRAP method (ferric reducing antioxidant potential)
- DPPH method
- TEAC method (Trolox equivalent antioxidant capacity)
- Determination of antioxidant potential by reduction with copper [52].

11.5.3 Other methods for determination of antioxidant potential

- TBARS method (thiobarbituric acid reactive substance): In the process of lipid peroxidation, lipid peroxides are formed with the subsequent formation of peroxyl radicals, and the entire process is followed by a decomposition phase in which aldehydes such as hexanal, malondialdehyde, and 4-hydroxynonenal are formed. This method is based on the detection of a stable pink-colored product formed in the reaction between aldehyde and thiobarbituric acid in the aqueous phase. The concentration of the reaction product is monitored spectrophotometrically and provides data on the strength of lipid peroxidation.
- CBT method (β-carotene bleaching test): This method is based on the loss of the β-carotene yellow color during the reaction with free radicals, which are formed by the oxidation process of linoleic acid. The presence of antioxidants slows down the process of β-carotene decolorization. The reaction is monitored spectrophotometrically [53].
- TOSC method (total oxidant scavenging capacity): This method enables determination of the antioxidant potential, specifically according to three oxidants: hydroxyl radical, peroxyl radical, and peroxynitrite. As a substrate that is oxidized, α-keto-γ-methylbutyric acid is used, which forms ethylene. The time of ethylene formation is monitored by the gas chromatography method, and the antioxidant potential is determined based on the antioxidant's ability to inhibit ethylene formation.
- PLC method (photo-chemiluminescence method): PLC is based on a thousand-fold acceleration of the oxidation reaction *in vitro* compared to normal conditions. This effect is achieved by optical excitation of a suitable photosensitizer, which results in the formation of superoxide radicals. The radical is detected with the chemiluminescent reagent luminol. Luminol acts as a photosensitizer, but it also participates in the reaction with radicals. The light emission is measured with the help of a luminometer. Light emission occurs as a result of the oxidation of luminol with the catalytic action of peroxidase. In the presence of antioxidants, the

oxidation of luminol is prevented and light emission is also inhibited. The duration of inhibition indicates the quantity of antioxidants present [54].
- Biosensor methods: These methods use the most common enzymatic electrodes based on superoxide dismutase, an enzyme used as a biosensor for the determination of superoxide radicals in aqueous and nonaqueous solutions for the determination of antioxidant potential [55].

11.6 Medicinal and aromatic plants as natural antioxidants

Aromatic and medicinal plants (AMPs) are plants with taste and smell qualities that are also utilized as medications because of their therapeutic properties. Since ancient times, people have utilized plants and their essential oils to treat certain medical conditions and enhance the flavor of food and drink. Their cultural and economic significance is demonstrated by their use to cover up offensive odors, draw attention from others, treat certain medical conditions, and benefit humans. The phenolic chemicals in AMPs' structure are linked to their antioxidant activity [56]. Flavonoids, phenolic acids, and phenolic terpenes are the most prevalent of these substances. By scavenging free radicals, forming compounds with metal ions (metal chelation), and preventing or lowering the generation of ROS, phenolic substances have an antioxidant effect [57, 58]. In order to stop free radicals from oxidizing lipids and other biological components, the compounds can supply hydrogen through hydroxyl groups in their aromatic rings. Plants' leaves, flowers, and woody parts are the primary sources of flavonoids and other phenolic chemicals. As a result, AMPs are frequently employed as essential oils or extracts made by extraction and distillation processes, or as medications made by drying sections of leaves and flowers. Since the chemical composition of aromatic plants varies, depending on many factors, their antioxidant effects will also vary (Figure 11.4) [59, 60].

Essential oils, which are made up of many chemical compounds, are volatile, aromatic, oily liquids that are extracted from plant materials such as leaves, roots, flowers, peels, bark, seeds, and twigs. These substances, which are secondary produced to shield plants from insects and microorganisms, have a potent odor. In addition to creating distinctive fragrances to draw pollinating insects, plants also create essential oils to protect themselves from unavoidable elements like sunshine, pollution, and hunger [61]. Medicinal plants continue to be a significant source of bioactive chemicals for drug development, many of which have served as the foundation for novel chemical structures in the food and pharmaceutical industries. The World Health Organization estimates that 80% of people worldwide still depend on herbal drugs, and few drugs are derived from medicinal plants.

Recently, there has been increasing interest in the therapeutic potential of medicinal plants as antioxidants in reducing tissue damage caused by free radicals. In addition

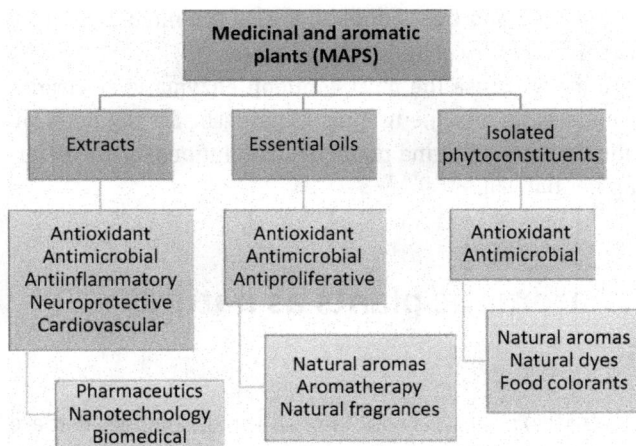

Figure 11.4: Few potential applications of medicinal and aromatic plants.

to the well-known and traditionally used natural antioxidants from tea, wine, fruits, vegetables, and spices, some natural antioxidants (e.g., rosemary and sage) are already commercially used either as antioxidant additives or as food supplements [62]. Many other plant species have been investigated for new antioxidants [63–65], but in general, there is a demand for more information on the antioxidant potential of different plant species. It is assumed that plants possess an antioxidant effect due to their phenolic compounds content [18]. In particular, despite the widespread use of wild plants as medicines, the literature contains few reports on the antioxidant activity and chemical composition of plants from different parts of the world. The relationship between total flavonoid content and total phenolic content and antioxidant activity is usually determined. In the long term, plant species (or their active ingredients) that have been found to have high levels of antioxidant activity *in vitro* may be valuable in the design of further studies to discover new treatment strategies for radical-induced disorders [66].

Spices as aromatic plants are very rich in antioxidants. Various metabolic products and their derivatives obtained from spices and aromatic plants have been identified as important antioxidants (Figure 11.5) [67]. A spice can be defined as a plant, the specific parts of which provide color and flavor, along with a stimulating odor, which is used in culinary and seasoning, as well as in cosmetics, fragrances, and medicines. These specific properties of herbs and spices have supported their application in functional foods for nutrients, bioactive compounds, disease prevention, and health promotion. The different parts of plants used as spices are rhizomes, leaves, buds, flowers, fruits, seeds, excretory products, and even tree bark [68]. Since a long time, plants have been used for almost all medical therapies until the development of synthetic drugs. Aromatic plants affect various systems of the body such as the cardiovascular, gastrointestinal, reproductive, and nervous systems [69]. All plant groups include common antioxidants, with a few exceptions. Spice and aromatic plants have been found to have some unique antioxidant chemicals. Rosmarinic acid is the dominant compound in some plants of the Lamiaceae family with four hydroxyl groups (catechol structures) in the structure, which is responsible for its antioxidant properties. Caffeic and gallic acids are also present in these plants and possess antioxidant activity due to the catechol structure [70].

Eugenol and its derivatives contain a phenolic group in the structure and have relatively lower antioxidant activity than other phenols with multiple hydroxyl groups. The phenolic group plays an important role in the free radical scavenging activity of eugenol. Eugenol, cumin aldehyde, curcumin, piperine, zingerone, and linalool have been reported as effective antioxidants. These compounds inhibit lipid peroxidation [71].

Recently, much attention has been focused on the development of less-toxic ethnomedicines and their potential application in the prophylaxis and treatment of various diseases. There is a lot of data on the antioxidant activity of essential oils extracted from plants of different species, and harvested in different places and different stages of plant development. Volatile essential oils and nonvolatile secondary metabolites

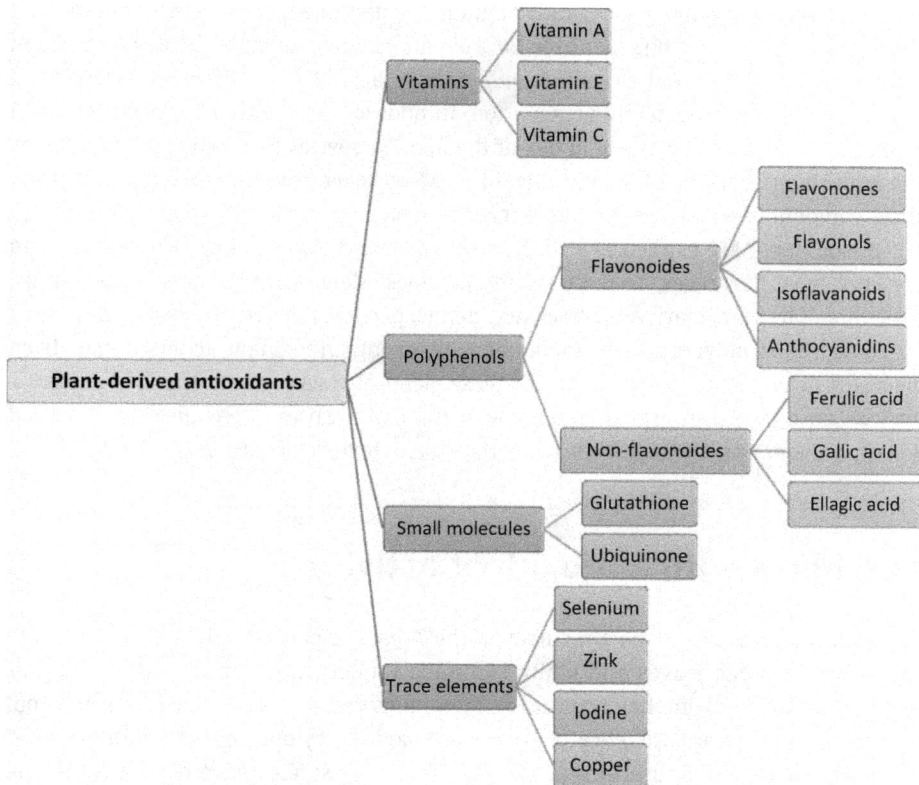

Figure 11.5: Examples of natural plant-derived antioxidants.

from plants have wide applications in food additives, flavorings and food preservation, folk medicine, and the fragrance industry [52]. Several reports have confirmed the antioxidant efficacy of plant-based essential oils *in vitro* and *in vivo* [72].

Chemical composition of these oils depends on several factors (age of the plant, part of the plant, developmental stage, growing site, harvest period, and chemotype), but the correlation between the biological activity (i.e., antioxidant activity) of essential oils and their chemical composition is often very complicated. Furthermore, due to the many different methods for *in vitro* assessment of antioxidant activity based on completely different mechanisms, the interpretation of the data is not straightforward. Consequently, the results on the antioxidant activity of essential oils from the same plant, reported in numerous studies, showed many variations [73, 74]. It is necessary to emphasize that there is no perfect system to assess the antioxidant activity of a single compound or a complex mixture. Differences in the analytical methods used and the measurement conditions may be responsible for such variations in the same samples [75].

Terpenes are the main constituents of essential oils extracted from medicinal plants, which are considered natural antioxidants. Essential oils of basil, cinnamon, cloves, nutmeg, oregano, and thyme possess antioxidant properties due to the presence of terpenes. Thymol and carvacrol are responsible for the antioxidant activity of the essential oils of *Thymus spathulifolius* and *Origanum vulgare* ssp., and *Melissa officinalis* essential oil shows free radical scavenging activity, too. In addition, isomenthone, 1,8-cineole, and menthone present in the essential oils of the *Mentha* species show antioxidant activity. The antioxidant capacity of the essential oil of *Melaleuca alternifolia* (tea tree) is a result of the compounds α-terpinene, γ-terpinene, and α-terpinolene [37, 76–78].

In addition, AMP extracts are widely used in most cultures to improve the taste and preserve food, beverages, cosmetics, and perfumes. Many MAP extracts (anise, fennel, basil, mint, tarragon, marjoram, rosemary, thyme, parsley, juniper, and bay leaf) serve as a rich source of polyphenolic compounds with strong antioxidant activity. Apart from their beneficial effects on human health, these plants also serve as natural food preservatives because they prevent oxidation, one of the main causes of chemical spoilage and deterioration of nutritional quality, color, flavor, and texture of various products [76].

11.7 MAPs with antioxidant activity

Angelica: *Angelica sinensis* is a member of the Apiaceae plant family. This hairy plant, which has fern-like leaves and white blooming umbels, was considered an "Angel's Herb". Chinese medicine has been using the plant's well-known yellowish-brown root for thousands of years. Angelica possesses antioxidant, cytoprotective, antimutagenic, antiproliferative, and antiseizure properties. Numerous studies have demonstrated the extremely strong antioxidant capacity of angelica root extracts. The antioxidant benefits of Angelica root extracts may generally be attributed to the presence of phenolic compounds, particularly ferulic acid and caffeic acid, as evidenced by the significant positive correlations found between total phenolic content and antioxidant potential [79]. According to *in vitro* antioxidant tests, at a concentration of 0.5 mg/mL of Angelica essential oil (EO), the scavenging ability of 2,2-azino-bis-3-ethyl-benzothiazoline-6-sulfonic acid (ABTS) and 1,1-diphenyl-2picrylhydrazyl (DPPH) for free radicals approached 100%. Furthermore, there was a concentration-dependent link between EO and the iron-reducing capacity. These findings show that Angelica EO has a strong antioxidant activity [80].

Anise: *Pimpinella anisum* is an annual herb belonging to the Apiaceae family of parsleys that is grown primarily for its aniseed fruits, which have a flavor similar to licorice. Home lands are Egypt, the eastern Mediterranean, southern Europe, the Middle East, North Africa, and America. Anise seed is a popular flavoring for many kinds of alcoholic beverages and is used extensively to flavor food in many parts of the world. According to reports, anise contains larvicidal, ovicidal, carminative, relaxing, antiviral, and antioxidant qualities. Flavonoids, terpenes, and essential oils are among the several substances that are known to be present in aniseed extracts. Several antioxidant tests, such as reducing power, free radical scavenging, superoxide anion radical scavenging, hydrogen peroxide scavenging, and metal chelating activities, were used to assess the antioxidant qualities of aniseed extracts. In the linoleic acid system, water and ethanol extracts at 20.00 µg/mL showed 99.1 and 77.5% inhibition of peroxidation, respectively, which was higher than the same dose of α-tocopherol (36.1%) [81]. The extract showed high antioxidant activity in another investigation using three methods: DPPH, ABTS, and iron-reducing power tests. With an IC_{50} value of 17.92 g/mL by the DPPH method, aniseed extracts demonstrated greater antioxidant activity than vitamin C as standard compound, which is considered to be caused by polyphenolic acids and flavonoids. Aniseed extract has encouraging antioxidant activity because of its high phenolic and flavonoid content [82].

Basil: *Ocimum basilicum*, also called great basil, is an aromatic herb of the family Lamiaceae. Although there are other types, the general name "basil" refers to the type commonly called Genovese basil or sweet basil. This sturdy, fragrant annual plant is grown extensively across Europe. Digestive, antifungal, antibacterial, antimelanoma, antimicrobial, radioprotective, anthelmintic, and antioxidant properties have been documented for basil. Numerous investigations looked into the various biological activities of basil essential oils and extracts. The study by Nadeem et al. (2022) [83] found that ethanol extracts had the highest concentration of plant secondary metabolites, such as total phenolic acid, flavonoids, and tannin content. In the DPPH, FRAP, and H_2O_2 tests, ethanol extracts likewise showed the best antioxidant activity. When compared to extracts from other locations, it was also found that the examined basil extracts had higher phenolic content (82.45 mg pyrocatechol equivalents (PE)/g) and antioxidant activity, as measured by the DPPH assay (IC_{50} 1.29 mg/mL). Using LC-ESI-MS/MS analysis, basil extracts were shown to be a prospective source of well-known medicinal and health-promoting substances such as umbelliferone, ellagic acid, rosmarinic acid, catechin, and liquiritigenin. According to the findings, basil is a powerful source of bioactive chemicals [83]. In another investigation, Minia basil extract demonstrated a high total phenolic content of 82.45 mg PE/g and radical scavenging activity with an IC_{50} value of 1.29 mg/mL. Antioxidant activity and the total phenolic content of basil extracts were shown to be highly correlated. There was a wide range in the essential oils' capacity to scavenge free radicals. Minia basil essential oil contained 41.3 mg PE/g of total phenolics and had strong DPPH radical scavenging activity, with an IC_{50} of 11.23 mg/mL [84]. Carbon dioxide was used to extract bioactive compounds from basil leaves in the study of Romano et al. [85]. The most effective technique was found to be extraction with supercritical CO_2 for two hours while utilizing 10% ethanol as a cosolvent. The extracts made using this method were tested for antioxidant activity, phenolic acid content, and volatile organic compounds. The ABTS assay revealed high antiradical activity with bergamotene (11–14%), linalool (35–27%), and caffeic acid (1.69–1.92 mg/g) contents that were significantly higher than the control. Besides enabling the production of extracts rich in bioactive compounds in an environmentally friendly way, supercritical CO_2 also reduced the need for ethanol and other solvents [85].

Bay/bay leaf: *Laurus nobilis*, sweet bay tree, an evergreen plant belonging to the Lauraceae family, is native to countries that border the Mediterranean. It can be used whole, dried, or fresh, in which case, it is taken out of the food before eating, or, less frequently, ground. Although the bay laurel (*Laurus nobilis*) is the most widely utilized source of bay leaves, they are used from a variety of species of this plant for their unique flavor and smell. The flavor and aroma of bay leaves are enhanced by the presence of essential oils that contain methyl eugenol, terpenes, and eucalyptol. In addition to its unique function as an insect repellent, bay leaf has been shown to possess antifungal, antibacterial, hypolipidemic, gastroprotective, digestive, antimicrobial, and antioxidant qualities. Using various antioxidant assays, numerous studies have documented the chemical composition and antioxidant properties of *Laurus nobilis* leaves. Terpenes including a-terpinol, a-terpinyl acetate, thymol, caryophyllene, selinene, farnesene, and cadinene. Eugenol and methyl eugenol, vitamin E, and sterols, are also known to be abundant in laurel. Extracts from bay leaves demonstrated strong radical scavenging activity, with an IC_{50} value of 1 mg/mL [86]. The antioxidant activity of bay leaf essential oil is lower than that of extracts. With IC_{50} values of 66.1 µg/mL, bay leaf essential oil demonstrated a scavenging effect on the DPPH radical. With an IC_{50} value of 35.6 µg/mL [87], the leaf EO exhibited the highest antioxidant activity in the β-carotene/linoleic acid system. The findings showed that *L. nobilis* EO has strong antioxidant activity, which may be because of its high 1,8-cineole content [87].

Clove: *Syzygium aromaticum* is an evergreen tropical tree from family Myrtaceae. It is indigenous to Indonesia and is frequently used as a flavoring, spice, or aroma in commercial goods like soaps, toothpaste, and makeup. Its increased polyphenol content is reported to be closely associated with its very strong antioxidant action. Antifungal, anesthetic, antiseptic, carminative, antispasmodic, antibacterial, antiviral, anti-inflammatory, and antioxidant qualities have all been documented for cloves. Eugenol and eugenyl acetate, the main components of clove buds' scent, exhibit antioxidant activity on par with that of natural antioxidants as α-tocopherol (vitamin E) [71]. The maximum antioxidant activity was demonstrated by TPC and TFC, which had contents of 247.61 and 141.70 mg/100 dry weight, respectively. Clove's antioxidant capacity was assessed by its total phenol content, total flavonoid content, ferric reducing antioxidant power (FRAP), and 2,2-diphenyl-1-picrylhydrazyl (DPPH). The maximum antioxidant activity in this investigation was demonstrated by the FRAP and DPPH methods (437.29 mg TE/100 g dry weight and 87.50%, respectively) [88]. Total phenolic content, free radical DPPH scavenging activity, hydrogen peroxide scavenging, and reducing power test were used in other studies to measure antioxidant activity. With a higher absorbance value at the highest concentration of methanolic extract (0.198 ± 0.001 A), ethanolic extract demonstrated the highest scavenging activity for free radicals (62.12%), followed by aqueous extract (48.32%). Additionally, methanolic extract demonstrated the highest scavenging activity for hydrogen peroxide (76.99 ± 0.09) and less reducing power properties [89]. In a study by Alfikri et al. [90], clove EO's antioxidant activity, which is highly significant economically, was examined. At various flowering phases, the DPPH scavenging activity, expressed as IC_{50} values, varied from 15.80 to 108.85 µg/mL. The findings showed that the best essential oil constituent and most effective source of natural antioxidants was clove during the flowering stage [90].

Hyssop: *Hyssopus officinalis* is a shrub from Lamiaceae family native to Southern Europe, and the Middle East. It has been utilized in traditional herbal medicine because of its antibacterial, carminative, stimulant, stomachic, expectorant, cough-relieving, and antioxidant properties. In the study by Fathiazad et al. [91], according to a description of the chemical composition of hyssop EO, the primary constituents were myrtenyl acetate, camphor, germacrene, and spathulenol. According to the same study, the total phenol content of the *n*-butanol and ethylacetate extracts of *Hyssopus officinalis* aerial parts were 246 mgGAE/g and 51 mgGAE/g, respectively. The DPPH radical scavenging experiment was also used to assess the extracts' antioxidant properties, and the IC_{50} values were 103×10^{-3} and 25×10^{-3} mg/mL, respectively. The largest concentration of phenolic components may have contributed to the best antioxidant activity of *n*-butanol extract [91]. The use of stirring, in conjunction with an ethanolic solvent, appears to be advantageous in optimizing the extraction of polyphenols from hyssop, leading to extracts with increased antioxidant activity, per the findings of Polaki et al. [92]. The antioxidant activity was 582.23 ± 16.88 µmol ascorbic acid equivalents (AAE)/g using the FRAP method and 343.75 ± 15.61 µmol AAE/g dry weight using the DPPH method. The total polyphenol content was 70.65 ± 2.76 mgGAE/g dry weight. The therapeutic potential of hyssop extract was further highlighted by the discovery of a wide spectrum of polyphenolic chemicals, such as rutin, *p*-coumaric acid, and caffeic acid, which are known for their medicinal and antioxidant properties [92]. According to Moulodi et al. 2018., the main constituents of *Hyssopus officinalis* essential oil were determined to be camphor (23.61%) and β-pinene (21.91%), with an IC_{50} of 11.22 μg/mL and a total phenolic content of 23.16 mgGAE/g of essential oil. The findings showed that hyssop essential oil has good antioxidant activity and can be utilized as a natural antioxidant in the food and pharmaceutical industries [93].

Juniper/*Juniperus*: *Juniperus communis* is an evergreen perennial shrub belonging to the cypress family Cupressaceae. There are between 50 and 67 species of junipers in North America, Africa, parts of Asia, and Central America, depending on the taxonomy. Both alcoholic and nonalcoholic beverages contain juniper extracts and essential oils, which have been shown to have anti-inflammatory, carminative, diuretic, fungicide, anticholinesterase, antibacterial, and antioxidant properties. Using GC-MS analysis of juniper essential oils, the authors Ennajar et al. (2009) identified 30 compounds, the most prevalent of which were α-pinene, δ-3-carene, and γ-cadinene. The primary polyphenols found in methanol, ethanol, ethyl acetate, and dichloromethane juniper extract were tannins, flavonoids, and anthocyanins. As a result, the extract from methanol leaves had a high antioxidant activity (IC_{50} value of 8.5 ± 0.3 mg/L), followed by the extract from ethanolic leaves (49.1 ± 0.6 mg/L). The highest activity for the berries was found in the ethanol extract (54 ± 1 mg/L), which was followed by the methanolic (642 ± 7 and 664 ± 7 mg/L, respectively) and ethyl acetate extract. The DPPH assay of juniper essential oils of leaves and berries showed radical-scavenging activity, with IC_{50} values of $5,364 \pm 121$ and $14,716 \pm 411$ mg/L, respectively. The high content of terpene hydrocarbons of essential oils was responsible for this low reactivity [94]. In a study by Miceli et al. [95], the authors compared the biological potential and flavonoid composition of berry methanol extracts from two Turkish juniperus varieties. There were 59.17 mg GAE/g extract of total polyphenols. By the HPLC analysis, 16 flavonoids were separated, with hypolaetin-7-pentoside and quercetin-hexoside being the principal constituents. Using several techniques, the *in vitro* antioxidant activity was assessed; the DPPH method yielded an IC_{50} of 0.63 mg/mL, whereas the TBA test yielded an IC_{50} of 4.44 μg/mL [95].

Mint: *Mentha piperita*, also called peppermint, is a hybrid species of mint, native to Europe and the Middle East. Peppermint is the most widely used of the more than 25 species in the genus *Mentha*. It belongs to the Lamiaceae family of perennial herbaceous plants, *Mentha*. It was discovered that the two components of peppermint essential oil, methone and isomenthone, were the most powerful radical scavenging compounds. According to reports, mint has cytoprotective, hepatoprotective, antibacterial, antiviral, anti-inflammatory, antiulcer, slightly anesthetic, antispasmodic, carminative, anthelmintic, and antioxidant properties. Polyphenols, flavonoids (such as luteolin and its derivative apigenin), flavanols (such as epicatechin and catechin), and coumarins (such as esculetin and scopoletin) are abundant in *Mentha* species. Mint essential oils are a primary emphasis in terms of formulations. Their primary constituents include alcohols, ketones, esters, ethers, and oxides, and they are either light yellow or greenish yellow in color. The primary components of essential oils made from various mint species are found to be menthol, menthone, isomenthone, menthyl acetate, linalool, linalyl acetate, pulegone, carvone, piperitenone oxide, and cis-piperitone epoxide. The researchers have effectively used a variety of *in vitro* antioxidant tests, including reducing power assays, ABTS suppression of linoleic acid peroxidation, and DPPH radical scavenging, to examine the antioxidant activity of mint. High concentrations of antioxidants, such as phenolic compounds, and vitamins that can postpone or prevent the oxidation of various molecules are found in many medicinal plants, including those in the genus *Mentha*. According to the numerous investigations on the antioxidant potential of *Mentha* species, *M. longifolia* was the most efficient of the nine species, exhibiting 88.6% antioxidant activity at a concentration of 100 µL/mL [96], which was higher than the 93.0% activity of ascorbic acid. Six *Mentha* species were found to have high antioxidant activity in an Algerian study, with the following findings for mint: Flavonoid content: 15.70 ± 0.10 mg RE/g DW, tannin content: 6.50 ± 0.41 mg CE/g DW, total phenolic content: 31.40 ± 0.80 mg GAE/g DW. The IC_{50} value for DPPH radical scavenging activity was 17.00 ± 0.88 µg/mL, while the IC_{50} value for the b-carotene bleaching assay was 516.00 ± 0.25 µg/mL. The correlation coefficients between phenolic contents and antioxidant activity of the six Algerian mints suggested that mint antioxidant activity is mostly correlated to tannins and polyphenols and less to flavonoids [97].

Nigella: *Nigella sativa*, also known as black caraway black or cumin, is an annual flowering plant from the family Ranunculaceae. The plant is native to the Eastern Europe and Western Asia (Cyprus, Turkey, Iran, and Iraq). High antioxidant activity and a high correlation with the total phenolic content characterize this herbaceous, erect annual herb. Antipyretic, antidiabetic, analgesic, carminative, diuretic, antineoplastic, antibacterial, anti-inflammatory, stimulant, expectorant, and anthelmintic qualities have all been recorded for nigella. Results of an analysis of the phenolic content and antioxidant activity of extracts made from Nigella seeds using Soxhlet and ultrasonic extraction procedures were reported in the work of Goga et al. [98]. According to the Folin-Ciocalteu tmethod, the total phenolic content ranged from 11.867 ± 0.338 to 31.148 ± 0.293 mgGAE/g. This ranged from $2.70 ± 0.22 \times 10^{-5}$ to $32.7 ± 1.31 \times 10^{-5}$ mgQE/g (quercetin equivalent) for total flavone and flavanol content. The DPPH and ABTS tests were used to assess the samples' radical scavenging activity. Samples varied from 3.01 ± 0.03 mg/mL to 12.04 ± 0.60 mg/mL in their capacity to reduce stable DPPH radicals, and in their ability to reduce stable ABTS radicals from 14.02 ± 0.62 mg/mL to 18.67 ± 1.54 mg/mL [98]. A wide range of bioactive components, such as flavonoids, phenols, steroids, triterpenoids, proteins, alkaloids, tannin, sesquiterpenoid hydrocarbons, monoterpenoid alcohol, and monoterpenoid ketone, were discovered during the analysis of the nigella essential oil, which was isolated as a pale-yellow liquid. Nigella EO's total flavonoid and phenolic content were measured at 442.25 µg QE/g and 641.23 µg GAE/g, respectively. At 1,000 µg/mL, nigella EO exhibited the highest percentage of inhibition (65.80%) when compared to the standard ascorbic acid (73.57%) [99]. In a different study, the antioxidant activity of nigella EO was examined using TLC screening techniques, which revealed that thymoquinone and its constituents carvacrol, t-anethole, and 4-terpineol demonstrated respectable radical scavenging properties. When examined for nonspecific hydrogen atom or electron donating activity using the DPPH assay, these four components and the essential oil showed varied antioxidant activity. Additionally, they were efficient OH radical scavengers in the deoxyribose degradation assay and the nonenzymatic lipid peroxidation in liposomes experiment [100].

Oregano: *Origanum vulgare* is a species of flowering plant from the Lamiaceae family. It is native to the Mediterranean region. The woody perennial oregano plant has white, pink, or light purple flowers. While its close relative *Origanum majorana* is known as sweet majoran, it is sometimes referred to as wild marjoram. A variety of biological properties, including carminative, antibacterial, antifungal, antiviral, anticancer, anti-inflammatory, hepatoprotective, and antioxidant, have been reported for oregano. The extracts' levels of sixteen bioactive phenolic compounds, total phenolic and total flavonoid content, and antioxidant activities (DPPH and FRAP tests) were all examined. With concentrations varying by subspecies, HPLC analyses revealed that rosmarinic acid (659.6–1646.9 mg/100 g dry weight (DW)) was by far the most prevalent constituent. It was followed by luteolin (46.5–345.4 mg/100 g DW), chicoric acid (36.3–212.5 mg/100 g DW), coumarin (65.7–193.9 mg/100 g DW), and quercetin (10.6–106.1 mg/100 g DW). Rosmarinic acid and antioxidant activity were shown to be significantly and favorably correlated ($r = 0.46$) [101]. According to Teixeira et al. [102], GC/MS research revealed that oregano essential oil includes phenolic chemicals and monoterpene hydrocarbons. Carvacrol, fenchyl alcohol, *p*-cymene, thymol, and terpinene were the main constituents. According to the reducing power analysis, the hot water extract had the highest antioxidant activity, followed by the ethanolic, cold water, and essential oil extracts. The same pattern of declining power analysis was observed in the DPPH free radical scavenging activity. While ethanolic extract (antioxidant activity index, AAI = 1.23), hot water extract (AAI = 3.16), and cold water extract (AAI = 0.55) were categorized as moderate, strong, and very strong antioxidants, respectively, oregano essential oil (AAI = 0.05) showed weak antioxidant activity. The authors came to the conclusion that Portuguese-origin oregano extracts and essential oils are excellent options to substitute industrially used synthetic compounds [102].

Rosemary: *Salvia rosmarinus* is a shrub with fragrant, evergreen, needle-like leaves and white, pink, purple, or bluish flowers. It is an aromatic plant from the Lamiaceae family, native to the Mediterranean region, but cultivated worldwide. Antimicrobial, antifungal, antiviral, antimicrobial, antiparasitic, antiproliferative, spasmolytic, anti-inflammatory, mildly analgesic, and antioxidant properties have all been documented. Among the most potent antioxidant components of rosemary are the cyclic diterpene diphenols, carnosolic acid, and carnosol, which are typically found in extracts. The polyphenolic profile of this plant is characterized by the presence of carnosic acid, carnosol, rosmarinic acid, and hesperidin as major components. According to reports, the primary constituents of rosemary essential oils are camphor, 1,8-cineole, α-pinene, *p*-cymene, and borneol [56]. The antioxidant properties of various solvent extracts and rosemary essential oil were assessed in the study of Al-Jaafreh 2024. The ethanol extract had the highest TPC 72.34 GAE mg/g and TFC 26.81 RE mg/g. Methanol extract demonstrated the strongest antioxidant activity in the NO radical scavenging assay (86.68 RE mg/g) and DPPH (138.3 GAE mg/g) assays, while the aqueous extract demonstrated the maximum activity in ABTS (125.33 TE mg/g). The authors observed a strong correlation between antioxidant activity and TPC, TFC, and TTC. They came to the conclusion that this is because most bioactive compounds, including flavonoids, polyphenols, and tannins, are found in more polar solvents, making these phytochemicals the primary source of the antioxidant properties [103].

Sage: *Salvia officinalis* (Figure 11.6), or common sage, is small evergreen subshrub used as an aromatic and culinary herb. Sage is a member of the Lamiaceae family and native to the Mediterranean region. Sage has been reported to have antiseptic, antimicrobial, anticancer, antiproliferative, antidiabetic, anti-inflammatory, hypolipidemic, memory-enhancing effects, and antioxidant properties. Studies on *Salvia officinalis* have demonstrated that certain phenolic antioxidants stop oxidative stress-induced cell death. Research has indicated that Salvia taxa have a large number of phenolic and flavonoid compounds. Extracts from the aerial parts of sage were shown to include a variety of phenolics, such as caffeic acid and chlorogenic acid, as well as flavonoids, such as apigenin and luteolin [7]. Three species of Salvia were compared by Pereira et al. [104]. The health benefits of their decoctions, specifically their antioxidant activity, were studied. The abundance of caffeic acid and its derivatives was closely associated with the sage decoctions' greater activity [104]. Yu et al. [105] reported on sage methanolic extracts' total phenols content of 50.89 ± 0.37 mgGAE/g DW, flavonoids of 43.92 ± 0.05 mgCATE/g

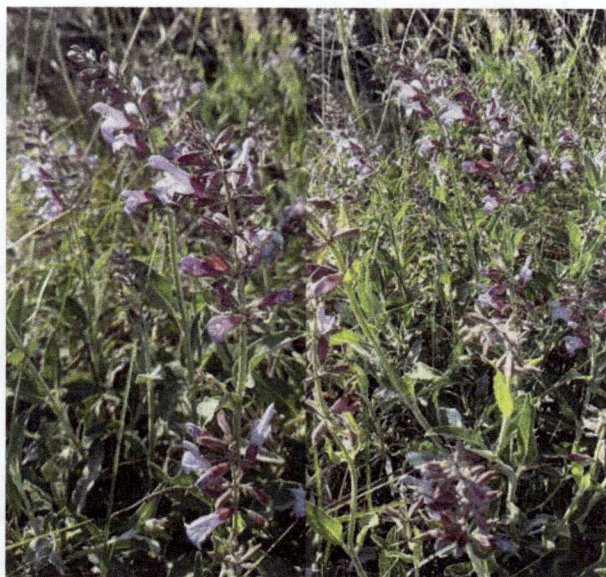

Figure 11.6: Flowering sage from Mostar, Bosnia and Herzegovina.

DW, tannins of 28.02 ± 1.40 mg ECE/g DW, ABTS 0.27 ± 0.01 mM Trolox/g DW, and DPPH 0.25 ± 0.00 mM Trolox/g DW [105].

Thyme: *Thymus vulgaris*, common or garden thyme is a species of flowering plant in the Lamiaceae family, native to southern Europe from the western Mediterranean to southern Italy. It is a perennial herbaceous shrub that bears clusters of purple or pink flowers in early summer along with tiny, very fragrant leaves. The plant's phenolic and flavonoid levels are linked to its potent antioxidant properties. Antibacterial, antimicrobial, antifungal, anti-inflammatory, expectorant, and spasmolytic qualities have all been documented for thyme. Several volatile chemicals were detected by GC-MS analysis, with thymol, carvacrol, geraniol, and *p*-cymene being the main constituents. Caffeic acid, quinic acid, *p*-coumaric acid, quercetin-7-*o*-glucoside, ferulic acid, carnosic acid, cinnamic acid, rosmarenic acid, apigenin, and naringenin were all detected in the methanolic extract by HPLC analysis. *T. vulgaris* had the highest levels of flavonoids and phenols (62.40 ± 0.03 mg TAE/g DW and 8.55 mg QE/g DW, respectively). Using the 1,1-diphenyl-2-picrylhydrazyl (DPPH) and reducing power test, the antioxidant activities of the samples were assessed, and the IC_{50} value was 289.3 µg/mL. The authors suggest that thyme is a good source of essential oil and flavonoids, which have a high level of antioxidant activity [106]. The total phenolic and flavonoid content of thyme extracts was ascertained in the Mokhtari et al. 2023 [107] study. The findings revealed a total flavonoid content of 3.87 mg QE/g DW and a total phenolic content of 8.89 mg GAE/g DW. When plant extracts were compared for their ability to scavenge DPPH radicals, thyme had an IC_{50} value of 69.39 µg/mL [107].

11.8 Conclusion

The era of herbal medicine is evident in both contemporary society and the scientific community. Numerous studies have demonstrated the biological benefits of MAPs, including their anti-inflammatory, antioxidant, immune-boosting, and antiaging properties, which guarantee longer and healthier life for humans. Presenting MAPs' scientific side requires analyzing their composition, medical uses, toxicity, physiological impacts, and converting their active ingredients into a useful product. Because they tend to enhance health and natural disease prevention, MAPs with high antioxidant content are highly sought after by consumers due to their greater safety and dependability. In addition to being well-known for their flavor and taste, these plants have paved the way for a new area of study where their antioxidant qualities help preserve food and improve health for customers, who are awaiting hard scientific evidence to support their interest in herbal therapy. Their natural antioxidants may provide an alternative to traditional treatments for oxidative stress, and these plant compounds have significant therapeutic effects with few side effects. They may also be good candidates for preventing free radical-induced diseases like diabetes, cancer, aging, and cardiovascular diseases. To further valorize MAPs, it would be crucial to increase the variety of parameters in research as well as the isolation, characterization, and identification of active chemical compounds.

References

[1] Bakkali, F., Averbeck, S., Averbeck, D. and Idaomar, M. (2008). Biological effects of essential oils – A review. Food and Chemical Toxicology: An International Journal Published for the British Industrial Biological Research Association, 46, 446–475.

[2] Buchbauer, G. (2010). Biological activities of essential oils. In: Khc, B. & Buchbauer, G. (eds) Handbook of Essential Oils: Science, Technology, and Applications, CRC Press, Taylor & Francis Group, Boca Raton, 1128.

[3] Karamaya, K. and Coşge Şenkal, B. (2022). The antioxidant capacities of leaf extracts from *Salvia viridis* L. Current Perspectives on Medicinal and Aromatic Plants, 5(2), 127–135.

[4] Mammadov, R. (2014). Tohumlu Bitkilerde Sekonder Metabolitler, Nobel Akademik Yayıncılık, Ankara, 15–129.

[5] Halliwell, B. and Gutteridge, J. M. C. (1990). A Role of free radicals and catalytic metal ions in human disease: An overview. Methods Enzymology, 186, 1–85.

[6] Kahkönen, M. P., Hopia, A. I., Vuorela, H. J., Rauha, J. P., Pihlaja, K., Kujala, T. S. and Heinonen, M. (1999). Antioxidant activity of plant extracts containing phenolic compounds. Journal of Agricultural and Food Chemistry, 47, 3954–3962.

[7] Rungsimakan, S. and Rowan, M. G. (2014). Terpenoids, flavonoids and caffeic acid derivatives from *Salvia viridis* L. var. blue jeans. Phytochemistry, 108, 177–188.

[8] Bendini, A., Cerretani, L., Pizzolante, L., Gallina-Toschi, T., Guzzo, F., Cedolo, F., Andereetta, F. and Levi, M. (2006). Phenol content related to antioxidant and antimicrobial activity of Passiflora spp. extracts. European Food Research & Technology, 223, 102–109.

[9] Baydar, H. (2013). Tıbbi Ve Aromatik Bitkiler Bilimi Ve Teknolojisi, Süleyman Demirel University Faculty of Agriculture Publication, Isparta, 23–141.

[10] Aqil, F., Ahmad, I. and Mehmood, Z. (2006). Antioxidant and free radical scavenging properties of twelve traditionally used Indian medicinal plants. Turkish Journal of Biology, 30, 77–183.

[11] Oraon, L., Jana, A., Prajapati, P. S. and Suvera, P. (2017). Application of medicinal and aromatic plants in functional dairy products – A review. Journal of Dairy, Veterinary & Animal Research, 5(3), 109–115.

[12] Viuda-Martos, M., Ruiz-Navajas, Y., Fernández-López, J. and Pérez-Álvarez, J. A. (2010). Spices as functional foods. Critical Reviews in Food Science and Nutrition, 51(1), 13–28.

[13] Sies, H. (2020). Oxidative stress: Concept and some practical aspects. Antioxidants, 9(9), 852.

[14] Rodrigo, R. (2009). Oxidative Stress and Antioxidants: Their Role in Human Disease. Nova biomedical books, New York.

[15] Čvorišćec, D. and Štrausova, Č. I. (2009). Medicinska Biokemija, Medicinska naklada, Zagreb, Croatia.

[16] Sun, W. and Shahrajabian, M. H. (2023). Therapeutic potential of phenolic compounds in medicinal plants-natural health products for human health. Molecules, 28(4), 1845.

[17] Kumpulainen, J. T. and Solonen, J. T. (1999). Natural Antioxidants and Anticarcinogens in Nutrition, Health and Disease, The Royal Society of Chemistry, Thomas Graham House, Science Park, Milton Road, Cambridge CB4 0W, UK.

[18] Cook, N.C. and Samman, S. (1996). Flavonoids-Chemistry, Metabolism, Cardioprotective Effects, and Dietary Sources. The Journal of Nutritional Biochemistry, 7, 66–76.

[19] Zampelas, A. and Micha, R. (2015). Antioxidants in Health and Disease, CRC Press Taylor & Francis Group, Boca Raton, USA.

[20] Charles, D. J. (2013). Antioxidant Properties of Spices, Herbs and Other Sources. Springer, New York.

[21] Fujii, J., Soma, Y. and Matsuda, Y. (2023). Biological action of singlet molecular oxygen from the standpoint of cell signaling, injury and death. Molecules, 28(10), 4085.

[22] Kahkeshani, N., Farzaei, F., Fotouhi, M., Alavi, S. S., Bahramsoltani, R., Naseri, R., Momtaz, S., Abbasabadi, Z., Rahimi, R., Farzaei, M. H. and Bishayee, A. (2019). Pharmacological effects of gallic acid in health and diseases: A mechanistic review. Iranian Journal of Basic Medical Sciences, 22(3), 225–237.

[23] Dvoršak, K. (2013). Magistrska Naloga: Superoksid Dismutazna Aktivnost Stabilnih Nitroksidnih Radikalov. Univerza v Ljubljani. Fakulteta za farmacijo, Ljubljana.

[24] Ayala, A., Muñoz, M. F. and Argüelles, S. (2014). Lipid peroxidation: Production, metabolism, and signaling mechanisms of malondialdehyde and 4-hydroxy-2-nonenal. Oxidative Medicine and Cellular Longevity, 2014, 360438.

[25] Salekeen, R., Haider, A. N., Akhter, F., Billah, M. M., Islam, M. E. and Didarul Islam, K. M. (2022). Lipid oxidation in pathophysiology of atherosclerosis: Current understanding and therapeutic strategies. Nternational Journal of Cardiology. Cardiovascular Risk and Prevention, 14, 200143.

[26] Cai, Z. and Yan, L. J. (2013). Protein oxidative modifications: beneficial roles in disease and health. Journal of Biochemical and Pharmacological Research, 1(1), 15–26.

[27] Xu, Z., Huang, J., Gao, M., Guo, G., Zeng, S., Chen, X., Wang, X., Gong, Z. and Yan, Y. (2021). Current perspectives on the clinical implications of oxidative RNA damage in aging research: Challenges and opportunities. Gero Science, 43(2), 487–505.

[28] Langseth, L. (1995). Oxidants, Antioxidants, and Disease Prevention. ILSI Europe, Brussels.

[29] Zakaria, N. A., Ibrahim, D., Sulaiman, S. F. and Supardy, N. A. (2011). Assessment of antioxidant activity, total phenolic content and in vitro toxicity of Malaysian red seaweed, Acanthophora spicifera. Journal of Chemical and Pharmaceutical Research, 3, 182–191.

[30] Halliwell, B. (1994). Free radicals and antioxidants free radicals, antioxidants, and human disease: Curiosity, cause, or consequence. Lancet, 344, 721–724.

[31] Wong, C. K., Ooi, V. E. and Ang, P. O. (2000). Protective effect of seaweeds against liver injury caused by carbon tetra chloride in rats. Chemosphere, 41, 173–176.

[32] Chen, K., Plumb, G. W., Bennett, R. N. and Bao, Y. (2005). Antioxidant activities of extracts from five anti-viral medicinal plants. Journal of Ethnopharmacology, 96, 201–205.

[33] Bjelakovic, G., Nikolova, D., Gluud, L. L., Simonetti, R. G. and Gluud, C. (2007). Mortality in randomized trials of antioxidant supplements for primary and secondary prevention: Systematic review and meta-analysis. Jama, 297, 842–857.

[34] Niwa, T., Doi, U., Kato, Y. and Osawa, T. (2001). Antioxidant properties of phenolic antioxidants isolated from corn steep liquor. Journal of Agricultural and Food Chemistry, 49, 177–182.

[35] Kahl, R. and Kappus, H. (1993). Toxicology of the synthetic antioxidants BHA and BHT in comparison with the natural antioxidants capacity and total vitamin E. Zeitschrift Für Lebensmittel-Untersuchung und–Forschung, 196, 329–338.

[36] Rietjens, I. M., Boersma, M. G., Haan, L., Spenkelink, B., Awad, H. M., Cnubben, N. H., Jj, V. Z., Woude, H., Alink, G. M. and Koeman, J. H. (2002). The pro-oxidant chemistry of the natural antioxidants vitamin C, vitamin E, carotenoids and flavonoids. Environmental Toxicology Pharmacology, 11, 321–333.

[37] Kim, D. O. and Lee, C. Y. (2004). Comprehensive study on vitamin C equivalent antioxidant capacity (VCEAC) of various polyphenolics in scavenging a free radical and its structural relationship. Critical Reviews in Food Science and Nutrition, 44, 253–273.

[38] Escarpa, A. and Gonzalez, M. C. (2001). An overview of analytical chemistry of phenolic compounds in foods. Critical Reviews in Analytical Chemistry, 31, 57–139.

[39]   Amić, D., Davidović-Amić, D., Bešlo, D. and Trinajstić, N. (2003). Structure-radical scavenging activity relationships of flavonoids. Croatica Chemica Acta, 76, 55–61.

[40]   Xiu-Qin, L., Chao, J., Yan-Yan, S., Min-Li, Y. and Xiao-Gang, C. (2009). Analysis of synthetic antioxidants and preservatives in edible vegetable oil by HPLC/TOF-MS. Food Chemistry, 113, 692–700.

[41]   Jacob, V. and Michael, A. (1999). Nutritional antioxidants: Mechanisms of action, analysis of activities and medical applications. Nutrition, 49, 1–7.

[42]   Tiwari, A. K. (2001). Imbalance in antioxidant defence and human diseases: Multiple approach of natural antioxidants therapy. Current Science, 81, 1179–1186.

[43]   Bostwick, D. G., Alexander, E. E., Singh, R., Shan, A., Qian, J., Santella, R. M., Oberley, L. W., Yan, T., Zhong, W., Jiang, X. and Oberley, T. D. (2000). Antioxidant enzyme expression and reactive oxygen species damage in prostatic intraepithelial neoplasia and cancer. Cancer, 89, 123–134.

[44]   Halliwell, B. (1991). Drug antioxidant effects. A basis for drug selection?. Drugs, 42, 569–605.

[45]   Shirwaikar, A., Rajendran, K. and Kumar, C. D. (2004). In vitro antioxidant studies of Annona squamosa Linn. leaves. Indian Journal of Experimental Biology, 42, 803–807.

[46]   SK, M., Jagdish, B. S. R., Kb, S. and Mk, U. (2003). In vitro evaluation of antioxidant properties of Cocos nucifera Linn. water. Nahrung, 47, 126–131.

[47]   Zheng, H., Yichen, X., Elisa, A. L. and Rusu, M. (2024). Vitamin C as Scavenger of reactive oxygen species during healing after Myocardial Infarction. International Journal of Molecular Sciences, 25(6), 3114.

[48]   Ali, S. S., Kasoju, N., Luthra, A., Singh, A., Sharanabasava, H., Sahu, A. and Bora, U. (2008). Indian medicinal herbs as sources of antioxidants. Food Research International, 41, 1–15.

[49]   Krinsky, N. I. (1992). Mechanism of action of biological antioxidants. Proceedings of the Society for Experimental Biology and Medicine Society for Experimental Biology and Medicine, 200, 248–254.

[50]   Nichenametla, S. N., Taruscio, T. G., Barney, D. L. and Exon, J. H. (2006). A review of the effects and mechanism of polyphenolics in cancer. Critical Reviews in Food Science and Nutrition, 46, 161–183.

[51]   Waheed Janabi, A. H., Kamboh, A. A., Saeed, M., Xiaoyu, L., BiBi, J., Majeed, F., Naveed, M., Mughal, M. J., Korejo, N. A., Kamboh, R., Alagawany, M. and Lv, H. (2020). Flavonoid-rich foods (FRF): A promising nutraceutical approach against lifespan-shortening diseases. Iranian Journal of Basic Medical Sciences, 23(2), 140–153.

[52]   Huang, D., Ou, B. and Prior, R. L. (2005). The chemistry behind antioxidant capacity assays. Journal of Agricultural and Food Chemistry, 53(6), 1841–1856.

[53]   Kulisic, T., Radonic, A., Katalinic, V. and Milos, M. (2004). Use different method for testing antioxidative activity of oregano esential oil. Food Chemistry, 85, 633–640.

[54]   Taso, R. and Deng, Z. (2004). Separation procedures for naturally occuring antioxidant phytochemicals. Journal of Chromatography B, 8(12), 85–89.

[55]   Campanella, L., Bonanni, A., Bellantoni, D., Finotti, E. and Tomassetti, M. (2004). Biosnsores for determination of total and natural antioxidant capacity of red and white wines: Comparison with other spectrophotometric and fluorimetric methods. Biosensors and Bioelectronics, 19, 641–651.

[56]   Nieto, G., Ros, G. and Castilla, J. (2018). Antioxidant and antimicrobial properties of rosemary (Rosmarinus officinalis L.): A review. Medicine, 5(98), 2–13.

[57]   Lopresti, L. A. (2017). Salvia (sage): A review of its potential cognitive-enhancing and protective effects. Drugs in R&D, 17, 53–64.

[58]   Cuvelier, M. E., Berset, C. and Richard, H. (1994). Antioxidant constituents in sage (Salvia officinalis). Journal of Agricultural and Food Chemistry, 42, 665–669.

[59]   Miraj, S., Kopaei, R. and Kiani, S. (2017). Melissa officinalis L: A review study with an antioxidant prospective. Journal of Evidence-Based Complementary & Alternative Medicine, 22(3), 385–394.

[60] Gutierrez, G. E. P., Salas, M. A. P., Lopez, L. N., Mendoza, M. S. C., Olivo, G. V. and Heredia, J. B. (2017). Flavonoids and phenolic acids from oregano: Occurrence. Biological Activity and Health Benefits, Plants, 7(2), 2–23.

[61] Burt, S. (2004). Essential oils: Their antibacterial properties and potential applications in foods-a review. International Journal of Food Microbiology, 94(3), 223–253.

[62] Schuler, P. (1990). Natural Antioxidants Exploited Commercially, Antioxidants, F. & Bjf, H. ed. Elsevier, London, 99–170.

[63] Chu, Y. (2000). Flavonoid content of several vegetables and their antioxidant activity. Journal of the Science of Food and Agriculture, 80, 561–566.

[64] Koleva, V. B. II, TA, L. J. P. H., De Groot, A. and Evstatieva, L. N. (2002). Screening of plant extracts for antioxidant activity: A comparative study on three testing methods. Phytochem Analysis, 13, 8–17.

[65] Mantle, D., Eddeb, F. and Pickering, A. T. (2000). Comparison of relative antioxidant activities of British medicinal plant species in vitro. Journal of Ethnopharmacology, 72, 47–51.

[66] Pourmorad, F., Hosseinimehr, S. J. and Shahabimajd, N. (2006). Antioxidant activity, phenol and flavonoid contents of some selected Iranian medicinal plants. African Journal of Biotechnology, 5(11), 1142–1145.

[67] Kessler, M., Ubeaud, G. and Jung, L. (2003). Anti- and pro-oxidant activity of rutin and quercetin derivatives. Journal of Pharmacy and Pharmacology, 55, 131–142.

[68] Williams P. G. (2006). Health benefits of herbs and spices: Public Health, Medical Journal of Australia, 185(4), S17–S18.

[69] Kochhar, K. P. (2008). Dietary spices in health and diseases (II). Indian Journal of Physiology and Pharmacology, 52(4), 327–354.

[70] Cuvelier, M. E., Richard, H. and Berset, C. (1996). Antioxidative activity and phenolic composition of pilot-plant and commercial extracts of sage and rosemary. Journal of the American Oil Chemists' Society, 73(5), 645–652.

[71] Lee, W. G. and Shibamoto, T. (2001). Antioxidant property of aroma extract isolated from clove buds. Food Chemistry, 74, 443–448.

[72] Hsu, F. L., Li, W. H., Yu, C. W., Hsieh, Y. C., Yang, Y. F., Liu, J. T., Shih, J., Chu, Y. J., Yen, P. L., Chang, S. T. and Liao, V. H. C. (2012). In vivo antioxidant activities of essential oils and their constituents from leaves of the Taiwanese Cinnamomum osmophloeum. Journal of Agricultural and Food Chemistry, 60(12), 3092–3097.

[73] Papageorgiou, V., Gardeli, C., Mallouchos, A., Papaioannou, M. and Komaitis, M. (2008). Variation of the chemical profile and antioxidant behavior of Rosmarinus officinalis L. and Salvia fruticosa Miller grown in Greece. Journal of Agricultural and Food Chemistry, 56(16), 7254–7264.

[74] Pizzale, L., Bortolomeazzi, R., Vichi, S., Uberegger, E. and Conte, L. S. (2002). Antioxidant activity of sage (Salvia officinalis and S fruticosa) and oregano (Origanum onites and O indercedens) extracts related to their phenolic compound content. Journal of Sol-Gel Science and Technology, 82, 1645.

[75] Kulisic, T., Radonic, A. and Milos, M. (2005). Antioxidant properties of thyme (Thymus vulgaris L.) and wild thyme (Thymus serpyllum L.) essential oils. Italian Journal of Food Science, 17, 315.

[76] Miguel, M. G. (2010). Antioxidant and anti-inflammatory activities of essential oils: A short review. Molecules, 15(12), 9252–9287.

[77] Mimica-Dukić, N., Božin, B., Soković, M., Mihajlović, B. and Matavulj, M. (2003). Antimicrobial and antioxidant activities of three Mentha species essential oils. Planta Medica, 69(05), 413–419.

[78] Sari, M., Biondi, D. M., Kaâbeche, M., Mandalari, G., D'Arrigo, M., Bisignano, G., Saija, A., Daquino, C. and Ruberto, G. (2006). Chemical composition, antimicrobial and antioxidant activities of the essential oil of several populations of Algerian Origanum glandulosum Desf. Flavour and Fragrance Journal, 21(6), 890–898.

[79]   Huang, S. H., Chen, C. C., Lin, C. M. and Chiang, B. H. (2008). Antioxidant and flavor properties of *Angelica sinensis* extracts as affected by processing. Journal of Food Composition and Analysis, 21(5), 402–409.

[80]   Penga, B., Zhoua, Z., Wanga, R., Fua, C., Wanga, W., Lia, Y., Qina, X., Zhangb, L. and Zhaoa, H. (2024). Analysis of components and antioxidant activity of *Angelica sinensis* essential oil (AEO) extracted from supercritical carbon dioxide. Journal of Food Bioactives, 26, 72–79.

[81]   Gülçin, I., Oktay, M., Kıreçcı, E. and Küfrevıoğlu, Ö. İ. (2003). Screening of antioxidant and antimicrobial activities of anise (*Pimpinella anisum* L.) seed extracts. Food Chemistry, 83(3), 371–382.

[82]   AlBalawi, A. N., Elmetwalli, A., Baraka, D. M., Alnagar, H. A., Alamri, E. S. and Hassan, M. G. (2023). Chemical constituents, antioxidant potential, and antimicrobial efficacy of *Pimpinella anisum* extracts against multidrug-resistant bacteria. Microorganisms, 11(4), 1024.

[83]   Nadeem, H. R., Akhtar, S., Sestili, P., Ismail, T., Neugart, S., Qamar, M. and Esatbeyoglu, T. (2022). Toxicity, antioxidant activity, and phytochemicals of Basil (*Ocimum basilicum* L.) leaves cultivated in Southern Punjab. Pakistan Foods, 11(9), 1239.

[84]   Ahmed, A. F., Attia, F. A. K., Liu, Z., Li, C., Wei, J. and Kang, W. (2019). Antioxidant activity and total phenolic content of essential oils and extracts of sweet basil (*Ocimum basilicum* L.) plants. Food Science and Human Wellness, 8(3), 299–305.

[85]   Romano, R., De Luca, L., Aiello, A., Pagano, R., Di Pierro, P., Pizzolongo, F. and Masi, P. (2022). Basil (*Ocimum basilicum* L.) leaves as a source of bioactive compounds. Foods, 11(20), 3212.

[86]   Conforti, F., Statti, G., Uzunov, D. and Menichin, F. (2006). Comparative chemical composition and antioxidant activities of wild and cultivated *Laurus nobilis* L. leaves and *Foeniculum vulgare* subsp. piperitum (Ucria) coutinho seeds. Biological & Pharmaceutical Bulletin, 29(10), 2056–2064.

[87]   Saab, A. M., Tundis, R., Loizzo, M. R., Lampronti, I., Borgatti, M., Gambari, R., Menichini, F., Esseily, F. and Menichini, F. (2012). Antioxidant and antiproliferative activity of *Laurus nobilis* L. (Lauraceae) leaves and seeds essential oils against K562 human chronic myelogenous leukaemia cells. Natural Products Research, 26(18), 1741–1745.

[88]   Mashkor, I. M. A. A. (2015). Evaluation of antioxidant activity of clove (*Syzygium aromaticum*. International Journal of Biological and Chemical Sciences, 13(1), 23–30.

[89]   Dahiru, N., Paliwal, R., Madungurum, M. A., Abubakar, A. S. and Abdullahi, B. (2022). Study on antioxidant property of *Syzygium aromaticum* (Clove). Journal of Microbiology and Biotechnology, 10(1), 13–16.

[90]   Alfikri, F. N., Pujiarti, R., Wibisono, M. G. and EB, H. (2020). Yield, quality, and antioxidant activity of clove (*Syzygium aromaticum* L.) bud oil at the different phenological stages in young and mature trees. Scientifica, 2020, Article ID 9701701.

[91]   Fathiazad, F., Mazandarani, M. and Hamedeyazdan, S. (2011). Phytochemical analysis and antioxidant activity of *Hyssopus officinalis* L. from Iran. Advanced Pharmaceutical Bulletin, 1(2), 63–67.

[92]   Polaki, S., Stamatelopoulou, V., Kotsou, K., Chatzimitakos, T., Athanasiadis, V., Bozinou, E. and Lalas, S. I. (2024). Exploring conventional and green extraction methods for enhancing the polyphenol yield and antioxidant activity of *Hyssopus officinalis* extracts. Plants, 13(15), 2105.

[93]   Moulodi, F., Khezerlou, A., Zolfaghari, H., Mohamadzadeh, A. and Alimoradi, F. (2018). Chemical Composition and antioxidant and antimicrobial properties of the essential oil of *Hyssopus officinalis* L. Journal of Kermanshah University of Medical Sciences, 22(4), 85256.

[94]   Ennajar, M., Bouajila, J., Lebrihi, A., Mathieu, F., Abderraba, M., Raies, A. and Romdhane, M. (2009). Chemical composition and antimicrobial and antioxidant activities of essential oils and various extracts of *Juniperus phoenicea* L. (Cupressacees). Journal of Food Science, 74(7), 364–371.

[95]   Miceli, N., Trovato, A., Dugo, P., Cacciola, F., Donato, P., Marino, A., Bellinghieri, V., La Barbera, T. M., Guvenc, A. E. and Taviano, M. F. (2009). comparative analysis of flavonoid profile, antioxidant and

antimicrobial activity of the berries of *Juniperus communis* L. var. communis and *Juniperus communis* L. var. saxatilis Pall. from Turkey. Journal of Agricultural and Food Chemistry, 57, 6570–6577.

[96]   Tafrihi, M., Imran, M., Tufail, T., Gondal, T. A., Caruso, G., Sharma, S., Sharma, R., Atanassova, M., Atanassov, L., Valere Tsouh Fokou, P. and Pezzani, R. (2021). The wonderful activities of the Genus Mentha: Not only antioxidant properties. Molecules, 26(4), 1118.

[97]   Benabdallah, A., Rahmoune, C., Boumendjel, M., Aissi, O. and Messaoud, C. (2016). Total phenolic content and antioxidant activity of six wild Mentha species (Lamiaceae) from northeast of Algeria. Asian Pacific Journal of Tropical Biomedicine, 6(9), 760–766.

[98]   Goga, A., Hasić, S., Š, B. and Ćavar, S. (2012). Phenolic Compounds and Antioxidant Activity of Extracts of *Nigella sativa* L. Bull Chem Technol Bosnia and Herzegovina, 39, 15–19.

[99]   Bhavikatti, S. K., Zainuddin, S. L. A., Ramli, R. B., Nadaf, S. J., Dandge, P. B., Khalate, M. and Karobari, M. I. (2024). Insights into the antioxidant, anti-inflammatory and anti-microbial potential of *Nigella sativa* essential oil against oral pathogens. Scientific Reports, 14, 11878.

[100]   Burits, M. and Bucar, F. (2000). Antioxidant activity of *Nigella sativa* essential oil. Phytotherapy Research: PTR, 14(5), 323–328.

[101]   Khorsand, G. J., Morshedloo, M. R., Mumivand, H., Bistgani, Z. E., Maggi, F. and Khademi, A. (2022). Natural diversity in phenolic components and antioxidant properties of oregano (*Origanum* vulgare L.) accessions, grown under the same conditions. Scientific Reports, 12, 5813.

[102]   Teixeira, B., Marques, A., Ramos, C., Serrano, C., Matos, O., Neng, N. R., Nogueira, J. M. F., Saraiva, J. A. and Nunes, M. L. (2013). Chemical composition and bioactivity of different oregano (*Origanum vulgare)* extracts and essential oil. Journal of Sol-Gel Science and Technology, 93(11), 2707–2714.

[103]   Al-Jaafreh, A. M. (2024). Evaluation of antioxidant activities of rosemary (*Rosmarinus officinalis* L.) essential oil and different types of solvent extractions. Biomedical & Pharmacology Journal, 17(1), 323–339.

[104]   Pereira, O. R., Catarino, M. D., Afonso, A. F., Silva, A. M. S. and Cardoso, S. M. (2018). *Salvia elegans, Salvia greggii* and *Salvia officinalis* decoctions: Antioxidant activities and inhibition of carbohydrate and lipid metabolic enzymes. Molecules, 23, 3169.

[105]   Yu, M., Gouvinhas, I., Rocha, J. and Barros Ana, I. R. N. A. (2021). Phytochemical and antioxidant analysis of medicinal and food plants towards bioactive food and pharmaceutical resources. Scientific Reports, 11, 10041.

[106]   Tohidi, B., Rahimmalek, M. and Arzani, A. (2017). Essential oil composition, total phenolic, flavonoid contents, and antioxidant activity of Thymus species collected from different regions of Iran. Food Chemistry, 220, 153–161.

[107]   Mokhtari R., Fard, K., Rezaei, M., Moftakharzadeh, M., Adel, S. and Amir, M. (2023). Antioxidant, antimicrobial activities, and characterization of phenolic compounds of Thyme (*Thymus vulgaris* L.), Sage (*Salvia officinalis* L.), and Thyme–Sage mixture extracts. Journal of Food Quality, 2023, 2602454.

Alema Dedić*, Hurija Džudžević-Čančar, Amra Alispahić,
and Emina Boškailo

Chapter 12
Medicinal and aromatic plants with antibacterial properties

Abstract: Medicinal and aromatic plants (MAPs) is the collective name for aromatic plants that belong to the category of medicinal plants. MAPs are becoming more popular around the world due to application in industries like the pharmaceutical business, healthcare products, cosmetics, organic food products, etc. Approximately 40% of newly approved medications over the past 20 years are made from natural ingredients, and most pharmaceutical corporations file patents on medical plants and their derivatives. These plants contain odorous volatile substances that exist in all their parts, including the root, wood, bark, stem, foliage, flower, and fruit, and are responsible for the distinctive fragrance. Extracts and essential oils are the most common applications of MAPs. Essential oils are complex volatile compounds, naturally synthesized by various parts of the plant during the secondary metabolism of plants, and have the ability to inhibit the growth of a wide range of pathogenic microorganisms. The knowledge of their medicinal qualities has been handed down by human societies. The aim of this article is to focus on the antibacterial activities of compounds from MAPs and the possible mechanisms involved in the inhibition of a variety of bacterial strains as their chemical potential. Plants hold great promise as a source of novel antibacterial agents due to a wide variety of chemically and structurally diverse secondary metabolites such as polyphenols, terpenoids, and alkaloids. Historical records and modern investigations highlight the importance of plant products in the treatment of various diseases caused by bacteria. Medicinal plant-derived compounds could provide novel, straightforward approaches against pathogenic bacteria.

Keywords: medicinal and aromatic plant, antibacterial activity, polyphenols, terpenoids, alkaloids

*Corresponding author: Alema Dedić**, Department of Chemistry in Pharmacy, University of Sarajevo-Faculty of Pharmacy, Zmaja od Bosne 8, Sarajevo 71000, Bosnia and Herzegovina, e-mail: alema.dedic@ffsa.unsa.ba
Hurija Džudžević-Čančar, Amra Alispahić, Department of Chemistry in Pharmacy, University of Sarajevo-Faculty of Pharmacy, Zmaja od Bosne 8, Sarajevo 71000, Bosnia and Herzegovina e-mail: alema.dedic@ffsa.unsa.ba
Emina Boškailo, Department of Ecology and Environmental Protection, Faculty of Social Sciences Dr. Milenko Brkić, Herzegovina University, Mostar 88000, Bosnia and Herzegovina; International Society of Engineering Science and Technology, Nottingham, UK

https://doi.org/10.1515/9783111469713-012

12.1 Introduction

Almost every culture has utilized medicinal plants as a source of treatment for different issues. In both developed and developing nations, ensuring the efficacy, safety, and quality of medical plants and herbal products has emerged as a crucial concern. Over the years, between human societies, the knowledge of their medicinal qualities has been carried down [1]. All around the world, medicinal plants have long been used to cure a wide range of ailments, including asthma, gastrointestinal disorders, skin disorders, respiratory and urinary problems, and cardiovascular and hepatic diseases. To live and thrive in their natural habitat, these plants create a range of physiologically active substances, including defenses against abiotic stresses brought on by temperature, water quality, nutrition and mineral availability, and insect pests. The species of the plant, the kind of soil, and the relationship between the plant and microbes all affect the physiologically active chemicals found in medicinal plants [2-4]. The chemical reactions of plant-associated microbial communities can also be significantly impacted by secondary metabolites produced by aromatic and medicinal plants [5, 6]. Aromatic plant species' biological traits are frequently attributed to active molecules generated during secondary metabolism. Essential oils, which are utilized for many different reasons around the world, including the treatment of infectious diseases, are mostly extracted from these components. Thanks to a growing number of papers on dangerous bacteria that are resistant to antibiotics, antimicrobial qualities of many plants that were once thought to be empirical have now been scientifically verified. In a variety of situations, plant-based products are able to regulate microbial development. The chemical composition of these plant antibacterials and the mechanisms underlying their capacity to suppress microbial growth in the particular context of disease therapy, either by themselves or in combination with traditional antibiotics, have been the subject of several research [7, 8].

Traditional medical methods have gained international attention in the last decade. According to current estimates, an important segment of human beings in many developing nations primarily depends on traditional healers and medicinal herbs to cover their basic medical needs. For historical and cultural reasons, herbal medicines have frequently maintained their popularity even when modern medicine is available in certain nations. Many times, medicinal plants are utilized as raw materials to extract the active compounds needed to make various medications. Plant-based compounds are used in blood thinners, antibiotics, antimalarial drugs, and laxatives. The World Health Organization (WHO) defines health as a condition of total physical, mental, and social wellbeing instead of merely being the absence of illness or disability [9]. For 75–80% of people worldwide, herbal medicine serves as their main source of healthcare. There are 28,187 species of plants that are utilized as medicines by humans, compared to an estimated 374,000 plants in total. Furthermore, the WHO has recognized over 20,000 species of medicinal plants as potential sources of new medications [10, 11]. The global market for herbal products is valued at over USD 62 billion, and it is projected to grow to USD 5 trillion by 2050 [12]. In more than 100 countries, regulations have been put in place regarding medicinal plants. Over 30,000 antibacterial compounds have been found in plants, and more than 1,340 species have been demonstrated to have particular antibacterial properties, and about 1,500 species are recognized for their flavor and fragrance. Furthermore, 74% of bioactive chemicals generated from plants have been found to be based on ethnomedicinal applications, and 14–28% of the most common plant species are thought to be therapeutic [13–15].

12.2 Definition, historical documents, and distribution related to the study of the usage of MAPs

The phrase "medicinal plants" refers to a wide variety of plants used in herbalism, some of which have therapeutic properties. In less-developed countries, more than 3.3 billion people frequently utilize herbal medicines, which are considered the "backbone" of traditional medicine [9, 16]. As the name implies, aromatic plants are those that exude fragrance. Many of them are only utilized in aromatherapy and other medical systems for therapeutic purposes. Since they belong to a unique category known to ethnobotanists as medicinal and aromatic plants, or MAP for short, aromatic plants are usually mentioned in conjunction with medicinal plants. Various writers have attempted to characterize medicinal and aromatic plants traditionally used since ages ago for medicinal purposes using various approaches [17].

Herbal medicine's historical relevance serves as an example of the long-standing connection between people and nature in the quest for health and wellbeing [18]. Many cultures from all over the world have recognized and made use of plants' healing properties throughout history. On a 5,000 year-old Sumerian clay slab from Nagpur, the earliest known written record of the utilization of medicinal herbs to create remedies was found. It had 12 drug preparation directions that cited more than 250 different plants, some of which were alkaloid, including mandrake, henbane, and poppies [19]. The Ebers Papyrus, written approximately 1550 BC, contains 800 prescriptions for 700 plant species and therapeutic cures, such as castor oil plants, pomegranates, aloe, garlic, onions, senna, coriander, figs, willows, junipers, and common centaury [20]. Treatments with plants, which are common in India, is mentioned in the Vedas, the country's sacred texts, around 2000 BC. Many of the spices and plants that are still used today come from India, including cloves, pepper, cinnamon, ginger, and sandalwood [21]. Around 2500 BC, Emperor Shen Nung wrote a book called "Pen T'Sao," which covered 365 dried sections of medicinal plants, many of which are still in use today. These consist of the big yellow gentian, ephedra, ginseng, jimson weed, Theae folium, Podophyllum, Rhei rhisoma, and camphor [22]. Sixty-three plant species from Minoan, Mycenaean, and Egyptian Assyrian pharmacotherapies were referenced in Homer's epics, The Iliad and The Odyssey, which were composed around 800 BC. Named after the Greek word *artemis*, which means "healthy," plants in the genus *Artemisia* were thought to maintain and regain their health [23]. Moreover, Pythagoras named the sea onion (*Scilla maritima*), mustard, and cabbage; Orpheus mentioned the fragrant garlic and hellebore; and Herodotus (c. 500 BC) recorded the castor oil plant. Hippocrates (459–370 BC) listed 300 medicinal plants based on their physiological action: wormwood and common centaury (*Centaurium umbellatum* Gilib) were used to treat intestinal parasites and fever; opium, henbane, fragrant hellebore, mandrake, sea onion, and garlic were used as narcotics; and celery, asparagus, sea onion, parsley, and garlic were used as diuretics [24, 25]. These are only a few historical facts on the recognition and understanding of MAPs as being essential to humanity.

The WHO reports indicate that 30% of the drugs sold worldwide contain substances derived from plant materials, including over 21,000 species. Throughout the world, aromatic and medicinal plants can be found in America, Europe, Africa, Australia, and South and Southeast Asia. More than 7,500 species, or half of India's native plant species, are employed in ethnomedicine there. India is one of the richest sources of MAPs; however, farmers have had limited success in utilizing these plants because they are unaware of their potential and profits. Approximately 6,000 species with therapeutic qualities are used in China. More than 5,000 plant species are used for therapeutic purposes in Africa. In Europe, there are at least 2,000 MAPs in use, two-thirds of which are native to the continent (1,200–1,300). Many of these plants are still being picked in their native environment [26]. Aromatic and therapeutic herbs are also exported in large quantities from Egypt and Turkey. South Asia's top exporters of MAPs are Japan, China, Hong Kong, Singapore, Korea, and Pakistan. In addition to these nations, Pakistan, Bangladesh, Afghanistan, and the Maldives recognized the value of this sector and are encouraging the commercial expansion of these facilities.

12.3 Antibacterial activity of MAPs

Antibacterial resistance has emerged as a result of the widely distributed, inappropriate, irregular, and indiscriminate use of antibiotics, rendering many routinely prescribed drugs useless [14, 27]. According to the WHO, this new tendency is alarming and may represent the most pressing problem confronting medical science. Consequently, there is an intensifying need to develop novel antibacterial agents that can stop the spread of antibiotic resistance and reduce the usage of antibiotics. Since almost 50% of modern medicines and nutraceuticals are natural compounds and their derivatives [28], this has prompted scientists to extract and identify novel bioactive molecules from plants that can combat microbial resistance [29, 30]. Numerous methods have been employed to utilize the nearly limitless supply of bioactive chemicals of medicinal plants as antibacterial agents [31]. However, there is still a lack of comprehensive research on the substances. To increase antibacterial activity against a variety of microorganisms, natural antibacterial compounds can work either alone or in conjunction with antibiotics [32]. Since many medicinal plants' antibacterial properties are still unknown, researchers are focusing more on finding novel, potent therapies that can be developed quickly [33].

12.4 Extracts and essential oils from MAPs as antibacterial agents

Medicinal and aromatic plants are considered potential sources to produce substances that might be used as a replacement for antibiotics to treat bacteria that are resistant to them because they are abundant in a wide range of biologically active compounds, which have been shown to have antibacterial properties *in vitro* [34].

In addition to their many biological qualities (antioxidant, antibacterial, anti-inflammatory, antifungal, and antiviral properties), MAP extracts and essential oils (EOs) have been screened worldwide as potential sources of new antibacterial compounds, alternatives to treat infectious diseases, and agents that help preserve food. The antibacterial chemicals found in medicinal plants may offer a substantial clinical benefit in the treatment of resistant microbial strains and may work differently from currently utilized antibacterials in inhibiting the growth of bacteria, fungi, and viruses. Since *Salmonella, Enterococcus, Staphylococcus aureus, Pseudomonas aeruginosa, Escherichia coli*, and *Shigella* are some of the most prevalent multidrug-resistant bacteria that are acquired in hospitals and the community, using plant extracts and essential oils (EOs) as possible antibacterial agents is crucial [35–37].

When used in conjunction with other medications, some of those active compounds can help bacteria overcome antibiotic resistance, which is a health issue caused by bacterial resistance to several antibiotics, even though they are not as effective as antibiotics alone. In addition to their natural antibacterial properties, some of those compounds can alter antibiotic resistance. The synergistic activity of the active components in medicinal plant extracts is also associated with the extent to which the extracts inhibit bacterial growth. The emergence of multi-target mechanisms, the presence of substances that can inhibit bacterial resistance mechanisms, and pharmacokinetic or physicochemical effects that improve bioavailability, solubility, and resorption rate, lessen toxicity, and mitigate side effects are some of the effects that contribute to the synergistic action [38, 39].

The production and use of EOs is increasing due to their multifunctional applications. Many industries, including pharmacy, medicine, and food preservation, use aromatic EOs [40]. EOs (volatile oils) are aromatic, oily liquids that are extracted from various plant parts, including leaves, fruits, flowers, wood, buds, twigs, bark, seeds, and roots (Figure 12.1) using the Clevenger device. It is expected that they will create new sources of antibacterial medications, particularly those that target bacteria. Aromatic oils have been categorized to have good, medium, or poor antibacterial efficacy and have been screened as potential sources of new antibacterial molecules. Furthermore, in reaction to external stress, aromatic oils may generate specific secondary metabolites to sustain their typical growth and development [41, 42].

Figure 12.1: Clevenger apparatus for extraction of essential oil.

The antibacterial properties of EOs from commonly consumed herbs, such as *Lavandula angustifolia, Lavandula latifolia, Citrus aurantium, Citrus limon, Satureja montana, Satureja hortensis, Hyssopus officinalis, Artemisia vulgaris, Clinopodium nepeta, Taxus baccata, Ocimum basilicum, Salvia officinalis, Origanum vulgare, Mentha spicata, Mentha piperita, Thymus vulgaris*, cinnamon, orange, and lemon, have been evaluated in many countries. Figure 12.2 shows some of the medicinal and aromatic plants collected from Bosnia and Herzegovina and Türkiye.

A number of chemicals influence drug resistance in different gram-negative bacterial species by focusing on efflux routes. The majority of research indicates that gram-positive bacteria are more resistant to EOs than gram-negative bacteria [43]. The hydrophobicity of EOs and their constituents is a key characteristic that enables them to interact with the lipids in the mitochondria and bacterial cell membrane, harming the cell structures and increasing their permeability.

Figure 12.2: (A) *Lavandula latifolia*; (B) *Clinopodium nepeta*; (C) *Taxus baccata*; (D) *Mentha piperita*; and (E) *Salvia officinalis* collected from Bosnia and Herzegovina and Türkiye.

12.5 Compounds of essential oils with antibacterial properties and their activity against a variety of bacterial strains

The main constituents of natural EOs characterized by a strong odor include two groups of different biosynthetic substances, which may determine the ability to fight against bacterial strains [44]. Terpenes and other low molecular weight aliphatic and aromatic compounds compose the majority of EOs. The most valuable compounds and the largest group of plant natural products are terpenes, which have a wide range of structural kinds. They could be categorized as monoterpenes (C10), sesquiterpenes (C15), and diterpenes (C20) based on the variety of their chemical structures. About 90% of EOs consist of monoterpenes, which represent almost all of their constituents. These typically have a nice odor and are volatile in nature [45, 46]. The number, quality, amount, and composition of molecules in the phytochemical profile of essential oils vary depending on the type of extraction, climate, soil composition, plant organ, age, and vegetative cycle time [47]. With differing outcomes, *in vitro* investigations have demonstrated active suppression of bacterial growth. When combined with other antibacterials, EOs can increase antibacterial efficacy and generate additive antibacterial action. The most evaluated components are limonene, pulegone, piperitenone oxide, cinnamaldehyde, geraniol, thymol, menthol, pinene, terpinene, carvacrol, linalool, etc. [48–50]. Clinical applications for essential oils and their components are limited. Some of these substances have been incorporated into topically applied creams, lotions, drops, or liposomal formulations for the treatment of skin conditions or for cosmetic purposes, while others have been employed in respiratory infection inhalation solutions [51].

Recent research has demonstrated the effectiveness of EOs as antibacterial activity restorers against resistant species and penetration enhancers for antiseptics. Antibiotic-resistant bacterial populations have been selected by the continuing application of antibiotics in hospital conditions. This stimulation of efflux pumps promotes multidrug resistance (MDR) [52]. Gram-negative bacteria such *Pseudomonas aeruginosa, Enterobacter* spp., *Escherichia coli*, and *Acinetobacter* are quickly becoming the most difficult to treat due to their nosocomial status, MDR phenotypes, and the fact that the few efflux pump inhibitors (EPIs) that are effective against them are toxic [53]. Finding EPIs that may effectively make MDR gram-negative bacteria vulnerable to antibiotics to which they are initially resistant is of utmost importance [54].

Finding EPIs in *Helichrysum italicum* EOs that are effective against the efflux processes of gram-negative bacteria was the aim of the study conducted by Lorenzi et al. [56]. According to additional research, using 2.5% of *H. italicum* essential oil reduces the minimum inhibitory concentration (MIC) of chloramphenicol for the Enterobacter aerogenes MDR strain EA27 by eight times, from 1,024 to 128 mg/L [52]. The remaining, less active plants reduce the MIC of chloramphenicol by two to four times. This EO was selected for research by Lorenzi et al. (2009) because it was able to reduce EA27's chloramphenicol resistance to a level equivalent to that of the control phenylalanine arginine-naphthylamide (PAN) [55]. According to this study, chemicals found in *H. italicum* essential oil target efflux pathways to change drug resistance in a variety of gram-negative bacterial species. The results show that EO reduces the chloramphenicol MIC for strains of *A. baumannii, P. aeruginosa*, and *E. aerogenes* isolates. Furthermore, for a strain of *E. aerogenes* (CM-64) that overproduced the tripartite efflux pump AcrAB-TolC, this EO reduces the MIC of chloramphenicol. Compounds not previously identified as modulators are present in the two most active fractions. Geraniol seemed to be the most potent of these substances at inhibiting efflux pathways. Curiously, the same geraniol that was initially tested for resistance to chloramphenicol was also found to be effective in inhibiting resistance to other clinically significant antibiotics, such as β-lactams and the fluoroquinolone norfloxacin. Additionally, an EPI activity ranking shows that geraniol is a more effective inhibitor of resistance in an acrAB mutant than PAN, suggesting that these two substances target different molecules. Together, these results provide a new supply of drugs that may help cure the condition, and geraniol may help us understand MDR in gram-negative bacteria that continue to pose a threat to public health [56].

12.6 Major groups with antibacterial activity from MAPs

Phenolics and polyphenols, terpenoids, and alkaloids are the main classes of antibacterial compounds produced by plants (Figure 12.3). Complex combinations of these groups are typically found in bioactive plant extracts, and when utilized together, they can have an even greater effect. These compounds are frequently used by plants as defensive mechanisms against insects, herbivores, and microbes. Some are responsible for plant pigment, while others, such as terpenoids, give plants their scents. Plant flavor is

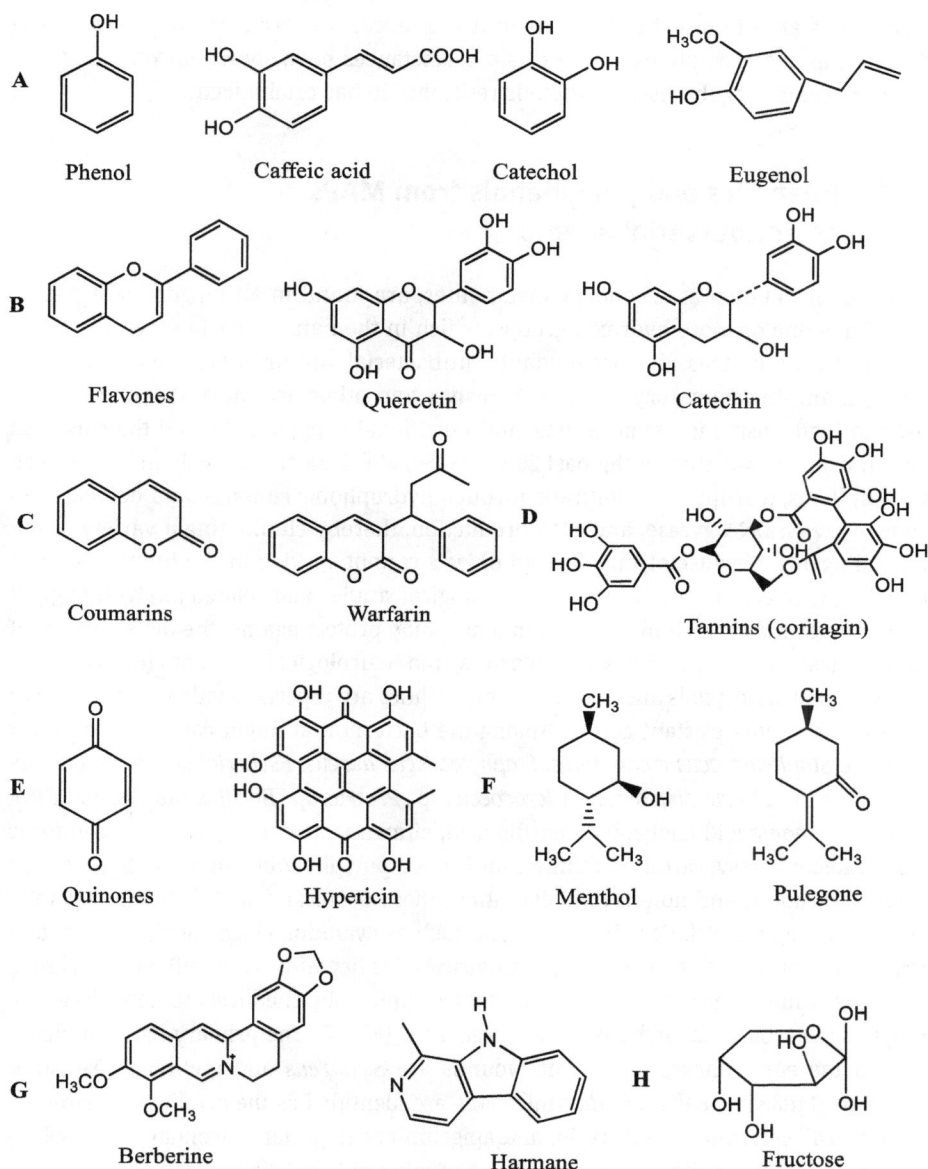

Figure 12.3: Structures of common antibacterial plant compounds: (A) phenols and phenolic acids; (B) flavonoids; (C) coumarins; (D) tannins; (E) quinones; (F) terpenoids; (G) alkaloids; and (H) sugars.

caused by a variety of components, and some of the similar herbs and spices that people use to season food also contain beneficial medical properties [57]. Despite the fact that many nations have previously approved synthetic antibacterial medicines, many researchers are interested in using natural substances that are derived from microorganisms, animals, or plants. These organic substances have shown encouraging outcomes in combating the rise of antibiotic resistance in bacterial infections [58].

12.6.1 Phenolics and polyphenols from MAPs as antibacterial agents

Polyphenols, known as secondary metabolites, are found in all kingdoms of plants. They have one or more hydroxyl groups, which in the natural world serve a number of biological purposes, like antioxidant, antibacterial, antiproliferative agents, antiallergic, anti-inflammatory, antihypertensive, and other activities. Polyphenols are used in food, cosmetics, medications, and nutritional supplements, and their use has grown dramatically during the past 20 years [59, 60]. Bioactive polyphenols enter cellular systems, disrupt the membrane through hydrophobic contacts, and decrease enzyme activity, DNA gyrase, and RNA production, thereby eliminating a variety of microbial agents. Because of this, foreign objects cannot survive in the human body or interfere with cellular processes. Epidemiological studies and related analyses suggest that long-term diets rich in plant polyphenols may protect against the development of cancer, heart disease, diabetes, osteoporosis, and neurological disorders [61–63].

Workers at hospitals and assisted living facilities are especially vulnerable to a large class of antibiotic-resistant germs. Among the bacteria that might cause issues in our lives are *Staphylococcus epidermidis*, *Staphylococcus aureus*, *Escherichia coli*, *Pseudomonas aeruginosa*, *Acinetobacter* sp., *Micrococcus* sp., *Proteus* sp., *Bacillus subtilis*, and *Klebsiella pneumoniae*. Phenolic acid, ferulic acid, cinnamic acid, sinapic acid, *p*-coumaric acid, catechin, resveratrol, curcumin, and other polyphenolic compounds (phenolic acids, flavonoids, and non-flavonoids) inhibit these bacteria, which is highly advantageous and helpful [64]. Certain substances, such as cyanidin, ellagic acid, luteolin, and resveratrol, may be able to kill dangerous viruses like hepatitis B and influenza and save our lives. You can use these more important polyphenolic chemicals to defend against fungi, viruses, bacteria, and other microorganisms [65–67]. The primary bacteria identified in the early phases of chronic wounds are *S. aureus* and methicillin-resistant *S. aureus* (MRSA); *E. coli* and other infections are identified as the condition progresses. Kaempferol, catechins, lutein, rutin, and apigenin are important secondary metabolites that aid in wound healing. Tannic acid has a number of beneficial properties, which also make it an effective compound for wound treatment [68, 69]. The numerous antibacterial benefits of polyphenols help fight off viruses, fungi, and bacteria. They can disrupt and interfere with cell membranes through quorum sensing; they can also chelate metal ions, block enzymes, generate reactive oxygen species (ROS), change the host immune response, and stop viruses from invading and growing. The bacteria species can be destroyed by polyphenols through these mechanisms. They can also improve resistance against microbiological infections by modifying the host immune response. The ability of polyphenols to degrade microbial cell membranes contributes to their antibacterial properties [70, 71]. Figure 12.4 presents polyphenols utilizing *in vitro* assays and their applications and possible antibacterial mechanisms [72].

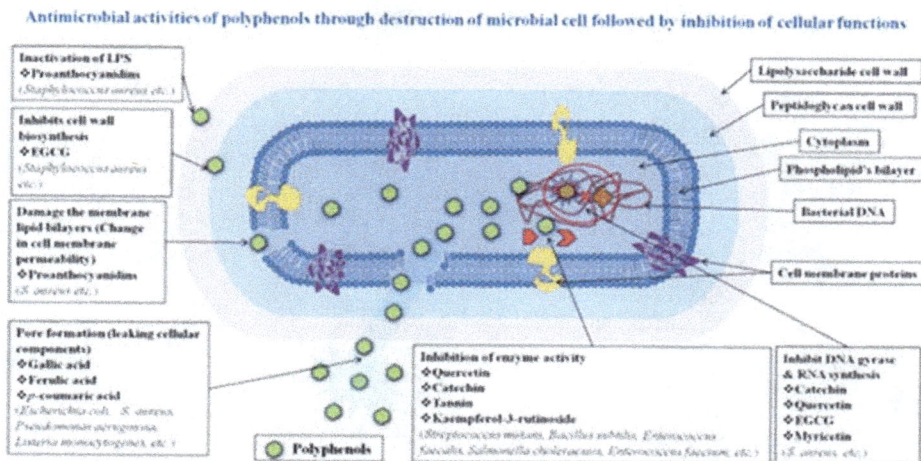

Figure 12.4: Antibacterial activity of polyphenols through inhibition of intracellular functions. Figure reused from open-access article reference [72].

To improve targeted and controlled release of polyphenols against microorganisms, it is necessary to make them more soluble [73]. In aqueous media, naturally occurring polyphenols that have been extracted from various plant parts (fruits, leaves, flowers, etc.) display reduced solubility. Because of this, it is required to turn them into salts in order to improve their solubility, which can be more beneficial in various dietary and medicinal applications [74, 75]. In order to discover novel biologically active compounds and increase the number of alternative raw materials for pharmaceutical and medical applications, future research should concentrate on wild or endangered species as well as medicinal and aromatic plant species. The development and targeting of drugs depend on an understanding of the underlying mechanisms of several well-known phenols, including the signaling routes and molecular processes by which they operate. Different phenolic components with a wide range of phytochemical characteristics can be obtained from a variety of aromatic and therapeutic plant kinds. Therefore, in order to find novel substances, future research should keep investigating other cultivars [76]. Table 12.1 presents some polyphenols with antibacterial properties.

Table 12.1: Summary of the antibacterial activity of some plant-derived polyphenols.

Polyphenols	Structure	Target microorganism	References
Catechins		*Helicobacter pylori* *Streptococcus mutans*	[77, 78]
Quercetin		*Pseudomonas aeruginosa* *Lactobacillus casei* var. *Shirota Proteus vulgaris* *Staphylococcus aureus* *Shigella flexneri* *Escherichia coli*	[79–81]
Resveratrol		*Helicobacter pylori* *Bacillus cereus* *Escherichia coli* *Staphylococcus aureus*	[82–84]
Curcumin		*Helicobacter pylori* *Staphylococcus aureus* *Escherichia coli* *Pseudomonas aeruginosa* MRSA strain	[85–88]
Naringenin		*Salmonella typhimurium* *Pseudomonas aeruginosa* *Klebsiella pneumoniae* *Escherichia coli* *Bacillus subtilis* *Staphylococcus aureus*	[89–91]
Apigenin		*Bacillus subtilis* *Pseudomonas aeruginosa* *Staphylococcus aureus* *Escherichia coli*	[92]
Luteolin		*Helicobacter pylori* *Escherichia coli* *Trueperella pyogenes* *Pseudomonas aeruginosa* *Staphylococcus aureus*	[93–95]

Table 12.1 (continued)

Polyphenols	Structure	Target microorganism	References
Daidzein		*Escherichia coli* *Acinetobacter baumannii* *Listeria monocytogenes* *Bacillus cereus* *Klebsiella pneumoniae* *Vibrio parahaemolyticus* *Pseudomonas aeruginosa* *Staphylococcus aureus* *Salmonella typhimurium* *Cronobacter sakazaki*	[96–98]
Ellagic acid		*Streptococcus mutans* *Helicobacter pylori*	[99–101]
Caffeic acid		*Pseudomonas aeruginosa* *Escherichia coli* *Staphylococcus aureus*	[102, 103]
Gallic acid		*Pseudomonas aeruginosa* *Staphylococcus aureus* *Klebsiella pneumoniae* *Escherichia coli* *Shigella flexneri* *Listeria monocytogenes*	[104–106]

12.6.2 Terpenoids from MAPs as antibacterial agents

One important source of naturally occurring bioactive compounds is terpenoids, sometimes referred to as isoprenoids. They include over 60,000 primary and secondary metabolites, such as monoterpenes (53%), diterpenoids (1%), sesquiterpenes (28%), and others (18%). Growth hormones, photosynthetic pigments, fragrance chemicals, and a variety of terpenoids (important metabolites) are produced by many plants [107]. The basic unit of terpenes is the isoprene unit (C_5H_8). It is the principal precursor and can undergo post-modification either in the cytosolic mevalonate (MVA) pathway or the plastid methyl erythritol phosphate (MEP) pathway. Because of their lipophilic qualities, terpenoids are currently one of the primary classes of antibacterial drugs that combat a wide range of microorganisms [108]. Previous research

has identified five primary pathways by which terpenoids exhibit antibacterial activity, which are:
1. Cell membrane destruction
2. Anti-quorum sensing (QS) action
3. Inhibition of ATP and its enzyme
4. Inhibition of protein synthesis
5. The synergistic effect

1. Cell membrane destruction: Terpenoids primarily destroy the bacterial cell membranes by using their lipophilicity. They have bactericidal or antibacterial activities by diffusing inward via the phospholipid bilayer of bacteria [109]. Since the integrity of the cell membrane is crucial to bacterial biological processes, terpenoids' damage to the membrane will impair the bacteria's basic physiological functions and lead to the loss of vital components like proteins and enzymes, which will ultimately result in the antibacterial effect [110]. Table 12.2 shows some terpenoids that inhibit the growth of microorganisms through this mechanism.

2. Intercellular communication is a function of the anti-quorum sensing (QS) system [108]. Bacteria use it as a communication tool to coordinate their interactions with other organisms, which is also the primary cause of antibiotic resistance [111]. The literature has presented and provided illustrations of the gram-positive and gram-negative bacteria's group sensing signal loop [112]. Research has demonstrated that the QS action between bacteria can be efficiently inhibited by a low quantity of cinnamon aldehyde [113]. QS can be efficiently inhibited by low quantities of carvacrol and thymol, which block the bacterial self-inducer acyl homoserine lactone (AHL) [114].

3. Inhibition of ATP and its enzyme: The main direct source of energy in living things, ATP is also necessary for microorganisms to maintain their regular functions. The antibacterial effect of terpenoids is carried out by rupturing the cell membrane, which results in a difference in the concentration of ATP inside and outside the cell [109]. For example, the terpenoids thymol and eugenol may have a fungicidal effect against *Candida albicans* by inhibiting H^+-ATPase, which targets the cell membrane and causes intracellular acidification and cell death [115]. In another study, the researchers used the MIC of carvacrol to treat the target infection. A luminometer (Biotek) was used to test the samples' levels of extracellular ATP. Based on absorbance analysis at 260 nm, this study discovered that carvacrol harmed the *E. coli* membrane and that potassium and ATP ions were also discharged [116].

4. Inhibition of protein synthesis: Protein synthesis is essential to bacterial physiological function. By preventing any step in the protein synthesis pathway, terpenoids, which are inhibitors of protein synthesis, may achieve an antibacterial impact. According to some research, cinnamaldehyde can lessen the binding and *in vitro* assembly reactions of the prokaryotic tubulin homolog FtsZ (filamenting temperature-

sensitive mutant Z)-type protein, which controls cell division. Additionally, by binding to FtsZ, preventing GTP hydrolysis, and disrupting the z-loop of cell dynamics, this chemical has antibacterial properties against bacteria [117]. The most current work includes calculations, biochemistry, and in vivo cell-based studies to confirm that cinnamaldehyde is a potential inhibitor of S. typhimurium (stFtsZ). Up to 70% of the activity and polymerization of stFtsZ GTPase are inhibited by it [118].

5. The synergistic effect: For instance, eugenol, carvacrol, and thymol have a synergistic antibacterial action because they may pass extracellular membranes. This is because they can either increase the number, size, and duration of holes that bind to membrane proteins for increased antibacterial activity, or allow eugenol to reach the cytoplasmic membrane [119].

Table 12.2: An overview of antibacterial properties of certain plant-derived terpenoids.

Terpenoids	Structure	Target microorganism	References
Limonene		*Acinetobacter baumannii* MRSA *Escherichia coli* *Candida albicans* *Salmonella enterica*	[120, 126]
Thymol		*Salmonella typhimurium* *Escherichia coli* *Brochothrix thermosphacta* *Staphylococcus aureus* *Pseudomonas fluorescens* *Pseudomonas fluorescens*	[114, 115, 119–121]
Carvacrol		*Salmonella typhimurium* *Escherichia coli* *Brochothrix thermosphacta* *Staphylococcus aureus* *Pseudomonas fluorescens* *Pseudomonas fluorescens*	[114, 116, 119–121]
Menthol		*Staphylococcus aureus* *Escherichia coli*	[122]

Table 12.2 (continued)

Terpenoids	Structure	Target microorganism	References
Eugenol		*Staphylococcus aureus* *Salmonella typhimurium* *Brochothrix thermosphacta* *Pseudomonas fluorescens* *Escherichia coli* *Pseudomonas aeruginosa* *Klebsiella pneumoniae*	[114, 115, 119, 121, 132]
Cinnamaldehyde		*Salmonella typhimurium* *Brochothrix thermosphacta* *Escherichia coli* *Staphylococcus aureus* *Pseudomonas fluorescens*	[113, 117, 118, 120]
1,8-Cineole		*Acinetobacter baumannii* *Candida albicans* MRSA *Escherichia coli*	[123]
(+)-α-Pinene		*Cryptococcus neoformans* *Rhizopus oryzae* *Salmonella enterica* *Staphylococcus aureus* *Escherichia coli* *Micrococcus luteus* MRSA	[124–126]
(+)-β-Pinene		*Rhizopus oryzae* *Cryptococcus neoformans* MRSA	[124]
α-Terpineol		*Escherichia coli* *Salmonella enterica* *Staphylococcus aureus*	[127]
Geraniol		*Salmonella enterica* *Salmonella enteritidis Klebsiella pneumoniae Escherichia coli* *Staphylococcus aureus* *Pseudomonas aeruginosa Enterococcus faecalis* *Listeria monocytogenes Proteus mirabilis*	[127–129]

Table 12.2 (continued)

Terpenoids	Structure	Target microorganism	References
Nerol		*Escherichia coli* *Staphylococcus aureus* *Salmonella enterica* *Streptococcus agalactiae* *Bacillus subtilis*	[127, 130, 131]
Linalool		*Staphylococcus aureus* *Klebsiella pneumoniae* *Salmonella enterica* *Pseudomonas fluorescens* *Pseudomonas aeruginosa* *Escherichia coli*	[127, 132, 133]
Myrcene		*Salmonella enterica* *Staphylococcus aureus* *Escherichia coli*	[127]

12.6.3 Alkaloids from MAPs as antibacterial agents

The structurally diverse class of nitrogen-containing organic compounds known as al-kaloids includes over 20,000 distinct compounds with a basic nitrogen atom that can exist as a primary amine (RNH_2), secondary amine (R_2NH), or tertiary amine (R_3N) [134]. Alkaloids can be categorized into two primary types based on their natural origin or chemical makeup. Alkaloids are classed as either heterocyclic or typical (also called true alkaloids) with nitrogen in the heterocycle or non-heterocyclic or atypical (also called protoalkaloids or biological amines) with nitrogen in the side chain. Because of its structural complexity, the second group can be further divided into 14 subgroups based on the ring structure, as shown in Figure 12.5 [135].

This natural group of compounds exhibits a variety of pharmacological actions [136–138]. Articles on the antibacterial properties of alkaloids produced from plants have become regular these days. They may be useful in treating resistant microbial strains and may be able to inhibit the growth of bacteria, viruses, protozoa, and fungi in a variety of ways [139]. As efflux pump inhibitors (EPIs), the majority of alkaloids exhibit antibacterial properties. For example, the bacterial and fungal efflux pumps can be competitively inhibited by quinolines, isoquinolines, monoterpene indoles, steroidal alkaloids, and protoberberines [140]. When coupled with ciprofloxacin, piperine, an alkaloid of the piperidine class, exhibits potent antibacterial activity against a range of bacterial strains and functions as an EPI in *S. aureus* [141, 142].

By blocking the synthesis and repair of nucleic acids, some alkaloids, such as berberine (an isoquinoline alkaloid), which is a potent DNA intercalator that accumulates

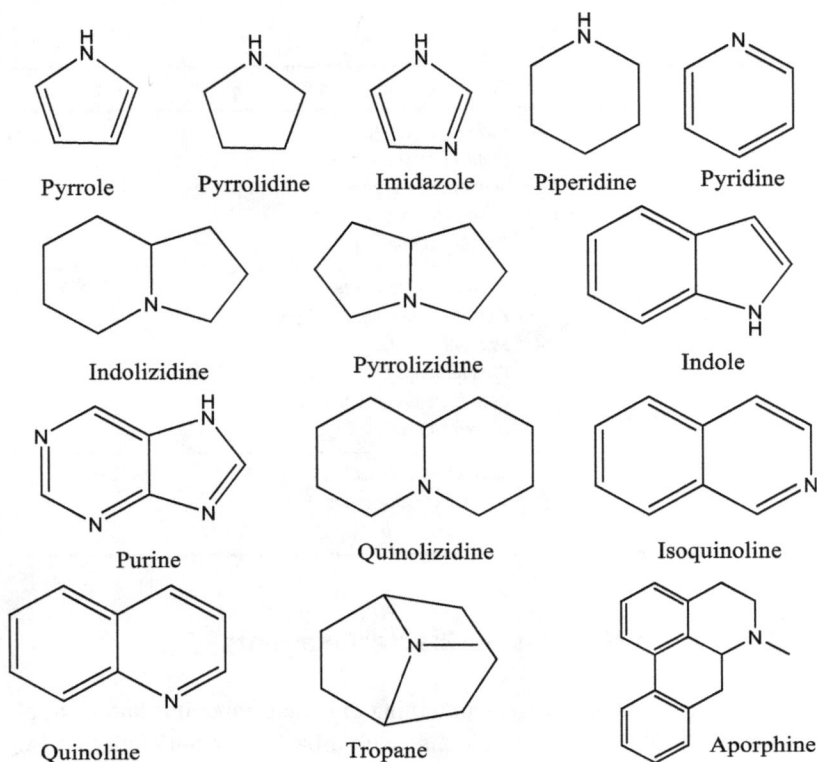

Figure 12.5: The 14 subgroups of alkaloids based on the ring structure.

under the influence of cell membrane potential, have antibacterial properties [143]. Additionally, by inhibiting the MexXY-OprM efflux pump system, berberine and the antibiotic carbapenem work together to resensitize imipenem-resistant *P. aeruginosa* [144, 145]. The isoquinoline alkaloid chelerythrine exhibits strong antibacterial activity against *S. aureus*, extended-spectrum β-lactamase *S. aureus* (ESBLs-SA), and MRSA by preventing cellular division and nucleic acid synthesis [146].

Certain alkaloids have an antibacterial effect via changing the permeability of the membrane. 8-hydroxyquinoline, for example, has antimicrobial properties against *Streptococcus pneumoniae, Haemophilus influenzae*, and *Staphylococcus aureus*. Its high lipophilicity allows it to penetrate bacterial cell membranes and reach its target site of action [147, 148].

A number of alkaloids have antibacterial qualities via preventing enzyme activity. By inhibiting the enzymatic activities of human DNA polymerases α and β as well as reverse transcriptases from HIV strains 1–2, michellamine B, a substance derived from the tropical plant *Ancistrocladus korupensis*, has shown anti-HIV action [149]. The benzophenanthridine alkaloid sanguinarine has antibacterial qualities by stopping the growth of microbes. It may disrupt Z-ring formation and stop cytokinesis in both gram-positive and gram-negative bacteria by blocking FtsZ binding [150]. By changing how FtsZ protofilaments bind, sanguinarine may also have a bacteriostatic effect [151]. Some significant alkaloids that prevent the growth of the different bacterial strains described in the publications are shown in Table 12.3.

Table 12.3: Summary of the antibacterial activity of some plant-derived alkaloids.

Alkaloids	Structure	Target microorganism	References
Berberine		*Escherichia coli* *Fusobacterium nucleatum* *Pseudomonas aeruginosa* *Micrococcus luteus* *Prevotella intermedia* *Bacillus subtilis* *Eberthella typhosa* MRSA	[142, 143, 150–152]
Reserpine		*Staphylococcus aureus* *Citrobacter freundii* *Enterococcus faecalis* *Escherichia coli Salmonella typhimurium*	[153, 154]
Piperine		*Pseudomonas aeruginosa* *Salmonella* sp. *Proteus vulgaris* *Bacillus subtilis* *Escherichia coli* *Staphylococcus aureus* *Klebsiella pneumoniae*	[142, 143, 155]
Chelerythrine		*Staphylococcus aureus* Spectrum β-lactamase *S. aureus* (ESBLs-SA) *Streptococcus agalactiae* *Escherichia coli* *Aeromonas hydrophila* MRSA	[156–158]
Roemerine		*Bacillus subtilis* *Escherichia coli* *Staphylococcus aureus*	[159–161]

Table 12.3 (continued)

Alkaloids	Structure	Target microorganism	References
Evodiamine		*Mycobacterium tuberculosis*	[162, 163]
Sanguinarine		*Klebsiella pneumoniae* *Pseudomonas aeruginosa* *Streptococcus pyogenes* MRSA	[148, 149]

12.7 Some medicinal and aromatic plants with antibacterial activity

12.7.1 *Prunus spinosa* L.

Prunus spinosa L. is a plant of the Rosaceae family, also referred to as blackthorn or sloe, which grows as a shrub on the slopes of wild, uncultivated terrain. Phenolics, alkaloids, terpenes, and sterols are among the powerful natural bioactive compounds found in this traditional medicinal plant, which has been used to handle a variety of illnesses. Blackthorn extracts have been found to contain the following polyphenolic compounds: kaempferol, quercetin, phenolic acids (caffeine and neochlorogenic derivatives), coumarin derivatives (umbelliferone, scopoletin, and esculetin), and anthocyanins, which are thought to be among the most potent natural antioxidants and antibacterial agents [164–166]. Fruits and leaves of blackthorn collected in Bosnia and Herzegovina are presented in Figure 12.6.

Because of their diuretic, spasmolytic, antibacterial, and antioxidant properties, all organ parts of blackthorn have therapeutic uses and are utilized for treating a wide range of disorders [167]. The fruit, for example, is used to manufacture tea, juice, and distillates utilized in the food industry, as well as several kinds of traditional jams and drinks [166]. Although polyphenolic compounds found in fruit extracts can significantly lessen the negative effects of free radicals and encourage the growth of pathogens in the body, extracts from blackthorn flowers are suggested for the treatment of urinary tract disorders, inflammation, and cardiovascular diseases [168].

In their study, Dedić et al. [165] documented the antibacterial property of ethanol extracts of blackthorn flowers, leaves, and fruits was tested against *Staphylococcus aureus*, *Bacillus subtilis*, *Escherichia coli*, *Pseudomonas aeruginosa*, *Salmonella enterica*, and *Enterococcus faecalis*, and antifungal property against *Candida albicans*. All investigated ex-

Figure 12.6: Fruits and leaves of *Prunus spinosa* L. from Sarajevo, Bosnia and Herzegovina.

tracts displayed effective antibacterial activity against those bacterial strains. These findings are in accordance with the study by Veličković et al. [168]. Extracts from *P. spinosa* may be utilized as additional sources of functional additives and may be a promising natural antibacterial agent that may be used to combat microbial resistance.

12.7.2 *Clinopodium nepeta* (L). Kuntze

Approximately 135 blooming species belong to the genus *Clinopodium nepeta* L. Kuntze, which is a part of the Lamiaceae family and is widely distributed over the Mediterranean, southern and southeastern Europe, Latin and North America, and even western Asia. This plant is also called *Calamintha, Satureja,* and *Thymus. Calamintha nepeta* (L.) Savi subsp. *nepeta* is the most frequently used synonym [169]. This genus is frequently abundant in essential oils and phenolic compounds, as well as consisting of flavonoids, alkaloids, terpenes, saponins, sterols, tannins, and glycosides. Due to possessing all these compounds, this aromatic and medicinal plant demonstrates numerous biological activities, including antioxidant, antibacterial, anti-inflammatory, antifungal, and antiviral [170–172]. In tea form, it has long been used to treat gastrointestinal disorders and reduce gas and cramps [173, 174]. While the EOs are used as a spice in Italian homes, they are also utilized as external compresses to treat hip pain [173, 175] and to reduce headaches, sleeplessness, and respiratory ailments [173, 176].

The study by [218] used GC and GC-MS to analyze an oil of *C. nepeta* (L.) Savi ssp. glandulosa made by hydrodistillation, where 36 components (98.4%) were identified.

Pulegone (37.5%), menthone (17.6%), piperitenone (15.0%), and piperitone (10.2%) were the EO's primary ingredients. The EOs antibacterial properties were examined against *Bacillus subtilis, Salmonella enteritidis, Aspergillus niger, Staphylococcus aureus, Escherichia coli*, and *Pseudomonas aeruginosa*. It was discovered that the microbes were vulnerable to the oil [177].

Figure 12.7 presents the aromatic plant *C. nepeta* L. Kuntze collected in the sub-Mediterranean area in Bosnia and Herzegovina.

Figure 12.7: *Clinopodium nepeta* L. Kuntze from Mostar, Bosnia, and Herzegovina.

Boškailo et al. [48, 169] reported that *C. nepeta* EOs collected in four areas in Bosnia and Herzegovina contained 42 compounds, including piperitenone oxide (60.2%), piperitenone (48.8%), and pulegone (44.8%) as the major compounds, followed by *p*-menthone, limonene, *cis*-piperitone oxide, and dihydrocarvyl acetate. These compounds could be a good source of antibacterial agents.

12.7.3 *Lavandula officinalis*

Lavender is a medicinal and aromatic plant belonging to the Lamiaceae family, which is valued mostly for its pleasing aroma. *Lavandula L.*, which comprises 41 species of flowering plants, has been used for a variety of uses since the times of ancient Greece and Rome. The flower and essential oil of lavender are used mostly in the toiletry and fragrance industries, aromatherapy, and folk medicine to treat a range of gastrointestinal and rheumatic disorders, depression, anxiety, and headaches [178, 179]. Lavender oil, derived from a number of plant species, is one of the most popular essential oils. The four primary species of lavender are *Lavandula latifolia, Lavandula angustifolia, Lavandula stoechas,* and *Lavandula × intermedia,* which is a sterile hybrid of *L. latifolia* and *L. angustifolia* [180]. *L. angustifolia,* commonly referred to as true lavender or commercial lavender, is the species that is most frequently grown among them. Monoterpenoids and sesquiterpenoids represent the majority of EO, with linalool and linalyl acetate being the most prevalent. Although other less common essential oil constituents (such as terpinen-4-ol, camphor, 1,8-cineole, carvacrol, lavandulyl acetate, and lavandol) have also been assessed and demonstrated synergistic effects alongside the main chemicals, the majority of studies have concentrated on the two primary constituents of most lavender EOs (linalyl acetate and linalool) [181–184]. The chemical composition of EO is extremely complicated and can vary greatly based on a number of variables, including the plant's morphological traits, processing methods, environmental circumstances, and cultivation area [185]. Furthermore, how EOs exhibit their biological function is influenced by their chemical composition [186].

Against both gram-positive and gram-negative bacteria, lavender oil demonstrated potent antibacterial activity [187, 188]. Linalyl acetate and linalool have been found to be strong antibacterial agents against pathogenic bacteria, including *E. coli* and *E. cloacae* [189, 190]. *L. angustifolia* Mill. and *L. latifolia* Vill., two lavender species grown in gardens in Sarajevo, were examined by Dudžević-Čančar and colleagues for their ability to fend off the fungus *C. albicans* and the bacterial strain *M. luteus.* According to the findings, the evaluated EOs exhibited potent antibacterial activity against *M. luteus* strains [179].

Lavender is now thriving and cultivated in botanical gardens and in private home gardens throughout Europe as well as in Bosnia and Herzegovina (Figure 12.8.).

The Lis-Balchin study presents that EO inhibits the growth of *S. enterica, A. hydrophila,* and *C. freundii* strains in disk diffusion tests, while the Danh et al. [194] study showed antibacterial property against *P. aeruginosa* and *E. faecalis* [192, 193]. The type of bacterium as well as the amount of active ingredients determines the antibacterial activity of plant EOs. Gram-negative bacteria are more resistant due to the hydrophilic lipopolysaccharides (LPS) in their membrane, which function as a barrier against hydrophobic and macromolecules [193, 194].

12.7.4 *Helichrysum italicum*

The perennial subshrub *Helichrysum italicum,* belonging to the genus *Helichrysum* and the family Asteraceae, has yellow flowers and grows in Mediterranean regions' alkaline, dry, sandy, and poor soil. Its choleretic, diuretic, and expectorant qualities

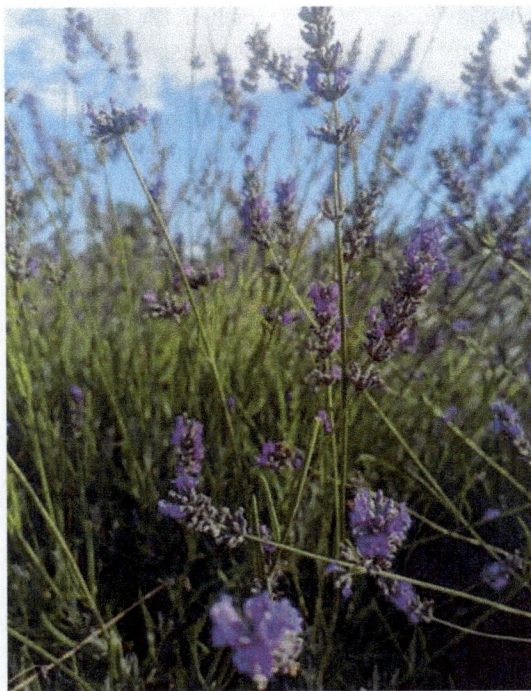

Figure 12.8: *Lavandula angustifolia* from Sarajevo Garden, Bosnia and Herzegovina.

have long been recognized in folk medicine [195, 196]. The unique EO composition and aroma of the *Helichrysum* species have attracted the interest of the pharmaceutical, cosmetic, and fragrance industries, which prompted new research on the topic. The commercial exploitation of wild *H. italicum* populations increased significantly in the Eastern European Mediterranean countries like Bosnia and Herzegovina and Croatia. Numerous pharmacological properties, including antioxidant, antibacterial, antiathero-sclerotic, antiproliferative, antidiabetic, neuroprotective, and anti-inflammatory properties, are present in EOs and extracts from this plant species [197–199].

Its blossoms and leaves are the parts most often used in Bosnia and Herzegovina (Figure 12.9), Spain, Portugal, and Italy to cure conditions like allergies, colds, coughs, issues of the liver, gallbladder, and skin, as well as inflammation, infections, and insomnia. A variety of scientific investigations have been carried out in recent decades to confirm some of the traditional uses and to identify additional possible uses for its extracts and isolated components. Also, it has been described as an antibacterial and anti-inflammatory agent *in vitro*. Its terpenoids, acetophenones, and phloroglucinols showed antifungal efficacy against *C. albicans*; flavonoids and phloroglucinols suppressed HIV and HSV, respectively; and its terpenes and flavonoids were efficient against bacteria such as *S. aureus* [197, 200–202].

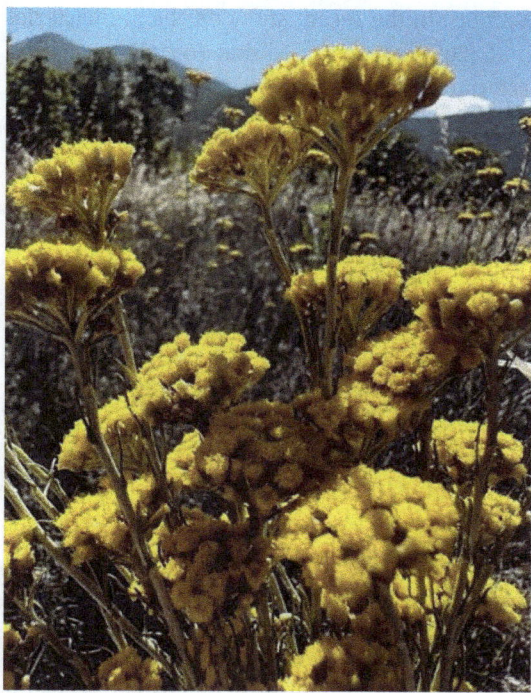

Figure 12.9: *Helichrysum italicum* from Mostar, Bosnia and Herzegovina.

Zheljazkov et al. [204] reported that *H. italicum* EO included 79 components, while *H. arenarium* EO contained 75 components. α-pinene (34.64–44.35%) and sabinene (10.63–11.1%) were the primary constituents of *H. arenarium* EO, confirming the population being studied as a novel chemical type. Originating in France, Bosnia and Herzegovina, and Corsica, the main constituents of *H. italicum's* EO were neryl acetate (4.04–14.87%) and β-himachalene (9.9–10.99%). Nonetheless, there were some differences in the EO profiles of *H. italicum* imported from the three aforementioned nations. *H. italicum* transplanted from France was dominated by neryl acetate, italicene, and α-guaiene (14.87%), D-limonene (5.23%), but plants brought from Bosnia and Herzegovina were dominated by α-pinene (13.74%), δ-cadinene (5.51%), β-caryophyllene (3.65%), α-cadinene (3.3%), and α-calacorene (1.63%). EOs from all three countries show antibacterial properties against the following bacterial strains: *E. faecalis*, *S. aureus* subs. *aureus*, *P. aeruginosa*, *S. pneumonia*, *Y. enterocolitica*, *S. enterica* subsp. *enterica*, *C. krusei*, and *C. tropicalis* [203].

H. italicum extracts also show antibacterial properties against a variety of bacterial strains. Ethanol extracts from *H. italicum*, *H. armenium*, *Gravolens*, and *plicatum* have been shown in recent research to be effective against *S. aureus* [204]. A few scientists [206–208] noted that gram-positive bacteria were sensitive to dichloromethane extract from *H. stoechas* and *H. aureonitens*. In the recent study on the antibacterial properties of plant extracts, Nostro et al. [209] demonstrated that *H. italicum* diethyl ether extract exhibited the best antibacterial activity against *S. aureus* [195, 208].

12.7.5 *Mentha piperita*

The genus *Mentha* (often called mint), belonging to the family Lamiaceae, includes a diverse group of 31 species and hybrids that differ widely in their biological characteristics. Mint is a perennial, potently fragrant medicinal herb that grows both wild and under cultivation in many countries across Europe and Asia. It is frequently used as a spice, an aroma component, in cosmetics, in the pharmaceutical industry, as well as in the form of tea, and hot or cold beverages. Because of its antioxidant potential, low toxicity, and high efficacy, the *Mentha* species has several health-promoting qualities, including antibacterial, anti-inflammatory, antidiabetic, and cardioprotective benefits [209–211]. The plant is aromatic and a stimulant and is used actually as a rub or liniment and internally as a tea, tincture, oil, or extract for allaying nausea, headaches, and vomiting. Mint oils are known to contain numerous monoterpenoids, with pulegone, D-limonene, piperitone, 1,8-cineole, piperitone oxide, menthone, piperitenone, menthol, β-caryophyllene, and carvone as predominating compounds. Due to this, it is among the most often used EOs in alcoholic liquors, mouthwash, cosmetics, medicines, food goods, and dental preparations. Nonetheless, chemogeographical diversity in the EO composition of the *Mentha* species has been noted, as well as some differences in the constituents of this oil from other nations [212, 213]. Figure 12.10 presents *Mentha* species from Türkiye.

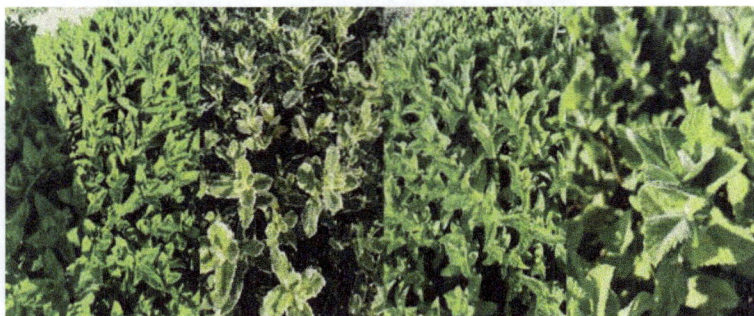

Figure 12.10: Different *Mentha* species from Isparta, Türkiye.

According to the chemical composition results of the study by Džudžević-Čančar et al. [180, 212], linalool (35.40%) is the main component, followed by linalyl acetate (28.60%), 1,8-cineole (6.00%), and geranyl acetate (2.60%), which is known as the linalool-linalyl acetate chemotype. This is in accordance with the previously published studies [211, 214].

Sujana et al. [216] investigated the antibacterial potential of six extracts (methanol, ethanol, chloroform, hexane, petroleum ether, and ethyl acetate) from the leaf, stem, and root of *M. piperita* against bacterial strains *S. aureus, S. pneumonia, E. coli, B. subtilis, K. pneumonia,* and *P. vulgaris* by the agar-well diffusion method. It was discovered that all leaf extracts had potent antibacterial qualities against these infections [216]. Antibacterial effects of spearmint oil from *M. spicata* var. *crispa*, peppermint oil from *Mentha piperita,* and cornmint oil from *M. arvensis* were tested against the various strains of *S. aureus,* including methicillin-resistant *S. aureus* (MRSA), in the research of Horváth and Koščová [217]. *M. spicata* var. *crispa* EO had the best resistance against *S. aureus* and also the MRSA strain [216].

12.8 Conclusion

Medicinal and aromatic plants, as well as the compounds they produce, known as phytochemicals, are in the focus of researchers because of their significant influence on human health. Many natural medical preparations that have been used for centuries to promote health and treat and prevent a variety of illnesses are mostly derived from plants. Plant-based natural substances have antifungal, antibacterial, anticarcinogenic, anti-inflammatory, antidiabetic, antiviral, antimutagenic, and antiallergic effects often linked to their antioxidant properties. Volatile compounds, commonly referred to as EOs and plant extracts, are one specific class of secondary metabolites found in plants. Because of their biological properties, these natural compounds have been utilized in numerous traditional medical systems throughout history since ancient times.

A growing worldwide issue that impacts people, animals, and the environment is antibacterial resistance. Public health is seriously threatened by the presence of pathogens that are resistant to antibiotics in food manufacturing systems. Finding novel compounds for therapeutic approaches against antibacterial resistance is vital, even though the problem is made worse by the absence of additional antibacterials. Numerous studies have demonstrated the great potential of MAPs for the development of novel antibacterial agents in the fight against microbes, which represent a major problem for the human body. To further validate these findings and support MAPs and their constituents as a crucial treatment tool for a number of illnesses, clinical trials are required.

References

[1] Samarth, R. M., Samarth, M. and Matsumoto, Y. (2017). medicinally important aromatic plants with radioprotective activity. Future Science OA, 3, 247–273.

[2] Tian, X. R., Feng, G. T., Ma, Z. Q., Xie, N., Zhang, J., Zhang, X., Zhang, X. and Tang, H.-F. (2014). Three new glycosides from the whole plant of *Clematis lasiandra Maxim* and their cytotoxicity. Phytochemistry Letter, 10, 168–2.

[3] Cushnie, T. P. T., Cushnie, B. and Lamb, A. J. (2014). Alkaloids: An overview of their antibacterial, antibiotic-enhancing and antivirulence activities. International Journal of Antimicrobial Agents, 44, 377–6.

[4] Vardhini, B. V. and Anjum, N. A. (2015). Brassinosteroids make plant life easier under abiotic stresses mainly by modulating major components of antioxidant defense system. Frontiers in Environmental Science, 2, 1–16.

[5] Chaparro, J. M., Badri, D. V. and Vivanco, J. M. (2014). Rhizosphere microbiome assemblage is affected by plant development. ISME Journal, 8, 790–793.

[6] Köberl, M., Schmidt, R., Ramadan, E. M., Bauer, R. and Berg, G. (2013). The microbiome of medicinal plants: Diversity and importance for plant growth, quality and health. Frontiers in Microbiology, 4, 1–9.

[7] Brenes, A. and Roura, E. (2010). Essential oils in poultry nutrition: Main effects and modes of action. Animal Feed Science and Technology, 158, 1–14.

[8] Greathead, H. (2003). Plants and plant extracts for improving animal productivity. Proceedings of the Nutrition Society, 62, 279–290.

[9] Singh, R. (2015). Medicinal Plants: A Review. Journal of Plant Sciences, 3, 50–55.

[10] Srinivasan, D., Nathan, S., Suresh, T. and Lakshmana, P. P. (2001). antimicrobial activity of certain Indian medicinal plants used in Folkloric medicine. Journal of Ethnopharmacology, 74, 217–220.

[11] Yadav, R. and Agarwala, M. (2011). Phytochemical analysis of some medicinal plants. The Journal of Phytological Research, 3, 10–14.

[12] Bhattacharya, R., Reddy, K. R. C. and Mishra, A. K. (2014). Export strategy of Ayurvedic products from India. International Journal of Ayurvedic Medicine, 5, 125–128.

[13] Pandey, A. and Kumar, S. (2013). Perspective on plant products as Antimicrobials agents: A review. Pharmacologia, 4, 469–480.

[14] WHO. (2014). Antimicrobial Resistance, World Health Organization, Geneva, Switzerland.

[15] Davidson-Hunt, I. (2000). Ecological Ethno Botany: Stumbling toward New Practices and Paradigms, Natural Resources Institute, University of Manitoba, Winnipeg, Manitoba.

[16] Akhtar, M. S., Degaga, B. and Azam, T. (2014). Antimicrobial activity of essential oils extracted from medicinal plants against the pathogenic microorganisms: A review. Issues in Biological Sciences and Pharmaceutical Research, 2, 1–7.

[17] Maiti, R., Rodriguez, H. G. and Kumari, A. (2016). Nutrient profile of native woody species and medicinal plants in Northeastern Mexico: A synthesis, Maiti et al. Journal of Bioprocessing and Biotechniques, 6, 1–8.

[18] Reis, H. T. (2001). Relationship experiences and emotional well-being. In: Carol D. Ryff, and Burton Singer (eds) Emotion, Social Relationships, and Health, Oxford University, New York, NY, USA, 57–86.

[19] Kelly, K. (2009). History of Medicine, Facts on file, New York.

[20] Tucakov, J. P. (1964). Beograd, Institute for Text Book Issuing in SR, Srbija, Belgrade, Serbia.

[21] Tucakov, J. (1971). Healing with Plants – Phytotherapy, Culture, Beograd.

[22] Wiart, C. (2006). Etnopharmacology of Medicinal Plants, Humana Press, New Jersey.

[23] Toplak, G. K. (2005). Domestic Medicinal Plants, Mozaic book, Zagreb.

[24] Gorunovic, M. and Lukic, P. (2001). Pharmacognosy, Beograd, Gorunovic M, Belgrade, Serbia.

[25] Petrovska, B. B. (2012). Historical review of medicinal plants usage. Pharmacognosy Reviews, 6, 1–5.

[26] Pandey, A. K., Kumar, P., Saxena, M. J. and Maurya, P. (2020). Distribution of aromatic plants in the world and their properties. In: Panagiota Florou-Paneri, Efterpi Christaki, Feed Additives, Academic Press, CEHTRA Chemical Consultants Pvt Ltd, Delhi, India, 89–114.

[27] Davies, J. and Davies, D. (2010). Origins and evolution of antibiotic resistance. Microbiology and Molecular Biology Reviews, 74, 417–433.

[28] Chavan, S. S., Damale, M. G., Shinde, D. B. and Sangshetti, J. N. (2018). Antibacterial and antifungal drugs from natural source: A review of clinical development. In: Atta-ur-Rahman, Shazia Anjum, Hesham El-Seedi (eds) Natural Products in Clinical Trials, Benthan Science Books, Sharjah, United Arab Emirates, 114–164.

[29] Tortorella, E., Tedesco, P., Palma Esposito, F., January, G., Fani, R., Jaspars, M. and De Pascale, D. (2018). Antibiotics from deep-sea microorganisms: Current discoveries and perspectives. Marine Drugs, 16, 355.

[30] Talib, W. H. (2011). Anticancer and antimicrobial potential of plant-derived natural products. In Rasooli, I. (ed.) Phytochemicals – Bioactivities and Impact on Health, IntechOpen, London, UK, 141–158.

[31] Lampinen, J. (2005). continuous antimicrobial susceptibility testing in drug discovery. International Journal of Drug Policy, 7, 1–3.

[32] Fazly Bazzaz, B. S., Khameneh, B., Zahedian Ostad, M. R. and Hosseinzadeh, H. (2018). *In vitro* evaluation of antibacterial activity of verbascoside, lemon verbena extract and caffeine in combination with gentamicin against drug-resistant *Staphylococcus aureus* and *Escherichia coli* clinical isolates. Avicenna Journal of Phytomedicine, 8, 246–253.

[33] Savoia, D. (2012). Plant-Derived antimicrobial compounds: Alternatives to antibiotics. Future Microbiol, 7, 979–990.

[34] Wagner, H. and Ulrich-Merzenich, G. (2009). Synergy research: Approaching a new generation of Phytopharmaceuticals. Phytomedicine, 16, 97–110.

[35] Shankar, S. R., Rangarajan, R., Sarada, D. V. L. and Kumar, C. S. (2010). Evaluation of antibacterial activity and phytochemical screening of *Wrightia Tinctoria* L. Pharmacognosy Journal, 2, 19–22.

[36] Alviano, D. S. and Alviano, C. S. (2009). Plant extracts: Search for new alternatives to treat microbial diseases. Current Pharmaceutical Biotechnology, 10, 106–121.

[37] Safaei-Ghomi, J. and Ahd, A. A. (2010). Antimicrobial and antifungal properties of the essential oil and methanol extracts of *Eucalyptus largiflorens and Eucalyptus intertexta*. Pharmacognosy Magazine, 6, 172–175.

[38] Ruddaraju, L. K., Pammi, S. V. N., Guntuku, G. S., Padavala, V. S. and Kolapalli, V. R. M. (2020). A review on anti-bacterials to combat resistance: From ancient era of plants and metals to present and future perspectives of green nano technological combinations. Asian Journal of Pharmaceutical Sciences, 15, 42–59.

[39] Almabruk, K. H., Dinh, L. K. and Philmus, B. (2018). Self-resistance of natural product producers: Past, present, and future focusing on self-resistant protein variants. American Chemical Society Chemical Biology, 13, 1426–1437.

[40] Singh, D., Gupta, R. and Saraf, S. A. (2012). Herbs-are they safe enough? an overview. Critical Reviews in Food Science and Nutrition, 52, 876–898.

[41] Bankole, M. A., Shittu, L. A., Ahmed, T. A., Bankole, M. N., Shittu, R. K., Terkula, K. and Ashiru, O. A. (2007). Synergistic antimicrobial activities of phytoestrogens in crude extracts of two sesame species against some common pathogenic microorganisms. African Journal of Traditional, Complementary and Alternative Medicines (AJTCAM), 4, 427–433.

[42] Shan, B., Cai, Y. Z., Brooks, J. D. and Corke, H. (2007). The *in vitro* antibacterial activity of dietary spice and medicinal herb extracts. International Journal of Food Microbiology, 117, 112–119.

[43] Devi, K. P., Nisha, S. A., Sakthivel, R. and Pandian, S. K. (2010). Eugenol (an essential oil of clove) acts as an antibacterial agent against *Salmonella typhi* by disrupting the cellular membrane. Journal of Ethnopharmacology, 130, 107–115.

[44] Pichersky, E., Noel, J. P. and Dudareva, N. (2006). Biosynthesis of plant volatiles: Nature's diversity and ingenuity. Science, 311, 808–811.

[45] Bakkali, F., Averbeck, S., Averbeck, D. and Idaomar, M. (2008). Biological effects of essential oils- A review. Food and Chemical Toxicology, 46, 446–475.

[46] Akthar, M. S., Degaga, B. and Azam, T. (2014). Antimicrobial activity of essential oils extracted from medicinal plants against the pathogenic microorganisms: A review. Issues in Biological Sciences and Pharmaceutical Research, 2, 001–7.

[47] Angioni, A., Barra, A., Coroneo, V., Dessi, S. and Cabras, P. (2006). Chemical composition, seasonal variability, and antifungal activity of Lavandula stoechas L. ssp. stoechas essential oils from stem/ leaves and flowers. Journal of Agricultural and Food Chemistry, 54, 4364–4370.

[48] Boškailo, E., Džudžević-Čančar, H., Dedić, A., Marijanović, Z., Alispahić, A., Čančar, I. F., Vidic, D. and Jerković, I. (2023). *Clinopodium nepeta* (L.) Kuntze from Bosnia and Herzegovina: Chemical characterisation of headspace and essential oil of fresh and dried samples. Records of Natural Products, 17(2), 300–311.

[49] Al-Bayati, F. A. (2009). Isolation and identification of antimicrobial compound from *Mentha longifolia* L. leaves grown wild in Iraq. Annals of Clinical Microbiology and Antimicrobials, 8, 2–6.

[50] Zomorodian, K., Saharkhiz, M. J., Rahimi, M. J., Bandegi, A., Shekarkhar, G., Bandegani, A., Pakshir, K. and Bazargani, A. (2011). Chemical composition and antimicrobial activities of the essential oils from three ecotypes of Zataria multiflora. Pharmacognosy Magazine, 7, 53–59.

[51] Van Vuuren, S. F., LC, D. T., Parry, A., Pillay, V. and Choonara, Y. E. (2010). Encapsulation of essential oils within a polymeric liposomal formulation for enhancement of antimicrobial efficacy. Natural Products Communications, 5, 1401–1408.

[52] Malléa, M., Chevalier, J., Bornet, C., Eyraud, A., Davin-Regli, A., Bollet, C. and Pagès, J. M. (1998). Porin alteration and active efflux: Two *in vivo* drug resistance strategies used by *Enterobacter aerogenes*. Microbiology, 144, 3003–3009.

[53] Blot, S., Depuydt, P., Vandewoude, K. and De Bacquer, D. (2007). Measuring the impact of multidrug resistance in nosocomial infection. Current Opinion in Infectious Diseases, Dis 20, 391–396.

[54] Mahamoud, A., Chevalier, J., Davin-Regli, A., Barbe, J. and Jm, P. (2006). Quinoline derivatives as promising inhibitors of antibiotic efflux pump in multidrug resistant Enterobacter aerogenes isolates. Current Drug Targets, 2006(7), 843–847.

[55] Lomovskaya, O., Warren, M. S., Lee, A., Galazzo, J., Fronko, R., Lee, M., Blais, J., Cho, D., Chamberland, S., Renau, T., Leger, R., Hecker, S., Watkins, W., Hoshino, K., Ishida, H. and Lee, V. J. (2001). Identification and characterization of inhibitors of multidrug resistance efflux pumps in *Pseudomonas aeruginosa*: Novel agents for combination therapy. Antimicrob Agents and Chemother, 45, 105–116.

[56] Lorenzi, V., Muselli, A., Bernardini, A. F., Berti, L., Pagès, J.-M., Amaral, L. and Bolla, J.-M. (2009). Geraniol restores antibiotic activities against multidrug-resistant isolates from gram-negative species. Antimicrob Agents and Chemother, 53, 2209–2211.

[57] MM, C. (1999). Plant products as antimicrobial agents. Clinical Microbiology Reviews, 12, 564–582.

[58] Fankam, A. G., Kuiate, J. R. and Kuete, V. (2014). Antibacterial activities of *Beilschmiedia obscura* and six other Cameroonian medicinal plants against multi-drug resistant Gram-negative phenotypes. BMC Complementary Alternative Healthcare and Medical, 14, 241–249.

[59] Abbas, M., Saeed, F., Anjum, F. M., Afzaal, M., Tufail, T., Bashir, M. S., Ishtiaq, A., Hussain, S. and Suleria, H. A. R. (2017). Natural polyphenols: An overview. International Journal of Food Properties, 20, 1689–1699.

[60] Bié, J., Sepodes, B., Fernandes, P. C. and Ribeiro, M. H. (2023). Polyphenols in health and disease: Gut microbiota, bioaccessibility, and bioavailability. Compounds, 3, 40–72.

[61] Rana, A., Samtiya, M., Dhewa, T., Mishra, V. and Aluko, R. E. (2022). Health benefits of polyphenols: A concise review. Journal of Food Biochemistry, 46, 1–24.

[62] Rahman, M. M., Rahaman, M. S., Islam, M. R., Rahman, F., Mithi, F. M., Alqahtani, T., Almikhlafi, M. A., Alghamdi, S. Q., Alruwaili, A. S. and Hossain, M. S. (2021). Role of phenolic compounds in human disease: Current knowledge and future prospects. Molecules, 27, 233–267.

[63] Minatel, I. O., Borges, C. V., Ferreira, M. I., Gomez, H. A. G., Chen, C.-Y. O. and Lima, G. P. P. (2017). Phenolic Compounds: Functional properties, impact of processing and bioavailability. Phenolic Compound Biochemistry, 8, 1–24.

[64] Marinaş, I. C., Chifiriuc, C., Oprea, E. and Lazăr, V. (2014). Antimicrobial and antioxidant activities of alcoholic extracts obtained from vegetative organs of *A. retroflexus*. Roumanian Archives of Microbiology and Immunology, 73, 35–2.

[65] Park, S. W., Kwon, M. J., Yoo, J. Y., Choi, H.-J. and Ahn, Y.-J. (2014). Antiviral activity and possible mode of action of ellagic acid identified in *Lagerstroemia speciosa* leaves toward human rhinoviruses. BioMed Central Complementary and Alternative Medicine, 14, 1–8.

[66] Bai, L., Nong, Y., Shi, Y., Liu, M., Yan, L., Shang, J., Huang, F., Lin, Y. and Tang, H. (2016). Luteolin inhibits hepatitis B virus replication through extracellular signal-regulated kinase-mediated down-regulation of hepatocyte nuclear factor 4αexpression. Molecular Pharmaceutics, 13, 568–7.

[67] Dinda, B., Dinda, M., Dinda, S., Ghosh, P. S. and Das, S. K. (2024). Anti-SARS-CoV-2, antioxidant and immunomodulatory potential of dietary flavanol quercetin: Focus on molecular targets and clinical efficacy. European Journal of Medicinal Chemistry Reports, 10, 1–35.

[68] Zulkefli, N., Che Zahari, C. N. M., Sayuti, N. H., Kamarudin, A. A., Saad, N., Hamezah, H. S., Bunawan, H., Baharum, S. N., Mediani, A. and Ahmed, Q. U. (2023). Flavonoids as potential wound-healing molecules: Emphasis on pathways perspective. International Journal of Molecular Sciences, 24, 1–29.

[69] Yupanqui Mieles, J., Vyas, C., Aslan, E., Humphreys, G., Diver, C. and Bartolo, P. (2022). Honey: An advanced antimicrobial and wound healing biomaterial for tissue engineering applications. Pharmaceutics, 14, 1–36.

[70] Mucha, P., Skoczyńska, A., Małecka, M., Hikisz, P. and Budzisz, E. (2021). Overview of the antioxidant and anti-inflammatory activities of selected plant compounds and their metal ions complexes. Molecules, 26, 2–52.

[71] Daglia, M. (2012). Polyphenols as antimicrobial agents. Current Opinion in Biotechnology, 23, 174–1.

[72] Mandal, M. K. and Domb, A. J. Antimicrobial activities of natural bioactive polyphenols. Pharmaceutics, 2024(16), 1–25.

[73] Liu, C., Dong, S., Wang, X., Xu, H., Yang, W. S. X., Jiang, X., Kan, M. and Xu, C. (2023). Research progress of polyphenols in nanoformulations for antibacterial application. Materials Today Bio, 21, 100729.

[74] Rajhard, S., Hladnik, L., Vicente, F. A., Srčič, S., Grilc, M. and Likozar, B. (2021). Solubility of luteolin and other polyphenolic compounds in water, nonpolar, polar aprotic and protic solvents by applying FTIR/HPLC. Processes, 9, 1952.

[75] Munin, A. and Edwards-Lévy, F. (2011). Encapsulation of natural polyphenolic compounds; a review. Pharmaceutics, 3, 793–829.

[76] Tungmunnithum, D., Thongboonyou, A., Pholboon, A. and Yangsabai, A. (2018). Flavonoids and other phenolic compounds from medicinal plants for pharmaceutical and medical aspects: An overview. Medicines, 5, 1–16.

[77] Kong, C., Zhang, H., Li, L. and Liu, Z. (2022). Effects of green tea extract epigallocatechin-3-gallate (EGCG) on oral disease-associated microbes: A review. Journal of Oral Microbiology, 14, 1–14.

[78] Bae, J., Kim, N., Shin, Y., Kim, S.-Y. and Kim, Y.-J. (2020). Activity of catechins and their applications. Biomed Dermatol, 4, 1–10.

[79] Veiko, A. G., Olchowik-Grabarek, E., Sekowski, S., Roszkowska, A., Lapshina, E. A., Dobrzynska, I., Zamaraeva, M. and Zavodnik, I. B. (2023). Antimicrobial activity of Quercetin, Naringenin and Catechin: Flavonoids inhibit *Staphylococcus aureus*-induced hemolysis and modify membranes of bacteria and Erythrocytes. Molecules, 28, 1–19.

[80] Jaisinghani, R. (2017). Antibacterial properties of quercetin. Microbiological Research, 8, 6877.

[81] Nguyen, T. L. A. and Bhattacharya, D. (2022). Antimicrobial activity of quercetin: An approach to its mechanistic principle. Molecules, 27, 2494–04.

[82] Roshani, M., Jafari, A., Loghman, A., Sheida, A. H., Taghavi, T., Tamehri Zadeh, S. S., Hamblin, M. R., Homayounfal, M. and Mirzaei, H. (2022). Applications of resveratrol in the treatment of gastrointestinal cancer. Biomedecine & Pharmacotherapy, 153, 113274.

[83] Mahady, G. B., Pendland, S. L. and Chadwick, L. R. (2003). Resveratrol and red wine extracts inhibit the growth of caga+strains of *Helicobacter Pylori in vitro*. Official Journal of the American College of Gastroenterology| ACG, 98, 1440–1441.

[84] Mattio, L. M., Catinella, G., Dallavalle, S. and Pinto, A. (2020). Stilbenoids: A natural arsenal against bacterial pathogens. Antibiotics, 9, 336.

[85] Ciuca, M. D. and Racovita, R. C. (2023). Curcumin: Overview of extraction methods, health benefits, and encapsulation and delivery using Microemulsions and Nanoemulsions. International Journal of Molecular Sciences, 24, 8874.

[86] Candra, A., Prasetyo, B. E. and Darge, H. F. (2023). Honey utilization in soursop leaves (*Annona muricata*) kombucha: Physicochemical, cytotoxicity, and antimicrobial activity. Biocatalysis and Agricultural Biotechnology, 52, 102815.

[87] Lüer, S., Troller, R. and Aebi, C. (20212). Antibacterial and anti-inflammatory kinetics of curcumin as a potential antimucositis agent in cancer patients. Nutrition and Cancer, 64, 975–981.

[88] Abdul Kadir, H., Hassandarvish, P., Tajik, H. and Abubakar, S. (2014). Zandi KA review on antibacterial, antiviral, and antifungal activity of curcumin. BioMed Research International, 2014, 186864.

[89] Duda-Madej, A., Stecko, J., Sobieraj, J., Szymańska, N. and Kozłowska, J. (2022). Naringenin and its derivatives – Health-promoting phytobiotic against resistant bacteria and fungi in humans. Antibiotics, 11, 1628.

[90] Yogesh, M. (2021). Biological evaluation of synthesized naringenin derivatives as antimicrobial agents. Anti-Infective Agents, 19, 192–199.

[91] Negm, W. A., El-aasr, M., Kame, A. A. and Elekhnawy, E. (2021). Investigation of the antibacterial activity and efflux pump inhibitory effect of Cycas Thouarsii r.Br. extract against *Klebsiella Pneumoniae* clinical isolates. Pharmaceuticals, 14, 756.

[92] Liu, R., Zhao, B., Wang, D. E., Yao, T. Y., Pang, L., Tu, Q., Ahmed, S. M., Liu, J. J. and Wang, J. Y. (2012). Nitrogen-containing apigenin analogs: Preparation and biological activity. Molecules, 17, 14748–14764.

[93] Qian, W., Liu, M., Fu, Y., Zhang, J., Liu, W., Li, J., Li, X., Li, Y. and Wang, T. (2020). Antimicrobial mechanism of luteolin against *Staphylococcus aureus* and Listeria monocytogenes and its antibiofilm properties. Microbial Pathogenesis, 142, 104056.

[94] Guo, Y., Liu, Y., Zhang, Z., Chen, M., Zhang, D., Tian, C., Liu, M. and Jiang, G. (2020). The antibacterial activity and mechanism of action of Luteolin against *Trueperella pyogenes*. Infection and Drug Resistance, 13, 1697–1711.

[95] Chaudhary, V., Sharma, N. and Ozogul, F. (2023). Luteolin: A flavone with myriads of bioactivities and food applications. Food Bioscience Volume, 52, 102366.

[96] Kim, M.-H., Han, J.-H. and Kim, S.-U. (2008). Isoflavone daidzein: Chemistry and bacterial metabolism. Journal of Applied Biological Chemistry, 51, 253–261.

[97] Zarei, M., Fazlara, A. and Mohammadi, S. (2014). Comparing the antimicrobial effectiveness of genistein and daidzein on some food-borne pathogen bacteries. Iranian Journal of Veterinary Medicine, 9, 55–2.

[98] GC, D. O. S., CC, V., AJ, L., De Sousacartágenes, M., AK, F., FR, D. N., RM, R., ER, P., MS, D. A., Fm, R. and De Andrade Monteiro, C. (2018). *Candida* infections and therapeutic strategies: Mechanisms of action for traditional and alternative agents. Frontiers in Microbiology, 9, 362855.

[99] Bell, C. and Hawthorne, S. (2008). Ellagic acid, pomegranate and prostate cancer – A mini review. Journal of Pharmacy and Pharmacology, 60, 139–144.

[100] Glazer, I., Masaphy, S., Marciano, P., Bar-Ilan, I., Holland, D., Kerem, Z. and Amir, R. (2012). Partial identification of antifungal compounds from *Punica granatum* peel extracts. Journal of Agricultural and Food Chemistry, 60, 4841–4848.

[101] An, J.-Y., Wang, L.-T., Lv, M.-J., Wang, J.-D., Cai, Z.-H., Wang, Y.-Q., Zhang, S., Yang, Q. and Fu, Y.-J. (2021). An efficiency strategy for extraction and recovery of ellagic acid from waste chestnut shell and its biological activity evaluation. Microchemical Journal, 160, 1–7.

[102] Kepa, M., Miklasinska-Majdanik, M., RD, W., Idzik, D., Korzeniowski, K., SmoleN-Dzirba, J. and Wdsik, T. J. (2018). Antimicrobial potential of Caffeic acid against *Staphylococcus aureus* clinical strains. BioMed Research International, 2018, 1–9.

[103] Lima, V. N., Oliveira-Tintino, D. M. C., Santos, S. E., Morais, P. L., Tintino, R. S., Freitas, T. S., Geraldo, S. Y., Pereira, L. S. R., Cruz, P. R., Menezes, R. A. I. and Coutinho, D. M. H. (2016). Antimicrobial and enhancement of the antibiotic activity by phenolic compounds: Gallic acid, caffeic acid and pyrogallol. Microbial Pathogenesis, 99, 56–61.

[104] Borges, A., Ferreira, C., Saavedra, M. J. and Simões, M. (2013). Antibacterial activity and mode of action of Ferulic and Gallic acids against Pathogenic Bacteria. Microbial Drug Resistance, 19, 256–265.

[105] Kang, J., Li, Q., Liu, L., Jin, W., Wang, J. and Sun, Y. (2018). The specific effect of gallic acid on *Escherichia coli* biofilm formation by regulating pgaABCD genes expression. Applied Microbiology and Biotechnology, 102, 1837–1846.

[106] Khorsandi, K., Keyvani-Ghamsari, S., Khatibi Shahidi, F., Hosseinzadeh, R. and Kanwal, S. (2021). A mechanistic perspective on targeting bacterial drug resistance with nanoparticles. Journal of Drug Targeting, 29, 941–959.

[107] Yamaguchi, T. (2022). Antibacterial effect of the combination of terpenoids. Archives of Microbiology, 204, 1–7.

[108] Sharma, A., Biharee, A., Kumar, A. and Jaitak, V. (2020). Antimicrobial terpenoids as a potential substitute in overcoming antimicrobial resistance. Current Drug Targets, 21, 1476–1494.

[109] Nazzaro, F., Fratianni, F., De Martino, L., Coppola, R. and De Feo, V. (2013). Effect of essential oils on pathogenic bacteria. Pharmaceuticals, 6, 1451–1474.

[110] Burt, S. A. and Reinders, R. D. (2003). Antibacterial activity of selected plant essential oils against Escherichia coli O157:H7. Letters in Applied Microbiology, 36, 162–167.

[111] Sharma, A., Biharee, A., Kumar, A. and Jaitak, V. (2020). Antimicrobial terpenoids as a potential substitute in overcoming antimicrobial resistance. Current Drug Targets, 21, 1476–1494.

[112] Mulat, M., Pandita, A. and Khan, F. (2019). Medicinal plant compounds for combating the multi-drug-resistant pathogenic bacteria: A Review. Current Pharmaceutical Biotechnology, 20, 183–186.

[113] Niu, C., Afre, S. and Gilbert, E. S. (2006). Subinhibitory concentrations of cinnamaldehyde interfere with quorum sensing. Letters in Applied Microbiology, 43, 489–4.

[114] Myszka, K., Schmidt, M. T., Majcher, M., Juzwa, W., Olkowicz, M. and Czaczyk, K. (2016). Inhibition of quorum sensing-related biofilm of *Pseudomonas fluorescens* KM121 by *Thymus vulgare* essential oil and its major bioactive compounds. International Biodeterioration & Biodegradation, 114, 252–259.

[115] Ahmad, A., Khan, A., Yousuf, S., Khan, L. A. and Manzoor, N. (2010). Proton translocating ATPase mediated fungicidal activity of eugenol and thymol. Fitoterapia, 81, 1157–2.

[116] Khan, I., Bahuguna, A., Shukla, S., Aziz, F., Chauhan, A. K., Ansari, M. B., Bajpai, V. K., Huh, Y. S. and Kang, S. C. (2020). Antimicrobial potential of the food-grade additive carvacrol against uropathogenic *E. coli* based on membrane depolarization, reactive oxygen species generation, and molecular docking analysis. Microbial Pathogenesis, 142, 104046.

[117] Domadia, P., Swarup, S., Bhunia, A., Sivaraman, J. and Dasgupta, D. (2007). Inhibition of bacterial cell division protein FtsZ by cinnamaldehyde. Biochemical Pharmacology, 74, 831–840.

[118] Naz, F., Kumar, M., Koley, T., Sharma, P., Haque, M. A., Kapil, A., Kumar, M., Kaur, P. and Ethayathulla, A. S. (2022). Screening of plant-based natural compounds as an inhibitor of FtsZ from *Salmonella typhi* using the computational, biochemical and *in vitro* cell-based studies. International Journal of Biological Macromolecules, 219, 428–7.

[119] Zhou, F., Ji, B., Zhang, H., Jiang, H., Yang, Z., Li, J., Li, J., Ren, Y. and Yan, W. (2007). Synergistic effect of thymol and carvacrol combined with chelators and organic acids against *Salmonella Typhimurium*. Journal of Food Protection, 70, 1704–1709.

[120] Di Pasqua, R., Betts, G., Hoskins, N., Edwards, M., Ercolini, D. and Mauriello, G. (2007). Membrane toxicity of antimicrobial compounds from Essential oils. Journal of Agricultural and Food Chemistry, 55, 4863–4870.

[121] Mittal, R. P., Rana, A. and Jaitak, V. (2019). Essential Oils: An impending substitute of synthetic antimicrobial agents to overcome antimicrobial resistance. Current Drug Targets, 20, 605–624.

[122] Turcheniuk, V., Raks, V., Issa, R., Cooper, I. R., Cragg, P. J., Jijie, R., Dumitrescu, N., Mikhalovska, L. I., Barras, A., Zaitsev, V., Boukherroub, R. and Szunerit, S. (2015). Antimicrobial activity of menthol modified nanodiamond particles. Diamond and Related Materials, 57, 2–8.

[123] Mulyaningsih, S., Sporer, F., Zimmermann, S., Reichling, J. and Wink, M. (2010). Synergistic properties of the terpenoids aromadendrene and 1,8-cineole from the essential oil of *Eucalyptus globulus* against antibiotic-susceptible and antibiotic-resistant pathogens. Phytomedicine, 17, 1061–1066.

[124] Silva, A. C. R., Lopes, P. M., Azevedo, M. M. B., Costa, D. C. M., Alviano, C. S. and Alviano, D. S. (2012). Biological Activities of a-Pinene and β-Pinene Enantiomers. Molecules, 17, 6305–6316.

[125] Dhar, P., Chan, P. Y., Cohen, D. T., Khawam, F., Gibbons, S., Snyder-Leiby, T., Dickstein, E., Rai, P. K. and Synthesis, W. G. (2014). Antimicrobial evaluation, and structure–activity relationship of α-Pinene Derivatives. Journal of Agricultural and Food Chemistry, 62, 3548–2.

[126] Wang, C. Y., Chen, Y. W. and Hou, C. Y. (2019). Antioxidant and antibacterial activity of seven predominant terpenoids. International Journal of Food Properties, 22, 230–238.

[127] Miladinovic, D. and Ilic, B. S. (2014). Kocic BD,.Miladinovic MD, An *in vitro* antibacterial study of savory essential oil and geraniol in combination with standard antimicrobials. Natural Product Communications, 9, 1629–2.

[128] Ilic, B. S., Kocit, B. D., Vojislav, M., Cvetkovit, O. and Miladinovit, D. L. (2014). An *in vitro* synergistic interaction of combinations of *Thymus glabrescens* essential oil and its main constituents with chloramphenicol. Scientific World, 2014, 1–12.

[129] Lorca, G., Ballestero, D., Langa, E. and Pino-Otín, M. R. (2024). Enhancing antibiotic efficacy with natural compounds: Synergistic activity of Tannic Acid and Nerol with commercial antibiotics against Pathogenic Bacteria. Plants, 13, 2717.

[130] Wang, Z., Yang, K., Chen, L., Yan, R., Qu, S., Li, Y., Liu, M., Zeng, H. and Tian, J. (2020). Activities of Nerol, a natural plant active ingredient, against *Candida albicans in vitro* and *in vivo*. Applied Microbiology and Biotechnology, 104, 5039–5052.

[131] Guo, F., Chen, Q., Chen, Q., Liang, Q., Zhang, M., Chen, W., Chen, H., Yonghuan, Y., Zhong, Q. and Chen, W. (2021). Antimicrobial activity and proposed action mechanism of Linalool against *Pseudomonas fluorescens*. Frontiers in Microbiology, 28, 1–11.

[132] Mahmoud Abd El-Baky, R. and Shawky Hashem, Z. (2016). Eugenol and linalool: Comparison of their antibacterial and antifungal activities. African Journal of Microbiology Research, 10, 1860–1872.

[133] Parthasarathy, A., Borrego, E. J., Savka, M. A., Dobson, R. C. J. and Hudson, A. O. (2021). Amino acid-derived defense metabolites from plants: A potential source to facilitate novel antimicrobial development. Journal of Biological Chemistry, 296, 100438.

[134] Othman, L., Sleiman, A. and Abdel-Massih, R. M. (2019). Antimicrobial activity of polyphenols and alkaloids in middle eastern plants. Frontiers in Microbiology, 10, 911.

[135] Cushnie, T. P. T., Cushnie, B. and Lamb, A. J. (2014). Alkaloids: An overview of their antibacterial, antibiotic-enhancing and antivirulence activities. International Journal of Antimicrobial Agents, 44, 377–6.

[136] Thawabteh, A., Juma, S., Bader, M., Karaman, D., Scrano, L., Bufo, S. A. and Karaman, R. (2019). The biological activity of natural alkaloids against herbivores, cancerous cells and pathogens. Toxins, 11, 656.

[137] Khameneh, B., Iranshahy, M., Ghandadi, M., Atashbeyk, D. G., Bazzaz, B. S. F. and Iranshahi, M. (2015). Investigation of the antibacterial activity and efflux pump inhibitory effect of co-loaded piperine and gentamicin nanoliposomes in methicillin-resistant Staphylococcus aureus. Drug Development and Industrial Pharmacy, 41, 989–4.

[138] Mabhiza, D., Chitemerere, T. and Mukanganyama, S. (2016). Antibacterial properties of Alkaloid extracts from *Callistemon citrinus* and *Vernonia adoensis* against *Staphylococcus aureus* and *Pseudomonas aeruginosa*. International Journal of Medicinal Chemistry, 2016, 6304163.

[139] Wink, M., Ashour, M. L. and El-Readi, M. Z. (2012). Secondary metabolites from plants inhibiting ABC transporters and reversing resistance of cancer cells and microbes to cytotoxic and antimicrobial agents. Frontiers in Microbiology, 3, 130.

[140] Hikal, D. M. (2018). Antibacterial activity of piperine and black pepper oil. Biosci Biotechnol Res Asia, 15, 877–0.

[141] Khan, I. A., Mirza, Z. M., Kumar, A., Verma, V. and Qazi, G. N. (2006). Piperine, a phytochemical potentiator of ciprofloxacin against *Staphylococcus aureus*. Antimicrob Agents and Chemother, 50, 810–812.

[142] Yi, Z. B., Yan, Y., Liang, Y. Z. and Bao, Z. (2007). Evaluation of the antimicrobial mode of berberine by LC/ESI-MS combined with principal component analysis. Journal of Pharmaceutical and Biomedical Analysis, 44, 301–304.

[143] Su, F. and Wang, J. (2018). Berberine inhibits the MexXY-OprM efflux pump to reverse imipenem resistance in a clinical carbapenem-resistant *Pseudomonas aeruginosa* isolate in a planktonic state. Experimental and Therapeutic Medicine, 15, 467–2.

[144] Laudadio, E., Cedraro, N., Mangiaterra, G., Citterio, B., Mobbili, G., Minnelli, C., Bizzaro, D., Biavasco, F. and Galeazzi, R. (2019). Natural alkaloid berberine activity against *Pseudomonas aeruginosa* MexXY-mediated aminoglycoside resistance: *In Silico* and *in vitro* Studies. Journal of Natural Products, 82, 1935–4.

[145] He, N., Wang, P., Wang, P., Ma, C. and Kang, W. (2018). Antibacterial mechanism of chelerythrine isolated from root of *Toddalia asiatica* (Linn) Lam. BMC Complementary and Alternative Medicine, 18, 261.

[146] Houdkova, M., Rondevaldova, J., Doskocil, I. and Kokoska, L. (2017). Evaluation of antibacterial potential and toxicity of plant volatile compounds using new broth microdilution volatilization method and modified MTT assay. Fitoterapia, 118, 56–2.

[147] Prachayasittikul, V., Prachayasittikul, S., Ruchirawat, S. and Prachayasittikul, V. (2013). 8-Hydroxyquinolines: A review of their metal chelating properties and medicinal applications. Drug Design, Development and Therapy, 7, 1157–1178.

[148] McMahon, J. B., Currens, M. J., Gulakowski, R. J., Buckheit, R. W., Lackman-Smith, C., Hallock, Y. F. and Boyd, M. R. (1995). Michellamine B, a novel plant alkaloid, inhibits human immunodeficiency virus-induced cell killing by at least two distinct mechanisms. Antimicrob Agents and Chemother, 39, 484–488.

[149] Hamoud, R., Reichling, J. and Wink, M. (2015). Synergistic antibacterial activity of the combination of the alkaloid sanguinarine with EDTA and the antibiotic streptomycin against multidrug resistant bacteria. Journal of Pharmacy and Pharmacology, 67, 264–3.

[150] Beuria, T. K., Santra, M. K. and Panda, D. (2005). Sanguinarine blocks cytokinesis in bacteria by inhibiting FtsZ assembly and bundling. Biochemistry, 44, 16584–3.

[151] Mujtaba, M. A., Akhter, M. H., Alam, M. S., Ali, M. D. and Hussain, A. (2022). An updated review on therapeutic potential and recent advances in drug delivery of berberine: Current status and future prospect. Current Pharmaceutical Biotechnology, 23, 60–71.

[152] Wen, S.-Q., Jeyakkumar, P., Avula, S. R., Ling, A., Zhang, L. and Zhou, C.-H. (2016). Discovery of novel berberine imidazoles as safe antimicrobial agents by down regulating ROS generation. Bioorganic & Medicinal Chemistry, 26, 2768–3.

[153] JA, P., AG, T., PA, N., DA, L., DA, S., VI, P., Bf, V., OV, E., MY, K., Paleskava, A. et al. (2023). Conjugates of Chloramphenicol Amine and Berberine as antimicrobial agents. Antibiotics, 12, 1–25.

[154] Mittal, R. P. and Jaitak, V. (2019). Plant-derived natural alkaloids as new antimicrobial and adjuvant agents in existing antimicrobial therapy. Current Drug Targets, 20, 1409–1433.

[155] Negi, J. S., Bisht, V. K., Bhandari, A. K., Bisht, D. S., Singh, P. and Singh, N. (2014). Quantification of reserpine content and antibacterial activity of Rauvolfia serpentina (L.) Benth. ex Kurz, Afr. Journal of Microbiology, 8, 162–166.

[156] Aldaly, Z. T. K. (2010). Antimicrobial activity of piperine purified from piper nigrum, ournal of basrah research. Science, 36, 54.

[157] He, N., Wang, P., Wang, P., Ma, C. and Kang, W. (2018). Antibacterial mechanism of chelerythrine isolated from root of *Toddalia asiatica* (Linn) Lam. BMC Complementary and Alternative Medicine, 18, 261.

[158] Xin, J., Pu, Q., Wang, R., Gu, Y., He, L., Du, X., Tang, G. and Han, D. (2024). Antibacterial activity and mechanism of chelerythrine against *Streptococcus agalactiae*. Frontiers in Veterinary Science, 11, 1–13.

[159] Miao, F., Yangab, X.-J., Zhou, L., Hub, H.-J., Zheng, F., Ding X-D, X.-D., Sun, D.-M., Zhou, C.-D. and Sun, W. (2011). Structural modification of sanguinarine and chelerythrine and their antibacterial activity. Natural Products Research, 25(9), 863–865.

[160] Yin, S. J., Rao, G. X., Wang, J., Luo, L. Y., He, G. H., Wang, C. Y., CY, M., XX, L., Hou, Z. and Xu, G. L. (2015). Roemerine improves the survival rate of septicemic BALB/c mice by increasing the cell membrane permeability of Staphylococcus aureus. PLoS ONE, 10, 1–13.

[161] Avci, F. G., Atas, B., Aksoy, C. S., Kurpejovic, E., Toplan, G. G., Gurer, C., Guillerminet, M., Orelle, C., Jault, J. M. and Akbulut, B. S. (2019). Repurposing bioactive aporphine alkaloids as efflux pump inhibitors. Fitoterapia, 139, 104371.

[162] Gokgoz, N. B. and Akbulut, B. S. (2015). Proteomics evidence for the activity of the Putative Antibacterial Plant Alkaloid (-)-Roemerine: Mainstreaming Omics-Guided drug discovery. OMICS A Journal of Integrative Biology, 19, 478–489.

[163] Guzman, J. D., Wube, A., Evangelopoulos, D., Gupta, A., Hufner, A., Basavannacharya, C., Rahman, M. M., Thomaschitz, C., Bauer, R., McHugh, T. D., Nobeli, I., Prieto, J. P., Gibbons, S., Bucar, F. and Bhakta, S. (2011). Interaction of N-methyl-2-alkenyl-4-quinolones with ATP-dependent MurE ligase of Mycobacterium tuberculosis: Antibacterial activity, molecular docking and inhibition kinetics. Journal of Antimicrobial Chemotherapy, 66, 1766–2.

[164] Hochfellner, C., Evangelopoulos, D., Zloh, M., Wube, A., Guzman, J. D., McHugh, T. D., Kunert, O., Bhakta, S. and Bucar, F. (2015). Antagonistic effects of indoloquinazoline alkaloids on antimycobacterial activity of evocarpine. Journal of Applied Microbiology, 118, 864–2.

[165] Dedić, A., Džudžević-Čančar, H., Alispahić, A., Tahirović, I. and Muratović, E. (2021). *In-vitro* antioxidant and antimicrobial activity of aerial parts of *Prunus spinosa* L. growing wild in Bosnia and Herzegovina. International Journal of Pharmaceutical Sciences Research, 12, 3643–3.

[166] Pinacho, R., Cavero, R. Y., Astiasarán, I., Ansorena, D. and Calvo, M. I. (2015). Phenolic compounds of blackthorn (*Prunus spinosa* L.) and influence of in-vitro digestion on their antioxidant capacity. Journal of Functional Foods, 19, 49–62.

[167] Shahidi, F., Vamadevan, V., Oh, W. Y. and Peng, H. (2019). Phenolic compounds in agri-food by-products, their bioavailability and health effects. Journal of Food Bioactives, 5, 57–119.

[168] Veličković, J. M., Kostić, D. A., Stojanović, G. S., Mitić, S. S., Mitić, M. N., Ranđelović, S. S. and Đorđević, A. S. (2014). Phenolic composition, antioxidant and antimicrobial activity of the extracts from *Prunus spinosa* L. fruit. Hemijska Industrija, 68, 297–03.

[169] Boškailo, E., Džudževič-Čančar, H., Dedić, A., Marijanović, Z., Alispahić, A., Čančar, I. F., Vidic, D. and Jerković, I. (2023). *Clinopodium nepeta* (L.) Kuntze from Bosnia and Herzegovina: Chemical Characterisation of Headspace and Essential Oil of Fresh and Dried Samples. Records of Natural Products, 17, 300–311.

[170] Bougandoura, N. and Bendimerad, N. (2013). Evaluation de l'activité antioxydante des extraits aqueux et méthanolique de *Satureja calamintha* ssp. Nepeta (L.) Briq. Nature Tech, 9, 14–19.

[171] Okach, D. O., Nyunja, A. R. O. and Opande, G. (2013). Phytochemical screening of some wild plants from Lamiaceae and their role in traditional medicine in Uriri District-Kenya. International Journal of Herbal Medicine, 1, 135–3.

[172] Boudjema, K., Bouanane, A., Gamgani, S., Djeziri, M., Mustapha, A. M. and Fazouane, F. (2018). Phytochemical profile and antimicrobial properties of volatile compounds of *Satureja calamintha* (L.) Scheel from northern Algeria, Trop. Journal of Pharmaceutical Research, 17, 857–4.

[173] Khodja, N. K., Boulekbache, L., Chegdani, F., Dahmani, K., Bennis, F. and Madani, K. (2018). Chemical composition and antioxidant activity of phenolic compounds and essential oils from Calamintha nepeta L. Journal of Complementary and Integrative Medicine, 15, 1–12.

[174] Mancini, E., De Martino, L., Malova, H. and De Feo, V. (2013). Chemical composition and biological activities of the essential oil from *Calamintha nepeta* plants from the wild in southern Italy. Natural Product Communications, 8, 139–2.

[175] Ristorcelli, D., Tom, F. and Casanova, J. (1996). Essential oils of *Calamintha nepeta* subsp. nepeta and subsp. glandulosa from Corsica (France). The Journal of Essential Oil Research, 8, 363–366.

[176] Božović, M. and Ragno, R. (2017). *Calamintha nepeta* (L.) Savi and its main essential oil constituent pulegone: Biological activities and chemistry. Molecules, 22, 290.

[177] Alan, S., Kürkçüoglu, M. and Baser, K. H. C. (2011). Composition of essential oils of *Calamintha nepeta* (L.) Savi subsp. nepeta and *Calamintha nepeta* (L.) Savi subsp. glandulosa (Req.) PW Ball. Asian Journal of Chemistry, 23, 2357–0.

[178] Kitic, D., Stojanovic, G., Palic, R. and Randjelovic, V. (2005). Chemical composition and microbial activity of the essential oil of Calamintha nepeta (L.) Savi ssp. nepeta var. subisodonda (Borb.) Hayek from Serbia. The Journal of Essential Oil Research, 17, 701–703.

[179] Walasek-Janusz, M., Grzegorczyk, A., Zalewski, D., Malm, A., Gajcy, S. and Gruszecki, R. (2022). Variation in the antimicrobial activity of essential oils from cultivars of *Lavandula angustifolia* and *L. × intermedia*. Agronomy, 12, 2955.

[180] Džudževič-Čančar, H., Alispahić, A., Dedić-Mahmutović, A., Dž, S., Uzunović, A., Boškailo, E. and Čančar, I. F. (8–11 September 2024), Essential oils of garden-growing lavender species: *In vitro* antimicrobial activity. 54th International Symposium on Essential Oils, Hunguest BAL Resort, Balatonalmádi, Hungary, 28.

[181] Cavanagh, H. M. and Wilkinson, J. M. (2002). Biological activities of lavender essential oil. Phytotherapy Research: PTR, 16, 301–308.

[182] Lis-Balchin, M. and Hart, S. (1999). Studies on the mode of action of the essential oil of lavender (*Lavandula angustifolia* P. Miller). Phytotherapy Research: PTR, 13, 540–542.

[183] Luo, W., Du, Z., Zheng, Y., Liang, X., Huang, G., Zhang, Q., Liu, Z., Zhang, K., Zheng, X., Lin, L. and Zhangt, L. (2019). Phytochemical composition and bioactivities of essential oils from six *Lamiaceae* species. Industrial Crops and Products, 133, 357–364.

[184] Bhalla, Y., Gupta, V. K. and Jaitak, V. (2013). Anticancer activity of essential oils: A review. Journal of Sol-Gel Science and Technology, 93, 3643–3653.

[185] Džudžević-Čančar, H., Dedić, A., Alispahić, A., Jerković, I., Boškailo, E., Stanojković, T., Marijanović, Z. and Kubat, N. (2022). Essential oil chemical profile of two different lavender species from Sarajevo gardens. Bulletin of the Chemists and Technologists of Bosnia and Herzegovina, CNP-02 74.

[186] Aprotosoaie, A. C., Gille, E., Trifan, A., Luca, V. S. and Miron, A. (2017). Essential oils of Lavandula genus: A systematic review of their chemistry. Phytochemistry Reviews, 16, 761–799.

[187] Bakkali, F., Averbeck, S., Averbeck, D. and Idaomar, M. (2008). Biological effects of essential oils-A review. Food and Chemical Toxicology, 46, 446–475.

[188] Sienkiewicz, M., Łysakowska, M., Cieæwierz, J., Denys, P. and Kowalczyk, E. (2011). Antibacterial activity of thyme and lavender essential oils. Medicinal Chemistry, 7, 674–689.

[189] Thosar, N., Basak, S., Bahadure, R. N. and Rajurkar, M. (2013). Antimicrobial efficacy of five essential oils against oral pathogens: An *in vitro* study. European Journal of Dentistry, 7, S71–S77.

[190] Sokoviæ, M., Glamoèlija, J., Marin, P. D., Brkiæ, D. and Van Griensven, L. J. (2010). Antibacterial effects of the essential oils of commonly consumed medicinal herbs using an *in vitro* model. Molecules, 15, 7532–7536.

[191] Jianu, C., Pop, G., Gruia, A. T. and Horhat, F. G. (2013). Chemical composition and antimicrobial activity of essential oils of lavender (*Lavandula angustifolia*) and lavandin (*Lavandula × intermedia*) grown in Western Romania. International Journal of Agriculture and Biology, 15, 772–776.

[192] Lis-Balchin, M. (2002). Lavender: The Genus *Lavandula*, Tylor and Francis, London, 174–175.

[193] Danh, L. T., Han, L. N., Triet, N. D. A., Zhao, J., Mammucari, R. and Foster, N. (2013). Comparison of chemical composition, antioxidant and antimicrobial activity of lavender (*Lavandula angustifolia* L.) essential oils extracted by supercritical CO_2, hexane and hydrodistillation. Food and Bioprocess Technology, 6, 3481–3489.

[194] Burt, S. (2004). Essential oils: Their antimicrobial properties and potential application in foods – A review. International Journal of Food Microbiology, 94, 223–253.

[195] Hyldgaard, M., Mygind, T. and Meyer, R. L. (2012). Essential oils in food preservation: Mode of action, synergies, and interactions with food matrix components. Frontiers in Microbiology, 3, 12.

[196] Nostro, A., Bisignano, G., Cannatelli, M. A., Crisafi, G., MP, G. and Alonzo, V. (2001). Effects of *Helichrysum italicum* extract on growth and enzymatic activity of *Staphylococcus aureus*. International Journal of Antimicrobial Agents, 17, 517–520.

[197] Galbany-Casals, M., Blanco-Moreno, J. M., Garcia-Jacas, N., Breitwieser, I. and Smissen, R. D. (2011). Genetic variation in Mediterranean *Helichrysum italicum* (*Asteraceae; Gnaphalieae*): Do disjunct populations of subsp. microphyllum have a common origin?. Plant Biology, 13, 678–687.

[198] Talić, S., Odak, I., Lukic, T., Brkljaca, M., Bevanda, A. M. and Lasić, A. (2021). Chemodiversity of *Helichrysum italicum* (Roth) G. Don subsp. *italicum* essential oils from Bosnia and Herzegovina. Fresenius Environmental Bulletin, 30, 2492–2502.

[199] Kramberger, K., Pražnikar Z, J., Baruca Arbeiter, A., Petelin, A., Bandelj, D. and Kenig, S. A. (2021). Comparative Study of the Antioxidative Effects of *Helichrysum italicum* and *Helichrysum arenarium* Infusions. Antioxidants, 10, 380.

[200] Rančić, A., Soković, M., Vukojević, J., Simić, A., Marin, P., Duletić-Laušević, S. and Djoković, D. (2005). Chemical composition and antimicrobial activities of essential oils of *Myrrhis odorata* (L.) Scop, *Hypericum perforatum* L and *Helichrysum arenarium* (L.) Moench. Journal of Essential Oil Research, 17, 341–345.

[201] Ninčević, T., Grdiša, M., Šatović, Z. and Jug-Dujaković, M. (2019). Helichrysum italicum (Roth) G. Don: Taxonomy, biological activity, biochemical and genetic diversity. Industrial Crops and Products, 138, 111487.

[202] Paolini, J., Desjobert, J. M., Costa, J., Bernardini, A. F., Castellini, C. B., Cioni, P. L., Guido Flamini, G. and Morelli, I. (2006). Composition of essential oils of *Helichrysum italicum* (Roth) G. Don fil subsp. *italicum* from Tuscan archipelago islands. Flavour and Fragrance Journal, 21, 805–808.

[203] Leonardi, M., Ambryszewska, K. E., Melai, B., Flamini, G., Cioni, P. L., Parri, F. and Pistelli, L. (2013). Essential oil composition of *Helichrysum italicum* (Roth) G.Don ssp. *italicum* from Elba Island (Tuscany, Italy). Chemistry & Biodiversity, 10, 343–355.

[204] Zheljazkov, V. D., Semerdjieva, I., Yankova-Tsvetkova, E., Astatkie, T., Stanev, S., Dincheva, I. and Kačániová, M. (2022). Chemical profile and antimicrobial activity of the essential oils of *Helichrysum arenarium* (L.) *Moench.* and *Helichrysum italicum* (Roth.). G. Don. Plants., 11, 951.

[205] Cosar, G. and Cubukcu, B. (1990). Antibacterial activity of *Helichrysum* species growing in Turkey. Fitoterapia, 61, 161–164.

[206] Rios, J. L., Recio, M. C. and Villar, A. (1991). Isolation and identification of the antibacterial compounds from *Helichrysum stoechas*. Journal of Ethnopharmacology, 33, 51–55.

[207] Meyer, J. J. M. and Afolayan, A. J. (1995). Antibacterial activity of *Helichrysum aureonitens* (Asteraceae). Journal of Ethnopharmacology, 47, 109–111.

[208] Afolayan, A. J. and Meyer, J. J. M. (1997). The antimicrobial activity of 3,5,7-trihydroxyflavone isolated from the shoots of *Helichrysum aureonitens*. Journal of Ethnopharmacology, 57, 177–1.

[209] Nostro, A., MP, G., D'Angelo, V., Marino, A. and Cannatelli, M. A. (2000). Extraction methods and bioautography for evaluation of medicinal plant antimicrobial activity. Letters in Applied Microbiology, 30, 379–4.

[210] Ali, M. A., Saleem, M., Ahmad, W., Parvez, M. and Yamdagni, R. (2002). A Chlorinated Monoterpene Ketone, Acylated-Sitosterol Glycosides and a Flavanone Glycoside from *Mentha longifolia* (Lamiaceae). Phytochemistry, 59(8) 2002 889–5.

[211] Tafrihi, M., Imran, M., Tufail, T., Gondal, T. A., Caruso, G., Sharma, S., Sharma, R., Atanassova, M., Atanassov, L., Valere Tsouh Fokou, P. et al. (2021). The wonderful activities of the Genus Mentha: not only Antioxidant properties. Molecules, 26, 1118.

[212] Džudžević-Čančar, H., Telci, I., Stanojković, T., Yavuz, M., Özek, T., Dedić-Mahmutović, A., Alispahić, A. and Yalçin, Ö. Ü. (8–11 September 2024), *Mentha gentilis* var. citrata essential oil phytochemical constituents and antiproliferative activity. 54th International Symposium on Essential Oils, Hunguest BAL Resort, Balatonalmádi, Hungary, 27.

[213] Mokaberinejad, R., Zafarghandi, N., Bioos, S., Dabaghian, H. F., Naseri, M., Kamalinejad, M., Amin, G., Ghobadi, A., Tansaz, M., Akhbari, A. and Hamiditabar, M. (2012). *Mentha longifolia* syrup in secondary amenorrhea: A double-blind, placebo-controlled, randomized trials. DARU, Journal of Pharmaceutical Sciences, 20, 1–8.

[214] Hajlaoui, H., Snoussi, M., Ben Jannet, H., Mighri, Z. and Bakhrouf, A. (2008). Comparison of chemical composition and antimicrobial activities of Mentha longifolia L. ssp.longifolia essential oil from two Tunisian localities (Gabes and Sidi Bouzid). Annals of Microbiology and Immunology, 58, 103–110.

[215] Al-Okbi, S. Y., Fadel, H. H. M. and Mohamed, D. A. (2015). Phytochemical constituents, Antioxidant and Anticancer activity of Mentha citrata and Mentha longifolia. Research Journal of Pharmaceutical, Biological and Chemical Sciences, 6, 739–751.

[216] Sujana, P., Sridhar, M. T., Josthna, P. and Naidu, C. V. (2013). Antibacterial activity and phytochemical analysis of Mentha piperita L. (peppermint) – An important multipurpose medicinal plant. American Journal of Plant Sciences, 4, 77–3.

[217] Horváth, P. and Koščová, J. (2017). *In vitro* antibacterial activity of *Mentha* essential oils against *Staphylococcus aureus*. Folia Veterinaria, 61(3), 71–77.

[218] Kitic, D., Jovanovic, T., Ristic, M., Palic, R., & Stojanovic, G. (2002). Chemical composition and antimicrobial activity of the essential oil of Calamintha nepeta (L.) Savi ssp. glandulosa (Req.) PW Ball from Montenegro. Journal of Essential Oil Research, 14(2), 150–152.

Part IV: **Uses of medicinal and aromatic plants in other areas**

Emina Boškailo*, Alema Dedić, Hurija Džudžević Čančar,
Amra Alispahić, and Kasapović Dejana

Chapter 13
Medicinal and aromatic plants used in cosmetics

Abstract: In order to promote sustainability and environmentally friendly practices, the chapter delves into medicinal and aromatic plant (MAP) extracts, their phytochemical, and their integration with novel technology aimed to promote applicability in cosmetic formulations. Huge progress is being made in replacing harnessing solvent with green substitutions, and increasing extraction process efficiency by manipulating parameters. Consequently, MAP extracts have complexed chemical profile and significant biological activities such as antioxidant, anti-inflammatory, antimicrobial, etc. MAPs are accelerating the revolution as environmentally benign materials with prominent usage in cosmetic industry. This chapter also emphasizes the significance of MAP extracts collaboration with nanoparticles (1–100 nm) as creative solutions being developed to leverage delivery of active ingredients, causing site-specificity, enhancing biocompatibility, or the drug-loading capacity. Therefore, the combination of MAP-derived nanoliposomes, nanocarriers such as ultradeformable vesicles have significant impact in upgrading the skin penetration of drugs and efficacy of anti-ageing performances of some metabolites. Tocoferol in transfersome (<100 nm), has great properties with entrapment potentiality of up to 90%. Such potential gives formulations another dimension, representing them as an eco-friendly source of materials for cosmetics.

Keywords: medicinal and aromatic plants, biodiversity, extraction, sustainability, nanotechnology, cosmetic products

*Corresponding author: Emina Boškailo, Department of Ecology and Environmental Protection, Faculty of Social Sciences Dr. Milenko Brkić, Herzegovina University, Mostar 88000, Bosnia and Herzegovina; International Society of Engineering Science and Technology, Nottingham, UK, e-mail: emina.zubovic@gmail.com; emina.boskailo@hercegovina.edu.ba
Alema Dedić, Hurija Džudžević Čančar, Amra Alispahić, Department of Chemistry in Pharmacy, University of Sarajevo-Faculty of Pharmacy, Sarajevo, Zmaja od Bosne 8 – Campus UNSA, Bosnia and Herzegovina
Kasapović Dejana, Department of Physics and Chemistry, Faculty of Engineering and Natural Sciences, University in Zenica, Zenica 72000, Bosnia and Herzegovina

https://doi.org/10.1515/9783111469713-013

13.1 Introduction

Since ancient times plants have served as primary care in protecting health. Nowadays, the prevalence of traditional use of plants is almost 80% in low-income countries. The world is rich in plant biodiversity that has broad of benefits for human health, so researchers have gained huge interest in plant investigation (qualitative and quantitative). There is increasing evidence for the significant value of medicinal and aromatic plants (MAPs) around the globe. Their use in numerous industry areas is influenced by their life-sustaining element content, role of oxygen provider, and sources of myriad of compounds beneficial for functioning of organisms. As the human population increases, intensity of wild plant use is high, but inappropriate ecosystem management and exploitations are a threat to medicinal plants [1]. Their medicinal potential has increased, especially in cosmetic industry for their functions on prevention and treatment in skin care. MAPs utilization has improved approaches in treatment of physical disorders, minimized synthetic antibiotics use, and prolonged life expectancy. As per the last estimation, nearly 223,300 seed plants have been investigated or known to humans, but only one-fifth are pharmacologically and chemically researched for novel drug discovery [2].

Antioxidant and antimicrobial potential of MAPs are high, due to polyphenols composition known as free radical scavengers, which prevent development of various diseases. The antioxidant activity they possess is responsible for their wide application in cosmetic industry. When humans are exposed to stress, they make very reactive oxygen-based species (ROS and RNS), and less non-enzymatic (e.g., vitamin E, tocopherols) and enzymatic antioxidants (e.g., superoxide dismutase). The pharmaceutical industry is dedicated to develop products based on MAPs, in particular based on their constituents, phenolics, terpenes, and alkaloids. Singh et al. [3] evaluated *Ajuga integrifolia* Buch.-Ham. leaf extracts (methanol, hexane, and water) for their chemical profile as well as antioxidant and antibacterial activity. Methanol extract showed the highest content of phenolic (196.16 ± 0.0083 mg GA equivalent/g) and flavonoid (222.77 ± 0.002 mg RU equivalent/g). DPPH test was used to examine the antioxidant activity and showed a minimum IC50 value. The highest inhibition zone (IZ) and minimum inhibitory concentration (MIC) was given in MRSA and β-lactam-resistant *E. coli*.

Unfortunately, concerns about synthetic products are growing due to their negative impact on human health, so preference for the innovative MAP-based cosmetics has peaked. Hence, in the European market, demand for MAP-based cosmetics has reached up to $4 billion in 2015, and keeps growing annually up to 11%. Rosemary essential oils and extracts are known for a plethora of biological potentials like antioxidant, anti-inflammatory, wound-healing, anti-wrinkle properties etc. These activities are related to various and versatile rosemary chemical contents (e.g., rosmarinic acid, carnosol, and carnosic acid). Hexane extract of *Rosmarinus officinalis* formulated in lipid nanocapsule-based mucoadhesive gel (particle size of 56.55–66.13 nm) showed antiaging potential. Collagen, elastin, and hyaluronic acid are crucial for the dermal extracellular matrix (ECM), and *Rosmarinus officinalis* hexane extract showed anti-collagenase activity of IC50 of 520.2 μg/mL. Moreover, in vitro anti-elastase (IC50 value of 57.6 μg/mL), and anti-hyaluronidase activities (448.1 μg/mL) improved antiaging potential of extracts [4].

The concept of One Health is aligned with UN sustainable development goals (SDGs) and a multidisciplinary approach including animal, environmental, and human health makes social approach correlated and ensures positive growth of medicinal products for human beings [5]. Before any cosmetic product gets on the market it needs to be proven for safety by assessment. Animal testing of cosmetics products is forbidden, so numerous alternatives such as 2D cell culture models or 3D human skin equivalent models are applied in order to investigate anti-inflammatory activity or skin irritation efficiency [6].

13.2 Medicinal and aromatic plant-derived extracts

13.2.1 Extraction techniques of MAPs

Natural products represent chemical constituents or phytochemicals isolated or produced from plants investigated for various purposes in numerous research fields. MAPs are a huge source of phytochemical extremely important for pharmaceutical industry, cosmetics, and nutraceuticals. MAP-based extracts like essential oils, concretes, resinoids, etc. consist of numerous chemical compounds groups (e.g., terpenoids, alkaloids, flavonoids, phenolics, and alcohols). They contribute to sensorial (taste and flavor) and functional (antioxidant, anti-cancer, and antimicrobial activities) preferences. Extraction as a simple method mainly implies selective solvent application (e.g., water, alcohol, and their mixture in various ratios) to dissolve chemical compounds for their separation and characterization. There is also a multitude of extraction techniques for production of natural products. The following parameters determine what type of extraction will be applied: the drug nature (quality of plant, its origin, climate conditions, harvest time, plant's organ, drying method, particle size, etc.), solvent type, costs and therapeutic values, concentration of the product, and stability of the drug (maceration, hydrodistillation, HD, Soxhlet extraction, SE). Such extraction approaches have pros and cons and are rapid, but mainly require organic and costly solvents.

Considering these drawbacks, innovative approaches are desirable such as ultrasound- and microwave-assisted extraction (UAH and MAH), supercritical fluid extraction (SFE), enzyme-assisted extraction (EAE), etc. Kırkıncı et al. [7] evaluated the effect of conventional and innovative extraction methods and revealed potential of HD and MAH on essential oil (EO) and wastewater (WW) yield and chemical profile of *Lavandula angustifolia* L. Accordingly, the potential for utilizing WW in numerous industrial applications is huge, for EOs and food, cosmetic, and health sectors. Both techniques were consistent in extracting major compounds by GC/MS (α-terpinolene (25.60% and 24.25% and (–)-borneol 19.55% and 19.37%), but the difference is visible through presence of minor compounds. MAH extract has higher phenolic and flavonoid content versus HD probably influenced by minimized thermal degradation and selective heating mechanisms. Moreover, MAH extract has shown better antioxidant activity. Naik et al. [8] used SE and SFE to extract *P. juliflora*. Regarding GC/MS analysis, 35 components were eluted by SFE and proved better antifungal activity. These studies highlight the importance of extraction methods in recovering phytochemicals as potential natural metabolites that may provide a positive impact on tested activities.

13.2.2 Influence of extraction operational parameters

Besides the impact of extraction techniques, other parameters might influence extracts: yield, composition, and results for tested activities are given in Table 13.1. For example, altitude influence (from 766 m to 1,387 m, with 100-m intervals) on *O. majorana* essential oil content, composition, and yield was evaluated. The best EO yield (6.50%) and linalool content (79.84%) were estimated by HD at the lowest altitudes (766 m). The highest yield of borneol, caryophyllene, linalool oxide, germacrene-D, *trans*-linalool oxide, a-humulene, and bicyclogermacrene compounds was achieved at 890 m altitude. Also, thymol and α-terpineol yield, as well as a-terpinene, terpinene-4-ol, *cis*-sabinene hydrate, and carvacrol were achieved at highest altitudes (1180 and 1,387 m). Climate parameters (temperature and its differences during the day and night, relative humidity, precipitation, day light hours, and intensity) change with altitude, and consequently affect EO yield and composition [9].

SC–CO_2 extraction was used to achieve yield of chia seed oil by manipulating different operational parameters (pressure, temperature, and particle size) and compared with SE results. The optimal SC–CO_2 parameters (335 bar, 45 °C, 100 – 400 μm, 24 s of grinding time) have contributed to the highest yield versus chia oil yield obtained by SE [10]. Puertolas et al. [11] investigated the impact of pulse electric field (PEF)-assisted extraction on olive seeds using the following parameters: electric field of 2 kV/cm and 65 J of energy. These conditions increased the yield up to 13.3%. Success of PEF is visible through content increase of compounds that belong to polyphenols (11.5%), phytosterols (9.9%) and total tocopherols (15%). Moreover, apple seed oil is obtained by SFE and SE considering optimization pressure, temperature, and CO_2 flow rate (10–30 MPa, 40–60 °C, 1–8 L/h) on the yield, antioxidant activity, and total phenolic content. The best SFE parameters were 24 MPa, 40 °C, 1 L/h of CO_2 flow rate, 140 min that obtained the highest yield 20.5 ± 1.5% (w/w) compared to SE yield of 22.5 ± 2.5% (w/w). Moreover, SFE contributed to higher linoleic acid content and better oxidative stability [12].

Vo et al. [13] optimized the UAE and MAE methods using a combination of three solvents (ethanol, acetone, and water) with an optimum ratio of 0.29:0.34:0.37, that achieved an appropriate polarity for recovering phenolics and flavonoids from passion fruit peels. The optimal UAE conditions were 28 mL/g of liquid-to-solid ratio (LSR), 608 W of ultrasonic power, and 63 °C for 20 min to achieve TPC of 39.38 mg GAE/g db and TFC of 25.79 mg RE/g db. The best MAE parameters were 26 mL/g of LSR and 606 W for 2 min to enable TPC and TFC recovery of 17.74 mg GAE/g db and 8.11 mg RE/g db, respectively. Similarly, 12 natural deep eutectic solvents were produced for recovering phenolics and terpenoids from *Abelmoschus sagittifolius* (Kurz) Merr roots; citric acid/glucose and lactic acid/glucose, with a molar ratio of 2:1 were the most suitable. For the highest terpenoid recovery at 69 ± 2 mg UA/g dw, the following conditions of UAE were used: 40 mL/g liquid-to-solid ratio, 40% water content, 30 °C, and 600 W ultrasonic power for 5 min. Regarding phenolics recovery, 150 W ultrasonic power was the most suitable at 9.56 ± 0.17 mg GAE/g dw. The conditions for MAE were 50 mL/g liquid-to-solid ratio, 20% water content, and 400 W microwave power for 2 min to achieve the maximum phenolics and terpenoids at 22.13 ± 0.75 mg GAE/g dw and 90 ± 1 mg UA/g dw, respectively [14].

13.3 MAPs in skin care products

13.3.1 MAPs as photoprotective agents against UV light and skin damage

Skin is often exposed to UV radiation and produced ROS is mainly responsible for oxidative damage and accumulation of oxidative products that leads to expression of elastase and collagenase. Moreover, skin aging is influenced by UV light and oxidative stress, which cause acute inflammatory reactions in the tissue. Numerous natural metabolites are strong antioxidants and show anti-inflammatory responses helpful in photoaging condition [15]. Abdallah et al. [16] prepared *R. damascena* emulgel (100 mg/g) and *R. damascena* oil nanoemulsion at 50 mg/g and 100 mg/g emulgel doses to investigate protective impact against UVB-influenced photoaging, measured antioxidant, antiwrinkle, and anti-inflammation parameters (CAT and SOD; MMP-9; TNF-α, and IL-6), etc. Moreover, total RNA isolated from rat dissected joints was used to evaluate the genes expression (JNK, ERK1/2, and p38 MAPK). Given results showed that *R. damascena* oil nanoemulsion achieved more significant antiaging potency versus *R. damascena* emulgel according to histological and biochemical evaluations. Such extraordinary activities might be linked with its chemical composition, geraniol, nerol, citronellol, phenyl ethyl alcohol, and linalool (29.2%, 23.4%, 16.34%, 4.96%, and 3.24%, respectively).

Strawberry (Fragaria × ananassa) rich in anthocyanins has strong anti-inflammatory potentials. Five pigments of anthocyanins were characterized from methanolic extracts and their potential to protect human dermal fibroblasts against UV-A radiation was evaluated by MTT cell growth assay and Comet assay. These extracts showed high photoprotective activity in fibroblasts emphasizing their ability to protect skin against the UV-A radiation negative effect [17]. Non-phototoxic chamazulene as major compound (38.92%) of *Artemisia sieversiana* Ehrhart ex Willd. EO in synergy with two UVB filters (ratio of

Table 13.1: Different extracts of medicinal and aromatic plants (MAPs) obtained using conventional and innovative extraction techniques along with operational parameters aimed at achieving MAP phytochemical profile and getting insight to discover potential natural metabolites that may provide versatile medicinal activities.

Plant species	Plant organ	Origin	Mass	Grounding type	Extraction	Solvent	Ratio (plant: solvent)	Extraction parameters	Yield (%)	Phytochemical profile	Medical benefit	Reference
Thymus vulgaris	Herb	Wild	150 g	Powder	SE	Hexane, chloroform, and methanol	1 g:5 mL	First hexane crude extract, then exposed to second chloroform and third extraction.	28.7 g, 19.2%; 35.1 g, 28.9%; 33.8 g, 39.2%	Flavonoids, phenolic molecules, aromatic compounds, saponins, tannins, iridoids, and quinones	Acne treatment, fragnance in perfumes	10.1038/s41598-024-71012-2
Ajuga integrifolia Buch.-Ham	Leaves	Wild	25 g	Powder	SE	Methanol, hexane, and water	–	25 g of powdered leaves extracted with 125 mL solvents at 55 °C, 50 °C, and 60 °C, 48 h. After that, viscous semisolid masses were evaporated.	3.98 g, 3.01 g, 2.71 g	Terpenes, phenolics, and alkaloids	Wound healing	10.1038/s41598-024-67133-3
Murraya paniculata	Leaves	Wild	–	Powder	SE	Hexane, acetone, chloroform, methanol and water	–	Soxhlet extraction with several solvents for 7–8 h and evaporated.	4.8, 3.7, 3.47, 1.761.07 g	Alkaloids, flavonoids, steroids, and tannins	Wound healing	10.1038/s41598-021-87404-7

Salvia officinalis L., Rosmarinus officinalis L., and Mentha piperita L.; Ruta graveolens L., Olea europaea L., Petroselinum crispum Mill., Punica granutum L.	Leaves; leaves and young stems	Botanical growing	40 mg	Powder	Agitating	Hydro-methanolic solution	MeOH: H₂O, 70:30, v/v	Agitating: 30 min, 200 rpm, RT.		Flavonoids, phenolic lignans and stilbenes; Rutin, psoralen, limonene, and pinene; oleuropein; pigenin, coumarins, myristicin; phenolic acid, flavonoids, tannins, amino acids, and alkaloidal	Anti-cancer, anti-microbial, anti-diabetes, and gastrointestinal diseases; tremors, paralysis, nervous disorders; cardioprotective activity; cyto-, gastro-, brain-, nephron-protective effects;	10.1038/s41598-021-89437-4
Lavandula angustifolia L.	Flowers	Field	100 g	Sieved to a particle size between 80 and 100 mesh	HD and microwave-assisted hydrodistillation (MHD)	Water	–	100 g in 500 mL water exposed do Clevenger-type HD; 100 g in 500 mL water exposed to microwave digestion (1,000 W): temp. increased to 100 °C in 10 min and maintained for 30 min.	–	Terpenoids (α-terpinolene and (–) borneol), phenolic	Aromatherapy, headaches, depression, and colds	[7]
Cinnamomum cassia	Leaves	Botanical growing	–	Sieved and powderd at particle size of 60 mesh	Ultrasonic	Methanol	–	the power set at 350 W, the frequency set at 35 kHz, the temperature set at 40 °C	–	Trans-Cinnamaldehyde	Anti-aging, hyperpigmentation, and acne treatment	10.1038/s41598-024-75189-4

(continued)

Table 13.1 (continued)

Plant species	Plant organ	Origin	Mass	Grounding type	Extraction	Solvent	Ratio (plant: solvent)	Extraction parameters	Yield (%)	Phytochemical profile	Medcial benefit	Reference
Taraxacum mongolicum Hand.-Mazz	Herb (leaf, stem, and root)	Wild	1 kg	Crushed	SE	Ethanol (conc. 60%)	20:1 (mL: g)	Extraction time (30, 40, 50, 60, and 70 min), extraction temperature (40, 50, 60 and 80 °C), and/or the ratio of liquid to material (20, 30, 40, 50 and 60:1 (v/w) mL/g) and ethanol concentration (40%, 50%, 60%, 70%, and 80% (v/w))	–	Flavonoids, triterpenes, sesquiterpenes, phenolic acids, sterols, and coumarins	Anti-aging and anti-cancer	10.1038/s41598-023-28775-x
Lepidium sativum	Seed	Market	2 kg	Coarse ground	Cold pressed, filtration, lyophilization	Water	2 kg seeds soaked in water (3 × 7 L)	at room temp. with frequent shaking for 12 h. The soaked seeds were cold pressed, filtered, and the filtrate was centrifuged at 3,000 rpm for 15 min. The supernatant was separated in a gel form and lyophilized. After 48 h. a lyophilized dry powder was produced (140 g).	140 g	Glucosinolates, phenolic acids derivatives, alkaloids, and flavonoids	Alopecia treatment	10.1038/s41598-023-33988-1

Species	Part	Source	Quantity	Form	Method	Extracts	Sample preparation	Conditions	Yield	Constituents	Properties	Reference
Colotropis procera	Aerial parts	Wild	266 g	Powder	Decoction and maceration	Aqueous and hydroalcoholic extracts	–	Decoction: 266 g of sample in deionized water at 60 °C for 3 h, with continuous stirring at 1,500 RPM. Hydro-alcoholic extract: 266 g macerated in 80% ethanol (1.5 L).	82 g (30%) and 65 g (24%)	Flavonoids, tannins, terpenoids, saponins, alkaloids, steroids, and cardiac glycosides	Antioxidant and skin anticancer properties	10.1038/s41598-024-76422w
P. juliflora	Leaves	Wild		Powder	SFE and SE	Ethanol and hexane	50 g of the P. juliflora leaf powder in 300 mL hexane	Pressure (200 bar) and temperature (450 °C) dynamic extraction time (90 min).	14.10 (93.37%) and 9.25 (61.25%) g/100 g	Phenol, 3,5-bis (dimethyl ethyl), benzene dicarboxylic acid, squalene; pentanoic acid, 5-hydroxy-2,4,-dibutyl phenyl ester, phytol, tetramethyl heptadecane, neophytadiene, and hexadecanal	Antimicrobial, antioxidant, and wound headlining	[8]

5:1) incorporated into sunscreen formulations reduces the usage of UVB filters up to 66%. Moreover, both have ability to minimize UVB-induced radiation cellular damage, which implies their possible application in the sunscreen products [18]. Also, the photo-protective effect of phenolics (oxyresveratrol and kuwanon O), obtained from the *Morus australis* (root) extract, in human primary epidermal keratinocytes was evaluated. Both phenolics were nontoxic to cells (conc. >10 and 0.5 μM). Oxyresveratrol increased cell via-bility at pretreatment at conc. 5 and 10 μM and attacked UVA- or H_2O_2-induced cellular ROS and also reduced UVA-improved nitrotyrosine. Kuwanon O also presented similar results with 0.25 μM and 0.5 μM, but without protection on cell survival after UVA irradi-ation. Both natural metabolites might be ingredients in cosmetic products aimed for skin photoprotection or the prevention of photocarcinogenesis in humans [19].

13.3.2 Regenerative and wound-healing properties of MAP-derived agents

Wound represents the disruption of the cellular and anatomic tissue layer due to sev-eral types of traumas (physical, thermal, microbial, etc.) or caused by immunological trauma. Accordingly, wound healing is a complicated process followed by repairing damaged tissue and depends on different phases such as inflammatory, proliferative, or remodeling. The final goal of wound healing is to reduce healing duration and mini-mize consequences such as scars. The healing process is dependent on physiological mechanisms like anti-inflammatory, antioxidant, and antimicrobial activities. Infected wounds are the main reason for wound-healing complications influencing efficiency of wound healing. It is well-known that bacteria are directly linked to infected wounds, so minimizing their load would be beneficial in the healing process. Many MAPs promote the skin's natural recovery processes and show potential in wound treatment. The em-ployment of MAP extracts in wound care is constantly increasing thanks to their wide spectra of compounds and physiological and pharmacological efficiency [20].

Numerous in vitro and in vivo methods might be employed in analyzing wound-healing performances of MAPs metabolites (Figure 13.1). In vivo artificial and tissue models, and others might be used according to factors taken into consideration. In vitro models are more robust, fast, and require less ethical considerations, and give detailed insight into biochemical and physiological processes that are influenced by the test agent/compound. It is important to emphasize that no animal tissue is an exact replica of human skin. Cream products based on *Kalanchoe pinnata* (KP leaves) extracts contain 0.15% of [quercetin 3-*O*-α-L-arabinopyranosyl-(1→2)- α-L-rhamnopyranoside]. Creams are used to investigate a rat excision model for 15 days, and on the 12th day, rat groups were treated with KP leaf-extract and its major flavonoid. The results have shown 95.3 ± 1.2% and 97.5 ± 0.8% of healing, respectively with significant re-epithelialization and denser collagen fibers that have huge importance in wound healing [21]. Gel formulated from *B. pinnatum* (leaves) aqueous extract (5%) was evaluated for treatment of back skin wounds in rats. The wounds reduction was followed by a reduction in inflammatory infiltrate and the levels of the IL-1β and TNF-α. The formulated gel exhibited phytochemical and biological stability for a month and showed quercetin 3-*O*-α-L-arabinopyranosyl -(1→2)-*O*-α-L-rhamnopyranoside (*B. pinnatum* compound) as significant chemical marker of MAP extracts and formulations containing *B. pinnatum* that might be used in evaluation of quality [22].

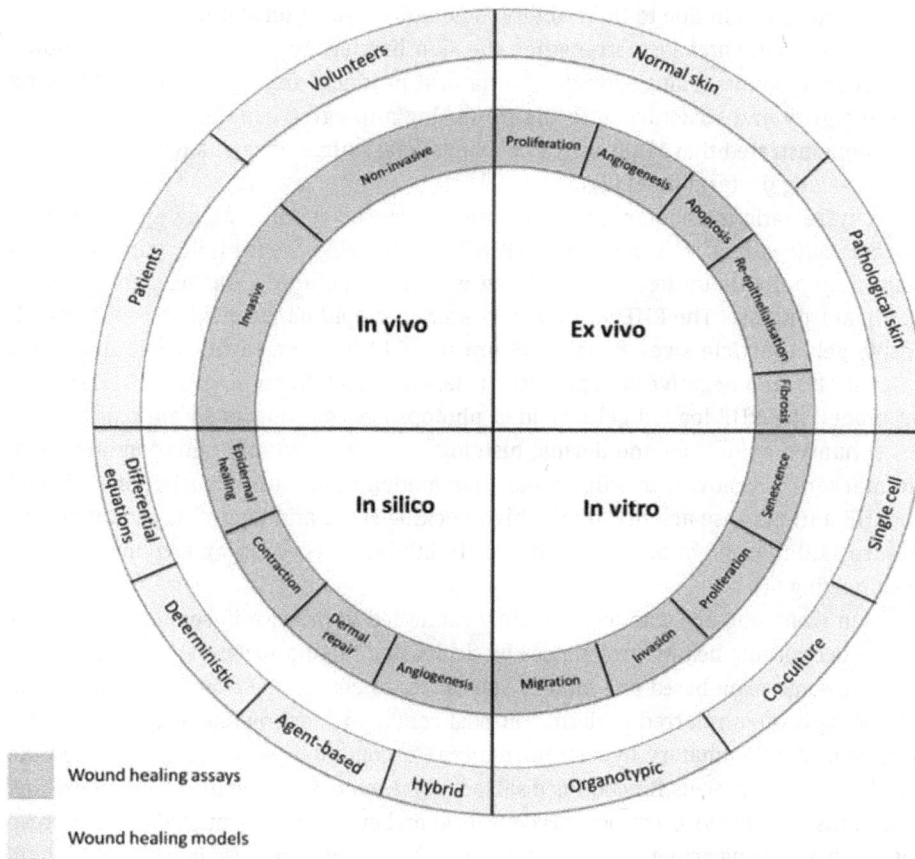

Figure 13.1: The proposed comprehensive scheme of several classes of human models of wound repair such as in silico, in vitro, ex vivo, in vivo, as well as applicable assays for various wound models [23].

13.3.3 MAPs as skin anti-aging and whitening agents

Today's skin care is very complex, and women in early ages (from 30 years) are showing signs of aging skin. The processes on skin are the most noticeable; the level of skin care products is highly in demand. The most age-related changes in the dermis include: (a) a decrease in the number and activity of fibroblasts, that are important in producing collagen, hyaluronic acid, and elastin, (b) breakdown of collagen fibers, increased cross-linking, and a decrease in skin resilience and stretchability, (c) alterations in elastin fibers, leading to clumping (elastosis), no elasticity, and wrinkle formation, and (d) a reduction in hyaluronic acid, resulting in skin that is less moisturized and less resilient. Botanical ingredients are commonly used in cosmetic products for dry and mature skin due to their ability to enhance hydration status of skin, decrease trans-epidermal water loss, strengthen the skin barrier, and prevent the breakdown of skin components. Plant extracts and natural products are valued for maintaining skin integrity and structure, offering promising anti-aging benefits. In vitro studies have demonstrated that MAPs might be a beneficial source of agents with huge potential anti-aging performances [24].

A systematic molecular modeling study explored the anti-aging effects of *Rosmarinus officinalis* L. hexane extract (RHE) by analyzing the inhibitory impact of its major components on key aging-related enzymes, including elastase, collagenase, and hyaluronidase. The RHE was incorporated into lipid nanocapsule-based mucoadhesive gels (particle sizes from 56.55 nm to 66.13 nm), uniform distribution (PDI 0.207–0.249), and negative zeta potential (−13.4 to − 15.6). In an in vivo UVB-irradiated rat model, the RHE-loaded gel provided photoprotection, improved antioxidant levels, enhanced epidermal and dermal histology, and reduced inflammation and wrinkle markers. Moreover, in silico molecular modeling identified verbenone, crucial for RHE anti-elastase activity, due to high docking score and favorable binding mode (Figure 13.2) that might be responsible for in inhibiting elastase, an enzyme involved in skin aging [4].

Skin whitening or lightening is mainly connected to practice in some ethnic groups for culture-specific beauty preferences by decreasing human melanin concentration or applying some plant-based or synthetic substances to change the skin tone or lighten it. Whitening is often referred to dermatological conditions, for treatment hyperpigmentation, or post-inflammatory hyperpigmentation, or conditions such as vitiligo when the skin loses pigment or its function is disabled [25]. A natural cosmetic product consisting of *Hibiscus cannabinus* L. extract was developed and evaluated on antioxidant, antityrosinase, and anti-aging activities. The results have shown antityrosinase potential on inhibition of monophenolase ($30.28 \pm 3.90\%$) and diphenolase ($11.40 \pm 0.29\%$) formation, and provided inhibition of collagenase ($36.41 \pm 0.54\%$) and elastase ($23.13 \pm 1.56\%$) [26]. Several MAPS and their aqueous extracts were exposed to the anti-aging activities, *Echinacea purpurea* J. Presl has shown the best inhibition of collagenase, elastase, and hyaluronidase activity (78.5 ± 0.0, 69.0 ± 1.4, and $64.2 \pm 0.3\%$). Also, *Morus alba* L. (leaves and steamed/roasted leaves) revealed the best anti-inflammatory potential, as they inhibited IL-6 and TNF-α secretion ($p < 0.05$). These aqueous extracts pose significant impacts on the skin

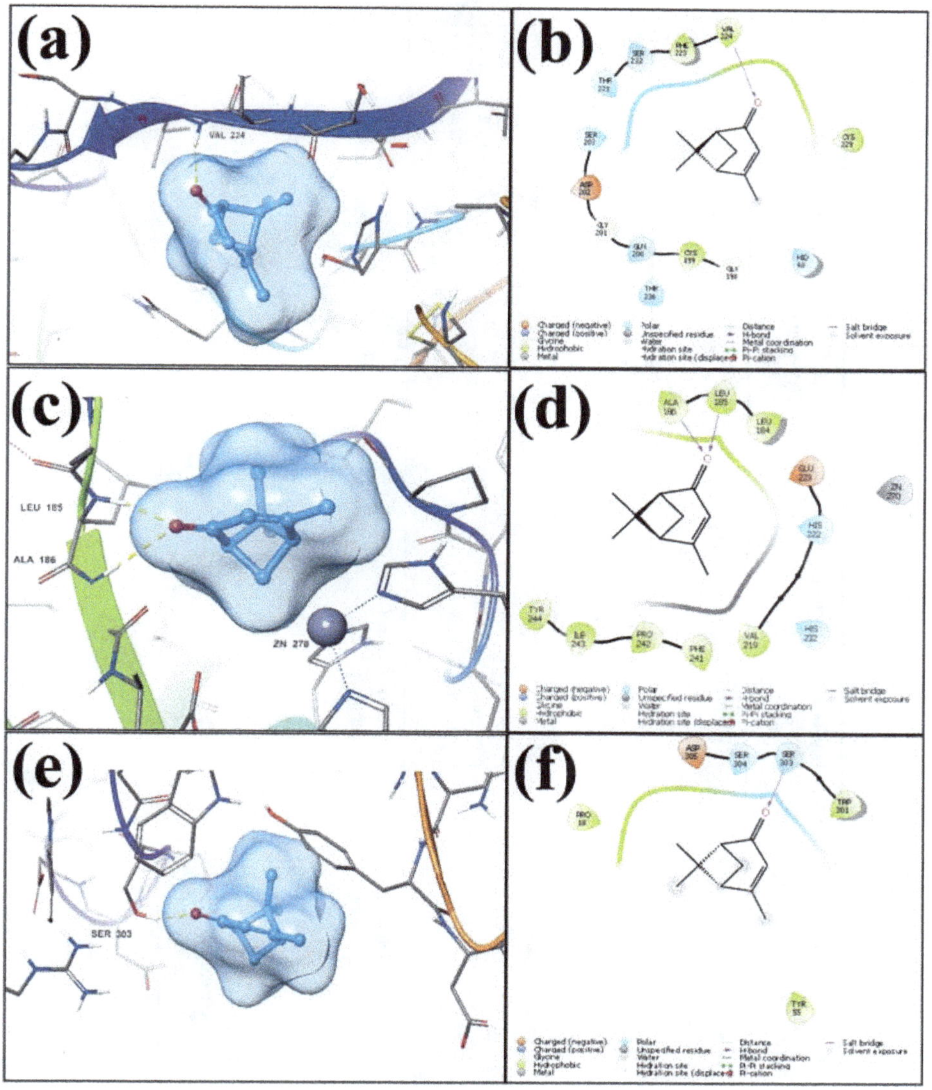

Figure 13.2: The representation of verbenone in the docking pose in elastase: (a, b) verbenone in 3D docking pose and 2D interaction diagram of in the binding site of elastase, (c, d) collagenase, and (e, f) hyaluronidase. Reproduced with permission from [4]. Copyright Nature ©2022.

and might be applied in cosmetic formulations like toners, facial mist, and facial serum [27]. A comprehensive summary of MAP extracts and their major constituents are given in Table 13.2, investigated by various in vitro, in vivo, and in silico methods applied in evaluation of their potential for cosmetic skin care formulations.

Table 13.2: Medicinal and aromatic plant extracts with their major constituents in maintaining skin care with huge benefits for skin aging, wound healing, and skin whitening.

Plant	Extract	Active compound	Model	Assay	Application area	Human cells	IC50	Cell viability (%)	Inhibition (%)	Reference
Centella asiatica	95% EtOH (1:3 w/v)	Asiaticoside (MF of 2.4%)	In vitro	Cytotoxicity and scratch assay test	Wound healing	HDF, HaCaT	0.19 and 100 µg/mL	80	–	[28]
Crocus sativus L.	Aqueous extrac	–, (0–320 µg/mL)	In vitro	Scratch assay test	Wound healing	HDF, HUVEC	–	At conc.10 µg/mL > 80	–	[29]
Solanum betaceum	MeOH	–, 500 – 1,800 µg/mL	In vitro	Trypan blue cytotoxicity assay, clonogenic assay	Skin aging, wound healing, proliferation of Ca9-22 cells	Ca9-22 oral carcinoma cells	Reduced the migration of Ca9-22 cells from 12.5% to 100%	From 6.9 to 82.1%	At conc. 100 µg/mL: Tyrosinase:50.4 ± 1.6 (seeds) and 30.7 ± 1.2 (pulp); Hyaluronidase: 20.2 ± 0.7 (seeds)	[30]
Borago officinalis	MeOh and MeOh-Aq	–, 100–1,000 µg/mL	In vitro	Inhibition of protein denaturation; determination of anti-collagenase and anti-elastase activity	Anti-aging properties	HaCaT, BJ	–	–	Collagenase: MeOH >44 and 48; Elastase: MeOh up to 40	[31]
Myracrodruon urundeuva	Aq	–, 2,000 µg/mL	In vitro	Inhibition of tyrosinase enzyme, antioxidant activity	Skin whitening	–	–	–	83.76, DPPH of 49.59	[32]

Plant	Extract	Compound / concentration	Study	Method	Application	Enzyme results		Other results	Ref.
Cyclopia sp	Aq, EtOH, Act, BtOH	Oleanolic acid –, 62.5–1,000 µg/ mL; 0–800 µg/ mL and 0–250 µg/mL	In vitro	Photoprotection (SPF estimation); antiaging activity (elastase and collagenase)	Wound healing and inhibition of the skin extracellular matrix enzymes	Elastase activity: BtOH of 666.27 ± 6.51 µg/mL; Act. of IC$_{50}$ 42.5 ± 1.05 µg/mL	–	SPF of 27.81 ± 0. 03	[33]
Sargassum thunbergii	EtOH	–, 0.0–99.5%	In vitro	Inhibition of TRP-1 and MMPs	Skin whitening and anti-wrinkling	–	–	RSA, TIA, and CIA of S. thunbergii extract of 86.5, 88.3, and 91.4	[34]
Rosmarinus officinalis L.	Hexane	Verbenone	In vivo, in vitro, in silico	Anti-elastase and anti-collagenase (fluorimetrically), anti-hyaluronidase activity (haluronidase inhibitor screening assay kit)	Skin aging and wound healing,	Elastase: 57.61 ± 2.93 Collagenase: 520.2 ± 26.5, hyaluronidase: 448.1 ± 22.8 µg/mL; at 10 µg/mL of 91.85 ± 5.1%	–	–	[4]

13.4 MAPs in hair cosmetics

13.4.1 MAPs in hair products

Nowadays, hair care and hairstyle are important to human society as they have huge impact on social life. Various hair conditions require different approaches due to the factors that that cause such conditions, like microbial infection, stress impact, etc. As MAP extract gain huge popularity for their benefits in hair care, researchers put in huge efforts to enhance cosmetics products intended for hair growth, reduce hair fall, for hair nourishing, reducing scalp, or even use their potential in dermatological conditions like seborrheic dermatitis [35]. This is supported by evidence of global market for organic products that reached almost $25 billion by 2025. Natural products are mainly produced from organic agriculture (95%) and do not contain chemical or synthetic additives. Essential oils (EOs) that do not contain constituents of animal origin are considered as vegan, without pesticide content or genetically modified organisms (GMOs), which is within the scope of sustainable production. Cosmetic dermatology has improved along with MAPs products popularity such as EOs due to their significant efficiency on the scalp and hair. The EOs metabolites easily penetrate in the scalp, and nourish and stimulate the deep hair follicles growth, supplement the nutrition, moisturize and strengthen the hair [36].

Plants like *Mentha piperita* L. and *Rosmarinus officinalis* L. are beneficial in enhancement of hair growth and prevent hair loss [37]. *M. piperita* L, EO (3.0%) has potential in the conservation of the dermal papillae vascularization, acts as a stimulant in hair growth, and is therapeutic for hair loss [38]. The clinical trials done on patients with alopecia, promoted *R. officinalis* L. consisting 3.87 mg 1,8-cineole/mL of the product; it is very effective against alopecia and decreased the scalp. *R. officinalis* L. EO antioxidant properties might be responsible for such effect as it generates free radicals [39]. Significant effects can be seen with a mixture of MAPs products in hair treatment. For example, the EO of *Melaleuca alternifolia, Rosa damascene, Citrus grandis*, and *Foeniculum vulgare* have moisturizing efficacy, and *Thymus vulgaris, Lavandula angustifolia*, and *Salvia sclarea* have benefits against alopecia and dandruff. Among various cosmetics products consisting Eos,e the most used are in products for various hair treatments [36].

13.4.2 MAPs in hair growth products

As hair loss is a huge issue, researchers pay great attention to find the right cosmetic products to prevent hair loss, by looking for novel approaches. So, MAP extracts might be beneficial due to their rich chemical profile. *Polygonum multiflorum* extract has revealed hair growth effects that might be related to anagen prolongation (Figure 13.3) [40], as well as 2,3,5,4′-Tetrahydroxystilbene-2-O-β-D-glucoside and emodin content [41]. Moreover, increased hair size and hair follicles number is proved by regulating β-catenin and sonic hedgehog expressions via topical and oral application, which is analyzed by histological methods of C57BL/6 mouse [42]. *Allium ascalonicum* L. methanolic extract contain-sphenolics such as rosmarinic, quercetin, and p-coumaric acids. The in shallot extract has hair growth potentiality due to the downregulation of the androgen gene expression (*SRD5A1* and *SRD5A2)* and the upregulation of the genes related with Wnt/β-catenin (*CTNNB1*), sonic hedgehog (*SHH, SMO,* and *GIL1*), and angiogenesis (*VEGF*) pathways [43].

Malva verticillata seed *n*-hexane extracts were identified with linoleic acid (LA) and oleic acid. LA treatment activated Wnt/β-catenin signaling and induced human follicles dermal papilla cells (HFDPCs) growth by increasing the expression of cell cycle proteins such as cyclin D1 and cyclin-dependent kinase 2. LA treatment also increased a vascular endothelial growth factor, insulin-like growth factor-1, and hepatocyte and keratinocyte growth factor, in a dose-dependent manner [44]. *Punica granatum* is associated with anti-inflammatory activity, and its usage in cosmetics (skin care, wrinkling care, pigmentation, etc.) is broad. Aqueous and alcohol *P.granatum* peel extracts have promoted growth of hair in albino mice with alopecia issues. These extracts enhanced hair growth up to 3% [45], showed effectivity in anti-dandruff and anti-lice performances, related to punicagranine 1 content as it has anti-inflammatory properties and inhibitory effects (IC50 of 22.8 ± 1.2 μm) [46].

13.5 MAPs in oral hygiene products

13.5.1 Formulations for toothpaste and mouthwash

Several antimicrobial agents have shown efficacy in mouthwashes available commercially and shown to be important in the treatment of oral disease. The oral flora in hospitalized and debilitated patients are subjected to an early gram-negative bacteria shift. So, the plaque might have a role as a reservoir for possible pathogens, as well as some resistant microbials, for infection at other body sites. Citrox® is based on MAPs soluble bioflavonoids, in particular derived from citrus fruits. Due to its hydroxylated phenolic structures it has potential against various microorganisms. Citrox® formulations in dilution range 0.007–8% v/v showed antimicrobial activity, with emphasis on the BC30 formulation that showed high inhibition activity of all bacterial species and several Candida spp. (*C. albicans, C. dubliniensi, C. glabratas, C. tropicalis:* MIC of 0.125% Citrox®, v/v; *C. krusei:* MIC of 0.03125% Citrox®, v/v, and *C. parapsilosis*: MIC of 0.5% Citrox®, v/v) analyzed by broth and biofilm assay at a concentration of 1% (v/v). These results showed potentiality of Citrox® as antimicrobial agents for oral care products such as mouthwash [47]. Also, eucalyptus oil presented prevention of the growth of oral microbials potentiality that affects creation of dental caries and endodontic infection. Diluted

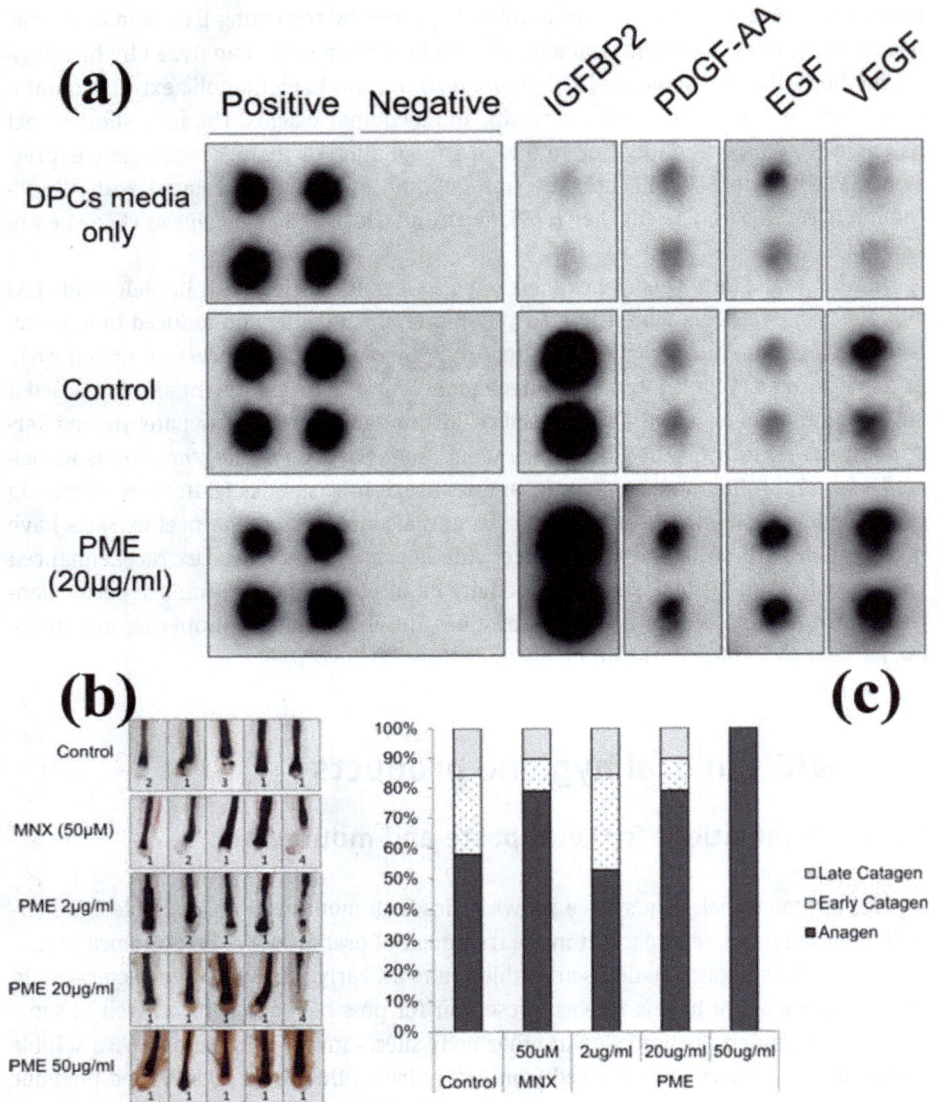

Figure 13.3: (a) Effect of PM extract (20 µg/mL) on 41 types of growth factors secreted from DPCs. Effect of PM extract (2 µg/mL, 20 µg/mL, and 50 µg/mL), and minoxidil (50 µM) on anagen elongation and catagen entry in human hair follicles (20 hair follicles/group) organ culture model was evaluated. After incubation of 6 days, hair follicle morphology was analyzed; (b) hair follicles images of for every experimental group; and (c) calculated ratio of anagen, early and late catagen. Reproduced with permission from [40]. Copyright Springer ©2020.

eucalyptus oil showed high total absorbance reduction against *S. mutans* and *E. faecalis* versus the control ($p \leq 0.001$) as shown in Figure 13.4. Moreover, both bacteria biofilms were reduced by nearly 60- and 30-fold for the biofilm measurement in comparison to the group with no EO ($p \leq 0.001$) [48]. Great antibacterial activity was seen in *Arnebia euchroma* (Royle) Johnst. root (AR) extract on *S. mutans* UA159 and anti-caries effect on rats with MIC, MBC, MBIC50, and MBRC50 of 1.0, 8.0, 4.0, 8.0 mg/mL, respectively [49]. *Aggregatibacter actinomycetemcomitans, Porphyromonas gingivalis, Tannerella forsythia,* and *Prevotella intermedia* known as periodontal pathogens were treated with *Nigella sativa* seeds EO and exhibited significant effect against these pathogens. These results present Nigella sativa seeds EO as a potent antimicrobial agent in the formulations for oral microbial pathogens [50].

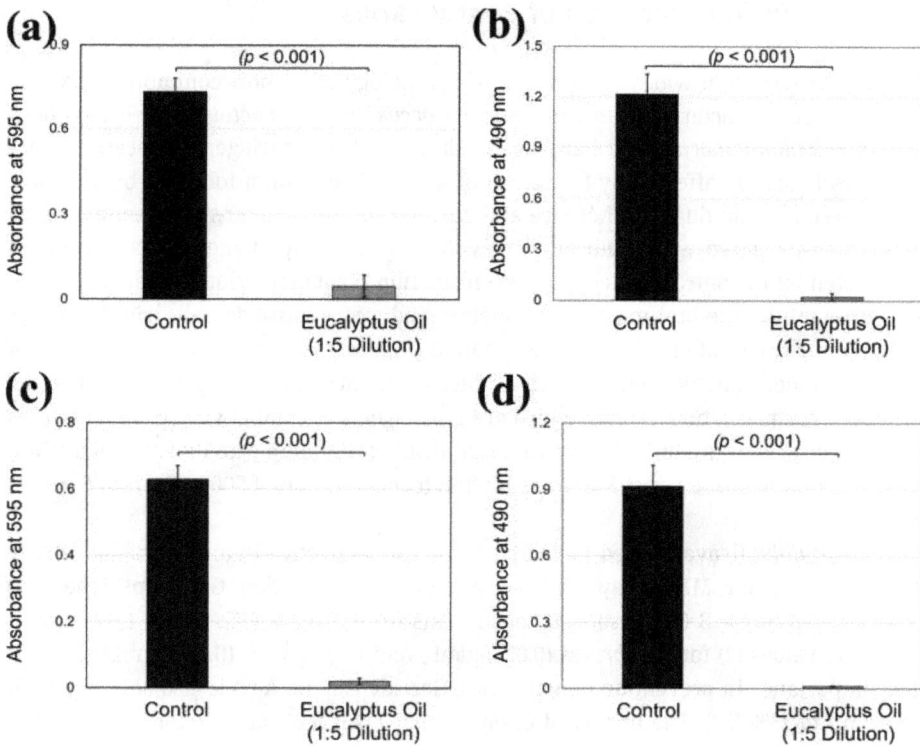

Figure 13.4: Essential oil of eucalyptus showed the antibacterial activities against Streptococcus mutans and Enterococcus faecalis with (a) and (c) reduced the total absorbance and (b) and (d) biofilm formation growth. Reproduced with permission from [48]. Copyright Nature ©2023.

Ethanol extract of *M. acuminata* was used in the formulation of toothpaste containing 1, 2, 3, 4, and 5% extract. It was also used to investigate the antibacterial activity of the toothpaste against *S. aureus* and *S. mutans*. The results revealed the best activity of toothpaste with 5% extract with an IZ of 19.30 ± 0.17 mm and 12.60 ± 0.52 mm against *S. aureus and S. mutans* versus the herbal toothpaste available in market [51]. The antimicrobial activities of the *Moringa oleifera* root extract were used in formulation of toothpaste and compared with commercial Oral-B toothpaste. Both were assessed against *S. mutans*, *Lactobacillus*, *S. aureus*, and *C. albicans*, and *M. oleifera*-based toothpaste showed better inhibitory potential against tested microbials [52]. Nine herbal toothpastes with pomegranate peel extract (0.2–1.8% w/w) and clove oil (1–1.4% w/w) showed promising in vitro antimicrobial activities against *C. albicans*, *S. mutans*, and *S. aureus*. Among all, formulations containing 1.4% w/w pomegranate peel extract and 1% w/w clove oil showed the best IZ against used strains with diameter 26, 27, and 25 mm [53].

13.5.2 MAPs in prevention of dental caries

Dental caries is often widespread in human pathology with most common worldwide incidence. Some bacterial strains like *Streptococcus mutans*, *Lactobacillus acidophilus*, and *Pseudomonas aeruginosa* have been related with the cariogenic processes. Dental caries is mainly affected by formation of a microbial biofilm followed by an acidic and anaerobic state due to adherence and colonization of *S. mutans* [54]. Antimicrobial films often prepared with chitosan films with *E. grandis* pyroligneous extracts might have potential in control strategy to prevent biofilm formation related to dental caries, as well as inhibit the oral microbial formation and may control dental caries by pH reducing impairment of enamel demineralization [55]. Turmeric also has broad spectra of noncurcuminoid phytochemicals such as curcumin, curcumenol, eugenol, zingiberene, curcumol, turmerin, turmerones, which are great agents to combat various oral diseases like dental caries. Curcumin has showed significant antibiofilm activity with sessile MIC concentration of 50% against *S. mutans* biofilm (concentration of 500 μM), from 5 min to 24 h [56].

   G. parvifolia (leaves)-based methanol extract was effective in antimicrobial activity proved by MIC and MBC assays with activity guide fractionation. Garcidepsidone A (1) and garcidepsidone B (2) as isolated constituents from *G. parvifolia* leaves also showed great MIC values (2) for *S. sobrinus* (0.02 mg/mL) and *P. gingivalis* (0.05 mg/mL) showing their importance in prevention dental caries. Garcidepsidone A (1) is also able to inhibit GTase up to 25% [57]. The impact of *C. myrrha* EO against *S. mutans* and *Lactobacillus* spp. involved in dental caries is evaluated as well. *S. mutans* bacteria were isolated from Khartoum Dental Teaching Hospital patients with dental caries and Lactobacillus spp. isolates from fermented milk. The results showed EOs (conc. of 100, 50, 25, and 12.5 mg/mL) had potential on *S. mutans* with IZ of 18.7 ± 0.6 mm and 14.00 mm by the well diffusion method and disc diffusion method, and with MBC of 3.125 mg/mL [58].

13.6 MAPs enhanced by sustainable materials in cosmetics

13.6.1 Nanotechnology in cosmetic formulations

Nanotechnology is an emerging technology with versatility of application, encouraged in cosmetic industry, due to nano-sized particles (1–100 nm). Several cosmetic products (shampoos, tonics, cleansers, etc.) use nanoparticles (NPs), mainly Au and Ag as they are the most controllable and simple to organize [59]. Numerous metabolites from MAP extracts such as polyphenols, flavonoids, alkaloids, terpenoids, quinones, and low molecular weight proteins are part of synthesis of NPs as source/s of reducing agent [60]. They have the ability to prolong action time by managing the delivery of active ingredients, causing site-specificity, enhancing biocompatibility or the drug-loading capacity. Such cosmetic products have more pros than traditional cosmetics due to nano size, big surface-to-volume ratio, and NPs incorporated in cosmetic formulations do not change their performances, upgrade their appearance, coverage, and adherence to the skin [61]. NPs like Ag has medical application thanks to its antimicrobial, antimycotic, cytotoxic, and antioxidant potentialities, and Ag-NPs have been employed in skin care products like cream and ointment for burns and open wounds [62]; they enhance antiaging effect, UV protection, skin penetration, stimulate hair and nail growth, strengthen their structure, enhance hydration power, etc. [61].

Aqueous extract of *Grewia optiva* leaves is employed as a reducing agent in the synthesis of Ag-NPs. The results have shown the important role of flavonoids and polyphenols in the fabrication of NPs. Extraordinary antibacterial potential of synthesized NPs has been revealed against *S. typhi*, and the impact of extract and NPs was analyzed on the hair growth of rabbit. The rate of hair growth was higher for NPs versus extract [60]. *Eucommia ulmoides* leaf aqueous extract (conc. 0.01%, 0.1%, 0.5%, and 1%) was used to synthesize AgNPs (10 mM, and diameter from 4 nm to 52 nm) an exhibited an absorption peak at 430 nm, and C, O, and Cl elements, which might enable absorption of flavonoids and phenolics on the AgNPs surface (Figure 13.5). The efficacy of AgNPs to suppress anti-TYR activity in both in vitro and cell and A375 cells, and diminish ROS formation in HaCat cells promote AgNPs as agents in preventing melanin development and improve their application in skin-whitening products [63]. *Cordyceps militaris* (CM) known for its wound-healing effect has been used for preparing its extract and SiO_2NP mixture (CM-SiO_2NP) to analyze its healing effects on burns induced wounds. Dissolved CM in SiO_2NP (roughly 20 nm) suspension was analyzed for cytotoxicity in human fibroblasts and also evaluated in treatment to burn-induced second-degree skin in mice. In small concentrations (0–160 µg/mL) CM and its CM-SiO_2NP mixture were not harmful to tested fibroblasts. Almost identical effects were shown by them on proliferating cells, contributing to wound healing, and skin recovering during the treatment period (7 days and 15 days). The wound closure rate, after seven days of treatment, in the CM group was faster (2.4 times) versus

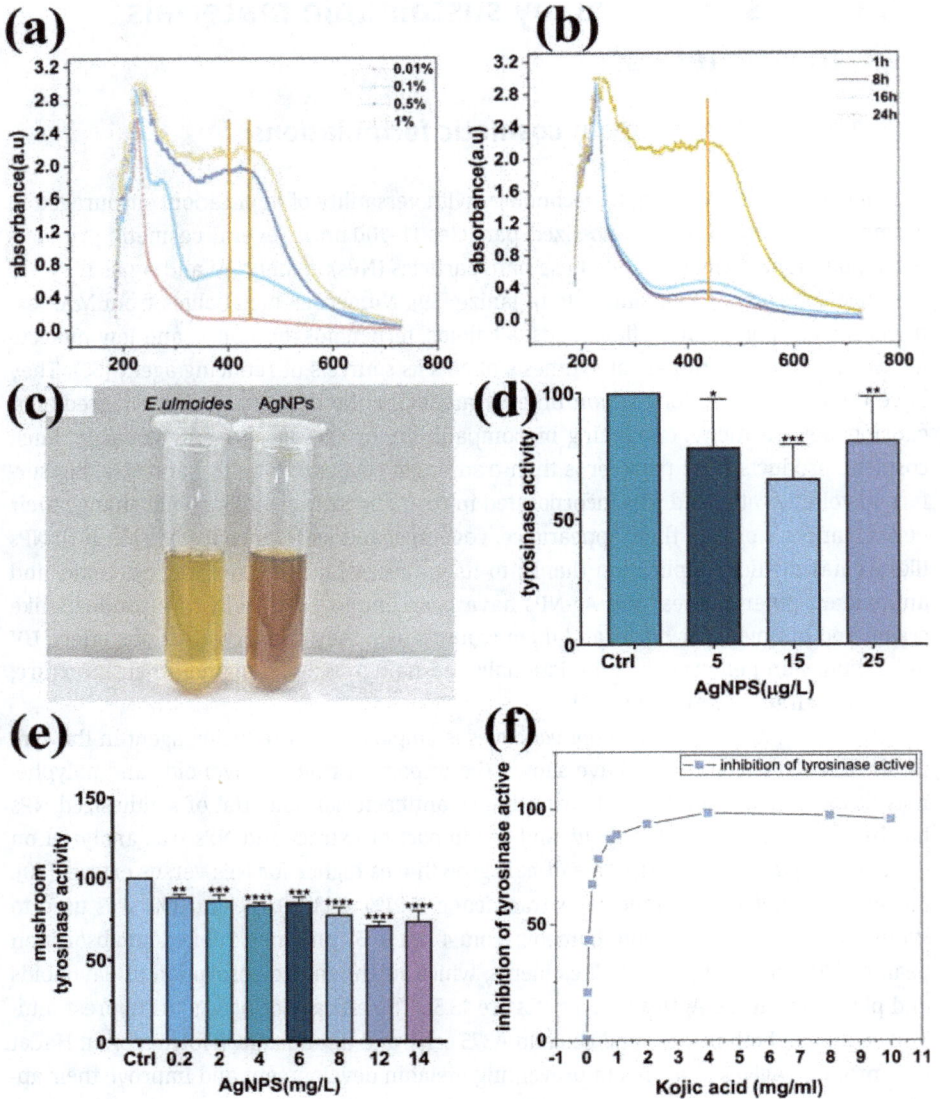

Figure 13.5: The representation of absorption spectrum of synthesized AgNPs from E. ulmoides leaf aqueous extract. (a) The adsorption spectrum of synthesized AgNPs (10 mM) and plant extract (conc. Of 0.01%, 0.1%, 0.5%, and 1%); (b) various reaction durations (1 h, 8 h, 16 h, and 24 h) between plant extract (1%) and AgNO$_3$ (10 mM); (c) visible solution color change due to generation of AgNPs; (d) anti-tyrosinase activity of AgNPs; (e) kojic acid standard curve for inhibiting tyrosinase activity; and (f) AgNPs anti-tyrosinase activity against A375 cells. Reproduced with permission from [63]. Copyright Elsevier ©2022.

sulfadiazine group and the untreated group (4.1 times faster). A complete wound closure is achieved after 15 days of treatment by CM-SiO$_2$NP treatment of burn-wound skin structure [64].

13.6.2 Innovative nanocarrier materials

Organic- and inorganic-based delivery systems have attracted scientists as potential and novel carriers in delivery systems for cosmetic applications. The biosynthesis of inorganic NPs based on the use of MAPs origin biomolecules extracted pathogens has gained huge interest of the researchers lately as the methods for their biosynthesis require safe and green solvents (e.g., water), room temperature conditions, or minimized heating. MAPs aqueous extracts were mainly used to synthesize different NPs composed of Ag, Au, Pd, ZnO, SiO$_2$, TiO$_2$, etc. [65]. AgNPs synthesis from *Brassica oleracea* was conducted to analyze its antibacterial, anticancer, and antioxidant properties. The ability of AgNPs is proved against numerous bacteria; they have strong antioxidant potential as well. The cytotoxicity in MCF-7 cells has increased up to 100 μg/mL, representing the maximum at concentration of the BO-AgNPs with IC50 of 55 μg/mL [66].

Nanocarriers (NCs) are able to easily penetrate through hair follicles, sebaceous glands, and sweat glands where they generate a depot impact and easy release the potential compounds. Hair follicles extend very deep in dermis (up to 1,000 μm) and create a 3-D contact surface with the skin, which offers sufficient absorption area for the potential constituents [67]. The permeability of NCs in hair follicles is mainly connected to their particle sizes. For example, lipid-based NCs of ≤ 640 nm might penetrate into hair follicles [68], but polymeric 300 nm-sized NPs might penetrate up to 300 μm deep into hair follicles [69]. Skin issues or/and diseases are mainly related to epidermal barrier dysfunction as they lead to skin dehydration, dryness, itchy, and chapped skin. NCs have efficacy with skin moisturizing components that moisturize and freshen the skin barrier, which can have impact on chronic dermatological issues, preventive or/and therapeutic impact on atopic dermatitis, eczema, and psoriasis [67].

Hyaluronic acid (HA) has beneficial effect on the skin, with moisturize and repair potential, but its large molecular weight restricts its penetration deep into the skin. It is reported that ethosomal carrier system encapsulates HA. Compared to PBS control, the system enhanced penetration of HA into the active epidermis and dermis by disrupting the dense structure of the stratum corneum. Concentrations of HA in skin were about 1,000-fold higher than those in blood [70]. Nanoemulsions (NE) developed from rice bran oil significantly moisturized skin without skin irritation. After 14 days of rice bran oil NEs treatment, skin was more moisturized, up to 38% and 30% in normal control volunteers and in patients with eczema issues, respectively [71]. Phenylethyl resorcinol (PR) is known as a whitening agent due to anti-tyrosinase activity. PR-loaded nanostructured lipid carriers were developed by the hot-melted ultrasonic method, and by their incorporation into nanostructured lipid carriers exhibited great physicochemical and photo stability. Also, they efficiently promoted the melanocyte's PR uptake in epidermis to inhibit migration of melanosome and provide whitening influence [71]. The comprehensive applicability of various MAP-based NCs is given in Table 13.3.

Table 13.3: Natural nanocarriers at different particle sizes synthesized from various MAPs being utilized due to their immense potential in cosmetic and dermatological areas in humans.

MAPs ingredient	Nanocarrier	Size (nm)	Effect	Application in cosmetic	References
Argan oil	Nanostructured lipid carrier	100–200	Prolonging contact time, facilitating local application, improving skin hydration	Moisturizer	[72]
Resveratrol	Solid lipid nanoparticle	≤200	Enhancing skin retention and, decreasing the cytotoxicity in high concentration	Whitening	[73]
Curcumin	Polymeric micelle	172.6	Enhancing the stability solubility, systemic bioavailability	Whitening	[74]
Lavender	Polymeric nanoparticle	301.8	Improving the physicochemical stability, upgrading and providing better the epidermal permeation, decreasing toxicity risks	Antiaging	[75]
Paeonol	Lipid liquid crystal	85	Improving the water stability, solubility, volatility reducing and irritation of the skin	Whitening	[76]
Phenylethyl resorcinol (4-(1-phenylethyl)1,3-benzenediol)	Nanostructured lipid carriers	57.9	Provided physicochemical stability and photostability, anti-tyrosinase activity	Whitening	[71]
Collagen peptides	Lipid-based nanocarrier	130	Better skin absorption biological safety, and significant impacts in anti-aging	Antiaging	[77]

13.6.3 Multifunctional nanocarriers in protection of the encapsulated ingredients

The ultradeformable vesicles have huge success in upgrading the skin penetration of drugs and efficacy of antiaging performances of some metabolites like tocopherol. Its preparation in transfersome, have great properties sized of less than 100 nm and entrapment potentiality of up to 90%. Also, in vitro assays and its distribution through the skin layer and have revealed its biocompatiblity with keratinocytes and fibroblasts, showing its protective effect against oxidative damage and the efficacy for wound healing [78]. NCs for coenzyme Q10 (CoQ10) delivery such as ethosomes [78, 79], transethosomes [80], are well-presented. CoQ10 is less soluble in water and it mainly restricts drug encapsulation effect in NCs, so the application of large amounts of lipid phase or ethanol might enhance its loading. Study by [78] significantly improved loading CoQ10 into protransfersomal emulgel observation during 28 days at low temperatures and contributed to the stability (particle sizes of 201.5 ± 6.1 nm, zeta potential of -11.26 ± 5.14 mV) and antiaging potentiality providing use as antiaging products.

Vitamin C is a dermocosmetic agents are related to whitening, anti-inflammatory, antioxidant, and antiaging effects on the skin. Its direct intro in cosmetic products is restricted by its low-fat solubility and dermal absorbability. So, delivery systems for stabilizing vitamin C for its application in cosmetics mainly use liposomes [81], chitosan-grafted cellulose nanocrystals [82], pectin [83], etc. Glycerosome represents a lipid vesicle containing edge activator (glycerol) to modify the fluidity of the bilayer structure, as well as upgrade transdermal delivery of quercetin and other active constituents. The quercetin deformable liposomes consisting phosphatidylcholine, cholesterol, and Tween 80 had a significant encapsulation potential due to particle size (132 ± 14 nm), increased elasticity (10.48 ± 0.71), and long-lasting drug release (up to 48 h) [84]. Liposome such as vesicles enriched with *Argania spinosa* oil might contribute skin hydrating and softening impacts as well as simplify the accumulation of natural molecules like allantoin in the skin and transdermal absorption. Conventional liposomes with incorporated allantoin and new argan oil with incorporated allantoin have enriched liposomes to enhance their local deposition in the skin. The argan oil-based formulations favored the allantoin accumulation in the dermis (~ 8.7 µg/cm^2), and permeation through the skin (~ 33 µg/cm^2) [85].

13.7 Conclusion

The future of MAPs holds promise for further development and improvement in cosmetic formulations. This chapter's main focus has been the role that MAPs play in addressing the transformative natural agents for development of various cosmetic formulations. Currently, MAPs stand as a testament for the remarkable potential they offer as they represent a huge source of phytochemical. MAPs serve in preparations if various extracts, and their chemical complexity is the reason for significant biological properties making them an attractive option for a rapidly cosmetic evolving field. Such upgraded MAP extracts developed by various sustainable approaches have a significant role in treatment of a broad spectra of medical disorders, such as dermatological. Extraction techniques, the drug nature (quality of plant, its origin, climate conditions, harvest time, plant's organ, drying method, particle size, etc.), solvent type, costs and therapeutic values, etc., have a big effect on the extract's potential. The results from latest studies underscore the exceptional capability of MAPs. Numerous natural metabolites are strong antioxidants and offer anti-inflammatory performances helpful in the treatment of oxidative stress-induced issues like photoaging, anti-elastase activity, the photoprotective effect of phenolics (e.g., oxyresveratrol and kuwanon O), improve hair growth and prevent hair loss, anti-dandruff, and anti-lice performances, and many more. Moreover, NPs have the potential to prolong action time by managing the delivery of active ingredients, causing site-specificity, enhancing biocompatibility, or the drug-loading capacity. Such potential has huge positive impact on cosmetic formulations due to nano size, big surface-to-volume ratio, etc. Further understanding of the concept of MAP extract mechanism and byproduct utilization, will help revolutionize cosmetic practices and contribute to a more sustainable future.

Abbreviations

MAPs	Medicinal and aromatic plants
ZI	Zone of inhibition
SDGs	Sustainable development goals
HD	Hydrodistillation
SE	Soxhlet extraction
WW	Wastewater
UAH	Ultrasound-assisted extraction
MAE	Microwave-assisted extraction
SFE	Supercritical fluid extraction
EAE	Enzyme-assisted extraction
EOs	Essential oils
NPs	Nanoparticles
NCs	Nanocarriers
NEs	Nanoemulsions
MIC	Minimum inhibitory concentration
MBC	Minimum bactericidal concentration
PR	Phenylethyl resorcinol

References

[1]   Sharafatmandrad, M. and Khosravi Mashizi, A. (2020). Ethnopharmacological study of native medicinal plants and the impact of pastoralism on their loss in arid to semiarid ecosystems of southeastern Iran. Scientific Reports, 10(1), 15526.

[2]   Ishtiaq, M. et al. (2024). Traditional ethnobotanical knowledge of important local plants in Sudhnoti, Azad Kashmir, Pakistan. Scientific Reports, 14(1), 22165.

[3]   Singh, H., Kumar, S. and Arya, A. (2024). Evaluation of antibacterial, antioxidant, and anti-inflammatory properties of GC/MS analysis of extracts of Ajuga. integrifolia Buch.-Ham. leaves. Scientific Reports, 14(1), 16754.

[4]   Ibrahim, N. et al. (2022). Rosmarinus officinalis L. hexane extract: Phytochemical analysis, nanoencapsulation, and in silico, in vitro, and in vivo anti-photoaging potential evaluation. Scientific Reports, 12(1), 13102.

[5]   Wirtu, S. F. et al. (2024). Isolation, characterization and antimicrobial activity study of Thymus vulgaris. Scientific Reports, 14(1), 21573.

[6]   Barthe, M. et al. (2021). Safety testing of cosmetic products: Overview of established methods and new approach methodologies (NAMs). Cosmetics, 8(2), 50.

[7]   Kırkıncı, S. et al. (2024). Evaluation of lavender essential oils and by-products using microwave hydrodistillation and conventional hydrodistillation. Scientific Reports, 14(1), 20922.

[8]   Naik, N. M. et al. (2023). Characterization of phyto-components with antimicrobial traits in supercritical carbon dioxide and soxhlet Prosopis juliflora leaves extract using GC-MS. Scientific Reports, 13(1), 4064.

[9]   Öner, E. K. and Yeşil, M. (2023). Effects of altitudes on secondary metabolite contents of Origanum majorana L. Scientific Reports, 13(1), 10765.

[10]   Ishak, I. et al. (2021). Optimization and characterization of chia seed (Salvia hispanica L.) oil extraction using supercritical carbon dioxide. Journal of CO2 Utilization, 45, 101430.

[11]   Puértolas, E., Alvarez-Sabatel, S. and Cruz, Z. (2016). Pulsed electric field: Groundbreaking technology for improving olive oil extraction. Inform, 27(3), 12–14.

[12]   Ferrentino, G. et al. (2020). Supercritical fluid extraction of oils from apple seeds: Process optimization, chemical characterization and comparison with a conventional solvent extraction. Innovative Food Science Emerging Technologies, 64, 102428.

[13]   Vo, T. P. et al. (2023). Optimizing ultrasonic-assisted and microwave-assisted extraction processes to recover phenolics and flavonoids from passion fruit peels. ACS Omega, 8(37), 33870–33882.

[14]   Vo, T. P. et al. (2023). Ultrasonic-assisted and microwave-assisted extraction of phenolics and terpenoids from abelmoschus sagittifolius (kurz) merr roots using natural deep eutectic solvents. ACS Omega, 8(32), 29704–29716.

[15]   Gendrisch, F. et al. (2021). Luteolin as a modulator of skin aging and inflammation. Biofactors, 47(2), 170–180.

[16]   Abdallah, H. M. et al. (2023). Taif rose oil ameliorates UVB-induced oxidative damage and skin photoaging in rats via modulation of MAPK and MMP signaling pathways. ACS Omega, 8(37), 33943–33954.

[17]   Giampieri, F. et al. (2012). Photoprotective potential of strawberry (Fragaria × ananassa) extract against UV-A irradiation damage on human fibroblasts. Journal of Agricultural Food Chemistry, 60(9), 2322–2327.

[18]   Zhou, Y. et al. (2023). Artemisia sieversiana Ehrhart ex willd. Essential oil and its main component, Chamazulene: Their photoprotective effect against UVB-induced cellular damage and potential as novel natural sunscreen additives. ACS Sustainable Chemistry Engineering, 11(50), 17675–17686.

[19] Hu, S., Chen, F. and Wang, M. (2015). Photoprotective effects of oxyresveratrol and Kuwanon O on DNA damage induced by UVA in human epidermal keratinocytes. Chemical Research in Toxicology, 28(3), 541–548.

[20] Maver, T. et al. (2015). A review of herbal medicines in wound healing. International Journal of Dermatology, 54(7), 740–751.

[21] Coutinho, M. A. S. et al. (2021). Wound healing cream formulated with Kalanchoe pinnata major flavonoid is as effective as the aqueous leaf extract cream in a rat model of excisional wound. Natural Product Research, 35(24), 6034–6039.

[22] Araújo, E. R. D. et al. (2023). Gel formulated with Bryophyllum pinnatum leaf extract promotes skin wound healing in vivo by increasing VEGF expression: A novel potential active ingredient for pharmaceuticals. Frontiers in Pharmacology, 13, 1104705.

[23] Ud-Din, S. and Bayat, A. (2017). Non-animal models of wound healing in cutaneous repair: In silico, in vitro, ex vivo, and in vivo models of wounds and scars in human skin. Wound Repair Regeneration, 25(2), 164–176.

[24] Michalak, M. (2023). Plant extracts as skin care and therapeutic agents. International Journal of Molecular Sciences, 24(20), 15444.

[25] Burger, P. et al. (2016). Skin whitening cosmetics: Feedback and challenges in the development of natural skin lighteners. Cosmetics, 3(4), 36.

[26] Sim, Y. Y. and Nyam, K. L. (2021). Application of Hibiscus cannabinus L.(kenaf) leaves extract as skin whitening and anti-aging agents in natural cosmetic prototype. Industrial Crops Products, 167, 113491.

[27] Chaiyana, W. et al. (2021). Herbal extracts as potential antioxidant, anti-aging, anti-inflammatory, and whitening cosmeceutical ingredients. Chemistry Biodiversity, 18(7), e2100245.

[28] Azis, H. et al. (2017). In vitro and in vivo wound healing studies of methanolic fraction of Centella asiatica extract. South African Journal of Botany, 108, 163–174.

[29] Soheilifar, M. H. et al. (2024). In vitro and in vivo evaluation of the diabetic wound healing properties of Saffron (Crocus Sativus L.) petals. Scientific Reports, 14(1), 19373.

[30] Huang, Y.-H. and Huang, C.-Y. (2024). Anti-skin aging and Cytotoxic effects of methanol-extracted solanum betaceum red fruit seed extract on Ca9-22 gingival carcinoma cells. Plants, 13(16), 2215.

[31] Michalak, M. et al. (2023). Phenolic profile and comparison of the antioxidant, anti-ageing, anti-inflammatory, and protective activities of Borago officinalis extracts on skin cells. Molecules, 28(2), 868.

[32] Sabóia Guerra Diógenes, É. et al. (2024). Evaluation of the skin whitening and antioxidant activity of Myracrodruon urundeuva extract (aroeira-do-sertão). Natural Product Research, 38(20), 3663–3668.

[33] Hering, A. et al. (2023). Photoprotection and antiaging activity of extracts from honeybush (Cyclopia sp.) – In vitro wound healing and inhibition of the skin extracellular matrix enzymes: Tyrosinase, collagenase, elastase and hyaluronidase. Pharmaceutics, 15(5), 1542.

[34] Gam, D.-H. et al. (2021). Effects of Sargassum thunbergii extract on skin whitening and anti-wrinkling through inhibition of TRP-1 and MMPs. Molecules, 26(23), 7381.

[35] Rajput, M. and Kumar, N. (2020). Medicinal plants: A potential source of novel bioactive compounds showing antimicrobial efficacy against pathogens infecting hair and scalp. Gene Reports, 21, 100879.

[36] Abelan, U. S. et al. (2022). Potential use of essential oils in cosmetic and dermatological hair products: A review. Journal of Cosmetic Dermatology, 21(4), 1407–1418.

[37] Carvalho, I. T., Estevinho, B. N. and Santos, L. (2016). Application of microencapsulated essential oils in cosmetic and personal healthcare products–a review. International Journal of Cosmetic Science, 38(2), 109–119.

[38] Oh, J. Y., Park, M. A. and Kim, Y. C. (2014). Peppermint oil promotes hair growth without toxic signs. Toxicological Research, 30, 297–304.

[39]   Panahi, Y. et al. (2015). Rosemary oil vs minoxidil 2% for the treatment of androgenetic alopecia: A randomized comparative trial. Skinmed, 13(1), 15–21.

[40]   Shin, J. Y. et al. (2020). Polygonum multiflorum extract support hair growth by elongating anagen phase and abrogating the effect of androgen in cultured human dermal papilla cells. BMC Complementary Medicine Therapies, 20, 1–12.

[41]   Chen, L. et al. (2018). Tetrahydroxystilbene glucoside effectively prevents apoptosis induced hair loss. BioMed Research International, 2018(1), 1380146.

[42]   Park, H.-J., Zhang, N. and Park, D. K. (2011). Topical application of Polygonum multiflorum extract induces hair growth of resting hair follicles through upregulating Shh and β-catenin expression in C57BL/6 mice. Journal of Ethnopharmacology, 135(2), 369–375.

[43]   Ruksiriwanich, W. et al. (2022). Phytochemical constitution, anti-inflammation, anti-androgen, and hair growth-promoting potential of shallot (Allium ascalonicum L.) extract. Plants, 11(11), 1499.

[44]   Ryu, H. S. et al. (2021). Activation of hair cell growth factors by linoleic acid in Malva verticillata seed. Molecules, 26(8), 2117.

[45]   Bhinge, S. D. et al. (2021). Screening of hair growth promoting activity of Punica granatum L.(pomegranate) leaves extracts and its potential to exhibit antidandruff and anti-lice effect. Heliyon, 7(4), e06903.

[46]   Sun, H.-Y. et al. (2019). Punicagranine, a new pyrrolizine alkaloid with anti-inflammatory activity from the peels of Punica granatum. Tetrahedron Letters, 60(18), 1231–1233.

[47]   Hooper, S. J. et al. (2011). Antimicrobial activity of Citrox® bioflavonoid preparations against oral microorganisms. British Dental Journal, 210(1), E22–E22.

[48]   Balhaddad, A. A. and AlSheikh, R. N. (2023). Effect of eucalyptus oil on Streptococcus mutans and Enterococcus faecalis growth. BDJ Open, 9(1), 26.

[49]   Wu, Z. et al. (2024). Inhibitory and preventive effects of Arnebia euchroma (Royle) Johnst. root extract on Streptococcus mutans and dental caries in rats. BDJ Open, 10(1), 15.

[50]   Bhavikatti, S. K. et al. (2024). Insights into the antioxidant, anti-inflammatory and anti-microbial potential of Nigella sativa essential oil against oral pathogens. Scientific Reports, 14(1), 11878.

[51]   Adeleye, O. A. et al. (2021). Physicochemical evaluation and antibacterial activity of Massularia acuminata herbal toothpaste. Turkish Journal of Pharmaceutical Sciences, 18(4), 476.

[52]   Amalunweze, A. and Ezumezu, C. (2022). Production of herbal toothpaste using Moringa root essential oil extract. International Journal of Advanced Biochemistry Research, 6, 49–51.

[53]   Chandakavathe, B. N., Kulkarni, R. G. and Dhadde, S. B. (2023). Formulation and assessment of In Vitro antimicrobial activity of herbal toothpaste. Proceedings of the National Academy of Sciences, India Section B: Biological Sciences, 93(2), 317–323.

[54]   Girisa, S. et al. (2021). From simple mouth cavities to complex oral mucosal disorders – Curcuminoids as a promising therapeutic approach. ACS Pharmacology Translational Science, 4(2), 647–665.

[55]   De Souza, J. L. S. et al. (2021). Antimicrobial and cytotoxic capacity of pyroligneous extracts films of Eucalyptus grandis and chitosan for oral applications. Scientific Reports, 11(1), 21531.

[56]   Li, B. et al. (2018). Curcumin as a promising antibacterial agent: Effects on metabolism and biofilm formation in S. mutans. BioMed Research International, 2018(1), 4508709.

[57]   Egra, S. et al. (2023). Garcidepsidone B from Garcinia parvifolia: Antimicrobial activities of the medicinal plants from East and North Kalimantan against dental caries and periodontal disease pathogen. Medicinal Chemistry Research, 32(8), 1658–1665.

[58]   Izzeldien, R. et al. (2023). Impact of Commiphora myrrha on Bacteria (Streptococcus mutans and Lactobacillus spp.) related to dental caries. International Journal of Dental Medicine, 12(2), 38–44.

[59]   Kumar, J. and Jaswal, S. (2021). Role of nanotechnology in the world of cosmetology: A review. Materials Today: Proceedings, 45, 3302–3306.

[60] Iftikhar, M. et al. (2020). Green synthesis of silver nanoparticles using Grewia optiva leaf aqueous extract and isolated compounds as reducing agent and their biological activities. Journal of Nanomaterials, 2020(1), 8949674.

[61] Gupta, V. et al. (2022). Nanotechnology in cosmetics and cosmeceuticals – A review of latest advancements. Gels, 8(3), 173.

[62] Khan, S. A., Shahid, S. and Lee, C.-S. (2020). Green synthesis of gold and silver nanoparticles using leaf extract of Clerodendrum inerme; characterization, antimicrobial, and antioxidant activities. Biomolecules, 10(6), 835.

[63] Xi, J. et al. (2022). Synthesis of silver nanoparticles using Eucommia ulmoides extract and their potential biological function in cosmetics. Heliyon, 8(8), e10021.

[64] Nguyen, L.-T.-T. et al. (2022). Effects of Cordyceps militaris extract and its mixture with silica nanoparticles on burn wound healing on mouse model. Journal of Drug Delivery Science Technology, 67, 102901.

[65] Kyriakoudi, A. et al. (2021). Innovative delivery systems loaded with plant bioactive ingredients: Formulation approaches and applications. Plants, 10(6), 1238.

[66] Ansar, S. et al. (2020). Eco friendly silver nanoparticles synthesis by Brassica oleracea and its antibacterial, anticancer and antioxidant properties. Scientific Reports, 10(1), 18564.

[67] Zhou, H. et al. (2021). Current advances of nanocarrier technology-based active cosmetic ingredients for beauty applications. Clinical, Cosmetic Investigational Dermatology, 14, 867–887.

[68] Lauterbach, A. and Müller-Goymann, C. C. (2015). Applications and limitations of lipid nanoparticles in dermal and transdermal drug delivery via the follicular route. European Journal of Pharmaceutics Biopharmaceutics, 97, 152–163.

[69] Główka, E. et al. (2014). Polymeric nanoparticles-embedded organogel for roxithromycin delivery to hair follicles. European Journal of Pharmaceutics Biopharmaceutics, 88(1), 75–84.

[70] Chen, M. et al. (2014). Topical delivery of hyaluronic acid into skin using SPACE-peptide carriers. Journal of Controlled Release, 173, 67–74.

[71] Bernardi, D. S. et al. (2011). Formation and stability of oil-in-water nanoemulsions containing rice bran oil: In vitro and in vivo assessments. Journal of Nanobiotechnology, 9, 1–9.

[72] Tichota, D. M. et al. (2014). Design, characterization, and clinical evaluation of argan oil nanostructured lipid carriers to improve skin hydration. International Journal of Nanomedicine, 9, 3855–3864.

[73] Rigon, R. B. et al. (2016). Skin delivery and in vitro biological evaluation of trans-resveratrol-loaded solid lipid nanoparticles for skin disorder therapies. Molecules, 21(1), 116.

[74] Cheng, Z. et al. (2020). Preparation and characterization of dissolving hyaluronic acid composite microneedles loaded micelles for delivery of curcumin. Drug Delivery Translational Research, 10, 1520–1530.

[75] Pereira, F. et al. (2015). Production and characterization of nanoparticles containing methanol extracts of Portuguese Lavenders. Measurement, 74, 170–177.

[76] Li, J.-C. et al. (2015). Self-assembled cubic liquid crystalline nanoparticles for transdermal delivery of paeonol. Medical Science Monitor: International Medical Journal of Experimental Clinical Research, 21, 3298.

[77] Han, S.-B. et al. (2021). Asterias pectinifera derived collagen peptide-encapsulating elastic nanoliposomes for the cosmetic application. Journal of Industrial Engineering Chemistry, 98, 289–297.

[78] Ayunin, Q. et al. (2022). Improving the anti-ageing activity of coenzyme Q10 through protransfersome-loaded emulgel. Scientific Reports, 12(1), 906.

[79] Sguizzato, M. et al. (2020). Ethosomes for coenzyme Q10 cutaneous administration: From design to 3D skin tissue evaluation. Antioxidants, 9(6), 485.

[80] El-Zaafarany, G. M. et al. (2021). Coenzyme Q10 phospholipidic vesicular formulations for treatment of androgenic alopecia: Ex vivo permeation and clinical appraisal. Expert Opinion on Drug Delivery, 18(10), 1513–1522.

[81] Jiao, Z. et al. (2018). Preparation and evaluation of a chitosan-coated antioxidant liposome containing vitamin C and folic acid. Journal of Microencapsulation, 35(3), 272–280.

[82] Akhlaghi, S. P., Berry, R. M. and Tam, K. C. (2015). Modified cellulose nanocrystal for vitamin C delivery. Aaps Pharmscitech, 16, 306–314.

[83] Zhou, W. et al. (2014). Storage stability and skin permeation of vitamin C liposomes improved by pectin coating. Colloids Surfaces B: Biointerfaces, 117, 330–337.

[84] Liu, D. et al. (2013). Quercetin deformable liposome: Preparation and efficacy against ultraviolet B induced skin damages in vitro and in vivo. Journal of Photochemistry Photobiology B: Biology, 127, 8–17.

[85] Manca, M. L. et al. (2016). Combination of argan oil and phospholipids for the development of an effective liposome-like formulation able to improve skin hydration and allantoin dermal delivery. International Journal of Pharmaceutics, 505(1–2), 204–211.

Binnur Bağci*, Gamze Tüzün, and Elyor Berdimurodov

Chapter 14
Edible medicinal and aromatic plants

Abstract: Edible medicinal and aromatic plants have historically played an important role in the treatment of diseases, and sometimes in the alleviation of their complications. In addition to their nutritional value, these plants also attract great attention due to their pharmaceutical and cosmetic importance. The active ingredients they contain are the main factors that reveal the therapeutic properties of these plants. These plants contain many bioactive compounds such as phenols, flavonoids, and terpenoids, and these serve the defense mechanisms of plants. Medicinal and aromatic plants are currently being researched in the fields of alternative medicine and nutrition because they provide important effects on health.

These plants have been traditionally used in the treatment of various diseases and these forms of use continue in many cultures today. Edible medicinal plants not only provide the body with the necessary nutrients, but also play a role in the treatment of diseases by offering pharmaceutical properties. Especially in recent years, due to the increase in chronic diseases and complications due to drugs, these plants have begun to attract more attention. In addition, since these plants contain bioactive compounds, they have great potential for the food and agricultural industries. In this chapter, edible medicinal aromatic plants and the medical conditions they are associated with are discussed in terms of their bioactive components.

Keywords: edible plants, culinary plants, medicinal and aromatic plants, bioactive compounds

14.1 Introduction

Historically, plants have served as the principal instrument for the treatment and, in certain instances, the mitigation of diseases and disease complications. Medicinal and aromatic plants are currently of significant interest due to their nutritional value, in

*Corresponding author: Binnur Bağci, Department of Nutrition and Dietetics, Faculty of Health, Sivas Cumhuriyet University Sciences, Sivas, Türkiye, e-mail: binnurkoksalbagci@gmail.com, https://orcid.org/0000-0003-1323-3359
Gamze Tüzün, Department of Chemistry, Faculty of Science, Cumhuriyet University, 58140, Sivas, Turkey
Elyor Berdimurodov, Chemical and Materials Engineering, New Uzbekistan University, 54 Mustaqillik Ave, Tashkent 100007, Uzbekistan; Faculty of Chemistry, National University of Uzbekistan, Tashkent, 100034, Uzbekistan

https://doi.org/10.1515/9783111469713-014

addition to their pharmacological and cosmetic relevance [1]. Medicinal and aromatic plants are used for medicinal purposes and typically possess an intense odor or flavor. Their therapeutic capabilities result from the active constituents in their chemical composition [2]. The compounds synthesized by these plants include phenols or their oxygen-substituted derivatives, tannins, flavonoids, tannins, saponins, terpenoids, anthocyanins, sterols, waxes, carotenoids, and fatty acids. It is known that they synthesize more than 12,000 aromatic substances, mostly secondary metabolites, as defense molecules against predation by microorganisms, insects, and herbivores. This number is thought to be a very small fraction of the amount of secondary metabolites claimed to be synthesized by medicinal aromatic plants. Consequently, medicinal and aromatic plants represent a promising and evolving field [3].

In recent years, the increased incidence of chronic illnesses (cancer, heart disease, diabetes, etc.) and various complications associated with pharmaceuticals have driven researchers to explore alternative solutions to these problems. Edible medicinal plants are recognized as significant resource of both nutrients and pharmaceuticals [4]. These plants not only provide adequate nutrients to supply the organic requirements of the body but also contain pharmacological substances that are used to treat a range of illnesses. They possess rich pharmaceutical components that can be employed both as therapeutic agents and as protective measures against disease risks.

Edible medicinal and aromatic plants occupy a research domain in disciplines such as ethnomedicine, which investigates the health-related beliefs and practices of societies, particularly traditional and culture-specific treatment methods; biomedicine, which examines the relationship between biological sciences and modern medicine, encompassing scientifically grounded medical applications; and traditional medicine, which explores health practices, beliefs, and treatments developed based on a society's cultural, historical, and local knowledge, passed down through generations.

The purpose of edible medicinal and aromatic plants, unlike other nutrients, is not solely to alleviate hunger and provide micro- and macronutrients to the body. Rather, they provide bioactive ingredients to promote physical and mental health and prevent diet-related disorders. [5].

Globally, these plants are an integral part of the cultural and traditional dietary practices of many societies. Researches have clearly demonstrated the critical role of healthy nutrition in preventing chronic diseases and its protective benefits. The primary objective of edible medicinal plants is to prevent chronic diseases and contribute to overall health. However, the effects of extracted compounds from certain conventionally used plants remain an active area of investigation.

From an ethnobotanical perspective, the rich ethnobotanical diversity offered by edible medicinal and aromatic plants is a situation that occurs in different ways all over the world. Each region of the world has its own methods and information regarding the utilization of these plants. For example, in countries with wide ecological and climatic diversity such as India, indigenous people have used wild plants for both nutritional and healing purposes for generations. Utilizing wild plants in the Mediterranean region, where Turkey is located, has a long history, and this tradition is an integral part of the regional culture [6]. In this region, historically, numerous wild plant species were commonly utilized as dietary supplements. Nevertheless, much of this knowledge now resides predominantly in the memories of older generations and faces the imminent risk of being lost within the next few decades [7]. In Turkey, the collection of wild plants, called "ot" is seen not only as a food source but also as a cultural heritage. This tradition is carried on all over the country and continues to be an important part of the culinary culture [6].

14.2 Bioactive compounds in edible medicinal and aromatic plants

Traditional folk medicine uses edible medicinal aromatic plants as they are a rich source of bioactive chemicals with a variety of medicinal uses, including antifungal, antimicrobial, insecticidal, antihyperglycemic, anti-inflammatory, and antioxidant functions, and positive influences on human health. Numerous articles in the literature have reported that edible medicinal aromatic plants are sources of polyphenols, natural pigments, essential oils, minerals, vitamins, and many phytochemicals [8]. Edible medicinal and aromatic plants have traditional usage patterns that have been integrated into lifestyles across generations in many cultures [9]. Moreover, they are frequently used to enhance the flavor of foods. They can also be used as natural colorants and flavorings or as substitutes of salt and sugar [10, 11]. Research on the effects of foods consumed in daily life on health is a popular field of study today. Edible plants and their bioactive compounds, with their different mechanisms of action in human nutrition, have an important place in this field. In addition, numerous studies and new techniques have provided new alternatives for the extraction of bioactive compounds in these plants. New applications such as the production, processing, and increased production of bioactive molecules in edible plants for the food and agricultural industries worldwide are of great interest [12].

Edible medicinal aromatic plants are rich in many phytochemical components with bioactive effects such as terpenes, alkaloids, glycosides, and polyphenols, in addition to their nutritional value, and are also valuable from a pharmacological point of view in Figure 14.1.

14.2.1 Polyphenols and edible medicinal and aromatic plants

Polyphenols, abundant in numerous plant species, represent a diverse family of structurally distinct compounds. They are named according to the number of phenol rings (aromatic rings with attached hydroxyl groups) and the nature of the substituents attached to these rings. Polyphenols play pivotal roles in plant protection mechanisms, offering protection against plant pathogens. This group encompasses compounds such as astringent tannins and hormone-mimicking phytoestrogens [10]. Polyphenols are categorized based on the structure of their carbon skeleton: phenolic acids character- ized by a benzene ring with one or more hydroxyl groups, flavonoids characterized by C6–C3–C6 skeleton (two aromatic rings connected by a three-carbon bridge), stil- benes characterized by C6–C2–C6 skeleton with two benzene rings connected by a two-carbon methylene bridge, and lignans characterized by C6–C3–C6 skeleton and derived from plant precursors through oxidative coupling, as shown in Figure 14.2.

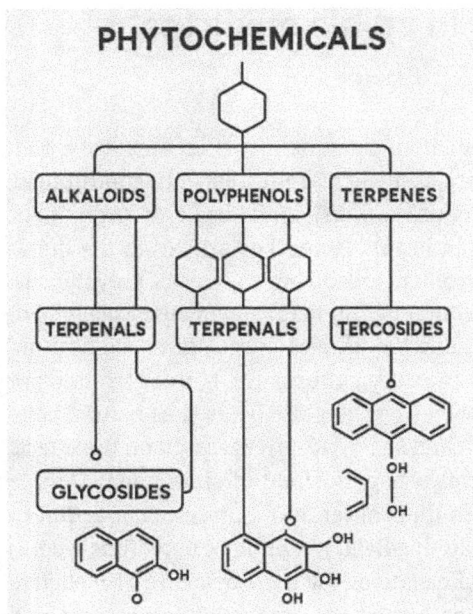

Figure 14.1: Classification of phytochemicals.

14.2.1.1 Phenolic acids

They are among the most abundant polyphenols in plants. The most commonly en- countered phenolic acids in edible medicinal aromatic plants are ferulic acid, caffeic acid, and gallic acid. Ferulic acid is found bound to hemicelluloses through ester bonds. It is present in artichokes, coffee, whole grains, wheat bran, strawberries, ap- ples, certain cereals, blackberries, raspberries, and mangoes [13, 14].

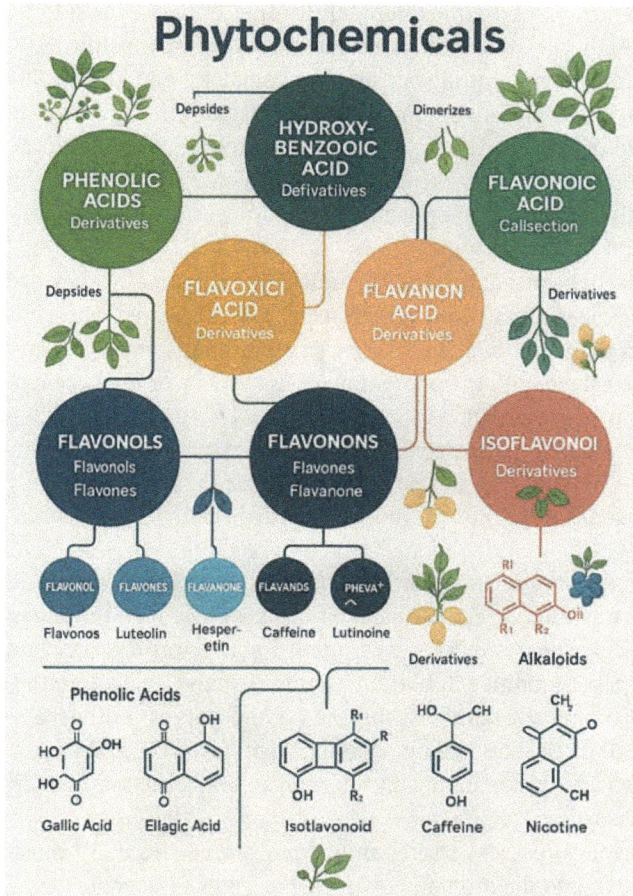

Figure 14.2: Main classes of polyphenols, their subclasses, example compounds, and their food sources.

14.2.1.2 Flavonoids

Among edible medicinal and aromatic plants, flavonoids are the most abundant polyphenol group. The Latin word "flavus," which means yellow, is where the word "flavonoid" originates. In addition to their medicinal aromatic nature, flavonoids are important functional nutrients for human physiology. They can be classified into several subgroups: flavanones, flavones, isoflavones, anthocyanidins, and flavonols, as shown in Figure 14.3.

Citrus fruits such as grapefruit, lemons, oranges, and pomelos are important sources of flavanones. [15–17]. The most commonly studied flavanones include naringenin, narirutin, and hesperetin, which exhibit antioxidant, anticancer, anti-atherosclerotic, and antiviral effects [18]. Both aglycone and glycoside forms of naringenin (5, 7, 4'-trihydroxyflavanone) are mostly present in sour oranges, tomatoes, and grapefruits. Another flavanone, narirutin, is also present in grapefruits. Naringenin-7-neohesperoside contributes to the bitter taste of grapefruits due to its glucose moiety [18]. Hesperetin, another widely distributed flavanone in citrus fruits, is less commonly found in its aglycone form, compared to its glycosides. It is abundant in sweet oranges, lemons, and tangerines [18].

Isoflavones are found in soybeans, natto, and soy milk [19]. Additionally, they are present in black beans, chickpeas, lima beans, green peas, alfalfa sprouts, and sunflower seeds. Genistein and daidzein, two isoflavones predominantly found in soybeans, are reported to be associated with the prevention of breast cancer and osteoporosis due to their estrogenic properties [20].

Due to their ability to interact with estrogen receptors, isoflavones can act as partial agonists or antagonists [21]. Daidzein is also involved in alcohol metabolism by inhibiting alcohol dehydrogenase and mitochondrial aldehyde dehydrogenase enzymes [22].

Anthocyanidins are a subgroup of flavonoids responsible for various colors in plants. They are found in fruits and vegetables such as blackberries, blood oranges, blueberries, blackcurrants, elderberries, grapes, red onions, plums, raspberries, and red cabbage [23]. Peonidin, pelargonidin, delphinidin, cyanidin, malvidin, and petunidin can be cited as some of the most commonly observed anthocyanins. Anthocyanidins exhibit anticancer, antidiabetic, antioxidant, cytotoxic, antibacterial, and antiangiogenic properties. They provide protection against cardiovascular diseases, obesity, diabetes, and neurological disorders [24].

Flavonols (3-hydroxyflavones) possess strong antioxidant properties and protect cells from oxidative damage. Due to the aromatic hydrocarbon rings in flavonoid compounds, they also protect against free radical damage. Flavonols are commonly found in onions, cabbage, broccoli, tomatoes, red lettuce, blackberries, grapes, apples, and tea [25]. The flavonols most extensively studied are quercetin, myricetin, and kaempferol.

Among these, quercetin (3,3',4',5,7-pentahydroxyflavone), the most well-known flavonol, protects low-density lipoproteins (LDLs) from oxidation due to its strong antioxidant capacity. It also demonstrates antioxidant and anti-inflammatory activity by protecting blood vessels [26]. Furthermore, it has antidiabetic properties by inhibiting aldose reductase, and activates thermogenesis and fat oxidation by suppressing norepinephrine [26, 27]. Quercetin is commonly found in onions, green tea, apples, grapes, and leafy vegetables. Kaempferol (3,4',5,7-tetrahydroxyflavone), another potent flavonol, is abundant in leeks, broccoli, cabbage, spinach, parsley, grapes, raspberries, and blackberries [28]. Kaempferol also exhibits antioxidant properties. This is achieved by inhibiting enzymes that produce reactive oxygen species (ROS) and inducing antioxidant enzymes to prevent lipid peroxidation. It also shows anti-atherosclerotic and anti-inflammatory properties and induces apoptosis while inhibiting angiogenesis and metastasis, thereby exhibiting anticancer effects [26].

Flavones, structurally differentiated from flavonols by an additional hydroxyl group at the carbon 3 position, are found in parsley, celery, chamomile, onions, wheat germ, tea, oranges, broccoli, carrots, peppers, and apples. Apigenin (4',5,7-trihydroxyflavone), a key bioactive component found in chamomile, is present in edible plants such as onions, tea, parsley, oranges, celery, corn, rice, and wheat germ. It is reported to have antioxidant, anticarcinogenic, anti-atherosclerotic, antiviral, and anti-inflammatory properties [29] Luteolin, another notable flavone, is found in cocoa carrots, peppermint, thyme, peppers, pomegranates, celery, artichokes, rosemary, lettuce, turnips, and cucumbers. Its antioxidant properties neutralize free radicals, bind metal ions, and prevent oxidative damage. Luteolin's anticancer properties limit cell growth and metastasis and support apoptosis mechanisms. It can cross the blood–brain barrier, making it significant in central nervous system diseases. Additionally, it reduces inflammation by protecting DNA, proteins, and lipids and has antibacterial, antiviral, and antifungal activities [30].

Flavanols are abundant in cocoa, beans, apricots, cherries, grapes, peaches, blackberries, apples, kiwis, green tea, and black tea. Their dietary intake is limited as they are primarily located in the peels and seeds of fruits and vegetables [31]. Catechins, a prominent subgroup of flavanols, are the building blocks of tannins. They are most commonly found in tea, cocoa, grapes, apricots, peaches, apples, blackberries, and red raspberries [32]. Flavanols increase nitric oxide (NO) levels in the bloodstream, leading to improved endothelial function and cardiovascular protection [32]. Catechins possess antioxidant activity, reducing oxidative stress-induced protein damage by scavenging free radicals [33]. Like other flavonoids, they exhibit anticancer, antidiabetic, and antimutagenic effects [32]. Catechins' antimicrobial properties make tea an effective defense against pathogens. Additionally, they may activate thermogenesis in synergy with caffeine [32, 34].

14.2.1.3 Stilbenes

Some stilbenes have been identified as phytoalexins, with their levels observed to increase in plants, following UV radiation exposure and infection, suggesting a role in defense mechanisms. Structurally, stilbenes consist of two aromatic rings (C6–C2–C6) connected by an ethylene bridge, as shown in Figure 14.4 and Figure 14.5. Among the various stilbene derivatives, resveratrol (3,4',5-trihydroxystilbene) is the most extensively studied and well-characterized compound. The presence of stilbenes is largely restricted to grapes, peanuts, and certain fruits [35, 36]. Polygonum cuspidatum, commonly known as Japanese knotweed, is a perennial herbaceous plant. Scientifically, it is also referred to as Fallopia japonica and belongs to the family Polygonaceae. This

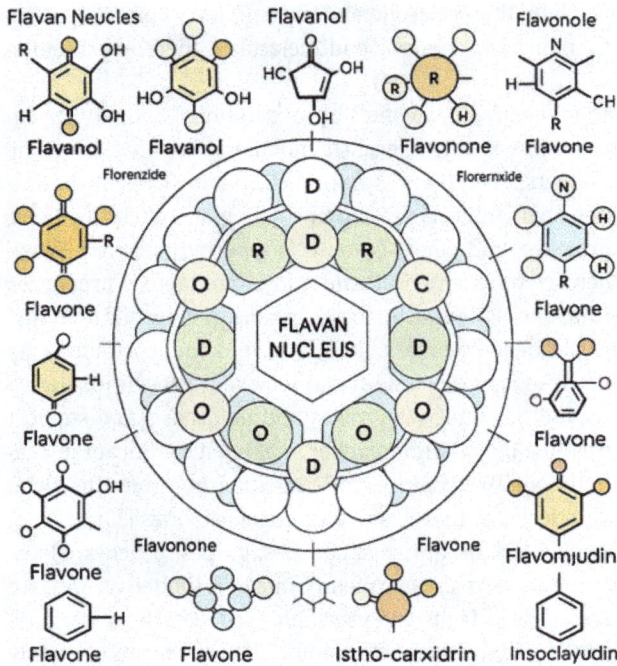

Figure 14.3: Basic flavan structure and chemical structures of flavonoid subgroups.

plant holds pharmacological significance, as its roots are rich in stilbene derivatives, like resveratrol. Due to these properties, it has been widely used in traditional Chinese medicine to address various health conditions [36].

14.2.1.4 Lignans

Lignans, due to their steroid-like chemical structures, are classified as phytoestrogens. They exhibit anti-inflammatory, antioxidant, anticancer, and cardioprotective properties [37, 38] Lignans are primarily found in seeds, particularly in sesame and flaxseeds, as well as in various cereals [39].

14.2.2 Alkaloids and edible medicinal and aromatic plants

Alkaloids are a broad category of chemical compounds that exist naturally. They usually have at least one nitrogen atom and frequently have simple characteristics. They usually have a bitter taste and demonstrate strong physiological effects in biological systems, as shown in Figure 14.4 [40].

Tea (*Camellia sinensis*) and coffee (*Coffea* spp.) are significant sources of purine alkaloids, primarily caffeine, and are also rich in theobromine and theophylline. The stimulating effects of tea and coffee are primarily attributed to caffeine's action as a central nervous system stimulant [41]. Numerous researches have documented tea's antidiabetic, antioxidant, and anti-obesity properties [42–44].

Cocoa (*Theobroma cacao*) contains theobromine, an alkaloid structurally similar to caffeine but with milder stimulant effects. Theobromine is responsible for the pleasurable effects of eating chocolate. Recently, theobromine has gained attention for its pharmacological effects. It acts as a vasodilator, improving blood flow and potentially reducing blood pressure. Its anti-inflammatory properties support vascular health and may lower the risk of atherosclerosis. Theobromine contributes to the prevention of cardiovascular diseases by inhibiting the oxidation of LDL cholesterol. Additionally, it promotes diuresis by increasing fluid excretion in the kidneys, making it beneficial for managing edema and hypertension. Theobromine also aids in reducing oxidative stress by scavenging free radicals, which is essential for slowing aging and preventing chronic diseases. Its metabolism-boosting effects increase energy expenditure, which may assist in weight management. Theobromine has been demonstrated to stimulate lipolysis and prevent fat cell differentiation [45–47].

Leguminous plants such as fenugreek (*Trigonella foenum-graecum*) and chickpea (*Cicer arietinum*) contain the alkaloid trigonelline. This compound exhibits hypoglycemic and antidiabetic properties and is notably abundant in fenugreek and pumpkin seeds [48–50].

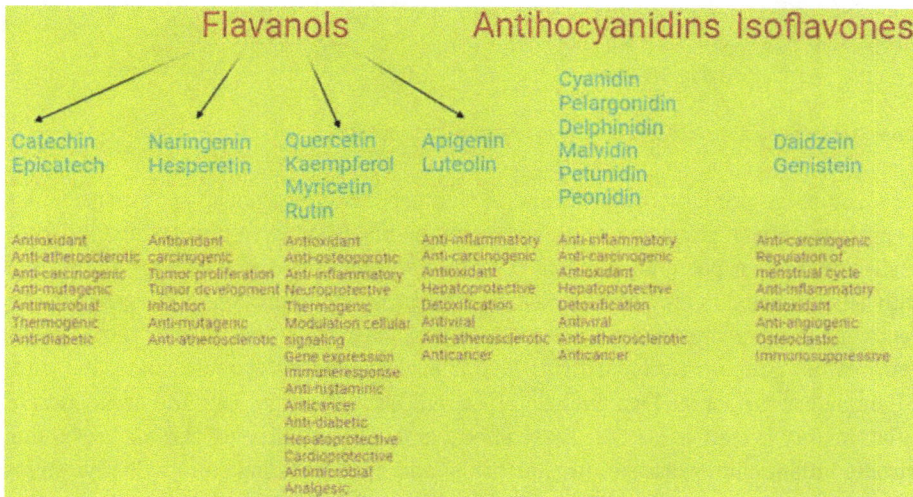

Figure 14.4: Biological activities of flavonoids.

Dihydroresveratrol

Resveratrol-3-O-β-D-glucuronide

Resveratrol-4-sulfate

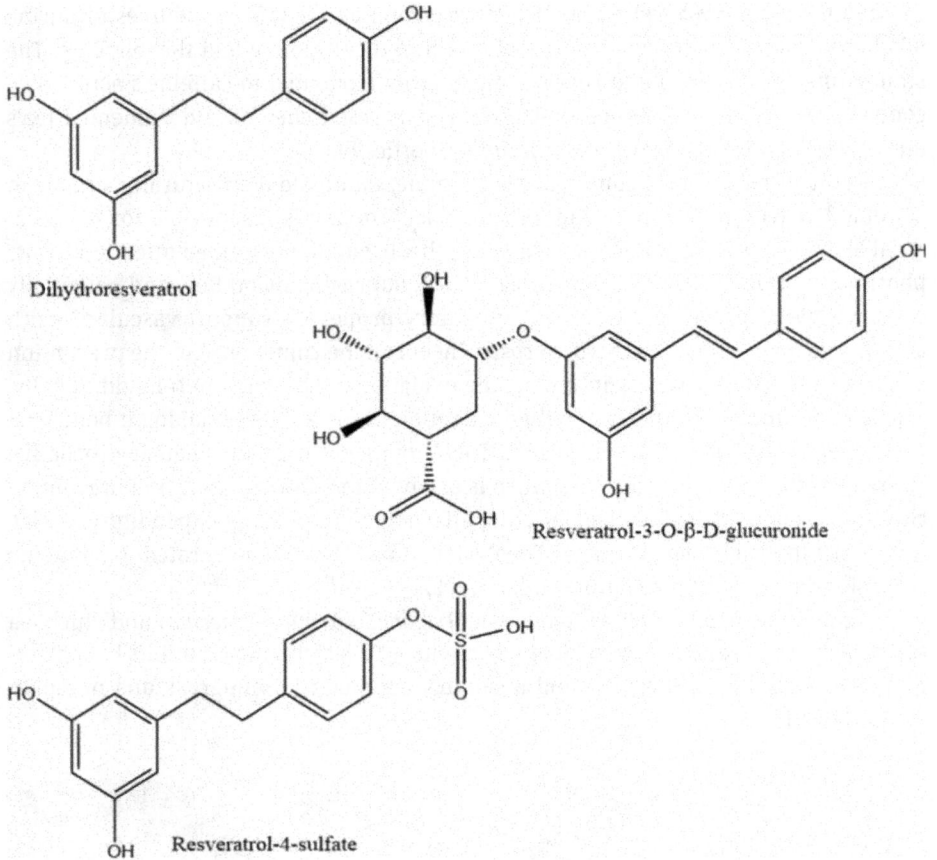

Figure 14.5: Chemical structures of the main resveratrol metabolites.

Capsaicin is the primary alkaloid found in chili peppers (*Capsicum* spp.), responsible for their characteristic pungency. Capsaicin exerts its biological effects by activating TRPV1 (transient receptor potential vanilloid 1) receptors on sensory nerve endings. Interaction with TRPV1 receptors temporarily blocks pain transmission, providing analgesic effects. Capsaicin-based creams and patches are used to manage conditions such as chronic pain, neuropathic pain, and osteoarthritis [51]. It also demonstrates antioxidant and anti-inflammatory properties by scavenging free radicals and suppressing the production of inflammatory cytokines [52, 53]. These effects make capsaicin beneficial for preventing chronic inflammatory diseases (e.g., arthritis) and conditions linked to oxidative stress [54]. Capsaicin improves blood flow through its vasodilatory effects, potentially lowering hypertension [55]. Moreover, it positively affects cholesterol metabolism by preventing LDL cholesterol oxidation, thereby reducing the risk of atherosclerosis [56]. Capsaicin also enhances energy expenditure and fat oxidation, aiding in weight management. Its thermogenic properties make it a promising compound for obesity treatment [57].

Plant/Source	Primary Alkoloids	Physiological effects
Tea (Camellia sinensis)	Caffeine Theobromine Theophylline	CNS stimulation (caffeine) - Antioxidant, Anti-diabetic, and anti-obesity effects
Coffee (Coffea spp.)	Caffeine	CNS stimulation, Increased alertness
Cocoa (Theobroma caco)	Theobromine	Mild stimulant effect- Vasodilation, improved blood flow -Anti-inflammatory properties -LDL cholesterol oxidantion - Diuretic effects - Suppports weight management by promoting lipolysis
Fenugreek (Trigonella foenum-graecum)	Trigonelline	Hypoglycemic and anti-diabetic effects
Chickpea (Cicer arietinum)	Trigonelline	Hypoglycemic and anti-diabetic effects
Chili pepper (Capsicum spp.)	Capsaicin	Analgesic effects via TRPV1 receptor activation -Anti-inflammatory and antioxidant properties - Vasodilation and improved blood flow - Supports cholesterol metabolism - Thermogenic and weight management properties

Figure 14.6: Alkaloids in edible plants and their health effects.

The alkaloids present in these edible plants offer significant nutritional and potential health benefits, yet they must be carefully evaluated for toxicity. Research on the bio-availability, metabolism, and physiological effects of alkaloids in the human body is crucial for understanding their positive and negative health impacts.

14.2.3 Terpenes and edible medicinal and aromatic plants

Although some microbes and insects are capable of producing terpenes, plants are the primary source of this family of chemicals. These compounds are characterized by simple hydrocarbon structures. Terpenes contribute to the characteristic aromas, flavors, and colors of plants, making them the principal constituents of essential oils [58]. Terpenes are structurally derived from isoprene units (C_5H_8), which serve as the building blocks for their synthesis. The structural diversity of terpenes arises primarily from the number of isoprene units they contain. Hemiterpenes consist of a single iso-prene unit (C_5), while monoterpenes, sesquiterpenes, diterpenes, triterpenes, and tetra-terpenes are composed of two (C_{10}), three (C_{15}), four (C_{20}), six (C_{30}), and eight (C_{40}) iso-prene units, respectively, as shown in Figure 14.7.

Monoterpenes are generally terpenes composed of two isoprene units and are found in many plants, as shown in Figure 14.8. These compounds contribute to the characteristic smells and flavors of plants. One example of a monoterpene is limonene, which is found in citrus fruits (especially lemons and oranges) and has a strong citrus aroma. Its anti-inflammatory, anticancer, antioxidant, and antimicrobial properties make it a potential therapeutic agent. Limonene has been observed to exhibit anti-inflammatory effects in the body. Studies have shown that limonene inhibits the production of pro-inflammatory cytokines and suppresses inflammation-related pathways. This property may contribute to the management of diseases such as arthritis, asthma, and other inflammatory conditions [59]. Limonene may also demonstrate potential anticancer properties against certain cancer types. Laboratory studies have shown that limonene inhibits tumor growth in breast, skin, and colon cancer cells and promotes apoptosis (programmed cell death) in cancer cells [60]. Therefore, limonene is thought to be a useful adjunct in cancer treatment, particularly in chemotherapy. Furthermore, limonene is known to be a powerful antioxidant. By neutralizing free radicals, it contributes to the reduction of oxidative stress. Since oxidative stress is a key factor in many chronic diseases, limonene's antioxidant effect may play an important role in preventing cardiovascular diseases, neurological disorders, and age-related conditions [61]. Limonene may also influence the central nervous system. Some clinical studies have observed limonene's effects on reducing stress and anxiety. It is believed that limonene modulates the levels of neurotransmitters such as serotonin and dopamine, improving mood and alleviating symptoms of depression [62].

Another monoterpene, menthol, is found in mint plants (*Mentha piperita, Mentha spicata, Mentha arvensis*, and *Mentha* × *piperita* var. *citrata*) and is a compound with significant medicinal benefits. Known for its cooling and relaxing effects, menthol is widely used in the pharmaceutical, cosmetic, and food industries. The pharmacological effects of menthol play an important role in various areas such as the digestive system, respiratory diseases, pain management, and psychological recovery. Menthol is considered an effective compound in pain management. This effect is related to menthol's ability to create a cooling sensation on the skin and to block pain signals neurologically. Menthol can inhibit pain perception by activating TRPM8 (transient receptor potential melastatin 8) channels, which are responsible for cold sensation. Several clinical studies have observed that topical preparations containing menthol provide relief in painful conditions such as headaches, muscle pain, and arthritis [63–65]. Menthol also exhibits antimicrobial properties and may assist in treating bacterial infections. Some studies have shown that menthol is effective against bacteria that cause respiratory tract infections. Additionally, menthol has antifungal properties and may be beneficial in treating fungal infections. Due to these antimicrobial properties, menthol is also used in the treatment of skin infections and in antiseptic preparations [66, 67].

The third monoterpene, pinene, is a volatile compound found in pine, black pine, and certain herbs (e.g., pine turpentine). Pinene can be found in edible plants such as spearmint (*Mentha spicata*) [68], rosemary (*Rosmarinus officinalis*) [69], sage (*Salvia officinalis*)

[70], and cumin (*Cuminum cyminum*) [71]. Pinene is recognized for its aromatic properties and biological activities, and has become a significant topic of research due to its pharmacological effects. Pinene exists in two isomers, α-pinene and β-pinene, each of which exhibits different biological effects. Pinene is particularly noted for its anti-inflammatory

Some Terpen structure

Acyclic monoterpenes

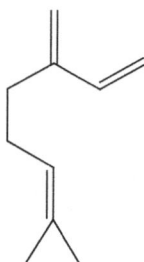

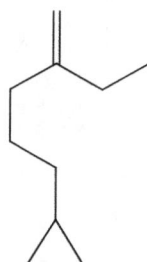

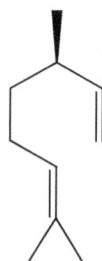

Myrcene   Ocimene   2,6-dimehtl   Citronelle
                                octane

Monocyclic monoterpenes

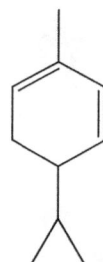

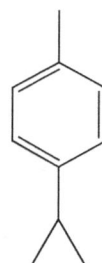

Menthane   Limonene   Alfa-Phellandrene   p-Cymene

Bicyclic monoterpenes

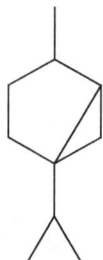

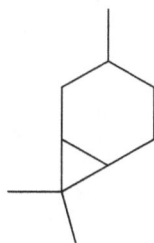

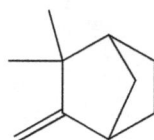

Thujane   Carane   Camphene   Alfa-Pinene   Fenchane

Figure 14.7: The chemical structures of terpenes.

Some Terpen structure

Acyclic sesquiterpenes

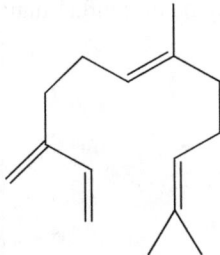

Farnesene

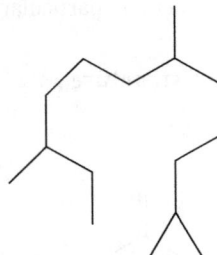

Farnesane

Monocyclic sesquiterpenes

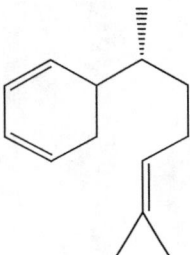

Menthane

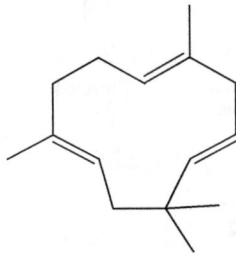

Alfa-humulene

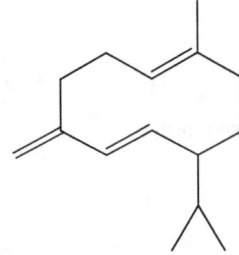

Germacrene D

Bicyclic sesquiterpenes

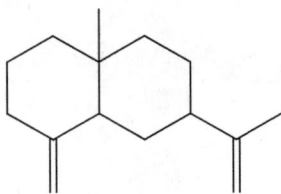

Seslinene

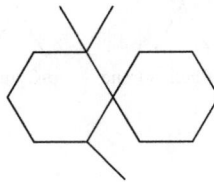

Chamigrane

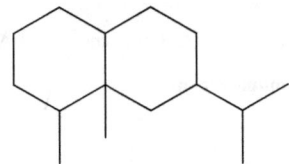

Eremophilane

Figure 14.7 (continued)

properties. Numerous studies have demonstrated that pinene reduces inflammation in the body. By inhibiting the production of molecules that trigger inflammatory responses, pinene may help manage inflammatory diseases such as rheumatoid arthritis, osteoarthritis, and other inflammatory conditions. Pinene also suppresses the production of pro-inflammatory cytokines, alleviating inflammation-related symptoms [72–74]. Moreover, pinene shows antimicrobial effects against various microorganisms. It has been reported to be effective against bacterial and fungal infections [75]. Pinene's antifungal properties suggest that it could be a potential therapeutic agent in treating fungal infections [76].

Linalool is a volatile terpene found primarily in lavender (*Lavandula angustifolia*) [77], basil (*Ocimum basilicum*) [78], thyme (*Thymus vulgaris*) [79], bay laurel (*Laurus nobilis*) [80], and coriander (*Coriandrum sativum*) [81]. Linalool is a fragrant compound frequently used in aromatherapy and has gained significant research attention due to its various pharmacological effects. Studies on the biological effects of linalool have highlighted its anti-inflammatory [82], analgesic [83], anxiolytic [84], and antioxidant [85] properties.

The last group of monoterpenes, terpinene, is found in plants such as thyme (*Thymus vulgaris* and *Origanum vulgare*) [86, 87] and tea tree (*Melaleuca alternifolia*) [88]. Terpinene is a monoterpene with various pharmacological properties and is an important component of essential oils used in healthcare. Research has demonstrated that terpinene exhibits antimicrobial, antioxidant, and anti-inflammatory activities [89–90].

Sesquiterpenes ($C_{15}H_{24}$) are terpenes composed of three isoprene units (C_5H_8) and are biologically active compounds primarily found in essential oils. These compounds are associated with the characteristic aromas and flavors of plants and have attracted attention due to their positive effects on health. Sesquiterpenes generally possess antimicrobial, anti-inflammatory, antioxidant, anticancer, and analgesic properties. Due to these properties, they stand out as potential therapeutic compounds in both traditional medicine and modern pharmaceutical research. Sesquiterpenes are found in ginger (*Zingiber officinale*) [91], cinnamon (*Cinnamomum verum*) [92], and turmeric (*Curcuma longa*) [93].

Diterpenes, which are predominantly found in resins, oils, and certain plant extracts, are abundant in edible plants. Rosemary (*Rosmarinus officinalis*) [94], sage (*Salvia officinalis*) [95], and coffee [96] are examples of plants that synthesize diterpenes. Diterpenes are significant compounds found in plants with various pharmacological effects. These compounds demonstrate a broad range of health impacts, particularly drawing attention for their antioxidant, anti-inflammatory, and anticancer potential [97].

Triterpenes are compounds with 30 carbon atoms belonging to the terpenoid class. These compounds are found in plants as secondary metabolites and exhibit various biological effects on health. Triterpenes generally offer a wide range of benefits, including antioxidant, anti-inflammatory, anticancer, antimicrobial, and cardiovascular protective effects. Olive (*Olea europaea*) contains triterpene components such as "oleuropein" [98]. Oleuropein is known for its antioxidant, anti-inflammatory, and antimicrobial properties. Cranberries (*Vaccinium* spp.) contain triterpenic acids and other bioactive compounds with antioxidant and anti-inflammatory characteristics [99]. Black cumin (*Nigella sativa*) contains triterpene components such as "thymoquinone." This compound can positively influence the immune system and has potential anticancer properties [100].

Monoterpene	Source Plants	Health Effects
Limonene	Orange (*Citrus sinensis*) Lemon (*Citrus limon*)	Anti-inflammatory, Anticancer, Antioxidant, Potential effects on the central nervous system; may reduce stress and anxiety.
Menthol	Peppermint (*Mentha piperita*) Spearmint (*Mentha spicata*) Cornmint (*Mentha arvensis*) Citrus mint (*Mentha x piperita var. citrata*)	Antimicrobial Antisepticanalgesic
Pinen	Spearmint (*Mentha spicata*) Rosemary (*Rosmarinus officinalis*) Sage (*Salvia officinalis*) Cumin (*Cuminum cyminum*)	Anti-inflammatory Antimicrobial
Linalool	Lavender (*Lavandula angustifolia*) Basil (*Ocimum basilicum*) Thyme (*Thymus vulgaris*) Laurel (*Laurus nobilis*) Coriander (*Coriandrum sativum*)	Anti-inflammatory. Analgesic AnxiolyticAntioxidant
Terpinene	Thyme (*Thymus vulgaris*) Tea Tree (*Melaleuca alternifolia*)	Antimicrobial Antioxidant Anti-inflammatory

Figure 14.8: Summary of the health effects of each monoterpene and the plants in which they are found.

Pumpkin (*Cucurbita pepo*) is rich in triterpenic compounds and steroids, which possess antioxidant properties and offers protective effects against cancer and heart diseases [101]. The triterpenes and flavonoids in St. John's wort (*Hypericum perforatum*) may have calming effects on the nervous system and can assist in treating depression [102]. Ginseng (*Panax ginseng*) contains triterpenoid compounds known as ginsenosides, which have energy-boosting, immune-strengthening, and anticancer effects [103].

Tetraterpenes, composed of 40 carbon atoms, are a class of terpenes commonly referred to as carotenoids. These compounds are widely present in plants and exhibit significant biological effects on health. Tetraterpenes are particularly recognized for their strong antioxidant properties and provitamin A activities. Carrot (*Daucus carota*), red pepper (*Capsicum annuum*), melon (*Cucumis melo*), mango (*Mangifera indica*), sweet potato (*Ipomoea batatas*), pumpkin (*Cucurbita* spp.), and apricot (*Prunus armeniaca*) are rich in beta-carotene and serve as important sources of vitamin A. Tomato (*Solanum lycopersicum*) and watermelon (*Citrullus lanatus*) stand out for their lycopene content. Lycopene is known as a potent antioxidant and is effective in cancer prevention and cardiovascular health support. Spinach (*Spinacia oleracea*) contains lutein and zeaxanthin, compounds that particularly support visual health [104–106].

14.2.4 Glycosides and edible medicinal and aromatic plants

Glycosides are organic compounds formed through a glycosidic bond between a sugar molecule (commonly glucose) and a functional group. Glycosides play significant roles as secondary metabolites in plants and are widely utilized in the pharmaceutical and food industries due to their diverse pharmacological effects. Tea (*Camellia sinensis*) is rich in flavonol glycosides, including myricetin glycosides, quercetin glycosides, and behenyl glycosides [107, 108]. Apples (*Malus domestica*) contain glycosylated flavonoids, specifically quercetin glycosides and phloretin glycosides [109]. Onions (*Allium cepa*) are known to contain delphinidin 3,5-diglycosides and quercetin 3-glycosides [110]. The edible parts of asparagus (*Asparagus officinalis*) contain quercetin triglycoside and isorhamnetin triglycoside [111].

References

[1]   Rasool, H. B. and Bassam, A. (2012). Medicinal plants (importance and uses). Pharmaceut Anal Acta, 3(10), 2153–2435.

[2]   Okigbo, R. N., Eme, U. E. and Ogbogu, S. (2008). Biodiversity and conservation of medicinal and aromatic plants in Africa. Biotechnology and Molecular Biology Reviews, 3(6), 127–134.

[3]   Ghorbanpour, M., Hadian, J., Nikabadi, S. and Varma, A. (2017). Importance of medicinal and aromatic plants in human life. Medicinal Plants and Environmental Challenges, Gewerbestrasse 11, 6330 Cham, Switzerland Springer, 1–23.

[4]   Ramadan, M. F. and Al-Ghamdi, A. (2012). Bioactive compounds and health-promoting properties of royal jelly: A review. Journal of Functional Foods, 4(1), 39–52.

[5]   Rivera, D., Obon, C., Inocencio, C., Heinrich, M., Verde, A., Fajardo, J. and Llorach, R. (2005). The ethnobotanical study of local Mediterranean food plants as medicinal resources in Southern Spain. Journal of Physiology and Pharmacology, Supplement 56(1), 97–114.

[6]   Kadioglu, Z., Yildiz, F., Kandemir, A., Cukadar, K., Kalkan, N. N., Vurgun, H. and Kaya, O. (2024). Preserving the richness of nature: Cultural and ecological importance of edible wild plants in Sivas. Genetic Resources and Crop Evolution, 71, 1–19.

[7]   Tardío, J., Pardo-de-santayana, M. and Morales, R. (2006). Ethnobotanical review of wild edible plants in Spain. Botanical Journal of the Linnean Society, 152(1), 27–71.

[8]   Awuchi, C. G. (2019). Medicinal plants: The medical, food, and nutritional biochemistry and uses. International Journal of Advanced Academic Research, 5(11), 220–241.

[9]   Charles, D. J. (2012). Antioxidant Properties of Spices, Herbs and Other Sources, Springer Science & Business Media, New york, Springer.

[10]   Opara, E. I. and Chohan, M. (2014). Culinary herbs and spices: Their bioactive properties, the contribution of polyphenols and the challenges in deducing their true health benefits. International Journal of Molecular Sciences, 15(10), 19183–19202, Ivanišová.

[11]   Ivanišová, E., Kačániová, M., Savitskaya, T. A. and Grinshpan, D. D. (2021). Medicinal Herbs: Important Source of Bioactive Compounds for Food Industry, Herbs and Spices – New Processing Technologies.

[12]   Mahmoud, E. A. and Elansary, H. O. (2024). Bioactive compounds, functional ingredients, antioxidants, and health benefits of edible plants. Frontiers in Plant Science, 15, 1420069.

[13]   Clifford, M. N. and Scalbert, A. (2000). Ellagitannins–nature, occurrence and dietary burden. Journal of the Science of Food and Agriculture, 80(7), 1118–1125.

[14]   Scalbert, A. and Williamson, G. (2000). Dietary intake and bioavailability of polyphenols. The Journal of Nutrition, 130(8), 2073S–2085S.

[15]   Guven, H., Arici, A. and Simsek, O. (2019). Flavonoids in our foods: A short review. Journal of Basic and Clinical Health Sciences, 3(2), 96–106.

[16] Manzoor, A., Dar, I. H., Bhat, S. A. and Ahmad, S. (2020). Flavonoids: Health benefits and their potential use in food systems. Functional Food Products and Sustainable Health, Singapore, Springer, 235–256.

[17] Manach, C., Scalbert, A., Morand, C., Rémésy, C. and Jiménez, L. (2004). Polyphenols: Food sources and bioavailability. The American Journal of Clinical Nutrition, 79(5), 727–747.

[18] Khan, M. K. and Dangles, O. (2014). A comprehensive review on flavanones, the major citrus polyphenols. Journal of Food Composition & Analysis, 33(1), 85–104.

[19] Reinli, K. and Block, G. (1996). Phytoestrogen content of foods – A compendium of literature values. Nutrition and Cancer, 26(2), 123–148.

[20] Adlercreutz, H. and Mazur, W. (1997). Phyto-oestrogens and Western diseases. Annals of Medicine, 29(2), 95–120.

[21] Cassidy, A., Bingham, S. and Setchell, K. D. (1994). Biological effects of a diet of soy protein rich in isoflavones on the menstrual cycle of premenopausal women. The American Journal of Clinical Nutrition, 60(3), 333–340.

[22] Hämäläinen, M., Nieminen, R., Vuorela, P., Heinonen, M. and Moilanen, E. (2007). Anti-inflammatory effects of flavonoids: Genistein, kaempferol, quercetin, and daidzein inhibit STAT-1 and NF-κB activations, whereas flavone, isorhamnetin, naringenin, and pelargonidin inhibit only NF-κB activation along with their inhibitory effect on iNOS expression and NO production in activated macrophages. Mediators of Inflammation, 2007(1), 045673.

[23] Lila, M. A. (2004). Anthocyanins and human health: An in vitro investigative approach. BioMed Research International, 2004(5), 306–313.

[24] Khoo, H. E., Azlan, A., Tang, S. T. and Lim, S. M. (2017). Anthocyanidins and anthocyanins: Colored pigments as food, pharmaceutical ingredients, and the potential health benefits. Food and Nutrition Research, 61(1), 1361779.

[25] Makris, D. P., Kallithraka, S. and Kefalas, P. (2006). Flavonols in grapes, grape products and wines: Burden, profile and influential parameters. Journal of Food Composition & Analysis, 19(5), 396–404.

[26] Lakhanpal, P. and Rai, D. K. (2007). Quercetin: A versatile flavonoid. Internet Journal of Medical Update, 2(2), 22–37.

[27] Erlund, I. (2004). Review of the flavonoids quercetin, hesperetin, and naringenin. Dietary sources, bioactivities, bioavailability, and epidemiology. Nutrition Research, 24(10), 851–874.

[28] M Calderon-Montano, J., Burgos-Morón, E., Pérez-Guerrero, C. and López-Lázaro, M. (2011). A review on the dietary flavonoid kaempferol. Mini Reviews in Medicinal Chemistry, 11(4), 298–344.

[29] Ali, F., Rahul, N., Jyoti, F. and Siddique, Y. H. (2017). Health functionality of apigenin: A review. International Journal of Food Properties, 20(6), 1197–1238.

[30] López-Lázaro, M. (2009). Distribution and biological activities of the flavonoid luteolin. Mini Reviews in Medicinal Chemistry, 9(1), 31–59.

[31] Pascual-Teresa, D., Moreno, D. A. and García-Viguera, C. (2010). Flavanols and anthocyanins in cardiovascular health: A review of current evidence. International Journal of Molecular Sciences, 11(4), 1679–1703.

[32] Gramza, A., Korczak, J. and Amarowicz, R. (2005). Tea polyphenols-their antioxidant properties and biological activity-a review. Polish journal of food and nutrition sciences, 14(55), 219–235.

[33] Vinson, J. A., Su, X., Zubik, L. and Bose, P. (2001). Phenol antioxidant quantity and quality in foods: Fruits. Journal of Agricultural and Food Chemistry, 49(11), 5315–5321.

[34] Samarghandian, S., Azimi-Nezhad, M. and Farkhondeh, T. (2017). Catechin treatment ameliorates diabetes and its complications in streptozotocin-induced diabetic rats. Dose-response, 15(1), 1559325817691158.

[35] Neveu, V., Perez-Jiménez, J., Vos, F., Crespy, V., Du Chaffaut, L., Mennen, L. and Scalbert, A. (2010). Phenol-Explorer: An online comprehensive database on polyphenol contents in foods. Database, 2010, 10, 1–9.

[36] El Khawand, T., Courtois, A., Valls, J., Richard, T. and Krisa, S. (2018). A review of dietary stilbenes: Sources and bioavailability. Phytochemistry Reviews, 17(5), 1007–1029.

[37] Ionkova, I. (2011). Anticancer lignans-from discovery to biotechnology. Mini Reviews in Medicinal Chemistry, 11(10), 843–856.

[38] Peterson, J., Dwyer, J., Adlercreutz, H., Scalbert, A., Jacques, P. and McCullough, M. L. (2010). Dietary lignans: Physiology and potential for cardiovascular disease risk reduction. Nutrition Reviews, 68(10), 571–603.

[39] Landete, J. M. (2012). Plant and mammalian lignans: A review of source, intake, metabolism, intestinal bacteria and health. Food Research International, 46(1), 410–424.

[40] Facchini, P. J. (2001). Alkaloid biosynthesis in plants: Biochemistry, cell biology, molecular regulation, and metabolic engineering applications. Annual Review of Plant Biology, 52(1), 29–66.

[41] Ashihara, H. and Crozier, A. (2001). Caffeine: A well known but little mentioned compound in plant science. Trends in Plant Science, 6(9), 407–413.

[42] Luca, V. S., Ana-Maria, S. T. A. N., Trifan, A., Miron, A. and Aprotosoaie, A. C. (2016). Catechins profile, caffeine content and antioxidant activity of camellia sinensis teas commercialized in Romania. The Medical-Surgical Journal, 120(2), 457–463.

[43] Xu, Y., Zhang, M., Wu, T., Dai, S., Xu, J. and Zhou, Z. (2015). The anti-obesity effect of green tea polysaccharides, polyphenols and caffeine in rats fed with a high-fat diet. Food & Function, 6(1), 296–303.

[44] Li, X., Liu, G. J., Zhang, W., Zhou, Y. L., Ling, T. J., Wan, X. C. and Bao, G. H. (2018). Novel flavoalkaloids from white tea with inhibitory activity against the formation of advanced glycation end products. Journal of Agricultural and Food Chemistry, 66(18), 4621–4629.

[45] Martínez-Pinilla, E., Oñatibia-Astibia, A. and Franco, R. (2015). The relevance of theobromine for the beneficial effects of cocoa consumption. Frontiers in Pharmacology, 6, 126866.

[46] Edo, G. I., Samuel, P. O., Oloni, G. O., Ezekiel, G. O., Onoharigho, F. O., Oghenegueke, O. and Igbodo, P. C. (2023). Review on the biological and bioactive components of cocoa (Theobroma cacao). Insight on food, health and nutrition. Natural Resources for Human Health, 3, 426–448.

[47] Baggott, M. J., Childs, E., Hart, A. B., De Bruin, E., Palmer, A. A., Wilkinson, J. E. and de Wit, H. (2013). Psychopharmacology of theobromine in healthy volunteers. Psychopharmacology, 228, 109–118.

[48] Zhou, J., Chan, L. and Zhou, S. (2012). Trigonelline: A plant alkaloid with therapeutic potential for diabetes and central nervous system disease. Current Medicinal Chemistry, 19(21), 3523–3531.

[49] Nguyen, V., Taine, E. G., Meng, D., Cui, T. and Tan, W. (2024). Pharmacological activities, therapeutic effects, and mechanistic actions of Trigonelline. International Journal of Molecular Sciences, 25(6), 3385.

[50] Farag, M. A., Porzel, A. and Wessjohann, L. A. (2015). Unraveling the active hypoglycemic agent trigonelline in Balanites aegyptiaca date fruit using metabolite fingerprinting by NMR. Journal of Pharmaceutical and Biomedical Analysis, 115, 383–387.

[51] Arora, V., Campbell, J. N. and Chung, M. K. (2021). Fight fire with fire: Neurobiology of capsaicin-induced analgesia for chronic pain. Pharmacology and Therapeutics, 220, 107743.

[52] Galano, A. and Martínez, A. (2012). Capsaicin, a tasty free radical scavenger: Mechanism of action and kinetics. The Journal of Physical Chemistry B, 116(3), 1200–1208.

[53] Li, J., Wang, H., Zhang, L., An, N., Ni, W., Gao, Q. and Yu, Y. (2021). Capsaicin affects macrophage anti-inflammatory activity via the MAPK and NF-κB signaling pathways. International Journal for Vitamin and Nutrition Research, 93 (4), 289–297.

[54] Guedes, V., Castro, J. P. and Brito, I. (2018). Topical capsaicin for pain in osteoarthritis: A literature review. Reumatología Clínica (English Edition), 14(1), 40–45.

[55] Roberts, R. G. D., Westerman, R. A., Widdop, R. E., Kotzmann, R. R. and Payne, R. (1992). Effects of capsaicin on cutaneous vasodilator responses in humans. Agents and Actions, 37, 53–59.

[56] Yang, S., Liu, L., Meng, L. and Hu, X. (2019). Capsaicin is beneficial to hyperlipidemia, oxidative stress, endothelial dysfunction, and atherosclerosis in Guinea pigs fed on a high-fat diet. Chemico-Biological Interactions, 297, 1–7.

[57] Irandoost, P., Lotfi Yagin, N., Namazi, N., Keshtkar, A., Farsi, F., Mesri Alamdari, N. and Vafa, M. (2021). The effect of Capsaicinoids or Capsinoids in red pepper on thermogenesis in healthy adults: A systematic review and meta-analysis. Phytotherapy Research, 35(3), 1358–1377.

[58] Masyita, A., Sari, R. M., Astuti, A. D., Yasir, B., Rumata, N. R., Emran, T. B. and Simal-Gandara, J. (2022). Terpenes and terpenoids as main bioactive compounds of essential oils, their roles in human health and potential application as natural food preservatives. Food Chemistry: X, 13, 100217.

[59] Retajczyk, M. and Wróblewska, A. (2018). Therapeutic applications of limonene. Pomeranian Journal of Life Sciences, 64(2), 51–57.

[60] de Vasconcelos, C. B., J., D. C., O., F., De Vasconcelos, C. M., Calixto, D., A., F., Santana, H. S., Almeida, I. B. and Serafini, M. R. (2021). Mechanism of action of limonene in tumor cells: A systematic review and meta-analysis. Current Pharmaceutical Design, 27(26), 2956–2965.

[61] Bacanlı, M., Başaran, A. A. and Başaran, N. (2015). The antioxidant and antigenotoxic properties of citrus phenolics limonene and naringin. Food and Chemical Toxicology, 81, 160–170.

[62] Hao, C. W., Lai, W. S., Ho, C. T. and Sheen, L. Y. (2013). Antidepressant-like effect of lemon essential oil is through a modulation in the levels of norepinephrine, dopamine, and serotonin in mice: Use of the tail suspension test. Journal of Functional Foods, 5(1), 370–379.

[63] Li, Z., Zhang, H., Wang, Y., Li, Y., Li, Q. and Zhang, L. (2022). The distinctive role of menthol in pain and analgesia: Mechanisms, practices, and advances. Frontiers in Molecular Neuroscience, 15, 1006908.

[64] McKemy, D. D. (2007). TRPM8: The cold and menthol receptor. TRP Ion Channel Function in Sensory Transduction and Cellular Signaling Cascades. Wolfgang B. Liedtke and Stefan Heller (eds.), CRC press, Boca Raton, Florida, United States.

[65] Pergolizzi, J. V. Jr, Taylor, R. Jr, LeQuang, J. A., Raffa, R. B. and Research Group, N. E. M. A. (2018). The role and mechanism of action of menthol in topical analgesic products. Journal of Clinical Pharmacy and Therapeutics, 43(3), 313–319.

[66] Kumar, A., Singh, S. P. and Chhokar, S. S. (2011). Antimicrobial activity of the major isolates of mentha oil and derivatives of menthol. Analytical Chemistry Letters, 1(1), 70–85.

[67] Suchodolski, J., Feder-Kubis, J. and Krasowska, A. (2017). Antifungal activity of ionic liquids based on (−)-menthol: A mechanism study. Microbiological Research, 197, 56–64.

[68] Buleandra, M., Oprea, E., Popa, D. E., David, I. G., Moldovan, Z., Mihai, I. and Badea, I. A. (2016). Comparative chemical analysis of Mentha piperita and M. spicata and a fast assessment of commercial peppermint teas. Natural Products Communications, 11(4), 1934578X1601100433.

[69] Wang, W., Wu, N., Zu, Y. G. and Fu, Y. J. (2008). Antioxidative activity of Rosmarinus officinalis L. essential oil compared to its main components. Food Chemistry, 108(3), 1019–1022.

[70] Novak, J., Marn, M. and Franz, C. M. (2006). An α-Pinene Chemotype in Salvia offcinalis L. (Lamiaceae). Journal of Essential Oil Research, 18(3), 239–241.

[71] Nadeem, M. and Riaz, A. (2012). Cumin (Cuminum cyminum) as a potential source of antioxidants. Pakistan Journal of Food Sciences, 22(2), 101–107.

[72] Kim, T., Song, B., Cho, K. S. and Lee, I. S. (2020). Therapeutic potential of volatile terpenes and terpenoids from forests for inflammatory diseases. International Journal of Molecular Sciences, 21(6), 2187.

[73] Del Prado-Audelo, M. L., Cortés, H., Caballero-Florán, I. H., González-Torres, M., Escutia-Guadarrama, L., Bernal-Chávez, S. A. and Leyva-Gómez, G. (2021). Therapeutic applications of terpenes on inflammatory diseases. Frontiers in Pharmacology, 12, 704197.

[74] Kim, D. S., Lee, H. J., Jeon, Y. D., Han, Y. H., Kee, J. Y., Kim, H. J. and Hong, S. H. (2015). Alpha-pinene exhibits anti-inflammatory activity through the suppression of MAPKs and the NF-κB pathway in mouse peritoneal macrophages. The American Journal of Chinese Medicine, 43(04), 731–742.

[75] Šarac, Z., Matejić, J. S., Stojanović-Radić, Z. Z., Veselinović, J. B., Džamić, A. M., Bojović, S. and Marin, P. D. (2014). Biological activity of Pinus nigra terpenes – Evaluation of FtsZ inhibition by selected compounds as contribution to their antimicrobial activity. Computers in Biology and Medicine, 54, 72–78.

[76] Ložienė, K., Švedienė, J., Paškevičius, A., Raudonienė, V., Sytar, O. and Kosyan, A. (2018). Influence of plant origin natural α-pinene with different enantiomeric composition on bacteria, yeasts and fungi. Fitoterapia, 127, 20–24.

[77] Carrasco, A., Martinez-Gutierrez, R., Tomas, V. and Tudela, J. (2016). Lavandula angustifolia and Lavandula latifolia essential oils from Spain: Aromatic profile and bioactivities. Planta Medica, 82(01/02), 163–170.

[78] Radulović, N. S., Blagojević, P. D. and Miltojević, A. B. (2013). α-Linalool–a marker compound of forged/synthetic sweet basil (Ocimum basilicum L.) essential oils. Journal of the Science of Food and Agriculture, 93(13), 3292–3303.

[79] Shabnum, S. and Wagay, M. G. (2011). Essential oil composition of Thymus vulgaris L. and their uses. Journal of Research and Development, 11, 83–94.

[80] Caputo, L., Nazzaro, F., Souza, L. F., Aliberti, L., De Martino, L., Fratianni, F. and De Feo, V. (2017). Laurus nobilis: Composition of essential oil and its biological activities. Molecules, 22(6), 930.

[81] Duman, A. D., Telci, I., Dayisoylu, K. S., Digrak, M., Demirtas, İ. and Alma, M. H. (2010). Evaluation of bioactivity of linalool-rich essential oils from Ocimum basilucum and Coriandrum sativum varieties. Natural Products Communications, 5(6), 1934578X1000500634.

[82] Peana, A. T., D'Aquila, P. S., Panin, F., Serra, G., Pippia, P. and Moretti, M. D. L. (2002). Anti-inflammatory activity of linalool and linalyl acetate constituents of essential oils. Phytomedicine, 9(8), 721–726.

[83] Peana, A. T. and Moretti, M. D. (2008). Linalool in essential plant oils: Pharmacological effects. In Botanical Medicine in Clinical Practice, R. R. Watson, Università degli Studi di Sassari Sassari Italy and V. R. Preedy (eds.), CAB International, Wallingford UK, 716–724.

[84] Cline, M., Taylor, J. E., Flores, J., Bracken, S., McCall, S. and Ceremuga, T. E. (2008). Investigation of the anxiolytic effects of linalool, a lavender extract, in the male Sprague-Dawley rat. AANA Journal, 76(1), 49–52.

[85] Jabir, M. S., Taha, A. A. and Sahib, U. I. (2018). Antioxidant activity of Linalool. Engineering Technology Journal, 36(1), 64–67.

[86] Satyal, P., Murray, B. L., McFeeters, R. L. and Setzer, W. N. (2016). Essential oil characterization of Thymus vulgaris from various geographical locations. Foods, 5(4), 70.

[87] Raina, A. P. and Negi, K. S. (2012). Essential oil composition of Origanum majorana and Origanum vulgare ssp. hirtum growing in India. Chemistry of Natural Compounds, 47, 1015–1017.

[88] Hammer, K. A., Carson, C. F. and Riley, T. V. (2012). Effects of Melaleuca alternifolia (tea tree) essential oil and the major monoterpene component terpinen-4-ol on the development of single-and multistep antibiotic resistance and antimicrobial susceptibility. Antimicrobial Agents and Chemotherapy, 56(2), 909–915.

[89] Sousa, L. G., Castro, J., Cavaleiro, C., Salgueiro, L., Tomás, M., Palmeira-Oliveira, R. and Cerca, N. (2022). Synergistic effects of carvacrol, α-terpinene, γ-terpinene, ρ-cymene and linalool against Gardnerella species. Scientific Reports, 12(1), 4417.

[90] De Oliveira Ramalho, T. R., de Oliveira, M. T. P., de Araujo Lima, A. L., Bezerra-Santos, C. R. and Piuvezam, M. R. (2015). Gamma-terpinene modulates acute inflammatory response in mice. Planta Medica, 81(14), 1248–1254.

[91] Kumar Gupta, S. and Sharma, A. (2014). Medicinal properties of Zingiber officinale Roscoe-A review. International Journal of Pharmacy and Biological Sciences, 9, 124–129.

[92] Wang, J., Su, B., Jiang, H., Cui, N., Yu, Z., Yang, Y. and Sun, Y. (2020). Traditional uses, phytochemistry and pharmacological activities of the genus Cinnamomum (Lauraceae): A review. Fitoterapia, 146, 104675.

[93] Ohshiro, M., Kuroyanagi, M. and Ueno, A. (1990). Structures of sesquiterpenes from Curcuma longa. Phytochemistry, 29(7), 2201–2205.

[94] Pertino, M. W. and Schmeda-Hirschmann, G. (2010). The corrected structure of rosmaridiphenol, a bioactive diterpene from Rosmarinus officinalis. Planta Medica, 76(06), 629–632.

[95] Li, L., Wei, S., Zhu, T., Xue, G., Xu, D., Wang, W. and Kong, L. (2019). Anti-inflammatory norabietane diterpenoids from the leaves of Salvia officinalis L. Journal of Functional Foods, 54, 154–163.

[96] Kurzrock, T. and Speer, K. (2001). Diterpenes and diterpene esters in coffee. Food Reviews International, 17(4), 433–450.

[97] Islam, M. T., Da Mata, A. M. O. F., De Aguiar, R. P. S., Paz, M. F. C. J., De Alencar, M. V. O. B., Ferreira, P. M. P. and De Carvalho Melo-cavalcante, A. A. (2016). Therapeutic potential of essential oils focusing on diterpenes. Phytotherapy Research, 30(9), 1420–1444.

[98] Sánchez-Quesada, C., López-Biedma, A., Warleta, F., Campos, M., Beltran, G. and Gaforio, J. J. (2013). Bioactive properties of the main triterpenes found in olives, virgin olive oil, and leaves of Olea europaea. Journal of Agricultural and Food Chemistry, 61(50), 12173–12182.

[99] Martău, G. A., Bernadette-Emőke, T., Odocheanu, R., Soporan, D. A., Bochiș, M., Simon, E. and Vodnar, D. C. (2023). Vaccinium species (Ericaceae): Phytochemistry and biological properties of medicinal plants. Molecules, 28(4), 1533.

[100] Gholamnezhad, Z., Havakhah, S. and Boskabady, M. H. (2016). Preclinical and clinical effects of Nigella sativa and its constituent, thymoquinone: A review. Journal of Ethnopharmacology, 190, 372–386.

[101] Tanaka, R., Kikuchi, T., Nakasuji, S., Ue, Y., Shuto, D., Igarashi, K. and Yamada, T. (2013). A Novel 3α-p-Nitrobenzoylmultiflora-7: 9 (11)-diene-29-benzoate and Two New Triterpenoids from the Seeds of Zucchini (Cucurbita pepo L). Molecules, 18(7), 7448–7459.

[102] Xiao, C. Y., Mu, Q. and Gibbons, S. (2020). The phytochemistry and pharmacology of Hypericum. Progress in the Chemistry of Organic Natural Products, 112, 85–182.

[103] Leung, K. W. and Wong, A. S. T. (2010). Pharmacology of ginsenosides: A literature review. Chinese Medicine, 5, 1–7.

[104] Imran, M., Ghorat, F., Ul-Haq, I., Ur-Rehman, H., Aslam, F., Heydari, M. and Rebezov, M. (2020). Lycopene as a natural antioxidant used to prevent human health disorders. Antioxidants, 9(8), 706.

[105] Ismail, J., Shebaby, W. N., Daher, J., Boulos, J. C., Taleb, R., Daher, C. F. and Mroueh, M. (2023). The wild Carrot (Daucus carota): A phytochemical and pharmacological review. Plants, 13(1), 93.

[106] Strugari, S. C., Butnariu, M. and Sokan-Adeaga, E. (2023). Bioactive compounds of vegetable origin. Biotechnology & Bioengineering, 4, 2766–2314.

[107] Balentine, D. A., Wiseman, S. A. and Bouwens, L. C. (1997). The chemistry of tea flavonoids. Critical Reviews in Food Science and Nutrition, 37, 693–704, doi: 10.1080/10408399709527797.

[108] Wang, H., Provan, G. J. and Helliwell, K. (2000). Tea flavonoids: Their functions, utilisation and analysis. Trends in Food Science and Technology, 11(4–5), 152–160.

[109] Patocka, J., Bhardwaj, K., Klimova, B., Nepovimova, E., Wu, Q., Landi, M. and Wu, W. (2020). Malus domestica: A review on nutritional features, chemical composition, traditional and medicinal value. Plants, 9(11), 1408.

[110] Teshika, J. D., Zakariyyah, A. M., Zaynab, T., Zengin, G., Rengasamy, K. R., Pandian, S. K. and Fawzi, M. M. (2019). Traditional and modern uses of onion bulb (Allium cepa L.): A systematic review. Critical Reviews in Food Science and Nutrition, 59(sup1), S39–S70.

[111] Fuentes-Alventosa, J. M., Jaramillo, S., Rodríguez-Gutiérrez, G., Cermeño, P., Espejo, J. A., Jiménez-Araujo, A. and Rodríguez-Arcos, R. (2008). Flavonoid profile of green asparagus genotypes. Journal of Agricultural and Food Chemistry, 56(16), 6977–6984.

Abdellatif Rafik* and Burak Tüzün

Chapter 15
The mysteries of Moroccan nature: aromatic plants and their therapeutic medicinal properties

Abstract: Morocco, with its rich and varied natural landscape, stands as a country of remarkable botanical diversity – a true testament to nature's generosity. The country is home to more than 4,200 plant species, of which over 800 are endemic, and approximately 600 are recognized for their medicinal and aromatic properties. This extraordinary wealth offers untapped potential, particularly in the areas of agriculture and healthcare, which are key to Morocco's sustainable development. To fully harness the benefits of this plant diversity, it is imperative to advance the field of ethnopharmacology, a discipline that focuses on understanding the medicinal and aromatic properties of plants, their variations, and the best practices for cultivation and utilization. By studying these plants, we can adapt them to Morocco's agricultural systems, optimizing their productivity and potential for enhancing the welfare of the population.

In this context, an ethnopharmacological study was undertaken in the Chaouia-Ouardigha region of Morocco. This area, known for its deep-rooted traditions and local knowledge, provided an ideal setting to explore the use of plants that have been part of the region's cultural heritage for generations. The study focused on four of the most widely known medicinal and aromatic plants, whose use is still prevalent in local herbal medicine. To gather meaningful data, a questionnaire was distributed to local herbalists, who are the custodians of this valuable knowledge. The results of this survey offer profound insights into the practices and customs surrounding the use of these plants. They reveal not only the depth of traditional knowledge passed down through generations but also the practical applications of these plants in the region's healthcare practices.

This ethnopharmacological data serves as an invaluable resource for understanding how these plants are used in local medicine and their potential role in modern pharmacology. It provides a foundation for further scientific exploration into the medicinal and aromatic properties of these plants. By documenting this local knowledge,

*Corresponding author: Abdellatif Rafik, Laboratory of Organic Chemistry, Catalysis and Environment Faculty of Sciences, Ibn Tofail University, Kenitra, Morocco, e-mail: rafik2013abdellatif@gmail.com, https://orcid.org/0000-0001-9617-052X
Burak Tüzün, Plant and Animal Production Department, Technical Sciences Vocational School of Sivas, Sivas Cumhuriyet University, Sivas, Turkey

https://doi.org/10.1515/9783111469713-015

the study also opens the door for future research and innovation in the field of natural medicine, with the potential to contribute to the development of new pharmaceutical products that could improve public health. Moreover, it highlights the importance of preserving this knowledge, as it holds the key not only to safeguarding local cultural practices but also to supporting sustainable agricultural and health practices in Morocco.

Keywords: medicinal and aromatic plants, ethnopharmacological study, Chaouia-Ouardigha, herbalists, pharmacy

15.1 Introduction

Aromatic and medicinal plants (AMPs) are a true natural heritage to cultivate, offering precious wealth for our wellbeing and our environment. Their cultivation is an enriching experience that connects humans to nature, with plants such as thyme, robust against winters, and mint, which diffuses its vitality in any space. In addition to their beauty and simplicity, they are full of benefits: thyme soothes the respiratory tract, rosemary promotes digestion and calms the mind, while chamomile and dandelion purify and soothe [1]. These versatile plants enhance our meals while taking care of our health, in the form of infusions, essential oils (EO), or ointments. Finally, they play a key role in the preservation of biodiversity, attracting pollinators and contributing to environmentally friendly practices. Growing these plants means adopting a lifestyle in harmony with nature [2, 3].

Medicinal and aromatic plants (MAPs), discreet jewels of nature, tell a story as old as humanity itself. These plants, bearers of healing secrets and captivating aromas, have long been at the heart of traditions, rituals, and care practices in all cultures of the world [4–6]. They embody a deep bond between man and his environment, testifying to human ingenuity in drawing on biodiversity to preserve health and enrich daily life.

Today, these plants are of paramount importance in addressing global challenges related to health, environment,, and sustainable development. Rich in natural active ingredients, they contribute not only to traditional medicines but also to modern research, paving the way for innovative solutions to meet the growing demand for healthcare and natural products [7–9]. However, this wealth is fragile. Intensive exploitation, loss of natural habitats, and climate change jeopardize their diversity and future.

Preserving these plant treasures is much more than protecting species; it is about safeguarding a universal heritage, a source of collective wellbeing, and a precious ecological balance. This requires a balanced approach, where conservation and sustainable use come together, and where traditional knowledge meets scientific advances to build a harmonious future.

The history of MAPs is closely intertwined with that of human civilizations. These plants, true treasures of nature, have always occupied a central place in both traditional and modern healthcare systems. Since the first discoveries of the curative virtues of plants, they have been valuable allies to humanity. It is estimated that at least 25% of modern medicines originate, directly or indirectly, from medicinal plants, which testifies to the wealth of traditional knowledge – often transmitted orally – and highlights the importance of their development through advances in modern technologies [10].

The growing recognition of these plants as essential natural resources continues to strengthen due to the spectacular increase in global demand for MAPs and their derivatives. This trend is also fueled by the diversification of the sectors in which these plants find their place: medicine, cosmetics, food, biotechnologies, and even sustainable agriculture. Each use, whether preventive or curative, contributes to enriching the economic and scientific fabric around these species, paving the way for new discoveries and applications in multiple fields [11]. In the heart of the Chaouia-Ouardigha region of Morocco, where nature and culture coexist in a delicate balance, our ethnobotanical survey sought to uncover and document the profound connections between local populations and the medicinal plants that surround them. This endeavor was more than a scientific exploration; it was a journey into the lives and traditions of a community that has long harmonized its existence with the natural world. The knowledge we sought to capture was not merely factual but deeply human, interwoven with the stories, practices, and beliefs that define this unique region. Our work began with a fundamental question: how do the inhabitants of this region interact with MAPs? To answer this, we set out to understand the frequency and contexts of their use in daily life. It was immediately clear that these plants are far more than mere resources; they are a cornerstone of the community's health, culture, and identity. Traditional medicine, practiced here for generations, draws upon the rich biodiversity of the region, blending natural remedies with ancestral knowledge passed down through the ages [12–15].

To delve deeper into this intricate relationship, we conducted an in-depth study. This involved engaging directly with the community, particularly with herbalists, who serve as custodians of a wealth of botanical wisdom. These interactions were complemented by careful observation of traditional practices, enabling us to witness firsthand how these plants are prepared and used in various contexts, from treating common ailments to addressing more complex health concerns. Through these efforts, we gained invaluable insights into the role of MAPs in the lives of the Chaouia-Ouardigha inhabitants. Amid the abundance of plants utilized in this region, four species stood out for their prominence and significance: *Mentha pulegium* (L.), *Dysphania ambrosioides*, *Ammi visnaga*, and *Ammodaucus leucotrichus* [16, 17]. Each of these plants is more than a biological entity; it is a symbol of the region's natural heritage and a vessel of cultural identity. Their frequent use underscores their importance in traditional medicine as well as their enduring presence in the collective memory of the community. These plants are not only celebrated for their therapeutic properties but also revered as carriers of ancestral knowledge, deeply rooted in the practices and beliefs of the people. Through this work, we aim to shine a light on the invaluable knowledge that these plants embody. Our research aspires to bridge the gap between tradition and modern science, ensuring that the rich legacy of medicinal plant use in the Chaouia-Ouardigha region is not only preserved but also appreciated and understood on a broader scale. By collaborating with local organizations, we hope to foster a sustainable approach to the use of aromatic plants in traditional medicine, one that respects the ecological balance of the region while honoring its cultural heritage. Ultimately, this work is a celebration of the profound relationship between people and nature. It is a testament to the resilience and wisdom of a community that has found health and harmony in its natural surroundings. As we document and share this knowledge, we are reminded of the importance of safeguarding these traditions, ensuring that they continue to thrive for generations to come.

15.2 Materials and methods

15.2.1 Study of the coherence and convergence of information

The study of the coherence and convergence of information constitutes a fundamental pillar in the process of validating the data collected. As part of our survey, we adopted a rigorous method of data confrontation, which allowed us to verify the reliability and relevance of the information gathered. This approach reflects a commitment to ensuring that the knowledge shared by informants, which represents ancestral practices and wisdom, is coherent and representative of a collective reality rather than merely the result of individual opinions. Information is considered coherent when it is reported in a similar manner by several informants from different localities. This strengthens the validity of the information by ensuring that it is not influenced by bias or personal interpretation, but is instead shared and recognized by various individuals from diverse contexts. In other words, the repetition of information across different witnesses and locations ensures its solidity and relevance within the local culture. Thus, when the same use or preparation of a plant is mentioned multiple times by informants located in different geographical areas, this information can be considered to have significant and universal value in the context of the survey.

On the other hand, information is considered divergent when it presents significant discrepancies in the following elements: the botanical identity of the plants, the symptoms or diseases for which they are used, the parts of the plant used, and the methods of preparation (decoction, infusion, etc.). These divergences, although they do not call into question the general validity of the survey, highlight local or individual variations that deserve special attention. They can be indicative of differences in cultural traditions, methods of knowledge transmission, or the ways in which information has adapted over time and across contexts. Thus, these divergences offer a wealth of analysis that allows us to deepen our understanding of local practices and better grasp the nuances that shape the use of medicinal plants in each community.

15.2.2 Sampling

The sampling we adopted in this study is based on a random approach, which made it possible to collect a representative range of the diversity of experiences and knowledge of the local population. This method was chosen with a view to ensuring the diversity of testimonies and capturing practices related to medicinal plants in their greatest variety. The dialog with the participants was conducted in Moroccan dialectal Arabic, a language that embodies the daily life and deep connection of the inhabitants with their cultural and plant heritage. This language, rich in its history and nuances, facilitated the exchange and understanding of local knowledge, often transmitted from generation to generation and sometimes tinged with subtleties that other languages could not have captured.

The information collected focused on several fundamental aspects: first, the profile of the respondents in order to understand their personal relationship with medicinal plants and their role in local society. We then explored the ethnopharmacological data concerning the chosen plants. This included collecting the specific uses of each plant, the parts of the plant used (whether leaves, roots, flowers, etc.), as well as the methods of preparation, whether decoctions, infusions, or other forms of transformation. We also collected information on the routes of administration of these remedies, whether oral, topical, or otherwise, and the potential adverse effects related to their consumption. All of this data was carefully noted on the questionnaires used during the survey. Once the survey was completed, the information was systematically entered and analyzed, thus making it possible to draw up a precise picture of local ethnopharmacological practices [18, 19]. This methodological approach not only made it possible to collect detailed information on the plants and their uses, but it also helped to enhance the value of this traditional knowledge. Our approach was not only to observe and record these uses but also to preserve them for future generations. This investigation thus aims to conserve and promote this rich plant and cultural heritage, which is both valuable for scientific development and fundamental for the health and wellbeing of local populations. By studying these plants, we hope to contribute to the preservation of traditional knowledge while reintegrating it into the framework of modern research that can enrich its understanding and application.

Since the dawn of time, humanity has always strived to shape its environment to meet its vital needs and improve its living conditions. The transition from hunter-gatherer to farmer, which began around 10,000 years ago at the end of the last ice age, is one of the most significant turning points in our history. In Mesopotamia, the cradle of this Neolithic revolution, the first farmers were able to take advantage of a favorable climate to initiate agricultural practices that would profoundly transform human societies. This domestication of plants, motivated by the search for subsistence and improved yields, was a foundational act, leading to the progressive selection of plant varieties over the generations.

Alongside this agricultural development, another essential quest began: that of understanding and exploiting the therapeutic properties of plants. From 4,200-year-old Sumerian clay tablets to Egyptian texts such as the Ebers Papyrus, we can trace the accumulated empirical knowledge that made plants a pillar of ancient medical practices. This knowledge has been passed down through the ages, right up to modern medicine, initiated by Hippocrates, where the study of medicinal plants has been enriched with solid scientific foundations. Today, this ancestral tradition is extended and amplified by technological advances, making it possible to manipulate the genome of plants to optimize their properties, as demonstrated by genetically modified organisms (GMOs).

However, in the era of modern science, herbal medicine and aromatherapy evoke ambivalent perceptions. Their "natural" nature is sometimes idealized, creating a simplistic opposition to synthetic drugs. However, their effectiveness is based, just like conventional treatments, on scientific evidence and rigorous studies. The totum of plants, this unique synergy of bioactive molecules, remains at the heart of therapeutic action mechanisms, illustrating the subtle interconnection between nature and science.

In addition, the cultivation and exploitation of MAPs are part of multiple issues affecting agronomy, economics, and pharmacology. In Occitanie, a region marked by the wine crisis, agricultural diversification offers an unprecedented opportunity. MAPs, due to the richness of local climates, have immense potential not only to supplement farmers' incomes but also to develop a sustainable local sector.

Finally, these efforts to revitalize the AMPs sector are part of a complex global context. Climate change, loss of biodiversity, and demographic pressure – which could lead the world population to exceed 10 billion people by 2050 – pose unprecedented challenges. In this context, MAPs not only represent an economic and therapeutic opportunity but also serve as a response to the quest for resilience in the face of environmental upheavals. Their preservation and enhancement are initiatives aimed at reconciling humans with their ecosystem by giving agriculture a role as the guardian of biodiversity and collective health [20, 22].

Thyme (*Thymus vulgaris*) is an AMP with multiple and ancient uses (Figure 15.1). Native to the Mediterranean regions, it is known for its antiseptic, anti-inflammatory, and antioxidant properties, thanks to active compounds such as thymol and carvacrol. These substances give thyme its therapeutic virtues, particularly for the relief of respiratory, digestive, and inflammatory disorders. Used since antiquity, it was employed by the Egyptians for mummification and by the Greeks to purify the air during ceremonies [23, 24]. In cooking, thyme is an essential herb in Mediterranean dishes,

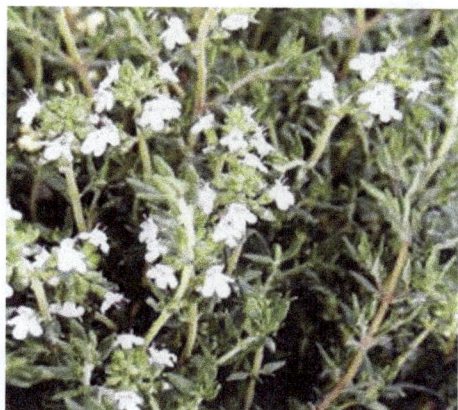

Figure 15.1: *Thymus vulgaris.*

enhancing the flavors of meats, vegetables, and marinades (Figure 15.2). It is also consumed in the form of herbal tea, often to relieve coughs or sore throats. In herbal medicine, its EO is used in aromatherapy for its purifying and relaxing effects. In addition, its ecological role is notable: it attracts pollinators and can protect other plants against certain parasites and diseases. Its cultivation is simple; thyme prefers well-drained soils and generous sunshine. Easy to maintain, it offers a sustainable contribution to cooking and natural health while beautifying gardens.

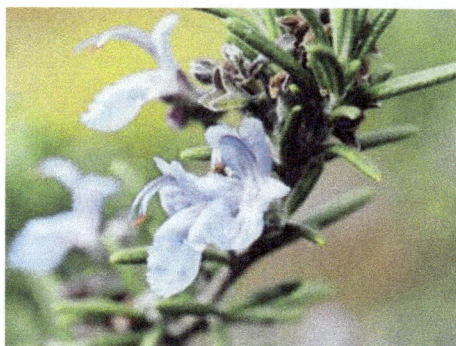

Figure 15.2: Rosemary (*Rosmarinus officinalis*).

Rosemary (*Salvia rosmarinus*), formerly known as *Rosmarinus officinalis*, embodies the very essence of the Mediterranean regions. This perennial shrub, belonging to the Lamiaceae family, is recognized for its medicinal and culinary properties, as well as its privileged place in cultural traditions. Its name, which reflects its natural environment on the Mediterranean shores, is often linked to memory and remembrance in the hearts of people.

15.2.3 Histoire et symbolism

Since ancient times, rosemary has been used for medicinal and ritual purposes. The Egyptians placed sprigs of rosemary in tombs to strengthen the souls of the deceased. During medieval epidemics, it was burned to purify the air. The plant is also famous for its role in the production of Hungarian Water, a preparation of alcohol and essences known for its rejuvenating effects in the fourteenth century. Shakespeare mentions it in Hamlet as a symbol of remembrance [25].

15.2.4 Medicinal properties

Rosemary is particularly rich in antioxidants, including rosmarinic acid, which protects cells against oxidative damage. It is also known to stimulate memory and cognitive functions, improve digestion and relieve mild gastrointestinal disorders, support liver function thanks to its detoxifying properties, and reduce fatigue while revitalizing the body. In herbal medicine, its EOs are used for their antiseptic, antifungal, and anti-inflammatory properties. However, precautions should be taken to avoid excessive use, especially in pregnant or breastfeeding women.

15.2.5 Culinary and practical uses

In cooking, rosemary brings a pronounced flavor, ideal for flavoring meats, vegetables, and sauces. Its leaves, often used fresh or dried, release an intense aroma. In the garden, this robust plant adapts well to well-drained and sunny soils, even resisting drought, although it is sensitive to excessive humidity [26]. Rosemary thus embodies a real bridge between ancient traditions and modern benefits, continuing to inspire uses and scientific research.

15.3 Generalities and characteristics

The species of lavender and lavandin (Figure 15.3) are identified by their genus name, *Lavandula*. Perennial shrubby plants with persistent foliage, lavender, and lavandin are heliophile species with very fragrant flowers, widely distributed in the south of France. The Latin name comes from lavare, which means "to wash" (its use dates back to the Roman era, when it was used to perfume laundry and baths). The optimal harvest period for extracting the EO is at full bloom, as the flowers concentrate most of the essence at that time.

Figure 15.3: Lavender and lavandin.

Since 2018 in France, and more particularly in Provence, a historically significant growing region for lavender and lavandin, a decrease in production has been observed due to several factors, including drought and high temperatures (which have a severe impact on plants by halting their development), as well as diseases and pests, including the gall midge. This change has resulted in a decrease in production and an upward trend in prices.

15.3.1 Lavender and lavandin: treasures of provençal nature

Lavender and lavandin, emblems of Provence, embody a centuries-old tradition associated with perfumery, natural medicine, and wellbeing. Although similar at first glance, these plants have botanical distinctions and unique uses.

15.3.2 Origin and history

Lavender, native to Mediterranean regions, has been used since antiquity. The Romans used it to perfume their baths and clothes. In the Middle Ages, its medicinal virtues were recognized: as a disinfectant and soothing agent, it was used to fight the plague and sanitize hospitals. Lavender became a key ingredient in perfumery in the eighteenth century, notably thanks to the city of Grasse, which developed its trade internationally [27].

15.3.3 Differences between lavender and lavandin

Fine lavender (*Lavandula angustifolia*), also called true lavender, grows naturally at high altitudes and produces a high-quality EO appreciated for its soothing, healing, and relaxing properties. It is often used in aromatherapy and skincare.

15.3.4 Herbs and traditions

A natural hybrid between fine lavender and spike lavender, it grows on lower ground. It is more productive in EO, but its scent is more intense and less subtle. Its oil, rich in camphor, is used mainly in the industry to perfume household and cosmetic products.

15.3.5 Benefits and uses

In health: Lavender EO is known for its calming, antistress, and antiseptic properties. Lavandin, which is more economical, is used for its decongestant and relaxing properties.

In perfumery: Lavender remains a staple of fragrances, thanks to its delicate floral notes, while lavandin is often preferred for its profitability.

In cooking and decoration: Lavender flowers bring a unique touch to desserts, while dried bouquets perfume interiors.

15.4 *Helichrysum italicum*

15.4.1 Choosing a nature-friendly herbal therapy

With its current lifestyle, humanity would need the equivalent of the resources of 1.7 Earths to sustainably meet its needs. Faced with the problem of resource depletion and the deadly relationship between humanity and its environment (Figure 15.4), in her book on Eco-responsible phytotherapy, the author emphasizes several key points. Firstly, the consideration of the capacities for self-healing, particularly the immune system's ability to function autonomously in many pathological cases (such as colds and viral infections), often does not require the use of basic therapy (except to alleviate symptoms), thereby contributing to spontaneous healing in a significant number of situations. Similarly, the dedicated role of the caregiver in the therapeutic relationship of trust established with the patient is highlighted, as the caregiver is keen to question the patient about potential imbalances in their lifestyle, which, as is often the case in modern societies, may be the root cause of pathological disorders linked to unhealthy habits and a consequent desynchronization of biological rhythms.

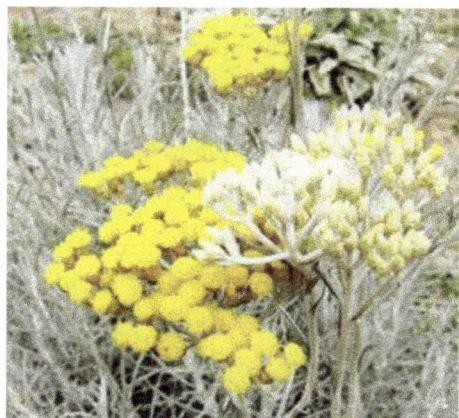

Figure 15.4: *Helichrysum italicum* dans son environnement naturel.

The first piece of advice will then be to return to a healthy way of life, including restful sleep, physical activity, stress management, and a diet aligned with individual metabolism. When medication is necessary, eco-responsible therapeutic practices should take into account distribution channels, in particular. Short supply chains, with traceable medicinal plant sources and simple plant extracts, are preferable to complex remedies with redundant activities [28]. We can thus find, in our latitudes, plants with benefits that are just as effective as those from another continent with the same indications.

Finally, consideration of the method of obtaining the therapeutic product and the preference for prioritizing a phototherapeutic use (which turns out to have fewer drawbacks related to production methods than aromatherapy with EOs) will be addressed. We will explore this further in the section devoted to the means used in the distillation of EOs.

15.5 Aromatic and medicinal plant (AMP) market

15.5.1 Agricultural rooms

Present at the departmental, regional, and national levels, they have been consular bodies since 1924, ensuring the representation of all economic agents in the agricultural, rural, and forestry sectors of France. Their role is both to represent the sector to public authorities and to assist farmers by supporting them in their development projects and improving their economic, social, and environmental performances. At the level of the plant active materials (PAM) sector, the Chamber of Agriculture of the

Occitanie region offers, for example, training in specialized techniques necessary for growing AMPs to farmers wishing to improve their skills as well as support in developing a commercial project based on technical and economic knowledge.

The cultivation of aromatic and medicinal plants is particularly relevant to farmers who wish to diversify their crops in order to generate additional income. However, this new activity first requires accepting a learning phase, which is essential for acquiring new knowledge, particularly understanding the requirements imposed by technical itineraries (working time, specific equipment). The chamber of agriculture specifically recommends considering, in the context of diversification into PAM, the establishment of at least two different species. This approach ensures a better distribution of the workload over the year (different harvest dates depending on the species cultivated) and helps mitigate certain risks, particularly climatic hazards that could impact one species more than another or certain economic risks linked to market fluctuations.

15.5.2 The national conservatory of perfume, medicinal, aromatic, and industrial plants (CNPMAI)

Founded in 1987, this nonprofit association is a valuable resource for producers, pickers, laboratories, and industrialists in the sector. Acting upstream of the production of PAM, it has set itself the mission of managing and promoting these plants while working to preserve and safeguard genetic and natural heritage. As an association governed by the 1901 law, it offers a catalog collection of plants and cultivars, providing producers with a valuable source of plant material adapted to their needs [29]. In addition, the label "Organic Agriculture" (AB) has been in force since 2002 at the Conservatoire National des Plantes à Parfum, Médicinales, Aromatiques et Industrielles (CNPMAI).

15.5.2.1 Cooperative societies

Cooperative societies are agricultural companies, positioning themselves in the collection of raw materials and playing the role of transactional intermediaries within the market, depending on the type of finished product: EO or dried plant. Their interest is, in particular, to group together significant quantities of PAM to offer a satisfactory market volume to buyers and thus to provide a certain market fluidity.

A cooperative is formed from the union of several producers who wish to pool their practices, particularly in terms of production and packaging. These activities, often demanding in terms of tools and investments, benefit from collective management that also facilitates the processing, storage, and marketing of products. By pooling their strengths, producers can negotiate more competitive purchase prices for their inputs, such as the supply of fertilizers. Cooperatives are generally managed by an elected producer and structured around a board of directors. Conversely, a private law company is managed by one or more associated individuals. These structures, like cooperatives, play a key role as supply platforms for buyers. They also enable rigorous quality management, particularly for the delivery of production batches that comply with standards and specifications, thanks to specific controls on the batches.

15.5.3 Local distilleries

An essential link in the production and processing chain, these distillation units specializing in AMPs allow, thanks to their geographical proximity, the management of harvests intended to be distilled in a very short time interval after mowing. The time factor is indeed very often a critical factor in the field of EOs; freshly harvested plants exposed to the open air can, in some cases, begin a fermentation process just a few hours after picking. A short time between transport and the distillation process will therefore limit this risk and guarantee a quality final product. Note here that drying the plant may, however, be necessary when the extraction of the EO is not immediately possible (if there is, e.g., a delay between harvesting and distillation), in which case the EO from the distilled dry plant matter will have a different yield and properties compared to a freshly harvested plant product. The large-scale distillation process requires specialized technical skills. Authorized distilleries can join the Union of Professionals of Perfume, Aromatic and Medicinal Plants through their departmental union.

15.5.4 Final product and quality controls

At the end of their growing cycle, the various parts of plants – flowers, leaves, stems, bark, or roots – are carefully harvested and transformed into a wide array of products, each tailored for specific purposes such as therapeutic, cosmetic, food, or even biocidal applications. This transformative process is guided by the necessity to ensure the highest quality and safety standards for the end consumers. A critical aspect of this process is the meticulous control of both the characteristics of the final products and the production intermediaries, as this guarantees their safety and efficacy. However, this requirement is particularly significant due to the inherent biological variability of the raw plant materials. The chemical composition and yield of the harvest can vary considerably depending on the timing within the plant's vegetative cycle. For instance, EO concentrations often peak during the flowering stage, making this period crucial for optimizing yield and quality. Similarly, other plant parts, like stems and leaves, are best harvested in the spring when they are rich in active compounds, whereas roots, which store nutrients and bioactive molecules, should ideally be collected in the fall. In addition, harvest timing is not only a matter of tradition but also a strategic decision rooted in both scientific observation and practical experience. By selecting the optimal harvest season for each plant part, producers can ensure the highest quality raw materials, which in turn support the development of safe and effective end products. This attention to detail underscores the importance of aligning agricultural practices with the biological rhythms of the plants, ultimately harmonizing the needs of producers, consumers, and the environment.

15.6 Herbal medicine

One of the operations necessary to obtain the final product for use in phytotherapy is the implementation of a drying and sorting process to optimize the conditions for preserving the plant, to maintain its properties between the time of harvest and its use, and to eliminate unwanted constituents (such as branches and foreign elements). In herbalism and on long circuits, it is generally leaves (more rarely flowers) that are sought. The majority of companies responsible for the outlets of the sector and for marketing expect the producer to provide a product that is already dried and sorted, with the price being directly dependent on the quality of the sorting carried out. This sorting does not necessarily require the use of high-tech equipment [30–33].

15.7 Aromatherapy

From the Greek word "aroma," meaning treatment or care, aromatherapy is a therapy that uses essences (natural secretions of certain plants), EOs, and hydrosols extracted from the aromatic parts of medicinal plants for therapeutic purposes. More specifically, an EO is composed of a complex mixture of plant origin, liquid at room temperature, volatile, fat-soluble, and grouping the alcohol-lipid-soluble compounds of the plant. The *European pharmacopoeia* 11th edition defines an EO as an odorous product, generally of complex composition, obtained from a plant drug, botanically defined, by steam distillation, dry distillation, or an appropriate mechanical process without heating. EOs are separated from the aqueous phase, if necessary, by a physical process that does not significantly modify their composition. EOs are considered herbal products under French regulations by the National Agency for the Safety of Medicines and Health Products (ANSM). Some of them can only be dispensed in pharmacies. In such cases, they must meet the pharmaceutical quality standards of the Pharmacopoeia (European or French). Where applicable, the exact scientific name of the EO, as well as its chemotype – and therefore its chemical composition – must be described.

EOs not covered by the pharmaceutical monopoly are available for sale over the counter and can be distributed through different channels (pharmacists, specialist stores). They must not mention a therapeutic indication since their composition is not guaranteed with regard to their potential therapeutic effect. The *European Pharmaco-poeia* specifies several extractive processes for obtaining EOs.

For therapeutic use: By steam distillation, the EO is produced by passing steam through the plant raw material within a distillation installation. This steam is of external origin, generated by water brought to a boil below the raw material, or by water brought to a boil in which the plant material is directly immersed (Figure 15.5). The water vapor then condenses, with the EO being separated from its aqueous part by decantation [34].

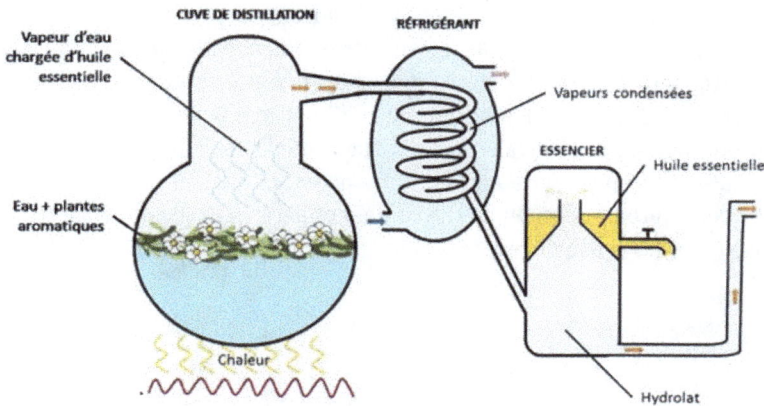

Figure 15.5: Schematic diagram of hydrodistillation.

Note that this extraction process requires heating conditions, which can involve certain limits for extracted constituents that are sensitive to heat (possible molecular re-arrangements of the structure, leading to modifications in the chemical nature of the EO compared to the original natural essence of the plant). In total, the result of the distillation will be an EO composed of a mixture of molecules naturally synthesized by the plant and molecules resulting from a rearrangement of structure due to the distillation process.

EOs can be obtained by different methods adapted to the nature of the raw material and the desired results. Dry distillation consists of heating the plant material, such as stems or bark, at a high temperature without adding water or water vapor. This process allows the EO to be extracted directly through the action of heat. Another method, called cold expression, is a mechanical process without heating. It is mainly used for citrus fruits, where the EO is extracted from the pericarp and then separated by a suitable physical process. Finally, rectification is a specific distillation process, generally carried out under vacuum. This step, which occurs after the initial extraction, allows residual water or other undesirable substances to be eliminated or the chemical composition of the EO to be modified to better meet specific quality or usage criteria. In general, it should be rightly considered that the process of obtaining an EO is a very energy-intensive process, with a low yield (very large quantities of fresh plant matter are used to obtain small quantities of EO) and often requires large volumes of water (particularly in the steam distillation process). Apart from the structural changes mentioned above, the nature of the solvent used (water) has certain limitations, and compounds may, for example, not dissolve in the aqueous fraction or be difficult to entrain within the tank due to a high molecular weight. For all these reasons, it is legitimate to question the relevance of the systematic use of aromatherapy. The distillation process requires a significant amount of energy to heat the plant material, as well as a large volume of water to produce the necessary steam. Other equally effective therapeutic alternatives can offer a more moderate consumption of resources. By nature, an EO is a powerful concentrate of active compounds, although it contains only a fraction of the total volatile molecules present in the plant's totum. These highly concentrated medicinal substances have intense pharmacological activity, requiring careful handling, especially in a pharmaceutical consulting setting. It is essential to prevent inappropriate use and limit the potential risks of toxicity, which must be carefully considered, especially in the context of personalized support if used orally. The dosage of an EO is usually expressed in standardized drops.

15.8 Production/collection

The cultivation and exploitation of AMPs have seen a significant increase over the past two decades both in developed and developing countries. Today, more than half of the world's production of EOs comes from these plants. For some countries, such as Morocco, domestic production is entirely export-oriented, making this sector a key economic pillar. In contrast, in countries such as China, India, or Indonesia, a significant proportion of production is consumed locally, reflecting growing domestic demand for these products. Wild AMPs, such as thyme, rosemary, or carob, grow naturally in large areas of national forests. Their harvesting and exploitation are supervised by public authorities responsible for water, forests, and collective land. These authorities oversee the marketing of these products, which is mainly conducted through tenders, thus ensuring organized and controlled management of this natural resource. Production, whether spontaneous or cultivated, is difficult to assess due to the lack of detailed regional statistics. However, export data provided by EACCE and the few studies available on the sector show that many of these herbs have a significant market. These figures 15.6 and 15.7 indicate growing demand and substantial economic potential for these products despite the lack of accurate information on local production.

The lack of reliable national and/or regional statistics, as well as the irregularity of production – mainly from spontaneous stands – explains this lack of information on the AMPs sector [35].

Figure 15.6: Sector for aromatic plants in the mountains.

Figure 15.7: Aromatic plant area on the depression.

The only available estimates are from the High Commissioner for Water, Forests, and Combating Desertification (HCEFLCD), which oversees the exploitation of forest areas where most wild MAP stands are located. In addition, EACCE, which is responsible for export figures, also provides data on this market, although limited to this specific aspect of the value chain [36].

15.9 Results and discussion

15.9.1 Production of PAM in dried leaves

This region is home to a rich diversity of plants, including rosemary, thyme, oregano, and carob, along with their derivatives. These natural treasures, however, face growing threats from unsustainable harvesting and collection practices, which jeopardize their future. Furthermore, the methods used for drying and cleaning these plants require significant improvement to preserve their quality and value. On the international stage, Moroccan rosemary has the potential to stand out as a competitive product. While Algeria presents a challenge in terms of pricing, other major suppliers, such as Spain and France, offer their products at significantly higher costs, leaving room for Morocco to carve its niche. As for other dried herbs like thyme and oregano, they originate from wild and natural stands scattered across various regions of the kingdom [37]. These plants not only embody the richness of Morocco's biodiversity but also hold promise for contributing to the sustainable development of local communities if managed wisely.

15.9.2 Evolution of Moroccan exports of essential oils

Tables 15.1 and 15.2 summarize the evolution of Moroccan exports of EOs and aromatic extracts from 1999 to 2003. They represent PAMs subjected to extraction (distillation or solvent extraction) and do not include products marketed fresh or after drying in the herbal, aromatic, and food markets.

Table 15.1: Exports by weight of essential oils (1999–2003).

Product	Poids in kg					
	1999	2000	2001	2002	2003	Moy
Rosemary essential oil	56,161	25,640	62,244	77,945	71,300	58,658
Myrtle essential oil	1,322	191	0	0	0	303
Eucalyptus essential oil	0	0	404	5	65	95
Other essential oils	549,173	270,414	389,845	465,893	307,793	396,624
Total HE (kg)	606,656	296,245	452,493	543,843	379,158	455,679

Of all the EOs, rosemary occupies the first place with an annual average of 58 tons. Note that the official statistics group "other essential oils" together with oils from more classic products such as oregano, bay leaf, and thyme.

Exports of EOs from AMPs have experienced significant growth. In fact, their value has doubled in the span of three years, from 56 million dirhams in 2000 to

Table 15.2: Exports in value of essential oils (1999–2003).

Product	Poids (kg)						Average price (Dh/kg)
	1999	2000	2001	2002	2003	Moyenne	
Rosemary essential oil	10,801	4,122	11,849	22,939	17,812	13,505	230.23
Myrtle essential oil	611	83	0	0	0	139	458.81
Eucalyptus essential oil	0	0	35	8	13	11	118.93
Other essential oils	82,459	51,805	92,954	102,809	94,655	84,936	214.15
Total HE (KDh)	93,872	56,010	104,838	125,756	112,480	98,591	216.36

112.4 million dirhams in 2003. This development reflects not only the increase in demand for these products but also the sector's continued adaptation to fluctuations in international markets. A thorough analysis of production trends and price variations of EOs between 1999 and 2003 allows us to grasp the subtleties of the dynamics that marked this period [38]. By studying these fluctuations, it becomes possible to better understand the forces that govern the supply and demand of these natural products. This analysis extends to a range of EOs, such as rosemary, myrtle, eucalyptus, as well as other oils, each with its own specificities. Each type of EO has its own characteristics that reveal a dual reality: the evolution of production, of course, but also the effects of price variations on the market. Indeed, these products, derived from the natural wealth and artisanal expertise of producers, are subject to the influences of global economic fluctuations while remaining anchored in traditional practices that have spanned the ages. Thus, this period of growth in EO exports reflects the adaptability of a sector that, while being part of ancestral production processes, must also deal with the contemporary challenges of the global market. The diversity of EOs and their unique properties continue to captivate interest, both for their therapeutic value and for their applications in various industries, ranging from cosmetics to the food industry [39].

15.9.3 Rosemary essential oil: a remarkable development

Rosemary EO experienced a particularly interesting development between 1999 and 2003. Its production recorded a substantial increase from 10,801 kg in 1999 to 17,812 kg in 2003, an increase of more than 60%. This growth not only reflects a growing demand for this oil but also likely sustained efforts to improve production, quality, and efficiency of harvesting methods. Rosemary, used for its medicinal and aromatic properties, has seen its popularity grow due to its many applications in the fields of herbal medicine, cosmetics, and even cooking.

However, this increase in production has not been accompanied by total price stability. The average price of rosemary EO has fluctuated over the years, reaching a peak of Dh230.23/kg in 2003, making it the most expensive oil among those studied. This price fluctuation may be linked to various factors, such as climatic conditions, fluctuating demand, or market strategies adopted by producers. Nevertheless, despite these variations, rosemary oil has maintained a high average price throughout the period studied at Dh230.23/kg. This reflects not only its leading position in terms of production but also the perceived value of its properties in the market.

15.9.4 Myrtle essential oil: low production but high price

As for myrtle EO, the analysis shows a different dynamic. Although production was relatively low, peaking at 611 kg in 1999, it then ceased in the following years. This low production volume can be attributed to several factors, such as specific growing conditions or a more restricted demand in the market for this particular oil. Despite this low production, myrtle oil had the highest average price among all the oils studied, with an average price of Dh458.81/kg. This disparity between low production and high price could be explained by the rarity of the oil and the specific properties it offers, particularly in the fields of traditional medicine and perfumery. The high price may also indicate that, although the demand for myrtle is more modest, those seeking this oil are willing to pay a premium price to obtain it. This dynamic illustrates how the rarity and perceived value of a product can directly influence its price, regardless of the volume of production. Over the entire period, the average price of myrtle EO remained stable at Dh458.81/kg, thus highlighting the uniqueness of this oil in the EOs market.

15.9.5 Eucalyptus essential oil: a limited market and a constant price

Eucalyptus EO has even more specific characteristics. Eucalyptus oil production remained extremely low between 2001 and 2003, fluctuating between 8 and 35 kg. This indicates that eucalyptus cultivation and oil distillation were not a priority for producers or that demand for the oil was insufficient to justify higher production volumes. In contrast, the average price of eucalyptus EO remained constant over the period at Dh118.93/kg, making it the cheapest oil among those studied. This price constancy could be related to low demand and limited supply, where producers were perhaps constrained in their ability to increase production, leading to price stability in a niche market. The low production volume and constant price may also suggest that eucalyptus oil did not experience increased demand during the study period. However, it remains a useful product in some sectors, such as aromatherapy and pharmaceutical manufacturing, justifying its continued presence in the market despite moderate production levels.

15.9.6 Other essential oils: stable production and fluctuating price

In addition, the group of other EOs, encompassing a variety of products, showed a different trend. The production of this category of oils was the most stable and highest over the years, ranging from 51,805 kg in 2000 to 102,809 kg in 2002. This high production suggests that EOs from this category are widely used and in demand, which may include oils such as lavender, basil, or chamomile, commonly used in perfumery, cosmetics, and traditional medicine. Despite high production, the average price of these oils remained relatively stable at around Dh214.15/kg, although fluctuations were noted. This could be due to the diversity of oils in this category, each of which has specific demands, slightly influencing their price according to market trends and consumer preferences. However, compared to the other oils studied, these oils remained more accessible and maintained a certain constancy in their price over the years.

15.10 Fluctuations and trends

Overall, total EO production peaked in 2002, with a total of 125,756 kg. This shows that, despite the fluctuations in production for some oils, the EO industry continued to grow. The average price for all EOs over the period was Dh216.36/kg, a figure that reflects a certain general stability in prices, although some oils, such as myrtle, maintained a much higher price due to their rarity and perceived value. Finally, the analysis of EO production and prices between 1999 and 2003 highlights some interesting market dynamics. While some oils, such as rosemary, have seen significant growth in production, others, such as eucalyptus, have shown limited production and consistent prices. Price fluctuations, particularly for rosemary oil and myrtle oil, are also indicative of the influence of demand, scarcity, and the perception of value of these oils in the market. This study provides valuable insight into the economic trends in the EOs sector and can help to better understand the factors that influence both production and pricing in this ever-changing industry.

15.10.1 Application of medicinal and aromatic plants

The Chaouia-Ouardigha region, located in central Morocco, is a land of great natural wealth, particularly in terms of plant biodiversity. This region, which benefits from a varied climate and soil, offers a terrain conducive to the growth of a wide variety of AMPs, used for centuries by local populations [40]. These plants, often at the heart of traditional practices rooted in ancestral customs, constitute a precious resource for the wellbeing of the inhabitants, both in terms of health and agriculture. The applications of MAPs in this region are multiple, ranging from traditional medicine to the modern cosmetics and pharmaceutical industries.

15.10.2 A plant treasure in the service of traditional health

AMPs occupy a central place in the traditional medicine of Chaouia-Ouardigha. The inhabitants of this region use a variety of plants to treat a wide range of ailments, ranging from digestive disorders to respiratory infections, including joint pain and skin conditions. A recent ethnobotanical census conducted among local herbalists revealed that four major plants are particularly used: pennyroyal (*Mentha pulegium*), ragweed (*Dysphania ambrosioides*), dill (*Ammi visnaga*), and mountain fennel (*Ammodaucus leucotrichus*). These plants, derived from ancestral knowledge, play an essential role in the daily management of health in this region. Pennyroyal is commonly used to treat digestive disorders, such as bloating and colic. It is also recognized for its antispasmodic and analgesic properties. According to the results of a survey conducted in 2023, 85% of local herbalists reported that they use this plant for its medicinal properties, particularly in the form of infusions. Similarly, ragweed is used in the treatment of respiratory conditions, such as coughs and bronchitis, thanks to its expectorant properties. Traditional practitioners in the region report that dill is often used to treat colic and urinary disorders, while *Ammodaucus leucotrichus* is mainly used for its anti-inflammatory and analgesic effects.

15.10.3 Traditional application and preparation methods

The methods of preparing these medicinal plants vary depending on the ailments to be treated, but infusions and decoctions remain the most common. A statistical study conducted among 50 herbalists in the region in 2023 revealed that 72% of respondents mainly use infusions, followed closely by decoctions (24%) and EOs (4%). Herbalists often collect fresh plants, drying them to preserve them before using them to prepare remedies. The use of EOs, although less widespread, is experiencing increasing interest due to demand from the cosmetics and natural health markets.

Local herbalists have also highlighted the uses of these plants for preventative purposes, such as boosting the immune system or improving digestion. For example, pennyroyal decoctions are often given to children to prevent stomach aches and aid digestion. These daily applications demonstrate a preventative approach that is deeply rooted in local culture.

15.11 Medicinal and aromatic plants: a potential for sustainable agriculture

In addition to their applications in traditional medicine, AMPs also represent an asset for sustainable agriculture in the Chaouia-Ouardigha region. The growth of national and international markets for EOs and plant extracts offers a considerable opportunity for local farmers. Indeed, several studies have shown that MAPs can be grown profitably while having a low environmental impact. The introduction of environmentally friendly agricultural practices, such as organic farming and agroecology, makes it possible to maximize the yield of these crops while preserving local biodiversity. According to a report published in 2022 by the National Institute for Agricultural Research (INRA), the cultivation of rosemary, lavender, and mint increased by 30% in the Chaouia-Ouardigha region between 2015 and 2020. This phenomenon is accompanied by a growing interest in the production of EOs due to the rising global demand for these products. In 2023, a market study estimated that the production of EOs in this region could generate additional income of around 50 million dirhams, thus contributing to the local economy. The results of surveys conducted among local producers show that nearly 60% of farmers consider the cultivation of medicinal plants to be a profitable activity. However, the lack of training on good agricultural practices and production management remains a major challenge for the expansion of these crops on a large scale. Awareness-raising and training programs in agroecology and sustainable natural resource management are therefore essential to enable farmers to take full advantage of the potential of AMPs.

15.11.1 A Promising future for medicinal and aromatic plants

The MAPs of the Chaouia-Ouardigha region represent an invaluable resource that, when exploited sustainably and scientifically, could have profound repercussions both economically and health-wise. Their use in traditional health practices is a precious heritage that deserves to be preserved and valued. In addition, medicinal plant agriculture offers promising prospects for the development of a sustainable and profitable agro-industrial sector, thus supporting local economies and contributing to improving the living conditions of rural populations [41, 42]. It is now crucial to invest in scientific research, education, and training of local stakeholders to guarantee the future of this plant wealth and take advantage of the opportunities it offers.

15.11.1.1 Amaghouss

However, this production encounters a significant limitation: the oasis suffers from a lack of sour pomegranate trees, which are essential for increasing the production volumes of amaghouss (Figure 15.8). Although this shrub is of particular interest for the manufacture of this remedy, it is unfortunately not very popular with local producers [43–45]. Even though they recognize its medicinal properties, the sour pomegranate tree is considered a harmful tree. Unlike its cousin, the sweet pomegranate, which is carefully cultivated in agricultural plots for its fruit production intended for consumption, the sour variety, which generally grows on the outskirts of fields, is often eliminated. Farmers, therefore, prefer to concentrate on crops that they consider more profitable and easier to manage.

Figure 15.8: Amaghouss (1/2 L).

However, despite the relatively accessible outlets at moderate prices (around Dh50/L) that amaghouss allows, this product does not yet seem attractive enough to justify its expansion. The reality of its production is indeed restrictive. To make 1 L of amaghouss, it takes no fewer than 7 L of sour pomegranate juice, a supply of firewood for a 24-h cooking time, and a significant workforce to supervise each step of this complex preparation. These material and human efforts make the production of amaghouss unprofitable for many producers, especially when one takes into account the work and time that must be devoted to it [46].

Despite these difficulties, there is a particular category of people who engage in the preparation of amaghouss: the wives of farmers. These women, often responsible for this task, devote a large part of their time to the preparation of this tradition based on sour pomegranates. This activity takes them hours, sometimes an entire day, and can seem secondary, even superfluous, in the eyes of their husbands, who view the production of honey or other agricultural products as more profitable priorities. However, these women play a central role in the transmission of this ancestral know-how.

In Tagmout, women carefully peel the sour pomegranates and crush them to extract the juice, which they then boil for about 24 h until it turns black and loses about 85% of its initial volume. The resulting liquid is a traditional folk remedy, renowned for its medicinal properties and widely used. Although it is only produced in a few specific regions such as Tagmout, Issafen, and Tinkhit, amaghouss has gained national recognition. Consumers come from all regions to obtain this precious liquid. Its reputation, based on its effectiveness and widely shared by word of mouth, has allowed demand to increase considerably. Amaghouss has thus become a prized product, the demand for which remains very high despite the difficulties of production.

15.11.1.2 Honey

Although the arid climate limits the expansion of beekeeping, this activity can find its place among the productions of mountain oases. Until now, actions aimed at developing honey production have remained limited. Beekeepers have considered grouping together to organize the sector, but above all to benefit from support and technical assistance. Although this idea has germinated in the minds of a few, it has not yet taken concrete form. It seems that the people concerned have not shown any real motivation to make this project a reality. Thus, only one beekeeper has been able to improve his techniques significantly in recent years, particularly thanks to a study trip organized by the DPA and the acquisition of "modern" hives at preferential prices.

15.11.1.3 Proposal for development actions

The development of honey production in mountain oases seems to be of great importance to us. This production, indeed, presents several advantages for local populations.

First of all, the climate at altitude is particularly favorable for beekeeping. Although droughts can threaten bees and limit production, the diversity and quantity of honey plants in these areas represent a real asset for quality production. In addition, the honey produced in these regions is recognized for its exceptional quality and is often considered a genuine remedy. Despite relatively modest quantities per order, overall demand remains very high. These elements explain the high price of honey while ensuring good market prospects. Even if production volumes were to increase, the means necessary for this activity are perfectly adapted to the realities of the Tagmout oases. The necessary financial investment is reasonable, and the permanent workforce, particularly that of women, is present and available to support this activity. In order to promote the expansion of this sector, farmers have expressed their needs and identified certain limitations [47]. Thus, it seems crucial to take into account three essential aspects to support the development of honey in the region: improve production by providing suitable technical support, promote the product by collectively structuring the profession, and preserve and promote the traditional know-how that makes this production rich.

15.11.1.4 Technical monitoring of beekeepers

Jusqu'à présent, l'apiculture n'a pas bénéficié d'un soutien technique important. Contrairement à d'autres régions du Maroc, la DPA a montré peu d'intérêt pour la production de miel. En réalité, Tagmout présente moins d'avantages que beaucoup d'autres régions du nord du pays pour une production de miel à grande échelle. Que le contexte dans lequel se trouve Tagmout lui permette ou non de bénéficier du soutien de la DPA, il apparaît pertinent d'organiser les producteurs (voir ci-dessous) afin d'améliorer les techniques apicoles locales. L'organisation des producteurs constitue en effet un moyen privilégié d'échanger des informations: savoirs, expériences, techniques, etc., permettant de tester et de diffuser des pratiques apicoles encore peu connues à Tagmout. Prenons l'exemple de l'élimination des futures reines excédentaires (pour limiter l'essaimage). Il s'agit d'une technique simple, sans investissement ni apprentissage complexe – consistant à écraser les larves des reines. Il n'est pas nécessaire de faire appel à un technicien pour diffuser une telle technique, car si un groupe professionnel local existe, cette méthode se propagera naturellement au sein de la communauté.

15.11.1.5 Support the organization of producers

Producers have clearly expressed their desire to organize the honey sector. We were all the more surprised by the interest they have shown in uniting – not only the "honey professionals" but also small producers with only one or two hives. In terms of development, we tend to focus on the economic aspect of such a grouping (the market for quality honey). However, it seems that the objective pursued by farmers is, above all, the improvement of their techniques. As mentioned above, the grouping of producers is the best way to create a discussion forum capable of circulating useful information to improve the management of hives. However, a group with an economic vocation will naturally be a driving force for the dissemination of techniques. It would be interesting, first of all, to clarify the expectations of beekeepers. The type of organization to be set up and its objectives can then be defined. It can be envisaged that this will be a group focused on the marketing of the product. The market is dynamic, and it is appropriate to promote this aspect as much as possible. This involves ensuring customer satisfaction while improving sales methods. The interested group will be able, if it wishes, to better target its customers, adapt part of its technical itinerary, and so on. But what seems most obvious is the promotion of the finished product [48]. Whether we think of a label or a designation of the Tagmout terroir, it is clearly essential to improve the packaging. A honey of this quality, often sold in small quantities, would certainly benefit from being sold in small pots. In the current context, it is very likely that it will be possible to sell 200 g pots labeled at around Dh100. A general box on the packaging of the products can be found in (Section 15.12).

15.12 Morocco, a traditional supplier to the world market

The AMP sector has experienced exponential growth in recent years, fueled by an ever-increasing international demand for these plants and their derivatives. This expansion is largely due to the rise in industrial demand, particularly from laboratories, cosmetics manufacturers, and the agro-industry, which are seeking to exploit the unique properties of AMPs in many areas. The applications of these plants have multiplied, ranging from pharmaceutical research to skin care products, food flavorings, and dietary supplements, thus strengthening their presence in the global market. Global AMP production is dominated by three major groups of countries, each with a specific role to play in this international value chain. First, developing countries with large domestic markets and diverse assets that allow them to position themselves as world leaders for certain species. Among them, China, India, and Indonesia are key players not only because of their rich biodiversity but also thanks to traditional agricultural systems that have proven themselves over time. Then, industrialized countries, which produce certain species in large quantities, thanks to cutting-edge technologies, particularly in the areas of processing and valorization of PAMs. These nations use modern processes to maximize the yield and quality of their products, thus meeting the demands of the international market while seeking to differentiate themselves through the sophistication of their production methods.

Finally, developing countries, often with low labor costs, are turning to the export of products from wild harvesting. Thanks to an abundant biomass, these nations exploit real economic potential by taking advantage of the richness of their natural flora. Their production capacities are often characterized by simpler but efficient methods, which utilize local resources without requiring heavy technological investments. Morocco has firmly established itself as a major player in the global market for PAMs, securing its position as the 12th largest exporter worldwide. This distinction is no small feat, reflecting both the country's rich natural biodiversity and its capacity to sustainably harness these resources for international trade. The role of Morocco in this market is crucial, with 75% of its production of PAMs dedicated to export, making it a key supplier to a variety of industries across the globe [49].

The distribution of Morocco's PAM exports is notably diverse, underscoring the versatility and wide range of applications these materials have. The largest share, approximately 60%, is directed toward the food sector, where PAMs are essential for the production of products such as carob, various spices, and aromatic substances. These ingredients not only enhance the flavor and texture of food but also add valuable nutritional and medicinal benefits. Carob, for instance, is a natural sweetener that has gained popularity as a healthier alternative to chocolate, particularly in the form of carob powder, which is also used in confectionery and baking. Morocco's aromatic spices, including cumin, saffron, and paprika, are highly sought after for their unique flavors and health benefits, contributing to the country's status as a leading exporter of high-quality food products.

The second-largest portion of Morocco's PAM exports, 35%, is allocated to the perfumery and cosmetics industries, where the natural and aromatic properties of PAMs are highly prized. Morocco's rich cultural heritage and the cultivation of aromatic plants such as roses, lavender, and argan trees have positioned the country as an important source of EOs and fragrances used in the production of perfumes and cosmetic products. Argan oil, in particular, is celebrated for its moisturizing and antiaging properties, making it a key ingredient in many skincare products. The demand for natural and organic beauty products has been growing globally, with consumers increasingly turning to ingredients like those provided by Morocco's PAM exports, (Tables 15.3 and 15.4) known for their purity and effectiveness. Morocco's standing in this market is bolstered by its expertise in cultivating these plant materials, which are often harvested using traditional, sustainable methods.

A smaller yet significant proportion – 5% – of Morocco's PAM exports is dedicated to medicinal applications. Many of the plants cultivated for PAM production possess therapeutic properties that are recognized and valued in the pharmaceutical industry. For centuries, Moroccan herbs and plants have been used in traditional medicine, and the country continues to be a supplier of natural remedies that contribute to modern healthcare. From the healing properties of argan oil, known for its antiinflammatory effects, to the medicinal benefits of certain spices and herbs, Morocco's natural resources are sought after for their ability to support health and wellbeing. These plant materials are used in various forms, including extracts, oils, and infusions, and they are incorporated into everything from over-the-counter medications to herbal supplements.

The diverse applications of PAMs reflect the broader trend toward sustainability and the growing demand for natural, plant-based alternatives across multiple industries. Morocco's ability to meet this demand, with products that are harvested in an environmentally responsible manner, ensures that the country will continue to play a significant role in the global supply chain for plant-based materials. Furthermore, Morocco's rich biodiversity and centuries-old agricultural practices contribute to the unique qualities of its PAMs, which are prized for their purity and potency. Finally, Morocco's position as the 12th largest exporter of PAMs is a testament to the country's ability to cultivate and supply a wide array of plant-based products that are essential to various industries worldwide. With a focus on the food sector, perfumery and cosmetics, and medicinal applications, Morocco continues to be a vital source of high-quality, natural materials. The country's role in the global market for PAMs is likely to expand further as more industries turn to sustainable, plant-based ingredients in response to growing consumer demand for natural products. Through its continued investment in sustainable agriculture and its deep-rooted knowledge of plant cultivation, Morocco will remain at the forefront of this dynamic and increasingly important global market.

Table 15.3: Evolution of PAM exports in volume between 2014 and 2019 (in tons).

2014	12,386
2019	46,000

Table 15.4: Evolution of exports of essential oils, extracts, and perfumes in volume between 2014 and 2019 (in tons).

2014	350
2019	687

Over the last decade, Moroccan exports of PAMs have experienced remarkable growth, underscoring the country's increasing prominence in the global trade of natural plant-based products. The volume of PAM exports surged from 12,386 ton in 2014 to an impressive 46,000 ton in 2019, reflecting a compound annual growth rate (CAGR) of 30%. This growth is a testament to Morocco's expanding role in supplying high-quality PAMs to various international markets, driven by both the demand for natural ingredients and the country's rich biodiversity and agricultural expertise.

A significant share of Morocco's exports of AMPs goes to the European Union (EU), which remains the leading destination for these products. Indeed, around 60% of Morocco's total AMP exports are absorbed by this region. This strong demand reflects the EU's growing preference for natural ingredients of plant origin, whether in the food, cosmetics, or pharmaceutical sectors [50]. This phenomenon is largely fueled by a change in consumer behavior, which increasingly favors sustainable and organic products. Moroccan exports include both raw materials such as carob, spices, and aromatic substances as well as processed products such as EOs and plant extracts. These products, appreciated for their quality, also benefit from Morocco's reputation for traditional and ecological production methods, which are perfectly in line with the EU's values of sustainability and respect for the environment. However, to better understand this dynamic, here is a summary table of the distribution of Moroccan PAM exports to the EU. The growth in the export of EOs, extracts, and perfumes mirrors the broader expansion of the PAM export market. Exports in this category have risen from 350 ton in 2014 to 687 ton in 2019, marking a CAGR of 14.5%. This increase highlights the global demand for natural aromatic products, especially as consumers seek out ingredients that are both effective and sustainably sourced. Morocco, known for its cultivation of aromatic plants, such as rose, lavender, and argan, has positioned itself as a leading supplier of EOs and perfumes, which are integral to the cosmetics and fragrance industries [51]. The country's capacity to supply these high-demand products, which are prized for their purity and therapeutic properties, reflects its deep expertise in the cultivation and extraction of plant-based materials.

Exported product	Proportion of exports (%)
Raw materials	40
Processed products	60
Main destination	60

This table illustrates the extent of the position occupied by Moroccan exports in the European market and highlights the most sought-after products as well as the characteristics that give them added value in this demanding market. Morocco, by focusing on quality and environmentally friendly methods, responds to European demand that is increasingly concerned about the origin and sustainability of the products it consumes.

This impressive growth trajectory of Moroccan PAM exports is not only a sign of the country's expanding footprint in international markets but also demonstrates the increasing value placed on natural, plant-based products globally. As the demand for sustainability and natural ingredients continues to rise, Morocco's ability to leverage its unique agricultural resources, combined with traditional practices and modern innovation, positions the country as a key supplier in this thriving market. The EU's continued absorption of these exports, coupled with the global expansion of the natural health and beauty sectors, suggests that Moroccan PAM exports will continue to see strong growth in the coming years.

15.13 Conclusion

In this study, herbal medicine is presented as a therapeutic technique that has existed for thousands of years, rooted in ancestral traditions, and passed down from generation to generation. Our ethnopharmacological study, conducted in the Chaouia-Ouardigh a region, highlighted the crucial importance of the use of AMPs by the local population. This work revealed the wealth of traditional knowledge in herbal care, knowledge that continues to play a key role in folk medicine. We chose to focus on four of the most frequently used plants in the local pharmacopoeia: *Mentha pulegium* (L.), *Dysphania ambrosioides*, *Ammi visnaga*, and *Ammodaucus leucotrichus*. These plant species, widely used in the treatment of digestive and respiratory conditions, as well as to relieve renal colic and act as antipyretics, demonstrate the effectiveness and diversity of remedies from nature, adapted to the needs of the population. Our results show that the leaves of the studied plants are the most commonly used parts, while infusions and decoctions are the most widespread preparation methods in local traditional medicine. It is clear that the use of medicinal plants remains a living and essential practice in the Grand Lac and Chaouia-Ouardigha region. This persistence of tradition, despite contemporary changes, is a testament to the depth of the relationship between humans and their natural environment. We hope that this study will contribute to enriching the already existing knowledge and will encourage more researchers to explore the exceptional biodiversity of this region, an infinite source of discoveries and innovations in the fields of natural sciences and health.

References

[1] Chaachouay, N., Belhaj, S., El Khomsi, M., Benkhnigue, O., & Zidane, L. (2024). Herbal remedies used to treat digestive system ailments by indigenous communities in the Rif region of northern Morocco. *Vegetos, 37*(1), 379–396.

[2] https://fertilisation-edu.fr/8-nutrition-des88plantes.html?start=10.

[3] Agriconomie – The partner site for farmers. NPK Fertilizers: https://www.agriconomie.com/.

[4] Rodrigues, L. et al. (2020). Ethnopharmacology and biodiversity conservation: A contemporary perspective. Frontiers in Pharmacology, 11, 231.

[5] Lange, D. et al. (2019). Traditional knowledge of medicinal plants and their modern-day uses. Journal of Ethnobiology and Ethnomedicine, 15(1), 68.

[6] Akinmoladun, A. F. et al. (2018). Exploring the relationship between aromatic plants and human health: From tradition to modernity. Phytotherapy Research, 32(5), 759–775.

[7] Rizvi, S. M. et al. (2022). Medicinal plants: A crucial asset for the future of sustainable healthcare and environment. Journal of Environmental Management, 303, 114089.

[8] Müller, A. and Köllner, T. G. (2021). Global challenges for the sustainable use of medicinal and aromatic plants. Sustainability, 13(2), 457.

[9] Zhao, Q. et al. (2019). Impact of climate change on the diversity and distribution of medicinal plants. Global Change Biology, 25(10), 3435–3445.

[10] Perfume, aromatic and medicinal plants market – Panorama 2021 FranceAgriMer; Apr 2023. https://www.franceagrimer.fr/content/download/71143/document/Marche_PPAM_Panorama_2021.

[11]   Marquet, V. and Viora, P. Soil characteristics and biomass production. SVT in high school. 2020.
       https://svtlyceedevienne.com/seconde/enjeux-contemporains/caracteristiques-des-sols-
       etproduction-de-biomasse/
[12]   Bellakhdar, J. (1997). Traditional Moroccan pharmacopoeia: Ancient arab medicine and popular
       knowledge. Editions Le Fennec, Casablanca/Ibis Press, Paris, 764 p.
[13]   Jiofack, T., Fokunang, C., Guedje, N. and Kemeuze, V. (2010). "Ethnobotanical uses of medicinal plants of
       two ethnic groups in the western region of cameroon." Journal of Ethnopharmacology, 133(1), 1–12.
[14]   Scherrer, A. M., Motti, R. and Weckerle, C. S. (2005). "Traditional plant use in Northern Tunisia:
       A comparison of medicinal and domestic use." Journal of Ethnopharmacology, 98(3), 345–357.
[15]   Elachouri, M. (2018). Ethnobotany/ethnopharmacology, and bioprospecting: Issues on knowledge
       and uses of medicinal plants by Moroccan people. In *natural products and drug discovery*
       (pp. 105–118). Elsevier.
[16]   Ghorbani, A., Esmaeilizadeh, M. and Alizadeh, M. (2017). "Pennyroyal (*Mentha pulegium* L.): A review
       of medicinal uses and toxicological effects." Journal of Ethnopharmacology, 213, 148–160.
[17]   M'hamed, H. and Eddouks, M. (2005). "Ethnobotanical survey on medicinal plants used by local
       populations in the oases of the Moroccan Sahara." Phytotherapy Research, 19(6), 459–464.
[18]   Albuquerque, U. P. and Hanazaki, N. (2009). "Five problems in current ethnobotanical research—
       and some suggestions for strengthening them." Human Ecology, 37(5), 653–661.
[19]   Eddouks, M., Maghrani, M., Lemhadri, A., Ouahidi, M. L. and Jouad, H. (2002).
       "Ethnopharmacological survey of medicinal plants used for the treatment of diabetes mellitus,
       hypertension and cardiac diseases in the southeast region of Morocco (Tafilalet)." Journal of
       Ethnopharmacology, 82(2–3), 97–103.
[20]   Bussmann, R. W. and Sharon, D. (2006). "Traditional medicinal plant use in Northern Peru: Tracking
       two thousand years of healing culture." Journal of Ethnobiology and Ethnomedicine, 2(1), 47.
[21]   FAO. (2010). "The state of the world's plant genetic resources for food and agriculture." Food and
       Agriculture Organization of the United Nations, Rome (http://www.fao.org/docrep/013/i1500e/
       i1500e.pdf). https://revuelautre.com/boutique/lautre-2015-vol-16-n2/
[22]   Tilman, D., Balzer, C., Hill, J. and Befort, B. L. (2011). "Global food demand and the sustainable
       intensification of agriculture." Proceedings of the National Academy of Sciences, 108(50), 20260–20264.
[23]   Granger, R. and Passet, J. Thymus vulgaris spontaneus of France: Chemical races and 91
       chemotaxonomy. Phytochemistry. 1 July 1973;12(7):1683–91.
[24]   Liaqat, I., Riaz, N., Saleem, Q. U. A., Tahir, H. M., Arshad, M., & Arshad, N. (2018). Toxicological
       evaluation of essential oils from some plants of Rutaceae family. *Evidence-Based Complementary and
       Alternative Medicine*, 2018(1), 4394687.
[25]   Grieve, M. (1931). A Modern Herbal. New York: Dover Publications.
[26]   National Agency for the Safety of Medicines (ANSM). Our Missions – Medicines Based on Plants and
       Essential Oils:https://ansm.sante.fr/qui-sommes-nous/notre-perimetre/les-medicaments/p/medica
       ments-abase-de-plantes-et-huiles-essentielles
[27]   Aouinti, F., Imelouane, B., Tahri, M., Wathelet, J. P., Amhamdi, H., & Elbachiri, A. (2014). New study of
       the essential oil, mineral composition and antibacterial activity of Pistacia lentiscus L. from Eastern
       Morocco. *Research on Chemical Intermediates*, 40, 2873–2886.
[28]   Bianchini, A., Santoni, F., Paolini, J., AF, B., Mouillot, D. and Costa, J. Partitioning therelative
       contributions of inorganic plant composition and soil characteristics to the quality of Helichrysum
       italicum subsp. italicum (Roth) G. Don fil. essential oil. Chemistry & Biodiversity. Juill
       2009;6(7):1014-33.
[29]   National Conservatory of Perfume, Medicinal, Aromatic and Industrial Plants (CNPMAI). (2023). The
       Conservatory: Mission and Actions. Consulted on [date of consultation], from https://www.cnpmai.net.
[30]   Schmidt, E. and Günther, C. (2001). "Drying and processing of medicinal plants." Journal of
       Medicinal Plant Research, 22(4), 250–258.

[31] Alaoui, S. B., Adan, A. M., Imani, Y., & Lahlou, O. (2018). Adaptation of Farmers to Climate Change: A Systems Approach to Cereal Production in Benslimane Region, Morocco. In *Applied Mathematics and Omics to Assess Crop Genetic Resources for Climate Change Adaptive Traits* (pp. 25–38). CRC Press.

[32] Monograph of the Chaouia-Ouardigha region. High Commission for Planning. Settat Regional Directorate. 6–7, 2008.

[33] El Rhaffari, L. and Zaid, A. Practice of Phytotherapy in the South-east of Morocco (Tafilalet): Empirical Knowledge for a Renovated Pharmacopoeia. Paris (FRA); Metz: IRD; SFE. ISBN 2-7099-1504-9. 293–318, 2002.

[34] Labadie, C., Cerutti, C., & Carlin, F. (2016). Fate and control of pathogenic and spoilage micro-organisms in orange blossom (Citrus aurantium) and rose flower (Rosa centifolia) hydrosols. *Journal of applied microbiology*, *121*(6), 1568–1579.

[35] Salhi, S., Fadli, M., Zidane, L. and Douira, A. (2010) Etudes floristique et ethnobotanique des plantes médicinales de la ville de Kénitra (Maroc). Lazaroa 31:133–146.

[36] Fadil, M., Farah, A., Haloui, T., & Rachiq, S. (2015). Ethnobotanical survey of plants operated by cooperatives and associations of the Meknes-Tafilalet aera in Morocco. *Phytothérapie*, 13, 19–30.

[37] Bnouham, M., Mekhfi, H., Legssyer, A. K. and Ziyyat, A. (2002) Medicinal plants used in the treatment of diabetes in Morocco. International Journal of Clinical Metabolism and Diabetes 10:33–50.

[38] Lahsissène, H., Kahouadji, A., Tijane, M. and Hseini, S. 2009. Catalogue of medicinal plants used in the region of Zaër (Western Morocco). Lejeunia, 186, 1–27.

[39] Mehdioui, R. and Kahouadji, A. 2007. Ethnobotanical Study among the Population Living near the Amsittène Forest: (Province of Essaouira). Bulletin of the Scientific Institute, Rabat, Life Sciences section, 2007, n 29, 11–20.

[40] HAJJAJI, A. (1990). Arboriculture, market gardening and cash crops in oasis areas. Oasis Agricultural Systems; Dollé, V., Toutain, G., Eds, 155–161.

[41] Essoubaiy, O., Gazzaz, B., Yahia, H. et al. Anthropogenetic study of the Arabic - speaking population of Chaouia Ouardigha (Morocco) based on autosomal STRs. *Egypt J Forensic Sci* 14, 17 (2024). https://doi.org/10.1186/s41935-024-00390-5

[42] Joshee, N., Dhekney, S. A., & Parajuli, P. (2019). Medicinal plants. *Switzerland*: Springer, Cham, 1–427.

[43] Boulmane, Z. and El Yassir, H. (2020). "Economic constraints of traditional Moroccan products: A case study of amaghouss production." Journal of Agricultural Economics, 12(3), 215–230.

[44] Benabdellah, A. and Kharbach, M. (2021). Traditional pomegranate processing practices in Morocco: Challenges and opportunities. Food Science Journal, 35(2), 123–135.

[45] Lachance, A., Moroccan aromas, essential oils. IDRC explores, the voice of southern research. Vol21, n2 July 93.

[46] Slikkerveer, L., Ethnobotanical knowledge systems and their potential for sustainable use of wild food and non-food plants. Option mediterrannénne n32.

[47] Heywood, V. H. (1999). The Mediterranean region a major centre of plant diversity. *Wild food and non-food plants: Information networking. Chania: CIHEAM-IAMC*, 5–13.

[48] Heywood, V. H. (2014). An overview of in situ conservation of plant species in the Mediterranean. *Flora Mediterranea*, *24*, 5–24.

[49] Greche, H., Belkheir, K., Hajjaji, N., Boukir, A., & Loukili, A. (2009). Organic essential oils from Morocco. Research on medicinal and aromatic plants, Proceedings of the International Congress, Hassan Greche & Abdeslam Ennabili (Ed.), 130–40.

[50] Bnouham, M., Merhfour, F. Z., Elachoui, M., Legssyer, A., Mekhfi, H., Lamnaouer, D., & Ziyyat, A. (2006). Toxic effects of some medicinal plants used in Moroccan traditional medicine. *Moroccan Journal of Biology*, *2*(3), 21–30.

[51] Charrouf, Z., & Guillaume, D. (2008). Argan oil: Occurrence, composition and impact on human health. *European Journal of Lipid Science and Technology*, *110*(7), 632–636.

Hurija Džudžević-Čančar*, Amra Alispahić, Emina Boškailo, and Alema Dedić

Chapter 16
The use of medicinal and aromatic plants in aromatherapy

Abstract: One of the fastest-growing alternative medical modalities is aromatherapy, which combines massage, counseling, and the pleasant scent of essential oils (EOs) and aromatic plant chemical compounds. Using EOs to improve health or well-being is called aromatherapy or EO therapy. Since ancient times, EOs have been utilized to enhance a person's mood or general health. The benefits of aromatherapy extend beyond mental relaxation. Stress and anxiety symptoms can be reduced by aromatherapy, which may indirectly enhance sleep. The inhalation or absorption of volatile compounds from aromatic plants may also aid in pain relief and body relaxation, as well as alleviate sadness, nausea, and discomfort, in addition to addressing other health issues, making it a useful adjunct to conventional medicine. Nevertheless, insufficient studies have been conducted to ascertain their efficacy in human health. This kind of treatment uses a variety of permutations and combinations to treat a wide range of conditions, including depression, indigestion, headaches, sleeplessness, respiratory issues, muscular discomfort, skin conditions, swollen joints, and troubles related to urinary problems, among others. EO inhalation activates the olfactory system, which includes the nose and the area of the brain linked to smell. Aromatherapy can be a very useful supplement to traditional medicine or utilized as a standalone treatment when done correctly and carefully.

Keywords: aromatherapy, essential oils, alternative medicine, health benefits

16.1 Introduction

There are many definitions for the term "aromatherapy." According to the Merriam-Webster dictionary, it is the inhalation or body application (such as by massage) of fragrant essential oils (EOs) (from flowers and fruits) for therapeutic purposes. More

*Corresponding author: Hurija Džudžević-Čančar, Department of Chemistry in Pharmacy, University of Sarajevo-Faculty of Pharmacy, Sarajevo, Bosnia and Herzegovina,
e-mail: hurija.dzudzevic-cancar@ffsa.unsa.ba
Amra Alispahić, Alema Dedić, Department of Chemistry in Pharmacy, University of Sarajevo-Faculty of Pharmacy, Sarajevo, Bosnia and Herzegovina
Emina Boškailo, Department of Ecology and Environmental Protection, Faculty of Social Sciences Dr. Milenko Brkić, Herzegovina University, Mostar, Bosnia and Herzegovina; International Society of Engineering Science and Technology, Nottingham, UK

https://doi.org/10.1515/9783111469713-016

broadly, aromatherapy is defined as the use of aroma to enhance feelings of well-being [1]. Aromatherapy is a holistic treatment that uses natural plant extracts in the form of EOs to promote human health, as well as mental and physical well-being. Aromatherapy is sometimes called "essential oil therapy." It is regarded as both a science and an art. The body and mind are dynamically affected by the scents experienced through EOs and aromatherapy. Another definition of aromatherapy is the methodical use of EOs to maintain and improve a person's physical, mental, and spiritual health. It seeks to integrate physiological, psychological, and spiritual processes to enhance an individual's natural healing process [2].

Aromatherapy has recently become increasingly well-known in the scientific and medical communities. The Persians are credited with distilling EOs in the tenth century, although the process may have existed for much longer. These natural substances were used for medicinal and religious purposes and were well-known for their psychological and physical advantages [3].

Aromatherapy is usually applied by inhalation or topically. EOs can be inhaled using a diffuser, spray, or dropper. They can also be added to a warm bath and inhaled. The component of the brain linked to scent, the olfactory system – which includes the nose and brain – is stimulated by inhaling EOs. After passing through the mouth or nose, molecules proceed to the lungs and then to other areas of the body. The limbic system, which is linked to emotions, is impacted by the chemicals as they enter the brain. Additionally, they affect breathing, memory, blood pressure, and heart rate [4]. In order to allow EOs to be absorbed via the skin, users can also apply them topically, for example, in massage oils, baths, and skin care products. Increasing circulation and absorption can be achieved by massaging the region where the oil is applied [4]. Proponents of aromatherapy claim that it can help with a variety of ailments, including pain, nausea, anxiety, depression and stress, insomnia, headaches, circulatory problems, menstrual problems, skin problems, and digestion. However, there is not enough research conducted to prove with certainty their effectiveness in human health [5].

There are a number of possible risks associated with using EOs for aromatherapy, such as allergic reactions, contact dermatitis, upset stomach if swallowed, breathing difficulties if EOs come into contact with the nose, chemical burns if EOs come into contact with the eyes, and an increased risk of sunburn if citrus EOs come into contact with the skin. It is true that using EOs directly on the skin is never a good idea. A carrier oil should always be used to dilute them. Because carrier oils are frequently made from nuts and seeds, consumers should inform their aromatherapist if they have any nut allergies when the oil is used with one. A person should cease using EOs immediately and avoid the aroma if they experience a new allergic reaction. Additionally, swallowing or ingesting EOs is not advised. Certain EOs can harm the kidneys or liver when consumed orally. They may also cause unanticipated changes in the gut and interact with other medications. However, it is crucial to remember that not all EOs are equally dangerous [6].

Aromatherapy may not be suitable for everyone. Chemical compounds in EOs can reduce the effectiveness of conventional medications or worsen an individual's existing health condition. When utilizing aromatherapy, individuals with epilepsy, respiratory disorders including asthma or chronic obstructive pulmonary disease, high blood pressure, or sensitive skin should exercise extra caution. Additionally, it is not advised for pregnant or nursing women to use aromatherapy because research has not demonstrated that it is safe for them [7].

A thorough medical history, lifestyle, and diet should be assessed by an aromatherapist. Treating the whole person is the goal of aromatherapy's holistic approach. Treatments are customized to meet each patient's unique physical and mental requirements. Based on these requirements, an aromatherapist might suggest a single oil or a blend. A massage therapist may utilize aromatherapy oils, but they are not the same as aromatherapists [8].

As of right now, the practice of aromatherapy and the production of aromatherapy goods are both unlicensed and unregulated. Since the FDA does not actively monitor aromatherapy goods, it can be challenging to determine whether a product is synthetic, tainted, or pure. Certain home and cosmetic products, such as lotions, makeup, and candles, contain ingredients that mimic EOs but are actually artificial scents. It is crucial to carefully follow the directions and seek expert guidance when using aromatherapy and its products [9–11].

EOs for aromatherapy come in about a hundred varieties. Clary sage, cypress, eucalyptus, fennel, geranium, ginger, helichrysum, lavender, lemon, lemongrass, mandarin, neroli, patchouli, peppermint, Roman chamomile, rose, rosemary, tea tree, and vetiver are a few of the well-known EOs [10, 11].

16.2 History of aromatherapy

EOs have been used in medicine, without a doubt, since ancient Egypt, through Greece and Rome, the Arab world, and Europe. The Greeks also had an impact on the development of aromatherapy; Megallus, a perfumer, employed myrrh as the primary ingredient in his fragrance, Megaleion. Before aromatherapy was given that name, Hippocrates, the "father of medicine," is thought to have utilized it for therapeutic purposes. The roots of aromatherapy are unclear, although it is said that the Egyptians used one of the earliest distillation tools to extract oils from plants like cinnamon, cloves, and cedarwood, which were subsequently used to embalm the deceased. However, it is thought that China is where the practice of using infused aromatic oils to elevate mood first emerged [12]. EOs have been used in the healing practices of the Ayurvedic system of traditional Indian medicine for at least 3,000 years. More than 700 medicinal plants, such as sandalwood, myrrh, cinnamon, and ginger, are mentioned in ancient Ayurvedic literature as being beneficial for healing. In addition to being used medicinally, aromatic herbs and oils were revered as natural elements and contributed to the region's philosophical and spiritual outlook [12].

Some Arab authors, already described in the history of phytotherapy, knew very precisely the medicinal use of EOs. However, bearing in mind the laboriousness of the distillation process, EOs were a medicine for a privileged class of people. They survived thanks to their obsession with perfumes and cosmetics. In the eighteenth and nineteenth centuries, industrial development occurred, which included improved distillation technology. EOs became more accessible and cheaper, but medical aromatherapy as we know it today began at the end of the nineteenth century with the isolation of active compounds from EOs, and aromatherapy experienced its explosion only in the twentieth century when everything changed [13]. René-Maurice Gattefossé, born in the nineteenth century in Lyon into a family that dealt in EOs, grew up surrounded by them. The company produced and sold EOs for the perfume industry. He was a chemist researching not only EOs but also synthetic fragrance substances. Gattefossé created the magazine *La Parfumerie Moderne* (Modern Perfumery), which was the bearer of modern scientific trends in the production and application of EOs. He encouraged lavender growers and also promoted new techniques in oil distillation and the chemistry of fragrant molecules. However, the term "aromatherapy" was first mentioned in the work *Aromathérapie – Les Huiles Essentielles Hormones Végétales* (1937) [14]. In the second edition of the book in 1944, René-Maurice explains that he initially wanted to give the name thymothérapie (thymotherapy), from the Greek word thymos or thumos (θυμός), which etymologically denotes the concept of strong, spiritual, lively, and symbolizes the light, volatile, and intense personality of EOs. However, he became aware that the name would easily be mistranslated (as "thyme" in French and English means the herb thyme). Therefore, he decided to coin a new word – aromatherapy. With this, aromatherapy officially entered history, and Gattefossé left a great legacy [14].

In the twentieth century, Jean Valnet emerged as an exceptionally talented military doctor, specializing in surgery, microbiology, hygiene, forensic pathology, colonial medicine, and occupational medicine. As a military doctor, he worked in Germany and Vietnam, and it was his work in Vietnam that proved crucial because there he began using EOs to accelerate the recovery of patients after operations and, according to him, achieved excellent results [15].

After his military service, he went to Paris to devote himself to further work in phytotherapy and aromatherapy. In 1964, he published the books *Les médecines différentes/Different medicines* and *Aromathérapie, traitement des maladies par les essences des plantes/Aromatherapy, treatment of diseases with plant essences* [16, 17]. His work may be partly controversial, but he undoubtedly combined traditional phytotherapy and then-modern trends in aromatherapy with great elegance, creating a foundation based on his clinical experience. These books are still read and cited today, if for no other reason than by those who study the development of modern aromatherapy and phytotherapy [16, 17]. For example, he emphasizes the use of niaouli oil in diseases of the respiratory system, which, due to its 1,8-cineole content, remains relevant today. Jean Valnet skillfully integrated phytotherapy and aromatherapy in his work. In his work *Aromatherapy*, without any reservations, he mentions tinctures and aqueous extracts of medicinal plants. He correlates some basic properties of molecules – such as acidity and basicity, dissociation constant, and electrical conductivity – with their effects. Dr. Jean Valnet worked closely with other experts. He actively promoted the use of aromatograms in selecting EOs for the treatment of infections. Together with Dr. Christian Duraffourd and Dr. Jean-Claude Lapraz, he published the book *Une médecine nouvelle. Phytothérapie et aromathérapie – comment guérir les maladies infectieuses par les plantes/ New Medicine. Phytotherapy and Aromatherapy – How to Treat Infectious Diseases with Medicinal Plants* [18].

Dr. Daniel Pénoël belongs to the generation of aromatherapists who introduced the concept of chemotype and its significance for the pharmacological action of EOs. He has been actively using EOs in practice since 1977, and today he operates in both Europe and the USA. Together with researcher Pierre Franchomme, he published the book *Aromathérapie Exactement* (*Exact Aromatherapy*). They passionately connect the electrical properties – namely, the basic physico-chemical properties of the molecules of EOs – with their pharmacological action, thus continuing the work of René-Maurice Gattefossé and Jean Valnet. At the end of the book, there are individual descriptions of over 100 EOs and a series of master preparations made from EOs for various ailments. Dr. Pénoël emphasized the importance of chemotypes: if the same species can produce EOs of different chemical compositions, then it is logical that the oils exhibit different effects. For example, the thyme chemotype thymol (phenol chemotype) is more irritating to the skin than the linalool chemotype (alcohol chemotype), but it also has a stronger antibacterial effect [19].

16.3 Essential oils

Naturally occurring volatile organic molecules, known as EOs (also referred to as ethereal or volatile oils), are produced in large quantities from the raw materials of plants or their organs, including flowers, seeds, buds, leaves, twigs, roots, bark, timber, and fruits. These oils can contain a blend of over 300 different chemical constituents. Chemically, these substances belong to several groups, including carbonyl compounds such as aldehydes, ketones, amides, and esters, as well as amines, alcohols, phenols, ethers, terpenes, and phenylpropanoids [20, 21].

Among plant extracts, EOs hold a unique position, probably because of their expense, aromatic qualities, and historical reputation as the "royal" plant extracts. Only a small amount of EO is made from extremely vast amounts of plant material. In certain cases, 100 kilograms of plant material are required to produce 1 kg of EO, and in other cases, tonnes are required. For every kilogram of rose EO, more than 4 ton of rose petals are typically required [22].

Most EOs are liquid at room temperature, but there are exceptions. Rose EO becomes solid at temperatures below 20 °C due to naturally occurring plant paraffins, and the EO of elecampane (*Inula helenium*) is a solid and melts only at elevated temperatures. Most EOs are slightly viscous but less viscous than vegetable oils and even water. However, some are very viscous, such as vetiver and patchouli EOs [23, 24]. Likewise, most EOs are less dense than water (relative density is less than 1) and therefore float on the surface of water. Only some EOs, such as German chamomile and clove, are denser than water. The volatility of EOs is their fundamental characteristic; that is, the molecules of the EO evaporate easily. This allows us to detect them with our sense of smell. However, volatility does not mean that EOs boil at temperatures below 100 °C. Most EO molecules boil only at temperatures above 150 °C [25].

Because they are lipophilic molecules, all EOs prefer nonpolar solvents over polar ones like water. Therefore, they are not soluble in water, but they dissolve well in vegetable oils and waxes, concentrated ethanol (90% ethanol and higher), diethyl ether, and similar solvents. EO molecules are mostly low molecular weight and are made up mainly of carbon molecules with 10–15 carbon atoms. Their high lipophilicity and low molecular weight give them biological characteristics, making them very easily absorbed through the skin, as well as generally through cell membranes, which includes easy absorption in the digestive system and rectum. Their volatility allows for easy inhalation, so for centuries, they were the only medicines that could be applied in this way [25–27].

The term "essential oils" can only refer to extracts that have been produced by steam distillation (the majority of EOs), pressing (citrus fruit oils), and direct heating of plant material without steam distillation (very uncommon, occasionally used for cinnamon bark oil). All other fragrant extracts obtained in a different manner are referred to by their manufacturing method, such as CO_2 extracts and n-hexane extracts, and may not be considered EOs. Only natural scents may be referred to as EOs, not synthetic blends or blends of synthetic and natural EOs. Since botanical traits greatly influence the chemical composition, all EOs must be characterized both botanically and chemically. This includes accurately identifying the species, potential subspecies, variety, and form. A chemically defined oil has a precisely defined proportion of individual molecules in the EO, usually expressed as a percentage (%) and determined by chemical analysis using gas chromatography. If necessary, chemical analysis also precisely defines the chemotype, which refers to the phenomenon where the same botanical material can yield EOs of different compositions depending on the place of growth [28].

16.4 Obtaining essential oils

EOs are usually obtained in two main ways: steam distillation and pressing. A very small number of oils are obtained by directly heating plant material without the presence of steam, and today such oils have almost completely replaced those obtained by steam distillation. These processes are based on the physical and chemical principles of EO production and the basics of technological implementation [29].

Why is steam distillation used? The boiling point of EOs is high and often exceeds 100 °C. This fact seems unusual because EOs appear to evaporate very easily. The laws of physics allow EOs, whose proportion in the mixture is relatively small, to be successfully distilled in sufficient quantities for meaningful production. Dalton's law theoretically explains how, in the process of steam distillation of two immiscible liquids, substances can be distilled far below their boiling point. This means that EOs can be distilled in the process at temperatures below 100 °C [29, 30].

16.4.1 Types of distillation

Regardless of the technological design of the distiller, each distillation device has these basic elements: a part containing water, where the water is heated until steam is released; a part containing plant material, through which the steam will pass; and a part where water vapor and EO are cooled and converted back into liquid form (condensed). In addition to the EO, a water layer saturated with EO is also formed, such as hydrolat or floral water [29].

Water distillation is the oldest method of distillation, originating in Mesopotamia. Plant material is literally boiled in water, and steam, together with the oil, emerges from this mass. Today, it is rarely used or not used at all because boiling in water can destroy the valuable ingredients of the EO. Hydrolysis of compounds occurs (decomposition under the influence of water), especially in EOs rich in esters (e.g., rose and lavender). The process itself is not well thermally controlled and can lead to local overheating of the plant material, resulting in a significant loss of the olfactory value of the EO (Figure 16.1).

Steam distillation is better and more commonly used. Above the boiling water layer is a mesh containing plant material. In this case, the water does not touch the plant material; only water vapor passes through it. Steam distillation is performed in such a way that there is a separate device, a steam generator, which is directed into a container with plant material. This is the best method of distillation because the plant material is in contact exclusively with steam, not water.

It is necessary to pay attention to the softness of the water (most modern devices contain a water softener), the temperature of the steam, the duration of the distillation, the pressure, the design and shape of the boilers, the material (stainless steel, copper), and the preparation of the plant material. Sometimes plants need to be left

Figure 16.1: Distillation using a modified Clevenger apparatus and the obtained essential oil.

untreated, such as lavender flowers, which have glands on their surface that allow the EOs to be easily released, and sometimes the plant material needs to be ground so that the water vapor can more efficiently penetrate the glands containing EO, as is the case with rosewood or cinnamon. Some plants are distilled strictly fresh, as is the case with lavender, while others can be distilled in a dry or semi-dry form [29, 30]. For EOs that contain a lot of water-soluble compounds, such as rose or cinnamon, the resulting hydrosol needs to be recirculated to improve the yield of EO and the olfactory properties of the oil. However, recirculation sometimes reduces the quality of the hydrosol. Therefore, for some plant species, such as rose, certain manufacturers specifically distill the plant material to obtain a high-quality hydrosol and separately to obtain EO. There is also distillation under reduced pressure. Long known in laboratory conditions, this is a relatively new method of obtaining commercial oils [31]. The theory of reduced pressure distillation is simple and is based on the already described Henry's law. Lowering the pressure also lowers the boiling point of water, so by sufficiently reducing the pressure, you can "force" water to boil at lower temperatures, even at room temperature. By releasing such cold steam under low pressure and condensing it at very low temperatures, EO and hydrolat are obtained with beautifully preserved delicate aromatic compounds and excellent olfactory properties. However simple it may be in theory, the technological implementation is not. Distillation boilers must have very thick steel walls to prevent the boiler from imploding, and the cooling device must use a different coolant than water, such as ethylene glycol, because the condensation temperature must often be well below 0 °C. This is why low-pressure distillation devices are much more expensive [31, 32].

Pressing – Citrus oil pressing is a much younger technique than distillation. Once manual and very laborious, today it is industrial and complex. It is not enough to simply squeeze a lemon; the quality of the oil depends on many factors, including the amount of water and temperature. Citrus fruits are pressed to extract their EOs. Anyone who peels lemons and oranges is aware that the oil is released in tiny droplets due to the physical separation of the fruit. Citrus fruits contain significant amounts of EO with a very pleasant aroma, but it was a great challenge to invent a technique that would collect sufficient quantities. In southern Italy, probably in the eighteenth century, the first method of extraction was developed, called sfumatura. Before sfumatura, citrus EOs were obtained by distillation for a long time [33]. Domenico Sestinio's 1776 work mentions the sfumatura technique. Culterera explains the sfumatura technique in an essay titled *Breve e patetica storia del commercio delle essenze agrumarie*, which was published in the journal *Industria Conserve* in 1954. Speaking nostalgically about the quality of today's industrially produced citrus oils compared to those obtained through the sfumatura technique, he said that we have descended "from the sublime to the simply good." Leading authors in the field agree that today's citrus EOs are not as good in olfactory terms as they used to be [33].

16.4.2 Post-production processing of essential oils

In aromatherapy, EOs that have not been specially processed after the distillation process itself, nor have their fractions of individual compounds been removed, are preferred. However, for some time, EOs were very often deterpenated in order to remove hydrocarbons. The reasons for removing hydrocarbons are multiple:

– Increasing the proportion of the fragrant fraction, as in the case of deterpenated citrus EOs, produces an oil enriched with oxygenated compounds (aldehydes, alcohols, esters, etc.) and is still widely used in perfumery and the food industry.
– Increasing the proportion of the active fraction, as in the case of lavender, where the proportion of linalool and linalyl acetate is increased, and eucalyptus globulus, where the proportion of 1,8-cineole is increased by removing α-pinene.
– Increasing the stability of the oil is important because hydrocarbons readily oxidize (increase in peroxide value) and polymerize (formation of precipitate) [34, 35].

Some oils are sometimes redistilled (distilled once more with steam) in order to enrich the EO with the main active compounds. This is the case with *Eucalyptus globulus*, in which the content of 1,8-cineole increases, and wintergreen (Gaultheria sp.), in which the content of methyl salicylate increases [36]. Distinct amounts of water are dissolved in several EOs. Water accelerates the oxidation of certain hydrocarbons and hydrolyzes esters, which is extremely harmful to EOs. Water has been found at the bottom of bottles of several EOs in retail establishments on multiple occasions. Water gradually separates and settles from EOs during storage, and the EO and water are combined while filling tiny packets. Since the EO and hydrolat must be separated by decanting, the maker must perform this operation with extreme caution. Anhydrous sodium sulfate can be used to dry EOs, if necessary, in order to eliminate any remaining moisture. Because sodium sulfate is inert and does not alter the oil's chemical or aromatic characteristics, such a method is allowed [36]. EO molecules react with other drying agents, such as sodium hydroxide, calcium chloride, or phosphorus pentoxide, and their usage is therefore prohibited. Due to their specific method of extraction by pressing, citrus oils can also be subjected to a winterization process, similar to the process used in processing vegetable oils. In the winterization process, the EO is kept at −20 °C for a week, during which waxes precipitate. The oil is carefully filtered, resulting in a pressed oil with a lower content of dry residue after evaporation. This prevents the formation of sediment during the storage of EOs. The winterization process is permitted for EOs used in aromatherapy. Steam-distilled oils do not undergo the winterization process, as they do not contain waxes. In the case of rose EO, winterization is prohibited because it contains paraffins that crystallize at low temperatures and are one of the characteristic compounds that prove the authenticity of EOs [37].

16.5 Classification of aromatherapy

Aromatherapy is often divided into two main categories based on how it is administered: topical application and inhalation. It encompasses a range of methods, including massage, psycho-aromatherapy, olfactory stimulation, cosmetics, and medicine.

In this therapy, certain EOs are used to create cosmetics for the skin, hair, body, and face. These products have multiple uses, such as cleansing, hydrating, drying, and toning. To promote healthy skin, EOs can be incorporated into facial products. A simple and effective way to enjoy cosmetic aromatherapy on a personal level is to use it for a foot bath or for the entire body. Similarly, a few drops of the appropriate oil can revitalize and rejuvenate [38, 39]. Topical aromatherapy is utilized in massage therapy, treatment, and cosmetics. Psycho-aromatherapy and olfactory stimulation are both included in inhalation aromatherapy. Figure 16.2 shows this in more detail. The

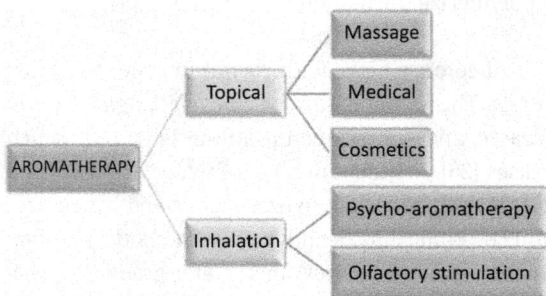

Figure 16.2: Aromatherapy: basic classification.

idea that aromatherapy only involves inhalation is among the most pervasive misconceptions about it. One of the primary methods in the broader field of aromatherapy is aromatherapy massage, which is also an effective strategy for reducing the discomfort of labor during pregnancy [40, 41]. Oral and rectal administration are two of the many methods through which EOs have been employed [42]. Applying pure vegetable oil containing jojoba, almond, or grape seed oil during an aromatherapy massage has been shown to produce remarkable benefits. In massage therapy, this is also known as "healing touch" [43, 44]. Cosmetics for the face, body, hair, and skin are made using specific EOs in this therapy. These products can be used for a variety of purposes, including toning, drying, moisturizing, and cleansing. EOs can be utilized in facial products to promote healthy skin. Using cosmetic aromatherapy for a foot bath or for the full body is an easy and efficient method to experience it on a personal level (Figure 16.3) [38].

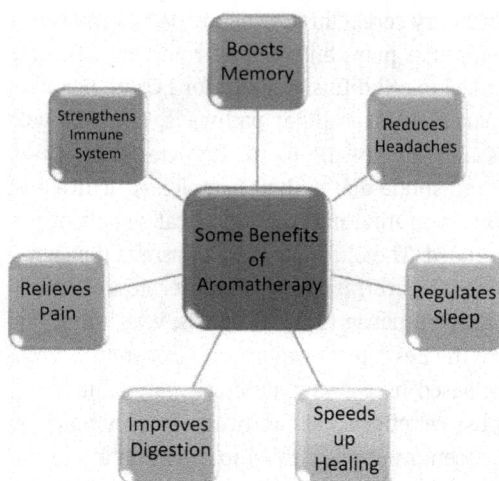

Figure 16.3: Specific benefits of aromatherapy.

The use of EOs to massage patients during surgery was pioneered by modern aromatherapist René-Maurice Gattefossé to explore medical aromatherapy and understand the effects of EOs on promoting and treating clinically proven medical conditions [44]. Olfactory aromatherapy, which involves inhaling EOs, has been demonstrated to enhance emotional health, calmness, relaxation, or physical renewal through the simple act of inhalation. Enjoyable scents that evoke memories are often associated with stress reduction. EOs serve as a supplement to medical treatment but cannot replace it [45]. In psycho-aromatherapy, these oils can be used to evoke particular feelings and moods, such as calmness, energy, or pleasant memories. Through infusion in the patient's room, the oils are directly inhaled as part of this therapy. Aromacology and psycho-aromatherapy study the effects of aroma, whether natural or manufactured. The scope of psycho-aromatherapy focuses on the study of natural EOs [46].

16.6 The way aromatherapy works

The therapeutic benefits of EOs for the body, mind, and soul have long been recognized. These fragrance molecules are incredibly powerful organic plant compounds that eliminate illness, fungi, viruses, and bacteria from the environment [47]. Numerous scientists have thoroughly established their multifaceted antibacterial, antiviral, anti-inflammatory, and immune system-boosting properties, which include effects on hormones, glands, emotions, circulation, relaxation, memory, and alertness [8, 10]. To understand their nature and significance in diseases and disorders, numerous human pilot projects and studies have been carried out [48]. Since their potency remains constant over time, these oils are renowned for having a certain energetic quality. These oils have stimulating qualities because of their structure, which is remarkably similar to that of real hormones [49]. The activity of EOs in the human body system involves the respiratory and olfactory pathways. EOs enter the brain through the respiratory cerebellum, which is part of the respiratory route. In this process, the trigeminal nerve plays an essential role. The olfactory nerve and olfactory bulb facilitate the three cell-to-cell diffusion paths for EOs in the olfactory pathway: the transcellular, paracellular, and intracellular pathways. The ability of the oils to penetrate the subcutaneous tissue is one of the main characteristics of this treatment. Their effects are also extensive and subtle due to their complex structure and chemical composition [50]. EOs function by integrating into the biological signals of the receptor cells in the nose when they are inhaled. The olfactory bulb transmits the signal to the brain's limbic and hypothalamic areas. Neurotransmitters like serotonin, endorphins, and others are released by the brain in response to these signals, which link our nerves and other bodily systems and ensure the desired alteration and relaxation. Serotonin, endorphins, and noradrenaline are released by calming, euphoric, and stimulating oils to produce the intended mental and physical effects [50, 51]. Aromatherapy employs scent therapy and EOs for conditioning. EO scent molecules travel to the brain and influence emotions. The emotional response in the brain can help with both psychological and physical issues. The nose and the skin are the primary routes of absorption. Oil absorption aids in disease prevention and the regulation of mental and bodily states (Table 16.1) [52].

Table 16.1: Essential oils used to treat common problems [53–55].

Condition	Essential oils
Anxiety, agitation, stress	*Angelica archangelica*/angelica, *Citrus aurantium* var. amara/petitgrain bigarade, *Citrus aurantium* var. amara/orange bigarade, *Citrus bergamia*/bergamot, *Citrus sinensis*/ sweet orange, *Cymbopogon martini*/palmarosa, *Eucalyptus staigeriana*/lemon-scented ironbark, *Lavandula angustifolia*/lavender, *Litsea cubeba*/may chang, *Ocimum basilicum*/basil, *Origanum majorana*/sweet marjoram, *Pelargonium graveolens*/ geranium, *Pogostemon cablin*/patchouli, *Valeriana officinalis*/valerian.

Table 16.1 (continued)

Condition	Essential oils
Fatigue	*Angelica archangelica*/angelica (nervous), *Cistus ladaniferus*/labdanum (chronic), *Citrus aurantium* var. amara/neroli bigarade, *Citrus paradisi*/grapefruit (exhaustion), *Coriandrum sativum*/coriander (including mental), *Cymbopogon nardus*/citronella, *Eucalyptus radiata*/black peppermint (chronic), *Eucalyptus smithii*/gully gum, *Juniperus communis* ram./juniper twig, *Mentha spicata*/spearmint (mental). *Pelargonium graveolens*/geranium (nervous), *Pinus sylvestris*/Scots pine, *Rosmarinus officinalis*/rosemary, *Salvia sclarea*/clary (nervous), *Zingiber officinale*/ginger.
Insomnia	*Angelica archangelica*/angelica, *Cananga odorata*/ylang-ylang, *Chamaemelum nobile*/Roman chamomile, *Citrus aurantium* var. amara/neroli bigarade, *Cistus ladaniferus*/labdanum, *Citrus bergamia*/bergamot, *Citrus limon*/lemon, *Citrus reticulata*/mandarin, *Citrus sinensis*/sweet orange, *Cuminum cyminum*/cumin, *Juniperus communis* fruct./juniper berry, *Lavandula angustifolia*/lavender, *Litsea cubeba*/may chang, *Melissa officinalis*/lemon balm, *Myrtus communis*/myrtle, *Ocimum basilicum*/basil (nervous), *Origanum majorana*/sweet marjoram, *Ravensara aromatica*/ravensara, *Thymus vulgaris*/sweet thyme, *Valeriana officinalis*/valerian.
Pain management	*Eucalyptus smithii* (gully gum), *Lavandula angustifolia* (lavender), *Matricaria recutita* (German chamomile), *Leptospermum scoparium* (manuka), *Origanum majorana* (sweet marjoram), *Pinus mugo* var. pumilio (dwarf pine), *Rosmarinus officinalis* ct. camphor (rosemary), *Zingiber officinale* (ginger).

In addition to olfactory stimulation, EOs primarily alter brain function through alveolar absorption. As a result, EO molecules can enter the bloodstream, penetrate the blood-brain barrier, and potentially interact with specific brain areas. The respiratory system facilitates gaseous exchange, and the diffusion process is the mechanism by which EOs enter the respiratory system. One possible route for the molecular transport of these volatile compounds into the bloodstream and brain remains alveolar diffusion. Lipophilic EO compounds have the ability to cross the blood-brain barrier and activate specific central nervous system affinity areas. Such activation could lead to positive psychological and physiological effects, helping to alleviate the symptoms of mood disorders [56].

EOs also stimulate the intestinal system. When applying EOs topically, the analgesic impact and sense of well-being are brought on by the rapid release of endorphins and certain painkillers. Studies show that applying lavender oil to the forehead triggers a complicated pharmaco-psychological cascade that produces a sense of relaxation. It was found that the therapeutic advantages of the Ayurvedic oil dripping treatment with lavender oil are due to the somato-autonomic reflex, which is initiated by thermo-sensors or pressure sensors via the trigeminal cranial nerve. According to reports, a complicated and cryptic pharmaco-psychological interplay created the physiological consequences [57]. Trigeminal nerve involvement in the transfer of medicinal substances from the nose to the brain has been shown in numerous investigations to be successful. Similarly, this pathway is expected for EOs; however, due to the complexity of EO compositions, the mechanistic participation is still unknown [58].

In order to be used topically, EOs must first penetrate the skin and dissolve in the lipid component of the skin's cell membrane. The chemical composition of an oil determines how deeply it penetrates the skin; for instance, oils such as jojoba, avocado, soybean, almond, and others can only penetrate the top layer of the epidermis, whereas oxygenated terpenes can penetrate deeper layers of the skin [59]. Certain oils are also used as penetration enhancers both internally and externally, depending on a number of mechanisms, including enhancing drug partitioning, dissolving the highly ordered intercellular (between corneocytes in the stratum corneum) lipid structure, and causing conformational modification by interacting with the intercellular protein domain [60]. Many volatile molecules with significant medicinal effects should ideally be present in EOs. The many ways that the EO molecules are delivered to the brain depend on their molecular sizes. As the formulation's size reduces, the inhalation rate and delivery success rise. Because of their uneven and variable sizes, these molecules do not enter the brain properly, which is a drawback of inhalational therapy (Table 16.2). However, the absorption of these EOs has been enhanced by the use of encapsulated nanoparticles, which are made possible by nanotechnology, making it a viable therapeutic endeavor for the future [56].

Table 16.2: The main active chemical components found in essential oils and their structures.

Essential oils/reference	Chemical compounds
Lavender (*Lavandula angustifolia* Mill., *Lavandula latifolia* Vill.) [61–64]	

4-terpineol Linalool 1,8-cineole Linalyl acetate

Ocimene Camphor Borneol

Table 16.2 (continued)

Essential oils/reference	Chemical compounds
Mint (*Mentha piperita*) [65–68]	

Linalool   Linalyl acetate   1,8-cineole   Limonene

Geranyl acetate   Menthol   Menthone Piperitone

Calamintha (*Clinopodium nepeta* L. *Kuntze*) [69, 70]

Dihydrocarvyl acetate Piperitenone oxide *cis*-piperitone oxide

Pulegone Limonene   Menthone Piperitenone

Clary sage (*Salvia sclarea* L.) [71]

Linalool   Linalyl acetate   Geranyl acetate

alpha-terpineol   Germacrene D

Table 16.2 (continued)

Essential oils/reference	Chemical compounds

Eucalyptus (*Eucalyptus globulus* Labill.)
[72, 73]

gama-terpinene   *p*-cimene   1,8-cineole   alpha-pinene

Aromadendrene   Globulol

Lemon (*Citrus limon* L.) [74, 75]

Limonene   Linalool   Geranial   Sabinen

Citronellal   Neral   gama-terpinene   alpha-pinene   beta-pinene

Ylang (*Cananga odorata*) [76, 77]

Linalool   Linalyl acetate   Germacrene D   Geranial

Geranyl acetate   beta-caryophyllene   Farnesene

16.7 Pharmacological testing of aromatherapy effects

Numerous EOs' pharmacological potentials have been examined. Many published studies have investigated the local antibacterial action of EOs, and most of them have found that EOs possess potent antimicrobial qualities [78]. While some EOs exhibit fungistatic and fungicidal effects on vaginal and oropharyngeal *Candida albicans*, others demonstrate antiviral properties that inhibit the reproduction of the herpes simplex virus [79, 80]. Research conducted in Europe and Japan has shown that rats exposed to different fragrances may become drowsy or excited, and their responses to pain and stress can change. A study on the sedative effects of EOs and other scent compounds (primarily specific chemical components of the EOs) on rat motility found that lavender oil (*Lavandula angustifolia* Mill.) in particular had a significant sedative effect. Compared to full EOs, several single-oil constituents exhibited similarly potent effects [81]. Variations in bioavailability are attributed to different levels of lipophilicity, with the most sedative effects produced by oils that are more lipophilic. Researchers also discovered significant amounts of volatile compounds in plasma following inhalation, indicating that direct pharmacological interactions, rather than indirect transmission through the central nervous system, are responsible for aromatherapy's benefits. Another study examined the impact of aromatherapy on rats' behavioral and immunological responses to unpleasant, stressful, or shocking stimuli. Rats exposed to pleasant scents during painful stimuli exhibited less pain-related behavior in two European investigations [82, 83].

There is not much research discussing aromatherapy as a specific cancer treatment in the peer-reviewed literature that has been published. The main topics of the research involving cancer patients include additional medical conditions, symptoms associated with treatments, the prevention of infections, as well as stress and anxiety levels as quality of life (QOL) metrics. The effectiveness of aromatherapy in certain cancer patient studies, which suggests that the items employed contain EOs, significantly restricts the ability of interested researchers and clinicians to replicate or compare studies or generate insightful meta-analyses of research findings. A thorough analysis of six studies that examined the use of aromatherapy massage to treat or prevent anxiety was released in 2000 [84]. The authors concluded that the research being conducted at the time was not thorough or reliable enough to demonstrate the efficacy of aromatherapy in treating anxiety, despite studies showing that aromatherapy massage has a slight, temporary anxiolytic effect. Research examining the effects of smells not specifically classified as aromatherapy was not included in this review, nor was research examining other impacts of aromatherapy (such as pain management). In another randomized controlled trial, 103 cancer patients were randomly assigned to receive massage using carrier oil (massage group) or massage using carrier oil plus EO of Roman chamomile (*Chamaemelum nobile* L.) (aromatherapy massage group) to examine the effects of massage and aromatherapy massage [85]. The aromatherapy massage group experienced a statistically significant improvement in symptoms (as deter-mined by the Rotterdam Symptom Checklist) and a decrease in anxiety (as determined by the State-Trait Anxiety Inventory) 2 weeks following the session. QOL, severe physical symptoms, severe psychological symptoms, and psychological well-being were the sub-scales with better results. Despite showing improvement, the massage-only group fell short of statistical significance. The experiences of patients referred to an aromatherapy service were also documented in research assessing the service after modifications were made following an initial trial at a UK cancer center [86]. Of the 89 patients initially re-ferred, 58 completed six aromatherapy sessions. At the conclusion of the research, there were notable improvements in anxiety and depression (as determined by the Hospital Anxiety and Depression Scale-HADS) when compared to measurements taken six sessions earlier. In Australia, a double-blind, randomized, placebo-controlled study examined how inhaled aromatherapy affected patients' anxiety levels following radiation treatment [87]. Three groups of 313 radiation therapy patients were randomly assigned to either nasal oil containing EOs, carrier oil alone, or pure EOs of cedar (*Cedrus atlantica*), lav-ender, and bergamot (*Citrus aurantium* L. ssp. bergamia). During radiation therapy, oils were inhaled by each of the three groups. Psychological outcomes (as determined by the Somatic and Psychological Health Report) and depression (as determined by the HADS) did not significantly differ between groups. The benefits of supplementary aro-matherapy on cancer patients' mood, QOL, and physical symptoms were investigated in a randomized, controlled pilot trial [88]. Using a standardized EO blend consisting of 1% lavender and chamomile in sweet almond oil, 46 patients were randomized to receive either traditional day care alone or day care plus weekly aromatherapy massages for 4 weeks. At baseline and at weekly intervals thereafter, patients self-rated their mood, QOL, and the severity of the two symptoms that affected them the most. Only 11 out of 23 patients (48%) in the aromatherapy group and 18 out of 23 patients (78%) in the control group completed all 4 weeks of the study. Both groups showed improvements in mood, symptoms, and patient-reported QOL, and none of these metrics showed statistically sig-nificant differences between the groups [89]. Some pharmacological effects are shown in Table 16.3.

Table 16.3: Populations, diseases, and/or symptoms that were the subject of research on essential oil-based therapies.

Disorder	Remarks regarding the study samples	Outcome category	Specific symptom or measurable result	References
Acne	Patients with mild to severe cases of facial acne vulgaris are being tested.	Physical	Skin lesion count, acne severity	[90]

Table 16.3 (continued)

Disorder	Remarks regarding the study samples	Outcome category	Specific symptom or measurable result	References
Anxiety, healthcare waiting spaces	Patients awaiting dental or surgical procedures were included in the sample.	Psychological	Anxiety	[91]
Anxiety, various populations	Clinically anxious individuals, patients in different medical contexts or circumstances (such as cancer, perioperative procedures, or colonoscopies), and healthy volunteers were included in the samples.	Psychological	Anxiety	[92]
Complications, hemodialysis	Persons with kidney diseases/malfunction	Psychological physical, global, sleep	Anxiety, depression, stress, fatigue, and sleep quality	[93]
Depressive symptoms, various populations	Out of eight qualified studies, only one focused on individuals who had anxiety or depression. The others recruited participants who were either healthy or had a variety of medical issues (e.g., mothers of children with ADHD, cancer, pregnancy, or postpartum).	Psychological	Depressive symptoms measured by various scales	[94]
Nausea/vomiting, postoperative	Research on individuals undergoing various types of surgery	Nausea/vomiting	Nausea severity	[95]
Pain, postoperative	Samples included various surgical procedures	Physical	Pain	[96]
Psychological symptoms, women aged 45+	Peri- and post-menopausal women	Psychological	Anxiety; depression	[97]
SBP/DBP, hypertension	Patients with hypertension or prehypertension	Physical	SBP, DBP	[98]
Sleep, various populations	The majority of the volunteers in the sample were healthy, although some had ischemic heart disease or sleeplessness.	Sleep	Sleep quality	[99]

Table 16.3 (continued)

Disorder	Remarks regarding the study samples	Outcome category	Specific symptom or measurable result	References
Various symptoms, dementia		Psychological global	Cognitive function, behavioral and psychological symptoms of dementia, and agitation	[100]

Abbreviations: SBP, systolic blood pressure; DBP, diastolic blood pressure; ADHD, attention-deficit/hyperactivity disorder.

16.8 Aromatherapy side effects

Other natural substances, such as EOs, may also have negative effects or be hazardous. Some EOs should never be used in aromatherapy because of their potential toxicity. Toxic oils can damage the kidneys, liver, and nervous system. These oils have more harmful effects when consumed. Additionally, some EOs used in aromatherapy may experience negative effects from prolonged exposure to direct sunlight. We should avoid applying oils like cumin, lemon, or orange to any part of our bodies that will be exposed to direct sunlight because of the risk of sunburn. One of the most common side effects of aromatherapy is the use of oils that might irritate the skin. Rashes, itching, and burning are just a few of the negative effects that oils with skin-irritating properties can cause, depending on how sensitive a person's skin is. Certain oils, such as lemon and peppermint, should never be applied directly, while others are safe to use when diluted appropriately. If EOs have the potential to irritate the skin when used in aromatherapy, caution is advised, and very low doses (about 1%) should be utilized [101].

16.9 Plants utilized in aromatherapy

Because they naturally emit EOs into the air when their leaves are crushed or touched, aromatic plants are excellent additions to outdoor gardens. They can be utilized in a garden for a variety of functions, such as cooking and making remedies at home. They are also a popular choice among gardeners due to their pleasant scent and potential health benefits. Due to the presence of EOs and volatile compounds in many plant materials, such as flowers, bark, stems, leaves, roots, and fruits, many plants have been used in aromatherapy. EO plants are multipurpose because they can be used as ingredients for food and beverages, medicines, cosmetics, and even as repellents. Because EOs and extracts contain significant levels of volatile, aromatic, and bioactive chemicals, they are highly valuable in a variety of sectors [102]. The pharmaceutical, food, cosmetic, agricultural, and health industries all depend heavily on these potent substances, which naturally possess antibacterial and antioxidant properties. This review provides a contemporary perspective on the use of EOs, plant extracts, and their constituents in food, agriculture, and cosmetic products, with a focus on their antibacterial, antifungal, herbicidal, and other properties. Additionally, EOs can be utilized in combination therapy, also known as polytherapy, which is a medical treatment that combines multiple approaches. Restoring the patient's physical and mental well-being enables them to resume their social lives. Middle-aged and older adults can benefit from aromatherapy as an adjuvant therapy to reduce stress and enhance their physical and mental well-being [104]. In this section, some of the plants used in aromatherapy will be presented.

16.9.1 Lavender

Lavender EO, usually obtained from *Lavandula angustifolia* Mill. (Figure 16.4) of the Lamiaceae family, is a multifunctional herbal EO. Lavender is a common household decor item and is often used as a wedding complement with dried flowers. Its well-known fragrance is utilized in cosmetics, balms, and soaps. Lavender is renowned for its relaxing, soothing, anti-inflammatory, pain-relieving, and anti-infective properties. It can balance and regulate the central nervous system and has significant calming effects on the nervous system. It can successfully alleviate symptoms of nervous system imbalance-related conditions such as bipolar disorder, anxiety, depression, headaches, migraines, and nervous sleeplessness. Additionally, it possesses sedative, relaxing, and circulatory system-calming properties. This plant's stunning purple blossoms provide a delightful scent that eases anxiety and encourages sleep. Its oil is frequently used in aromatherapy to promote relaxation and reduce stress [105, 106].

Figure 16.4: *Lavandula angustifolia* Mill.

16.9.2 Mint

The genus Mentha, sometimes referred to as peppermint (*Mentha piperita*) (Figure 16.5) or simply mint, is a member of the taxonomic family Lamiaceae and is found in temperate regions of the world. A variety of substances found in mint are categorized as either nonessential or mint EO. A living space can be revitalized by adding a few drops of mint oil to a diffuser. Mint is a cooling herb with a pleasant scent that is frequently used. The strong scent of mint leaves relieves headaches, increases focus, and reduces fatigue. It is widely used in teas and EOs to revitalize the body and soul. Mint is easy to cultivate in garden beds or containers and has a refreshing, energizing taste. Its leaves can be used in many different types of food and beverages. In addition to being delicious and used as a spice or tea, mint oil has medicinal properties that help with headaches, digestion, and enhancing energy levels. It can also help alleviate digestive discomfort, soothe muscle aches and pains, and repel insects. Mint EO is a mixture of volatile metabolites with anti-inflammatory, antibacterial, antiviral, scolicidal, immunomodulatory, anticancer, neuroprotective, antifatigue, and antioxidant properties. It is mostly composed of menthol, menthone, neomenthol, and iso-menthone. There is growing evidence that this EO may have hypoglycemic and hypolipidemic effects, in addition to pharmacologically protecting the respiratory, skin, gastrointestinal, liver, kidney, brain, and neurological systems [107, 108].

Figure 16.5: Mentha piperita.

16.9.3 Chamomile

One of the most valued therapeutic plants in the world is chamomile (*Matricaria chamomilla* L.). It is usually called chamomile or chamomile, German chamomile, Hungarian chamomile (kamilla), wild chamomile, blue chamomile, or scented mayweed. Chamomile is an annual plant of the family Asteraceae. In order to treat a wide range of illnesses, the dried flower heads and EO are used as mouthwashes, tisanes, baths, and inhalations [109]. Inflammatory diseases, bacterial infections, skin and mucosal lesions (oral cavity, gastrointestinal tract, and respiratory tract), gastrointestinal tract spasms and ulcers, sleeplessness, and nervousness are among the therapeutic indications [109]. Many traditional medical traditions have made extensive use of *Matricaria chamomilla* L. In order to assess the many activities that *M. chamomilla* exhibits, such as antibacterial, antioxidant, anti-inflammatory, antiulcer, hypoglycemic, hypolipidemic, cardioprotective, hepatoprotective, neuroprotective, nephroprotective, antispasmodic, wound healing, and anticancer properties, as well as medical conditions like anxiety, sleep deprivation, and depression, numerous preclinical and clinical studies have been carried out. There are various chemotypes of *Matricaria chamomilla* L. EO, which is frequently utilized for both medicinal and cosmetic purposes. Because it is mild and calming, chamomile EO is perfect for encouraging relaxation, lowering anxiety, and enhancing the quality of sleep. Having a soothing cup of chamomile tea before bed is a great way to improve the quality of your sleep [110].

16.9.4 Eucalyptus

In the Myrtaceae family, the genus *Eucalyptus* consists of about 700 species of flowering plants. A small number of eucalyptus species are shrubs, while the majority are trees, frequently mallees. They are referred to as "gum trees" or eucalypts, along with a number of other genera. *Eucalyptus globulus* trees, which are native to Australia, are now planted for their therapeutic qualities all over the world. The oval-shaped leaves of the tree are used to make eucalyptus oil, which gives them their medicinal properties. The EO is extracted by crushing, drying, and distilling the leaves. Before the oil can be utilized as a complementary alternative medicine or in aromatherapy, it needs to be diluted after extraction. Purifying, cleansing, clarifying, and immune-boosting, eucalyptus EO is perfect for use on skin, in aromatherapy, and as a fabric refresher and surface cleaner. Approximately 500 different types of eucalyptus are used to make EOs. Although some types have subtle aroma differences, they all have similar medicinal properties and a distinctive fresh, camphoraceous scent. Because of its calming, invigorating, and antibacterial qualities, eucalyptus EO is a common component in massage blends, balms, inhalers, and oral hygiene products. Eucalyptus EO is known for its respiratory benefits. Inhaling eucalyptus oil provides relief from respiratory issues, acting as a decongestant, which is especially beneficial during cold and flu seasons. It helps with coughing, chest cleansing, wound disinfection, and other health issues. A spa-like atmosphere can be created at home by adding a few drops of eucalyptus oil to your shower for a refreshing and invigorating experience. It can enhance respiratory health, ease sinusitis and bronchitis symptoms, and assist in eliminating congestion. Additionally, eucalyptus oil works well to keep mosquitoes away and to improve mental clarity [111–113].

16.9.5 Citrus

One of the primary fruit crops grown in tropical and subtropical climates across the world is citrus, members of the Rutaceae (or Citrus) family (Figure 16.6). The most commercially important citrus species are sweet oranges (*Citrus sinensis* L. Osbeck) and tangerines (*Citrus unshiu* Marc., *Citrus nobilis* Lour., *Citrus deliciosa* Ten., *Citrus reticulata* Blanco, and their hybrids) (more than 80%), followed by lemons (*Citrus limon* L. Burm. f.), limes (*Citrus aurantifolia* Christm. Swing.), and grapefruits (*Citrus paradisi* Macf.). Although they are mostly cultivated for fresh consumption, the food sector also uses them to make fruit juice. Additionally, the waste materials produced after their industrial processing are a source of significant bioactive substances that could be used in manufactured meals, animal feed, and medical treatment [114]. Citrus EO is a complex blend of over 400 chemicals that has been proven to be effective in aromatherapy, fragrances, medicines, food and beverage color enhancers, and aromatic infusions for personal healthcare. Citrus EOs have a pleasant aroma and provide calming, uplifting, cheering, and relaxing qualities. They are used in aromatherapy either in diffusion sprays for homes and cars or in massage oils. The diffusion improves relaxation from stress and anxiety, lifts the spirits, and increases both physical and emotional vitality. The rinds or peels of different citrus fruits are used to make citrus EOs. In addition to their energizing and revitalizing aromas, citrus EOs are well-known for their many health advantages and versatile applications. Citrus EOs are a popular choice for many people and a traditional inclusion in many EO blends because of their various benefits, which range from aromatherapy to skincare and household cleaning. Oils like lemon EO, sweet orange EO, and grapefruit EO are very helpful at elevating mood and lowering stress because of their energetic and uplifting aromas. Citrus EOs have the ability to stimulate and revitalize the senses, encouraging relaxation and a sense of well-being. A cheerful and lively mood can be created by simply adding a few drops of citrus oil or a citrus oil combination to an EO diffuser or by directly inhaling their aroma. The EO's qualities include a fresh and aromatic aroma as well as a calming, soothing, and invigorating effect. In addition to sterilizing and healing wounds, it can aid in reducing skin irritation. It has the ability to regulate the autonomic nervous system and alleviate the symptoms of stress, anxiety, and depression [115]. Mixtures of volatiles and EOs derived from plants are used to symbolize their "essence" or odoriferous components. They have been used as flavoring elements in cosmetics and fragrances since ancient times because of their therapeutic properties. Cold-pressed citrus oils are known to produce phototoxic reactions because they include photoactive furocoumarins (psoralens), a family of naturally occurring plant components having a coumarin basic structure linked to a furan ring [116].

Figure 16.6: *Citrus limon* L.

16.9.6 Rosemary

Salvia rosmarinus (Figure 16.7), often referred to as rosemary, belongs to the Lamiaceae family. It is a shrub with aromatic, evergreen, needle-like leaves and flowers
that can be white, pink, purple, or blue, and is known for its distinct aroma. It is native to Portugal, Spain, and the Mediterranean region. In addition to its culinary applications, rosemary oil is thought to improve mental clarity, memory, and focus. The
delightful scent of rosemary enhances memory, sharpens focus, and reduces fatigue. Incorporating the refreshing scent of rosemary into the workplace can increase concentration and enhance productivity. It is frequently used in aromatherapy to stimulate the
senses and improve mental clarity. Beyond its fragrant appeal, rosemary has antioxidant
properties and is used in cooking because of its amazing flavor and potential health benefits. This resilient plant thrives in gardens and containers, enabling ease of maintenance.
The highest-quality EO of rosemary is extracted from the plant's blooming tops and is
well-known for its calming, pain-relieving, and energizing effects. When used in aromatherapy, rosemary oil boosts mental activity, promotes insight and clarity, eases weari-
ness, supports respiratory function, and lowers stress and nervous tension. By strengthening concentration, it helps people become more attentive, dispel negative emotions,
and retain more knowledge. In addition to increasing appetite, the aroma of rosemary
EO is believed to lower the amount of negative stress hormones released during
stressful situations. It also relieves congestion in the throat and nose by clearing the
respiratory tract and strengthens the immune system by promoting internal antioxidant activity, which combats diseases caused by free radicals. Rosemary EO, when diluted and applied topically, has been shown to boost hair growth, reduce pain, decrease inflammation, alleviate headaches, enhance immunity, and condition hair to
make it feel and look healthy. The purifying qualities of rosemary oil, when applied
during a massage, can promote a healthy digestive system, alleviate cramps, bloating,

and flatulence, and ease constipation. This oil promotes circulation through massage, improving the body's ability to absorb nutrients from meals [117, 118].

Figure 16.7: Salvia rosmarinus.

16.9.7 Frankincense

Frankincense is obtained from the resin of the Boswellia tree, which is native to the Middle East and Africa. Boswellia plants, belonging to the Burseraceae family, yield frankincense, sometimes called olibanum or myrrh, an aromatic resin used in incense and perfumes. *Boswellia sacra* (syn. *Boswellia bhaw-dajiana*, syn. *Boswellia carteri*), *Boswellia frereana*, *Boswellia serrata*, and *Boswellia papyrifera* are among the species of Boswellia that provide real frankincense [119]. Depending on when it is harvested, different grades of resin are available from each species. For quality, the resin is hand-sorted and has a rich history in spiritual rituals. In addition to its spiritual significance, frankincense oil has therapeutic uses and is well-known for its anti-inflammatory qualities and skincare applications. Frankincense is one of more than 90 types of EOs that are frequently used in aromatherapy. Despite not being one of the most popular oils, frankincense may be beneficial for human health. It can also be found as an extract or resin and is occasionally offered for sale as a nutritional supplement or as a component of skin care or other products. It is frequently characterized as woody, toasty, and rich, with the addition of spicy, fruity, or lemony notes. Frankincense has been utilized for thousands of years and has long been a part of Chinese, Indian, and Islamic medicine [119]. Research shows that frankincense naturally exhibits many activities, such as anti-inflammatory, antimicrobial, analgesic, and anticancer properties, and it may have various effects on the immune system. This characteristic is referred to as immune-modulating. *Boswellia serrata* is an Indian tree that produces special compounds with strong anti-inflammatory and potentially anticancer effects. Two of the most beneficial extracts from Boswellia trees are terpenes and boswellic acids, which have strong anti-inflammatory properties. Frankincense offers numerous advantages, such as lowering stress and negative emotions, strengthening the immune system and preventing disease, preventing cancer and its side effects, protecting the skin and preventing signs of aging, improving memory, balancing hormones and enhancing fertility, promoting sleep, easing digestion, reducing inflammation, and alleviating pain [120].

16.10 Conclusion

We may conclude that aromatherapy is gaining popularity among people nowadays as a means of reducing stress and evoking specific feelings. In the realm of aromatherapy, EOs are crucial. In addition to alleviating the symptoms of diseases, the use of perfumes revitalizes the body as a whole. Aromatherapy regulates the body, mind, and soul's elevation for a new phase of life. This treatment can be used for both the acute and chronic stages of a condition, and it is also preventive. The pharmaceutical industry is looking for natural, alternative, and eco-friendly treatments for illnesses linked to metabolism and infections. By utilizing EOs, it might be possible to improve the bioavailability and reaction rate of medications. These volatile oils may work in concert with medications used to treat problems of the central nervous system if they are well-researched. Furthermore, there is disagreement on when the plant has the highest concentration of volatile oil with various chemical components. As long as safety and quality are considered, EOs can be a helpful nonmedical alternative or used in conjunction with traditional treatments for some medical disorders. The scientific community's preference for complementary and alternative medicine has raised hopes that EOs can alleviate the negative effects of contemporary medicine. Aromatherapy is neither a science nor a medicine that should be used to treat medical concerns, even if most people report short-term advantages from massages, baths, and candles. Not every type of aromatherapy is beneficial to our health. Customers should use caution before utilizing aromatherapy. Certain oils may have negative effects on people with specific medical conditions, including pregnant women. The study of aromatherapy is still in its infancy. More research is needed to reach a scientific conclusion about the use and effects of aromatherapy.

References

[1]   https://www.merriam-webster.com/dictionary/aromatherapy
[2]   Mojay, G. What Is Aromatherapy?, National Association for Holistic Aromatherapy, https://naha.org/explore-aromatherapy/about-aromatherapy/what-is-aromatherapy/.
[3]   Hedaoo S.A., Chandurkar P.A. (2019) A Review on Aromatherapy. 8(7), 635–651.

[4]   Vora, L. K., Gholap, A. D., Hatvate, N. T., Naren, P., Khan, S., Chavda, V. P., Balar, P. C., Gandhi, J. and Khatri, D. K. (2024). Essential oils for clinical aromatherapy: A comprehensive review. Journal of Ethnopharmacology, 330, 118180.

[5]   Libster, M. M. (2015). Evolution of aromatherapy. In: Buckle, J. (ed) Clinical Aromatherapy Essential Oils in Healthcare, 3rd edition, Elsevier, St Louis (MO), 2–14.

[6]   Krishna, A., Tiwari, R. and Kumar, S. (2000). Aromatherapy-an alternative health care through essential oils. Journal of Medicinal and Aromatic Plant Sciences, 22, 798–804.

[7]   Worwood, V. A. (2000). Aromatherapy for the Healthy Child: More than 300 Natural, Non-toxic, and Fragrant Essential Oil Blends, New World Library, Novato.

[8]   Svoboda, K. P. and Deans, S. G. (1995). Biological activities of essential oils from selected aromatic plants. Acta Horticulturae, 390, 203–209.

[9]   Dunning, T. (2013). Aromatherapy: Overview, safety and quality issues. OA Altern Med, 1(1), 6.

[10]   Esposito, E. R., Bystrek, M. V. and Klein, J. S. (2014). An elective course in aromatherapy science. American Journal of Pharmaceutical Education, 78(4), 79.

[11]   Schiller, C. and Schiller, D. (1994). 500 Formulas for Aromatherapy: Mixing Essential Oils for Every Use, Sterling Publications, USA.

[12]   Elshafie, H. S. and Camele, I. (2017). An overview of the biological effects of some Mediterranean essential oils on human health. BioMed Research International, 2017, 9268468.

[13]   Aromatherapy history. http://www.aromatherapy.com/history.html

[14]   Gattefossé, R. M. (1937). AromathéRapie; les huiles essentielles hormones végétales. Librairie Des Sciences, Girardot & Cie, 187 pages.

[15]   Valnet, J. (1990). The Practice of Aromatherapy: A Classic Compendium of Plant Medicines & Their Healing Properties, Healing Arts Press, Rochester, USA, 280 pages.

[16]   Valnet, J., Bruno, F., Mahé, A., Le Prestre, C., Lambert, G. and Ménétrier, J. (1964). Les Médecines Différentes, éd, Planète, Paris.

[17]   Valnet, J. (1964). L'Aromathérapie Ou Aromathérapie, Traitement Des Maladies Par Les Essences Des Plantes, Paris, éd. Le Livre de Poche N°7885, Paris.

[18]   Valnet J. Duraffourd C, Lapraz JC, (1978) Une médecine nouvelle. Phytothérapie et aromathérapie - comment guérir les maladies infectieuses par les plantes, éd. Presses de la Renaissance, Paris, France.

[19]   Franchomme P, Jollois R., Pénoël D. (2001) L'aromatherapie exactement: encyclopédie de l'utilisation thérapeutique des huiles essentielles : fondements, démonstration, illustration et applications d'une science médicale naturelle, Published by Roger Jollois, Paris, France, 490 pages.

[20]   Lillehei, A. S. and Halcon, L. L. (2014). A review of the effect of inhaled essential oils on sleep. Journal of Alternative and Complementary Medicine, 20(6), 441–451.

[21]   Morone-Fortunato, I., Montemurro, C., Ruta, C., Perrini, R., Sabetta, W., Blanco, A., Lorusso, E. and Avato, P. (2010). Essential oils, genetic relationships and in vitro establishment of *Helichrysum italicum* (Roth) G. Don ssp. italicum from wild Mediterranean germplasm. Industrial Crops and Products, 32, 639–649.

[22]   Ahmadi, L., Mirza, M. and Shahmir, F. (2002). The volatile constituents of *Artemisia marschaliana* sprengel and its secretory elements. Flavour and Fragrance Journal, 17, 141–143.

[23]   Ciccarelli, D., Garbari, F. and Pagni, A. M. (2008). The flower of *Myrtus communis* (Myrtaceae): Secretory Structures, Unicellular Papillae, and Their Ecological Role, Flora, Flora - Morphology, Distribution, Functional Ecology of Plants, 203, 85–93.

[24]   Martín, A., Varona, S., Navarrete, A. and Cocero, M. J. (2010). Encapsulation and co-precipitation processes with supercritical fluids: Applications with essential oils. The Open Chemical Engineering Journal, 4, 31–41.

[25] Bezić, N., Šamanić, I., Dunkić, V., Besendorfer, V. and Puizina, J. (2009). Essential oil composition and Internal Transcribed Spacer (ITS) sequence variability of four South-Croatian satureja species (Lamiaceae). Molecules, 14, 925–938.

[26] Skold, M., Hagvall, L. and Karlberg, A. T. (2008). Autoxidation of linalyl acetate, the main compound of lavender oil, creates potent contact allergens. Contact Dermatitis, 58, 9–14.

[27] Gupta, V., Mittal, P., Bansal, P., Khokra, S. L. and Kaushik, D. (2010). Pharmacological potential of *Matricaria recutita*. International Journal of Pharmaceutical Science and Research, 2, 12–16.

[28] Tabanca, N., Demirci, B., Ozek, T., Kirimer, N., Baser, K. H. C., Bedir, E., Khan, I. A. and Wedge, D. E. (2006). Gas chromatographic–mass spectrometric analysis of essential oils from Pimpinella species gathered from central and Northern Turkey. Journal of Chromatography A, 1117, 194–205.

[29] Bowles, E. J. (2003). The Chemistry of Aromatherapeutic Oils, 3rd Edition, Griffin Press, New York, USA.

[30] Surburg, H. and Panten, J. (2006). Common Fragrance and Flavor Materials. Preparation, Properties and Uses, 5th Ed, WILEY-VCH, Weinheim.

[31] Miguel, M. G. (2010). Antioxidant and anti-inflammatory activities of essential oils. Molecules, 15, 9252–9287.

[32] Andrade, E. H. A., Alves, C. N., Guimarães, E. F., Carreira, L. M. M. and Maia, J. G. S. (2011). Variability in essential oil composition of *Piper dilatatum* L.C. Biochemical Systematics and Ecology, 39, 669–675.

[33] Dugo, G. and Di Giacomo, A., Eds. (2002). Citrus: The Genus Citrus, 1st ed, CRC Press, Taylor & Francis Group, London, UK.

[34] Ganem, F., Mattedi, S., Rodríguez, O., Rodil, E. and Soto, A. (2020). Deterpenation of citrus essential oil with 1-ethyl-3-methylimidazolium acetate: A comparison of unit operations. Separation and Purification Technology, 250, 117208.

[35] Fernández-Marín, R., Mujtaba, M., Cansaran-Duman, D., Ben Salha, G., Andrés Sánchez, M. Á., Labidi, J. and Fernandes, S. C. M. (2021). Effect of deterpenated *Origanum majorana* L. essential oil on the physicochemical and biological properties of Chitosan/β-Chitin Nanofibers Nanocomposite Films. Polymers (Basel), 13(9), 1507.

[36] Wu, Y. and Cheng, X. (2017). Aromatherapy, 3rd ed, New Wenjing, New Taipei City, Taiwan.

[37] Liu Y, Zhang J, Lam SP, Yu MW, Li SX, Zhou J, Chan JW, Chan NY, Li AM, Wing YK. (2016) Help-seeking behaviors for insomnia in Hong Kong Chinese: a community-based study. Sleep Medicine, 21, 106–113.

[38] Ziosi, P., Manfredini, S., Vertuani, S., Ruscetta, V., Radice, M. and Sacchetti, G. (2010). Evaluating essential oils in cosmetics: Antioxidant capacity and functionality. Cosmet Toilet, 125, 32–40.

[39] Ali, B., Al-Wabel, N. A., Shams, S., Ahamad, A., Khan, S. A. and Anwar, F. (2015). Essential oils used in aromatherapy: A systemic review. Asian Pacific Journal of Tropical Biomedicine, 5, 601–611.

[40] Mueller, S. M. and Grunwald, M. (2021). Effects, side effects and contraindications of relaxation massage during pregnancy: A systematic review of randomized controlled trials. Journal of Clinical Medicine, 10(2021), 3485.

[41] Horrigan, B., Lewis, S., Abrams, D. I. and Pechura, C. (2012). Integrative medicine in America; how integrative medicine is being practiced in clinical centers across the United States. Global Advances in Health and Medicine, 1, 18–94.

[42] Soden, K., Vincent, K., Craske, S., Lucas, C. and Ashley, S. (2004). A randomized controlled trial of aromatherapy massage in a hospice setting. Palliative Medicine, 18, 87–92.

[43] Chang, S. Y. (2008). Effects of aroma hand massage on pain, state anxiety and depression in hospice patients with terminal cancer. Taehan Kanho Hakhoe Chi, 38, 493–502.

[44] Maeda, K., Ito, T. and Shioda, S. (2012). Medical aromatherapy practice in Japan. Essence, 10, 14–16.

[45] Maxwell-Hudson, C. (1995). Aromatherapy Massage Book, Dorling Kindersley, London.

[46] Perry, N. and Perry, E. (2006). Aromatherapy in the management of psychiatric disorders clinical and neuropharmacological perspectives. CNS Drugs, 20, 257–280.

[47] Baratta, M. T., Dorman, H. J. D., Dean, S. G., Figueiredo, C., Barroro, J. G. and Ruberto, G. (1998). Antimicrobial and antioxidant property of some commercial essential oils. Flavour and Fragrance Journal, 13, 235–244.

[48] Liu, S. H., Lin, T. H. and Chang, K. M. (2013). The physical effects of aroma- therapy in alleviating work-related stress on elementary school teachers in Taiwan. Evid Based Complement Altern Med, 2013, 853809.

[49] Colgate, S. M. and Molyneux, R. J. (1933). Bioactive Natural Products Detection, Isolation and Structural Determination, CRC Press, Florida.

[50] Krishna, A., Tiwari, R. and Kumar, S. (2000). Aromatherapy-an alternative health care through essential oils. Journal of Medicinal and Aromatic Plant Sciences, 22, 798–804.

[51] Buchbauer, G. and Jirovetz, L. (1994). Aromatherapy-use of fragrances and essential oils as medicaments. Flavour and Fragrance Journal, 9, 217–222.

[52] Alan, R. H. (2005). Aromatherapy for Pain Relief. In Weiner's Pain Management-A Practical Guide for Clinicians, CRC Press, Boca Raton, FL, USA.

[53] Buckle, J. (2007). Literature review: Should nursing take aromatherapy more seriously?. British Journal of Nursing, 16, 116–120.

[54] Price, S. and Price, L. (2011). Aromatherapy for Health Professionals, 4th ed, Elsevier Churchill Livingstone, New York.

[55] Varney, E. and Buckle, J. (2013). Effect of inhaled essential oils on mental exhaustion and moderate burnout: A small pilot study. Journal of Alternative and Complementary Medicine (New York, NY), 19, 69–71.

[56] Fung Timothy, K. H., Lau, B. W. M., Ngai, S. P. C. and Tsang, H. W. H. (2021). Therapeutic effect and mechanisms of essential oils in mood disorders: Interaction between the nervous and respiratory systems. International Journal of Molecular Sciences, 22, 4844.

[57] Dobetsberger, C. and Buchbauer, G. (2011). Actions of essential oils on the central nervous system: An updated review. Essential oil effects on the CNS. Flavour and Fragrance Journal, 26, 300.

[58] Sanna, M. D., Les, F., Lopez, V. and Galeotti, N. (2019). Lavender (*Lavandula angustifolia* mill.) essential oil alleviates neuropathic pain in mice with spared nerve injury. Frontiers in Pharmacology, 10, 472.

[59] De Andrade, S. F., Rijo, P., Rocha, C., Zhu, L. and Rodrigues, L. M. (2021). Characterizing the mechanism of action of essential oils on skin homeostasis-data from sonographic imaging, epidermal water dynamics, and skin biomechanics. Insight Cosmetics, 8(2021), 36.

[60] Herman, A. and Herman, A. P. (2015). Essential oils and their constituents as skin penetration enhancer for transdermal drug delivery: A review. Journal of Pharmacy and Pharmacology, 67, 473–485.

[61] Caprari, C., Fantasma, F., Divino, F., Bucci, A., Iorizzi, M., Naclerio, G., Ranalli, G. and Chemical Profile, S. G. (2021). *In vitro* biological activity and comparison of essential oils from fresh and dried flowers of *Lavandula angustifolia* L. Molecules, 26, 5317.

[62] Masoumeh, B. N., Seyed, A. S., Attieh, N., Fatemeh, E. and Fahimeh-Sadat, G. K. (2016). The effects of aromatherapy with lavender essential oil on fatigue levels in haemodialysis patients: A randomized clinical trial. Complementary Therapies in Clinical Practice, 22, 33–37.

[63] Džudžević-Čančar, H., Alispahić, A., Dedić-Mahmutović, A., Dž, S., Uzunović, A., Boškailo, E. and Čančar, I. F. (2024). Essential oils of garden-growing lavender species: *In vitro* antimicrobial activity. In: 54th International Symposium on Essential Oils - Book of Abstracts, Balatonalmádi, Hungary, 28.

[64] Džudžević-Čančar, H., Dedić, A., Alispahić, A., Jerković, I., Boškailo, E., Stanojković, T., Marijanović, Z. and Kubat, N. (2022 CNP). Essential oil chemical profile of two different lavender species from Sarajevo gardens. Bulletin of the Chemists and Technologists of Bosnia and Herzegovina, 02, 74.

[65] Džudžević-Čančar, H., Telci, I., Stanojković, T., Yavuz, M., Özek, T., Dedić-Mahmutović, A., Alispahić, A. and Yalçin, Ö. Ü. (2024). *Mentha gentilis* var. citrata essential oil phytochemical constituents and antiproliferative activity. In: 54th International Symposium on Essential Oils - Book of Abstracts, Balatonalmádi, Hungary, 27.

[66] Mokaberinejad, R., Zafarghandi, N., Bioos, S., Dabaghian, H. F., Naseri, M., Kamalinejad, M., Amin, G., Ghobadi, A., Tansaz, M., Akhbari, A. and Hamiditabar, M. (2012). *Mentha longifolia* syrup in secondary amenorrhea: A double-blind, placebo-controlled, randomized trials. DARU Journal of Pharmaceutical Sciences, 20(1), 97–104.

[67] Hajlaoui, H., Snoussi, M., Ben Jannet, H., Mighri, Z. and Bakhrouf, A. (2008). Comparison of chemical composition and antimicrobial activities of *Mentha longifolia* L. ssp. longifolia essential oil from two Tunisian localities (Gabes and Sidi Bouzid). Annals of Microbiology and Immunology, 58, 103–110.

[68] Al-Okbi, S. Y., Fadel, H. H. M. and Mohamed, D. A. (2015). Phytochemical constituents, antioxidant and anticancer activity of *Mentha citrata* and *Mentha longifolia*. Research Journal of Pharmaceutical, Biological and Chemical Sciences, 6, 739–751.

[69] Boškailo, E., Džudžević-Čančar, H., Dedić, A., Marijanović, Z., Alispahić, A., Čančar, I. F., Vidic, D. and Jerković, I. (2023). *Clinopodium nepeta*. In: (L.) Kuntze from Bosnia and Herzegovina: Chemical Characterisation of Headspace and Essential Oil of Fresh and Dried Samples, Records of Natural Products, 17(2), 300–311.

[70] Kitic, D., Stojanovic, G., Palic, R. and Randjelovic, V. (2005). Chemical composition and microbial activity of the essential oil of *Calamintha nepeta* (L.) Savi ssp. nepeta var. subisodonda (Borb.) Hayek from Serbia. He Journal of Essential Oil Research, 17, 701–703.

[71] Acimovic, M. G., Loncar, B., Ljj, V., DJ, L. L. P., Ljujic, J. P., Miljkovic, A. R. and Vujisic, L. V. (2022). Comparison of volatile compounds from clary sage (*Salvia sclarea* L.) verticillasters essential oil and hydrolate. Journal of Essential Oil-Bearing Plants, 25, 555–570.

[72] Jerbi, A., Derbali, A., Elfeki, A. and Kammoun, M. (2017). Essential oil composition and biological activities of *Eucalyptus globulus* leaves extracts from Tunisia. Journal of Essential Oil-Bearing Plants, 20, 438–448.

[73] Si Said, Z. B.-O., Haddadi-Guemghar, H., Boulekbache-Makhlouf, L., Rigou, P., Remini, H., Adjaoud, A., Khoudja, N. K. and Madani, K. (2016). Essential oils composition, antibacterial and antioxidant activities of hydrodistillated extract of *Eucalyptus globulus* fruits. Industrial Crops and Products, 89, 167–175.

[74] Moses, S. O., Opeyemi, N. A., Isiaka, A. O., William, N. S., Rukayat, O., Akintayo, L. O., Oladipupo, A. L. and Guido, F. (2018). Chemical composition of *Citrus limon* (L.) Osbeck growing in southwestern Nigeria: Essential oil chemo types of both peel and leaf of lemon. American Journal of Essential Oils and Natural Products, 6, 36–40.

[75] Aguilar-Hernandez, M. G., Sanchez-Bravo, P., Hernandez, F., Carbonell-Barrachina, A. A., Pastor-Perez, J. J. and Legua, P. (2020). Determination of the volatile profile of lemon peel oils as affected by rootstock. Foods, 9, 241.

[76] Mrani, S. A., Zejli, H., Azzouni, D., Fadili, D., Alanazi, M. M., Hassane, S. O. S., Sabbahi, R., Kabra, A., Moussaoui, A. E., Hammouti, B. and Taleb, M. (2024). Chemical composition, antioxidant, antibacterial, and hemolytic properties of ylang-ylang (*Cananga odorata*) essential oil: Potential therapeutic applications in dermatology. Pharmaceuticals, 17, 1376.

[77] Zhang, N., Zhang, L., Feng, L. and Yao, L. (2018). Cananga odorata essential oil reverses the anxiety induced by 1-(3-chlorophenyl) piperazine through regulating the MAPK pathway and serotonin system in mice. Journal of Ethnopharmacology, 219, 23–30.

[78] Aridoğan, B. C., Baydar, H., Kaya, S., Demirci, M., Ozbaşar, D. and Mumcu, E. (2002). Antimicrobial activity and chemical composition of some essential oils. Archives of Pharmacal Research, 25(6), 860–864.

[79] Minami, M., Kita, M., Nakaya, T., Yamamoto, T., Kuriyama, H. and Imanishi, J. (2003). The inhibitory effect of essential oils on herpes simplex virus type-1 replication *in vitro*. Microbiology & Immunology, 47(9), 681–684.

[80] D'Auria, F. D., Tecca, M., Strippoli, V., Salvatore, G., Battinelli, L. and Mazzanti, G. (2005). Antifungal activity of *Lavandula angustifolia* essential oil against *Candida albicans* yeast and mycelial form. Medical Mycology, 43(5), 391–396.

[81] Buchbauer, G., Jirovetz, L., Jäger, W., Plank, C. and Dietrich, H. (1993). Fragrance compounds and essential oils with sedative effects upon inhalation. Journal Pharmaceutical Sciences, 82(6), 660–664.

[82] Aloisi, A. M., Ceccarelli, I., Masi, F. and Scaramuzzino, A. (2002). Effects of the essential oil from citrus lemon in male and female rats exposed to a persistent painful stimulation. Behavioural Brain Research, 136(1), 127–135.

[83] Jahangeer, A. C., Mellier, D. and Caston, J. (1997). Influence of olfactory stimulation on nociceptive behavior in mice. Physiology & Behavior, 62(2), 359–366.

[84] Cooke, B. and Ernst, E. (2000). Aromatherapy: A systematic review. The British Journal of General Practice, 50(455), 493–496.

[85] Wilkinson, S., Aldridge, J., Salmon, I., Cain, E. and Wilson, B. (1999). An evaluation of aromatherapy massage in palliative care. Palliative Medicine, 13(5), 409–417.

[86] Kite, S. M., Maher, E. J., Anderson, K., Young, T., Young, J., Wood, J., Howells, N. and Bradburn, J. (1998). Development of an aromatherapy service at a Cancer Centre. Journal of Clinical Oncology, 12(3), 171–180.

[87] Graham, P. H., Browne, L., Cox, H. and Graham, J. (2003). Inhalation aromatherapy during radiotherapy: Results of a placebo-controlled double-blind randomized trial. Journal of Clinical Oncology, 21(12), 2372–2376.

[88] Wilcock, A., Manderson, C., Weller, R., Walker, G., Carr, D., Carey, A. M., Broadhurst, D., Mew, J. and Ernst, E. (2004). Does aromatherapy massage benefit patients with cancer attending a specialist palliative care day centre? Palliative Medicine, 18(4), 287–290.

[89] Blackburn, L., Achor, S., Allen, B., Bauchmire, N., Dunnington, D., Klisovic, R. B., Naber, S. J., Roblee, K., Samczak, A., Tomlinson-Pinkham, K. and Chipps, E. (2017). The effect of aromatherapy on insomnia and other common symptoms among patients with acute leukemia. Oncology Nursing Forum, 44(4), E185–E193.

[90] Cao, H., Yang, G., Wang, Y., Liu, J. P., Smith, C. A., Luo, H. and Liu, Y. (2015). Complementary therapies for acne vulgaris. The Cochrane Database of Systematic Reviews, 1(1), CD009436.

[91] Biddiss, E., Knibbe, T. J. and McPherson, A. (2014). The effectiveness of interventions aimed at reducing anxiety in health care waiting spaces: A systematic review of randomized and nonrandomized trials. Anesthesia Ans Analgesia, 119(2), 433–448.

[92] Perry, R., Terry, R., Watson, L. K. and Ernst, E. (2012). Is lavender an anxiolytic drug? A systematic review of randomised clinical trials. Phytomedicine, 19(8–9), 825–835.

[93] Bouya, S., Ahmadidarehsima, S., Badakhsh, M., Balouchi, A. and Koochakzai, M. (2018). Effect of aromatherapy interventions on hemodialysis complications: A systematic review. Complementary Therapies in Clinical Practice, 32, 130–138.

[94] Sanchez-Vidana, D. I., Ngai, S. P., He, W., Chow, J. K., Lau, B. W. and Tsang, H. W. (2017). The effectiveness of aromatherapy for depressive symptoms: A systematic review. Journal of Evidence-Based Complementary & Alternative Medicine, 2017, 1–21.

[95] Chen, S. F., Wang, C. H., Chan, P. T., Chiang, H. W., Hu, T. M., Tam, K. W. and Loh, E. W. (2019). Labour pain control by aromatherapy: A meta-analysis of randomised controlled trials. Women and Birth: Journal of the Australian College of Midwives, 32(4), 327–335.

[96] Dimitriou, V., Mavridou, P., Manataki, A. and Damigos, D. (2017). The use of aromatherapy for postoperative pain management: A systematic review of randomized controlled trials. Journal of Perianesthesia Nursing, 32(6), 530–541.

[97] Babakhanian, M., Ghazanfarpour, M., Kargarfard, L., Roozbeh, N., Darvish, L., Khadivzadeh, T. and Dizavandi, F. R. (2018). Effect of aromatherapy on the treatment of psychological symptoms in postmenopausal and elderly women: A systematic review and meta-analysis. Journal of Menopausal Medicine, 24(2), 127–132.

[98] Hur, M. H., Lee, M. S., Kim, C. and Ernst, E. (2012). Aromatherapy for treatment of hypertension: A systematic review. Journal of Evaluation in Clinical Practice, 18(1), 37–41.

[99] Lillehei, A. S. and Halcon, L. L. (2014). A systematic review of the effect of inhaled essential oils on sleep. Journal of Alternative and Complementary Medicine (New York, NY), 20(6), 441–451.

[100] Fung, J. K., Tsang, H. W. and Chung, R. C. (2012). A systematic review of the use of aromatherapy in treatment of behavioral problems in dementia. Geriatrics & Gerontology International, 12(3), 372–382.

[101] Posadzki, P., Alotaibi, A. and Ernst, E. (2012). Adverse effects of aromatherapy: A systematic review of case reports and case series. The International Journal of Risk & Safety in Medicine, 24(3), 147–161.

[102] Samadi, S., Asgari Lajayer, B., Moghiseh, E. and Rodríguez-Couto, S. (2021). Effect of carbon nanomaterials on cell toxicity, biomass production, nutritional and active compound accumulation in plants. Environmental Technology & Innovation, 21, 101323.

[103] Aćimović, M., Šovljanski, O., Šeregelj, V., Pezo, L., Zheljazkov, V. D., Ljujić, J., Tomić, A., Ćetković, G., Čanadanović-Brunet, J. and Miljković, A. (2022). Chemical composition, antioxidant, and antimicrobial activity of *Dracocephalum moldavica* L. essential oil and hydrolate. Plants, 11, 941.

[104] Hong, Y., Chen, M. and Li, P. (2001). Clinical application of integrated traditional Chinese and western medicine in cancer patients. Chinese Journal of Integrative Medicine, 3, 39–42.

[105] Barut, M., Tansı, L. S. and Karaman, S. (2022). Essential oil composition of lavender (*Lavandula angustifolia* Mill.) at various plantation ages and growth stages in the Mediterranean Region. Turkish Journal of Agriculture – Food Science and Technology, 10(4), 746–753.

[106] Prusinowska, R. and Śmigielski, K. B. (2014). Composition, biological properties and therapeutic effects of lavender (*Lavandula angustifolia* L.). A review. Herba Polonica, 60(2), 56–66.

[107] Zhao, H., Ren, S., Yang, H., Tang, S., Guo, C., Liu, M., Tao, Q., Ming, T. and Xu, H. (2022). Peppermint essential oil: Its phytochemistry, biological activity, pharmacological effect and application. Biomedicine and Pharmacotherapy, 154, 113559.

[108] Hudz, N., Kobylinska, L., Pokajewicz, K., Sedláčková V, H., Fedin, R., Voloshyn, M., Myskiv, I., Brindza, J., Wieczorek, P. P. and Lipok, J. (2023). *Mentha piperita*: Essential oil and extracts, their biological activities, and perspectives on the development of new medicinal and cosmetic products. Molecules, 28(21), 7444.

[109] Alvarado-García, P. A., Soto-Vásquez, M. R., Rodrigo-Villanueva, E. M., Gavidia-Valencia, J. G., Rodríguez, N. M., Rengifo-Penadillos, R. A. et al. (2024). Chamomile (*Matricaria chamomilla* L.) essential oil and its potential against stress, anxiety, and sleep quality. Journal of Young Pharmacists, 16(1), 100–107.

[110] Höferl, M., Wanner, J., Tabanca, N., Ali, A., Gochev, V., Schmidt, E., Kaul, V. K., Singh, V. and Jirovetz, L. (2020). Biological activity of *Matricaria chamomilla* essential oils of various chemotypes. Planta Medica, 07(03), e114–e121.

[111] Ben Marzoug, H. N., Romdhane, M., Lebrihi, A., Mathieu, F., Couderc, F., Abderraba, M., Khouja, M. L. and Bouajila, J. (2011). *Eucalyptus oleosa* essential oils: Chemical composition and antimicrobial and antioxidant activities of the oils from different plant parts (stems, leaves, flowers and fruits). Molecules, 16(2), 1695–1709.

[112] Lee, K. G. and Shibamoto, T. (2001). Antioxidant activities of volatile components isolated from Eucalyptus species. Journal of the Science of Food and Agriculture, 81, 1573–1597.

[113] Bignell, C. M., Dunlop, P. J., Brophy, J. J. and Jackson, J. F. (1994). Volatile leaf oils of some south-Western and southern Australian species of the genus Eucalyptus. part V. subgenus symphyomyrtus, section bisectaria, series oleosae. Flavour and Fragrance Journal, 10, 313–317.

[114] Agarwal, P., Sebghatollahi, Z., Kamal, M., Dhyani, A., Shrivastava, A., Singh, K. K., Sinha, M., Mahato, N., Mishra, A. K. and Baek, K. H. (2022). Citrus essential oils in aromatherapy: Therapeutic effects and mechanisms. Antioxidants (Basel), 11(12), 2374.

[115] Palazzolo, E., Laudicina, V. A. and Germanà, M. A. (2013). Current and potential use of citrus essential oils. Current Organic Chemistry, 17, 3042–3049.

[116] Tisserand, R. and Balocs, T. (1995). Essential Oil Safety. A Guide Health Care Professional, Elsevier Health Sciences, Edinburgh.

[117] Ghasemzadeh Rahbardar, M. and Hosseinzadeh, H. (2020). Therapeutic effects of rosemary (*Rosmarinus officinalis* L.) and its active constituents on nervous system disorders. Iranian Journal of Basic Medical Sciences, 23(9), 1100–1112.

[118] De Oliveira, J. R., Camargo, S. E. A. and de Oliveira, L. D. (2019). *Rosmarinus officinalis* L. (rosemary) as therapeutic and prophylactic agent. Journal of Biomedical Science, 26(5), 1–22.

[119] Al-Yasiry, A. R. M. and Kiczorowska, B. (2016). Frankincense – Therapeutic properties. Postepy Hig Med Dosw, 70, 380–391.

[120] Abbood, S. M., Kadhim, S. M., AlEthari, A. Y. H., AL-Qaisia, Z. H. and Mohammed, M. T. (2022). Review on frankincense essential oils: Chemical composition and biological activities. Misan Journal of Academic Studies, 21(44), 332–345.

Alireza Mirzaei* and Jamshid Fooladi

Chapter 17
Medicinal and aromatic plants with antioxidant properties

Abstract: In recent years, medicinal plants have shown significant effects on many diseases. The latest research on the medicinal and therapeutic effectiveness of these plants has demonstrated that about 92% of medicinal plants contribute to the treatment of three important groups of diseases: digestive, respiratory, and nervous system disorders. Additionally, these plants contain powerful compounds that can heal and relieve pain. On the other hand, the active substances in these plants can also prevent common and significant diseases such as Alzheimer's, cancer, diabetes, and cardiovascular diseases. The rarity of some medicines is attributed to the limited availability of these valuable plant resources. Fortunately, extensive research has been conducted on both the propagation and breeding of medicinal plants, as well as on understanding the effective compounds and essential oils derived from these plants. This research has laid the foundation for identifying many secondary metabolites, particularly effective antioxidants, in the treatment of diseases. It should also be noted that medicinal and aromatic plants are rich sources of biological compounds that neutralize free radicals and possess anti-inflammatory and antimicrobial properties, which help counteract harmful metabolic processes in the human body. This research specifically examines the antioxidant activity of medicinal and aromatic plants and provides valuable insights for researchers in this field. It is hoped that this body of research will lead to the recognition and discovery of effective natural antioxidants for therapeutic purposes by presenting a comprehensive plan for the propagation and cultivation of medicinal plants.

Keywords: aromatic plants, antioxidant activity, free radicals, antimicrobial activity

17.1 Introduction

Free radicals of internal or external origin are among the most important factors in the emergence of pathological conditions in inflammatory diseases, diabetes, cardiac and cerebral ischemia, cancer, immune deficiency, and aging in humans [1–4]. Numerous

*Corresponding author: Alireza Mirzaei, Department of Biotechnology, Faculty of Biological Sciences, Alzahra University, Tehran, Iran, e-mail: amirzaei25@gmail.com, https://orcid.org/0000-0002-7765-4948
Jamshid Fooladi, Department of Biotechnology, Faculty of Biological Sciences, Alzahra University, Tehran, Iran, e-mail: jfooladi@alzahra.ac.ir

https://doi.org/10.1515/9783111469713-017

experiments have shown that free radicals possess a reactive electron in their outermost shell. This structure creates an unstable state, which ultimately forces the resulting energy to react with nearby biochemical molecules [5, 6]. Antioxidants are compounds that, at a low concentration compared to the substrate, can reduce the speed of oxidative reactions through different mechanisms. Numerous internal and external antioxidant compounds, both natural and industrial, have been introduced to treat or prevent diseases related to free radicals. On the other hand, it has been proven that many industrial antioxidants used as preservatives in the food industry have side effects. Although the amount of consumption of these compounds as preservatives does not cause side effects, researchers have increasingly focused on finding antioxidants of natural origin [7, 8]. The antioxidant activity of several plant derivatives has been examined in the form of extracts, phenolic derivatives, and flavonoids, and in many cases, derivatives much more active than known compounds such as vitamins C, E, and beta-carotene have been identified. These derivatives are found in different parts of the plant, including leaves, stems, fruits, roots, and seeds. It has been proven that these types of derivatives and plants with antioxidant activity reduce the risk of degenerative diseases such as cancer, Parkinson's, arthritis, Alzheimer's, osteoporosis, and diseases related to joints and the spine, and they have a preventive effect against oxidative stress [9–11].

17.2 Free radicals

Body cells are constantly facing metabolic risks and chemical, biological, and physical threats. Imbalances in these reactions lead to many changes in the body. Oxidation-reduction reactions, biological changes, intracellular electron changes, and membrane reactions are all examples of such imbalances [12]. In general, free radicals are divided into three main groups: free radicals with oxygen, nitrogen, and sulfur. The reactive oxygen species (ROS) are the most well-known ROS. Reactive nitrogen species (RNS) are the most well-known active nitrogen species. Reactive sulfur species (RSS) are the most well-known active sulfur species [13]. Figure 17.1 shows the different types of free radicals.

Numerous experiments have shown that free radicals possess a reactive electron in their outermost shell. This structure creates an unstable state, ultimately forcing the resulting energy to interact with nearby biochemical molecules. The important characteristic shared among all of them is that these molecules have a strong tendency to take electrons from other molecules. This tendency causes the structure and function of molecules to change. In some cases, these changes are beneficial and reduce cellular threats. However, in other cases, these molecules can be dangerous, to the extent that they can even alter the coding and structure of DNA [14–16].

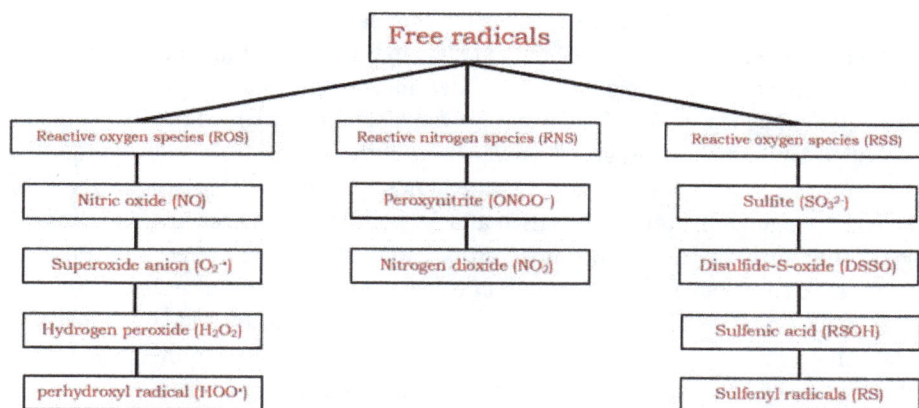

Figure 17.1: Types of free radicals.

17.3 Assessment of inhibition of free radicals

Measuring the level of inhibition of 2,2-diphenyl-1-picrylhydrazyl (DPPH) free radicals is one of the reliable, accurate, easy, and cost-effective methods with high reproducibility, which is used to examine the antioxidant activity of plant extracts under laboratory conditions. The DPPH stable radical scavenging model is widely used to evaluate the free radical scavenging ability in various samples. DPPH is one of the free and stable hydrophilic radicals with a dark purple color, whose maximum absorption is in the range of 515–517 nm. When receiving an electron from reducing compounds such as phenols, this radical turns into a colorless hydrazine form, and this structural change is associated with a decrease in absorption [17–19]. The compounds that can perform this action are proposed as antioxidants. Inhibition of free radicals is one of the most well-known mechanisms through which antioxidant compounds can inhibit the oxidation of fats.

17.4 Antioxidants

In general, the most important antioxidation system against oxidative stress is the presence of antioxidants. Therefore, if we look at the matter from a general point of view, we will realize that any substance that leads to the inhibition of oxidative stress, even on a small scale, is considered a type of antioxidant. Different molecules have antioxidant properties, but their two general groups are endogenous and exogenous antioxidants. In detail, it can be said that imbalance is one of the important parameters in oxidation. As we see in the cell membrane, this imbalance leads to oxidation and, in some cases, the leakage of molecules and elements in the opposite direction. Thus, one of the important tasks of antioxidants in cells is to maintain cell balance. In other classifications, antioxidants are divided into two groups: chain-breaking and preventive antioxidants. The first group prevents oxidation by breaking the chain of molecules, while the second group prevents the formation of molecules that lead to oxidation. Table 17.2 lists the active antioxidants of these two groups. As given in Table 17.1, flavonoids prevent oxidative attacks both by chain breaking and preventively. Therefore, the use of flavonoids in nutrition can be very effective. In recent years, researchers have named a group of diseases called radical diseases. These diseases include lung diseases, brain diseases, reproductive diseases, and circulatory diseases. The common factor between these diseases, which has led to their naming, is the presence of free radicals and abnormal oxidations in the body. This research has shown that the use of antioxidants is the most effective way to prevent these diseases [20–22].

Table 17.1: Types of antioxidants based on their mode of action.

Antioxidants that disrupt the molecular chain associated with oxidation	Antioxidants that prevent the formation of oxidation reactions
Albumin	Metallothionein
Bilirubin	Ceruloplasmin
Catalase	Selenium
Vitamins C and E	Ethylenediamine tetraacetate
Uric acid	Diethylenetriamine pentaacetate
Lipoic acid	Transferrin
Glutathione	Ferritin
Flavonoids	Myoglobin
Superoxide dismutase	Flavonoids
Glutathione peroxidase	

17.5 The role of antioxidants in the human body

The human body is constantly producing and consuming antioxidants and free radicals. Antioxidants and free radicals are opposites of each other. When the liver fights toxins in the body, it produces antioxidants. On the other hand, white blood cells create free radicals in the body to fight viral diseases and repair damaged cells. The properties of antioxidants in the body are such that they can help ensure human health. An imbalance between the levels of antioxidants and free radicals can lead to disease in the body. Another point that should be noted is that the use of medicinal plants is much more effective than the use of a single type of antioxidant because they contain several types of antioxidants.

The most important role of antioxidants is to neutralize free radical molecules. Free radicals can damage cell membranes, DNA, and other parts of the cell. On the other hand, the reduction of antioxidant levels in the body causes signs of aging and disease symptoms. As we know, medicinal plants are a rich source of natural antioxidants that can be effective in the treatment of many diseases. As a result, by consuming these plants, you can increase the amount of natural antioxidants in the body [23–26]. In the following, we will discuss in detail the most important medicinal plants with antioxidant properties.

17.6 Natural antioxidants found in medicinal and aromatic plants

Biological compounds found in nature have been used for many years as valuable resources for treatment. According to a report by the World Health Organization, approximately 80% of the world's population currently relies on these medicinal sources. Furthermore, in many countries, the discovery of medicine, as well as the livelihood and wellbeing of the population, depends on these medicinal plants. The history of the use of these plants indicates that, since around 5000 BC in the Middle East region, they have been utilized not only as medicine but also for their aroma and flavor in food. Today, herbal medicines are employed to address a wide range of diseases due to their soothing effects and antioxidant, anti-inflammatory, and antiseptic properties. These potential therapeutic effects are attributed to the chemical structures, physiological functions, and phytochemical properties of medicinal plants [27, 28].

17.7 Natural antioxidants

As shown in Figure 17.2, the natural antioxidants found in medicinal plants are divided into two general groups: enzymatic and nonenzymatic. In general, there are two classifications regarding antioxidants. In the first type, antioxidants are divided into two groups: soluble in water and soluble in fat. Water-soluble antioxidants react more with body fluids. Blood serum, plasma, and all types of extracellular fluids are among the fluids with which antioxidants are combined. Meanwhile, the antioxidants that are combined with fat molecules in the body are mostly responsible for protecting the cell membrane. It should also be noted that, in general, antioxidants with lower molecular weight combine more with body fluids and are sent to different tissues and organs through the blood [29–32].

In another division, antioxidants are categorized into natural and synthetic groups. The production of synthetic antioxidants is primarily due to the instability of natural antioxidants. These artificial antioxidants are also used to stabilize fats and oils. Butylated hydroxytoluene is a synthetic antioxidant formulated to protect petroleum against oxidative gumming. Additionally, ethoxyquin is another artificial antioxidant that is widely used in the feed industry. Natural antioxidants are found predominantly in sources such as herbs and vegetables with dark green leaves [33–36].

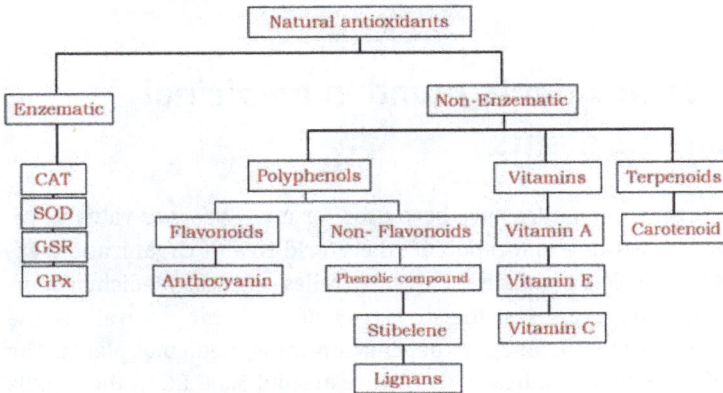

Figure 17.2: General classification of natural antioxidants.

17.7.1 Enzymatic antioxidants

Enzymatic antioxidants are endogenous antioxidants that neutralize ROS and RNS. In general, the unpaired electrons of these antioxidants are highly reactive and lead to the neutralization of harmful metabolites in the human body. These radicals contribute to detoxification through concentrated enzyme groups and thus prevent harmful oxidation in the body, even at low concentrations. The most important of these antioxidants are catalase, superoxide dismutase (SOD), glutathione reductase, and glutathione peroxidase [37–39].

17.7.1.1 Catalase (CAT)

Catalase, with the chemical formula $C_9H_{10}O_3$ is one of the natural antioxidants found in medicinal plants. As we know, oxygenated water H_2O_2 is one of the by-products of cellular reactions. This substance is highly toxic to cells and needs to be eliminated in some way. Certain immune cells use oxygenated water to kill microbes. Catalase is an effective enzyme involved in the decomposition of hydrogen peroxide. It ultimately facilitates the breakdown of hydrogen peroxide into water and oxygen. Catalase is abundant in the storage tissues of medicinal plants [40, 41].

17.7.1.2 Superoxide dismutase (SOD)

SOD, with the chemical formula $C_{10}H_5NO_4$ is one of the active antioxidants that leads to the inactivation of free radicals. This antioxidant neutralizes the harmful effects of O_2^-, OH^-, HOO^-, and H_2O_2. On the other hand, SOD can have a positive impact on other antioxidants by preventing the effect of free radicals on them. GP_X is one of the types of antioxidants that can remain active in the cellular environment for a long time as a result of SOD activity.

SOD is an active enzyme that is activated in response to inflammation, lipid metabolism, and various types of oxidative stress, reacting to these conditions. Controlling and improving diabetes through signaling pathways and lipid metabolism, preventing the aging process of cells by reducing skin inflammation, treating burns and skin darkening, improving physiological functions, and reducing liver toxicity are among the positive effects of SOD. On the other hand, this antioxidant plays a significant role in breeding processes and agricultural biotechnology. In this context, interaction with various elements and ions, such as copper and manganese, leads to increased resistance to environmental stresses such as cold, heat, and drought stress in plants. Increasing resistance to cold in potatoes and enhancing drought stress resistance in wheat are successful examples of the application of this antioxidant. However, one of the challenges in the activity of this antioxidant is the issue of permeability through the cell membrane. In general, the membrane acts as a barrier against the penetration of SOD, which has reduced the activity of this antioxidant in various studies. Although researchers believe that establishing a balance can address this issue, the problem has not yet been completely resolved, and research on this matter is still ongoing [42–44].

17.7.1.3 Glutathione reductase or glutathione-disulfide reductase (GSR)

Glutathione reductase has been identified in bacteria, yeast, pea, wheat germ, and also in most medicinal plants. Glutathione reductase is a vital antioxidant enzyme responsible for maintaining the antioxidant molecule GSH. The most common antioxidant in the brain is glutathione, which is found in millimolar (mM) concentrations in many cells. Glutathione reductase plays an important role in gene regulation. Glutathione, in its reduced form, plays a key role in the cellular control of ROS. Additionally, glutathione is a reductase flavoprotein. It plays a role in many enzymatic and metabolic processes of the body. Furthermore, this antioxidant is needed in the production of niacin, the activity of the adrenal gland, and the formation of red blood cells [45].

17.7.1.4 Glutathione peroxidase (GPx)

Glutathione peroxidase is another important enzymatic antioxidant. Recent research shows that the level of antioxidant activity of glutathione peroxidase and SOD is interdependent. Usually, these two chemical compounds have a positive effect on each other, leading to the strengthening and prolongation of antioxidant effects. It should be noted that if elements such as copper, zinc, manganese, and selenium are present in the environment, the antioxidant activity of glutathione peroxidase increases.

Glutathione peroxidase has a significant effect on the treatment of type 1 and type 2 diabetes. By controlling and regulating blood sugar, the activity of enzymes and hemoglobin greatly reduces the complications caused by diabetes.

Glutathione peroxidase has several isozymes encoded by different genes. These isozymes are located in different parts of the cell and have different substrate specificities. Glutathione peroxidase activity is measured using a colorimetric method at a wavelength of 412 nm. So far, eight different isoforms of glutathione peroxidase have been identified in humans.

Cellular glutathione peroxidase is present in all tissues, but various diseases affect its activity and levels. Therefore, measuring glutathione peroxidase enzyme activity can serve as a diagnostic tool for several diseases. Glutathione peroxidase is associated with diseases such as diabetes, liver disorders, prostate cancer, autism, Alzheimer's, and pregnancy-related complications. Considering the importance of glutathione peroxidase enzyme function in the body's antioxidant activity and its role in various diseases, the need for further studies and research is still evident [46–48].

17.7.2 Nonenzymatic antioxidants

17.7.2.1 Polyphenols

Polyphenols are important compounds that have significant therapeutic potential. The treatment of cardiovascular diseases is one of the most notable properties of these chemical compounds. Generally, the healing properties of these antioxidants and their biological activity are related to their structure. In this regard, the orientation and placement of molecules, especially hydroxybenzoic and hydroxycinnamic acids, are very important. Nowadays, research has shown that these compounds play a crucial and key role in the treatment process, and the treatment of heart diseases, cancer, and cholesterol is highly dependent on these compounds. So far, about 8,000 different polyphenols have been identified. In general, polyphenols can be divided into two categories: flavonoid and nonflavonoid polyphenols. Approximately 60% of polyphenols are flavonoids. One of the plants that is rich in flavonoids is the medicinal plant *Asplenium nidus*. The results of research by Jarial et al. in 2018, conducted through GC/MS, showed that the medicinal plant *Asplenium nidus* contains a large amount of flavonoids. In addition, this plant exhibits anticancer and anti-inflammatory antioxidant effects. It also has a unique potential for preventing chemotherapy-related side effects [49]. Table 17.2 highlights other important medicinal plants with polyphenols and flavonoids.

Table 17.2: Medicinal plants with flavonoid compounds.

Aromatic plants	Flavonoids (average ± standard deviation)
Quercus persica	241.1 ± 1.0
Rosa foetida	125.1 ± 1.7
Zataria multiflora	128.1 ± 1.5
Teucrium polium	121.1 ± 2.6
Matricaria recutita	147.2 ± 1.9
Berberis integerrima	126.1 ± 1.8

17.7.2.2 Anthocyanins

One of the most important flavonoid polyphenols is anthocyanin, with the chemical formula $C_{15}H_{11}O^+$. Figure 17.3 shows the chemical structure of anthocyanin. Anthocyanin causes the formation of red and blue colors in plant organs such as fruits, flowers, and leaves. Its antioxidant activity is well-known and is involved in various biological activities, including reducing the risk of heart diseases, diabetes, arthritis, and cancer. However, its stability and biological characteristics depend on its chemical structure. The diversity of anthocyanin is due to the number and position of hydroxyl and methoxyl groups in the main anthocyanidin skeleton, the type, number, and position of sugars attached to it, as well as the degree of sugar acylation and the type of acylation. The intensity and type of color of anthocyanin are influenced by hydroxyl and methoxyl groups. If hydroxyl groups are dominant, the color tends to be blue, whereas if methoxyl groups are dominant, the color tends to be red [50, 51]. The effect of scavenging ROS varies among anthocyanidins. For example, delphinidin has the highest activity against superoxide anion and is the most effective against pelargonidin hydroxyl radical [52]. Anthocyanin and pigments derived from it are important factors in assessing the quality of fruits and vegetables in terms of microbial and viral contamination. Today, health experts use this antioxidant to evaluate the quality and health of products. The higher the amount of this antioxidant, the higher the quality of the product. Additionally, this antioxidant also protects products against external factors, such as UV rays. Considering that the pigments caused by anthocyanin ensure the health of the product, many countries, including the USA, the European Union, Japan, Australia, Saudi Arabia, the UAE, Chile, Colombia, and Iran, use anthocyanin to certify food products [53–55].

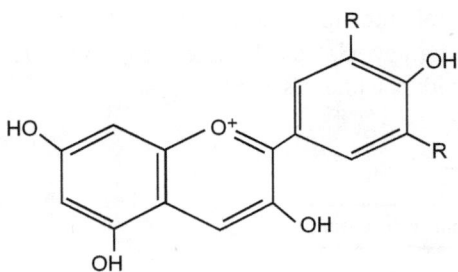

Figure 17.3: Chemical structure of anthocyanins.

Plants such as saffron petals, blackberries, raspberries, black rice, black soybeans, and other plant parts that have red, blue, purple, and black colors contain large amounts of anthocyanins. Additionally, anthocyanins are highly sensitive to environmental factors such as pH, light, temperature, concentration, and even the presence of other chemical compounds in the environment such as oxygen, metal ions, and enzymes, and they undergo changes as a result. Therefore, extensive research is currently being conducted on the stabilization of these compounds [56].

17.7.2.3 Phenolic compounds

Phenolic compounds are phytochemicals with the chemical formula C_6H_5OH, which act as antioxidants in plant tissues (Figure 17.4). Phenolic compounds are a large group of active plant glycosidic compounds that contribute to certain physiological functions in plants. These polymeric and monomeric compounds are present in most plants and provide a distinctive aroma and color to them. For example, in pepper, the spicy taste is due to the presence of these compounds. Rice, wheat, rye, and triticale are rich sources of phenolic compounds.

Another group of these compounds is also necessary for plant development and photosynthesis. Additionally, the vital pigments in these compounds are important for flower coloring and help attract pollinating insects.

Generally, these compounds are present in the cell wall or cell vacuole, and chemically, they typically have an aromatic ring. These compounds cause certain activities in plants, such as stimulating enzymes and destroying bacteria and free radicals, which are among the important functions of these antioxidant compounds. Many of these antioxidants protect against diseases; for example, a group of these compounds reduces the adhesion of blood platelets or helps lower blood pressure, thereby improving cardiovascular health.

In recent years, many advances have been made in the extraction and identification of phenolic compounds. However, the important point in this case is to pay special attention to the type of solvent and the extraction temperature. Nowadays, solvents such as water, alcohol, and acetone are used to extract these compounds.

Phenolic acids, lignans, and stilbenes are among the main and important polyphenols found naturally in medicinal plants [57, 58].

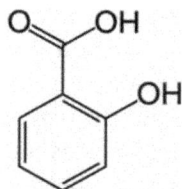

Figure 17.4: Chemical structure of phenolic acid.

17.7.2.4 Stilbenes

Stilbenes are allelochemicals whose chemical formula is $C_{14}H_{12}$. Figure 17.5 shows the chemical structure of these polyphenols. In general, stilbenes are synthesized by medicinal plants, especially grapes, peanuts, rhubarb, and berries, to defend themselves in stressful situations. Currently, these antioxidants are used in the treatment of medicinal plants and agriculture. As we know, in inflammatory conditions, cytokines released from invading agents such as bacteria or chemicals destroy healthy cells. In this case, stilbene reduces inflammation factors by eliminating external agents. In general, stilbene reduces inflammation symptoms by affecting the NF-κB, MAPK, and JAK/STAT pathways. Additionally, these antioxidants have many antiproliferative properties that contribute to the homeostatic maintenance of cells.

Stilbenes have diverse effects on functional cells. They modulate many biological reactions and also affect numerous pathways and molecular interactions. These properties have led to these antioxidants being used as leading agents in drug preparation [59].

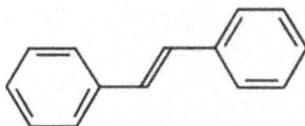

Figure 17.5: Chemical structure of stilbenes.

17.7.2.5 Lignans

Lignans are a group of antioxidants with the chemical formula $C_{22}H_{22}O_8$ that are found in medicinal plants (Figure 17.6). These compounds are mainly from the family of phytoestrogens. Plant phytoestrogens are similar to the hormone estrogen and possess antioxidant properties.

Lignans are also found in plants such as oilseeds, grains, vegetables, and fruits in large amounts and contribute to beneficial biological properties in the human body.

These compounds have strong antioxidant properties and prevent the oxidation of fats and the production of harmful substances such as free radicals and free fatty acids. Most plant lignans in food are converted into enterolactone and enterodiol by microflora in the upper intestine, where they play a protective role. Some studies show a direct relationship between the level of lignans in the body and a reduced risk of diseases such as breast cancer, ovarian cancer, osteoporosis, and cardiovascular diseases. Other studies have demonstrated the effect of lignans in reducing the risk of thyroid and endometrial cancers. Among edible plant sources, flaxseed contains the highest amount of lignans. Intact or powdered flaxseeds, as well as capsules containing flaxseed oil, are excellent sources of essential fatty acids and fiber in the human diet. Other major sources of lignans include pumpkin seeds, sesame seeds, rye, soybeans, broccoli, and beans. Fruits and vegetables, cereal husks, and tea are also among the foods with moderate to good lignan levels [60–63].

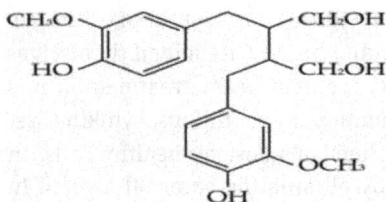

Figure 17.6: Chemical structure of lignans.

17.7.2.6 Vitamins

Antioxidant vitamins were identified by scientists in the 1980s. During this decade, the effects of vitamins on chronic diseases and cancer were determined. Further investigations revealed the role of vitamins in relation to eye diseases. In the following years, it was discovered that vitamins combat threats related to chemicals and free radicals. In general, vitamins such as A, E, and C are the most potent vitamins with antioxidant properties [64].

17.7.2.6.1 Vitamin A
Vitamin A should be obtained through diet as well as the purchase and consumption of vitamin A tablets. Most medicinal plants, green and yellow vegetables, dairy products, and fruits are sources of vitamin A. In the body, vitamin A exists in the forms of retinol, retinal, and retinoic acid. This vitamin is stored as a carotenoid provitamin in the liver, kidneys, and fat tissues. Vitamin A, with its antioxidant properties, can be effective in preventing various diseases, especially vision-related diseases [65].

17.7.2.6.2 Vitamin E

Vitamin E is a fat-soluble vitamin and a strong, effective antioxidant. Vitamin E comes in many forms, but α-tocopherol is the only form used by the body. Its main role is to act as an antioxidant and prevent free radicals from damaging the body. This vitamin also strengthens the function of the immune system and prevents the formation of clots in the arteries of the heart. Vitamin E, like vitamin A, is effective in strengthening vision. It is also effective in improving conditions such as fatty liver, Alzheimer's disease, and diabetes. Vitamin E is found in medicinal plants such as canola, camelina, almonds, avocados, leafy greens, peanuts, and spinach [65–67].

17.7.2.6.3 Vitamin C

In general, the concentration of vitamin C varies across different tissues, and from a biological point of view, and it can be said that its levels depend on factors such as consumption, renal excretion, and the biological conditions of the tissues. Vitamin C has a significant effect in neutralizing infections. Many human and animal infections are treated by this compound, and numerous bacteria and viruses are suppressed by this antioxidant. Daily and regular consumption of this vitamin can enhance its antioxidant effects. In general, this vitamin is found in many plant sources. Medicinal plants, citrus fruits, and especially thyme are rich in vitamin C. In addition to its many biological effects, this vitamin is also present in plasma and extracellular fluids, where it can prevent the invasion of pathogens in these areas. Recent research has shown that gene effects, as well as transcription and translation, may be involved in the regulation of this vitamin, although further studies are required to confirm this [68].

17.7.2.7 Terpenoids

Terpenoids, also known as isoprenoids, are often considered the main components of essential oils in medicinal plants. This is why terpenoids are regarded as the primary compounds in medicinal plants. They exhibit a wide spectrum of antioxidant and pharmacological activities and are reported in most medicinal plants, grains, and legumes, where they are recognized as effective compounds for eliminating free radicals. In developing countries, the average consumption of terpenoids is approximately 250 mg/day, whereas in Mediterranean countries, where diets are based on olive oil, daily consumption reaches 400 mg/day. In general, terpenoids undergo two general biosynthetic pathways: the mevalonate (MVA) pathway and the methylerythritol phosphate (MEP), which extend from the initial stages to the production of isopentenyl diphosphate and exhibit significant diversity. Carotenoids are the most important terpenoids found in medicinal plants. The identification of these antioxidants enables the commercial development of medicinal plants based on specific metabolites, as well as advancements in metabolite engineering and molecular modification of medicinal plants (Figure 17.7) [69].

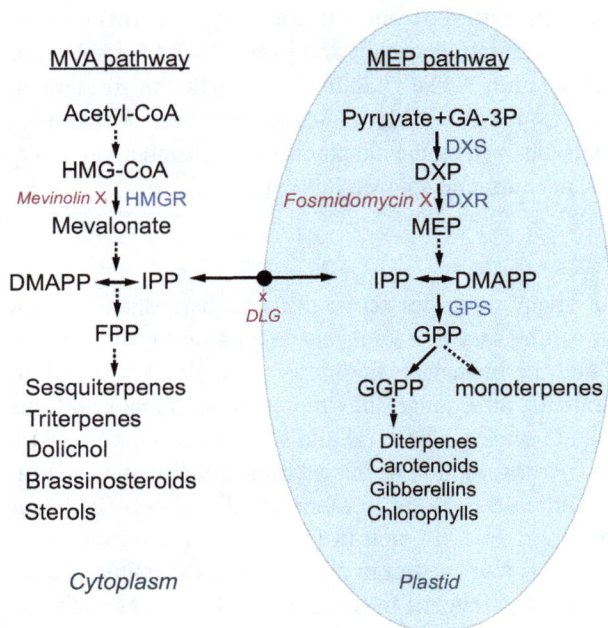

MVA pathway

Acetyl-CoA

HMG-CoA

Mevinolin X ↓ HMGR

Mevalonate

DMAPP ←→ IPP ←———

FPP

Sesquiterpenes
Triterpenes
Dolichol
Brassinosteroids
Sterols

Cytoplasm

MEP pathway

Pyruvate+GA-3P

↓ DXS

DXP

Fosmidomycin X ↓ DXR

MEP

IPP ←→ DMAPP

↓ GPS

GPP

GGPP   monoterpenes

Diterpenes
Carotenoids
Gibberellins
Chlorophylls

Plastid

X DLG

Figure 17.7: Mevalonate pathway and methylerythritol phosphate pathway in terpenoid synthesis [69].

17.7.2.8 Carotenoids

Carotenoids are one of the most important terpenoids. Today, many carotenoids are extracted from natural sources, and a group of them also act as precursors to vitamins. Carotenoids trigger many cellular responses in the body. For example, increasing cellular immunity, protecting cells, and eliminating the harmful effects of mutations and damage to the cell nucleus are among the effects of carotenoids. In general, it can be said that carotenoids are molecules that are strongly influenced by the surrounding molecules. However, chemical effects and their reactions to free radicals occur under the influence of proteins and cell membranes. In general, it can be said that the more effective antioxidant effect of these molecules depends on the precise location and proper orientation of the molecule. Carotenoids have beneficial biological and pathological functions in human health. However, their maximum use should be achieved by carefully checking their solubility, molecular weight, and combination with other molecules. Lutein and zeaxanthin are the most important carotenoids found in medicinal plants [70–72].

17.8 Recent experiments and studies on antioxidants

Various studies show that free radicals cause many diseases by destroying vital molecules and sensitive tissues. This problem requires a coherent defense network to address these factors. By creating a defense line, antioxidants can establish a coherent network and counteract these molecules. However, it should be noted that antioxidants may also be strengthened or weakened by interactions with other molecules. As mentioned earlier, antioxidants may influence each other, generally in positive ways, which can lead to the strengthening of the defense network. However, these cases require further investigation, which can be achieved through numerous tests. Elbalola et al. in 2024 reviewed the extracts of medicinal and aromatic plants in the Asteraceae and Lamiaceae families. Their results showed that there are approximately 160 different chemicals in these extracts, which can be classified into 36 phytochemical classes. Most of these compounds include terpenoids, benzene and its derivatives, naphthalenes, fatty acyls, and phenols [73]. Esra also investigated the antioxidant and chemical activity of *Polygonum aviculare* in 2024. The results of this research showed that increasing the concentration of the extract has a direct effect on enhancing the antioxidant activity of this *Polygonum aviculare* plant. Additionally, the antiproliferative study revealed that the aqueous extract exhibited more active antiproliferative properties than the methanolic extract [74]. Samira et al. demonstrated in 2024 that active antioxidant peptides extracted from plant proteins are involved in the treatment of oxidation-related diseases, in addition to their biotechnological applications. The results of Samira's research indicated that these molecules have constructive interactions that play a role in modulating the catabolism of fatty acids and regulating blood glucose levels [75]. Ghudhaib et al. also investigated the antioxidant, antifungal, and antibacterial properties of marjoram in 2024. Their findings showed that marjoram plant extract is effective in the treatment of diabetes and liver diseases, with its antioxidant properties increasing at higher doses. It was also found that marjoram contains numerous flavonoids as well as alkaloid, terpenoid, tannin, and saponin compounds [76]. A study conducted by Bouloumpasi et al. in 2024 revealed that the antibacterial activity of rosemary and sage plants is higher than that of mint and oregano, while the antioxidant activity of oregano was significantly higher than that of the other plants studied. In the next phase of the study, rosemary and sage also demonstrated the highest antioxidant activity [77]. In 2024, Baibuch et al. investigated the antioxidant properties of roses. Their results showed that this medicinal plant, like other medicinal plants, contains strong antioxidants such as phenols, flavonoids, and anthocyanins [78]. On the other hand, Kebert et al. examined the antimicrobial and antibacterial properties of the rose plant. Their findings indicated that this plant can inhibit free radicals and the growth of *E. coli* bacteria. It was also discovered that the rose plant possesses potential antioxidant properties in addition to containing secondary metabolites [79]. Ricardo and colleagues in 2024 also demonstrated that medicinal and aromatic plants such as thyme and cloves have unique medicinal properties. The results of their research showed that these plants contain specific biochemical compounds with significant antioxidant and antimicrobial properties. In general, their antioxidant properties reduce the oxidation of lipids and proteins [80]. Mokhtari et al. in 2023 investigated the antioxidant properties of thyme (*Thymus vulgaris* L.) and sage (*Salvia officinalis* L.). The results of GC-MS analysis in methanolic extract revealed that thymol, apigenin, rosmarinic acid, and carvacrol were the most prominent phenolic compounds in thyme and sage extracts. Additionally, their findings showed that the amount of antioxidant compounds such as phenols and flavonoids in the mixture of the two plants is higher than the amount of antioxidants in each plant individually [81].

17.9 The most important medicinal and aromatic plants, rich in antioxidants

17.9.1 Carnation dianthus

Dianthus is one of the fragrant perennial plants, which is usually highly valued because of the properties found in its flowers. Dianthus is among the medicinal plants with the highest antioxidant levels, which protect the health of the body, and the consumption of its extract prevents the growth of cancer cells, especially breast cancer. Additionally, the antioxidant properties of this plant in eliminating free radicals make the body resistant to diseases such as Parkinson's and Alzheimer's [82]. Just half a teaspoon of dianthus contains more antioxidants than half a cup of blueberries. In traditional medicine, this plant is used to treat many diseases, including bronchitis, asthma, tuberculosis, nausea and diarrhea, sore throat, gum infections, and digestive diseases [83]. In 2024, Wang et al. investigated the antioxidant and antitumor effects of dianthus. The results of this research showed that this plant contains flavonoids such as methyl folate. Furthermore, their research revealed that this plant contains large amounts of antioxidants in its roots, stems, leaves, and flowers. On the other hand, studies have shown that phytochemical compounds are distributed throughout the anatomical parts of dianthus, which may have antimutation and antibacterial properties in addition to antioxidant properties [84].

17.9.2 Oregano (*Origanum vulgare*)

Origanum vulgare belongs to the Lamiaceae family, and its medicinal and therapeutic role in many diseases has been recognized since ancient times. *Origanum vulgare* is a plant rich in antioxidant compounds that help treat chronic diseases such as cancer and heart disease by neutralizing the effects of free radicals. This medicinal plant has anti-inflammatory properties, and according to some sources, it is useful for weight loss and for inhibiting the growth of *E. coli* bacteria and certain Staphylococcal infections. In research conducted by Ferreira-Anta et al. in 2024, the antioxidant properties of *Origanum vulgare* were investigated using an ultrasound device, which revealed that this plant contains important antioxidants such as peroxidase, catalase, and polyphenol oxidase [85]. Additionally, in research conducted by Tarkesh Esfahani et al. in 2024, methyl jasmonate treatment was applied to the plant at four levels. The results of this research showed that 96 h after methyl jasmonate treatment, the amount of antioxidants in *Origanum vulgare* increased significantly, with foliar spray application proving to be more effective [86]. In 2024, Latifeh et al. investigated the effects of nanotechnology and nanoparticles (AgNPs) on increasing the amount of antioxidants in *Origanum vulgare*. This research demonstrated that temperature, time, and the extraction solvent are critical factors influencing the amount of antioxidants produced. However, in general, silver nanoparticles were found to increase the amount of antioxidants, particularly phenolic compounds, in the *Origanum vulgare* plant [87].

17.9.3 Ginger (*Zingiber officinale*)

Zingiber officinale is a famous spice and one of the most useful medicinal plants, considered one of the best antioxidant foods. This plant has been used as a strong medicinal remedy in Southeast Asia since ancient times. In general, many essential oils and biological compounds present in this plant lead to a wide range of reactions in the body, ultimately strengthening the immune system. *Zingiber officinale* prevents Alzheimer's disease, helps reduce inflammation and joint pain in patients with arthritis, and reduces the risk of cancer [88]. Ayustaningwarno et al. investigated the antioxidant properties of *Zingiber officinale* in 2024. The results of their investigations showed that the antioxidant properties of *Zingiber officinale* are related to gingerol, paradol, shogaol, and zingerone. It was also found that the mechanism of activation of antioxidants occurs through the Nrf2 signaling pathway. In general, it was found that *Zingiber officinale* and its active compounds have strong antioxidant properties that reduce inflammation by eliminating free radicals [89]. Additionally, Mustafa et al., in 2023, investigated the antioxidant properties of *Zingiber officinale* during the drying process. Their results showed that there is a direct relationship between total phenol content, total flavonoid content, and antioxidant activity in *Zingiber officinale*. They also demonstrated that during the drying process, the use of ethanol and natural sunlight plays an effective role in preserving as many plant antioxidants as possible [90].

17.9.4 Cinnamon (*Cinnamomum verum*)

Cinnamon is another popular and famous spice that, in addition to its strong antioxidant properties, helps stabilize blood sugar levels, control diabetes, and prevent cancer. Consumption of cinnamon is also useful for increasing insulin sensitivity and reducing cholesterol and inflammation. So far, more than 300 types of cinnamon have been identified, many of which possess antioxidant properties. Recent research has shown that the antioxidant, anti-inflammatory, and anticancer properties of cinnamon are mostly due to the presence of compounds such as proanthocyanidins A and B, cinnamic acid, and kaempferol [91]. Al-Mijalli et al. in 2023 investigated the antioxidant properties of *Cinnamomum verum* through gas chromatography-mass spectrometry (GC/MS). Their results showed that cinnamon leaves have a wide range of antioxidants. These antioxidants contribute to resistance against antimicrobial agents, antifungal agents, and a wide range of Gram-positive and Gram-negative bacteria, molds, and yeasts, including *Micrococcus luteus*, *Bacillus subtilis*, *Bacillus cereus*, *E. coli*, *Klebsiella aerogenes*, and *Salmonella enterica* [92]. In 2015, Pandey investigated the methanolic and aqueous extracts of *Cinnamomum verum* leaves to evaluate their antioxidant and analgesic properties. Their results showed that the extract of the *Cinnamomum verum* plant exhibits high antioxidant and analgesic activity by inhibiting free radicals [93].

17.9.5 Turmeric (*Curcuma longa*)

Turmeric is one of the widely used antioxidant spices that is commonly incorporated into many foods and has been renowned for its extraordinary properties for a long time. Turmeric protects liver health against various toxins, including alcohol, and the antioxidants present in it help prevent Alzheimer's disease, blood cancer, and reduce blood cholesterol levels [94]. *Curcuma longa* is rich in antioxidant compounds and polyphenols. The active ingredient in turmeric rhizome is curcumin, which inhibits NF-kappaB activity and contributes to antioxidant and anti-inflammatory properties. Research has also shown that this compound can activate other antioxidants as well [95, 96].

17.9.6 Mustard (*Sinapis arvensis*)

Sinapis arvensis antioxidant seeds exhibit strong antioxidant activity that is beneficial for various diseases, ranging from allergies to arthritis and joint pain. *Sinapis arvensis* is also recognized as an analgesic, antibacterial, expectorant, and diuretic herb that enhances digestion, reduces mucus, and protects the body against free radicals and cancer cells [97, 98]. The seeds of *Sinapis arvensis* contain potent phenolic compounds and glucosinolates, which contribute to antibacterial and antimicrobial resistance by inhibiting free radicals. Furthermore, in addition to the seeds, the leaves and flowers of *Sinapis arvensis* possess strong antioxidant compounds. Kaempferol, hydroxycinnamic acid, and catechin are other types of phenolic compounds found in mustard, which promote antioxidant and antibacterial properties through the inhibition of free radicals [99].

17.9.7 Rosemary (*Rosmarinus officinalis*)

Rosmarinus officinalis is an evergreen plant. Rosemary leaves are native to coastal areas in Southern Europe and Northern Africa and are now cultivated worldwide. *Rosmarinus officinalis* also has many medicinal properties that have increased its popularity. Since ancient times, rosemary has been used in traditional medicine to treat various ailments, including digestive problems, infections, and to improve skin and hair health. Rosemary is a powerful antioxidant and anti-inflammatory plant that helps strengthen the body's immune system, improves blood flow, and helps neutralize the effects of free radicals. Phenolic diterpenes present in *Rosmarinus officinalis* promote liver detoxification and are also useful in the treatment of lung and prostate diseases [100–102].

17.9.8 Saffron (*Crocus sativus*)

Saffron contains a variety of powerful antioxidant compounds that protect cells from free radicals and oxidative stress. Crocin, crocetin, carotene, safranal, and kaempferol are among the notable antioxidants in saffron. Saffron has antidepressant properties and helps reduce inflammation, appetite, and weight. Crocin and crocetin are carotenoid pigments responsible for its red color. These pigments protect brain cells and have strong anti-inflammatory effects. In general, saffron's antioxidant compounds reduce the risk of heart disease, improve age-related vision problems in adults, and also have strong anticancer properties [103, 104]. Boskabady et al. demonstrated in 2016 that the antioxidants in saffron contribute to effective therapeutic properties in nervous disorders, vascular diseases, asthma, bronchitis, colds, fever, and diabetes [105].

17.9.9 Sage (*Salvia officinalis*)

Salvia is a plant belonging to the Lamiaceae family. This plant is closely related to rosemary. Sage is a medicinal plant, and its leaves are mostly used. Its antioxidant compounds include anthocyanins, diterpenoids, phenolic acids, rosmarinic acid, and flavonoids. *Salvia officinalis* contains more than 160 different antioxidant polyphenolic compounds that reduce the risk of cancer and help improve brain function and memory. *Salvia officinalis* is also useful for reducing bad cholesterol and increasing good cholesterol. Sage also has strong antioxidant vitamins such as vitamins A, C, and

E. Additionally, by eliminating free radicals, this plant contributes to resistance against fatty liver and reduces fat and liver inflammation [106, 107]. Furthermore, the antioxidants in this plant provide resistance against oral and dental diseases, lung diseases, herpes simplex, and viral diseases. This plant deactivates these viruses by occupying their active sites and creating a disinfecting effect. Moreover, this plant contains active antioxidants, which, with its anti-inflammatory and pain-relieving properties, contribute to resistance and improvement in the treatment of joint pain, rickets in children, rheumatism, and varicose veins [108, 109].

17.9.10 Stevia (*Stevia rebaudiana*)

Stevia has received significant attention in the pharmaceutical and therapeutic industries due to its antioxidant properties, very low calorie content, high resistance to heat, and superior sweetening ability compared to sucrose. Additionally, studies on stevia plant extract have highlighted its antiviral, antimicrobial, antiparasitic, antibacterial, antitumor properties, and its highly effective ability to prevent diabetes, making it a natural sweetener without calories. The sweetening properties of stevia are primarily attributed to the glycosides it contains, including stevioside and rebaudioside A. Various studies have confirmed the antioxidant capacity of this plant and demonstrated that the antioxidant activity of stevia leaf extract is associated with the scavenging of free radical electrons and superoxides. These antioxidants ultimately reduce the risk of chronic diseases such as cancer, heart disease, and Alzheimer's disease [110, 111]. One of the key antioxidant compounds in stevia is kaempferol, which has been shown to reduce the incidence of pancreatic cancer [112]. Additionally, apigenin, isoquercitrin, caffeic acid, quercetin, and luteolin are other antioxidant compounds in stevia that contribute to resistance against cardiovascular diseases, stroke, and cancer [113, 114]. Miocene and chlorogenic acid are among the anti-inflammatory, antimicrobial, and antifungal antioxidants in stevia that provide resistance to inflammatory bowel diseases as well as cardiovascular diseases [115, 116].

17.9.11 Dracocephalum

Due to its strong antioxidant properties, this medicinal plant prevents the growth and spread of cancer cells in the body and is useful for the treatment of neurological disorders caused by excessive consumption of manganese. Dracocephalum also has anti-inflammatory properties that help relieve pain [117, 118]. Chlorogenic acid and isoorientin are two important antioxidant compounds in Dracocephalum that contribute to its anti-inflammatory properties [119]. Additionally, phenolic compounds such as rosmarinic acid and tannin present in this plant help prevent cardiovascular diseases, cancer, and osteoporosis [120]. Rhamnocitrin, rhamnazine, and isoquercitrin are other flavonoid antioxidants that exhibit anticancer, anti-inflammatory, and antiviral properties [121].

17.9.12 Paprika (*Capsicum annuum*)

Dry pepper powder, or paprika, is another antioxidant spice that supports heart and blood vessel health while also helping to prevent cancer. The flavonoids in paprika combat free radicals and are beneficial for detoxifying the body as well as reducing chronic pain and inflammation in diseases such as arthritis [122–124].

Garlic (*Allium sativum*), coriander (*Coriandrum sativum*), cumin (*Cuminum cyminum*), fenugreek (*Trigonella foenum-graecum*), and basil (*Ocimum basilicum*) are also important medicinal and aromatic plants that contain many flavonoid and carotenoid compounds. The antioxidants present in these plants contribute to resistance against cardiovascular diseases, diabetes, cancer, Alzheimer's, osteoporosis, and arthritis, which are very common [125].

17.9.13 Rapeseed (*Brassica napus*)

Rapeseed has different varieties, and these varieties contain many antioxidant compounds. Rapeseed contains magnesium, calcium, copper, iron, SOD, thiamine (vitamin B_1), and riboflavin (vitamin B_2). In general, *Brassica campestris*, *Brassica juncea*, *Brassica rapa*, and *Brassica napus* contain the highest levels of antioxidants, including riboflavin and niacin. Research has shown that riboflavin content increases gradually during the germination period. The antioxidant compounds in rapeseed increase during the growth period and reach a constant level. It should be noted that in *Brassica napus*, increasing the amount of a compound causes a decrease in antioxidant compounds. For example, increasing the amount of copper in the plant causes a reduction in antioxidant compounds, especially in the aerial parts. For this reason, the correct consumption of nutrients and maintaining a proper balance are essential to obtain the maximum amount of antioxidant compounds. In general, obtaining the maximum antioxidant compounds in rapeseed oil is a necessity and is highly effective in promoting human health [126–128].

17.10 Conclusion

Medicinal and aromatic plants are rich in natural antioxidants that can neutralize the activity of free radicals. Therefore, it is recommended to use these plants to prevent common diseases such as cardiovascular diseases, cancer, Alzheimer's, Parkinson's, and diabetes. Increasing scientific attention to this research is important in several ways. First, it provides the basis for obtaining new information and antioxidant compounds. Additionally, the discovery of new natural antioxidants can lead to the production of valuable products for human health. Finally, considering that the consumption of antioxidants enhances the immune system's resistance, reduces the aging process of cells, and aids in detoxification within the body, more information can help improve food consumption patterns.

References

[1]   Kumar, S. and Pandey, A. (2015). Free radicals: Health implications and their mitigation by herbals. British Journal of Medicine and Medical Research, 7(6), 438–457.

[2]   Mustafa, Y. F. (2024). Harmful free radicals in aging: A narrative review of their detrimental effects on health. Indian Journal of Clinical Biochemistry, 39(2), 154–167.

[3]   Hassan, H. A., Ahmed, H. S. and Hassan, D. F. (2024). Free radicals and oxidative stress: Mechanisms and therapeutic targets. Human Antibodies, 32(4), 151–167.

[4]   Sindhi, V., Gupta, V., Sharma, K., Bhatnagar, S., Kumari, R. and Dhaka, N. (2013). Potential applications of antioxidants–A review. Journal of Pharmacy Research, 7(9), 828–835.

[5]   Pisoschi, A. M. and Pop, A. (2015). The role of antioxidants in the chemistry of oxidative stress: A review. European Journal of Medicinal Chemistry, 97, 55–74.

[6]   Kahrizi, D. and Mohammadi, S. (2023). Anticancer, antimicrobial, cardioprotective, and neuroprotective activities of luteolin: A systematic-narrative mini-review. Nano Micro Biosystems, 2(2), 1–9.

[7]   Eruygur, N., Ucar, E., Akpulat, H. A., Shahsavari, K., Safavi, S. M. and Kahrizi, D. (2019). In vitro antioxidant assessment, screening of enzyme inhibitory activities of methanol and water extracts and gene expression in *Hypericum lydium*. Molecular Biology Reports, 46(2), 2121–2129.

[8]   Shahdadi, F., Khorasani, S., Salehi-Sardoei, A., Fallahnajmabadi, F., Fazeli-Nasab, B. and Sayyed, R. Z. (2023). GC-MS profiling of *Pistachio vera* L., and effect of antioxidant and antimicrobial compounds of it's essential oil compared to chemical counterparts. Scientific Reports, 13(1), 21694.

[9]   Fazeli-Nasab, B., Shahraki-Mojahed, L., Jahantigh, M. and Dahmardeh, N. (2022). *Scrophularia striata*: Therapeutic and healing properties with an emphasis on oxidative stress and gastric ulcer treatment. Gene, Cell and Tissue, 9(4).

[10]   Babarabi, M., Sardoei, A. S., Dhanalakshmi, K., Malathi, G., Sayyed, R. Z., Sunita, K. . . . and Fazeli-Nasab, B. (2024). Triacontanol: The role player in *Polianthes tuberosa* var. Pearl response to under natural conditions. Biocatalysis and Agricultural Biotechnology, 103228.

[11]   Houldsworth, A. (2024). Role of oxidative stress in neurodegenerative disorders: A review of reactive oxygen species and prevention by antioxidants. Brain Communications, 6(1), fcad356.

[12]   Naviaux, R. K. (2014). Metabolic features of the cell danger response. Mitochondrion, 16, 7–17.

[13]   Zhang, R., Zimmerman, A. R., Zhang, R., Li, P., Zheng, Y. and Gao, B. (2024). Persistent free radicals generated from a range of biochars and their physiological effects on wheat seedlings. Science of the Total Environment, 908, 168260.

[14]   Mathew, B. B., Tiwari, A. and Jatawa, S. K. (2011). Free radicals and antioxidants: A review. Journal of Pharmacy Research, 4(12), 4340–4343.

[15]   Yang, S., Li, Y., Chen, L., Wang, H., Shang, L., He, P. . . . and Ding, G. (2023). Fabrication of carbon-based quantum dots via a "bottom-up" approach: Topology, chirality, and free radical processes in "Building Blocks". Small, 19(31), 2205957.

[16] Qiu, Y., Zhang, T. and Zhang, P. (2023). Fate and environmental behaviors of microplastics through the lens of free radical. Journal of Hazardous Materials, 453, 131401.

[17] Anbessa, B., Lulekal, E., Hymete, A., Debella, A., Debebe, E., Abebe, A. and Degu, S. (2024). Ethnomedicine, antibacterial activity, antioxidant potential and phytochemical screening of selected medicinal plants in Dibatie district, Metekel zone, western Ethiopia. BMC Complementary Medicine and Therapies, 24(1), 199.

[18] El Kamari, F., El Omari, H., El-Mouhdi, K., Chlouchi, A., Harmouzi, A., Lhilali, I. . . . and Ousaaid, D. (2024). Effects of different solvents on the total phenol content, total flavonoid content, antioxidant, and antifungal activities of *Micromeria graeca* L. from middle atlas of Morocco. Biochemistry Research International, 2024(1), 9027997.

[19] Zhou, K. and Yu, L. (2004). Effects of extraction solvent on wheat bran antioxidant activity estimation. LWT-Food Science and Technology, 37(7), 717–721.

[20] Somogyi, A., Rosta, K., Pusztai, P., Tulassay, Z. and Nagy, G. (2007). Antioxidant measurements. Physiological Measurement, 28(4), R41.

[21] Houldsworth, A. (2024). Role of oxidative stress in neurodegenerative disorders: A review of reactive oxygen species and prevention by antioxidants. Brain Communications, 6(1), fcad356.

[22] Rogers, S., Witz, G., Anwar, M., Hiatt, M. and Hegyi, T. (2000). Antioxidant capacity and oxygen radical diseases in the preterm newborn. Archives of Pediatrics & Adolescent Medicine, 154(6), 544–548.

[23] Chrysargyris, A., Petrovic, J. D., Tomou, E. M., Kyriakou, K., Xylia, P., Kotsoni, A. . . . and Tzortzakis, N. (2024). Phytochemical profiles and biological activities of plant extracts from aromatic plants cultivated in cyprus. Biology, 13(1), 45.

[24] Muscolo, A., Mariateresa, O., Giulio, T. and Mariateresa, R. (2024). Oxidative stress: The role of antioxidant phytochemicals in the prevention and treatment of diseases. International Journal of Molecular Sciences, 25(6), 3264.

[25] Halliwell, B. (2024). Understanding mechanisms of antioxidant action in health and disease. Nature Reviews Molecular Cell Biology, 25(1), 13–33.

[26] Sen, S. and Chakraborty, R. (2011). The role of antioxidants in human health. In Oxidative Stress: Diagnostics, Prevention, and Therapy, American Chemical Society, 1–37.

[27] Yadav, A., Kumari, R., Yadav, A., Mishra, J. P., Srivatva, S. and Prabha, S. (2016). Antioxidants and its functions in human body - A review. Research in Environment and Life Sciences, 9(11), 1328–1331.

[28] Jamshidi-Kia, F., Lorigooini, Z. and Amini-Khoei, H. (2017). Medicinal plants: Past history and future perspective. Journal of Herbmed Pharmacology, 7(1), 1–7.

[29] Lazzarino, G., Listorti, I., Bilotta, G., Capozzolo, T., Amorini, A. M., Longo, S. . . . and Bilotta, P. (2019). Water-and fat-soluble antioxidants in human seminal plasma and serum of fertile males. Antioxidants, 8(4), 96.

[30] Atta, E. M., Mohamed, N. H. and Abdelgawad, A. A. (2017). Antioxidants: An overview on the natural and synthetic types. European Chemical Bulletin, 6(8), 365–375.

[31] Gulcin, İ. (2020). Antioxidants and antioxidant methods: An updated overview. Archives of Toxicology, 94(3), 651–715.

[32] Hamid, A. A., Aiyelaagbe, O. O., Usman, L. A., Ameen, O. M. and Lawal, A. (2010). Antioxidants: Its medicinal and pharmacological applications. African Journal of Pure and Applied Chemistry, 4(8), 142–151.

[33] Ram, V. and Salkuti, S. R. (2023). An overview of major synthetic fuels. Energies, 16(6), 2834.

[34] Shakir, R. M. (2014). Synthesis and Antioxidant Properties of Some New di-*tert*-butylphenol Derivatives Bearing Heterocyclic Ring, Doctoral dissertation University of Malaya, Malaysia.

[35] Kumar, S., Sharma, S. and Vasudeva, N. (2017). Review on antioxidants and evaluation procedures. Chinese Journal of Integrative Medicine, 1–12.

[36] Dang, Y., Li, Z. and Yu, F. (2024). Recent advances in astaxanthin as an antioxidant in food applications. Antioxidants, 13(7), 879.

[37] Mirończuk-Chodakowska, I., Witkowska, A. M. and Zujko, M. E. (2018). Endogenous non-enzymatic antioxidants in the human body. Advances in Medical Sciences, 63(1), 68–78.

[38] Lubrano, V. and Balzan, S. (2015). Enzymatic antioxidant system in vascular inflammation and coronary artery disease. World Journal of Experimental Medicine, 5(4), 218.

[39] Silva, B. R. and Silva, J. R. (2023). Mechanisms of action of non-enzymatic antioxidants to control oxidative stress during in vitro follicle growth, oocyte maturation, and embryo development. Animal Reproduction Science, 249, 107186.

[40] Bian, Y. and Chang, T. M. S. (2023). Nanobiotechnological basis of an oxygen carrier with enhanced carbonic anhydrase for CO2 transport and enhanced catalase and superoxide dismutase for antioxidant function. Frontiers in Bioengineering and Biotechnology, 11, 1188399.

[41] Ighodaro, O. M. and Akinloye, O. A. (2018). First line defence antioxidants-superoxide dismutase (SOD), catalase (CAT) and glutathione peroxidase (GPX): Their fundamental role in the entire antioxidant defence grid. Alexandria Journal of Medicine, 54(4), 287–293.

[42] Hemerková, P. and Vališ, M. (2021). Role of oxidative stress in the pathogenesis of amyotrophic lateral sclerosis: Antioxidant metalloenzymes and therapeutic strategies. Biomolecules, 11(3), 437.

[43] Zhao, H., Li, W., Zhao, X., Li, X., Yang, D., Ren, H. and Zhou, Y. (2017). Cu/Zn superoxide dismutase (SOD) and catalase (CAT) response to crude oil exposure in the polychaete Perinereis aibuhitensis. Environmental Science and Pollution Research, 24, 616–627.

[44] Azadmanesh, J. and Borgstahl, G. E. (2018). A review of the catalytic mechanism of human manganese superoxide dismutase. Antioxidants, 7(2), 25.

[45] Kocaoğlu, E., Talaz, O., Çavdar, H., Şentürk, M., Supuran, C. T. and Ekinci, D. (2019). Determination of the inhibitory effects of N-methylpyrrole derivatives on glutathione reductase enzyme. Journal of Enzyme Inhibition and Medicinal Chemistry, 34(1), 51–54.

[46] Jeremic, A., Vasiljevic, M., Mikovic, Z., Bukumiric, Z., Simic, P., Stanisavljevic, T. . . . and Djukic, T. (2025). Oxidative homeostasis in follicular fluid and embryo quality – A pilot study. International Journal of Molecular Sciences, 26(1), 388.

[47] Pei, J., Pan, X., Wei, G. and Hua, Y. (2023). Research progress of glutathione peroxidase family (GPX) in redoxidation. Frontiers in Pharmacology, 14, 1147414.

[48] Gusti, A. M., Qusti, S. Y., Alshammari, E. M., Toraih, E. A. and Fawzy, M. S. (2021). Antioxidants-related superoxide dismutase (SOD), catalase (CAT), glutathione peroxidase (GPX), glutathione-S-transferase (GST), and nitric oxide synthase (NOS) gene variants analysis in an obese population: A preliminary case-control study. Antioxidants, 10(4), 595.

[49] Jarial, R., Thakur, S., Sakinah, M., Zularisam, A. W., Sharad, A., Kanwar, S. S. and Singh, L. (2018). Potent anticancer, antioxidant and antibacterial activities of isolated flavonoids from Asplenium nidus. Journal of King Saud University-Science, 30(2), 185–192.

[50] Horbowicz, M., Kosson, R., Grzesiuk, A. and Dębski, H. (2008). Anthocyanins of fruits and vegetables-their occurrence, analysis and role in human nutrition. Journal of Fruit and Ornamental Plant Research, 68(1), 5–22.

[51] Pinto, T., Aires, A., Cosme, F., Bacelar, E., Morais, M. C., Oliveira, I. . . . and Gonçalves, B. (2021). Bioactive (poly) phenols, volatile compounds from vegetables, medicinal and aromatic plants. Foods, 10(1), 106.

[52] Ullah, R., Khan, M., Shah, S. A., Saeed, K. and Kim, M. O. (2019). Natural antioxidant anthocyanins – A hidden therapeutic candidate in metabolic disorders with major focus in neurodegeneration. Nutrients, 11(6), 1195.

[53] Mahanta, S. K. and Challa, S. R. (2022). Phytochemicals as pro-oxidants in cancer: Therapeutic implications. Handbook of Oxidative Stress in Cancer: Therapeutic Aspects, 611–619.

[54] LaFountain, A. M. and Yuan, Y. W. (2021). Repressors of anthocyanin biosynthesis. New Phytologist, 231(3), 933–949.

[55] Jarial, R., Thakur, S., Sakinah, M., Zularisam, A. W., Sharad, A., Kanwar, S. S. and Singh, L. (2018). Potent anticancer, antioxidant and antibacterial activities of isolated flavonoids from *Asplenium nidus*. Journal of King Saud University-Science, 30(2), 185–192.

[56] Diaconeasa, Z., Ştirbu, I., Xiao, J., Leopold, N., Ayvaz, Z., Danciu, C. . . . and Socaciu, C. (2020). Anthocyanins, vibrant color pigments, and their role in skin cancer prevention. Biomedicines, 8(9), 336.

[57] Metsämuuronen, S. and Sirén, H. (2019). Bioactive phenolic compounds, metabolism and properties: A review on valuable chemical compounds in Scots pine and Norway spruce. Phytochemistry Reviews, 18, 623–664.

[58] Kumar, N. and Goel, N. (2019). Phenolic acids: Natural versatile molecules with promising therapeutic applications. Biotechnology Reports, 24, e00370.

[59] Al-Khayri, J. M., Mascarenhas, R., Harish, H. M., Gowda, Y., Lakshmaiah, V. V., Nagella, P. . . . and Rezk, A. A. S. (2023). Stilbenes, a versatile class of natural metabolites for inflammation – An overview. Molecules, 28(9), 3786.

[60] El-Razek, A., Mohamed, T. A., Abdel-Halim, S., Bata, S. M. and Kubacy, T. M. (2023). Comprehensive NMR reassignments of lignans derived from *Commiphora myrrha*. Egyptian Journal of Chemistry, 66(12), 45–57.

[61] Brito, A. F. and Zang, Y. (2018). A review of lignan metabolism, milk enterolactone concentration, and antioxidant status of dairy cows fed flaxseed. Molecules, 24(1), 41.

[62] Soleymani, S., Habtemariam, S., Rahimi, R. and Nabavi, S. M. (2020). The what and who of dietary lignans in human health: Special focus on prooxidant and antioxidant effects. Trends in Food Science & Technology, 106, 382–390.

[63] Polat Kose, L. and Gulcin, İ. (2021). Evaluation of the antioxidant and antiradical properties of some phyto and mammalian lignans. Molecules, 26(23), 7099.

[64] Niki, E. and Noguchi, N. (2004). Dynamics of antioxidant action of vitamin E. Accounts of Chemical Research, 37(1), 45–51.

[65] Didier, A. J., Stiene, J., Fang, L., Watkins, D., Dworkin, L. D. and Creeden, J. F. (2023). Antioxidant and anti-tumor effects of dietary vitamins A, C, and E. Antioxidants, 12(3), 632.

[66] Baeza-Jiménez, R., López-Martínez, L. X., García-Varela, R. and García, H. S. (2017). Lipids in fruits and vegetables: Chemistry and biological activities. Fruit and Vegetable Phytochemicals: Chemistry and Human Health, 2nd Edition, 423–450.

[67] Carr, A. C. and Maggini, S. (2017). Vitamin C and immune function. Nutrients, 9(11), 1211.

[68] Ghafarifarsani, H., Hoseinifar, S. H., Javahery, S. and Van Doan, H. (2022). Effects of dietary vitamin C, thyme essential oil, and quercetin on the immunological and antioxidant status of common carp (*Cyprinus carpio*). Aquaculture, 553, 738053.

[69] Angeli, L., Robatscher, P. and Chitarrini, G. (2021). Volatile organic compounds in apples: From biosynthesis to compound identification. Laimburg Journal, 3.

[70] Ma, L. and Lin, X. M. (2010). Effects of lutein and zeaxanthin on aspects of eye health. Journal of the Science of Food and Agriculture, 90(1), 2–12.

[71] Semmar, N. (2024). Secondary Metabolites in Plant Stress Adaptation: Analytic Space of Secondary Metabolites, Springer Nature.

[72] Hoshino, Y. (2024). Terpenoids and membrane dynamics evolution. Frontiers in Ecology and Evolution, 12, 1345733.

[73] Elbalola, A. A. and Abbas, Z. K. (2024). Chemotaxonomy, antibacterial and antioxidant activities of selected aromatic plants from Tabuk region-KSA. Heliyon, 10(1).

[74] Uçar, E. (2024). Polygonum aviculare L.'s biological activities: Investigating its Anti-Proliferative, Antioxidant, chemical properties supported by molecular docking study. Inorganic Chemistry Communications, 162, 112228.

[75] Samira, O., Laila, B., Moussa, N. A., Mohamed, I., Devkota, K., Abdelhakim, B. . . . and Said, G. (2024). Recent advances in the extraction of bioactive compounds from plant matrices and their use as potential antioxidants for vegetable oils enrichment. Journal of Food Composition and Analysis, 105995.

[76] Ghudhaib, K. K. and Khaleel, F. M. (2024). Evaluation of antioxidants, antibacterial and antidiabetic activities of aqua-alcoholic marjoram extract. Baghdad Science Journal.

[77] Bouloumpasi, E., Hatzikamari, M., Christaki, S., Lazaridou, A., Chatzopoulou, P., Biliaderis, C. G. and Irakli, M. (2024). Assessment of antioxidant and antibacterial potential of phenolic extracts from post-distillation solid residues of oregano, rosemary, sage, lemon balm, and spearmint. Processes, 12(1), 140.

[78] Baibuch, S. Y., Schelegueda, L. I., Bonifazi, E., Cabrera, G., Mondragón Portocarrero, A. C., Franco, C. M. . . . and Campos, C. A. (2024). Argentinian rose petals as a source of antioxidant and antimicrobial compounds. Foods, 13(7), 977.

[79] Kebert, M., Rašeta, M., Kostić, S., Vuksanović, V., Božanić Tanjga, B., Ilić, O. and Orlović, S. (2024). Metabolically tailored selection of ornamental rose cultivars through polyamine profiling, osmolyte quantification and evaluation of antioxidant activities. Horticulturae, 10(4), 401.

[80] Ricardo-Rodrigues, S., Rouxinol, M. I., Agulheiro-Santos, A. C., Potes, M. E., Laranjo, M. and Elias, M. (2024). The antioxidant and antibacterial potential of thyme and clove essential oils for meat preservation – An overview. Applied Biosciences, 3(1), 87–101.

[81] Mokhtari, R., Kazemi Fard, M., Rezaei, M., Moftakharzadeh, S. A. and Mohseni, A. (2023). Antioxidant, antimicrobial activities, and characterization of phenolic compounds of Thyme (*Thymus vulgaris* L.), Sage (*Salvia officinalis* L.), and Thyme–Sage mixture extracts. Journal of Food Quality, 2023(1), 2602454.

[82] Yun, B. R., Yang, H. J., Weon, J. B., Lee, J., Eom, M. R. and Ma, C. J. (2016). Neuroprotective properties of compounds extracted from *Dianthus superbus* L. against glutamate-induced cell death in HT22 cells. Pharmacognosy Magazine, 12(46), 109.

[83] Al-Snafi, A. E. (2017). Chemical contents and medical importance of *Dianthus caryophyllus* - A review. IOSR Journal of Pharmacy, 7(3), 61–71.

[84] Wang, M., Shen, Q., Pang, J., Mao, Y., Li, X., Tao, Y. . . . and Zhou, X. (2024). Study on chemical constituents and antioxidant activities of *Dianthus caryophyllus* L. Frontiers in Plant Science, 15, 1438967.

[85] Ferreira-Anta, T., Torres, M. D., Mourelle, L., Legido, J. L., Domínguez, H. and Flórez-Fernández, N. (2024). Ecofriendly cascade extraction of antioxidants from *Origanum vulgare*: Morphological and rheological behavior of microparticles formulations. Journal of Industrial and Engineering Chemistry.

[86] Tarkesh Esfahani, S., Pourgholamian, F. and Saravi, A. T. (2024). Time and concentration dependent changes in antioxidant and photosynthetic capacity of *Origanum vulgare* L. (Lamiaceae) in response to exogenous methyl jasmonate. Russian Journal of Plant Physiology, 71(3), 87.

[87] Latifeh, P., Helena, T., Neda, F., Sina, S. M., Svetlana, R. N. and Jelena, P. D. (2024). Antioxidant potential of green synthesized AGNPs, essential oil, and extracts of *Origanum vulgare* L. leaves. Genetics & Applications.

[88] Sharifi-Rad, M., Varoni, E. M., Salehi, B., Sharifi-Rad, J., Matthews, K. R., Ayatollahi, S. A. . . . and Rigano, D. (2017). Plants of the genus Zingiber as a source of bioactive phytochemicals: From tradition to pharmacy. Molecules, 22(12), 2145.

[89]   Ayustaningwarno, F., Anjani, G., Ayu, A. M. and Fogliano, V. (2024). A critical review of Ginger's (*Zingiber officinale*) antioxidant, anti-inflammatory, and immunomodulatory activities. Frontiers in Nutrition, 11, 1364836.

[90]   Mustafa, I. and Chin, N. L. (2023). Antioxidant properties of dried ginger (*Zingiber officinale* Roscoe) var. Bentong. Foods, 12(1), 178.

[91]   Pagliari, S., Forcella, M., Lonati, E., Sacco, G., Romaniello, F., Rovellini, P. . . . and Bruni, I. (2023). Antioxidant and anti-inflammatory effect of cinnamon (*Cinnamomum verum* J. Presl) bark extract after in vitro digestion simulation. Foods, 12(3), 452.

[92]   Al-Mijalli, S. H., Mrabti, H. N., El Hachlafi, N., El Kamili, T., Elbouzidi, A., Abdallah, E. M. . . . and Chahdi, F. O. (2023). Integrated analysis of antimicrobial, antioxidant, and phytochemical properties of *Cinnamomum verum*: A comprehensive In vitro and In silico study. Biochemical Systematics and Ecology, 110, 104700.

[93]   Pandey, M. and Chandra, D. R. (2015). Evaluation of ethanol and aqueous extracts of *Cinnamomum verum* leaf galls for potential antioxidant and analgesic activity. Indian Journal of Pharmaceutical Sciences, 77(2), 243.

[94]   Ramsewak, R. S., DeWitt, D. L. and Nair, M. G. (2000). Cytotoxicity, antioxidant and anti-inflammatory activities of curcumins I–III from Curcuma longa. Phytomedicine, 7(4), 303–308.

[95]   Memarzia, A., Khazdair, M. R., Behrouz, S., Gholamnezhad, Z., Jafarnezhad, M., Saadat, S. and Boskabady, M. H. (2021). Experimental and clinical reports on anti-inflammatory, antioxidant, and immunomodulatory effects of Curcuma longa and curcumin, an updated and comprehensive review. BioFactors, 47(3), 311–350.

[96]   Boroumand, N., Samarghandian, S. and Hashemy, S. I. (2018). Immunomodulatory, anti-inflammatory, and antioxidant effects of curcumin. Journal of Herbmed Pharmacology, 7(4), 211–219.

[97]   Dang, R., Guan, H. and Wang, C. (2023). Sinapis Semen: A review on phytochemistry, pharmacology, toxicity, analytical methods and pharmacokinetics. Frontiers in Pharmacology, 14, 1113583.

[98]   Nayak, G. R., Hegde, S., Shenoy, S., Parida, A. and Kg, M. R. (2024). Evaluation of anti-epileptic effect of *Sinapis alba* using Maximal Electroshock Seizure Model. Biomedical and Pharmacology Journal, 17(1), 153–161.

[99]   Başyiğit, B., Alaşalvar, H., Doğan, N., Doğan, C., Berktaş, S. and Çam, M. (2020). Wild mustard (*Sinapis arvensis*) parts: Compositional analysis, antioxidant capacity and determination of individual phenolic fractions by LC–ESI–MS/MS. Journal of Food Measurement and Characterization, 14, 1671–1681.

[100] Li Pomi, F., Papa, V., Borgia, F., Vaccaro, M., Allegra, A., Cicero, N. and Gangemi, S. (2023). *Rosmarinus officinalis* and skin: Antioxidant activity and possible therapeutical role in cutaneous diseases. Antioxidants, 12(3), 680.

[101] Nieto, G., Ros, G. and Castillo, J. (2018). Antioxidant and antimicrobial properties of rosemary (*Rosmarinus officinalis* L.): A review. Medicines, 5(3), 98.

[102] Hendel, N., Sarri, D., Sarri, M., Napoli, E., Palumbo Piccionello, A. and Ruberto, G. (2024). Phytochemical analysis and antioxidant and antifungal activities of powders, methanol extracts, and essential oils from *Rosmarinus officinalis* L. and Thymus ciliatus Desf. Benth. International Journal of Molecular Sciences, 25(14), 7989.

[103] Marrone, G., Urciuoli, S., Di Lauro, M., Cornali, K., Montalto, G., Masci, C. . . . and Noce, A. (2024). Saffron (*Crocus sativus* L.) and its by-products: Healthy effects in internal medicine. Nutrients, 16(14), 2319.

[104] Hernández, F., Amanpour, A., Soltani, M., Lipan, L., García Garví, J. M., Carbonell-Barrachina, Á. A. and Sendra Nadal, E. (2024). Comparative study on nutraceutical andsensorial characteristics of saffron (*Crocus sativus* L.) cultivated in Iran, Spain, and Türkiye.

[105] Boskabady, M. H. and Farkhondeh, T. (2016). Antiinflammatory, antioxidant, and immunomodulatory effects of *Crocus sativus* L. and its main constituents. Phytotherapy Research, 30(7), 1072–1094.

[106] Garcia, C. S., Menti, C., Lambert, A. P. F., Barcellos, T., Moura, S., Calloni, C. . . . and Henriques, J. A. (2016). Pharmacological perspectives from Brazilian salvia officinalis (Lamiaceae): Antioxidant, and antitumor in mammalian cells. Anais da Academia Brasileira de Ciências, 88, 281–292.

[107] Miraj, S. and Kiani, S. (2016). A review study of therapeutic effects of *Salvia officinalis* L. Der Pharmacia Lettre, 8(6), 299–303.

[108] Amer, A. A., Kassem, S. H. and Hussein, M. A. (2024). Chemical composition, antioxidant, cytotoxic, antiviral, and lung-protective activities of *Salvia officinalis* L. ethanol extract herb growing in Sinai, Egypt. Beni-Suef University Journal of Basic and Applied Sciences, 13(1), 39.

[109] Schnitzler, P., Nolkemper, S., Stintzing, F. C. and Reichling, J. (2008). Comparative in vitro study on the anti-herpetic effect of phytochemically characterized aqueous and ethanolic extracts of *Salvia officinalis* grown at two different locations. Phytomedicine, 15(1–2), 62–70.

[110] Rahmani, A. F., Sakandari, M. N., Mirzaei, A. and Uçar, E. (2024). Investigating the physiological responses and the expression of effective genes in Steviol Glycosides production in Stevia (*Stevia rebaudiana*). Agrotechniques in Industrial Crops, 4(3), 126–133.

[111] Shivanna, N., Naika, M., Khanum, F. and Kaul, V. K. (2013). Antioxidant, anti-diabetic and renal protective properties of *Stevia rebaudiana*. Journal of Diabetes and its Complications, 27(2), 103–113.

[112] Sharangi, A. B. and Bhutia, P. H. (2016). Stevia: Medicinal miracles and therapeutic magic. International Journal of Crop Science and Technology, 2(2).

[113] Ghanta, S., Banerjee, A., Poddar, A. and Chattopadhyay, S. (2007). Oxidative DNA damage preventive activity and antioxidant potential of *Stevia rebaudiana* (Bertoni) Bertoni, a natural sweetener. Journal of Agricultural and Food Chemistry, 55(26), 10962–10967.

[114] Zhao, L., Wang, X., Xu, M., Lian, Y., Wang, C., Yang, H. and Mehmood, A. (2018). Dietary intervention with Stevia residue extracts alleviates impaired glucose regulation in mice. Journal of Food Biochemistry, 42(6), e12651.

[115] Myint, K. Z., Wu, K., Xia, Y., Fan, Y., Shen, J., Zhang, P. and Gu, J. (2020). Polyphenols from *Stevia rebaudiana* (Bertoni) leaves and their functional properties. Journal of Food Science, 85(2), 240–248.

[116] Yoon, S., Hong, S. J., Jo, S. M., Jeong, H., Cho, S., Lee, Y. B. and Shin, E. C. (2023). Development of Enzymatically Modified Stevia Analysis Using HPLC.

[117] Aćimović, M., Šovljanski, O., Šeregelj, V., Pezo, L., Zheljazkov, V. D., Ljujić, J. . . . and Vujisić, L. (2022). Chemical composition, antioxidant, and antimicrobial activity of *Dracocephalum moldavica* L. essential oil and hydrolate. Plants, 11(7), 941.

[118] Kazempour, M., Shahangian, S. S. and Sariri, R. (2024). *Dracocephalum kotschyi*: Inhibition of critical enzyme relevant to type-2 diabetes, essential oil composition, bactericidal and anti-oxidant activity. Caspian Journal of Environmental Sciences, 22(2), 289–303.

[119] Che Zain, M. S., Osman, M. F., Lee, S. Y. and Shaari, K. (2021). UHPLC-UV/PDA method validation for simultaneous quantification of luteolin and apigenin derivatives from *Elaeis guineensis* leaf extracts: An application for antioxidant herbal preparation. Molecules, 26(4), 1084.

[120] Yu, H., Chen, Z., Chen, H. and Wang, Z. (2024). *Dracocephalum moldavica* L. extract ameliorates hypertension through modulating the interaction between miRNAs and Gut Microbiota in 2K1C Rats. Natural Product Communications, 19(7), 1934578X241261020.

[121] Lobach, A. R., Schmidt, F., Fedrizzi, D. and Müller, S. (2024). Toxicological safety evaluation of an aqueous lemon balm (*Melissa officinalis*) extract. Food and Chemical Toxicology, 187, 114565.

[122] Loizzo, M. R., Pugliese, A., Bonesi, M., Menichini, F. and Tundis, R. (2015). Evaluation of chemical profile and antioxidant activity of twenty cultivars from *Capsicum annuum*, *Capsicum baccatum*, *Capsicum chacoense* and *Capsicum chinense*: A comparison between fresh and processed peppers. LWT-Food Science and Technology, 64(2), 623–631.

[123] Shotorbani, N. Y., Jamei, R. and Heidari, R. (2013). Antioxidant activities of two sweet pepper *Capsicum annuum* L. varieties phenolic extracts and the effects of thermal treatment. Avicenna Journal of Phytomedicine, 3(1), 25.

[124] Choi, M. H., Kim, M. H. and Han, Y. S. (2023). Physicochemical properties and antioxidant activity of colored peppers (*Capsicum annuum* L.). Food Science and Biotechnology, 32(2), 209–219.

[125] Ozdal, T., Tomas, M., Toydemir, G., Kamiloglu, S. and Capanoglu, E. (2021). Introduction to nutraceuticals, medicinal foods, and herbs. In: Aromatic Herbs in Food, Academic Press, 1–34.

[126] Zieliński, H., Frias, J., Piskuła, M. K., Kozłowska, H. and Vidal-Valverde, C. (2006). The effect of germination process on the superoxide dismutase-like activity and thiamine, riboflavin and mineral contents of rapeseeds. Food Chemistry, 99(3), 516–520.

[127] Jeon, J., Jang, Y., Shin, Y., Lee, S. H., Choi, Y. M. and Choung, M. G. (2022). Contents of thiamine, riboflavin, and niacin in sprouted vegetables. Korean Society of Food Science and Nutrition Conference Proceedings, 423–423.

[128] Feigl, G., Kumar, D., Lehotai, N., Pető, A., Molnár, Á., Rácz, É. . . . and Laskay, G. (2015). Comparing the effects of excess copper in the leaves of *Brassica juncea* (L. Czern) and *Brassica napus* (L.) seedlings: Growth inhibition, oxidative stress and photosynthetic damage. Acta Biologica Hungarica, 66(2), 205–221.

Ines Sifaoui*, Atteneri López-Arencibia, María Reyes-Batlle,
José E. Piñero*, and Jacob Lorenzo-Morales*

Chapter 18
Medicinal and aromatic plants with anti-parasitic properties

Abstract: According to WHO, parasitic diseases, such as Malaria, leishmaniasis, Trypanosomiasis, Helminthiasis among other, cause significant morbidity worldwide, affecting millions of people annually particularly in rural areas of low-income countries. The chemotherapy of those infection has been weakened by the appearance of resistance form, the inconstant efficacy between strains or species, the toxicity, and requirement of long courses of administration. Consequently, there has been a marked increase in the need to identify new molecules with antiparasitic activities. In this field, Medicinal and Aromatic Plants (MAP) constitute a great mine as they have been used since the prehistoric times to treat a wide range of human infections. Recently, the use of bioactive compounds from MAP have increased exponentially in the pharmaceutical industry and the food sector. The present chapter provides a comprehensive overview of MAP that has been cited in the treatment of parasitic infections.

*Corresponding author: Ines Sifaoui**, Instituto Universitario de Enfermedades Tropicales y Salud Pública de Canarias, Universidad de La Laguna, Tenerife, Islas Canarias, La Laguna, Spain; Consorcio Centro de Investigación Biomédica en Red de Enfermedades Infecciosas (CIBERINFEC), Instituto de Salud Carlos III, Madrid, Spain; Departamento de Obstetricia y Ginecología, Pediatría, Medicina Preventiva y Salud Pública, Toxicología, Medicina Legal y Forense, y Parasitología, Universidad de La Laguna, Tenerife, Spain, e-mail: isifaoui@ull.edu.es
*Corresponding author: José E. Piñero**, Instituto Universitario de Enfermedades Tropicales y Salud Pública de Canarias, Universidad de La Laguna, Tenerife, Islas Canarias, La Laguna, Spain; Consorcio Centro de Investigación Biomédica en Red de Enfermedades Infecciosas (CIBERINFEC), Instituto de Salud Carlos III, Madrid, Spain; Departamento de Obstetricia y Ginecología, Pediatría, Medicina Preventiva y Salud Pública, Toxicología, Medicina Legal y Forense, y Parasitología, Universidad de La Laguna, Tenerife, Spain, e-mail: jpinero@ull.edu.es
*Corresponding author: Jacob Lorenzo-Morales**, Instituto Universitario de Enfermedades Tropicales y Salud Pública de Canarias, Universidad de La Laguna, Tenerife, Islas Canarias, La Laguna, Spain; Consorcio Centro de Investigación Biomédica en Red de Enfermedades Infecciosas (CIBERINFEC), Instituto de Salud Carlos III, Madrid, Spain; Departamento de Obstetricia y Ginecología, Pediatría, Medicina Preventiva y Salud Pública, Toxicología, Medicina Legal y Forense, y Parasitología, Universidad de La Laguna, Tenerife, Spain, e-mail: jmlorenz@ull.edu.es
Atteneri López-Arencibia, María Reyes-Batlle, Instituto Universitario de Enfermedades Tropicales y Salud Pública de Canarias, Universidad de La Laguna, Tenerife, Islas Canarias, La Laguna, Spain; Consorcio Centro de Investigación Biomédica en Red de Enfermedades Infecciosas (CIBERINFEC), Instituto de Salud Carlos III, Madrid, Spain; Departamento de Obstetricia y Ginecología, Pediatría, Medicina Preventiva y Salud Pública, Toxicología, Medicina Legal y Forense, y Parasitología, Universidad de La Laguna, Tenerife, Spain

https://doi.org/10.1515/9783111469713-018

Keywords: Medicinal and aromatic plants, parasitic infections, protozoa, Helminths

18.1 Introduction

According to the World Health Organization (WHO), infectious illnesses are diseases caused by a multitude of organisms such as viruses, bacteria, fungi, and parasites. Distinct from most viral and bacterial infections, parasitic diseases are usually chronic. They can last for months or even years [1].

Among the most important parasitic infections, we cite leishmaniasis, malaria, toxoplasmosis, trichomoniasis, trypanosomiasis, amebiasis, giardiasis, and helminthiasis [2]. Three of the cited diseases account for the highest death toll among all neglected tropical diseases including leishmaniasis and American and African trypanosomiasis.

Notwithstanding the considerable measures that have been taken to control and eradicate Neglected Tropical Diseases (NTDs), they remain an important global health problem, especially in developing countries. Thanks to the heightened awareness of the disease's severity and the implementation of mass drug administration programs, some of these diseases are now under control. The majority of NTDs outlined above have been classified as emerging and re-emerging diseases, presenting unique challenges in terms of treatment. The absence of an alternative chemotherapeutic modality for treatment represents a pressing need for immediate attention. Many of these issues are caused by complications arising from chemotherapeutic treatments. It is evident that the treatment of these infections has been hindered by several factors, such as resistance, variability in efficacy between strains or species, toxicity, parenteral administration, the necessity for lengthy courses of treatment, adverse effects, and cost considerations.

The utilization of medicinal and aromatic plants (MAPs) as a therapeutic modality for the treatment of various diseases has witnessed a notable surge recently. Due to the limitations and inaccessibility of modern healthcare in low- and middle-income countries, the population still depends on traditional medicine for primary health care [3, 4]. Since time immemorial, medicinal plants have been used by local populations as traditional medicines to treat a wide range of human ailments. In the present chapter, an overview was conducted to summarize MAP used to treat most parasitic infections.

18.2 Medicinal plants used for the treatment of malaria

Malaria is among the world's oldest and most lethal parasitic diseases. The latest World Malaria Report indicates that 249 million cases of malaria have been reported globally, with 94% of cases concentrated in Africa [5]. Transmitted by an infected female Anopheles mosquito, the causative agent of this protozoan disease is the genus *Plasmodium*. To date, five species have been identified in humans: *P. falciparum*, *P. vivax*, *P. ovale*, *P. malariae*, and *P. knowlesi* with *P. falciparum* being the deadliest species and the most widespread on the African continent. The principal mode of transmission of *Plasmodium* from human to human is through the bite of the *Anopheles gambiae* mosquito. However, vertical transmission via the transplacental route and transfusion-transmitted malaria have also been previously described [6, 7]. The malaria life cycle is complex and comprises various stages. Researchers often simplify the life cycle into two phases: a nonsexual stage in humans and a sexual stage in the vector. First, the parasite carrier bites a human and injects the parasite in the form of sporozoites. These sporozoites travel through the bloodstream to the liver, where they divide. The resulting parasites, called merozoites, burst from the liver cells to infect red blood cells and continue to multiply (Figure 18.1). These are ingested by a new mosquito when it bites the infected human, and the cycle begins again [8].

The clinical manifestations of malaria are highly variable. Fever develops within 30 days post-infection. It may be accompanied by weakness, headache, myalgia, vomiting, diarrhea, and/or cough. Fever with chills, cold sweats, and profuse sweating may occur cyclically due to the different phases of the parasite cycle [9, 10]. More serious symptoms may occur such as difficulty breathing, bleeding, jaundice, extreme tiredness, and convulsions. In some patients, infected erythrocytes can obstruct the blood vessels that supply the brain, which can ultimately result in vascular occlusion and, in severe cases, may prove fatal [11]. In zones where malaria is highly endemic, a proportion of people may be

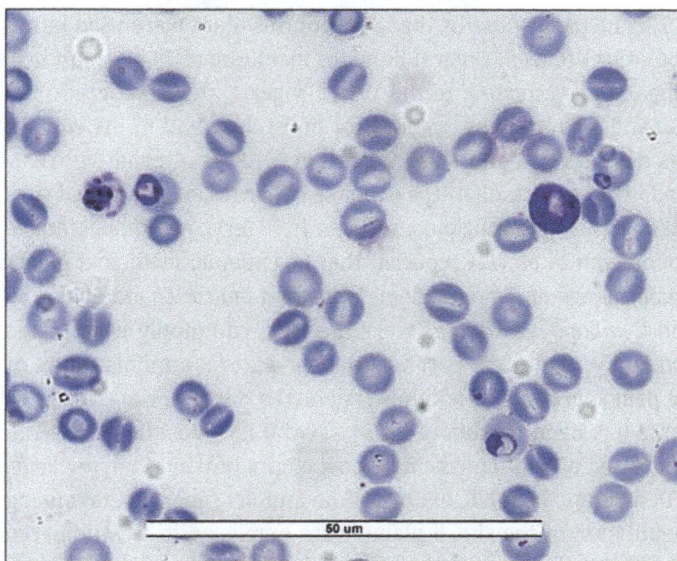

Figure 18.1: *P. berghei* blood smear showing vegetative forms and schizogony stages from Johannes Lieder's *General Parasitology, 3900*. The image was captured using the echo revolution microscope with a 100× objective.

asymptomatic. In fact, after multiple infections, the immune system may develop the ability to tolerate the presence of the parasite in the body [12].

The choice of antimalarial therapy depends on several factors, including the specific type and severity of the infection, the age of the patient, their underlying health status, the likelihood of their susceptibility to antimalarial drugs, and the cost and availability of such drugs [13]. Antimalarial therapy is based on the use of chloroquine to treat nonsevere *Plasmodium* spp. infections in sensitive areas. It is highly effective, well-tolerated, and economical [9]. In the case of chloroquine-resistant *Plasmodium falciparum*, the treatment is much more complex. Depending on the area of infection, in the United States, the treatment would be based on the use of mefloquine alone, or quinine plus doxycycline or pyrimethamine/sulfadoxine (Fansidar^R). In other regions, the treatment is primarily based on the use of halofantrine, artemisinin (qinghaosu) derivatives, and clindamycin [9, 13]. To combat drug-resistant malaria, the development of new antimalarial drugs that are both safe and economical is of the utmost importance.

In this context, Willcox and Bodeker [14] have documented that more than 1,200 plant species, originating from 160 distinct families, have been employed for the purpose of malaria and fever treatment [14]. Nowadays, artemisinin is used as an antimalarial agent, while formerly, the leaves of *Artemisia annua* were used to reduce fever by Chinese herbalists in AD 340 [15, 16]. *Parthenocissus tricuspidata (Sieb. and Zucc.) Planch* leaves have been stated by Il Hong Son et al. [17] to present antimalarial activity against *Plasmodium falciparum* [17]. The authors reported that the activity is mainly caused by the presence of stilbene derivatives, namely the pieceid-(1 → 6)-β-D-glucopyranoside (IC$_{50}$ = 5.3 ± 1.6 μM). Actually, the leaves of this plant were used in the past by South Asian populations to treat several illnesses including arthritis, jaundice, insect bites, and neuralgia [18]. According to the WHO, Nigeria, the Democratic Republic of Congo, Uganda, and Mozambique account for more than half of the world's malaria fatalities [5]. As developing countries, the use of MAP as alternative treatments is considered mandatory. As a result, these endemic countries are renowned sites of folk medicine, cultural diversity, and religious practices. In Nigeria, an ethnobotanical study conducted by Evbuomwan et al. [19] reported that 62 endemic medicinal plants were identified as antimalarial agents. Among them, *Mangifera indica, Enantia chlorantha, Alstonia boonei,* and *Cymbopogon citratus* were the most commonly used plants [19]. While the most frequently used plant part was the leaves, oral administration of water decoction was the principal route of administration. The antimalarial efficacy of the aforementioned plants has been corroborated by several in vivo studies [20–22]. Chiribagula et al. [23] conducted an ethnomedicinal knowledge study in Mampa village, Haut-Katanga province, Democratic Republic of Congo. The authors reported that among 38 plants used by the population of the site, 5 plants were employed to treat malaria, with *Crossopteryx febrifuga* being the most cited species [23]. In another study conducted by Manya et al. [24] in Bukavu and Uvira, the authors highlighted 45 plant species used as antimalarial agents. The most frequently cited plants were *Artemisia annua* L., belonging to the Asteraceae family, and *Carica papaya* L., belonging to the Caricaceae family, with a 34% citation rate for each. Oral administration of leaf decoction was the most cited route of administration [24]. In Uganda, ethnomedicinal studies have mentioned *Vernonia amygdalina, Aloe vera, Azadirachta indica, Chamaecrista nigricans, Aloe nobilis, Warburgia ugandensis, Abrus precatorius, Kedrostis foetidissima, Senna occidentalis*, and *Mangifera indica* as the most commonly used plants to treat malaria and its symptoms. According to previous works, the main route of administration was oral, and leaf decoction and maceration were the most commonly used plant preparations [25, 26]. In Mozambique, several plants were used to treat malaria symptoms. *Momordica balsamina*, usually recognized as balsam apple, is an endemic plant of tropical and subtropical regions of Africa, Asia, and Australia and is frequently employed as a means of alleviating symptoms of vomiting, which are thought to be linked to bile and fever [27, 28]. *Spirostachys africana*, an indigenous plant of Southern Africa, was frequently cited to alleviate headaches [29, 30]. *Bridelia cathartica* is a well-known folk medicinal plant used in sub-Saharan Africa for headache pain relief. The antiplasmodial effect of *Bridelia cathartica* was confirmed by Jurg et al. [31]. The authors reported that a concentration of 0.05 μg/mL of ethanolic and aqueous extracts from the root part was able to reduce the growth of *Plasmodium falciparum* schizonts by 50% [31].

18.3 Medicinal plants used for treatment of leishmaniasis

Leishmaniasis is transmitted by the bite of a female phlebotomine sand fly and is caused by protist parasites of the genus *Leishmania*. *Leishmania* is a dimorphic protozoan parasite (having two morphological forms) carried by the females of small insects of the genus Phlebotomine. At this stage, the parasite is in its promastigote form

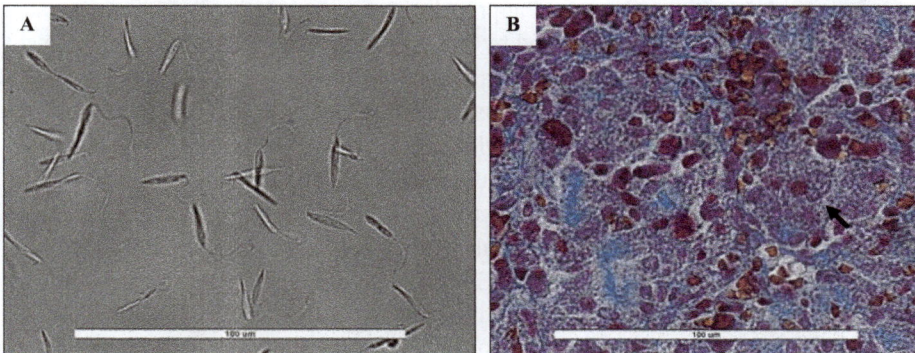

Figure 18.2: *Leishmania donovani*, promastigote form (A) and intracellular form, from Johannes Lieder's *General Parasitology, 3900*. The images were taken using the echo revolution microscope with a 60× objective.

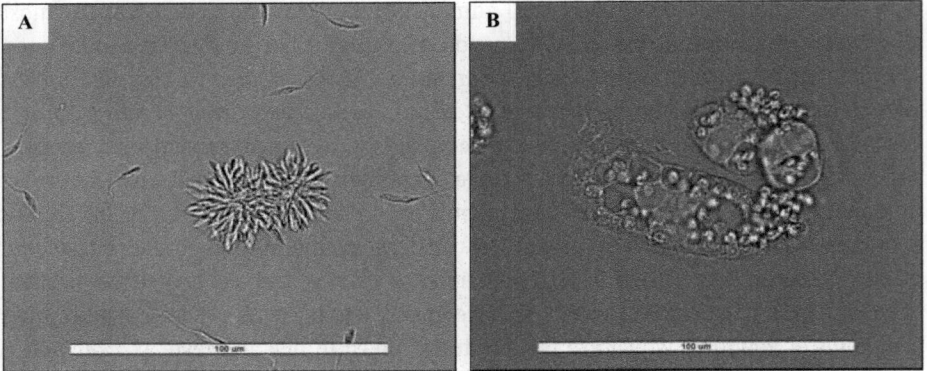

Figure 18.3: *Leishmania amazonensis*, promastigote form (A) and intramacrophagic form (B). The images were taken using the echo revolution microscope with a 60× objective.

(Figures 18.2A and 18.3A), which is mobile. Once injected into a mammalian host, it transforms inside macrophages into rounded intracellular cells called amastigotes (Figures 18.2B and 18.3B).

Depending on the *Leishmania* species, the infection occurs in three forms: cutaneous leishmaniasis (CL), mucocutaneous leishmaniasis (MCL), and visceral leishmaniasis (VL) [32]. Although CL is a non-life-threatening form, it can cause lifelong scars, ulcerated lesions, severe disability, and stigma. As reported by the WHO in 2023, the annual incidence of new cases ranges from 600,000 to 1 million. Furthermore, the majority (approximately 95%) of CL cases are concentrated in the Americas, the Mediterranean basin, the Middle East, and Central Asia. The MCL form has the potential to cause significant damage to the mucous membranes of the nose, mouth, and throat, resulting in partial or complete destruction. A review of case studies revealed that Bolivia, Brazil, Ethiopia, and Peru collectively account for over 90% of cases of MCL [33]. The most severe clinical form of leishmaniasis is VL. In the absence of appropriate intervention, the disease can be fatal. The current manifestation is described by irregular spikes in body temperature, weight loss, damage to internal organs, including the spleen and liver, and the induction of anemia in the bone marrow [33, 34]. According to the WHO, in 2020 nearly 13,000 cases of VL occurred. Ninety percent of worldwide reported cases happen in seven countries: Brazil, Ethiopia, India, Kenya, Somalia, South Sudan, and Sudan [35, 36]. Pentavalent antimonial has been the standard treatment for VL since the 1940s. However, in cases where this treatment proves ineffective, alternative options such as amphotericin B deoxycholate and liposomal amphotericin B can be employed. To date, Miltefosine, an alkyl phospholipid molecule, is the only oral drug currently available for the treatment of VL [35, 37]. However, these drugs are noticeably toxic to patients and susceptible to the development of drug resistance. It is therefore imperative to identify different natural and safe sources of leishmanicidal molecules. In recent times, there has been a marked increase in the quest for new active antileishmanial molecules, particularly those of plant origin. Digging into the folk medicine knowledge of leishmaniasis-endemic countries could boost drug development against the present illness. Persian folk medicine represents one of the most ancient forms of traditional medicine. As a country with a high prevalence of leishmaniasis, Iran encompasses a range of efficacious techniques for the management of CL. In this context, an assessment of the plants used in Iranian traditional medicine for the treatment of leishmaniasis was conducted by Bahmani et al. [38]. The authors compiled the application of plants in phytomedicine with their confirmed activity against the parasite [38]. The Asteraceae family, including *Artemisia* and *Marigold,* constitutes the most native and active plants against *Leishmania* spp. [38]. In an additional review, the authors reported the utilization of a variety of medicinal plants, including *Aristolochia* sp., Alumroot, *Ferula ammoniacum, Commiphora mukul, Rhamphospermum nigrum,* and honey soaked in wheat oil or vinegar as potential treatments for CL ulcers [39]. *Juniperus excelsa* M. *Bieb* was mentioned as being used in cases of resistant ulcer forms [40]. In 2017, a study conducted by Mahmoodreza et al. proved that the petroleum ether extract of *Juniperus excelsa* M. Bieb could completely inhibit the growth of the promastigote stage of *Leishmania major* at a concentration of 2.5 mg/mL [41].

In the Amazonia region, an endemic site for various *Leishmania* species, including *L. amazonensis, L. braziliensis,* and *L. guyanensis,* traditional antileishmanial phytomedicines are numerous and widely used [42]. Most of the reported remedies are used to treat both cutaneous and mucocutaneous forms. In this context, Copaiba oil has been used to treat anti-inflammatory symptoms, ulcers, and scars caused by leishmaniasis [43]. An extensive review on medicinal plants used in the Amazonia region related to leishmaniasis has been conducted by Odonne et al. [42]. The authors reported that topical application of medicinal plants was the most commonly used administration route.

Inga, Vismia, Citrus, Piper, Jatropha, Spondias, Manihot, Anacardium, Carapa, and *Solanum* were the most cited genera of plants. Various studies have confirmed the in vitro effects of the mentioned plants:

- Monzote et al. [44] have confirmed the leishmanicidal effect of *Inga sierrae* (Fabaceae) extract against the promastigote stage of *L. amazonensis* with an IC_{50} of 25.7 µg/mL [44].
- Diel et al. [45] have confirmed the antileishmanial activity of *Vismia* species extracts against *L. amazonensis,* with the *V. guianensis* extract exhibiting the lowest IC_{50} (4.3 µg/mL); indeed, this species is the most commonly used for treating CL [45].
- Garcia et al. [46] have reported the leishmanicidal activity of *Citrus sinensis* leaf organic extracts, with the hexane extract being the most active against *L. amazonensis,* showing an IC_{50} of as organic leaf extract of *Citrus sinensis* 25.91 ± 4.87 µg/mL [46].
- Bosquiroli et al. [47] have reported that the essential oil of *P. angustifolium* could inhibit the amastigote stage of *L. infantum,* the causative agent of CL and VL, with an IC_{50} of 1.43 µg/mL [47].
- Hermoso et al. [48] conducted a bio-guided fractionation against *Leishmania braziliensis* using the extract of *Piper elongatum* (Piperaceae), which is traditionally

used by indigenous people of Cusco, Perú to treat ulcers, and they obtained two dihydrochalcones endowed with leishmanicidal properties [48].

In addition to the aforementioned plants, the olive tree (*Olea europaea*, Oleaceae) has historically provided significant economic and dietary benefits to the Mediterranean region [49]. Indeed, olive leaf extracts (OLEs) have also been employed by local populations in traditional medicine to treat conditions such as fever and malaria. In 2014, we conducted a bio-guided fractionation of OLE against the promastigote stage of *L. infantum* and *L. amazonensis*. Two anti-leishmanicidal triterpenic acids were isolated, including oleanolic and maslinic acids [50, 51].

Apart from the mentioned plants, our research group has been conducting work on the leishmanicidal effect of medicinal plants. In recent years, we have reported the antileishmanial activity of various medicinal plants, including *Matricaria recutita* L. [52], *Inula viscosa* [53], *Maytenus chiapensis* [54], *Pituranthos battandieri* [55], *Zinowiewia integerrima, Maytenus segoviarum, Quetzalia ilicina, Euonymus enantiophyllus*, and *Wimmeria cyclocarpa*, among others [56]. Table 18.1 shows further examples of extracts with antileishmanial activity.

Table 18.1: Examples of plant extracts with antileishmanial activity.

Botanical name	Tested species	Preparation	References
Thymus hirtus sp. *algeriensis*	*L. major* *L. infantum*	Essential oil	[57]
Ruta chalepensis	*L. infantum*		
Vanillosmopsis arborea	*L. amazonensis*	Essential oil	[58]
Aspidosperma ramiflorum Muell	*L. amazonensis*	Alkaloid extract	[59]
Aniba canelilla H.B.K.	*L. braziliensis* *L. amazonensis*	Ethyl acetate extract	[60]
Cardiopetalum calophyllum Schldl	*L. amazonensis* *L. braziliensis* *L. donovani*	Alkaloid fraction	
Duguetia spixiana Mart.	*L. amazonensis* *L. braziliensis* *L. donovani*	Alkaloid fraction	
Serjania tenuifolia Radlk	*L. braziliensis* *L. amazonensis*	Ethanolic extract	
Ampelocera edentula Kulm	*L. amazonensis*	Ethanolic extract	
Hura crepitans, Bambusa vulgaris, Simarouba glauca	*L. amazonensis*	Hydroethanol extract	[61]

Table 18.1 (continued)

Botanical name	Tested species	Preparation	References
Artemisia herba-alba Asso.	*L. tropica*	Aqueous extract	[62]
Thymus capitellatus	*L. infantum* *L. tropica* *L. major*	Essential oil	[63]
Vitis vinifera L.	*L. infantum*	Ethanolic extract	[64]
Inula montana L.	*L. infantum*	Methanolic extract	[65]
Chenopodium ambrosioides	*L. amazonensis*	Essential oil	[66]
Yucca filamentosa L.	*L. amazonensis*	Ethanolic extract	[67]
Lantana ukambensis	*L. donovani*	Methanolic extract	[68]
Salvia officinalis	*L. major*	Hydroalcoholic extract	[69]

18.4 Medicinal plants used for the treatment of human trypanosomiasis

Human trypanosomiasis is an NTD caused by various species of the hemoflagellate parasite *Trypanosoma*. Two main pathologies are associated with the infection, which depend on the infecting species: sleeping sickness (or African trypanosomiasis) and Chagas disease (or *American trypanosomiasis* (AT) in humans). In terms of geographical distribution, trypanosomiasis is endemic in certain regions of both Africa and Latin America, with implications for both human and animal health.

18.4.1 American trypanosomiasis

AT, also referred to as Chagas disease, is a vector-borne zoonosis caused by a flagellate protozoan parasite, *Trypanosoma cruzi* (WHO). In endemic countries, in Latin America, the disease is generally transmitted via the triatomine bug (kissing bug), though there are other modes of infection, including blood transfusion, organ transplant, oral, sexual, and congenital transmission [70]. The life cycle of *Trypanosoma cruzi* comprises two hosts and four developmental stages. The cycle commences with the infection of the mammalian host by the metacyclic trypomastigotes, which are found in the excreta of the blood-feeding triatomine vector. This occurs either through the bite wound of the vector or via various mucous membranes.

Upon entering both phagocytic and nonphagocytic nucleated cells, the parasite differentiates into small, round, nonmotile cells known as amastigotes. In this phase, the parasite proliferates by binary fission until the cells are filled and begin to elongate, reacquiring the flagella and undergoing a transformation into trypomastigotes. The intense movement of trypomastigotes results in the lysis of the host cell. Upon release, the parasites can invade neighboring cells or enter the lymphatic system and spread throughout the body. When ingested by the vector, the trypomastigotes present in the bloodstream are transformed into epimastigotes in the midgut of the vector, migrate to the hindgut, and differentiate into metacyclic trypomastigotes. This process marks the beginning of a new cycle (Figure 18.4) [71].

In cases of missed or late diagnosis, this infection can become life-threatening. In undeveloped countries, it constitutes a global public health burden [72]. According to the WHO, 6–7 million people worldwide are estimated to be infected with *T. cruzi*, mainly in Latin America, and approximately 75 million people are at risk of infection.

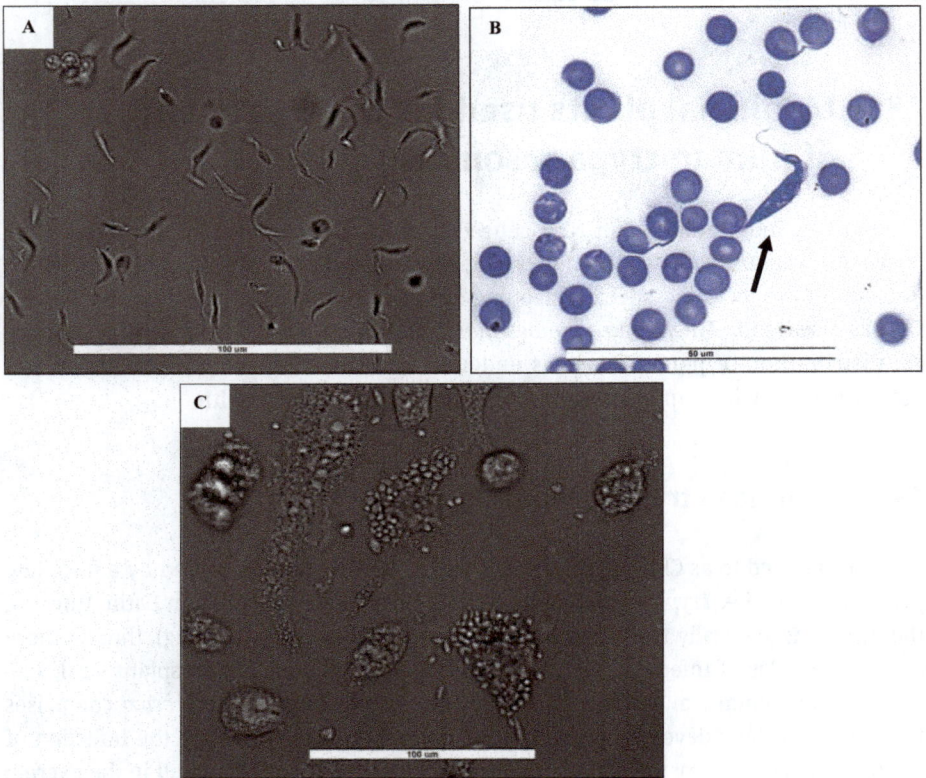

Figure 18.4: *Trypanosoma cruzi* spp., epimastigote form (A), trypomastigote form (B) from Johannes Lieder's *General Parasitology, 3900*, and intra-macrophagic form (C). The images were taken using the echo revolution microscope, with a 60× objective for A and C, and a 100× objective for B.

An estimated 12,000 deaths associated with the disease occur annually [73, 74]. Globalization and the immigration of unaware hosts could increase the occurrence of Chagas disease in nonendemic areas [73]. The current therapeutic approach for AT is limited to the administration of two pharmacological agents: benznidazole and nifurtimox [75].

Although the cited drugs are highly effective when administered at the beginning of the infection, as in congenital infection, their efficacy decreases in the chronic stage [76]. Moreover, the adverse effects associated with the long-term administration of both drugs remain a matter of concern [77, 78]. Chagas disease remains prevalent in underdeveloped and poor countries, where medicinal plants constitute the first line of treatment. Among the endemic regions for Chagas disease, Salvador is considered one of the most affected, with an estimated prevalence of 1.3–3.7% [79]. As a result, the region is characterized by an extensive use of folk medicine, especially in rural communities. In this context, the in vitro trypanocidal effects of various Salvadoran plant extracts on the epimastigote forms of *T. cruzi* have been investigated by Castillo et al. [80]. Among the 38 evaluated species, four methanolic extracts were found to inhibit parasite growth, including *P. jacquemontianum*, *P. lacunosum*, *T. havanensis*, and *P. pseudo-pereskiifolia* [80]. Our group has worked on the evaluation of five Salvadoran *Celastraceae* species extracts against the epimastigote stage of *T. cruzi*. Among the tested extracts, seven obtained from three species, including *Zinowiewia integerrima*, *Maytenus segoviarum*, and *Quetzalia ilicina* presented potent anti-trypanosomal activity. In addition, hexane-ethanol extracts from *Z. integerrima* and *M. segoviarum* and acetonic extracts from *Z. integerrima* and *Q. ilicina* exhibited an IC_{50} lower than the reference drug benznidazole (IC_{50} < 1.81 µg/mL) [54]. Nekoei et al. [81] conducted a review of natural trypanocidal products isolated from plants worldwide. The authors listed 165 plants used globally against trypanosomiasis [81]. Most of the cited plants with activity against *Trypanosoma cruzi* were from Latin America. Among the effective plants, we focus on those presenting a potent trypanocidal effect with an IC_{50} ≤ 5 µg/mL, including *Iochroma arborescens*, *Anthemis tinctoria*, *Casearia sylvestris*, *Connarus suberosus*, *Maianthemum paludicola*, *Psidium laruotteanum Cambess*, *Scoparia dulcis* L., and *Tagetes caracasana Humb.*, among others.

18.4.2 African trypanosomiasis

Human AT is an NTD prevalent in sub-Saharan Africa, where the vector tsetse fly is present. The infection is caused by a flagellate protozoan, *Trypanosoma brucei*. Depending on the species, researchers distinguish two forms of the illness: the slowly progressing form (gambiense HAT), caused by *T. brucei gambiense*, prevalent in western and central Africa; and the more rapidly progressive form (rhodesiense HAT), caused by *T. brucei rhodesiense*, widespread in eastern and southern Africa [82].

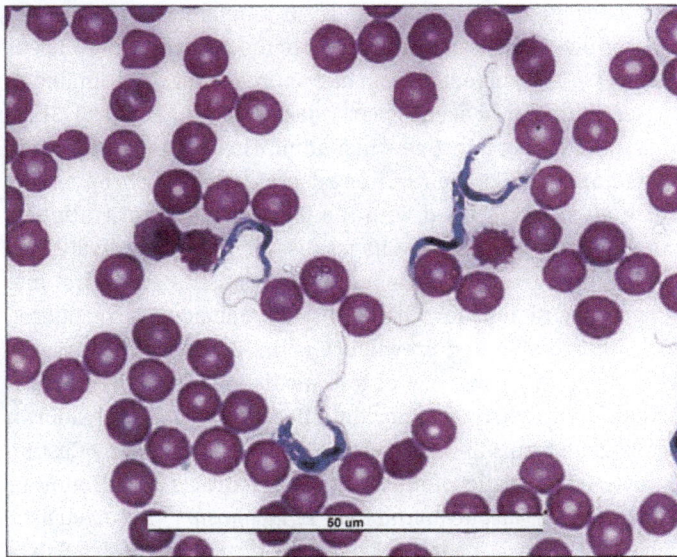

Figure 18.5: *Trypanosoma gambiense*, Central African sleeping sickness, blood smear from Johannes Lieder's *General Parasitology, 3900*. The image was taken using the echo revolution microscope with a 100× objective.

Following the bite of a mammalian host, an infected vector injects metacyclic trypomastigotes into the skin tissue. These trypanosomes enter the bloodstream and undergo a conversion process, becoming pleomorphic bloodstream trypomastigotes (Figure 18.5). They then multiply in the bloodstream and are carried to further sites, including the cerebrospinal fluid, the interstitial space of the lymph nodes, the spleen, and the brain [83]. During the first phase of the disease, parasites in the blood cause nonspecific symptoms, such as fever, headache, fatigue, and inflammation of the lymph nodes, making diagnosis difficult. Left untreated, the parasites will certainly invade the central nervous system (CNS). When the CNS becomes involved, disturbances in the sleep/wake cycle appear. This deterioration of the nervous system is always fatal if left untreated [84]. The type of treatment depends on the parasite species and the stage of the disease, but the earlier the diagnosis, the better the prospects of recovery. The drugs used in the first phase have few side effects and are relatively easy to administer: Pentamidine for *T. brucei gambiense* [85] and Suramin for *T. brucei rhodesiense* [86]. Both cited drugs are unable to cross the blood-brain barrier, so they are ineffective during the second phase [87]. In the case of infection progression to the CNS, the only available therapeutic option is the trivalent arsenical melarsoprol. However, melarsoprol therapy is associated with a number of adverse effects, with reactive arsenical encephalopathy being the most severe side effect [88]. Since 2009, the combination of nifurtimox and eflornithine (NECT) has been recommended for advanced chronic forms of *T. brucei gambiense* [88]. Recently, the WHO has recommended the use of fexinidazole instead of suramin and melarsoprol in patients aged 6 years or older. The oral administration and efficacy of fexinidazole are considered a significant advance in the treatment of rhodesiense human African trypanosomiasis [89]. This disease has a significant impact on both human and animal health, particularly in areas of Africa where there is limited access to modern healthcare systems. It has been identified as a major challenge affecting both human health and livestock productivity [90]. As in most poor countries, medicinal plants constitute the first line of treatment to combat infectious diseases. Thus, various authors have reported the ethnobotanical uses to manage African trypanosomiasis. Among them, Ibrahim et al. [90] conducted an ethnobotanical review on anti-*Trypanosoma* medicinal plants used in the African continent. The authors detailed that most of the reported plants were from West African countries, essentially Nigeria, Burkina Faso, Mali, and Ivory Coast. In the second place, almost 32% of the cited plants were from East African countries such as Tanzania, Ethiopia, and Uganda. African mahogany, *African custard-apple*, gum arabic tree, and Indian lilac plants were reported to exhibit the highest trypanocidal activity [90]. Ibrahim et al. (2013) proved in vivo the efficacy of the phenolic-rich fraction of *Khaya senegalensis* to eliminate *T. brucei* in infected mice at a dose of 300 mg/kg body weight [91]. Aqueous extract from *Annona senegalensis* has also been reported to eradicate in vivo *T. b. brucei* in infected mice at a dose of 200 mg/kg body weight per day. In vitro assays on the anti-*T. brucei* activity of essential oils from various Cameroonian medicinal plants were conducted by Ngahang Kamte et al. [92]. The authors demonstrated that the essential oil from *Aframomum danielli* exhibited the highest trypanocidal activity, followed by *Echinops giganteus* and *Azadirachta indica*, with IC_{50} values of 7.65 ± 1.1, 10.50 ± 1.7, and 15.21 ± 0.97 μg/mL, respectively [92]. The recent review conducted by Nekoei et al. [93] on trypanocidal activities detected that most of the cited plants from Africa were investigated against *T. brucei* subspecies. Among the 24 plants reported for their activity against *T. brucei*, four species were mentioned to provide various molecules endowed with anti-*Trypanosoma* effects: *Tanacetum cinerariifolium*, *Keetia leucantha*, *Tecla trichocarpa*, and *Terminalia avicennioides* [93].

18.5 Medicinal plants used for the treatment of giardiasis

Giardiasis (or lambliasis) is an infection caused by a flagellate intestinal protozoan known as *Giardia lamblia* that occurs worldwide and is the most widespread intestinal protozoal infection in the United States. *Giardia* can develop an external shell (called a cyst), which allows the parasite to survive for prolonged periods outside a host organism (e.g., in a lake or stream) and makes it less sensitive to chlorine (e.g., in swimming pools). These cysts are excreted in the feces and are infectious (Figure 18.6) [94].

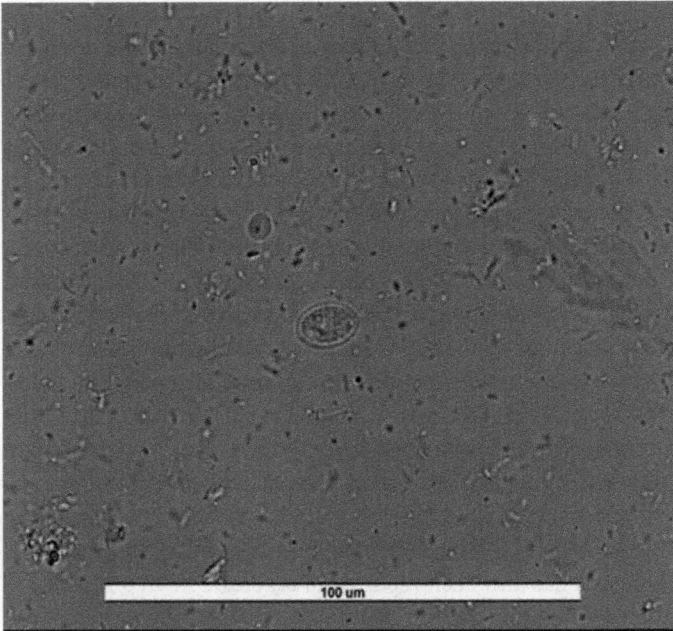

Figure 18.6: Cyst of *Giardia lamblia* from a stool sample. The image was taken using the echo revolution microscope with a 60× objective.

The protozoan genus *Giardia* is a common contaminant of freshwater, including lakes and rivers. The use of inefficient filtration techniques in municipal drinking water systems contributes to the outbreak of epidemics. Most people become infected by drinking contaminated water. Nevertheless, the infection can also be transmitted via ingestion of contaminated foodstuffs or through contact with the feces of an infected individual, which may occur between children or sexual partners [95]. The life cycle of *G. lamblia* can be divided into two distinct stages. This has enabled it to adapt and survive in a range of disparate and inhospitable environments. The cyst form responsible for the transmission of giardiasis is capable of enduring for months in fresh water at 4 °C, while the trophozoite stage colonizes the human small intestine and is the cause of disease [96]. Symptoms of giardiasis usually appear 1–3 weeks after ingestion of viable cysts. Most cases of giardiasis are asymptomatic; in others, the clinical manifestations consist of a variety of nonspecific symptoms such as pasty stools, diarrhea, malaise, abdominal cramps, meteorism, bloating, and fat malabsorption [97]. Other common symptoms are anorexia and weight loss [98]. In the absence of treatment, giardiasis either resolves spontaneously or becomes chronic, characterized by worsening digestive symptoms [99]. As treatment, nitroimidazoles, including metronidazole, tinidazole, ornidazole, and secnidazole, constitute the first line to treat various protozoan infections, including giardiasis [97]. In Iranian phytomedicine, various plants have been cited as anti-giardiasis agents, including French lavender, asafetida, feverfew, few-flowered garlic, Jerusalem oak goosefoot, ajowan, garlic, sweet wormwood, shallot, blue mint bush, Avishan-e-Shirazi, *Tasmanian bluegum, Mexican oregano,* and pomegranate [100]. A systematic review on medicinal plants used to treat giardiasis has been carried out by Alnomasy et al. [101]. The authors reported that 19 plant families have been cited to exhibit anti-Giardia activity in vitro and in vivo, with the Lamiaceae, Asteraceae, and Apiaceae families being the most commonly used. The authors observed that essential oils and aqueous extracts were the most common extract types used to treat *Giardia* infection, and that the most commonly used plant part to prepare the crude extract was the leaf. Azadbakht et al. [102] tested the effect of *Allium sativum, A. sieberi, Z. multiflora, C. botrys,* and *E. globulus* essential oils on *G. lamblia* cysts. Among the tested species, the essential oil from *E. globulus* was the most active extract, able to reduce almost 80% of *G. lamblia* cysts at a concentration of 0.2 μg/mL [102]. In a study conducted in Mexico, the effects on *G. lamblia* of 27 plants used in Mexican phytomedicine on *G. lamblia* were evaluated. Among the tested methanolic extracts, 15 presented an IC_{50} lower than 100 μg/mL, with *D. contrajerva, S. villosa,* and *R. chalepensis* being the most active species, showing an $IC_{50} < 38$ μg/mL [103].

18.6 Medicinal plants used for the treatment of amebiasis

Amebiasis is caused by a protozoan amoebozoa called *Entamoeba histolytica,* which is specific to humans. *E. histolytica* presents two life stages: trophozoite and cyst. The cycle starts with the ingestion of mature cysts, which germinate in the small intestine to excyst into the trophozoite stage. Herein, the cells reach the large intestine, where they multiply and encyst again. It is in this more resistant cyst form that *E. histolytica* is shed in the stools and is likely to contaminate other people [104]. Generally infecting humans, *E. histolytica* is usually found in areas with poor sanitation. The parasite is transmitted via the fecal-oral route, generally through the ingestion of contaminated water or food or by dirty hands or objects soiled with stools containing ameba cysts [105]. According to the WHO, it is estimated that approximately 90% of carriers are asymptomatic. In regions where amebiasis is not endemic, the WHO recommends that asymptomatic carriers should also be treated with contact medications in order to reduce the risk of transmission [106]. It is worth noting that in a small number of infected people, the parasite has the potential to penetrate the lining of the colon. When this occurs, the amoebae can cause the intestinal form of amebiasis, also known as amoebic dysentery. The clinical symptoms of amebiasis may progress from an asymptomatic stage to a severe stage, which may result in the development of amoebic colitis and/or amoebic liver abscesses [107]. The treatment of acute amebiasis comprises the administration of broad-spectrum antiparasitic drugs and contact amebicides that act locally within the lumen of the digestive tract. These include metronidazole and tinidazole. Additionally, drugs that act against cysts can be employed, such as paromomycin and iodoquinol [106]. The infection is endemic in poor countries where the only available health treatment often relies on the use of medicinal plants. Most plant species exploited for the treatment of gastrointestinal disorders have been reported based on their use in the management of diarrhea and dysentery rather than as specific treatments for identified pathogens [108]. A review conducted by Nezaratizade et al. [109] reported a total of 24 plant families with anti-*E. histolytica* effects. Among the cited plants, various extracts present outstanding activity, including *Elaeodendron trichotomum*, *Virgillia oroboides*, *Syzygium cordatum*, *Artemisia sieberi*, *Pimpinella anisum*, and *Kigelia pinnata* with an $IC_{50} < 1$ µg/mL [109]. McGaw et al. [110] studied the anti-amoebic activity of various South African plants on *E. histolytica*. Among the tested species, two ethanolic extracts stand out: *A. calamus* and *S. birrea* with IC_{50} values of 0.3125 and 1.25 µg/mL, respectively [110].

18.7 Medicinal plants used for the treatment of cryptosporidiosis

Cryptosporidiosis is a disease caused by a parasite of the genus *Cryptosporidium*, transmitted via the fecal-oral route. It is considered one of the most common waterborne diseases and a significant cause of waterborne disease outbreaks, particularly in countries with inadequate sanitation.

Cryptosporidium has a life cycle that can be subdivided into six groups based on the developmental phases. It begins with the ingestion of the oocyst, which then undergoes excystation into infectious sporozoites within the gastrointestinal tract. This marks the beginning of the second stage of asexual proliferation, known as merogony. The third stage, gametogony, is defined by the formation of micro- and macrogametes, which are subsequently fertilized in the fourth phase. The fifth stage is characterized by the development of oocyst walls, which confer environmental tolerance and facilitate the transmission of the infection from one host to the next. In the sixth and final stage, sporogony, infectious sporozoites are formed (Figure 18.7) [111]. The disease can manifest in any age group. In immunocompetent individuals, the infection is typically asymptomatic or manifests as a common gastrointestinal infection (gastroenteritis). Symptoms may emerge 5–21 days following ingestion including diarrhea, abdominal cramps, fever, and nausea. These symptoms typically resolve within a few days or weeks. However, in immunocompromised patients, such as those with AIDS, the infection can be life-threatening, causing severe diarrhea accompanied by fever. In such cases, the patient may require prolonged parenteral nutrition [112]. The actual infection has a much greater impact on livestock, as it affects young animals such as calves, resulting in significant economic losses due to factors such as morbidity, growth retardation, and the costs associated with treatment [113]. Most immuno-

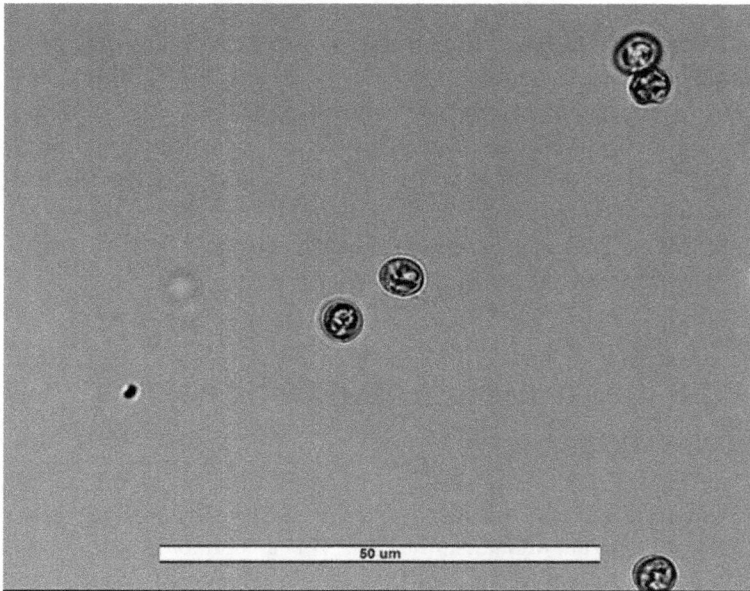

Figure 18.7: Thick-walled sporulated oocyst stage of *Cryptosporidium* spp. from a stool sample. The image was taken using the echo revolution microscope, with a 100× objective.

competent individuals will recuperate without the need for intervention. However, the illness may be prolonged and associated with worsening nutritional status in children. In this case, hydration supportive therapy should be included to substitute fluids and electrolytes, including sodium, potassium, bicarbonate, and glucose. It is important to note that patients with compromised immune systems may develop severe cryptosporidiosis, which requires careful management. This may include antimotility drugs, fluids and electrolytes, antiretroviral therapy, and often antiparasitic drugs. A limited number of drugs have demonstrated efficacy in vitro. Halofuginone has been approved for the prophylaxis and metaphylaxis of disease in animals in Europe. In humans, only nitazoxanide has been approved by the Food and Drug Administration for the treatment of cryptosporidiosis in patients aged ≥1 year [114]. The medicinal plants used to treat cryptosporidiosis aim to address intestinal disorders, particularly diarrhea, abdominal cramps, fever, and nausea, and/or to reduce oocyst shedding, thereby preventing new contamination. In this field, in vivo studies in calves and mice have demonstrated the effect of *A. sativum* extract in reducing or eliminating oocyst shedding while alleviating inflammation symptoms [115, 116]. *Anethum graveolens* was found to reduce oocyst shedding and led to the clearance of the *Cryptosporidium* antigen from the intestinal tissue of Swiss albino male mice [117]. El-Hamed et al. [118] demonstrated that a combination of *Ficus carica* and *Olea europaea* extracts led to a decrease in oocyst shedding and an improvement in histopathological lesions as well as hepatic and renal functions in immunosuppressed mice [118]. A recent study conducted by Abd El Wahab et al. [119] revealed that *Citrus sinensis* peel ethanolic extract could reduce the number of oocysts shed in immunocompetent mice, with an improvement in clinical manifestations [119]. Various in vitro studies have reported the anti-cryptosporidial effects of medicinal plant extracts. Among them, Ahmed et al, (2023) demonstrated that the phenolic fraction of Artemisia judaica L. showed a potent anti-cryptosporidial effect in vitro, with an IC_{50} of 31.6 µg/ mL against C. parvum oocysts. Meanwhile, in vivo, the phenolic-enriched fraction of Artemisia judaica L. was found to protect mice from infection [120].

18.8 Medicinal plants used for the treatment of toxoplasmosis

Toxoplasmosis is an infectious disease that results from the transmission of a protozoan parasite, *Toxoplasma gondii,* from animals to humans. It is a common disease that is frequently unrecognized, as most infected individuals are asymptomatic. Belonging to the phylum Apicomplexa, *Toxoplasma gondii* is an obligate intracellular protozoon. The life cycle of the parasite presents three life stages: sporozoites, tachyzoites, and bradyzoites [121]. Cats serve as the definitive host for the asexual reproduction (schizogony) and sexual reproduction (gametogony) of the parasite in the epithelial cells of the small intestine, resulting in the production of unsporulated oocysts [122].

Once shed in feces by cats, the oocysts transform into infective sporulated oocysts (Figure 18.8). Cats can become reinfected by ingesting the sporulated oocysts. Infection of intermediate hosts, including birds, animals bred for consumption, and vegetables, occurs when they ingest contaminated materials. Once ingested, the oocysts undergo a transformation into tachyzoites, which disseminate throughout the body and form tissue cysts in neural, ocular, and muscular tissues (Figure 18.9). Infection in humans can occur as a result of ingesting undercooked food or water contaminated with feline feces or through contact with a cat's litter.

The present disease can lead to more serious consequences, particularly in pregnant women, where it may result in miscarriage (occurring in approximately 3% of cases) or stillbirth. Infants infected with the parasite often exhibit severe symptoms of congenital abnormalities, including neurological disorders and ocular lesions [123]. Several drugs are currently available for the treatment of toxoplasmosis. The most prescribed combination of drugs is pyrimethamine with sulfonamide. Folinic acid is typically administered in conjunction with pyrimethamine with the objective of preventing bone marrow suppression [122].

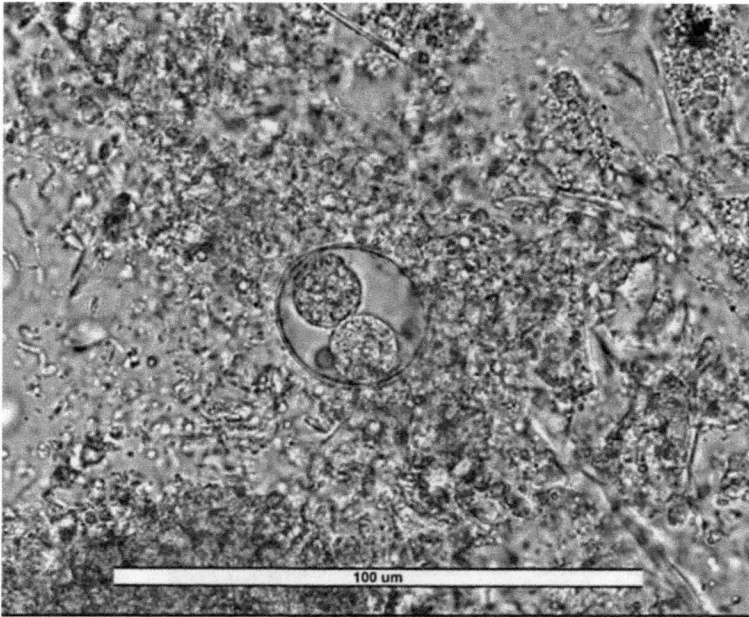

Figure 18.8: Sporulated oocyst of *Toxoplasma gondii* from a stool sample. The image was taken using the echo revolution microscope, with a 60× objective.

Folk knowledge about phytomedicine plays a pivotal role in the discovery of novel natural products with anti-toxoplasmosis properties. A systematic review of medicinal plants with activity toward *Toxoplasma gondii* has been conducted by Cheraghipour et al. [124]. The authors cited various extracts endowed with such effects, including *A. annua, Z. officinale, Sophora flavescens Aiton*, and *E. longifolia* [124].

From Egypt, Elazab et al. (2021) studied the anti-toxoplasmosis effects of various Egyptian plants, in the form of oils or extracts, such as chamomile, bay laurel, bitter apple, camphor tree, frankincense tree, lemon balm, West Indian lemon grass, marjoram, watercress, common wheat, sesame, rosemary, citronella grass, clove, jojoba, and basil. Among the cited plants, the authors demonstrated that methanolic extracts *of C. colocynthis, L. nobilis*, and *M. chamomilla* as well as oils from *C. citratus, O. majorana*, and *C. nardus* exhibited the highest anti-*Toxoplasma* activities (IC$_{50}$ < 35 µg/mL) [125]. From South Korea, ethanolic extracts from *S. flavescens, S. acutum, P. koreana, U. macrocarpa*, and *Torilis japonica* were tested for their ability to inhibit *T. gondii* proliferation. The authors showed that, at a concentration of 156 ng/mL, both *T. japonica* and *S. flavescens* were able to inhibit 99% of the cell culture [126]. In a study published by Leesombun et al. [127], three species of Piperaceae plants – *Piper betle, P. sarmentosum*, and *Piper nigrum* – were tested against *Toxoplasma gondii*. The authors found that the parasite was highly inhibited by *P. betle* extract [127].

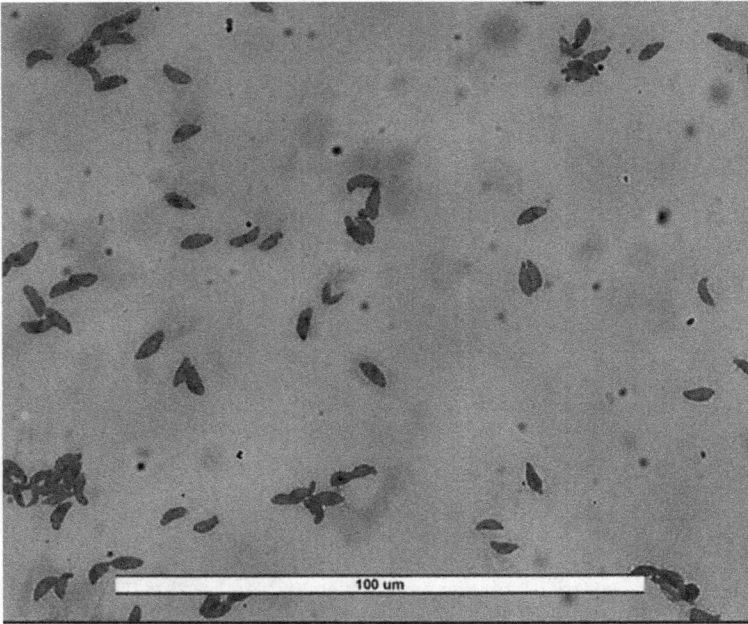

Figure 18.9: Tachyzoites of *Toxoplasma gondii* from Johannes Lieder's *General Parasitology, 3900*. The image was taken using the echo revolution microscope, with a 60× objective.

18.9 Medicinal plants used for the treatment of free-living ameba infections

Recently, in addition to the widespread prevalence of neglected tropical diseases, we are also witnessing an increase in the incidence of other diseases caused by protozoa, among other organisms. These diseases are categorized as Emerging and Re-Emerging Infectious Diseases. The primary causes of these infections include new or evolved species or strains of infectious agents, their spread to new populations or areas, or even the reemergence of previously controlled infections. Free-living amebic (FLA) infections represent a category of emerging and re-emerging diseases. These protozoan parasites are ubiquitous and have been isolated from various environmental sources, including water, air, and soil. Although most FLAs are nonpathogenic, free-living ameba from the genera *Acanthamoeba* spp., *Balamuthia mandrillaris*, *Naegleria fowleri*, and *Sappinia diploidea* can act as opportunistic pathogens for humans and animals. The diseases they cause are characterized by a relatively high mortality rate. These four genera have been associated with fatal CNS infections and other serious diseases in humans [128].

18.9.1 Infection caused by *Naegleria fowleri*

Known as the "Brain eating amoeba," *Naegleria fowleri* is a protozoan parasite and the causative agent of primary amoebic meningoencephalitis. It is frequently isolated from untreated or contaminated warm water sources, including swimming pools and natural hot springs. This parasite primarily affects healthy children and young adults, particularly following activities that involve nasal exposure to contaminated water, such as splashing. The mortality rate of this disease exceeds 95–97% of registered cases, reflecting the rapid and fulminant nature of the infection [129]. The life cycle of this protozoan comprises three distinct stages, contingent on environmental conditions: the trophozoite (Figure 18.10), flagellate, and cyst stages [130]. Once the ameba gains access to the olfactory nerve via the nasal cavity, it migrates and penetrates the cribriform plate, thereby gaining entry to the CNS. The infection causes widespread parenchymal inflammation and hemorrhagic necrosis [131].

The symptoms of PAM are not specific and manifest between 1 and 9 days following exposure. In the initial stages of infection, these symptoms include seizures, fever, and a frontal headache. As the infection progresses, patients may experience hallucinations, paralysis, and ultimately coma [131]. The lack of established PAM therapy protocols presents a significant challenge in the treatment of the infection. Presently, ex-

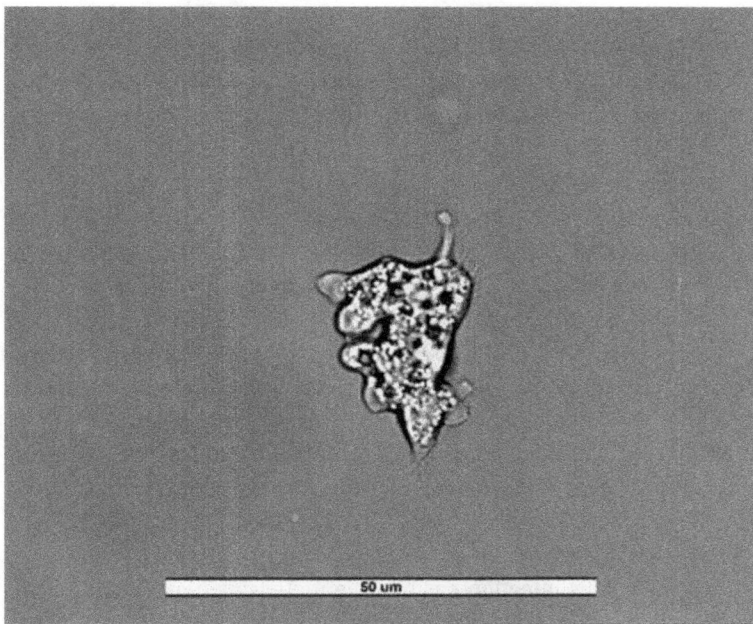

Figure 18.10: Trophozoite of *Naegleria fowleri* (ATCC 30808) in axenic culture. The image was taken using the echo revolution microscope with a 100× objective.

perimental therapy options include Amphotericin B (Amp B) and Miltefosine, either alone or in combination with other drugs (rifampin, azithromycin, or azoles) or with hypothermia [132]. As a result, novel therapeutic molecules for PAM are needed since the current options present undesired toxic side effects. In the literature, only a few papers have been published on the use of medicinal plants against *Naegleria fowleri*. In 2020, we published the activity of the ethanolic extract of *Inula viscosa* leaves against the trophozoite stage of *N. fowleri* with an IC_{50} of 17.89 μg/mL [133]. Compounds from plants have been cited for their activity against *N. fowleri* including nordihydroguaiaretic acid and methylnordihydroguaiaretic acid, with IC_{50} values of 37 and 38 μM, respectively. Both acids were isolated from the methanolic extract of *L. tridentata* [134]. Siddiqui et al. [135] reported that ursolic acid, betulinic acid, and betulin, isolated from a mixture of *S. triloba* and *R. yaundensis* extracts, were able to inhibit parasite growth by 19%, 52%, and 63% at a concentration of 100 μg/mL [135]. Huong Giang et al. [136] tested the amoebicidal activity of 18 flavonoids against *Naegleria fowleri*. Only four compounds showed activity: demethoxycurcumin, kaempferol, resveratrol, and silybin (A + B) [136].

18.9.2 Infection caused by *Acanthamoeba* spp

Acanthamoeba is widely distributed as a protozoan parasite that can be found in a variety of environmental and clinical sources. It has two distinct life-cycle stages: the vegetative form, which is known as a trophozoite (Figure 18.11), and a cyst form that is highly resistant. Depending on the point of access to the human body, the affected organ or tissue, and the immunological state of the patient, *Acanthamoeba* can invade the CNS, leading to granulomatous amebic encephalitis (GAE). It can also form skin ulcers or invade the cornea, causing *Acanthamoeba keratitis* (AK). In immunocompromised patients, several studies have described a general amebic infection disseminated to different organs [137].

In the case of GAE, it is reported to be a chronic disease with a prodromal period that apparently lasts for several weeks to months. The initial indications and symptoms may include headache, photophobia, nausea, personality changes, and tonic-clonic seizures. It is important to emphasize that the present infection is usually fatal, and the treatment is based on individual patient cases. The treatment used in these cases is a combination of certain antiparasitic molecules such as sulfadiazine, fluconazole, pentamidine, ketoconazole, hexadecylphosphocholine, and amphotericin B [137, 138]. As for AK, the infection is marked by an invasive infection of the cornea that can result in blindness. Recently, this infection has become increasingly prevalent, particularly in developed countries, due to the increased use of contact lenses [139]. AK typically manifests with symptoms that are nonspecific and may resemble those of herpetic keratitis. These include intense pain, photophobia, red eyes, and reduced visual acuity. The prevailing therapeutic strategy for AK entails the topical administration of

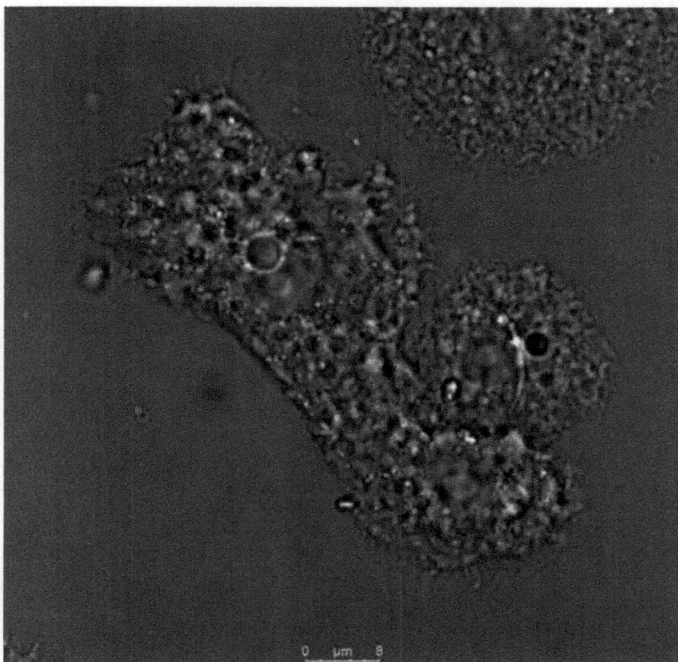

Figure 18.11: Trophozoite of *Acanthamoeba* spp. in axenic culture. The image was taken using the confocal microscope Leica SPE, with an oil objective of 63×.

antimicrobials, encompassing a range of combinations including propamidine isethionate and neomycin or biguanides [140]. Considering the resistance observed at the cyst stage, the variable efficacy of treatment among Acanthamoeba genotypes, and the emergence of resistance to drugs, including those previously mentioned, researchers have identified a pressing need to prioritize the identification of novel compounds as a means of treating Acanthamoeba infections. Consequently, there has been a notable shift in research focus toward natural-origin compounds, predominantly derived from plants and herbs. A review was conducted on the application of medicinal plants against *Acanthamoeba* infections by Niyyati et al. [141]. The authors cited various plant extracts as amoebicidal agents such as *Trigonella foenum-graecum, Origanum syriacum, Inula oculuschristi* (L.), and *Melissa officinalis* among others [141]. Recently, a second review was conducted by Chegeni et al. [142], and the authors reported a total of 110 species of plants studied for their amoebicidal effects. The most active species belong to the genera *Citrus* (six species), *Allium* (five species), *Peucedanum, Piper, Lippia,* and *Olea* (four species) [142]. Our group has been working for a decade on the amoebicidal effects of medicinal plants, including Tunisian *Thymus capitatus, Maltese half-blood (Citrus sinensis), chamomile (Matricaria recutita* L.), *Rubus ulmifolius Schott,* and *Olea europaea* [143–147]. In the same field, a study conducted by Mitsuwan et al.

[148] investigated the anti-acanthamoeba effect of Cambodian plant extracts on *Acanthamoeba triangularis*. The authors demonstrated that among the evaluated 39 plant extracts, two extracts were active against both trophozoites and cysts: *Annona muricata* and *Clausena trifoliata* [148]. Table 18.2 provides more examples of plants with anti-amoebic activity.

Table 18.2: Examples of plant extracts with anti-amoebic activity.

Botanical name	Tested species	Preparation	References
Peganum harmala	*Acanthamoeba triangularis*	Ethanolic extract	[149]
Pulicaria inuloides	*Acanthamoeba castellanii*	Chloroform extract	[150]
Eryngium planum	*Acanthamoeba castellanii*	Ethanolic extract	[151]
Origanum syriacum *Origanum laevigatum*	*Acanthamoeba castellanii*	Methanolic extract	[152]
P. harmala *M. azedarach* *R. communi*	*Acanthamoeba castellanii*	Methanolic extract	[153]
Arachis hypogaea L. *Curcuma longa* L. *Pancratium maritimum* L.	*Acanthamoeba castellanii*	Ethanolic extract	[154]
Teucrium polium *Teucrium chamaedrys*	*Acanthamoeba castellanii*	Methanolic extract	[155]
Pastinaca armenia *Inula oculus-Christi*	*Acanthamoeba castellanii*	Aqueous extract	[156]
Peucedanum caucasicum *Peucedanum palimbioides* *Peucedanum chryseum* *Peucedanum longibracteolatum*	*Acanthamoeba castellanii*	Methanolic extract	[157]
Pouzolzia indica	*Acanthamoeba spp*	Methanolic extract	[158]
Salvia staminea *Salvia caespitosa*	*Acanthamoeba castellanii*	Methanolic extract	[159]

18.10 Medicinal plants used for the treatment of helminthiases

Helminthiases are infectious diseases caused by parasitic worms [160]. They constitute the most important class of neglected tropical diseases, affecting a minimum of one billion individuals across the globe, with a disproportionate impact on resource-poor regions with constrained disease surveillance capabilities [161]. The present infection constitutes a health, economic, and social burden to developing countries as it affects both humans and livestock. According to Hotez et al. [160], the most widespread helminthiases are the intestinal helminthiases, mainly ascariasis, trichuriasis, and hookworm. These are followed by schistosomiasis and lymphatic filariasis [160]. There are three principal groups of helminths responsible for helminthiasis: nematodes (roundworms), cestodes (tapeworms), and trematodes (flukes). There are several potential routes of infection for these parasitic worms, including ingestion of contaminated food or water, penetration of the skin, and transmission via vectors [162]. Individuals afflicted with mild infections (characterized by a low worm burden) tend to be asymptomatic. However, more severe infections can manifest in a multitude of ways, including intestinal distress (diarrhea and abdominal discomfort), malnourishment, weakness, and, in the case of children, growth retardation and decreased physical development [163, 164]. According to the WHO, the treatment of helminthiases is based on the use of albendazole and mebendazole. As with most parasitic infections, there are emerging issues of drug resistance and an urgent need for new, safe, and effective anthelmintic molecules [165]. Medicinal plants have constituted a pivotal element in the struggle against helminth infections in humans and livestock of indigenous communities across the globe. In Ethiopia, extracts from *Hagenia abyssinica*, leaves of *Myrsine africana*, *Rhus glabra*, *Jasminum abyssinicum*, *Rhus vulgaris*, and *Acokanthera schimperi* have been used to treat helminth infections in humans and animals [166]. The antiparasitic effects of the cited plants have been demonstrated by several authors. Hirpa E (2015) attributed the anthelmintic activity of *Hagenia abyssinica* against various parasites, including *Schistosoma mansoni*, *Clonorchis sinensis*, *Fasciola hepatica*, and *Echinostoma caproni*, to the nonpolar fraction. A study from Ethiopia examined the effect of polar extracts from *Myrsine africana*, *Rhus glabra*, *Jasminum abyssinicum*, *Rhus vulgaris*, *Acokanthera schimperi*, and *Foeniculum vulgare* against *Haemonchus contortus*. Among the tested plants, polar extracts of *Foeniculum vulgare*, at a concentration of 1 mg/mL, were able to totally inhibit egg hatching of the studied nematode, while methanolic extracts of *Rhus glabra* were the most effective in inhibiting larval development [167]. Essential oils were also reported to exhibit anthelmintic effects. Panda et al. [168] conducted an extensive review of the use of essential oils and their components as anthelmintic treatments [168]. Among the standout listed essential oils, the essential oil of *Baccharis dracunculifolia* was able to inhibit *Schistosoma mansoni* adult worms at a concentration of 10 µg/mL [169], while the essential oil of *Bunium persicum* (Boiss) was found to inactivate 100% of protoscoleces at a concentration of 25 µg/mL [170].

18.11 Conclusions

In conclusion, a considerable number of MAP have been demonstrated to exhibit high antiparasitic activities. In 2000, the WHO stated that 80% of the global population relies on traditional and folk medicines for the treatment of their ailments. Traditional and folk medicines are a vast and invaluable repository of knowledge, skills, and practices. They are based on the use of local ingredients, such as medicinal plants, to prevent, treat, or improve physical and mental illnesses. Yet, the extinction of traditional medicinal knowledge is occurring at an accelerated pace, largely because of cultural transformation and the reduction in accessibility to natural medicinal resources in both urban and rural settings. Each population should conduct an inventory of local medicinal plants used as antiprotozoals to serve as a resource for future drug discovery.

18.12 Funding

The work was funded by the Consorcio Centro de Investigación Biomédica (CIBER) de Enfermedades Infecciosas (CIBERINFEC); Instituto de Salud Carlos III, 28006 Madrid, Spain (CB21/13/00100); Ministerio de Sanidad, Spain; and Project 2022CLISA26 "PROTCAN" from Fundación Cajacanarias/Fundación Bancaria la Caixa. IS was funded by the programa postdoctoral ULL-Fundación Bancaria La Caixa-Fundación Bancaria Cajacanarias 2024.

References

[1] Seed, J. R. (1996). Protozoa: Pathogenesis and defenses. In: Baron, S. (ed) Medical Microbiology, The University of Texas Medical Branch at Galveston, Galveston (TX).

[2] Scholar, E. (2007). Diseases caused by Protozoa. In: Enna, S. J. & Bylund, D. B. (eds) xPharm: The Comprehensive Pharmacology Reference, Elsevier, New York, 1–3.

[3] Heirangkhongjam, M. D. and Ngaseppam, I. S. (2018). Traditional medicinal uses and pharmacological properties of *Rhus chinensis* Mill.: A systematic review. European Journal of Integrative Medicine, 21, 43–49. doi: 10.1016/j.eujim.2018.06.011. Available online: https://www.scien cedirect.com/science/article/pii/S1876382018303846.

[4] Sato, A. (2012). Revealing the popularity of traditional medicine in light of multiple recourses and outcome measurements from a user's perspective in Ghana. Health Policy and Planning, 27, 625–637. doi: 10.1093/heapol/czs010. Available online: https://doi.org/10.1093/heapol/czs010 (accessed on Nov 28, 2024).

[5] World Health Organization. (2023). *World malaria report 2023*; World Health Organization.

[6] Ouédraogo, A., Tiono, A. B., Diarra, A., Bougouma, E. C. C., Nébié, I., Konaté, A. T. and Sirima, S. B. (2012). Transplacental transmission of *Plasmodium falciparum* in a highly malaria endemic area of Burkina Faso. Journal of Tropical Medicine, 2012, 109705. doi: 10.1155/2012/109705. Available online: https://doi.org/10.1155/2012/109705.

[7] Verra, F., Angheben, A., Martello, E., Giorli, G., Perandin, F. and Bisoffi, Z. (2018). A systematic review of transfusion-transmitted malaria in non-endemic areas. Malaria Journal, 17, 36. doi: 10.1186/s12936-018-2181-0. Available online: https://doi.org/10.1186/s12936-018-2181-0.

[8] Thera, M. A. and Plowe, C. V. (2012). Vaccines for malaria: How close are we? Annual Review of Medicine, 63, 345–357. doi: 10.1146/annurev-med-022411-192402. Available online: https://www.an nualreviews.org/content/journals/10.1146/annurev-med-022411-192402 (accessed on Nov 28, 2024).

[9] Crutcher, J. M. and Hoffman, S. L. (1996). Malaria. In: Baron, S. (ed) Medical Microbiology, University of Texas Medical Branch at Galveston, Galveston (TX).

[10] Commons, R. J., Rajasekhar, M., Allen, E. N., Yilma, D., Chotsiri, P., Abreha, T., Adam, I., Awab, G. R., Barber, B. E., Brasil, L. W., Chu, C. S., Cui, L., Edler, P., Margarete do Socorro, M., Gonzalez-Ceron, L., Grigg, M. J., Hamid, M. M. A., Hwang, J., Karunajeewa, H., Lacerda, M. V. G., Ladeia-Andrade, S., Leslie, T., Longley, R. J., Monteiro, W. M., Pasaribu, A. P., Poespoprodjo, J. R., Richmond, C. L., Rijal, K. R., Taylor, W. R. J., Thanh, P. V., Thriemer, K., Vieira, J. L. F., White, N. J., Zuluaga-Idarraga, L. M., Workman, L. J., Tarning, J., Stepniewska, K., Guerin, P. J., Simpson, J. A., Barnes, K. I. and Price, R. N. (2024). Primaquine for uncomplicated *Plasmodium vivax* malaria in children younger than 15 years: A systematic review and individual patient data meta-analysis. Lancet Child and Adolescent Health, 8, 798–808. doi: 10.1016/S2352-4642(24)00210-4. Available online: https://pubmed.ncbi.nlm. nih.gov/39332427/ (accessed on Nov 28, 2024).

[11] Okagu, I. U., Aguchem, R. N., Ezema, C. A., Ezeorba, T. P. C., Eje, O. E. and Ndefo, J. C. (2022). Molecular mechanisms of hematological and biochemical alterations in malaria: A review. Molecular and Biochemical Parasitology, 247, 111446. doi: 10.1016/j.molbiopara.2021.111446. Available online: https://www.sciencedirect.com/science/article/pii/S0166685121000931 (accessed on Nov 28, 2024).

[12] Minassian, A. and Cowan, R. (2023). Understanding how the immune system responds to repeated malaria infections.

[13] White, N. J. (1996). The treatment of malaria. New England Journal of Medicine, 335, 800–806. doi: 10.1056/NEJM199609123351107. Available online: https://pubmed.ncbi.nlm.nih.gov/8703186/ (accessed on Nov 28, 2024).

[14] Willcox, M. L. and Bodeker, G. (2004). Traditional herbal medicines for malaria. BMJ, 329, 1156–1159. doi: 10.1136/bmj.329.7475.1156. Available online: https://pubmed.ncbi.nlm.nih.gov/15539672/ (accessed on Nov 28, 2024).

[15] Noronha, M., Pawar, V., Prajapati, A. and Subramanian, R. B. (2020). A literature review on traditional herbal medicines for malaria. South African Journal of Botany, 128, 292–303. doi: 10.1016/j.sajb.2019.11.017. Available online: https://www.sciencedirect.com/science/article/pii/ S0254629919311755 (accessed on Nov 28, 2024).

[16] Meshnick, S. R., Taylor, T. E. and Kamchonwongpaisan, S. (1996). Artemisinin and the Antimalarial Endoperoxides: from Herbal Remedy to Targeted Chemotherapy.

[17] Son, I. H., Chung, I., Lee, S. and Moon, H. (2007). Antiplasmodial activity of novel stilbene derivatives isolated from *Parthenocissus tricuspidata* from South Korea. Parasitology Research, 101, 237–241. doi: 10.1007/s00436-006-0454-y. Available online: https://doi.org/10.1007/s00436-006-0454-y (accessed on Nov 28, 2024).

[18] Hwang, H. K., Sung, H. K., Wang, W. K. and Kim, I. H. (1995). Medicinal plants of Korea. Yakhak Hoechi, 39, 289.

[19] Evbuomwan, I. O., Stephen Adeyemi, O. and Oluba, O. M. (2023). Indigenous medicinal plants used in folk medicine for malaria treatment in Kwara State, Nigeria: An ethnobotanical study. BMC Complementary Medicine and Therapies, 23, 324.

[20] Adebajo, A. C., Odediran, S. A., Aliyu, F. A., Nwafor, P. A., Nwoko, N. T. and Umana, U. S. (2014). In vivo antiplasmodial potentials of the combinations of four Nigerian antimalarial plants. Molecules, 19, 13136–13146. doi: 10.3390/molecules190913136. Available online: https://www.ncbi.nlm.nih.gov/ pmc/articles/PMC6271372/ (accessed on Nov 28, 2024).

[21] Abdulai, S. I., Ishola, A. A. and Bewaji, C. O. (2023). Antimalarial activities of a therapeutic combination of *Azadirachta indica, Mangifera indica* and *Morinda lucida* leaves: A molecular view of

its activity on *Plasmodium falciparum* proteins. Acta Parasitologica, 68, 659–675. doi: 10.1007/s11686-023-00698-7. Available online: https://pubmed.ncbi.nlm.nih.gov/37474844/ (accessed on Nov 28, 2024).

[22] Onyishi, G. C., Nwosu, G. C. and Eyo, J. E. (2020). In vivo studies on the biochemical indices of *Plasmodium berghei* infected mice treated with *Alstonia boonei* leaf and root extracts. African Health Sciences, 20, 1698–1709. doi: 10.4314/ahs.v20i4.21. Available online: https://pubmed.ncbi.nlm.nih.gov/34394229/ (accessed on Nov 28, 2024).

[23] Chiribagula, B. V., Salvius, B. A., Martin, B. B. and Baptiste, L. S. J. Ethnomedical knowledge of plants used in traditional medicine in Mampa village, Haut-Katanga province, Democratic Republic of Congo. https://www.researchsquare.com/article/rs-5116022/v1 (accessed Nov 28, 2024).

[24] Manya, M. H., Keymeulen, F., Ngezahayo, J., Bakari, A. S., Kalonda, M. E., Kahumba, B. J., Duez, P., Stévigny, C. and Lumbu, S. J. (2020). Antimalarial herbal remedies of Bukavu and Uvira areas in DR Congo: An ethnobotanical survey. Journal of Ethnopharmacology, 249, 112422. doi: 10.1016/j.jep.2019.112422. Available online: https://pubmed.ncbi.nlm.nih.gov/31765762/ (accessed on Nov 28, 2024).

[25] Gumisiriza, H., Olet, E. A., Mukasa, P., Lejju, J. B. and Omara, T. (2023). Ethnomedicinal plants used for malaria treatment in Rukungiri District, Western Uganda. Tropical Medicine and Health, 51, 49. doi: 10.1186/s41182-023-00541-9. Available online: https://www.ncbi.nlm.nih.gov/pmc/articles/PMC10466780/ (accessed on Nov 28, 2024).

[26] Tabuti, J. R. S., Obakiro, S. B., Nabatanzi, A., Anywar, G., Nambejja, C., Mutyaba, M. R., Omara, T. and Waako, P. (2023). Medicinal plants used for treatment of malaria by indigenous communities of Tororo District, Eastern Uganda. Tropical Medicine and Health, 51, 34. doi: 10.1186/s41182-023-00526-8. Available online: https://doi.org/10.1186/s41182-023-00526-8 (accessed on Nov 28, 2024).

[27] Thiaw, M., Samb, I., Genva, M., Gaye, M. L. and Fauconnier, M. (2023). *Momordica balsamina* L.: A plant with multiple therapeutic and nutritional potential – A review. Nutraceuticals, 3, 556–573. doi: 10.3390/nutraceuticals3040040. Available online: https://www.mdpi.com/1661-3821/3/4/40 (accessed on Nov 28, 2024).

[28] Bandeira, S. O., Gaspar, F. and Pagula, F. P. (2001). African ethnobotany and healthcare: Emphasis on Mozambique. Pharmaceutical Biology, 39, 70–73. doi: 10.1076/phbi.39.s1.70.0002. Available online: https://doi.org/10.1076/phbi.39.s1.70.0002 (accessed on Nov 28, 2024).

[29] Singh, K. and Baijnath, H. (2024). Micromorphology, and preliminary phytochemical screening of *Spirostacys africana* L. Leaf and Bark. Student's Journal of Health Research Africa, 5, 9–9. doi: 10.51168/sjhrafrica.v5i6.1194. Available online: https://sjhresearchafrica.org/index.php/public-html/article/view/1194 (accessed on Nov 28, 2024).

[30] Bandeira, S. O., Gaspar, F. and Pagula, F. P. (2001). African ethnobotany and healthcare: Emphasis on Mozambique. Pharmaceutical Biology, 39, 70–73. doi: 10.1076/phbi.39.s1.70.0002. Available online: https://doi.org/10.1076/phbi.39.s1.70.0002 (accessed on Nov 28, 2024).

[31] Jurg, A., Tomás, T. and Pividal, J. (1991). Antimalarial activity of some plant remedies in use in Marracuene, southern Mozambique. Journal of Ethnopharmacology, 33, 79–83. doi: 10.1016/0378-8741(91)90165-A. Available online: https://www.sciencedirect.com/science/article/pii/037887419190165A (accessed on Nov 28, 2024).

[32] Bero, J., Hannaert, V., Chataigné, G., Hérent, M. and Quetin-Leclercq, J. (2011). *In vitro* antitrypanosomal and antileishmanial activity of plants used in Benin in traditional medicine and bio-guided fractionation of the most active extract. Journal of Ethnopharmacology, 137, 998–1002. doi: 10.1016/j.jep.2011.07.022. Available online: https://www.sciencedirect.com/science/article/pii/S0378874111004983 (accessed on Nov 28, 2024).

[33] WHO Leishmaniasis. https://www.who.int/news-room/fact-sheets/detail/leishmaniasis.

[34] De Vries, H. J. C. and Schallig, H. D. (2022). Cutaneous leishmaniasis: A 2022 updated narrative review into diagnosis and management developments. American Journal of Clinical Dermatology,

23, 823–840. doi: 10.1007/s40257-022-00726-8. Available online: https://pubmed.ncbi.nlm.nih.gov/36103050/ (accessed on Nov 28, 2024).

[35] Scarpini, S., Dondi, A., Totaro, C., Biagi, C., Melchionda, F., Zama, D., Pierantoni, L., Gennari, M., Campagna, C., Prete, A. and Lanari, M. (2022). Visceral leishmaniasis: Epidemiology, diagnosis, and treatment regimens in different geographical areas with a focus on pediatrics. Microorganisms, 10, 1887. doi: 10.3390/microorganisms10101887. Available online: https://www.mdpi.com/2076-2607/10/10/1887 (accessed on Nov 28, 2024).

[36] World Health Organization (2021). Organisation mondiale de la Santé Global leishmaniasis surveillance: 2019–2020, a baseline for the 2030 roadmap – Surveillance mondiale de la leishmaniose: 2019–2020, une période de référence pour la feuille de route à l'horizon 2030. Weekly Epidemiological Record = Relevé épidémiologique hebdomadaire, 96, 401–419. Available online: https://iris.who.int/handle/10665/344795.

[37] Kayser, O., Kiderlen, A. F., Laatsch, H. and Croft, S. L. (2000). In vitro leishmanicidal activity of monomeric and dimeric naphthoquinones. Acta Tropica, 77, 307–314. doi: 10.1016/S0001-706X(00)00161-3. Available online: https://www.sciencedirect.com/science/article/pii/S0001706X00001613 (accessed on Nov 28, 2024).

[38] Bahmani, M., Saki, K., Ezatpour, B., Shahsavari, S., Eftekhari, Z., Jelodari, M., Rafieian-Kopaei, M. and Sepahvand, R. (2015). Leishmaniosis phytotherapy: Review of plants used in Iranian traditional medicine on leishmaniasis. Asian Pacific Journal of Tropical Biomedicine, 5, 695–701. doi: 10.1016/j.apjtb.2015.05.018. Available online: https://www.sciencedirect.com/science/article/pii/S2221169115001537 (accessed on Nov 28, 2024).

[39] Parvizi, M. M., Ghahartars, M., Jowkar, Z., Saki, N., Kamgar, M., Hosseinpour, P., Zare, H. and Aslani, F. S. (2022). Association of non-melanoma skin cancer with temperament from the perspective of traditional Persian medicine: A case-control study. Iranian Journal of Medical Sciences, 47, 477.

[40] Parvizi, M. M., Handjani, F., Moein, M., Hatam, G., Nimrouzi, M., Hassanzadeh, J., Hamidizadeh, N., Khorrami, H. R. and Zarshenas, M. M. (2017). Efficacy of cryotherapy plus topical *Juniperus excelsa* M. Bieb cream versus cryotherapy plus placebo in the treatment of Old World cutaneous leishmaniasis: A triple-blind randomized controlled clinical trial. PLoS Neglected Tropical Diseases, 11, e0005957. doi: 10.1371/journal.pntd.0005957. Available online: https://www.ncbi.nlm.nih.gov/pmc/articles/PMC5655399/ (accessed on Nov 28, 2024).

[41] Moein, M., Hatam, G., Taghavi-Moghadam, R. and Zarshenas, M. M. (2017). Antileishmanial activities of Greek Juniper (*Juniperus excelsa* M.Bieb.) against Leishmania major Promastigotes. Journal of Evidence-Based Complementary and Alternative Medicine, 22, 31–36. doi: 10.1177/2156587215623435. Available online: https://doi.org/10.1177/2156587215623435 (accessed on Nov 28, 2024).

[42] Odonne, G., Houël, E., Bourdy, G. and Stien, D. (2017). Treating leishmaniasis in Amazonia: A review of ethnomedicinal concepts and pharmaco-chemical analysis of traditional treatments to inspire modern phytotherapies. Journal of Ethnopharmacology, 199, 211–230. doi: 10.1016/j.jep.2017.01.048. Available online: https://www.sciencedirect.com/science/article/pii/S0378874116315884 (accessed on Nov 28, 2024).

[43] de Albuquerque, K. C. O., da Veiga, A. D. S. S., Silva, J. V. D. S. E., Brigido, H. P. C., Ferreira, E. P. D. R., Costa, E. V. S., Marinho, A. M. D. R., Percário, S. and Dolabela, M. F. (2017). Brazilian Amazon traditional medicine and the treatment of difficult to heal leishmaniasis wounds with Copaifera. Evidence Based Complementary and Alternative Medicine, 2017, 8350320. doi: 10.1155/2017/8350320. Available online: https://pubmed.ncbi.nlm.nih.gov/28194218/ (accessed on Nov 28, 2024).

[44] Monzote, L., Piñón, A. and Setzer, W. N. (2014). Antileishmanial potential of tropical rainforest plant extracts. Medicines, 1, 32–55. doi: 10.3390/medicines1010032. Available online: https://www.mdpi.com/2305-6320/1/1/32 (accessed on Nov 28, 2024).

[45] Diel, K. A. P., Santana Filho, P. C., Pitol Silveira, P., Ribeiro, R. L., Teixeira, P. C., Rodrigues Júnior, L. C., Marinho, L. C., Romão, P. R. T. and von Poser, G. L. (2024). Antiprotozoal potential of *Vismia* species (Hypericaceae), medicinal plants used to fight cutaneous leishmaniasis. Journal of Ethnopharmacology, 328, 118028. doi: 10.1016/j.jep.2024.118028. Available online: https://www.scien cedirect.com/science/article/pii/S0378874124003271 (accessed on Nov 28, 2024).

[46] Garcia, A. R., Amaral, A. C. F., Azevedo, M. M. B., Corte-Real, S., Lopes, R. C., Alviano, C. S., Pinheiro, A. S., Vermelho, A. B. and Rodrigues, I. A. (2017). Cytotoxicity and anti-*Leishmania amazonensis* activity of *Citrus sinensis* leaf extracts. Pharmaceutical Biology, 55, 1780–1786. doi: 10.1080/13880209.2017.1325380. Available online: https://doi.org/10.1080/13880209.2017.1325380 (accessed on Nov 28, 2024).

[47] Bosquiroli, L. S. S., Demarque, D. P., Rizk, Y. S., Cunha, M. C., Marques, M. C. S., Matos, M. D. F. C., Kadri, M. C. T., Carollo, C. A. and Arruda, C. C. P. (2015). *In vitro* anti-*Leishmania infantum* activity of essential oil from *Piper angustifolium*. Revista Brasileira de Farmacognosia, 25, 124–128. doi: 10.1016/j.bjp.2015.03.008. Available online: https://www.sciencedirect.com/science/article/pii/S0102695X15000666 (accessed on Nov 28, 2024).

[48] Hermoso, A., Jiménez, I. A., Mamani, Z. A., Bazzocchi, I. L., Piñero, J. E., Ravelo, A. G. and Valladares, B. (2003). Antileishmanial activities of dihydrochalcones from *piper elongatum* and synthetic related compounds. Structural requirements for activity. Bioorganic and Medicinal Chemistry, 11, 3975–3980. doi: 10.1016/S0968-0896(03)00406-1. Available online: https://www.sciencedirect.com/science/article/pii/S0968089603004061 (accessed on Nov 28, 2024).

[49] Japón-Luján, R. and Luque de Castro, M. D. (2006). Superheated liquid extraction of oleuropein and related biophenols from olive leaves. Journal of Chromatography A, 1136, 185–191. doi: 10.1016/j.chroma.2006.09.081. Available online: https://www.sciencedirect.com/science/article/pii/S002196730601867X (accessed on Nov 28, 2024).

[50] Sifaoui, I., López-Arencibia, A., Martín-Navarro, C. M., Chammem, N., Reyes-Batlle, M., Mejri, M., Lorenzo-Morales, J., Abderabba, M. and Piñero, J. E. (2014). Activity of olive leaf extracts against the promastigote stage of *Leishmania* species and their correlation with the antioxidant activity. Experimental Parasitology, 141, 106–111. doi: 10.1016/j.exppara.2014.03.002. Available online: https://www.sciencedirect.com/science/article/pii/S0014489414000368 (accessed on Nov 28, 2024).

[51] Sifaoui, I., López-Arencibia, A., Martín-Navarro, C. M., Ticona, J. C., Reyes-Batlle, M., Mejri, M., Jiménez, A. I., Lopez-Bazzocchi, I., Valladares, B., Lorenzo-Morales, J., Abderabba, M. and Piñero, J. E. (2014). In vitro effects of triterpenic acids from olive leaf extracts on the mitochondrial membrane potential of promastigote stage of *Leishmania* spp. Phytomedicine, 21, 1689–1694. doi: 10.1016/j.phymed.2014.08.004. Available online: https://pubmed.ncbi.nlm.nih.gov/25442278/ (accessed on Nov 28, 2024).

[52] Hajaji, S., Sifaoui, I., López-Arencibia, A., Reyes-Batlle, M., Jiménez, I. A., Bazzocchi, I. L., Valladares, B., Akkari, H., Lorenzo-Morales, J. and Piñero, J. E. (2018). Leishmanicidal activity of α-bisabolol from Tunisian chamomile essential oil. Parasitology Research, 117, 2855–2867. doi: 10.1007/s00436-018-5975-7. Available online: https://doi.org/10.1007/s00436-018-5975-7 (accessed on Nov 28, 2024).

[53] Zeouk, I., Sifaoui, I., López-Arencibia, A., Reyes-Batlle, M., Bethencourt-Estrella, C. J., Bazzocchi, I. L., Bekhti, K., Lorenzo-Morales, J., Jiménez, I. A. and Piñero, J. E. (2020). Sesquiterpenoids and flavonoids from *Inula viscosa* induce programmed cell death in kinetoplastids. Biomedicine and Pharmacotherapy, 130, 110518. doi: 10.1016/j.biopha.2020.110518. Available online: https://pubmed.ncbi.nlm.nih.gov/32674017/ (accessed on Nov 28, 2024).

[54] Núñez Marvin, J., Martínez Morena, L., López-Arencibia, A., Bethencourt-Estrella, C. J., San Nicolás-Hernández, D., Jiménez Ignacio, A., Lorenzo-Morales, J., Piñero José, E. and Bazzocchi Isabel, L. (2021). In vitro susceptibility of kinetoplastids to celastroloids from *Maytenus chiapensis*. Antimicrobial Agents and Chemotherapy, 65. doi: 10.1128/aac.02236-20,10.1128/aac.02236-20. Available online: https://doi.org/10.1128/aac.02236-20.

[55]   Mennai, I., Sifaoui, I., Esseid, C., López-Arencibia, A., Reyes-Batlle, M., Benayache, F., Benayache, S., Bazzocchi, I. L., Lorenzo-Morales, J., Piñero, J. E. and Jiménez, I. A. (2021). Bio-guided isolation of leishmanicidal and trypanocidal constituents from *Pituranthos battandieri* aerial parts. Parasitology International, 82, 102300. doi: 10.1016/j.parint.2021.102300. Available online: https://www.sciencedir ect.com/science/article/pii/S1383576921000192 (accessed on Nov 28, 2024).

[56]   Núñez, M. J., Martínez, M. L., Castillo, U. G., Flores, K. C., Menjívar, J., López-Arencibia, A., Bethencourt-Estrella, C. J., Jiménez, I. A., Piñero, J. E., Lorenzo-Morales, J. and Bazzocchi, I. L. (2024). *Salvadoran celastraceae* species as a source of antikinetoplastid quinonemethide triterpenoids. Plants, 13, 360. doi: 10.3390/plants13030360. Available online: https://www.mdpi.com/2223-7747/13/ 3/360 (accessed on Nov 28, 2024).

[57]   Ahmed, S. B. H., Sghaier, R. M., Guesmi, F., Kaabi, B., Mejri, M., Attia, H., Laouini, D. and Smaali, I. (2011). Evaluation of antileishmanial, cytotoxic and antioxidant activities of essential oils extracted from plants issued from the leishmaniasis-endemic region of Sned (Tunisia). Natural Products Research, 25, 1195–1201. doi: 10.1080/14786419.2010.534097. Available online: https://pubmed.ncbi. nlm.nih.gov/21740286/ (accessed on Nov 28, 2024).

[58]   Colares, A. V., Almeida-Souza, F., Taniwaki, N. N., Souza, C. D. S. F., da Costa, J. G. M., Calabrese, K. D. S. and Abreu-Silva, A. L. (2013). In vitro antileishmanial activity of essential oil of *Vanillosmopsis arborea* (Asteraceae) baker. Evidence Based Complementary and Alternative Medicine, 2013, 727042. doi: 10.1155/2013/727042. Available online: https://pubmed.ncbi.nlm.nih.gov/23935675/ (accessed on Nov 28, 2024).

[59]   Cunha, A. D. C., Chierrito, T. P. C., Machado, G. M. D. C., Leon, L. L. P., Da Silva, C. C., Tanaka, J. C., de Souza, L. M., Gonçalves, R. A. C. and de Oliveira, A. J. B. (2012). Anti-leishmanial activity of alkaloidal extracts obtained from different organs of *Aspidosperma ramiflorum*. Phytomedicine, 19, 413–417. doi: 10.1016/j.phymed.2011.12.004. Available online: https://pubmed.ncbi.nlm.nih.gov/22326547/ (accessed on Nov 28, 2024).

[60]   Fournet, A., Barrios, A. A. and Muñoz, V. (1994). Leishmanicidal and trypanocidal activities of Bolivian medicinal plants. Journal of Ethnopharmacology, 41, 19–37. doi: 10.1016/0378-8741(94) 90054-x. Available online: https://pubmed.ncbi.nlm.nih.gov/8170156/ (accessed on Nov 28, 2024).

[61]   García, M., Monzote, L., Scull, R. and Herrera, P. (2012). Activity of Cuban plants extracts against *Leishmania amazonensis*. ISRN Pharmacology, 2012, 104540. doi: 10.5402/2012/104540. Available online: https://pubmed.ncbi.nlm.nih.gov/22530133/ (accessed on Nov 28, 2024).

[62]   Hatimi, S., Boudouma, M., Bichichi, M., Chaib, N. and Idrissi, N. G. (2001). In vitro evaluation of antileishmania activity of *Artemisia herba* alba Asso. Bulletin de la Societe de Pathologie Exotique, 94, 29–31. Available online: https://pubmed.ncbi.nlm.nih.gov/11346978/ (accessed on Nov 28, 2024).

[63]   Machado, M., Dinis, A. M., Santos-Rosa, M., Alves, V., Salgueiro, L., Cavaleiro, C. and Sousa, M. C. (2014). Activity of *Thymus capitellatus* volatile extract, 1,8-cineole and borneol against Leishmania species. Veterinary Parasitology, 200, 39–49. doi: 10.1016/j.vetpar.2013.11.016. Available online: https://pubmed.ncbi.nlm.nih.gov/24365244/ (accessed on Nov 28, 2024).

[64]   Mansour, R., Haouas, N., Ben Kahla-Nakbi, A., Hammami, S., Mighri, Z., Mhenni, F. and Babba, H. (2013). The effect of *Vitis vinifera* L. leaves extract on *Leishmania infantum*. Iranian Journal of Pharmaceutical Research, 12, 349–355. Available online: https://www.ncbi.nlm.nih.gov/pmc/articles/ PMC3813254/ (accessed on Nov 28, 2024).

[65]   Martín, T., Villaescusa, L., Gasquet, M., Delmas, F., Bartolomé, C., Díaz-Lanza, A. M., Ollivier, E. and Balansard, G. (1998). Screening for protozoocidal activity of Spanish plants. Pharmaceutical Biology, 36, 56–62. doi: 10.1076/phbi.36.1.56.4627. Available online: https://doi.org/10.1076/phbi.36.1.56.4627 (accessed on Nov 28, 2024).

[66]   Monzote, L., García, M., Pastor, J., Gil, L., Scull, R., Maes, L., Cos, P. and Gille, L. (2014). Essential oil from *Chenopodium ambrosioides* and main components: Activity against Leishmania, their mitochondria and other microorganisms. Experimental Parasitology, 136, 20–26. doi: 10.1016/j.

exppara.2013.10.007. Available online: https://pubmed.ncbi.nlm.nih.gov/24184772/ (accessed on Nov 28, 2024).

[67]   Plock, A., Sokolowska-Köhler, W. and Presber, W. (2001). Application of flow cytometry and microscopical methods to characterize the effect of herbal drugs on *Leishmania* spp. Experimental Parasitology, 97, 141–153. doi: 10.1006/expr.2001.4598. Available online: https://www.sciencedirect.com/science/article/pii/S0014489401945989 (accessed on Nov 28, 2024).

[68]   Sawadogo, W. R., Le Douaron, G., Maciuk, A., Bories, C., Loiseau, P. M., Figadère, B., Guissou, I. P. and Nacoulma, O. G. (2012). In vitro antileishmanial and antitrypanosomal activities of five medicinal plants from Burkina Faso. Parasitology Research, 110, 1779–1783. doi: 10.1007/s00436-011-2699-3. Available online: https://pubmed.ncbi.nlm.nih.gov/22037827/ (accessed on Nov 28, 2024).

[69]   Serakta, M., Djerrou, Z., Mansour-Djaalab, H., Kahlouche-Riachi, F., Hamimed, S., Trifa, W., Belkhiri, A., Edikra, N. and Hamdi Pacha, Y. (2013). Antileishmanial activity of some plants growing in Algeria: *Juglans regia*, *Lawsonia inermis* and *Salvia officinalis*. African Journal of Traditional, Complementary and Alternative Medicines, 10, 427–430. doi: 10.4314/ajtcam.v10i3.7. Available online: https://pubmed.ncbi.nlm.nih.gov/24146470/ (accessed on Nov 28, 2024).

[70]   Velázquez-Ramírez, D. D., de Léon, P., Adalberto, A. and Ochoa-Díaz-López, H. (2022). Review of *American trypanosomiasis* in southern Mexico highlights opportunity for surveillance research to advance control through the one health approach. Frontiers in Public Health, 10. doi: 10.3389/fpubh.2022.838949. Available online: https://www.frontiersin.org/journals/public-health/articles/10.3389/fpubh.2022.838949/full (accessed on Nov 28, 2024).

[71]   Martín-Escolano, J., Marín, C., Rosales, M. J., Tsaousis, A. D., Medina-Carmona, E. and Martín-Escolano, R. (2022). An updated view of the *Trypanosoma cruzi* life cycle: Intervention points for an effective treatment. ACS Infectious Diseases, 8, 1107–1115. doi: 10.1021/acsinfecdis.2c00123. Available online: https://doi.org/10.1021/acsinfecdis.2c00123 (accessed on Nov 28, 2024).

[72]   Andrade, M. V., Noronha, K. V. M. D. S., Souza, A. D., Motta-Santos, A. S., Braga, P. E. F., Bracarense, H., Miranda, M. C. C. D., Nascimento, B. R., Molina, I., Martins-Melo, F. R., Perel, P., Geissbühler, Y., Quijano, M., Machado, I. E. and Ribeiro, A. L. P. (2023). The economic burden of Chagas disease: A systematic review. PLOS Neglected Tropical Diseases, 17, e0011757. doi: 10.1371/journal.pntd.0011757. Available online: https://journals.plos.org/plosntds/article?id=10.1371/journal.pntd.0011757 (accessed on Nov 28, 2024).

[73]   Gómez-Ochoa, S. A., Rojas, L. Z., Echeverría, L. E., Muka, T. and Franco, O. H. (2022). Global, regional, and national trends of Chagas disease from 1990 to 2019: Comprehensive analysis of the global burden of disease study. Global Heart, 17, 59. doi: 10.5334/gh.1150. Available online: https://pubmed.ncbi.nlm.nih.gov/36051318/ (accessed on Nov 28, 2024).

[74]   WHO Chagas disease (also known as *American trypanosomiasis*). https://www.who.int/news-room/fact-sheets/detail/chagas-disease-(american-trypanosomiasis).

[75]   Pérez-Molina, J. A., Crespillo-Andújar, C., Bosch-Nicolau, P. and Molina, I. (2020). Trypanocidal treatment of Chagas disease. Enfermedades Infecciosas y Microbiologia Clinica (English Ed), S0213–2. doi: 10.1016/j.eimc.2020.04.011. Available online: https://pubmed.ncbi.nlm.nih.gov/32527494/ (accessed on Nov 28, 2024).

[76]   Gabaldón-Figueira, J. C., Martinez-Peinado, N., Escabia, E., Ros-Lucas, A., Chatelain, E., Scandale, I., Gascon, J., Pinazo, M. and Alonso-Padilla, J. (2023). State-of-the-art in the drug discovery pathway for Chagas disease: A framework for drug development and target validation. RRTM, 14, 1–19. doi: 10.2147/RRTM.S415273. Available online: https://www.dovepress.com/state-of-the-art-in-the-drug-discovery-pathway-for-chagas-disease-a-fr-peer-reviewed-fulltext-article-RRTM (accessed on Nov 28, 2024).

[77]   Jackson, Y., Alirol, E., Getaz, L., Wolff, H., Combescure, C. and Chappuis, F. (2010). Tolerance and safety of nifurtimox in patients with chronic Chagas disease. Clinical Infectious Diseases, 51,

e69–e75. doi: 10.1086/656917. Available online: https://doi.org/10.1086/656917 (accessed on Nov 28, 2024).

[78] Vázquez, C., García-Vázquez, E., Carrilero, B., Simón, M., Franco, F., Iborra, M. A., Gil-Gallardo, L. J. and Segovia, M. (2023). Tolerance and adherence of patients with chronic Chagas disease treated with benznidazole. Revista da Sociedade Brasileira de Medicina Tropical, 56, e0384–86822023000100305. doi: 10.1590/0037-8682-0384-2022. Available online: https://pubmed. ncbi.nlm.nih.gov/36700605/ (accessed on Nov 28, 2024).

[79] Beltrami, M., Grande, R., Giacomelli, A., Sabaini, F., Biondo, L., Longo, M., Grosso, S., Oreni, L., Fadelli, S., Galimberti, L., Ridolfo, A. L. and Antinori, S. (2023). Chagas disease prevalence among migrants from El Salvador in Milan: A cross-sectional study of an often-overlooked population. Infectious Diseases (London), 55, 559–566. doi: 10.1080/23744235.2023.2222817. Available online: https://pubmed.ncbi.nlm.nih.gov/37317783/ (accessed on Nov 28, 2024).

[80] Castillo, U. G., Komatsu, A., Martínez, M. L., Menjívar, J., Núñez, M. J., Uekusa, Y., Narukawa, Y., Kiuchi, F. and Nakajima-Shimada, J. (2022). Anti-trypanosomal screening of *Salvadoran flora*. Journal of Natural Medicines, 76, 259–267. doi: 10.1007/s11418-021-01562-6. Available online: https://doi.org/ 10.1007/s11418-021-01562-6 (accessed on Nov 28, 2024).

[81] Nekoei, S., Khamesipour, F., Habtemariam, S., de Souza, W., Mohammadi Pour, P. and Hosseini, S. R. (2022). The anti-Trypanosoma activities of medicinal plants: A systematic review of the literature. Veterinary Medicine and Science, 8, 2738–2772. doi: 10.1002/vms3.912. Available online: https://pubmed.ncbi.nlm.nih.gov/36037401/ (accessed on Nov 28, 2024).

[82] Büscher, P., Cecchi, G., Jamonneau, V. and Priotto, G. (2017). Human African trypanosomiasis. The Lancet, 390, 2397–2409. doi: 10.1016/S0140-6736(17)31510-6. Available online: https://www.science direct.com/science/article/pii/S0140673617315106 (accessed on Nov 28, 2024).

[83] Berkowitz, F. E. (2012). Trypanosoma species (Trypanosomiasis). In: Long, S. S. (ed) Principles and Practice of Pediatric Infectious Diseases (Fourth Edition), Elsevier, London, 1319–1326.e2.

[84] Carpio, A., Romo, M. L., Parkhouse, R. M. E., Short, B. and Dua, T. (2016). Parasitic diseases of the central nervous system: Lessons for clinicians and policy makers. Expert Review of Neurotherapeutics, 16, 401–414. doi: 10.1586/14737175.2016.1155454. Available online: https://doi. org/10.1586/14737175.2016.1155454 (accessed on Nov 28, 2024).

[85] Doua, F., Miezan, T. W., Sanon Singaro, J. R., Boa Yapo, F. and Baltz, T. (1996). The efficacy of pentamidine in the treatment of early-late stage *Trypanosoma brucei* gambiense trypanosomiasis. Americal Journal of Tropical Medicine and Hygiene, 55, 586–588. doi: 10.4269/ajtmh.1996.55.586. Available online: https://pubmed.ncbi.nlm.nih.gov/9025682/ (accessed on Nov 28, 2024).

[86] Albisetti, A., Hälg, S., Zoltner, M., Mäser, P. and Wiedemar, N. (2023). Suramin action in African trypanosomes involves a RuvB-like DNA helicase. International Journal for Parasitology: Drugs and Drug Resistance, 23, 44–53. doi: 10.1016/j.ijpddr.2023.09.003. Available online: https://www.science direct.com/science/article/pii/S2211320723000301 (accessed on Nov 28, 2024).

[87] Jennings, F. W., Rodgers, J., Bradley, B., Gettinby, G., Kennedy, P. G. E. and Murray, M. (2002). Human *African trypanosomiasis*: Potential therapeutic benefits of an alternative suramin and melarsoprol regimen. Parasitology International, 51, 381–388. doi: 10.1016/S1383-5769(02)00044-2. Available online: https://www.sciencedirect.com/science/article/pii/S1383576902000442 (accessed on Nov 28, 2024).

[88] Álvarez-Rodríguez, A., Jin, B., Radwanska, M. and Magez, S. (2022). Recent progress in diagnosis and treatment of Human African Trypanosomiasis has made the elimination of this disease a realistic target by 2030. Frontier Medicine, 9. doi: 10.3389/fmed.2022.1037094. Available online: https://www.frontiersin.org/journals/medicine/articles/10.3389/fmed.2022.1037094/full (accessed on Nov 28, 2024).

[89] Lindner, A. K., Lejon, V., Barrett, M. P., Blumberg, L., Bukachi, S. A., Chancey, R. J., Edielu, A., Matemba, L., Mesha, T., Mwanakasale, V., Pasi, C., Phiri, T., Seixas, J., Akl, E. A., Probyn, K.,

Villanueva, G., Simarro, P. P., Kadima Ebeja, A., Franco, J. R. and Priotto, G. (2024). New WHO guidelines for treating rhodesiense human *African trypanosomiasis*: Expanded indications for fexinidazole and pentamidine. The Lancet Infectious Diseases, S1473–4. doi: 10.1016/S1473-3099(24) 00581-4. Available online: https://pubmed.ncbi.nlm.nih.gov/39389073/ (accessed on Nov 28, 2024).

[90] Ibrahim, M. A., Mohammed, A., Isah, M. B. and Aliyu, A. B. (2014). Anti-trypanosomal activity of African medicinal plants: A review update. Journal of Ethnopharmacology, 154, 26–54. doi: 10.1016/j. jep.2014.04.012. Available online: https://www.sciencedirect.com/science/article/pii/ S0378874114002876 (accessed on Nov 28, 2024).

[91] Ibrahim, M. A., Musa, A. M., Aliyu, A. B., Mayaki, H. S., Gideon, A. and Islam, M. S. (2013). Phenolics-rich fraction of *Khaya senegalensis* stem bark: Antitrypanosomal activity and amelioration of some parasite-induced pathological changes. Pharmaceutical Biology, 51, 906–913. doi: 10.3109/ 13880209.2013.771191. Available online: https://pubmed.ncbi.nlm.nih.gov/23627467/ (accessed on Nov 28, 2024).

[92] Kamte, S. L. N., Ranjbarian, F., Campagnaro, G. D., Nya, P. C. B., Mbuntcha, H., Woguem, V., Womeni, H. M., Ta, L. A., Giordani, C., Barboni, L., Benelli, G., Cappellacci, L., Hofer, A., Petrelli, R. and Maggi, F. (2017). *Trypanosoma brucei* inhibition by essential oils from medicinal and aromatic plants traditionally used in Cameroon (*Azadirachta indica, Aframomum melegueta, Aframomum daniellii, Clausena anisata, Dichrostachys cinerea* and *Echinops giganteus*). International Journal of Environmental Research and Public Health, 14, 737. doi: 10.3390/ijerph14070737. Available online: https://www.mdpi.com/1660-4601/14/7/737 (accessed on Nov 28, 2024).

[93] Nekoei, S., Khamesipour, F., Habtemariam, S., de Souza, W., Mohammadi Pour, P. and Hosseini, S. R. (2022). The anti-Trypanosoma activities of medicinal plants: A systematic review of the literature. Veterinary Medicine and Science, 8, 2738–2772.

[94] Gutiérrez, L. and Bartelt, L. (2024). Current understanding of *Giardia lamblia* and pathogenesis of stunting and cognitive deficits in children from low- and middle-income countries. Current Tropical Medicine Reports, 11, 28–39. doi: 10.1007/s40475-024-00314-2. Available online: https://doi.org/10. 1007/s40475-024-00314-2 (accessed on Nov 28, 2024).

[95] Adam Rodney, D. (2021). *Giardia duodenalis*: Biology and pathogenesis. Clinical Microbiology Reviews, 34, 24. doi: 10.1128/CMR.00024-19. Available online: https://doi.org/10.1128/CMR.00024-19.

[96] Birkeland, S. R., Preheim, S. P., Davids, B. J., Cipriano, M. J., Palm, D., Reiner, D. S., Svärd, S. G., Gillin, F. D. and McArthur, A. G. (2010). Transcriptome analyses of the *Giardia lamblia* life cycle. Molecular and Biochemical Parasitology, 174, 62–65. doi: 10.1016/j.molbiopara.2010.05.010. Available online: https://www.sciencedirect.com/science/article/pii/S0166685110001337 (accessed on Nov 28, 2024).

[97] Gardner Timothy, B. and Hill David, R. (2001). Treatment of giardiasis. Clinical Microbiology Reviews, 14, 114–128. doi: 10.1128/cmr.14.1.114–128.2001. Available online: https://doi.org/10.1128/cmr.14.1. 114-128.2001.

[98] Gardner, T. B. and Hill, D. R. (2001). Treatment of giardiasis. Clinical Microbiology Reviews, 14, 114–128. doi: 10.1128/CMR.14.1.114–128.2001. Available online: https://www.ncbi.nlm.nih.gov/pmc/ar ticles/PMC88965/ (accessed on Nov 28, 2024).

[99] Granados, C. E., Reveiz, L., Uribe, L. G. and Criollo, C. P. (2012). Drugs for treating giardiasis. Cochrane Database of Systematic Reviews. doi: 10.1002/14651858.CD007787.pub2. Available online: https://www.cochranelibrary.com/cdsr/doi/10.1002/14651858.CD007787.pub2/full (accessed on Nov 28, 2024).

[100] Nazer, M. R., Abbaszadeh, S., Anbari, K. and Shams, M. (2019). A review of the most important medicinal herbs affecting giardiasis. Journal of Herbmed Pharmacology, 8, 78–84. doi: 10.15171/ jhp.2019.13. Available online: https://herbmedpharmacol.com/Article/jhp-8358 (accessed on Nov 28, 2024).

[101] Alnomasy, S., Al-Awsi, G. R. L., Raziani, Y., Albalawi, A. E., Alanazi, A. D., Niazi, M. and Mahmoudvand, H. (2021). Systematic review on medicinal plants used for the treatment of *Giardia* infection. Saudi

Journal of Biological Sciences, 28, 5391–5402. doi: 10.1016/j.sjbs.2021.05.069. Available online: https://www.sciencedirect.com/science/article/pii/S1319562X21004472 (accessed on Nov 28, 2024).

[102] Azadbakht, M., Chabra, A., Saeedi Akbarabadi, A., Motazedian, M. H., Monadi, T. and Akbari, F. (2020). Anti-parasitic activity of some medicinal plants essential oils on *Giardia lamblia* and *Entamoeba histolytica*, in vitro. Research Journal of Pharmacognosy, 7, 41–47. doi: 10.22127/rjp.2019.168142.1462. Available online: https://www.rjpharmacognosy.ir/article_96944.html (accessed on Nov 28, 2024).

[103] Calzada, F., Yépez-Mulia, L. and Aguilar, A. (2006). *In vitro* susceptibility of *Entamoeba histolytica* and *Giardia lamblia* to plants used in Mexican traditional medicine for the treatment of gastrointestinal disorders. Journal of Ethnopharmacology, 108, 367–370. doi: 10.1016/j.jep.2006.05.025. Available online: https://www.sciencedirect.com/science/article/pii/S037887410600287X (accessed on Nov 28, 2024).

[104] Pacheco-Yépez, J., Martínez-Castillo, M., Cruz-Baquero, A., Serrano-Luna, J. and Shibayama, M. (2017). Role of cytokines and reactive oxygen species in the amebic liver abscess produced by *Entamoeba histolytica*. In: Muriel, P. (ed) Liver Pathophysiology, Academic Press, Boston, 187–197.

[105] Conlan, J. and Lal, A. (2015). 5 – Socioeconomic burden of foodborne parasites. In: Gajadhar, A. A. (ed) Foodborne Parasites in the Food Supply Web, Woodhead Publishing, Oxford, 75–98.

[106] Morán, P., Serrano-Vázquez, A., Rojas-Velázquez, L., González, E., Pérez-Juárez, H., Hernández, E. G., Padilla, M. D. L. A., Zaragoza, M. E., Portillo-Bobadilla, T., Ramiro, M. and Ximénez, C. (2023). Amoebiasis: Advances in diagnosis, treatment, immunology features and the interaction with the intestinal ecosystem. International Journal of Molecular Sciences, 24, 11755. doi: 10.3390/ijms241411755. Available online: https://www.mdpi.com/1422-0067/24/14/11755 (accessed on Nov 28, 2024).

[107] Castellanos-Castro, S., Bolaños, J. and Orozco, E. (2020). Lipids in *Entamoeba histolytica*: Host-dependence and virulence factors. Frontiers in Cellular and Infection Microbiology, 10. doi: 10.3389/fcimb.2020.00075. Available online: https://www.frontiersin.org/journals/cellular-and-infection-microbiology/articles/10.3389/fcimb.2020.00075/full (accessed on Nov 28, 2024).

[108] Cock, I. E., Selesho, M. I. and Van Vuuren, S. F. (2018). A review of the traditional use of southern African medicinal plants for the treatment of selected parasite infections affecting humans. Journal of Ethnopharmacology, 220, 250–264. doi: 10.1016/j.jep.2018.04.001. Available online: https://pubmed.ncbi.nlm.nih.gov/29621583/ (accessed on Nov 28, 2024).

[109] Nezaratizade, S., Hashemi, N., Ommi, D., Orhan, I. E. and Khamesipour, F. (2021). A systematic review of anti-*Entamoeba histolytica* activity of medicinal plants published in the last 20 years. Parasitology, 148, 672–684. doi: 10.1017/S0031182021000172. Available online: https://pubmed.ncbi.nlm.nih.gov/33536098/ (accessed on Nov 28, 2024).

[110] McGaw, L. J., Jäger, A. K. and van Staden, J. (2000). Antibacterial, anthelmintic and anti-amoebic activity in South African medicinal plants. Journal of Ethnopharmacology, 72, 247–263. doi: 10.1016/S0378-8741(00)00269-5. Available online: https://www.sciencedirect.com/science/article/pii/S0378874100002695 (accessed on Nov 28, 2024).

[111] Dhal, A. K., Panda, C., Yun, S. and Mahapatra, R. K. (2022). An update on Cryptosporidium biology and therapeutic avenues. Journal of Parasitic Diseases, 46, 923–939. doi: 10.1007/s12639-022-01510-5. Available online: https://doi.org/10.1007/s12639-022-01510-5 (accessed on Nov 28, 2024).

[112] Farthing, M. J. (2000). Clinical aspects of human cryptosporidiosis. Contributions to Microbiology, 6, 50–74. doi: 10.1159/000060368. Available online: https://pubmed.ncbi.nlm.nih.gov/10943507/ (accessed on Nov 28, 2024).

[113] Helmy, Y. A., El-Adawy, H. and Abdelwhab, E. M. (2017). A comprehensive review of common bacterial, parasitic and viral zoonoses at the human-animal interface in Egypt. Pathogens, 6, 33. doi: 10.3390/pathogens6030033. Available online: https://www.mdpi.com/2076-0817/6/3/33 (accessed on Nov 28, 2024).

[114] Helmy, Y. A. and Hafez, H. M. (2022). Cryptosporidiosis: From prevention to treatment, a narrative review. Microorganisms, 10, 2456. doi: 10.3390/microorganisms10122456. Available online: https://www.mdpi.com/2076-2607/10/12/2456 (accessed on Nov 28, 2024).

[115] Gaafar, M. R. (2012). Efficacy of *Allium sativum* (garlic) against experimental cryptosporidiosis. Alexandria Journal of Medicine, 48, 59–66. doi: 10.1016/j.ajme.2011.12.003. Available online: https://doi.org/10.1016/j.ajme.2011.12.003 (accessed on Nov 28, 2024).

[116] Aboelsoued, D. Control of Cryptosporidiosis in Buffalo Calves Using Garlic (*Allium sativum*) and Nitazoxanide with Special Reference to Some Biochemical Parameters. Available online: https://www.academia.edu/39644208/Control_of_Cryptosporidiosis_in_Buffalo_Calves_Using_Gar lic_Allium_sativum_and_Nitazoxanide_with_Special_Reference_to_Some_Biochemical_Parameters (accessed on Nov 28, 2024).

[117] Gaber, M., Galal, L. A. A., Farrag, H. M. M., Badary, D. M., Alkhalil, S. S. and Elossily, N. (2022). The effects of commercially available *Syzygium aromaticum*, *Anethum graveolens*, *Lactobacillus acidophilus* LB, and zinc as alternatives therapy in experimental mice challenged with *Cryptosporidium parvum*. Infection and Drug Resistance, 15, 171–182. doi: 10.2147/IDR.S345789. Available online: https://pubmed.ncbi.nlm.nih.gov/35087280/ (accessed on Nov 28, 2024).

[118] Abd El-Hamed, W. F., Yousef, N. S., Mazrou, Y. S. A., Elkholy, W. A. E. S., El-Refaiy, A. I., Elfeky, F. A., Albadrani, M., El-Tokhy, A. I. and Abdelaal, K. (2021). Anticryptosporidium efficacy of *Olea europaea* and *Ficus carica* leaves extract in immunocompromised mice associated with biochemical characters and antioxidative system. Cells, 10, 2419. doi: 10.3390/cells10092419. Available online: https://www. mdpi.com/2073-4409/10/9/2419 (accessed on Nov 28, 2024).

[119] Abd El Wahab, W. M., Shaapan, R. M., El-Naggar, E. B., Ahmed, M. M., Owis, A. I. and Ali, M. I. (2022). Anti-*Cryptosporidium* efficacy of *Citrus sinensis* peel extract: Histopathological and ultrastructural experimental study. Experimental Parasitology, 243, 108412. doi: 10.1016/j. exppara.2022.108412. Available online: https://www.sciencedirect.com/science/article/pii/ S0014489422002065 (accessed on Nov 28, 2024).

[120] Ahmed, S. A., Eltamany, E. E., Nafie, M. S., Elhady, S. S., Karanis, P. and Mokhtar, A. B. (2023). Anti-Cryptosporidium parvum activity of *Artemisia judaica* L. and its fractions: In vitro and in vivo assays. Frontiers in Microbiology, 14. doi: 10.3389/fmicb.2023.1193810. Available online: https://www.frontier sin.org/journals/microbiology/articles/10.3389/fmicb.2023.1193810/full (accessed on Nov 28, 2024).

[121] Dubey, J. P., Lindsay, D. S. and Speer, C. A. (1998). Structures of *Toxoplasma gondii* Tachyzoites, bradyzoites, and sporozoites and biology and development of tissue cysts. Clinical Microbiology Reviews, 11, 267–299. doi: 10.1128/cmr.11.2.267. Available online: https://doi.org/10.1128/cmr.11.2.267.

[122] Halonen, S. K. and Weiss, L. M. (2013). Chapter 8 – Toxoplasmosis. In: Garcia, H. H., Tanowitz, H. B. & Del Brutto, O. H. (eds) Handbook of Clinical Neurology, Vol. 114, Elsevier, 125–145.

[123] Bollani, L., Auriti, C., Achille, C., Garofoli, F., De Rose, D. U., Meroni, V., Salvatori, G. and Tzialla, C. (2022). Congenital Toxoplasmosis: The state of the art. Frontiers in Pediatrics, 10. doi: 10.3389/ fped.2022.894573. Available online: https://www.frontiersin.org/journals/pediatrics/articles/10. 3389/fped.2022.894573/full (accessed on Nov 28, 2024).

[124] Cheraghipour, K., Masoori, L., Ezzatpour, B., Roozbehani, M., Sheikhian, A., Malekara, V., Niazi, M., Mardanshah, O., Moradpour, K. and Mahmoudvand, H. (2021). The experimental role of medicinal plants in treatment of *Toxoplasma gondii* infection: A systematic review. Acta Parasitologica, 66, 303–328. doi: 10.1007/s11686-020-00300-4. Available online: https://pubmed.ncbi.nlm.nih.gov/ 33159263/ (accessed on Nov 28, 2024).

[125] Elazab, S. T., Soliman, A. F. and Nishikawa, Y. (2021). Effect of some plant extracts from Egyptian herbal plants against *Toxoplasma gondii* tachyzoites *in vitro*. Journal of Veterinary Medical Science, 83, 100–107. doi: 10.1292/jvms.20-0458.

[126] Youn, H. J., Lakritz, J., Kim, D. Y., Rottinghaus, G. E. and Marsh, A. E. (2003). Anti-protozoal efficacy of medicinal herb extracts against *Toxoplasma gondii* and *Neospora caninum*. Veterinary Parasitology,

116, 7–14. doi: 10.1016/s0304-4017(03)00154-7. Available online: https://pubmed.ncbi.nlm.nih.gov/14519322/ (accessed on Nov 28, 2024).

[127] Leesombun, A., Boonmasawai, S., Shimoda, N. and Nishikawa, Y. (2016). Effects of extracts from Thai Piperaceae plants against infection with *Toxoplasma gondii*. Plos One, 11, e0156116. doi: 10.1371/journal.pone.0156116. Available online: https://journals.plos.org/plosone/article?id=10.1371/journal.pone.0156116 (accessed on Nov 28, 2024).

[128] Visvesvara, G. S., Moura, H. and Schuster, F. L. (2007). Pathogenic and opportunistic free-living amoebae: *Acanthamoeba* spp., *Balamuthia mandrillaris*, *Naegleria fowleri*, and *Sappinia diploidea*. FEMS Immunology and Medical Microbiology, 50, 1–26. doi: 10.1111/j.1574-695X.2007.00232.x. Available online: https://pubmed.ncbi.nlm.nih.gov/17428307/ (accessed on Nov 28, 2024).

[129] Rizo-Liendo, A., Sifaoui, I., Reyes-Batlle, M., Chiboub, O., Rodríguez-Expósito, R. L., Bethencourt-Estrella, C. J., San Nicolás-Hernández, D., Hendiger, E. B., López-Arencibia, A. and Rocha-Cabrera, P. (2019). In vitro activity of statins against *Naegleria fowleri*. Pathogens, 8, 122.

[130] Arberas-Jiménez, I., Rizo-Liendo, A., Ines, S., Chao-Pellicer, J., Piñero José, E. and Lorenzo-Morales, J. (2022). A fluorometric assay for the in vitro evaluation of activity against *Naegleria fowleri* cysts. Microbiology Spectrum, 10, 515. doi: 10.1128/spectrum.00515-22. Available online: https://doi.org/10.1128/spectrum.00515-22.

[131] Rizo-Liendo, A., Arberas-Jiménez, I., Sifaoui, I., Gkolfi, D., Santana, Y., Cotos, L., Tejedor, D., García-Tellado, F., Piñero, J. E. and Lorenzo-Morales, J. (2021). The therapeutic potential of novel isobenzofuranones against *Naegleria fowleri*. International Journal for Parasitology: Drugs and Drug Resistance, 17, 139–149. doi: 10.1016/j.ijpddr.2021.09.004. Available online: https://www.sciencedirect.com/science/article/pii/S2211320721000476 (accessed on Nov 28, 2024).

[132] Rizo-Liendo, A., Sifaoui, I., Cartuche, L., Arberas-Jiménez, I., Reyes-Batlle, M., Fernández, J. J., Piñero, J. E., Díaz-Marrero, A. R. and Lorenzo-Morales, J. (2020). Evaluation of indolocarbazoles from *Streptomyces sanyensis* as a novel source of therapeutic agents against the brain-eating amoeba Naegleria fowleri. Microorganisms, 8, 789. doi: 10.3390/microorganisms8050789. Available online: https://www.mdpi.com/2076-2607/8/5/789 (accessed on Nov 28, 2024).

[133] Zeouk, I., Sifaoui, I., Rizo-Liendo, A., Arberas-Jiménez, I., Reyes-Batlle, M., L. Bazzocchi, I., Bekhti, K., E. Piñero, J., Jiménez, I. A. and Lorenzo-Morales, J. (2021). Exploring the anti-infective value of inuloxin A isolated from inula viscosa against the brain-eating amoeba (*Naegleria fowleri*) by activation of programmed cell death. ACS Chemical Neuroscience Journal, 12, 195–202. doi: 10.1021/acschemneuro.0c00685. Available online: https://doi.org/10.1021/acschemneuro.0c00685 (accessed on Nov 28, 2024).

[134] Bashyal, B., Li, L., Bains, T., Debnath, A. and LaBarbera, D. V. (2017). *Larrea tridentata*: A novel source for anti-parasitic agents active against *Entamoeba histolytica*, *Giardia lamblia* and *Naegleria fowleri*. PLoS Neglected Tropical Diseases, 11, e0005832. doi: 10.1371/journal.pntd.0005832. Available online: https://www.ncbi.nlm.nih.gov/pmc/articles/PMC5565192/ (accessed on Nov 28, 2024).

[135] Siddiqui, R., Boghossian, A., Khatoon, B., Kawish, M., Alharbi, A. M., Shah, M. R., Alfahemi, H. and Khan, N. A. (2022). Antiamoebic properties of metabolites against *Naegleria fowleri* and *Balamuthia mandrillaris*. Antibiotics, 11, 539. doi: 10.3390/antibiotics11050539. Available online: https://www.mdpi.com/2079-6382/11/5/539 (accessed on Nov 28, 2024).

[136] Lê, H. G., Võ, T. C., Kang, J., Nguyễn, T. H., Hwang, B., Oh, Y. and Na, B. (2023). Antiamoebic activities of flavonoids against pathogenic free-living amoebae, *Naegleria fowleri* and *Acanthamoeba* species. Parasites, Hosts and Diseases, 61, 449–454. doi: 10.3347/PHD.23078. Available online: https://www.ncbi.nlm.nih.gov/pmc/articles/PMC10693969/ (accessed on Nov 28, 2024).

[137] Schuster, F. L. and Visvesvara, G. S. (2004). Free-living amoebae as opportunistic and non-opportunistic pathogens of humans and animals. International Journal for Parasitology, 34, 1001–1027. doi: 10.1016/j.ijpara.2004.06.004. Available online: https://pubmed.ncbi.nlm.nih.gov/15313128/ (accessed on Nov 28, 2024).

[138] Khan, N. A. (2006). Acanthamoeba: Biology and increasing importance in human health. FEMS Microbiology Reviews, 30, 564–595. doi: 10.1111/j.1574-6976.2006.00023.x. Available online: https://pubmed.ncbi.nlm.nih.gov/16774587/ (accessed on Nov 28, 2024).

[139] Lorenzo-Morales, J., Khan, N. A. and Walochnik, J. (2015). An update on *Acanthamoeba keratitis*: Diagnosis, pathogenesis and treatment. Parasite, 22, 10. doi: 10.1051/parasite/2015010. Available online: https://www.parasite-journal.org/articles/parasite/abs/2015/01/parasite140120/para site140120.html (accessed on Nov 28, 2024).

[140] Sifaoui, I., Díaz-Rodríguez, P., Rodríguez-Expósito, R. L., Reyes-Batlle, M., López-Arencibia, A., Salazar Villatoro, L., Castelan-Ramírez, I., Omaña-Molina, M., Oliva, A., Piñero, J. E. and Lorenzo-Morales, J. (2022). Pitavastatin loaded nanoparticles: A suitable ophthalmic treatment for *Acanthamoeba keratitis* inducing cell death and autophagy in *Acanthamoeba polyphaga*. European Journal of Pharmaceutics and Biopharmaceutics, 180, 11–22. doi: 10.1016/j.ejpb.2022.09.020. Available online: https://www.sciencedirect.com/science/article/pii/S0939641122002156 (accessed on Nov 28, 2024).

[141] Niyyati, M., Dodangeh, S. and Lorenzo-Morales, J. (2016). A review of the current research trends in the application of medicinal plants as a source for novel therapeutic agents against *Acanthamoeba* infections. Iranian Journal of Pharmaceutical Sciences, 15, 893–900. Available online: https://www.webofscience.com/wos/woscc/full-record/WOS:000391797900029 (accessed on Nov 28, 2024).

[142] Chegeni, T. N., Fakhar, M., Ghaffarifar, F. and Saberi, R. (2020). Medicinal plants with anti-Acanthamoeba activity: A systematic review. Infectious Disorders – Drug Targets, 20, 620–650. doi: 10.2174/1871526519666190716095849. Available online: https://www.eurekaselect.com/article/99642 (accessed on Nov 28, 2024).

[143] Saoudi, S., Sifaoui, I., Chammem, N., Reyes-Batlle, M., López-Arencibia, A., Pacheco-Fernández, I., Pino, V., Hamdi, M., Jiménez, I. A., Bazzocchi, I. L., Piñero, J. E. and Lorenzo-Morales, J. (2017). Anti-*Acanthamoeba* activity of Tunisian *Thymus capitatus* essential oil and organic extracts. Experimental Parasitology, 183, 231–235. doi: 10.1016/j.exppara.2017.09.014. Available online: https://www.science direct.com/science/article/pii/S001448941730317X (accessed on Nov 28, 2024).

[144] Zouaghi, G., Najar, A., Chiboub, O., Sifaoui, I., Abderrabba, M. and Lorenzo Morales, J. (2017). The effect of viroid infection of citrus trees on the amoebicidal activity of 'Maltese half-blood' (*Citrus sinensis*) against trophozoite stage of *Acanthamoeba castellanii* Neff. Experimental Parasitology, 183, 182–186. doi: 10.1016/j.exppara.2017.09.006. Available online: https://www.webofscience.com/api/ gateway?GWVersion=2&SrcAuth=DOISource&SrcApp=WOS&KeyAID=10.1016%2Fj.exppara.2017.09. 006&DestApp=DOI&SrcAppSID=EUW1ED0FE5g71Iecp26kEYn0bzxhw&SrcJTitle=EXPERIMENTAL+PARA SITOLOGY&DestDOIRegistrantName=Elsevier (accessed on Nov 28, 2024).

[145] Hajaji, S., Sifaoui, I., López-Arencibia, A., Reyes-Batlle, M., Valladares, B., Pinero, J. E., Lorenzo-Morales, J. and Akkari, H. (2017). Amoebicidal activity of α-bisabolol, the main sesquiterpene in chamomile (*Matricaria recutita* L.) essential oil against the trophozoite stage of *Acanthamoeba castellani* Neff. Acta Parasitologica, 62, 290–295. doi: 10.1515/ap-2017-0036. Available online: https://doi.org/10.1515/ap-2017-0036 (accessed on Nov 28, 2024).

[146] Hajaji, S., Jabri, M., Sifaoui, I., Lopez-Arencibia, A., Reyes-Batlle, M., B'chir, F., Valladares, B., Pinero, J. E., Lorenzo-Morales, J. and Akkari, H. (2017). Amoebicidal, antimicrobial and *in vitro* ROS scavenging activities of Tunisian *Rubus ulmifolius* Schott, methanolic extract. Experimental Parasitology, 183, 224–230. doi: 10.1016/j.exppara.2017.09.013. Available online: https://www.webofs cience.com/api/gateway?GWVersion=2&SrcAuth=DOISource&SrcApp=WOS&KeyAID=10.1016%2Fj.ex ppara.2017.09.013&DestApp=DOI&SrcAppSID=EUW1ED0FE5g71Iecp26kEYn0bzxhw&SrcJTitle=EXPERI MENTAL+PARASITOLOGY&DestDOIRegistrantName=Elsevier (accessed on Nov 28, 2024).

[147] Sifaoui, I., López-Arencibia, A., Martín-Navarro, C. M., Chammem, N., Mejri, M., Lorenzo-Morales, J., Abderabba, M. and Piñero, J. E. (2013). Activity assessment of Tunisian olive leaf extracts against the trophozoite stage of Acanthamoeba. Parasitology Research, 112, 2825–2829. doi: 10.1007/s00436-

013-3453-9. Available online: https://pubmed.ncbi.nlm.nih.gov/23681194/ (accessed on Nov 28, 2024).

[148] Mitsuwan, W., Sin, C., Keo, S., Sangkanu, S., de Lourdes Pereira, M., Jimoh, T. O., Salibay, C. C., Nawaz, M., Norouzi, R., Siyadatpanah, A., Wiart, C., Wilairatana, P., Mutombo, P. N. and Nissapatorn, V. (2021). Potential anti-Acanthamoeba and anti-adhesion activities of *Annona muricata* and *Combretum trifoliatum* extracts and their synergistic effects in combination with chlorhexidine against *Acanthamoeba triangularis* trophozoites and cysts. Heliyon, 7, e06976. doi: 10.1016/j. heliyon.2021.e06976. Available online: https://pubmed.ncbi.nlm.nih.gov/34027178/ (accessed on Nov 28, 2024).

[149] Boonhok, R., Sangkanu, S., Chuprom, J., Srisuphanunt, M., Norouzi, R., Siyadatpanah, A., Mirzaei, F., Mitsuwan, W., Wisessombat, S., de Lourdes Pereira, M., Rahmatullah, M., Wilairatana, P., Wiart, C., Ling, L. C., Dolma, K. G. and Nissapatorn, V. (2021). *Peganum harmala* extract has antiamoebic activity to *Acanthamoeba triangularis* trophozoites and changes expression of autophagy-related genes. Pathogens, 10, 842. doi: 10.3390/pathogens10070842. Available online: https://www.mdpi. com/2076-0817/10/7/842 (accessed on Nov 28, 2024).

[150] Fadel, H., Sifaoui, I., López-Arencibia, A., Reyes-Batlle, M., Hajaji, S., Chiboub, O., Jiménez, I. A., Bazzocchi, I. L., Lorenzo-Morales, J., Benayache, S. and Piñero, J. E. (2018). Assessment of the antiprotozoal activity of *Pulicaria inuloides* extracts, an Algerian medicinal plant: Leishmanicidal bioguided fractionation. Parasitology Research, 117, 531–537. doi: 10.1007/s00436-017-5731-4. Available online: https://doi.org/10.1007/s00436-017-5731-4 (accessed on Nov 28, 2024).

[151] Derda, M., Hadaś, E. and Thiem, B. (2009). Plant extracts as natural amoebicidal agents. Parasitology Research, 104, 705–708. doi: 10.1007/s00436-008-1277-9. Available online: https://pubmed.ncbi.nlm.nih.gov/19050923/ (accessed on Nov 28, 2024).

[152] Degerli, S., Tepe, B., Celiksoz, A., Berk, S. and Malatyali, E. (2012). In vitro amoebicidal activity of *Origanum syriacum* and *Origanum laevigatum* on *Acanthamoeba castellanii* cysts and trophozoites. Experimental Parasitology, 131, 20–24. doi: 10.1016/j.exppara.2012.02.020. Available online: https://pubmed.ncbi.nlm.nih.gov/22417972/ (accessed on Nov 28, 2024).

[153] Shoaib, H. M., Muazzam, A. G., Mir, A., Jung, S. and Matin, A. (2013). Evaluation of inhibitory potential of some selective methanolic plants extracts on biological characteristics of *Acanthamoeba castellanii* using human corneal epithelial cells in vitro. Parasitology Research, 112, 1179–1188. doi: 10.1007/s00436-012-3249-3. Available online: https://pubmed.ncbi.nlm.nih.gov/23306385/ (accessed on Nov 28, 2024).

[154] El-Sayed, N. M., Ismail, K. A., Ahmed, S. A. and Hetta, M. H. (2012). In vitro amoebicidal activity of ethanol extracts of *Arachis hypogaea* L., *Curcuma longa* L. and *Pancratium maritimum* L. on *Acanthamoeba castellanii* cysts. Parasitology Research, 110, 1985–1992. doi: 10.1007/s00436-011-2727-3. Available online: https://pubmed.ncbi.nlm.nih.gov/22146994/ (accessed on Nov 28, 2024).

[155] Tepe, B., Malatyali, E., Degerli, S. and Berk, S. (2012). In vitro amoebicidal activities of *Teucrium polium* and *T. chamaedrys* on *Acanthamoeba castellanii* trophozoites and cysts. Parasitology Research, 110, 1773–1778. doi: 10.1007/s00436-011-2698-4. Available online: https://pubmed.ncbi.nlm.nih.gov/22037826/ (accessed on Nov 28, 2024).

[156] Degerli, S., Berk, S., Malatyali, E. and Tepe, B. (2012). Screening of the in vitro amoebicidal activities of *Pastinaca armenea* (Fisch. & C.A.Mey.) and *Inula oculus*-christi (L.) on *Acanthamoeba castellanii* cysts and trophozoites. Parasitology Research, 110, 565–570. doi: 10.1007/s00436-011-2524-z. Available online: https://pubmed.ncbi.nlm.nih.gov/21735149/ (accessed on Nov 28, 2024).

[157] Malatyali, E., Tepe, B., Degerli, S., Berk, S. and Akpulat, H. A. (2012). In vitro amoebicidal activity of four Peucedanum species on *Acanthamoeba castellanii* cysts and trophozoites. Parasitology Research, 110, 167–174. doi: 10.1007/s00436-011-2466-5. Available online: https://pubmed.ncbi.nlm. nih.gov/21626154/ (accessed on Nov 28, 2024).

[158] Roongruangchai, K., Kummalue, T., Sookkua, T. and Roongruangchai, J. (2010). Comparison of *Pouzolzia indica* methanolic extract and Virkon against cysts of *Acanthamoeba* spp. Southeast Asian

Journal of Tropical Medicine and Public Health, 41, 776–784. Available online: https://pubmed.ncbi.
nlm.nih.gov/21073052/ (accessed on Nov 28, 2024).

[159] Goze, I., Alim, A., Dag, S., Tepe, B. and Polat, Z. A. (2009). In vitro amoebicidal activity of Salvia
staminea and *Salvia caespitosa* on *Acanthamoeba castellanii* and their cytotoxic potentials on corneal
cells. Journal of Ocular Pharmacology and Therapeutics, 25, 293–298. doi: 10.1089/jop.2008.0132.
Available online: https://pubmed.ncbi.nlm.nih.gov/19450152/ (accessed on Nov 28, 2024).

[160] Hotez, P. J., Brindley, P. J., Bethony, J. M., King, C. H., Pearce, E. J. and Jacobson, J. (2008). Helminth
infections: The great neglected tropical diseases. Journal of Clinical Investigation, 118, 1311–1321.
doi: 10.1172/JCI34261. Available online: https://www.jci.org/articles/view/34261 (accessed on
Nov 28, 2024).

[161] Schluth, C. G., Standley, C. J., Bansal, S. and Carlson, C. J. (2023). Spatial parasitology and the
unmapped human helminthiases. Parasitology, 150, 1–9. doi: 10.1017/S0031182023000045. Available
online: https://pubmed.ncbi.nlm.nih.gov/36632014/ (accessed on Nov 29, 2024).

[162] Khan, J. S., Provencher, J. F., Forbes, M. R., Mallory, M. L., Lebarbenchon, C. and McCoy, K. D. (2019).
Chapter One – Parasites of seabirds: A survey of effects and ecological implications. In: Sheppard,
C. (ed) Advances in Marine Biology, Vol. 82, Academic Press, 1–50.

[163] Fauziah, N., Ar-Rizqi, M., Hana, S., Patahuddin, N. M. and Diptyanusa, A. (2022). Stunting as a risk
factor of soil-transmitted Helminthiasis in children: A literature review. Interdisciplinary
Perspectives on Infectious Diseases, 2022, 8929025. doi: 10.1155/2022/8929025. Available online:
https://doi.org/10.1155/2022/8929025.

[164] WHO Soil-transmitted helminth infections. https://www.who.int/news-room/fact-sheets/detail/soil-
transmitted-helminth-infections.

[165] Chai, J., Jung, B. and Hong, S. (2021). Albendazole and mebendazole as anti-parasitic and anti-cancer
agents: An update. Korean Journal of Parasitology, 59, 189–225. doi: 10.3347/kjp.2021.59.3.189.
Available online: https://pubmed.ncbi.nlm.nih.gov/34218593/ (accessed on Nov 29, 2024).

[166] Hirpa, E. (2015). Advanced review on anthelmintic medicinal plants. Available online: https://www.
academia.edu/24835859/Advanced_Review_on_Anthelmintic_Medicinal_Plants (accessed on
Nov 29, 2024).

[167] Getachew, S., Ibrahim, N. and Eguale, T. (2014). In vitro evaluation of anthelmintic activities of crude
extracts of selected medicinal plants against *Haemonchus contortus* in Alemgena Wereda. Ethiopia.

[168] Panda, S. K., Daemen, M., Sahoo, G. and Luyten, W. (2022). Essential oils as novel anthelmintic drug
candidates. Molecules, 27, 8327. doi: 10.3390/molecules27238327. Available online: https://www.
mdpi.com/1420-3049/27/23/8327 (accessed on Nov 29, 2024).

[169] Parreira, N., Magalhães, L., Morais, D., Caixeta, S., De Sousa, J., Bastos, J., Cunha, W., Silva, M.,
Nanayakkara, N., Rodrigues, V. and da Silva Filho, A. (2010). Antiprotozoal, schistosomicidal, and
antimicrobial activities of the essential oil from the leaves of *Baccharis dracunculifolia*. Chemistry and
Biodiversity, 7, 993–1001. doi: 10.1002/cbdv.200900292. Available online: https://doi.org/10.1002/
cbdv.200900292.

[170] Mahmoudvand, H., Tavakoli Oliaei, R., Mirbadie, S. R., Kheirandish, F., Tavakoli Kareshk, A.,
Ezatpour, B. and Mahmoudvand, H. (2016). Efficacy and safety of *Bunium persicum* (Boiss) to
inactivate protoscoleces during hydatid cyst operations. Surgical Infection (Larchmt), 17, 713–719.
doi: 10.1089/sur.2016.010. Available online: https://pubmed.ncbi.nlm.nih.gov/27501060/ (accessed
on Nov 29, 2024).

Danial Kahrizi* and Masoumeh Khanahmadi

Chapter 19
Medicinal and aromatic plants used in personal care products

Abstract: This chapter delves into the significant role of medicinal and aromatic plants (MAPs) in personal care products, reflecting a growing consumer preference for natural and organic ingredients. As awareness of the potential adverse effects of synthetic chemicals increases, there is a resurgence in the use of MAPs, which have been valued for their therapeutic properties throughout history. This work explores the historical applications of these plants, tracing their use from ancient civilizations, such as the Egyptians and Greeks, to contemporary practices in skincare, haircare, and aromatherapy. The chapter highlights key medicinal plants, including aloe vera, chamomile, tea tree oil, rosehip oil, and camelina oil, emphasizing their unique benefits and applications in personal care formulations. Each plant's bioactive compounds are examined for their contributions to skin health and overall well-being. Furthermore, the text discusses the profound impact of incorporating MAPs into cosmetic products on the cosmetics industry, showcasing the expanding market for botanicals and the increasing consumer demand for plant-based formulations. However, the integration of MAPs into personal care products is not without challenges. The book addresses critical issues such as sustainable harvesting practices to prevent resource depletion, consumer skepticism regarding the efficacy of natural versus synthetic ingredients, and navigating complex regulatory frameworks that govern botanical ingredient use. By providing a comprehensive overview of the historical context, current applications, benefits, challenges, and future prospects surrounding MAPs in personal care products, this book serves as an essential resource for both consumers and manufacturers. It aims to guide informed choices in personal care while promoting a deeper understanding of how these remarkable plants can enhance health and beauty in a sustainable manner.

Keywords: natural ingredients, organic cosmetics, skincare, haircare, essential oils, plant-based ingredients

*Corresponding author: Danial Kahrizi, Department of Biotechnology, Faculty of Agriculture, Tarbiat Modares University, Tehran, Iran, e-mail: dkahrizi@modares.ac.ir, https://orcid.org/0000-0002-1717-6075
Masoumeh Khanahmadi, Academic Center for Education, Culture and Research (ACECR), Kermanshah, Iran, e-mail: chem_khanahmadi@yahoo.com, https://orcid.org/0000-0003-0181-0868

https://doi.org/10.1515/9783111469713-019

19.1 Introduction

The use of medicinal and aromatic plants (MAPs) in personal care products has gained significant traction in recent years, driven by a consumer shift toward natural and organic ingredients. This interest stems from the therapeutic properties these plants offer, alongside their aromatic qualities, which enhance the sensory experience of personal care products. This document will explore the historical and modern applications of these plants, their benefits, impacts on the industry, and challenges faced [1, 2].

The trend toward incorporating MAPs into personal care formulations is not merely a passing fad; it reflects a deeper cultural shift toward holistic wellness and sustainability. The resurgence of interest in traditional herbal remedies can be traced back to ancient civilizations, where plants were revered for their healing properties. For example, ancient Egyptians utilized oils from plants such as myrrh and frankincense for both medicinal and cosmetic purposes, laying the groundwork for modern practices. Today, many consumers are rediscovering these age-old practices, seeking products that align with their values of health, wellness, and environmental responsibility [3, 4].

Moreover, the benefits of using MAPs extend beyond mere esthetics. Many botanical ingredients possess bioactive compounds that can provide significant health benefits. For instance, *aloe vera* is well-known for its soothing properties, making it a staple in skincare products aimed at treating sunburns and other skin irritations. Similarly, tea tree oil has established itself as an effective treatment for acne due to its antimicrobial properties. The incorporation of these natural ingredients not only enhances product efficacy but also appeals to consumers who prioritize safety and sustainability in their purchasing decisions [5, 6].

In addition to skincare applications, MAPs are increasingly being recognized for their roles in haircare and aromatherapy. Ingredients such as rosemary and lavender are celebrated not only for their pleasant scents but also for their ability to promote hair growth and relaxation, respectively. The versatility of these plants allows manufacturers to create multifunctional products that cater to diverse consumer needs while minimizing the use of synthetic additives [7].

The impact of this trend on the cosmetics industry is profound. The global market for botanicals has expanded significantly, with estimates suggesting it was valued at approximately $108 billion in 2015, encompassing sectors such as herbal medicines, cosmetics, and functional foods [7].

Countries like Germany and France lead the European market, emphasizing the growing acceptance and demand for plant-based formulations. This shift is supported by scientific research, which continues to validate the efficacy of various botanical ingredients through clinical studies [8, 9].

However, while the benefits are substantial, there are challenges associated with sourcing and utilizing MAPs sustainably. Overharvesting poses a threat to biodiversity; thus, it is crucial for companies to engage in responsible sourcing practices that support local communities and preserve ecosystems. Additionally, regulatory frameworks governing the use of botanical ingredients can vary widely between regions, creating barriers to market entry for new products [10].

In conclusion, the integration of MAPs into personal care products signifies a transformative movement toward natural beauty solutions that resonate with contemporary consumer values. As this trend continues to evolve, it will be essential for industry stakeholders to address sustainability concerns while harnessing the full potential of these remarkable plants. This document aims to provide a comprehensive overview of the historical context, current applications, benefits, challenges, and future prospects surrounding MAPs in personal care products. [11]

The aim of this chapter is to provide a comprehensive overview of the role and significance of MAPs in personal care products, highlighting their historical usage, current applications, and the benefits they offer, while also addressing the challenges faced by the industry in terms of sustainability and regulation. This exploration will serve as a foundation for understanding how these natural ingredients can be effectively integrated into modern formulations, ultimately guiding both consumers and manufacturers toward more informed choices in personal care.

19.2 Historical context

Historically, civilizations such as the Egyptians, Greeks, and Romans utilized various plant extracts for skincare and therapeutic purposes. For instance, Cleopatra is said to have used rose and almond oils in her beauty rituals. In contemporary times, there is a resurgence in the incorporation of these botanical ingredients into cosmetics due to growing consumer demand for natural products [12, 13].

Historically, civilizations such as the Egyptians, Greeks, and Romans utilized various plant extracts for skincare and therapeutic purposes, establishing a rich tradition of herbal medicine that has persisted through the ages. The Egyptians, for instance, are well-documented for their extensive use of aromatic plants in both daily life and religious practices. They employed ingredients like frankincense and myrrh, which were not only used in embalming but also in perfumes and cosmetics. Cleopatra, renowned for her beauty, is said to have used rose and almond oils in her beauty rituals, highlighting the importance placed on natural ingredients even in ancient times. Additionally, the use of honey and milk as skin moisturizers and cleansers further underscores the sophisticated understanding of skincare that existed among ancient Egyptians [3].

In ancient Greece, the medicinal properties of plants were systematically studied and documented by figures such as Hippocrates, who is often referred to as the "Father of Medicine." He emphasized the importance of diet and herbal remedies in maintaining health. His student, Aristotle, also contributed to the understanding of plants' medicinal properties. Theophrastus, known as the "Father of Botany," wrote extensively about plants and their uses, laying the groundwork for future botanical studies. His works included descriptions of over 500 plant species, many of which are still recognized for their therapeutic benefits today [14].

The Romans adopted many Greek practices but expanded on them by integrating their own local flora into medicinal use. They utilized plants like lavender for its calming properties and rosemary for memory enhancement. The Roman physician Galen further advanced herbal medicine by developing complex formulations that combined multiple plant extracts to treat various ailments [15].

As we move into the Middle Ages, herbal medicine continued to flourish, particularly within monastic communities, where monks meticulously cultivated medicinal gardens. These gardens were filled with a variety of herbs used for healing purposes. The knowledge of these plants was preserved in manuscripts, which became crucial during a time when much scientific knowledge was lost due to societal upheaval [16].

The Renaissance marked another pivotal moment in the history of medicinal plants. This era saw a revival of interest in classical texts and an increase in botanical exploration. Herbalists began to document their findings more rigorously, leading to significant advancements in pharmacognosy – the study of medicines derived from natural sources. The publication of "De Materia Medica" by Dioscorides, a Greek physician in the first century A.D., became a cornerstone text that influenced herbal medicine for centuries [17].

In contemporary times, there has been a resurgence in the incorporation of botanical ingredients into cosmetics due to growing consumer demand for natural products. This shift is driven by increasing awareness of the potential side effects associated with synthetic chemicals commonly found in personal care products. Consumers are now more inclined to seek products containing natural ingredients that not only enhance beauty but also promote overall well-being [11].

The modern market has responded to this demand by incorporating a wide range of MAPs into personal care formulations. Ingredients such as aloe vera, known for its soothing properties; tea tree oil, celebrated for its antimicrobial effects; and shea butter, valued for its moisturizing capabilities, have become staples in many cosmetic lines. Furthermore, brands are increasingly emphasizing transparency regarding ingredient sourcing and formulation processes, appealing to ethically minded consumers [18].

This historical context illustrates that the relationship between humans and medicinal plants is deeply rooted in our quest for health and beauty. As we continue to explore the benefits of these natural ingredients, it is essential to recognize their historical significance and how they have shaped modern practices within the personal care industry. Understanding this lineage not only enriches our appreciation for these plants but also informs future innovations that prioritize sustainability and efficacy in personal care products.

Table 19.1 summarizes the historical uses of various MAPs across different civilizations. This table highlights the diverse applications of these plants in ancient cultures, illustrating how they were integrated into daily life for both medicinal and cosmetic purposes.

Table 19.1: Historical uses of some medicinal plants across different civilizations.

Civilization	Plant used	Purpose
Ancient Egypt	Myrrh	Embalming and skincare
Ancient Greece	Rosemary	Memory enhancement and hair health
Ancient Rome	Lavender	Calming effects and perfumes
Middle Ages	Chamomile	Healing skin irritations
Renaissance	Aloe vera	Skin healing and moisturizing

19.3 Common medicinal and aromatic plants in personal care

19.3.1 Skincare

Medicinal plants are renowned for their various skin benefits, and their incorporation into personal care products has become increasingly popular. Here, we explore some of the most notable plants used in skincare formulations.

See Figures 19.1–19.5 for a visual representation of key MAPs used in personal care products. These figures feature a diverse array of MAPs, including Aloe vera, chamomile, tea tree, rosehip, and camelina. These plants are essential components in the formulation of personal care products, each offering unique therapeutic properties that enhance skin health and overall well-being.

19.3.1.1 *Aloe vera*

Aloe vera is one of the most widely recognized medicinal plants for skincare, celebrated for its extensive therapeutic properties. The gel extracted from its leaves is rich in vitamins A, C, E, and B12, which contribute to its healing effects. *Aloe vera* is particularly known for its soothing and hydrating properties, making it an ideal ingredient for treating sunburns, minor burns, and skin irritations. Studies have shown that aloe vera can significantly speed up wound healing by promoting collagen synthesis and increasing the moisture content of the skin. The polysaccharides found in aloe vera help bind moisture to the skin, enhancing its elasticity and reducing the appearance of fine lines and wrinkles [18, 19].

Furthermore, aloe vera possesses anti-inflammatory properties that can alleviate redness and swelling associated with various skin conditions, including acne and eczema. Its ability to inhibit the production of inflammatory cytokines makes it a valuable ingredient in formulations aimed at sensitive skin. Aloe vera also contains antiseptic agents such as salicylic acid and lupeol, which provide antimicrobial benefits that help combat acne-causing bacteria [20, 21]. Regular application of aloe vera gel can lead to improved skin texture and overall health, making it a staple in many skincare routines [22]. Figure 19.1 shows an aloe vera field.

Figure 19.1: An aloe vera field.

19.3.1.2 Chamomile

Chamomile (*Matricaria chamomilla*) is another prominent medicinal plant valued for its calming effects on the skin. The flowers of the chamomile plant contain flavonoids and essential oils that exhibit significant anti-inflammatory and antioxidant properties. Chamomile is particularly effective in soothing irritated skin, reducing redness, and calming conditions like eczema or dermatitis. Its gentle nature makes it suitable for all skin types, including sensitive skin [23].

Chamomile extracts are commonly used in skincare products such as creams, lotions, and serums. These extracts help promote healing by encouraging the regeneration of skin cells while also providing a protective barrier against environmental stressors. The antioxidant properties of chamomile help neutralize free radicals, which can contribute to premature aging [24].

Additionally, chamomile's natural fragrance offers a soothing aroma that enhances the sensory experience of personal care products [25]. Figure 19.2 shows chamomile plants.

19.3.1.3 Tea tree oil

Tea tree (*Melaleuca alternifolia*) oil, derived from the leaves of the *M. alternifolia* tree, is renowned for its powerful antimicrobial properties. It has been extensively studied

Figure 19.2: Chamomile plants.

for its effectiveness against acne due to its ability to inhibit the growth of acne-causing bacteria. Tea tree oil works by penetrating the pores and reducing inflammation while preventing future breakouts. Its antiseptic qualities also make it useful for treating minor cuts and wounds [26].

In addition to its acne-fighting abilities, tea tree oil has been shown to soothe irritated skin conditions such as psoriasis and eczema. Its anti-inflammatory effects can help alleviate itching and redness associated with these conditions [27].

Many personal care products now incorporate tea tree oil as a key ingredient due to its versatility and efficacy in promoting clearer skin. A number of plant bushes are shown in Figure 19.3.

19.3.1.4 Rosehip oil

Rosehip oil, extracted from the seeds of wild rose bushes, is highly regarded for its anti-aging properties. Rich in essential fatty acids and vitamins A and C, rosehip oil helps improve skin texture by promoting cell regeneration and reducing the appearance of scars and fine lines. Its antioxidant content protects the skin from oxidative stress caused by environmental factors such as pollution and UV radiation [28].

Rosehip oil is often included in formulations aimed at improving skin tone and texture due to its ability to enhance collagen production. This makes it a popular choice for those looking to achieve a more youthful appearance. Additionally, rosehip oil has moisturizing properties that help maintain hydration levels in the skin without clogging pores.

Its lightweight consistency allows for easy absorption, making it suitable for both daytime and nighttime use [29]. A picture of this plant can be seen in Figure 19.4.

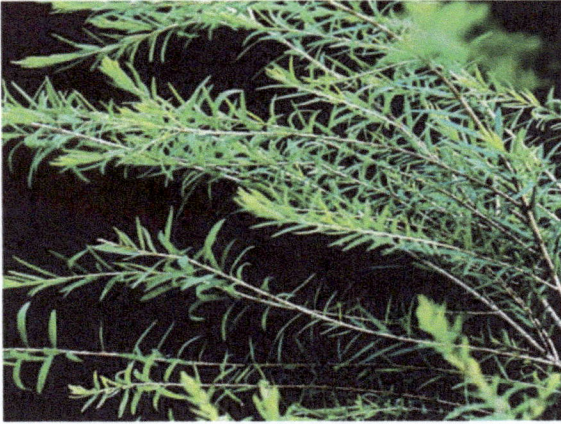

Figure 19.3: Tea tree bush.

Figure 19.4: A rose hip plant.

19.3.1.5 Camelina oil

Camelina oil, derived from the seeds of the *Camelina sativa* plant, is an emerging star in the realm of skincare due to its impressive profile of beneficial properties. Often referred to as the "gold of pleasure," this oil has been used for centuries, particularly in Europe, and is now gaining recognition globally for its multifaceted benefits [30].

One of the standout features of camelina oil is its rich content of essential fatty acids, particularly omega-3 and omega-6 fatty acids. These fatty acids play a crucial role in maintaining skin health by forming a lipid barrier that locks in moisture and reduces transepidermal water loss. This barrier function is vital for keeping the skin hydrated and preventing dryness, which can lead to irritation and premature aging. The high levels of alpha-linolenic acid, an omega-3 fatty acid, contribute to its emollient properties, making it an excellent choice for those with dry or sensitive skin [31].

In addition to its moisturizing benefits, camelina oil is also known for its antioxidant properties. It is rich in vitamin E, which helps protect the skin from oxidative stress caused by environmental factors such as UV radiation and pollution. By neutralizing free radicals, camelina oil can help slow down the signs of aging and maintain a youthful appearance. Regular use can lead to improved skin elasticity and a reduction in the appearance of fine lines and wrinkles [32].

The soothing and anti-inflammatory effects of camelina oil make it particularly beneficial for sensitive or irritated skin. It has been shown to reduce redness and inflammation, making it suitable for conditions such as eczema and rosacea. The oil's ability to promote cellular regeneration further aids in repairing damaged skin and improving overall skin texture. This regenerative property is particularly valuable in restorative treatments, such as those aimed at minimizing the appearance of stretch marks or scars [33].

Camelina oil's versatility extends beyond skincare; it can be incorporated into various cosmetic formulations, including creams, lotions, balms, and even haircare products. Its lightweight texture allows for quick absorption without leaving a greasy residue, making it ideal for daily use. Additionally, camelina oil can enhance the sensory experience of products by extending the aroma of essential oils when used in natural bath salts or massage oils [34]. Figure 19.5 shows a camelina field.

Figure 19.5: A camelina field.

Table 19.2 provides a comprehensive overview of the key MAPs discussed in this chapter, along with their notable properties and common applications in personal care products. This table serves as a quick reference for understanding how these plants contribute to skincare, haircare, and aromatherapy, illustrating their therapeutic benefits and versatility.

Table 19.2: Some medicinal plants along with their key properties and common uses in personal care products.

Plant name	Key properties	Common uses
Aloe vera	Soothing and hydrating	Sunburn relief and skin moisturizer
Chamomile	Anti-inflammatory and calming	Skin-soothing creams and serums
Tea tree oil	Antimicrobial and anti-inflammatory	Acne treatments and antiseptic
Rosehip oil	Antioxidant and antiaging	Antiaging creams and serums
Camelina oil	Moisturizing and antioxidant	Skin hydrators and restorative oils

19.3.2 Haircare

Several plants are utilized in haircare formulations, each offering unique benefits that cater to various hair and scalp conditions. The following plants are particularly noteworthy.

19.3.2.1 Rosemary

Rosemary (*Rosmarinus officinalis*) has a long-standing reputation in both culinary and medicinal applications, but its role in haircare is particularly significant. Traditionally, rosemary has been used as a natural remedy to stimulate hair growth and improve overall scalp health. The active compounds in rosemary, such as caffeic acid and rosmarinic acid, are known for their potent antioxidant properties, which help protect hair follicles from oxidative stress caused by free radicals. This protection is crucial in preventing premature hair loss and maintaining healthy hair [35].

Research has shown that rosemary oil can enhance blood circulation to the scalp, which is vital for nourishing hair follicles and promoting hair growth. Improved circulation ensures that essential nutrients reach the hair roots, thereby supporting stronger and healthier hair strands. Additionally, rosemary oil has antimicrobial properties that can help combat dandruff and other scalp conditions, making it an effective treatment for maintaining a clean and balanced scalp environment [36].

Incorporating rosemary into haircare products, such as shampoos and conditioners, not only provides these benefits but also imparts a refreshing scent that enhances the overall user experience. Regular use of rosemary-infused products can lead to thicker, fuller hair over time, making it a popular choice among those seeking natural solutions for hair vitality (Figure 19.6) [37].

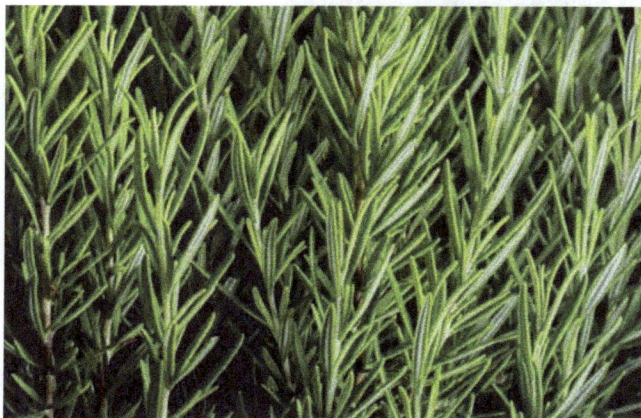

Figure 19.6: Rosemary (*Rosmarinus officinalis*).

19.3.2.2 Hibiscus

Hibiscus (*Hibiscus rosa-sinensis*) is another powerful botanical used in haircare formulations. Known for its vibrant flowers, hibiscus is rich in vitamins A and C, amino acids, and antioxidants, all of which contribute to its effectiveness in promoting hair health. One of the primary benefits of hibiscus is its ability to stimulate hair regrowth. The plant contains natural acids that help exfoliate the scalp, removing dead skin cells and promoting a healthy environment for new hair growth [38].

Hibiscus extracts are also known to enhance the shine and luster of hair. The amino acids present in hibiscus help strengthen the hair shaft, reducing breakage and split ends. This makes hibiscus an excellent ingredient for those with dry or damaged hair. Additionally, hibiscus has been traditionally used to combat premature graying due to its ability to condition the hair and improve its overall texture [39].

Moreover, hibiscus can be used as a natural conditioner. When applied as a paste or infused oil, it helps detangle hair while providing moisture and nourishment. Its soothing properties also make it beneficial for sensitive scalps, alleviating irritation and promoting overall scalp health (Figure 19.7) [40].

19.3.2.3 Neem

Neem (*Azadirachta indica*), often referred to as Indian lilac, is a remarkable plant known for its extensive medicinal properties. In the realm of haircare, neem is partic-

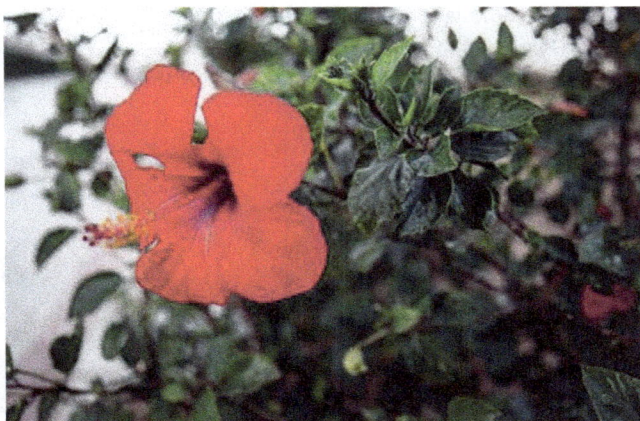

Figure 19.7: Hibiscus (*Hibiscus rosa-sinensis*).

ularly valued for its effectiveness in addressing various scalp conditions such as dandruff, psoriasis, and other inflammatory issues. Neem's antifungal and antibacterial properties make it an ideal ingredient for treating dandruff by combating the underlying causes of flakiness and irritation [41].

The active compounds in neem leaves, such as azadirachtin, have been shown to inhibit the growth of fungi and bacteria that can contribute to scalp problems. Regular use of neem-infused shampoos or oils can help maintain a healthy scalp environment while promoting optimal conditions for hair growth. Furthermore, neem oil acts as a natural moisturizer that nourishes the scalp without clogging pores [42].

In addition to its anti-dandruff benefits, neem is known for strengthening hair follicles and reducing hair fall due to breakage or thinning. Its rich nutrient profile includes essential fatty acids that help improve the overall health of both the scalp and hair strands. Neem's ability to balance oil production on the scalp makes it suitable for all hair types – whether oily or dry (Figure 19.8) [43].

In summary, incorporating plants like rosemary, hibiscus, and neem into haircare formulations offers numerous advantages for maintaining healthy hair and scalp conditions. These botanical ingredients not only provide effective treatments but also align with the growing consumer preference for natural and sustainable beauty solutions. As awareness of these benefits continues to rise, more individuals are turning to herbal remedies to enhance their haircare routines naturally.

19.3.3 Aromatherapy

Essential oils derived from aromatic plants play a crucial role in aromatherapy, a holistic healing practice that utilizes the natural fragrances of plants to enhance physi-

Figure 19.8: Leaves and fruits of the neem tree.

cal, emotional, and spiritual well-being. Aromatherapy harnesses the power of these concentrated plant extracts to promote relaxation, alleviate stress, and improve overall health [44].

The use of essential oils in aromatherapy provides a holistic approach to wellness by leveraging the natural benefits of aromatic plants. Each essential oil brings unique therapeutic properties that can enhance emotional well-being, promote relaxation, improve physical health, or even support skincare routines. As interest in natural remedies continues to grow, understanding how these oils work can empower individuals to incorporate them into their daily lives effectively – whether through diffusion, topical application, or inclusion in personal care products [45].

Here are some of the most commonly used essential oils in aromatherapy, along with their benefits:

19.3.3.1 Lavender oil

Lavender (*Lavandula angustifolia*) oil is one of the most popular essential oils in aromatherapy, renowned for its calming and soothing effects [46]. It is frequently used in bath products, massage oils, and diffusers. The scent of lavender has been shown to reduce anxiety and promote relaxation, making it an excellent choice for those dealing with stress or insomnia. Studies have indicated that inhaling lavender essential oil can lower heart rate and blood pressure, contributing to a sense of tranquility [47].

In addition to its calming properties, lavender oil is also known for its antiseptic and anti-inflammatory benefits. This makes it useful for treating minor burns, cuts, and insect bites when diluted with a carrier oil. Its versatility allows it to be incorporated into skincare products as well, where it can help balance oily skin and reduce acne (Figure 19.9) [48, 49].

Figure 19.9: Lavender (*Lavandula angustifolia*).

19.3.3.2 Sandalwood oil

Sandalwood (*Santalum album*) oil is prized for its distinctive sweet and woody aroma. It has been used for centuries in traditional medicine and spiritual practices. In aromatherapy, sandalwood oil is known for its grounding effects, helping to promote mental clarity and emotional stability. It is often used during meditation to enhance focus and create a peaceful atmosphere [27].

Beyond its emotional benefits, sandalwood oil possesses anti-inflammatory properties that can help soothe irritated skin conditions such as eczema or psoriasis. Its moisturizing qualities make it beneficial in formulations aimed at dry skin. Sandalwood oil can also be blended with other essential oils to create complex fragrances that enhance personal care products (Figure 19.10) [50].

19.3.3.3 Ylang oil

Extracted from the flowers of the ylang-ylang (*Cananga odorata*) tree, ylang-ylang oil is celebrated for its sweet, floral scent. It is often used to reduce stress and anxiety while promoting feelings of joy and relaxation. Additionally, ylang-ylang oil can help balance oil production in the skin, making it beneficial for both dry and oily skin types (Figure 19.11) [51].

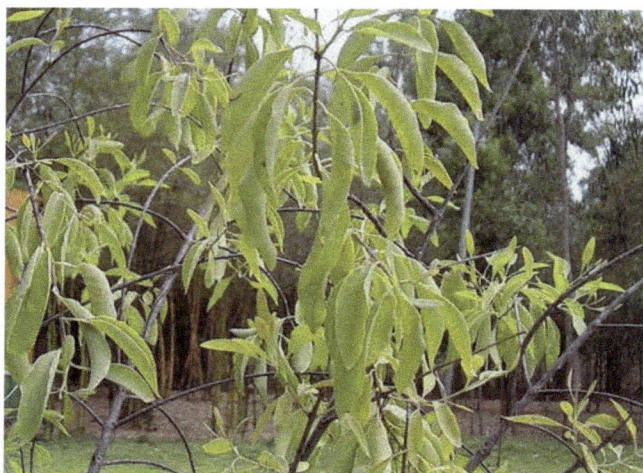

Figure 19.10: Sandalwood (*Santalum album*).

Figure 19.11: Ylang-ylang (*Cananga odorata*).

19.3.3.4 Peppermint oil

Known for its refreshing aroma, peppermint (*Mentha piperita*) oil invigorates the senses and enhances mental clarity. It is commonly used to alleviate headaches and migraines when inhaled or applied topically with a carrier oil. Peppermint oil also has cooling properties that can soothe sore muscles when added to massage oils (Figure 19.12) [52].

Figure 19.12: Peppermint (*Mentha piperita*).

19.3.3.5 Tea tree oil

Derived from the leaves of the *Melaleuca alternifolia* tree, tea tree oil is renowned for its powerful antimicrobial properties. It is often used in skincare products to treat acne due to its ability to combat bacteria that cause breakouts. In aromatherapy, tea tree oil can help purify the air and promote respiratory health when diffused (Figure 19.3) [26].

19.3.3.6 Bergamot oil

Extracted from the peel of the bergamot orange (*Citrus bergamia*), this essential oil has a fresh, citrusy scent that uplifts mood while reducing feelings of stress and anxiety. Bergamot oil is also known for its skin benefits; it can help treat oily skin and improve overall complexion when diluted properly (Figure 19.13) [53].

19.3.3.7 Frankincense oil

Frankincense (*Boswellia serrata*) has been used for thousands of years in religious ceremonies and traditional medicine. In aromatherapy, frankincense promotes deep relaxation and spiritual awareness. It has anti-inflammatory properties that can benefit skin health by reducing redness and irritation (Figure 19.14) [54].

Figure 19.13: Bergamot orange (*Citrus bergamia*).

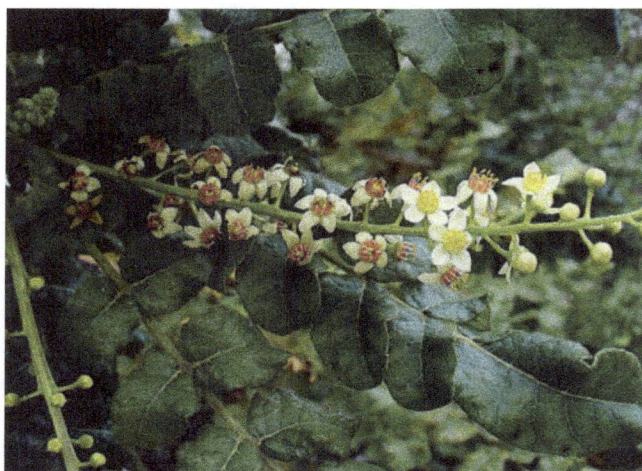

Figure 19.14: Frankincense (*Boswellia serrata*).

19.3.3.8 Geranium oil

Geranium (*Pelargonium graveolens*) oil offers a sweet, floral aroma that helps balance emotions while promoting feelings of calmness and peace. It is also known for its ability to regulate sebum production in the skin, making it beneficial for both dry and oily skin types (Figure 19.15) [55].

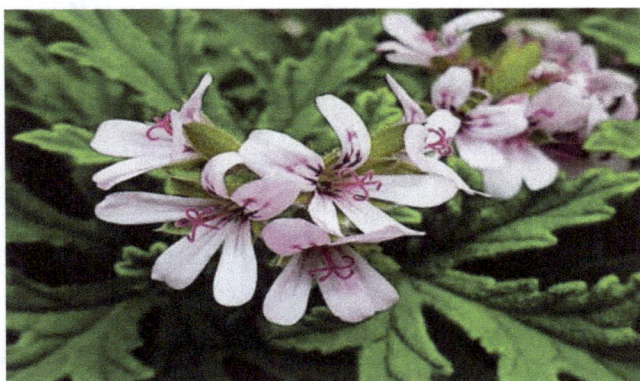

Figure 19.15: Geranium (*Pelargonium graveolens*).

19.3.3.9 Lemon oil

Extracted from lemon peels (*Citrus limon*), lemon oil has a bright and uplifting scent that energizes the mind while enhancing mood. It is often used in cleaning products due to its antibacterial properties but is also effective in skincare formulations for brightening dull skin (Figure 19.16) [45].

Figure 19.16: Lemon peels (*Citrus limon*).

19.3.3.10 Rosemary oil

Rosemary (*Rosmarinus officinalis*) oil not only enhances cognitive function but also improves concentration and memory retention when inhaled or diffused. Its invigorating scent makes it popular in haircare formulations as well, due to its ability to stimulate hair growth (Figure 19.6) [36, 43].

19.3.3.11 Jasmine oil

Jasmine (*Jasminum grandiflorum*) essential oil is renowned for its exotic floral fragrance, which promotes relaxation while enhancing feelings of romance and intimacy. It is often used in massage oils due to its ability to nourish dry skin (Figure 19.17) [56].

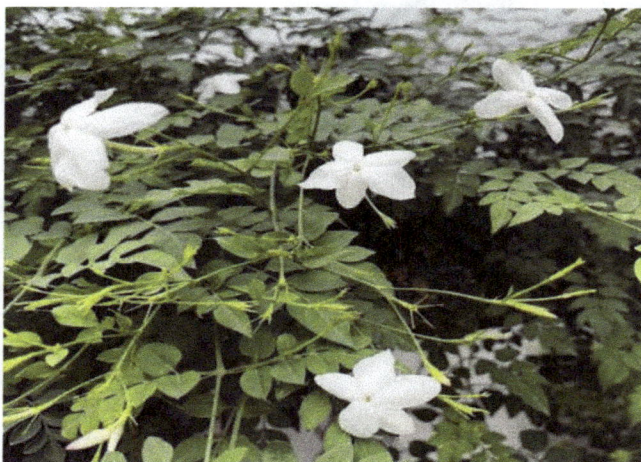

Figure 19.17: Jasmine (*Jasminum grandiflorum*).

19.3.3.12 Clary sage oil

Clary sage (*Salvia sclarea*) offers a sweet, herbaceous aroma that helps alleviate stress and anxiety while promoting hormonal balance – making it particularly beneficial during menstruation or menopause (Figure 19.18) [57].

19.3.4 Benefits of medicinal and aromatic plants

The utilization of MAPs in personal care and wellness products offers a multitude of benefits that extend beyond mere esthetics. These plants have long been revered for

Figure 19.18: Clary sage (*Salvia sclarea*).

their therapeutic properties, and their incorporation into various products reflects a growing consumer preference for natural alternatives [58]. Below are some of the key benefits associated with the use of MAPs:

19.3.4.1 Consumer health and safety

Many consumers perceive plant-based products as safer alternatives to synthetic chemicals, significantly reducing the risks of allergies and irritation. The shift toward natural ingredients is driven by increasing awareness of the potential side effects associated with synthetic additives commonly found in personal care products. Numerous studies have highlighted that essential oils and extracts derived from medicinal plants possess antimicrobial, anti-inflammatory, and antioxidant properties, which can enhance skin health without the adverse effects often linked to synthetic compounds [59].

For example, essential oils like tea tree oil have been shown to effectively combat acne-causing bacteria while being less irritating than many conventional treatments. Similarly, aloe vera is widely recognized for its soothing properties, making it a preferred choice for individuals with sensitive skin or conditions such as eczema. As consumers become more informed about ingredient safety, they are increasingly opting for products that feature MAPs, viewing them as not only effective but also gentle on the skin [60].

19.3.4.2 Environmental sustainability

The trend toward natural ingredients supports sustainable farming practices and ethical sourcing, promoting biodiversity within ecosystems. MAPs are often cultivated using organic farming methods that avoid harmful pesticides and fertilizers, thereby reducing environmental impact. This sustainable approach not only benefits the planet but also supports local communities involved in the cultivation of these plants [61].

Moreover, the cultivation of MAPs can contribute to the preservation of traditional agricultural practices and biodiversity. By prioritizing native plant species, farmers can help maintain local ecosystems while providing consumers with high-quality natural products. The increasing demand for sustainably sourced ingredients encourages companies to adopt eco-friendly practices, further enhancing their commitment to environmental stewardship [61].

The use of MAPs also aligns with the principles of green chemistry, which emphasize the design of chemical products and processes that reduce or eliminate hazardous substances. By replacing synthetic chemicals with natural alternatives derived from plants, companies can minimize their ecological footprint while delivering effective personal care solutions [62].

19.3.4.3 Economic opportunities

The demand for MAPs creates significant economic benefits for farmers and communities involved in their cultivation. As consumers increasingly seek out natural products, there is a corresponding rise in market opportunities for growers of MAPs. This trend not only supports local economies but also fosters job creation in rural areas where these plants are cultivated [63, 64].

Farmers who cultivate MAPs can benefit from higher profit margins compared to traditional crops due to the growing interest in organic and specialty products. Additionally, many communities are turning to agro-tourism by offering educational experiences related to herbal medicine and sustainable farming practices. This diversification can provide additional income streams for families relying on agriculture [65, 66].

Furthermore, as research continues to uncover new applications for MAPs in various industries – including cosmetics, food production, and pharmaceuticals – the potential for economic growth expands significantly. Companies that invest in sustainable sourcing practices can differentiate themselves in the marketplace, attracting environmentally conscious consumers who are willing to pay a premium for ethically produced goods [67].

In summary, the benefits of MAPs extend far beyond their immediate applications in personal care products. They offer consumers safer alternatives to synthetic chemicals while promoting environmental sustainability and providing economic opportunities for local communities. As awareness regarding the advantages of MAPs continues to grow, both consumers and manufacturers are likely to embrace these natural ingredients even more fervently in the years to come. By supporting sustainable practices and prioritizing health-conscious choices, we can foster a more holistic approach to beauty and wellness that respects both people and the planet.

19.3.5 Challenges

The integration of MAPs into personal care products represents a significant evolution in the cosmetics industry. As consumers increasingly seek natural alternatives, understanding the historical context, benefits, and challenges associated with these plants is essential for future developments. Despite the numerous benefits associated with the use of MAPs in personal care products, the industry faces several significant challenges that must be addressed to ensure sustainable growth and consumer trust. These challenges encompass environmental, economic, and regulatory aspects, each of which plays a critical role in shaping the future of the natural cosmetics market [68].

19.3.5.1 Ensuring sustainable harvesting practices

One of the most pressing challenges is ensuring sustainable harvesting practices to prevent the depletion of natural resources. Many MAPs are harvested from wild populations, which can lead to overexploitation if not managed properly. Unsustainable harvesting can result in a decline in plant populations, threatening biodiversity and disrupting local ecosystems [69].

To combat this issue, it is essential for companies to adopt sustainable sourcing practices that prioritize the long-term health of plant species and their habitats. This includes implementing guidelines for responsible harvesting, such as only taking a certain percentage of a plant population or utilizing cultivation methods that allow for regeneration. Additionally, certification programs like fair trade and organic certifications can help consumers identify products sourced from environmentally responsible suppliers [70].

The establishment of community-based management systems can also play a vital role in promoting sustainability. By involving local communities in the cultivation and harvesting processes, companies can ensure that traditional knowledge is respected while providing economic incentives for conservation efforts. This collaborative approach not only protects plant species but also empowers local populations, fostering a sense of stewardship over their natural resources [71].

19.3.5.2 Addressing consumer skepticism

Another significant challenge is addressing consumer skepticism regarding the efficacy of natural versus synthetic ingredients. While there is a growing demand for natural products, some consumers remain doubtful about their effectiveness compared to well-established synthetic alternatives. This skepticism can stem from various factors, including misinformation, a lack of scientific evidence supporting the benefits of certain botanical ingredients, and past negative experiences with natural products [72].

To overcome this challenge, it is crucial for manufacturers to invest in research and development that validates the efficacy of their products. Conducting clinical trials and publishing findings in reputable journals can help build consumer confidence in natural ingredients. Additionally, transparent communication about ingredient sourcing, formulation processes, and product benefits can enhance trust between brands and consumers [73].

Education plays a vital role in dispelling myths surrounding natural cosmetics. Companies can engage consumers through informative marketing campaigns that highlight the science behind their products while emphasizing the safety and effectiveness of botanical ingredients. Collaborating with dermatologists, herbalists, and other experts can further lend credibility to claims made about natural formulations [74].

19.3.5.3 Navigating regulatory frameworks

Navigating regulatory frameworks that govern the use of botanical ingredients in cosmetics presents another challenge for the industry. Regulations regarding the safety and efficacy of cosmetic products vary significantly between countries and regions, creating complexities for manufacturers seeking to market their products globally. In some cases, stringent regulations may limit the use of certain botanicals or require extensive documentation to prove safety [75].

This regulatory landscape can pose barriers to entry for smaller companies or those looking to innovate with new plant-based formulations. To address this challenge, it is essential for industry stakeholders to advocate for clear and consistent regulations that support the safe use of MAPs while fostering innovation within the sector [76].

Collaboration between regulatory bodies and industry representatives can help establish guidelines that are both practical and protective of consumer health. Furthermore, ongoing dialog with researchers and scientists can ensure that regulations are informed by the latest scientific findings regarding botanical safety and efficacy [77].

19.3.5.4 Market competition

The increasing popularity of natural cosmetics has led to heightened competition within the market. While this can drive innovation and improve product quality, it also poses challenges for brands striving to differentiate themselves in a crowded marketplace. Many companies may resort to "greenwashing," where they exaggerate or misrepresent their environmental claims to attract consumers [78].

To combat this issue, brands must focus on authenticity and transparency in their marketing efforts. Clearly communicating sourcing practices, ingredient lists, and product benefits can help build trust with consumers who are increasingly discerning about their purchases. Establishing strong brand values centered on sustainability and ethical practices will also resonate with consumers who prioritize these aspects in their buying decisions [79].

In conclusion, while the incorporation of MAPs into personal care products offers substantial benefits, several challenges must be addressed to ensure sustainable growth within the industry. By focusing on sustainable harvesting practices, building consumer trust through education and research, navigating complex regulatory environments, and maintaining authenticity amidst competition, stakeholders can work together to create a thriving market for natural cosmetics that respects both people and the planet. Overcoming these challenges will ultimately pave the way for a more responsible approach to beauty that prioritizes health, sustainability, and ethical practices [80].

19.4 Conclusion

The exploration of MAPs used in personal care products reveals a rich tapestry of history, benefits, and challenges that shape the current landscape of the cosmetics industry. As consumers increasingly gravitate toward natural and organic alternatives, the significance of MAPs has surged, reflecting a broader cultural shift toward holistic wellness and sustainability. This chapter highlights the multifaceted roles that these plants play in skincare, haircare, and aromatherapy, showcasing their therapeutic properties and the positive impact they can have on health and well-being.

Historically, the use of MAPs dates back thousands of years, with ancient civilizations recognizing their value for both medicinal and cosmetic purposes. From the soothing properties of aloe vera to the calming effects of lavender oil, these plants have been integral to beauty rituals across cultures. The resurgence of interest in traditional herbal remedies is not merely a trend; it is a testament to humanity's enduring connection with nature and its resources. As we navigate modern life, the wisdom embedded in these ancient practices offers valuable insights into creating effective and safe personal care products.

The benefits of incorporating MAPs into personal care formulations are manifold. They provide consumers with safer alternatives to synthetic chemicals, reducing the risk of allergies and skin irritations. Furthermore, the sustainable cultivation of these plants promotes environmental stewardship by supporting biodiversity and ethical sourcing practices. This alignment with consumer values fosters a sense of trust and loyalty, as customers increasingly seek brands that prioritize health, sustainability, and community engagement.

However, the journey toward widespread adoption of MAPs is not without its challenges. Ensuring sustainable harvesting practices is crucial to prevent the depletion of natural resources and protect biodiversity. Additionally, addressing consumer skepticism regarding the efficacy of natural ingredients compared to synthetic ones is essential for building confidence in these products. Regulatory frameworks also pose significant hurdles that can complicate the marketing and formulation processes for companies looking to innovate with botanical ingredients.

To overcome these challenges, collaboration among stakeholders – including farmers, manufacturers, researchers, and regulatory bodies – is essential. By working together to establish sustainable practices, validate product efficacy through scientific research, and create clear regulatory guidelines, the industry can foster an environment conducive to growth and innovation. Education plays a pivotal role in this process: informing consumers about the benefits of MAPs and dispelling myths surrounding natural products will empower them to make informed choices.

Looking ahead, the future of MAPs in personal care products appears promising. As scientific research continues to unveil new applications for these botanical ingredients, opportunities for innovation abound. The integration of MAPs into cutting-edge formulations – ranging from skincare to haircare – not only enhances product efficacy but also aligns with consumer demands for authenticity and transparency.

In conclusion, MAPs represent a vital component of the personal care industry that bridges tradition with modernity. Their rich history, coupled with their proven benefits, positions them as essential ingredients for creating effective and sustainable beauty solutions. By embracing these plants responsibly and ethically, we can cultivate a more holistic approach to personal care that honors our connection to nature while promoting health and well-being for all. As we move forward in this dynamic landscape, it is imperative that we remain committed to sustainability, education, and innovation – ensuring that the legacy of MAPs continues to thrive in our daily lives.

References

[1]   Dini, I. and Laneri, S. (2019). Nutricosmetics: A brief overview. Phytotherapy Research: PTR, 33(12),
      3054–3063. doi: 10.1002/ptr.6494.
[2]   Rahmati E, Khoshtaghaza MH, Banakar A, Ebadi MT (2022). Decontamination technologies for
      medicinal and aromatic plants: A review. Food Science & Nutrition, 10(3):784–799. doi:10.1002/
      fsn3.2707.
[3]   Cao, B., Wei, X. C., Xu, X. R., Zhang, H. Z., Luo, C. H., Feng, B., Xu, R. C., Zhao, S. Y., Du, X. J., Han,
      L. and Zhang, D. K. (2019). Seeing the unseen of the combination of two natural resins.
      Frankincense and Myrrh: Changes in Chemical Constituents and Pharmacological Activities.
      Molecules (Basel, Switzerland), 24(17), doi: 10.3390/molecules24173076.
[4]   Morikawa T, Matsuda H, Yoshikawa M. (2017). A Review of Anti-inflammatory Terpenoids from the
      Incense Gum Resins Frankincense and Myrrh. Journal of Oleo Science, 66(8):805–814. doi:10.5650/
      jos.ess16149.
[5]   Li, S., Liu, Y., Wu, Y., Ren, L., Lu, Y., Yamaguchi, S., Lu, Q., Hu, C., Li, D. and Jiang, N. (2024). An
      outlook on platinum-based active ingredients for dermatologic and skincare applications.
      Nanomaterials (Basel, Switzerland), 14(15). doi: 10.3390/nano14151303.
[6]   Molsaghi, M., Moieni, A. and Kahrizi, D. (2014). Efficient protocol for rapid aloe vera
      micropropagation. Pharmaceutical Biology, 52(6), 735–739.
[7]   Cadar, R.-L., Amuza, A., Dumitras, D. E., Mihai, M. and Pocol, C. B. (2021). Analysing clusters of
      consumers who use medicinal and aromatic plant products. Sustainability, 13(15), 8648.
[8]   Kolekar, Y. S., Tamboli, F. A., More, H. N., Mulani, S. A. and Mali, N. P. (2021). Medicinal plants used
      in cosmetics for skin and hair care. International Journal of Pharmaceutical Chemistry and Analysis,
      8(2), 36–40.
[9]   Srivastava, S. K. and Singh, N. K. (2020). General overview of medicinal and aromatic plants:
      A. Journal of Medicinal Plants, 8(5), 91–93.
[10] Rahimi, A., Karimipour Fard, H. and Mohamadzadeh, H. (2024). A review of secondary metabolites
      and inhibitory effect of medicinal plants on some plant pathogens. Journal of Medicinal Plants and
      By-Products, doi: 10.22034/jmpb.2024.364494.1643.
[11] Alnuqaydan, A. M. (2024). The dark side of beauty: An in-depth analysis of the health hazards and
      toxicological impact of synthetic cosmetics and personal care products Frontiers in Public Health, 12,
      1439027. doi: 10.3389/fpubh.2024.1439027.
[12] Nasiri, A., Fallah, S., Sadeghpour, A. and Barani-Beiranvand, H. (2024). Essential oil profile in
      different parts of *Echinophora cinerea* (Boiss.). Agrotechniques in Industrial Crops, 4(2), 98–105. doi:
      10.22126/atic.2023.9492.1108.
[13] Nayak, M. and Ligade, V. S. (2021). History of Cosmetic in Egypt, India, and China. Journal of
      Cosmetic Science, 72(4), 432–441.
[14] Memariani, Z., Hashemimehr, M. and Mohammadi, F. (2024). Ibn Wāfid Andalusi, a medieval
      physician, pharmacist, and botanist, with a look at his most important work Al-Adwiyah Al-
      Mufradah. Journal of Medical Biography, 9677720241273608. doi: 10.1177/09677720241273608.
[15] Shah, S. A., Iqbal, W., Sheraz, M., Javed, B., Zehra, S. S., Abbas, H., Hussain, W., Sarwer, A. and
      Mashwani, Z. U. (2021). Ethnopharmacological study of medicinal plants in Bajwat Wildlife
      Sanctuary, District Sialkot, Punjab Province of Pakistan Evidence-based Complementary and
      Alternative Medicine: eCAM, 2021, 5547987. doi: 10.1155/2021/5547987.
[16] Menale, B., De Castro, O., Cascone, C. and Muoio, R. (2016). Ethnobotanical investigation on
      medicinal plants in the Vesuvio National Park (Campania, Southern Italy) Journal of
      Ethnopharmacology, 192, 320–349. doi: 10.1016/j.jep.2016.07.049.
[17] Zhao, Z., Guo, P. and Brand, E. (2018). A concise classification of bencao (materia medica) Chinese
      Medicine, 13, 18. doi: 10.1186/s13020-018-0176-y.

[18] Varpe, B. D., Kulkarni, A. A. and Mali, A. S. (2021). Aloe vera Compositions used for medicinal applications: A patent review (2013-till 2020). Recent Patents on Food, Nutrition & Agriculture, 12(2), 104–111. doi: 10.2174/2212798411999201228192616.

[19] Tanaka, M., Misawa, E., Yamauchi, K., Abe, F. and Ishizaki, C. (2015). Effects of plant sterols derived from Aloe vera gel on human dermal fibroblasts in vitro and on skin condition in Japanese women Clinical, Cosmetic and Investigational Dermatology, 8, 95–104. doi: 10.2147/ccid.s75441.

[20] Surjushe, A., Vasani, R. and Saple, D. G. (2008). Aloe vera: A short review. Indian Journal of Dermatology, 53(4), 163–166. doi: 10.4103/0019-5154.44785.

[21] Chelu, M., Musuc, A. M., Popa, M. and Calderon Moreno, J. (2023). Aloe vera-based hydrogels for wound healing: Properties and therapeutic effects. Gels (Basel, Switzerland, 9(7). doi: 10.3390/gels9070539.

[22] Lotfizadeh, V., Mollaei, S. and Hazrati, S. (2023). Biological Activities of Aloin-rich Extracts Obtained from Aloe vera (L.) Burm.f. Journal of Medicinal Plants and By-Products, 12(3), 275–281. doi: 10.22092/jmpb.2021.355897.1395.

[23] Sepp, J., Koshovyi, O., Jakstas, V., Žvikas, V., Botsula, I., Kireyev, I., Tsemenko, K., Kukhtenko, O., Kogermann, K., Heinämäki, J. and Raal, A. (2024). Phytochemical, Technological, and pharmacological study on the galenic dry extracts prepared from German Chamomile (*Matricaria chamomilla* L.) Flowers. Plants (Basel, Switzerland), 13(3), doi: 10.3390/plants13030350.

[24] Jurek, J. M. and Neymann, V. (2024). The role of the ImmunatuRNA® complex in promoting skin immunity and its regenerative abilities: Implications for antiaging skincare. Journal of Cosmetic Dermatology, 23(4), 1429–1445. doi: 10.1111/jocd.16131.

[25] Vilela, A., Ferreira, R., Nunes, F. and Correia, E. (2020). Creation and Acceptability of a Fragrance with a Characteristic Tawny Port Wine-Like Aroma. Foods (Basel, Switzerland), 9(9), doi: 10.3390/foods9091244.

[26] Kairey, L., Agnew, T., Bowles, E. J., Barkla, B. J., Wardle, J. and Lauche, R. (2023). Efficacy and safety of *Melaleuca alternifolia* (tea tree) oil for human health - A systematic review of randomized controlled trials Frontiers in Pharmacology, 14, 1116077. doi: 10.3389/fphar.2023.1116077.

[27] Moy, R. L. and Levenson, C. (2017). Sandalwood Album Oil as a Botanical Therapeutic in Dermatology. The Journal of Clinical and Aesthetic Dermatology, 10(10), 34–39.

[28] Negrean, O. R., Farcas, A. C., Nemes, S. A., Cic, D. E. and Socaci, S. A. (2024). Recent advances and insights into the bioactive properties and applications of *Rosa canina* L. and its by-products. Heliyon, 10(9), e30816. doi: 10.1016/j.heliyon.2024.e30816.

[29] Kovács, A., Péter-Héderi, D., Perei, K., Budai-Szűcs, M., Léber, A., Gácsi, A., Csányi, E. and Berkó, S. (2020). Effects of formulation excipients on skin barrier function in creams used in pediatric care. Pharmaceutics, 12(8), doi: 10.3390/pharmaceutics12080729.

[30] Eskandarzadeh, M., Janmohammadi, M., Sabaghnia, N. and Kheshtpaz, N. (2024). Exogenous application of growth-stimulating substances alleviated the effects of water-deficit stress on the spring *Camelina sativa*. Agrotechniques in Industrial Crops, 5(1), 70–80. doi: 10.22126/atic.2024.10928.1155.

[31] Fereidooni, L., Tahmasebi, Z., Kahrizi, D., Safari, H. and Arminian, A. (2024). Evaluation of drought resistance of Camelina (*Camelina sativa* L.) doubled haploid lines in the climate conditions of Kermanshah Province. Agrotechniques in Industrial Crops, 4(3), 134–146. doi: 10.22126/atic.2023.9570.1111.

[32] Teimoori, N., Ghobadi, M. and Kahrizi, D. (2023). Improving the growth characteristics and grain production of Camelina (*Camelina sativa* L.) under salinity stress by silicon foliar application. Agrotechniques in Industrial Crops, 3(1), 1–13. doi: 10.22126/atic.2023.8681.1081.

[33] Veronese, S., Picelli, A., Zoccatelli, A., Amuso, D., Amore, R., Smania, N., Frisone, A., Sbarbati, A. and Scarano, A. (2024). Morphological characterization of two dermal and hypodermal alterations in an

adult man: Surgical scar vs. stretch mark. Journal of Ultrasound, 27(4), 857–862. doi: 10.1007/s40477-024-00956-y.

[34] Ekiert, H., Klimek-Szczykutowicz, M., Rzepiela, A., Klin, P. and Szopa, A. (2022). Artemisia Species with High Biological Values as a Potential Source of Medicinal and Cosmetic Raw Materials. Molecules (Basel, Switzerland), 27(19), doi: 10.3390/molecules27196427.

[35] Nabavi, S. F., Tenore, G. C., Daglia, M., Tundis, R., Loizzo, M. R. and Nabavi, S. M. (2015). The cellular protective effects of rosmarinic acid: From bench to bedside. Current Neurovascular Research, 12(1), 98–105. doi: 10.2174/1567202612666150109113638.

[36] Khosravi, A., Pourmoslemi, S. and Moradkhani, S. (2024). Exploring the chemical landscape and biological potentials of *Rosmarinus officinalis* essential Oil: A GC analysis approach. Journal of Medicinal Plants and By-Products, 13(4), 1137–1148. doi: 10.22034/jmpb.2024.365089.1663.

[37] Tawfik, M., Rodriguez-Homs, L. G., Alexander, T., Patterson, S., Okoye, G. and Atwater, A. R. (2021). Allergen content of best-selling ethnic versus nonethnic shampoos, conditioners, and styling products. Dermatitis: Contact, atopic, occupational, drug, 32(2), 101–110. doi: 10.1097/der.0000000000000668.

[38] Mehrnia, M., Filizadeh, Y. and Naji, A. (2024). Evaluation the Effects of shade and Humic Acid on the Eco-Physiological Traits of Roselle (*Hibiscus sabdariffa* L.) under Different Irrigation Regimes. Journal of Medicinal Plants and By-Products, 13(4), 1026–1036. doi: 10.22034/jmpb.2024.364076.1625.

[39] Dadkhah, A. (2024). Effect of Seed Priming and Soil Application of Humic Acid on Growth and Yield of Roselle (*Hibiscus sabdariffa* L.). Journal of Medicinal Plants and By-Products, 13(3), 513–518. doi: 10.22034/jmpb.2024.362229.1556.

[40] Merja, A., Patel, N., Patel, M., Patnaik, S., Ahmed, A. and Maulekhi, S. (2024). Safety and efficacy of REGENDIL™ infused hair growth promoting product in adult human subject having hair fall complaints (alopecia). Journal of Cosmetic Dermatology, 23(3), 938–948. doi: 10.1111/jocd.16084.

[41] Lakshmi, T., Krishnan, V., Rajendran, R. and Madhusudhanan, N. (2015). *Azadirachta indica*: A herbal panacea in dentistry – An update. Pharmacognosy Reviews, 9(17), 41–44. doi: 10.4103/0973-7847.156337.

[42] Tiple, R. H., Jamane, S. R. and Khobragade, D. S. (2024). Antifungal Activity of Neem Leaf Extract With *Eucalyptus citriodora* Oil and *Cymbopogon martini* Oil Against *Tinea capitis*: An In-Vitro Evaluation. Cureus, 16(5), e59671. doi: 10.7759/cureus.59671.

[43] Hashem, M. M., Attia, D., Hashem, Y. A., Hendy, M. S., AbdelBasset, S., Adel, F. and Salama, M. M. (2024). Rosemary and neem: An insight into their combined anti-dandruff and anti-hair loss efficacy. Scientific Reports, 14(1), 7780. doi: 10.1038/s41598-024-57838-w.

[44] Vora, L. K., Gholap, A. D., Hatvate, N. T., Naren, P., Khan, S., Chavda, V. P., Balar, P. C., Gandhi, J. and Khatri, D. K. (2024). Essential oils for clinical aromatherapy: A comprehensive review Journal of Ethnopharmacology, 330, 118180. doi: 10.1016/j.jep.2024.118180.

[45] Agarwal, P., Sebghatollahi, Z., Kamal, M., Dhyani, A., Shrivastava, A., Singh, K. K., Sinha, M., Mahato, N., Mishra, A. K. and Baek, K. H. (2022). Citrus essential oils in aromatherapy: Therapeutic effects and mechanisms. Antioxidants (Basel, Switzerland), 11(12), doi: 10.3390/antiox11122374.

[46] Sayafi, M., Pourmoslemi, S. and Moradkhani, S. (2024). Exploring the chemical composition and bioactivity of *Lavandula angustifolia* Essential Oil: A GC analysis approach. Journal of Medicinal Plants and By-Products: -, doi: 10.22034/jmpb.2024.364861.1655.

[47] Usta, C., Tanyeri-Bayraktar, B. and Bayraktar, S. (2021). Pain control with lavender oil in premature infants: A double-blind randomized controlled study. Journal of Alternative and Complementary Medicine (New York, NY), 27(2), 136–141. doi: 10.1089/acm.2020.0327.

[48] Sarfaraz, S., Asgharzadeh, A. and Zabihi, H. (2024). Enhancing lavender germination and growth: An assessment of biofertilizer seed priming on physiological and biochemical attributes under water stress. Journal of Medicinal Plants and By-Products, 13(4), 1037–1054. doi: 10.22034/jmpb.2024.363735.1614.

[49] Donadu, M., Usai, D., Pinna, A., Porcu, T., Mazzarello, V., Fiamma, M., Marchetti, M., Cannas, S., Delogu, G., Zanetti, S. and Molicotti, P. (2018). In vitro activity of hybrid lavender essential oils against multidrug resistant strains of *Pseudomonas aeruginosa*. Journal of Infection in Developing Countries, 12(1), 9–14. doi: 10.3855/jidc.9920.

[50] Da Silva Ja, T., Kher, M. M., Soner, D., Page, T., Zhang, X., Nataraj, M. and Ma, G. (2016). Sandalwood: Basic biology, tissue culture, and genetic transformation. Planta, 243(4), 847–887. doi: 10.1007/s00425-015-2452-8.

[51] Tan, L. T., Lee, L. H., Yin, W. F., Chan, C. K., Abdul Kadir, H., Chan, K. G. and Goh, B. H. (2015). Traditional Uses, Phytochemistry, and Bioactivities of *Cananga odorata* (Ylang-Ylang) Evidence-based Complementary and Alternative Medicine: eCAM, 2015, 896314. doi: 10.1155/2015/896314.

[52] Nemati Lafmejani, Z., Jafari, A. A., Moradi, P. and Ladan Moghadam, A. (2021). Application of Chelate and Nano-Chelate Zinc Micronutrient Onmorpho-physiological Traits and Essential Oil Compounds of Peppermint (*Mentha piperita* L.). Journal of Medicinal Plants and By-Products, 10(Special), 21–28. doi: 10.22092/jmpb.2020.125992.1109.

[53] Navarra, M., Mannucci, C., Delbò, M. and Calapai, G. (2015). Citrus bergamia essential oil: From basic research to clinical application Frontiers in Pharmacology, 6, 36. doi: 10.3389/fphar.2015.00036.

[54] Wu, Y. R., Xiong, W., Dong, Y. J., Chen, X., Zhong, Y. Y., He, X. L., Wang, Y. J., Lin, Q. F., Tian, X. F. and Zhou, Q. (2024). Chemical Constituents and Pharmacological Properties of Frankincense: Implications for Anticancer Therapy. Chinese Journal of Integrative Medicine, 30(8), 759–767. doi: 10.1007/s11655-024-4105-x.

[55] Santos, F. N. D., Fonseca, L. M., Jansen-Alves, C., Crizel, R. L., Pires, J. B., Kroning, I. S., De Souza, J. F., Fajardo, A. R., Lopes, G. V., Dias, A. R. G. and Zavareze, E. D. R. (2024). Antimicrobial activity of geranium (*Pelargonium graveolens*) essential oil and its encapsulation in carioca bean starch ultrafine fibers by electrospinning. International Journal of Biological Macromolecules, 265(Pt 1), 130953. doi: 10.1016/j.ijbiomac.2024.130953.

[56] Bera, P., Kotamreddy, J. N., Samanta, T., Maiti, S. and Mitra, A. (2015). Inter-specific variation in headspace scent volatiles composition of four commercially cultivated jasmine flowers. Natural Products Research, 29(14), 1328–1335. doi: 10.1080/14786419.2014.1000319.

[57] Yang, H. J., Kim, K. Y., Kang, P., Lee, H. S. and Seol, G. H. (2014). Effects of *Salvia sclarea* on chronic immobilization stress induced endothelial dysfunction in rats BMC Complementary and Alternative Medicine, 14, 396. doi: 10.1186/1472-6882-14-396.

[58] Zulkipli, I. N., David, S. R., Rajabalaya, R. and Idris, A. (2015). Medicinal Plants: A Potential Source of Compounds for Targeting Cell Division Drug Target Insights, 9, 9–19. doi: 10.4137/dti.s24946.

[59] Šojić, B., Milošević, S., Savanović, D., Zeković, Z., Tomović, V. and Pavlić, B. (2023). Isolation, bioactive potential, and application of essential oils and terpenoid-rich extracts as effective antioxidant and antimicrobial agents in meat and meat products. Molecules (Basel, Switzerland), 28(5), doi: 10.3390/molecules28052293.

[60] Mansoor, K., Aburjai, T., Al-Mamoori, F. and Schmidt, M. (2023). Plants with cosmetic uses. Phytotherapy Research: PTR, 37(12), 5755–5768. doi: 10.1002/ptr.8019.

[61] Jiang, L., Chen, Y., Wang, X., Guo, W., Bi, Y., Zhang, C., Wang, J. and Li, M. (2022). New insights explain that organic agriculture as sustainable agriculture enhances the sustainable development of medicinal plants Frontiers in Plant Science, 13, 959810. doi: 10.3389/fpls.2022.959810.

[62] Chrysargyris, A., Petrovic, J. D., Tomou, E. M., Kyriakou, K., Xylia, P., Kotsoni, A., Gkretsi, V., Miltiadous, P., Skaltsa, H., Soković, M. D. and Tzortzakis, N. (2024). Phytochemical profiles and biological activities of plant extracts from aromatic plants cultivated in Cyprus. Biology, 13(1), doi: 10.3390/biology13010045.

[63] Awasthi, P. and Adhikari, S. P. (2024). Blooming markets: The economic dynamics of the Nepal rose trade. Heliyon, 10(15), e35585. doi: 10.1016/j.heliyon.2024.e35585.

[64] Hatami, S., Hatami Badrbani, M. and Kahrizi, D. (2024). Data mining approach in the agricultural industry, medicinal plants (case study): A review. Journal of Medicinal Plants and By-Products, 13(2), 247–256.

[65] Mishra, B., Gyawali, B. R., Paudel, K. P., Poudyal, N. C., Simon, M. F., Dasgupta, S. and Antonious, G. (2018). Adoption of Sustainable Agriculture Practices among Farmers in Kentucky, USA. Environmental Management, 62(6), 1060–1072. doi: 10.1007/s00267-018-1109-3.

[66] Pashaei, M., Fayçal, Z., Kahrizi, D. and Ercisli, S. (2024). Medicinal plants and natural substances for poultry health: A review. The Journal of Poultry Sciences and Avian Diseases, 2(2), 36–49.

[67] Dashtian, K., Kamalabadi, M., Ghoorchian, A., Ganjali, M. R. and Rahimi-Nasrabadi, M. (2024). Integrated supercritical fluid extraction of essential oils Journal of Chromatography A, 1733, 465240. doi: 10.1016/j.chroma.2024.465240.

[68] Liu, J. K. (2022). Natural products in cosmetics. Nat Prod Bioprospecting, 12(1), 40. doi: 10.1007/s13659-022-00363-y.

[69] Brendler, T., Brinckmann, J. A. and Schippmann, U. (2018). Sustainable supply, a foundation for natural product development: The case of Indian frankincense (*Boswellia serrata* Roxb. ex Colebr.) Journal of Ethnopharmacology, 225, 279–286. doi: 10.1016/j.jep.2018.07.017.

[70] Azhar, B., Saadun, N., Prideaux, M. and Lindenmayer, D. B. (2017). The global palm oil sector must change to save biodiversity and improve food security in the tropics. Journal of Environmental Management, 203(Pt 1), 457–466. doi: 10.1016/j.jenvman.2017.08.021.

[71] Fajinmi, O. O., Olarewaju, O. O. and Van Staden, J. (2023). Propagation of medicinal plants for sustainable livelihoods, economic development, and biodiversity conservation in South Africa. Plants (Basel, Switzerland), 12(5), doi: 10.3390/plants12051174.

[72] Sang, S. H., Akowuah, G. A., Liew, K. B., Lee, S. K., Keng, J. W., Lee, S. K., Yon, J. A., Tan, C. S. and Chew, Y. L. (2023). Natural alternatives from your garden for hair care: Revisiting the benefits of tropical herbs. Heliyon, 9(11), e21876. doi: 10.1016/j.heliyon.2023.e21876.

[73] Chen, S. W., Lee, K. Y. and Hsieh, C. M. (2021). Determinants of Consumers' Trust in Biotech Brands and Purchase Intentions towards the Cord Blood Products. International Journal of Environmental Research and Public Health, 18(21), doi: 10.3390/ijerph182111574.

[74] AR, B. (2015). Herbal medicinal products versus botanical-food supplements in the European market: State of art and perspectives. Natural Products Communications, 10(1), 125–131.

[75] Bilia, A. R. and Costa, M. D. C. (2021). Medicinal plants and their preparations in the European market: Why has the harmonization failed? The cases of St. John's wort, valerian, ginkgo, ginseng, and green tea Phytomedicine International Journal of Phytotherapy & Phytopharmacology, 81, 153421. doi: 10.1016/j.phymed.2020.153421.

[76] Sato, K., Kodama, K. and Sengoku, S. (2023). Optimizing the Relationship between Regulation and Innovation in Dietary Supplements: A Case Study of Food with Function Claims in Japan. Nutrients, 15(2), doi: 10.3390/nu15020476.

[77] Gabrielson, J. P. and WFt, W. (2015). Technical decision-making with higher order structure data: Starting a new dialogue. Journal of Pharmaceutical Sciences, 104(4), 1240–1245. doi: 10.1002/jps.24393.

[78] Tong, Y., Lau, Y. W. and Binti Ngalim, S. M. (2024). Do pilot zones for green finance reform and innovation avoid ESG greenwashing? Evidence from China. Heliyon, 10(13), e33710. doi: 10.1016/j.heliyon.2024.e33710.

[79] Bornkessel, S., Smetana, S. and Heinz, V. (2019). Nutritional sustainability inside-marketing sustainability as an inherent ingredient Frontiers in Nutrition, 6, 84. doi: 10.3389/fnut.2019.00084.

[80] Dini, I. (2024). "Edible Beauty": The evolution of environmentally friendly cosmetics and packaging. Antioxidants (Basel, Switzerland), 13(6), doi: 10.3390/antiox13060742.

Dilara Ülger Özbek* and Elyor Berdimurodov

Chapter 20
Genotoxic effects of medicinal and aromatic plants

Abstract: Since the beginning of human history, medicinal and aromatic plants, which have been recognized as a source of healing in nature, have been used in various areas such as food, medicine, clothing, cosmetics, and spices. Thanks to the bioactive molecules they contain, medicinal and aromatic plants are effective in the treatment of many diseases and complaints. These metabolites are utilized as herbal drug targets. Many diseases can be treated by leveraging the antioxidant, anticancer, antifungal, antimicrobial, and anti-inflammatory properties of their bioactive molecules. Today, the increasing costs and side effects of synthetic drugs have encouraged people to cultivate and use medicinal and aromatic plants. Recent research, however, has shown that some plant extracts may have genotoxic effects, especially when used in high quantities or under specific experimental conditions. Additionally, many folkloric plants have been found to possess DNA-damaging effects. Hence, assessing the safety profile of medicinal herbs before their approval for use is essential. However, given the fact that the chemical composition of these plants is not well understood, their uncontrolled use and the secondary metabolites synthesized due to the environmental conditions in which they grow can also be harmful. These harmful effects include poisonous, genotoxic, cytotoxic, and mutagenic properties, which may lead to negative outcomes.

A genotoxin is a biochemical or agent that can cause DNA or chromosomal damage. Such damage in a germ cell has the potential to cause a heritable altered trait. DNA damage in a somatic cell may result in a somatic mutation, which may lead to malignant transformation (cancer). Genotoxicity is the capacity to induce chromosomal abnormalities, DNA strand breakage, or mutation. Genotoxicity is an especially insidious toxicity that may result in carcinoma development years after exposure; it can arise from multiple compounds, with or without metabolic activation. In this section, the genotoxic effects of medicinal aromatic plants will be discussed.

Keywords: genotoxicity, medicinal and aromatic plants, genotoxic plants

---

*Corresponding author: Dr. Dilara Ülger Özbek**, Advanced Technology Research and Application Centre, Sivas Cumhuriyet University, Sivas, Turkey, e mail: dilaraulger@cumhuriyet.edu.tr, https://orcid.org/0000-0002-6834-020X
Elyor Berdimurodov, Chemical and Materials Engineering, New Uzbekistan University, 54 Mustaqillik Ave, Tashkent 100007, Uzbekistan; Faculty of Chemistry, National University of Uzbekistan, Tashkent 100034, Uzbekistan

https://doi.org/10.1515/9783111469713-020

20.1 Introduction

There has been a long-standing link between humans and plants. Plants provide not only food but also non-food industrial items. In addition to being utilized in medicine and as pharmaceuticals by the general public, plants are also employed in landscaping, paint manufacturing, insecticides, food, spices, drinks, fragrances, and cosmetics. The widespread cultivation of these crops is usually done to produce niche goods or high-quality chemicals [1]. Another such category of plants is medicinal plants, which are an amazing treasure trove of chemical compounds with a wide range of uses in human health and wellness. The World Health Organization (WHO) defines a medicinal plant as a plant of any kind that, whether whole or in one or more of its organs, either contains compounds that can be used to make beneficial medication or has medicinal properties that can be used directly for therapeutic purposes or have an impact on the health of humans [2]. These plants, either wholly or in parts, including roots, stems, leaves, stem bark, fruits, or seeds, can be used to manage or treat diseases. Additionally, they can be applied in a variety of ways, including teas made from various plant species, tinctures, food additives, and nasal or oral applications [3].

Over time, health issues have also given rise to a variety of plant applications. Together with contemporary medical understanding, the use of plants has been described using techniques known as complementary and traditional medicine. One of the methods, called traditional and complementary medicine, is the phytotherapy method [4].

The Greek terms "Phyton" (plant) and "Therapeia" (treatment) are combined to produce the phrase phytotherapy. It is well recognized that utilizing plant parts such as leaves, seeds, blossoms, or roots can help prevent and treat illnesses [5]. A remarkable number of contemporary medications have also been identified from these natural plant species, in part due to their usage in traditional medicine. Phytochemicals, biologically active compounds, or active ingredients are terms frequently used to describe these chemical components or therapeutically active non-nutrient compounds found in these plants [5]. Many phytochemicals are essential medications that are being utilized all over the world to treat a wide range of dangerous illnesses. Since medicinal plants contain a wide range of phytochemicals, just one of them can treat a number of disorders and diseases. Traditional medicinal plants and their components used in alternative medicine have been shown to heal and have beneficial effects on a variety of diseases, including asthma, fever, intestinal problems, urinary tract infections, digestive system issues, and skin infections [6].

The therapeutic use of medicinal plants is significantly influenced by a nation's degree of development. Despite the fact that 80% of people in poorer nations utilize plants for medicinal purposes, this percentage drops dramatically in developed nations like Germany (40–50%), the USA (42%), Australia (48%), and France (49%). In underdeveloped nations, the rate is 95% [7]. The usage of medicinal plants is expanding rapidly across the world due to an increase in demand for herbal medications, alternative medicines, and secondary metabolites [8].

The origins of medicinal plants used for therapy trace back to antiquity. The reasons why medicinal plants have been preferred more in recent years can be listed as the increasing ineffectiveness of modern synthetic drugs, bacterial resistance, rising costs, and side effects [9]. These herbs are combined with the oldest traditional remedies, such as those from China, Egypt, and Ayurveda [10]. Over time, medicinal plants have gained prominence in the health-care system and as an alternative source of health-beneficial medications.

Medicinal plants have therapeutic capacities and are used in the treatment of many diseases and complaints, providing raw materials for numerous traditional and modern medicines [11]. Bioactive molecules obtained from medicinal aromatic plants are the primary metabolites used in treatment. With their therapeutic properties, they offer many benefits, such as preventing and treating diseases, improving symptoms, minimizing chemotherapy side effects, and regulating the physical and mental state of the body [12].

It has been estimated that there are between 50,000 and 1,000,000 plant species across the globe, and that 4,000 herbal medicines have been made from approximately 20,000 species and are administered to patients. Of the roughly 11,000 plant species that may be produced in Turkey, only 500 have been reported to be suitable for therapeutic purposes [13, 14]. According to the WHO, 21,000 plant species can be used to make medicines [15].

Turkey has a rich flora and is home to many plant species. Approximately 500 plant taxa are used in alternative medicine [16]. These plants are utilized both fresh and dried. In Turkey, medicinal and aromatic plants are mostly harvested in the Aegean, Mediterranean, Southeastern Anatolia, Marmara, and Eastern Black Sea areas. Despite its great plant variety, Turkey has been unable to achieve significant levels of medicinal and aromatic plant exports. Anise, bay leaf, poppy, and thyme are the most exported plants, with black pepper ranking highest in imports [17]. Plants frequently used in the treatment of some diseases in Turkey are listed in Table 20.1.

Table 20.1: Medicinal and aromatic plants commonly used in Turkey [18].

Disease/ disorder	Medicinal plants
Kidney disease	*Helichrysum arenarium, Equisetum, Elymus repens*
Dyspepsia	*Pimpinella anisum, Anethum graveolens, Alpinia officinarum, Elettaria cardamomum, Cuminum cyminum, Anthemis, Foeniculum vulgare, Pimenta racemosa, Zingiber officinale*
Hemorrhoids	*Achillea millefolium, Rosa canina, Thuja, Sambucus ebulus* L., *Zingiber officinale*
Constipation	*Linum usitatissimum, Foeniculum vulgare, Cassia*

Table 20.1 (continued)

Disease/ disorder	Medicinal plants
Heart diseases	*Crataegus Orientalis, Viscum album*
Cancer diseases	*Urtica, Capsicum annuum, Viscum album*
Liver disease	*Cynara scolymus, Cichorium endivia, Lycopodium clavatum, Curcuma longa*
Menopause	*Salvia officinalis, Achillea millefolium, Syzygium aromaticum, Anthemis, Cinnamomum verum*
Stomach bleeding	*Achillea millefolium, Rosa canina, Rhus*
Stomach disorders	*Acorus calamus, Mentha, Zingiber officinale*
Prostate	*Acorus calamus, Urtica, Camellia sinensis, Zingiber officinale*
Rheumatism	*Pimpinella anisum, Equisetum, Salvia rosmarinus, Syzygium aromaticum, Thymus, Lavandula, Melissa officinalis, Anthemis*
Gall bladder	*Helichrysum arenarium, Achillea millefolium, Cichorium endivia, Artemisia absinthium, Curcuma longa*
Cold, cough	*Juniperus sabina, Malva parviflora, Tilia, Echinacea, Eucalyptus*
Stress, depression	*Hypericum perforatum, Humulus lupulus, Anthemis, Matricaria, Foeniculum vulgare*
Sleep disturbance	*Pimpinella anisum, Primula veris, Valeriana officinalis, Foeniculum vulgare, Humulus lupulus*
Exhaustion	*Salvia officinalis, Salvia rosmarinus, Thymus, Rosa canina, Elettaria*
High cholesterol	*Momordica charantia, Camellia sinensis, Thymus, Rosa canina*
High glucose	*Momordica charantia, Prunus mahaleb, Cinnamomum verum*

20.2 Physiology of medicinal and aromatic plants

Plant physiology is the branch of science that studies the physical and biochemical functions of plants, their systems, and the effects of the environment and genetics on plant life span. Metabolic functions can vary from the plant's youth to adulthood. This variability can affect the plant's metabolite levels, synthesis, and degradation. Metabolism is the totality of the many chemical processes and reactions that take place in each plant cell at any one time. The primary and secondary products of metabolism are called metabolites [19]. The categorization of plant metabolites is shown in Figure 20.1.

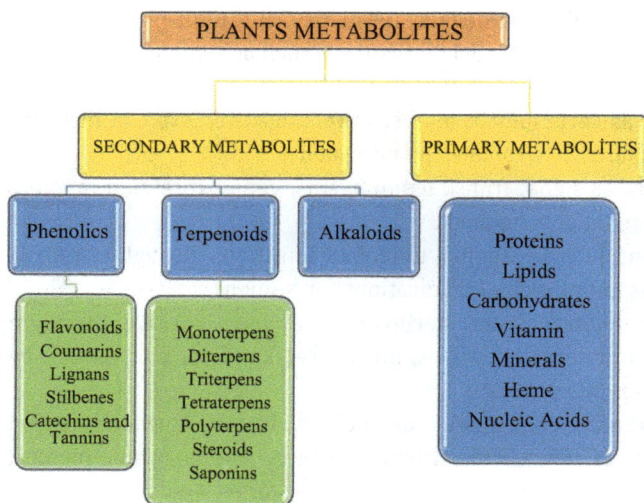

Figure 20.1: Classification of plant metabolites.

20.2.1 Primary metabolites

Primary metabolites of medicinal and aromatic plants are organic compounds produced by plants to maintain their basic life functions. These metabolites play an important role in vital processes such as plant growth, development, and energy production. Primary metabolites are generally common in all living organisms and are synthesized under the control of the genetic code. The main categories, according to their functions, are amino acids, lipids, carbohydrates, heme, chlorophyll, vitamins, minerals, nucleic acids, and organic acids [20].

20.2.2 Secondary metabolites

Although medicinal and aromatic plant secondary products have been defined over the years as invisible chemicals with vital biochemical functions that play a role in the construction and maintenance of plant cells, recent research has shown that these metabolites play an important role in the ecophysiology of plants. Secondary metabolites are primarily formed under various stress circumstances, resulting in a diverse set of molecules that are not found in all species [21]. They are employed for certain objectives, such as defense or to satisfy specific ecological demands. Humans utilize the secondary metabolites generated in medicinal plants for a variety of health benefits. Here is a list of secondary metabolite characteristics [22]:

- Serve as both a defense against pathogens, herbivory, and inter-plant competition, as well as an attractant for beneficial organisms such as symbionts and pollinators.
- Abiotic stressors, such as variations in temperature, moisture, levels of light, UV exposure, and mineral nutrients, can also be mitigated.
- At the level of the cells, as signal transduction agents, gene expression modulators, and regulators of plant development.
- If not highly toxic, plant defenses against microbial infections through cytotoxicity may be beneficial as antimicrobial medications for humans.
- Neurotoxin activity in defense against herbivores may have positive effects on people, such as antidepressants, sedatives, muscle relaxants, or anesthetics, by acting on the central nervous system.
- They may work by imitating endogenous metabolites, ligands, hormones, signal transduction molecules, or neurotransmitters, and thus have favorable medical effects on humans due to similarities in their possible target areas (e.g., central nervous system, endocrine system, etc.).

The primary distinction that separates medicinal plants from others is their physiological action. The plant species' features influence the synthesis, breakdown, and accumulation of secondary metabolites [23].

With their diverse variety of pharmacological effects, including anti-inflammatory, antibacterial, anticancer, antioxidant, and antifungal qualities, medicinal plants are clearly important for human health. These bioactive secondary metabolites play a crucial role in plants' ability to combat a variety of biotic and abiotic stressors, including serving as chemical barriers against predators, pathogens, and allelopathic agents [24]. Oxidation, reduction, substitution, and condensation reactions are the primary chemical processes that produce bioactive metabolites in medicinal plants. A wide range of these phytochemicals, including phenolic compounds, flavonoids, alkaloids, tannins, and terpenes, are synthesized and accumulated by medicinal plants and are used therapeutically or as substrates for the production of beneficial medications [25]. Medicinal plants' secondary metabolites can be broadly categorized into three groups: alkaloids, terpenoids, and phenolic compounds. More than 200,000 known secondary metabolites derived from plants are classified as nitrogen or nonnitrogen substances based on their chemical structure, content, function, and synthesis. Alkaloids are nitrogen-containing secondary metabolites, whereas nonnitrogen secondary metabolites include terpenoids, steroids, and phenolics generated via the malonic, shikimic, and mevalonic acid biosynthetic pathways. The biosynthesis of phenolic compounds (8,000) occurs via the shikimic acid or malonate/acetate route [26]. The alkaloids (12,000) contain one or more nitrogen atoms derived from amino acids.

20.2.2.1 Phenolic compounds

Phenolic compounds are low molecular mass compounds that have at least one hydroxyl (OH) group in the aromatic chain as an additive. These substances can have a single ring, a large complex, or multiple fused rings. The R group in Figure 20.2 indicates the different types of phenolic chemicals. The malonate pathway and shikimic acid pathway are the production pathways of phenolics. The elimination of ammonia and the transformation of phenylalanine into cinnamic acid are the most crucial steps in this process [27]. Phenolic chemicals play vital roles in a plant's defensive systems, growth, development, and reproduction. Plant phenolics oxidize and turn brown when they come into contact with air. Their ability to combine with proteins to form complexes and generate intermediate products that block certain enzymes is another important characteristic [28].

Polyphenols are phenols that include a high number of phenolic rings and OH groups. These chemical compounds possess antiviral, antibacterial, anti-inflammatory, sedative, antioxidant, and wound-healing properties. According to a study on COVID-19, a recent infectious and incurable illness, turmeric's curcumin phenolic plays a powerful role in combating the virus by activating the protease enzyme [29].

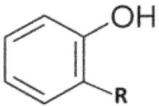

Figure 20.2: Phenolic compound.

Phenolic compounds are divided into groups among themselves, and they are listed below.

20.2.2.1.1 Flavonoids

Flavonoids include about 6,000 species and are the largest class of phenolic chemicals. They are low molecular weight compounds with a chemical structure that includes one oxygenated heterocyclic chain and two aromatic rings (C6-C3-C6). They are hydrophilic and abundantly distributed in plants. Flavonoids significantly influence the color, flavor, odor, and nutritional content of plant foods. Due to their beneficial effects on human health, they are the subject of extensive scientific investigation. Flavonoids are also categorized into six groups, as shown in Figure 20.3. The importance of flavonoids for human health is outlined below [30]:

1. **Antioxidant defense:** Flavonoids and anthocyanins are useful in avoiding oxidative stress-related chronic illnesses.
2. **Support for the cardiovascular system:** Inhibits the oxidation of LDL cholesterol and improves blood vessel flexibility.

3. **Protection of the nerves:** Anthocyanins have the potential to delay the progression of neurodegenerative disorders, including Parkinson's and Alzheimer's.
4. **Potential for preventing cancer:** Through the suppression of cellular mutations, they may prevent the formation of tumors.

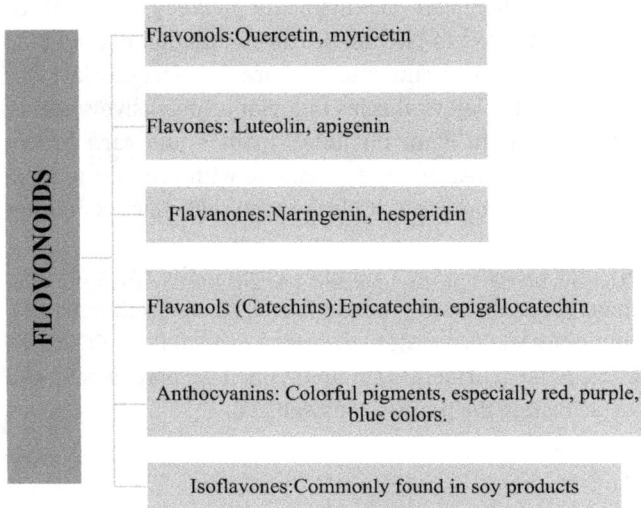

Figure 20.3: Classification of flavonoids.

20.2.2.1.2 Coumarins

Simple phenolic compounds found in nature are called coumarins. They have a benzopyranone ring (a mix of pyrone and benzene rings) in their fundamental chemical structure. They are categorized based on the various functional groups (R groups) that have been added to the benzopyran ring, which is their fundamental structure. The primary coumarin types are [31]:

Easy coumarins: Substances like umbelliferone and coumarin.

Furanocoumarins: Psoralen, bergapten, and a furan ring are all parts of their structure.

Pyranocoumarins: They have a pyrano ring called selina in their structure.

Dicoumarols: Dicoumarol is an anticoagulant substance that inhibits blood coagulation.

Plant metabolism and interactions with the environment are significantly influenced by coumarins. By preventing other plants from growing near certain plants, shielding them from UV radiation, bacteria, and herbivores, and providing plants with their scent, they produce allelopathic effects [31].

The chemicals known as coumarins are of interest because of their potential medical uses. Overconsumption of coumarin can cause toxicity to the liver. Consequently, it is advised to consume coumarin-containing foods (e.g., cinnamon) in moderation. The following is a list of the consequences for human health.

1. **Effect of anticoagulants**: The active components of medications frequently used to prevent blood clotting are coumarin derivatives, such as warfarin.
2. **The activity of antimicrobials:** They may work well against fungi, viruses, and bacteria.
3. **Inflammatory reduction:** Through the reduction of inflammatory processes, they assist the immune system.
4. **Properties of antioxidants**: Neutralize free radicals and lower oxidative stress.
5. **Capability to prevent cancer:** Certain coumarin derivatives have the ability to stop cancer cells from growing.
6. **Impact on nervous system function**: It has calming and anticonvulsant qualities that support the nervous system [32].

20.2.2.1.3 Lignans

Lignans are phenolic secondary metabolites derived from phenylpropanoid metabolism. The molecular structure is created by the bonding of two phenylpropanoids (C6-C3). They are present in glycoside form and are transformed into free form during digestion. Lignans serve several functions in plants, including defense, physiological control, and adaptability to external challenges. In the human colon, certain lignans are converted into phytoestrogens like enterolactone and enterodiol. Because of their estrogenic, anti-inflammatory, anti-carcinogenic, and antioxidant properties, they are extremely important for human health. Additionally, lignans are categorized as follows:

Sesamin and sesamolin: Sesame seeds contain them.

Secoisolariciresinol: The most prevalent lignan found in flaxseed.

Matairesinol: The forerunner of other lignans.

Pinoresinol and lariciresinol: Numerous plant sources include them [33].

20.2.2.1.4 Stilbenes

Stilbenes are phenolic chemicals that are naturally present in plants and are generated from phenylpropanoids. A basic skeleton joins two aromatic rings with a vinyl group (–CH=CH–) to form their chemical structure. Stilbene varieties differ depending on how this fundamental structure is modified. Highlighted stilbenes are:

– **Resveratrol:** It has been studied the most. It may be found in plants like Japanese knotweed, red wine, and grape skins. The biological activity of *trans*-resveratrol is higher than that of its *cis*-isomers.
– **Pinosylvin:** Pine tree resin contains it and it possesses antibacterial and antifungal properties.

- **Pterostilbene**: A resveratrol derivative that can be found in blueberries and other fruits. It increases bioavailability [34].

Stilbenes are often found either glycosylated or free. They are involved in plant defense, responding to stress, and metabolism. Their beneficial impacts on human health, particularly their anti-inflammatory and antioxidant qualities, draw significant attention. Here is a list of how they affect human health:

- **Antioxidant activity:** By scavenging free radicals, stilbenes reduce oxidative stress in cells. This characteristic is associated with antiaging benefits.
- **The cardiovascular system**: Resveratrol reduces LDL cholesterol oxidation and improves blood vessel flexibility. It helps control blood pressure and lowers the risk of cardiovascular disease.
- **Anticancer properties:** They regulate the cell cycle and induce apoptosis. These properties have been demonstrated to be very beneficial in treating breast, colon, and prostate cancer.
- **Anti-inflammatory effect:** Enzymes involved in the regulation of inflammation are inhibited by stilbenes.
- **Neurologically protective**: It protects against neurodegenerative illnesses, including Alzheimer's and Parkinson's diseases.
- **Metabolic impacts:** Pterostilbene may lower the risk of metabolic syndrome by controlling blood sugar and cholesterol levels. [35]

20.2.2.1.5 Catechins and tannins

Catechins are a type of flavonoid that are primarily formed through dimerization into procyanidins or polymerization into oligomeric procyanidins. Flavonoids have a fundamental structure of C6–C3–C6, whereas catechins consist of two aromatic rings and a heterocyclic ring containing one oxygen atom. Catechins include stereoisomers such as catechin and epicatechin, which may vary in their biological functions. For example, epigallocatechin gallate has a more complex structure, including a gallic acid group, and is considered more significant in terms of health benefits [36].

Tannins, better known as tannic acid, are hydrophilic phenolic compounds generated by the condensation of flavan derivatives or the polymerization of quinone units. Aside from lignans, they are another phenolic component that aids in plant defense. Condensed tannins, which are mostly present in woody plants, are generated by the bonding of flavonoid units. Condensed tannins react with powerful acids and hydrolyze to anthocyanidins, resulting in "proanthocyanidins" [37]. The grouping of tannins is shown in Figure 20.4.

Tannins have incredible medicinal effects. A wide range of conditions can be treated by them, including microbial infections, snake bites, tonsillitis, cardiovascular diseases, burns, injuries, gastrointestinal issues, heavy metal exposure, inflammation, inhibition of platelet aggregation, and therapy for tumors [38].

Tannins

Hydrolyzable tannins
- Gallotannins
- Ellagitannins

Condensed tannins
- Prproanthocyanidin B1
- Proanthocyanidin B2

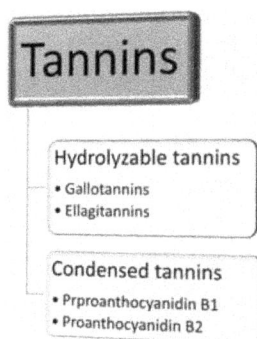

Figure 20.4: Grouping of tannins.

20.2.2.2 Alkaloids

Alkaloids, which are essentially amino acid derivatives, are bioactive secondary metabolites that include at least one nitrogen atom in their heterocyclic structure. Over 12,000 natural alkaloids have been identified in plants, and their fundamental structures include a basic amine group. In general, alkaloids are toxic compounds, but when used in certain doses, they exhibit therapeutic properties. They serve as analgesics (pain relievers), antibiotics, antimalarials, and cardiovascular regulators in humans, and they have a profound effect on the neurological system. Alkaloid chemicals are classified into distinct groups based on their precursors and heterocyclic content. Figure 20.5 illustrates these groupings [39].

20.2.2.2.1 True alkaloids
True alkaloids, formed from amino acid precursors and including a heterocyclic nitrogen, include atropine, nicotine, and morphine. Aside from the heterocyclic nitrogen ring, they may also include terpene or peptide residues [40].

20.2.2.2.2 Proto-alkaloids
These are other alkaloid types that are amino acid derivatives but do not contain a heterocyclic nitrogen. These include mescaline, adrenaline, and ephedrine [40].

20.2.2.2.3 Pseudo-alkaloids
These alkaloids originate from terpene and steroid sources and include caffeine, theobromine, theacrine, and theophylline [40].

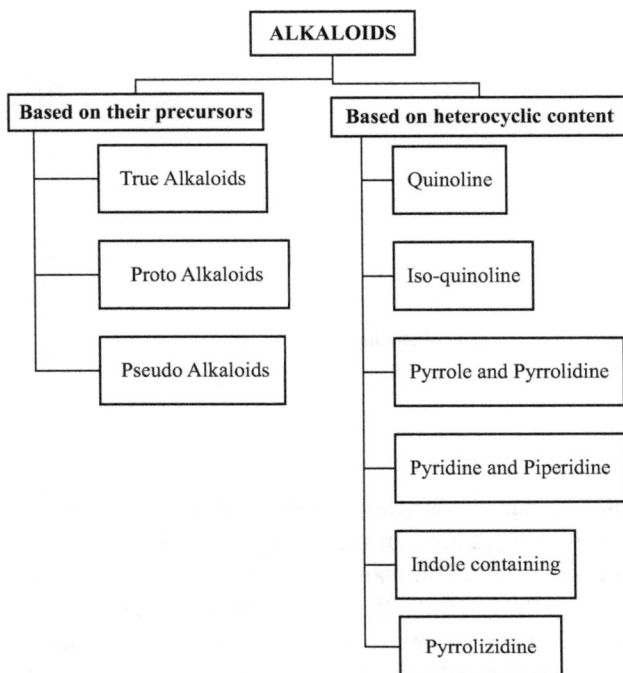

Figure 20.5: The categorization of alkaloids.

20.2.2.3 Terpenoids

Terpenoids are a class of fundamental secondary metabolites recognized for their volatile qualities, which contribute to the pleasant scents and biological activities of medicinal and fragrant plants. Terpenoids, which are made up of isoprene units (C_5H_8), can contain functional groups including alcohol, aldehyde, ketone, and acid. They are produced by the mevalonate and methylerythritol phosphate pathways. In plants, these compounds function as hormones (gibberellins, abscisic acid), pigments (phytol, carotenoids), and cell wall structural components (phytosterol). Terpenes exhibit cytotoxic activity against bacteria, fungi, insects, and diseases [41]. Terpenes are classified based on the number of isoprene units (C5). The classification of terpenes is shown in Figure 20.6.

20.2.2.3.1 Monoterpenes

Initially separated from turpentine, they are terpenoids composed of two isoprene units (C10). These chemicals are often extracted as essential oils in angiosperm species and as resins in gymnosperm species. Pinenes are among the most prevalent monoterpenes produced by trees, including pine, spruce, and fir [41].

20.2.2.3.2 Diterpenes

Diterpenes are terpenoids generated by combining two isoprene units with a 20-carbon backbone. They are made from intermediate products such as phytyl diphosphate and geranylgeranyl diphosphate. Taxol (paclitaxel), derived from the yew tree (*Taxus* sp.), is a potent anti-tumor drug used in cancer treatments. Another example is forskolin, which is derived from Coleus forskohlii, a plant in the coconut-family with pharmacological properties, including blood pressure management [42].

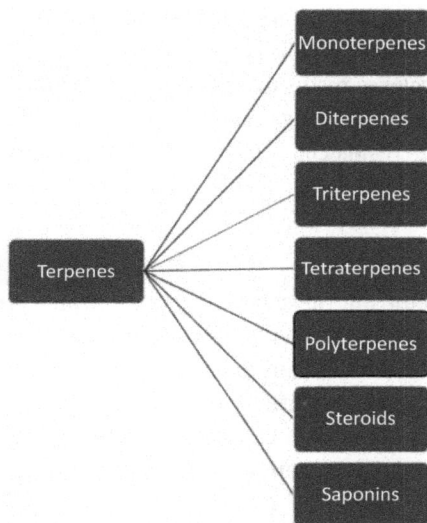

Figure 20.6: Types of terpenes.

20.2.2.3.3 Triterpenes

Triterpenes with 30 carbon atoms are formed by the head-to-head coupling of two 15-carbon chains, resulting in three isoprene segments joined in a head-to-tail configuration. Brassinosteroids, phytosterol membrane components, several phytoalexins, various poisons, oleanolic acid, and surface wax factors are all examples of terpenoids [43].

20.2.2.3.4 Tetraterpenes

Tetraterpenes are lipophilic molecules made up of eight isoprene units, each containing 40 carbon atoms. They function as catalysts for the synthesis of pigments, which are often yellow, orange, red, or green. Among tetraterpenes, carotenoids are the most widely known. These coloring agents serve two purposes in plants: they aid photosynthesis and color flowers and fruits [43].

20.2.2.3.5 Polyterpenes

Polyterpenes are high molecular weight compounds containing 100 or more isoprene units. They are produced by mixing the chemicals isopentenyl diphosphate and dimethylallyl diphosphate. Important types and examples are listed below:

- **Natural rubber (*cis*-1,4-polyisoprene):** Rubber trees and other plants produce this substance. It is employed in the automobile, shoe, and medical industries because of its elastic structure.
- **Gutta-percha (*trans*-1,4-polyisoprene):** Produced from plants such as the *Palaquium* species, it serves as an electrical insulator as well as a filler material in dentistry.
- **Resins and waxes:** Pine and other plants release resins that include polyterpenic chemicals. These chemicals help preserve and defend against microbes.
- **Polyterpenoid resins:** Gums and aromatic resins come from a variety of plant species [44].

20.2.2.3.6 Steroids

Lipid derivatives known as steroids are derived from the sterane skeleton, which is a four-ring structure composed of one cyclopentane ring and three cyclohexane rings Figure 20.7. They perform specific functions as hormones, coenzymes, and provitamins. Sterols are plant steroids that have undergone C3 hydroxylation. Cardenolides and bufadienolides are two subgroups of plant steroids, along with cardiac glycosides, ecdysterols, sterols, pregnane derivatives, steroid saponins, and steroid alkaloids [45].

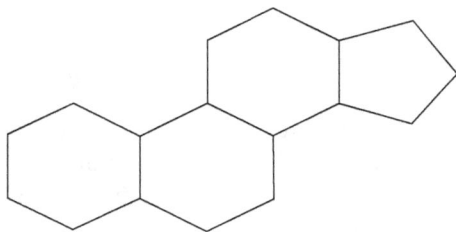

Figure 20.7: A basic steroid ring.

20.2.2.3.7 Saponins

Saponins are composed of a triterpenoid skeleton known as a lipophilic aglycone (sapogenin) and a glycoside structure with hydrophilic sugar groups. They received their name from the froth that forms when shaken with water. This trait is attributable to the properties of surface-active agents (surfactants). Excessive and high-dose saponin ingestion lowers dietary protein digestion, causes gastrointestinal discomfort, and can degrade red blood cells [46].

20.3 Genotoxicity

A genotoxin is a substance or agent capable of causing DNA or chromosomal damage. Mutations in DNA may arise spontaneously or as a result of chemical action by agents of either endogenous or exogenous origin. Such damage to a germ cell can produce a heritable characteristic (germline mutation). DNA damage in a somatic cell can result in somatic changes, which can progress to a malignant mutation (cancer) [47]. The term "genotoxicity" describes the theoretically deleterious effects on genetic material caused by the induction of irreversible, transmissible alterations in the quantity or structure of the genetic material [48]. Genotoxicity, broadly defined as "damage to the genome," is a distinct and important type of toxicity, as specific genotoxic events are considered hallmarks of cancer [49]. As a result, a compound's genotoxicity is frequently investigated in order to better understand its carcinogenic mechanism. The search for compounds with genotoxic properties is critical for establishing whether a novel molecule is carcinogenic and/or mutagenic. This approach makes at least a significant contribution to the identification of dangerous substances.

20.4 Determination tests of medicinal and aromatic plants' genotoxicity

There are various protocols and analysis methods known as "genotoxic tests" to assess the safety of any suspected medicinal aromatic plants or products. Standard tests and methods have been established for genotoxicity testing, which aim to gather information regarding all forms of mutations and DNA breaks, including gene mutations, structural chromosome disorders (clastogenicity), and numerical chromosomal abnormalities [50]. These include the in vivo or in vitro comet assay (CA), micronucleus test (MN), Ames test, Salmonella/microsome analysis, bacterial reverse mutation analysis (BRM), Allium cepa test (ACT), chromosomal aberration analysis (ChA), and somatic mutation and recombination test (SMART). The Vitotox test (VA), γH2AX intracellular Western test (WA), mouse lymphoma tk test (MLtk), and lysogenic induction test (LI) are also utilized as test procedures [51].

The Ames test is an effective method for assessing the genotoxicity of complex herbal mixtures and predicting the genotoxic effects of herbal usage in humans. The responses of different Salmonella strains used in this test can help identify the types of genotoxic compounds prevalent in plants [52].

The selection of the test to be used in genotoxicological studies is important. Genotoxicity should be evaluated using several methods together because no single test can determine total genotoxicity [53]. For this reason, when deciding on test selection, a holistic decision is made involving plant extracts and solvents. Table 20.2 presents data on various genotoxicity tests conducted on extracts from certain plants, along with their corresponding results [54].

Table 20.2: Genotoxicity tests conducted on various plant extracts and their results.

Plant name	Extraction method	Method	Genotoxic effect
Melaleuca alternifolia	Steam distillation	Comet	Genotoxic
Curcuma longa	Water extraction	Micronucleus	Anti-genotoxic
Matricaria chamomilla	Alcohol extraction	Chromosome aberration	Mild genotoxic
Thymus vulgaris	Steam distillation	Ames	May be genotoxic

20.5 Genotoxic effects of medicinal and aromatic plants

Medicines derived from medicinal and aromatic plants are used worldwide and have recently gained renewed interest in developed countries. The main arguments for utilizing them are patient dissatisfaction with the success rate and/or safety of traditional natural medicines, pleasure with therapeutic outcomes, the satisfying sense of being involved in the selection of therapeutic means, and the (mis)belief that medicinal and aromatic plants are inherently "natural" and thus safe [55]. Since this assumption is quite incorrect, emphasis has been placed on toxicological studies on herbal medicines, and despite the benefits of herbal medicines, the literature contains reports of toxicity [56]. Medicinal and aromatic plants show their health benefits through their secondary metabolites and their products. However, these secondary metabolites do not always provide a curative effect and can be harmful due to their toxic effects. In order to take precautions against adverse conditions such as genotoxicity and carcinogenicity, concentration tests, animal studies, model organism experiments, and cell and tissue experiments are very useful.

According to studies, certain plants often used in traditional medicine may contain genotoxic compounds. These chemicals often react with cellular macromolecules, including DNA, resulting in cellular toxicity and/or genotoxicity [57]. The combined effect of any harmful chemical can cause a variety of chromosomal abnormalities, including chromatid breaks, isochromatid breaks, gaps, chromosomal fragments, exchanges, and sister chromatid unions, even if the DNA structure changes [58]. Such DNA disorders may result in the formation of diseases, susceptibility to diseases, increased morbidity/mortality, changes in hereditary characteristics, and reproductive disorders [59].

A great deal of research has been conducted examining the effects of medicinal and aromatic plants on human tissue and DNA. It has been reported that some phytochemicals produce free oxygen radicals and that excessive production of these harmful radicals causes cell death by damaging DNA and biomolecules [60]. Oxidative damage rates caused by free radicals are highest in animals, and they have generally been shown to have short life spans. However, in humans, genetic diseases resulting from spontaneous DNA damage or inadequate DNA repair often manifest as premature aging. Additionally, exposure to external factors such as ultraviolet light and cigarette smoke contributes to a decrease in lifespan. Therefore, protecting the cell against DNA damage and increasing DNA repair efficiency allows for an extension of lifespan. Antioxidant compounds such as vitamins C and E, contained in medicinal and aromatic plants, can transform into oxidant precursor molecules in the presence of redox-active molecules such as iron and under certain conditions. Therefore, it is important to evaluate the genotoxic properties of plants [61].

Secondary metabolites are bioactive compounds eliciting pharmacological or toxicological effects in humans and animals. A monoterpene with numerous biological properties (antibacterial, antifungal, acaricidal, antileishmanial, antinociceptive, antioxidant, anticancer, antiviral, and insecticidal activities), γ-terpinene (1-Isopropyl-4-methyl-1,4-cyclohexadiene) is widely present in the essential oils of aromatic and medicinal plants, including *Thymus vulgaris* L., *Eucalyptus camaldulensis Dehnh.*, *Nigella sativa* L., *Cuminum cyminum* L., *Majorana hortensis Moench*, *Protium icicariba* (DC.) *Marchand*, *Citrus deliciosa Tenore*, *Origanum onites* L., *Melissa officinalis* L., *Satureja thymbra* L., and *Pistacia khinjuk Stocks* [62]. In a study, it was reported that γ-terpinene had cytotoxic and genotoxic effects on insect hemocytes and that the vital pathways of the insect were damaged as a result of the harm inflicted on hemocytes, which have many functions [63].

Estragole, a secondary metabolite present in plants such as basil (*Ocimum basilicum*), fennel (*Foeniculum vulgare*), *Tagetes*, and *Artemisia,* is transformed into an electrophilic and carcinogenic carbocation through the decomposition of 1'-sulfoxestragole. The bioactivation of estragole to this carcinogenic form involves cytochrome P450 (CYP) and sulfotransferases, which convert 1'-hydroxyestragole to 1'-sulfoxestragole. The bioactivation of estragole from 1'-hydroxyestragole to 1'-sulfoxestragole is mediated by cytochrome P450 (CYP) and sulfotransferases. Various phenolic compounds exhibit a chemoprotective effect by increasing the activity of CYPs and a cancer-inducing effect by decreasing it [63].

Proteins and DNA undergo conformational changes and mutations when alkylated. Reactives such as epoxides, pyrrolizidine alkaloids, aristolochic acids, cycasin, and furanocoumarins may alkylate DNA. Additionally, furanocoumarins induce intercalations that result in DNA alkylation and frameshift mutations [64].

In a study conducted on *M. officinalis*, it was reported that the phenolic substances in the plant, especially caffeic acid derivatives, possess genotoxic and mutagenic properties and have carcinogenic potential [65]. Water extracts of *M. officinalis* exhibited the highest genotoxic properties among all extracts.

Table 20.3: Secondary metabolites that are converted into genotoxic and carcinogenic active compounds [66].

Secondary metabolite	Proximate metabolite	Active metabolite
Cycasin		Methylazoxymethanol
Ranunculin		Protoanemonin
Elemicin	1'-Hyrdoxyelemicin	1'-Sulfooxyelemicin
Methyleugenol	1'-Hydroxymethyleugenol	1'-Sulfooxymethyleugenol
Myristicin	1'-Hydroxymyristicin	1'-Sulfooxymyristicin
Apiol, β-asarone	1'-Hydroxy metabolites	1'-Sulfooxy metabolites
Estragole	1'-Hyrdoxyestragole	1'-Sulfooxyestragole
Safrole	1'-Hyrdoxysafrole	1'-Sulfooxysafrole
Pyrrolizidine alkaloids	Dehydropyrrolizidine alkaloids (pyrroles)	6,7-Dihydro-7-hydroxy-1 hydroxymethyl 5H-pyrrolizine
Aristolochic acid	Aristolochic acid I-II	Aristolactam-nitrenium ion
Quercetin	ortho-Quinone	Quinone and quinone methides
Lucidin-3-O-primiveroside		Lucidin

The alkanylbenzene group's submetabolites – safrole, estragole, apiol, elemicin, beta-asarone, trans-anethole, myristicin, and methyleugenol – are used in a variety of foods and fragrances, mostly to enhance food and drink flavor or for therapeutic purposes Table 20.3. For medicinal applications, they are used to treat cervical cancer and possess antibacterial, anti-inflammatory, and antidiabetic properties. Plants containing alkanylbenzene groups include basil, black pepper, anise, fennel, dill, tarragon, elephant apple, and gazelle grass. If not utilized properly, this particular category of metabolites has been shown to induce hepatocarcinoma and tumor formation, alter DNA structure, and cause genotoxic effects [67].

Aristolochic acid I and aristolochic acid II are two distinct carcinogenic metabolites found in aristolochic acid. This secondary metabolite is mostly present in the structures of worm killer, birth weed, maternity weed, and pelican flower. Its anti-inflammatory, anti-repellent, diuretic, analgesic, and insect-repelling qualities make it a useful health aid. According to reports, nephritis, renal pelvis, and ureter malignancies might result from overdosing and negligent usage [68].

Pyrrolizidine alkaloids are mostly found in the Boraginaceae, Compositae, and Leguminosae plant families. Capillary and artery bleeding, diabetes, rheumatism, asthma, skin irritations, wounds, burns, injuries, inflammation therapy, circulation improvement, antibacterial and anticancer action, and blood purification are all common applications. They include subcarcinogenic products such as riddelliine, senecionine, lasiocarpine, senkirkine, and monocrotaline. Excessive and unregulated usage results in genotoxic activity, particularly for riddelliine and lasiocarpine metabolites, while other components have hepatotoxic, carcinogenic, and tumor-inducing effects [69].

Monoterpenes are present in herbs such as swamp cedar, sage, California worm-wood, marigold, false cypress, tansy, mint species, juniper, and spearmint, and they are degraded into carcinogenic sub-metabolites, including Thujone and Pulegone. Herbs are utilized for their aromatizing, flavor-enhancing, bioantimutagenic, anthelmintic, tonic, stimulant, abortive, acaricidal, convulsant, antiulcerogenic, and anti-inflammatory activ-ities. However, when used excessively and unmanaged, they have been found to be gen-otoxic and to influence the development of gland cancers [70].

Cycasin is a type of alkaloidal glycoside present in the Dionysian chestnut and palm species, which are members of the cycad family. It has several medicinal uses. Numerous animal studies have demonstrated its mutagenic, genotoxic, and carcino-genic properties [64].

Furanocoumarins serve a purpose in treating psoriasis, kidney and bladder issues, as well as their sub-metabolites psoralen, 5-methoxypsoralen, and 8-methoxypsoralen. Plants such as gazelle, parsley, wild celery, rue, rabbitwort, and foam bush contain them. They penetrate the DNA structure and create mono-adducts when exposed to sunlight. This has genotoxic consequences. They are mutagenic, phototoxic, and may cause cancer if exposed to the skin over time [71].

Quercetin, a phenolic molecule, is classified as an antioxidant flavonoid, al-though it behaves as a prooxidant when covalently coupled to cellular DNA and pro-teins. It is recommended for its atherosclerosis, cardiovascular disease, anticarcino-genic, anti-inflammatory, antiaggregant, and vasodilator properties. Various fruits, vegetables, and therapeutic aromatic herbs contain quercetin. Excessive usage has resulted in genotoxic consequences, such as DNA breakage in rats [72].

20.5.1 Genotoxic effects of medicinal aromatic plants widely used in different regions

Coptis chinensis Franch is preferred in alternative Chinese medicine to treat dysen-tery, diarrhea, vomiting, and inflammatory symptoms. One of its main metabolites, the alkaloid berberine, has been reported to have genotoxic activity [73].

In African countries, many medicinal plants are used in the treatment of malaria, which is especially common. In addition to these plants, the hazardous qualities of *Acanthospermum hispidum*, also known as bristly starbur, have been studied. Both in vivo and in vitro studies have revealed that the plant contains genotoxic proper-ties [74].

Dipteryx alata is a plant that originates exclusively in the Brazilian Cerrado and is used medicinally. In one study, a hydroalcoholic *D. alata* bark extract was tested for cytotoxic and genotoxic properties [75].

Acacia nilotica, commonly referred to as kikar, black babul, and gum arabic tree, is a cost-effective plant that serves as fodder and fuel, and all of its parts are utilized for healing a variety of diseases, including as painkillers and treatments for chest diseases, fever, dysentery, gastrointestinal problems, sclerosis, smallpox, leprosy, menstrual problems, and as a sedative. It contains phenolics, flavonoids, anthocyanins, and saponins, which can be utilized for the treatment of the disorders indicated [76]. However, a study found that *A. nilotica* has a dose-dependent genotoxic effect [77].

It has been reported that all hydroalcoholic extracts of pomegranate fruit (*Punica granatum* L.), which are used in the treatment of various diseases in alternative medicine in America, Asia, Africa, and Europe, are genotoxic [78].

Mentha piperita essential oil is used in alternative medicine to treat minor ailments. It has properties such as alleviating muscular pain, nerve discomfort, itching, and providing a pleasant aroma. High doses of this oil may cause mucosal irritation and heartburn. Research on *Mentha piperita* essential oil found that it mildly induces sister chromatid exchange and is genotoxic according to the SMART test [79].

The medicinal plant known as *Stachys annua* L. is used as an antipyretic in Turkey [80]. The *Nepeta nuda* plant, endemic to Turkey, is used for its antiseptic, diuretic, and expectorant effects [81]. In a study investigating the genotoxic and anti-genotoxic properties of these plants, it was determined that *N. nuda* and *S. annua* created genotoxic effects on plasmid DNA [82].

In Ethiopia, medicinal and aromatic herbs are utilized as treatments, although little is known about their potential genotoxic consequences. For this reason, hydroalcoholic extracts of medicinal plants such as *Glinus lotoides, Plumbago zeylanica, Rumex steudelii,* and *Thymus schimperi* were evaluated for genotoxicity using the comet method [83]. The major components of these plants are generally saponins, flavonoids, coumarins, plumbagin, glucosides, phytoestrogens, and polyphenols. Table 20.4 shows the treatment purposes of the plants used.

Table 20.4: Traditional medicinal plants of Ethiopia and their medicinal uses.

Plant name	Medical use
Glinus lotoides	Tapeworm infestation treatment
Plumbago zeylanica	Treat malaria and topical as well as different microbial infections
Rumex steudelii	antifertility agent
Thymus schimperi	Hypertension, fungal infections, and bacterial infections

The current collaboration shows that *G. lotoides, P. zeylanica, R. steudelii,* and *T. schimperi* extracts produce substantial DNA damage in mouse lymphoma cells without inducing cytotoxicity, and that the compounds in these extracts may directly interact with DNA and be genotoxic [83].

In Turkish flora, *Helichrysum pamphylicum* and *Helichrysum noeanum* species are endemic medicinal aromatic plants that are represented by 26 taxa. In Anatolia, they are referred to as the eternal bloom, unfading flower, or golden herb. They are consumed as herbal tea to treat asthma, diarrhea, jaundice, stomachaches, urogenital problems, and kidney stones [84]. The genotoxic effects of the species *Helichrysum sanguineum*, *Helichrysum pamphylicum*, *Helichrysum orientale*, and *Helichrysum noeanum*, which are utilized as traditional medicine in Turkey, were investigated. According to the results of the study, the *Helichrysum orientale* species is not genotoxic and can be used safely. However, due to their genotoxic effects, *Helichrysum sanguineum*, *Helichrysum noeanum*, and particularly *Helichrysum pamphylicum* should not be taken in excessive dosages. These plant extracts may raise micronucleus rates, cause chromosomal damage, and delay mitosis, all of which might be carcinogenic [85].

Species of the genus *Artemisia*, widely used in Serbia, have garnered attention in recent years for their protective effects against SARS-CoV and COVID-19. These plants are recognized for their therapeutic potential in the management of various diseases. A study investigated the genotoxic effects of acetone and water extracts of two subspecies, *Artemisia vulgaris* L. and *Artemisia alba Turra*, on peripheral blood lymphocytes. The findings indicated that both acetone and water extracts of *A. vulgaris* exhibit genotoxic properties, whereas only the acetone extract of *A. alba* was reported to possess genotoxic effects. According to conventional medicine, *A. alba* water extract may be used for therapy; however, excessive dosages should not be utilized [86].

Traditional medicine often uses the Chinese herb *Fritillaria cirrhosa* D. Don, which has a high alkaloid content. Its antihistaminic, anti-inflammatory, and analgesic effects are utilized. Research on *Fritillaria cirrhosa* water extract found that it had genotoxic properties due to the alkaloids, terpenoids, and glycoside phytochemicals it contains [87].

20.6 Conclusion

The use of medicinal and aromatic plants, which have been utilized for treatment, nutrition, cosmetics, and many other purposes for centuries, has increased rapidly in recent years due to economic factors and the side effects of rapidly growing industrialization. However, the plants that nature offers, which are excellent for addressing many diseases and health problems, may not always be beneficial. The metabolites and toxic compounds they contain may become harmful to human health due to changes in the vegetation in which they grow, environmental conditions, climate, and other factors. Exceeding certain dosages and frequent, intensive use can render medicinal and aromatic plants genotoxic and carcinogenic. It is recommended that the medicinal and aromatic plants we use for their natural benefits be thoroughly researched before use and consumed carefully in appropriate doses. Otherwise, it is inevitable to encounter difficult-to-predict damages, such as genotoxicity. Furthermore, it is unavoidable that these genotoxic effects will be passed on from generation to generation.

References

[1]   Barata, A. M. et al. (2016). Conservation and sustainable uses of medicinal and aromatic plants genetic resources on the worldwide for human welfare. Industrial Crops and Products, 88, 8–11.

[2]   Farnsworth, N. R. and Soejarto, D. D. (1991). Global importance of medicinal plants. The Conservation of Medicinal Plants, 26(26), 25–51.

[3]   Adekunle, A. and Adekunle, O. (2009). Preliminary assessment of antimicrobial properties of aqueous extract of plants against infectious diseases. Biology and Medicine, 1(3), 20–24.

[4]   Dağlar, N. and Dağdeviren, H. N. (2018). Geleneksel ve tamamlayıcı tıp uygulamalarında fitoterapinin yeri. Eurasian Journal of Family Medicine, 7(3), 73–77.

[5]   Parildar, H., Serter, R. and Yesilada, E. (2011). Diabetes mellitus and phytotherapy in Turkey. JPMA-Journal of the Pakistan Medical Association, 61(11), 1116.

[6]   Saganuwan, A. S. (2010). Some medicinal plants of Arabian Peninsula. Journal of Medicinal Plant Research, 4(9), 766–788.

[7]   Acıbuca, V. and Budak, D. B. (2018). Dünya'da ve Türkiye'de tıbbi ve aromatik bitkilerin yeri ve önemi. Çukurova Tarım Ve Gıda Bilimleri Dergisi, 33(1), 37–44.

[8]   Chen, S. et al. (2010). Validation of the ITS2 region as a novel DNA barcode for identifying medicinal plant species. PloS One, 5(1), e8613.

[9]   Smolinski, M. S., Hamburg, M. A. and Lederberg, J. (2003). Microbial Threats to Health. Emergence, Detection and Response. National Academies Press at: http://www.nap.edu/catalog/10636.html Washington D.C.

[10] Nahar, L. and Sarker, S. D. (2019). Chemistry for Pharmacy Students: General, Organic and Natural Product Chemistry. John Wiley & Sons, UK.

[11] Oksman-Caldentey, K.-M. (2007). Tropane and nicotine alkaloid biosynthesis-novel approaches towards biotechnological production of plant-derived pharmaceuticals. Current Pharmaceutical Biotechnology, 8(4), 203–210.

[12] Canter, P. H., Thomas, H. and Ernst, E. (2005). Bringing medicinal plants into cultivation: Opportunities and challenges for biotechnology. TRENDS in Biotechnology, 23(4), 180–185.

[13] Baytop, T. (1999). Türkiye'de Bitkiler Ile Tedavi: Geçmişte Ve Bugün. Nobel Tıp Kitabevleri, İstanbul, TURKEY.

[14] Nohutçu, L., Tunçtürk, M. and Tunçtürk, R. (2019). Yabani bitkiler ve sürdürülebilirlik. Yüzüncü Yıl Üniversitesi Fen Bilimleri Enstitüsü Dergisi, 24(2), 142–151.

[15] Başaran, A. (2012). Ülkemizdeki bitkisel ilaçlar ve ürünlerde yasal durum. Missed (27–28), 22–26.

[16] Türkan, Ş. et al. (2006). Ordu ili ve çevresinde yetişen bazı bitkilerin etnobotanik özellikleri. Süleyman Demirel Üniversitesi Fen Bilimleri Enstitüsü Dergisi, 10(2), 162–166.

[17] Bayraktar, Ö. V., Öztürk, G. and Arslan, D. (2017). Türkiye'de bazı tıbbi ve aromatik bitkilerin üretimi ve pazarlamasındaki gelişmelerin değerlendirilmesi. Tarla Bitkileri Merkez Araştırma Enstitüsü Dergisi, 26(2), 216–229.

[18] Göktaş, Ö. and Gıdık, B. (2019). Tıbbi ve aromatik bitkilerin kullanım alanları. Bayburt Üniversitesi Fen Bilimleri Dergisi, 2(1), 145–151.

[19] Stevens, R. (1957). Botany: An Introduction to Plant Science. American Institute of Biological Sciences Circulation, AIBS, 1313 Dolley . . . .

[20] Wink, M. (1999). Functions of Plant Secondary Metabolites and Their Exploitation in Biotechnology. Vol. 3, Taylor & Francis, UK.

[21] Cseke, L. et al. Natural Products from Plants. Second edition, CRC, London, New York, 551. Daborn, PJ, Lumb, C., Boey, A., Wong, W. & Batterham, P.(2007). Evaluating the insecticide resistance potential of eight *Drosophila melanogaster* cytochrome P450 genes by transgenic over-expression. Insect Biochemistry and Molecular Biology, 1999. 37(5), 512–519.

[22] Briskin, D. P. (2000). Medicinal plants and phytomedicines. Linking plant biochemistry and physiology to human health. Plant Physiology, 124(2), 507–514.

[23] Hegnauer, R. (1986). Phytochemistry and plant taxonomy – An essay on the chemotaxonomy of higher plants. Phytochemistry, 25(7), 1519–1535.

[24] Naikoo, M. I. et al. (2019). Role and regulation of plants phenolics in abiotic stress tolerance: An overview. Plant Signaling Molecules, 157–168

[25] Edeoga, H. O., Okwu, D. and Mbaebie, B. (2005). Phytochemical constituents of some Nigerian medicinal plants. African Journal of Biotechnology, 4(7), 685–688.

[26] Poiroux-Gonord, F. et al. (2010). Health benefits of vitamins and secondary metabolites of fruits and vegetables and prospects to increase their concentrations by agronomic approaches. Journal of Agricultural and Food Chemistry, 58(23), 12065–12082.

[27] Taiz, L. and Zeiger, E. (2002). Plant Physiology. Sinauer associates.

[28] Karakaya, S. (2004). Bioavailability of phenolic compounds. Critical Reviews in Food Science and Nutrition, 44(6), 453–464.

[29] Mohammadi, N. and Shaghaghi, N. (2020). Inhibitory effect of eight secondary metabolites from conventional medicinal plants on COVID_19 virus protease by molecular docking analysis.

[30] Middleton, E., Kandaswami, C. and Theoharides, T. C. (2000). The effects of plant flavonoids on mammalian cells: Implications for inflammation, heart disease, and cancer. Pharmacological Reviews, 52(4), 673–751.

[31] Riveiro, M. E. et al. (2010). Coumarins: Old compounds with novel promising therapeutic perspectives. Current Medicinal Chemistry, 17(13), 1325–1338.

[32] Borges, F. et al. (2005). Simple coumarins and analogues in medicinal chemistry: Occurrence, synthesis and biological activity. Current Medicinal Chemistry, 12(8), 887–916.

[33] Adlercreutz, H. (2007). Lignans and human health. Critical Reviews in Clinical Laboratory Sciences, 44(5–6), 483–525.

[34] Baur, J. A. and Sinclair, D. A. (2006). Therapeutic potential of resveratrol: The in vivo evidence. Nature Reviews Drug Discovery, 5(6), 493–506.

[35] Frémont, L. (2000). Biological effects of resveratrol. Life Sciences, 66(8), 663–673.

[36] Bravo, L. (1998). Polyphenols: Chemistry, dietary sources, metabolism, and nutritional significance. Nutrition Reviews, 56(11), 317–333.

[37] Ky, I., Le Floch, A., Zeng, L., Pechamat, L., Jourdes, M., & Teissedre, P. L. (2016). Tannins; Caballero, B., Finglas, PM, Toldrá, F., Eds. Academic Press: Oxford, UK.

[38] Ghosh, D. (2015). Tannins from foods to combat diseases. International Journal of Pharmaceutical Sciences Review and Research, 4(5), 40–44.

[39] Thirumurugan, D. et al. (2018). An introductory chapter: Secondary metabolites. Secondary Metabolites-sources and Applications, 1, 13.

[40] Rajput, A., Sharma, R. and Bharti, R. (2022). Pharmacological activities and toxicities of alkaloids on human health. Materials Today: Proceedings, 48, 1407–1415.

[41] Tiring, G., Satar, S. and Özkaya, O. (2021). Sekonder metabolitler. Bursa Uludağ Üniversitesi Ziraat Fakültesi Dergisi, 35(1), 203–215.

[42] Van Wyk, B.-E. and Wink, M. (2015). Phytomedicines, Herbal Drugs & Plant Poisons. Briza Publications, London, UK.

[43] Ludwiczuk, A., Skalicka-Woźniak, K. and Georgiev, M. (2017). Terpenoids, in Pharmacognosy. Elsevier, Amsterdam, 233–266.

[44] Tang, C. et al. (2016). The rubber tree genome reveals new insights into rubber production and species adaptation. Nature Plants, 2(6), 1–10.

[45] Aftab, T. and Hakeem, K. R. (2020). Medicinal and Aromatic Plants. Expanding Their Horizons through Omics. Elsevier, Amsterdam, The Netherlands.

[46] Hostettmann, K. and Marston, A. (1995). Chemistry and Pharmacology of Natural Products. Vol. 548, Cambridge University Press, Cambridge, UK.

[47] Phillips, D. H. and Arlt, V. M. (2009). Genotoxicity: Damage to DNA and its consequences. Molecular, Clinical and Environmental Toxicology: Volume 1: Molecular Toxicology, Springer, Germany, 87–110.

[48] Cavalcanti, B. et al. (2010). Structure–mutagenicity relationship of kaurenoic acid from Xylopia sericeae (Annonaceae). Mutation Research/Genetic Toxicology and Environmental Mutagenesis, 701(2), 153–163.

[49] Ellinger-Ziegelbauer, H. et al. (2009). Application of toxicogenomics to study mechanisms of genotoxicity and carcinogenicity. Toxicology Letters, 186(1), 36–44.

[50] Kirkland, D. et al. (2005). Evaluation of the ability of a battery of three in vitro genotoxicity tests to discriminate rodent carcinogens and non-carcinogens: I. Sensitivity, specificity and relative predictivity. Mutation Research/Genetic Toxicology and Environmental Mutagenesis, 584(1–2), 1–256.

[51] Guideline, I. H. T. (2011). Guidance on genotoxicity testing and data interpretation for pharmaceuticals intended for human use S2 (R1). in International Conference on Harmonisation of Technical Requirements for Registration of Pharmaceuticals for Human Use, Rockville, Maryland, US.

[52] Maron, D. M. and Ames, B. N. (1983). Revised methods for the Salmonella mutagenicity test. Mutation Research/Environmental Mutagenesis and Related Subjects, 113(3–4), 173–215.

[53] Saravanan, V., Murugan, S. and Kumaravel, T. (2020). Genotoxicity studies with an ethanolic extract of Kalanchoe pinnata leaves. Mutation Research/Genetic Toxicology and Environmental Mutagenesis, 856, 503229.

[54] Kawasaki, K. et al. (2015). A hot water extract of Curcuma longa inhibits adhesion molecule protein expression and monocyte adhesion to TNF-α-stimulated human endothelial cells. Bioscience, Biotechnology, and Biochemistry, 79(10), 1654–1659.

[55] Huxtable, R. J. (1990). The harmful potential of herbal and other plant products. Drug Safety, 5(Suppl 1), 126–136.

[56] Cosyns, J.-P. et al. (1999). Urothelial lesions in Chinese-herb nephropathy. American Journal of Kidney Diseases, 33(6), 1011–1017.

[57] Rietjens, I. M. et al. (2005). Molecular mechanisms of toxicity of important food-borne phytotoxins. Molecular Nutrition & Food Research, 49(2), 131–158.

[58] Arpita, B. and Deep, S. A. (2022). Genotoxicity induced by medicinal plants. Bulletin of the National Research Centre, 46(1), 1–11.

[59] Carnesoltas Lázaro, D. et al. (2010). Genotoxic assessment of aqueous extract of Rhizophora mangle L.(mangle rojo) by spermatozoa head assay. Revista Cubana de Plantas Medicinales, 15(1), 0–0.

[60] Ahmad, S. (1992). Biochemical defence of pro-oxidant plant allelochemicals by herbivorous insects. Biochemical Systematics and Ecology, 20(4), 269–296.

[61] Friedberg, E. C. et al. (2005). DNA Repair and Mutagenesis. American Society for Microbiology Press. Washington D.C, USA.

[62] Moreira, R. C. et al. (2022). Health properties of dietary monoterpenes. Biomolecules from Natural Sources: Advances and Applications, volume 1, UK, 362–389

[63] Diksha, et al. (2023). Growth inhibitory, immunosuppressive, cytotoxic, and genotoxic effects of γ-terpinene on Zeugodacus cucurbitae (Coquillett)(Diptera: Tephritidae). Scientific Reports, 13(1), 16472.

[64] Wink, M. and Van Wyk, B.-E. (2008). Mind-altering and Poisonous Plants of the World. Timber Press, South Africa.

[65] Alves, A. et al. (2009). Genotoxic and mutagenic effects of Melissa officinalis (Erva Cidreira) extracts. The Open Toxicology Journal, 3(1), 58–69.

[66] Prinsloo, G., Nogemane, N. and Street, R. (2018). The use of plants containing genotoxic carcinogens as foods and medicine. Food and Chemical Toxicology, 116, 27–39.

[67] Alajlouni, A. M. et al. (2017). Determination and risk assessment of naturally occurring genotoxic and carcinogenic alkenylbenzenes in nutmeg-based plant food supplements. Journal of Applied Toxicology, 37(10), 1254–1264.

[68] Dhouioui, M. et al. (2016). Seasonal changes in essential oil composition of Aristolochia longa L. ssp. paucinervis Batt.(Aristolochiaceae) roots and its antimicrobial activity. Industrial Crops and Products, 83, 301–306.

[69] Chen, T., Mei, N. and Fu, P. P. (2010). Genotoxicity of pyrrolizidine alkaloids. Journal of Applied Toxicology: An International Journal, 30(3), 183–196.

[70] Akkol, E. K. et al. (2015). Thuja occidentalis L. and its active compound, α-thujone: Promising effects in the treatment of polycystic ovary syndrome without inducing osteoporosis. Journal of Ethnopharmacology, 168, 25–30.

[71] Yockey, O. P. et al. (2017). Mechanism of error-free DNA replication past lucidin-derived DNA damage by human DNA polymerase κ. Chemical Research in Toxicology, 30(11), 2023–2032.

[72] Erlund, I. (2004). Review of the flavonoids quercetin, hesperetin, and naringenin. Dietary sources, bioactivities, bioavailability, and epidemiology. Nutrition Research, 24(10), 851–874.

[73] Zhang, Q. et al. (2011). Preventive effect of Coptis chinensis and berberine on intestinal injury in rats challenged with lipopolysaccharides. Food and Chemical Toxicology, 49(1), 61–69.

[74] Adukpo, S. et al. (2020). Research article antiplasmodial and genotoxic study of selected ghanaian medicinal plants.

[75] Esteves-Pedro, N. M. et al. (2011). Implementation of the three Rs in the human hazard assessment of Brazilian medicinal plants: An evaluation of the cytotoxic and genotoxic potentials of Dipteryx alata Vogel. Alternatives to Laboratory Animals, 39(2), 189–196.

[76] Ali, A. et al. (2012). Acacia nilotica: A plant of multipurpose medicinal uses. Journal of Medicinal Plants Research, 6(9), 1492–1496.

[77] Diab, K. A. et al. (2022). Evaluation of the cytotoxic, anticancer, and genotoxic activities of Acacia nilotica flowers and their effects on N-methyl-N-nitrosourea-induced genotoxicity in mice. Molecular Biology Reports, 49(9), 8439–8448.

[78] Chidambara Murthy, K. et al. (2004). Study on wound healing activity of Punica granatum peel. Journal of Medicinal Food, 7(2), 256–259.

[79] Romero-Jiménez, M. et al. (2005). Genotoxicity and anti-genotoxic of some traditional medicinal herbs. Mutation Research/Genetic Toxicology and Environmental Mutagenesis, 585(1–2), 147–155.

[80] Güner, A. et al. Güner, A., Aslan, S., Ekim, T. & Vural, M. (2012). Türkiye Bitkileri Listesi (Damarlı Bitkiler). Nezahat Gökyiğit Botanik Bahçesi ve Flora Araştırmaları Derneği Yayını. İstanbul, İstanbul.

[81] Teber, İ. and Bursal, E., Nepeta nuda subsp. albiflora Bitki Ekstrelerinin Antioksidan Aaktivitelerinin Farklı in vitro Biyoanalitik Metotlar ile Belirlenmesi.

[82] Ayar, A. et al. (2018). Genotoxic and antigenotoxic effects of some plant species of lamiaceae family. European Journal of Science and Technology, 14, 348–352.

[83] Demma, J., Engidawork, E. and Hellman, B. (2009). Potential genotoxicity of plant extracts used in Ethiopian traditional medicine. Journal of Ethnopharmacology, 122(1), 136–142.

[84] Baytop, T. (1994). Türkçe Bitki Adları Sözlüğü. Vol. 578, Turk Dil Kurumu, Ankara, TURKEY.

[85] Erolu, E. H. et al. (2010). In vitro genotoxic effects of four Helichrysum species in human lymphocytes cultures. Biological Research, 43(2), 177–182.

[86] Radović Jakovljević, M. et al. (2022). Comparative study of the genotoxic activity of Artemisia vulgaris L. and Artemisia alba Turra extracts in vitro. Drug and Chemical Toxicology, 45(4), 1915–1922.

[87] Guo, X. et al. (2020). Aqueous extract of bulbus Fritillaria cirrhosa induces cytokinesis failure by blocking furrow ingression in human colon epithelial NCM460 cells. Mutation Research/Genetic Toxicology and Environmental Mutagenesis, 850, 503147.

Yeter Çilesiz*, Uzma Nadeem, and Tolga Karaköy

Chapter 21
Applicability of start codon targeted (SCoT) polymorphism markers in determining genetic diversity in medicinal and aromatic plants

Abstract: Medicinal and aromatic plants hold critical importance in agricultural, industrial, and pharmaceutical sectors globally due to their diverse applications and significant economic value. These plants possess remarkable potential in terms of their adaptability to environmental stresses, diverse secondary metabolite profiles, and their role in biodiversity conservation. Identifying genetic diversity in medicinal and aromatic plants is not only crucial for ensuring their sustainable utilization but also facilitates the development of effective conservation strategies and the preservation of genetic resources for future generations. This chapter comprehensively discusses the applicability of start codon targeted (SCoT) polymorphism markers in assessing genetic diversity in medicinal and aromatic plants. SCoT markers, which target coding regions surrounding translation start codons, are recognized as a method capable of detecting polymorphisms. This approach has gained widespread usage in recent years due to its high reproducibility and the generation of meaningful molecular data for genetic diversity studies. In this chapter, the fundamental principles of SCoT markers, their role in molecular biology, and a comparative analysis with other molecular marker techniques are elaborated upon, along with case studies showcasing their applications in medicinal and aromatic plants. Furthermore, the advantages of SCoT markers in genetic diversity analyses and their technical limitations are critically examined. This information serves as a valuable resource for researchers in molecular biology and genetics, as well as scientists focusing on medicinal and aromatic plants.

Keywords: genetic diversity, SCoT markers, medicinal and aromatic plants

*Corresponding author: Assist. Prof. Dr. Yeter Çilesiz, Department of Field Crops Faculty of Agricultural Sciences and Technology, Sivas Science and Technology University, 58140 Sivas, Turkey, e-mail: yetercilesiz_mbg@hotmail.com, https://orcid.org/0000-0002-4313-352X
Uzma Nadeem, Department of Environmental Studies, Mata Sundri College for Women, University of Delhi, Delhi 110002, India
Tolga Karaköy, Department of Plant Protection, Faculty of Agricultural Sciences and Technology, Sivas Science and Technology University, 58140 Sivas, Turkey

https://doi.org/10.1515/9783111469713-021

21.1 The importance and conservation of genetic diversity

Medicinal and aromatic plants are important in ecological systems as well as in the fields of economics and science due to their biological and chemical richness. These plants are the main source of raw ingredients for a variety of goods, such as food preservatives, cosmetic formulations, pharmaceutical drug research, aromatherapy, and nutritional supplements.

Significantly, secondary metabolites from these plants have anti-inflammatory, anticancer, antimicrobial, and antioxidant qualities, which make them essential parts of the medical field. However, increasing demand and environmental factors pose threats to the genetic diversity of medicinal and aromatic plants, highlighting the necessity of their conservation. The most traded medicinal and aromatic plants in the world are coffee, sesame, garlic, red pepper, allspice, black pepper, green tea, mustard seed, poppy seed, ginger, salep, and cumin [1–3] Figure 21.1.

A   B   C

Figure 21.1: Some of the most traded medicinal and aromatic plants in the world: (A) sesame, (B) garlic, and (C) black pepper [4].

Genetic diversity refers to the degree of variation in the genetic material within a population or species. This variation determines the capacity of individuals to respond to environmental stressors. High genetic diversity in medicinal and aromatic plants is crucial for their adaptation to diverse ecological conditions, the development of resistance to diseases, and their utility in targeted breeding programs. Therefore, a comprehensive understanding of the genetic structures of these plants is essential for conserving natural populations and developing sustainable production strategies [5].

Deciphering the intricacies of biological diversity is made possible by the molecular techniques employed to detect and analyze genetic diversity. Specifically, molecular marker technologies enable the detection of variations in an organism's DNA with high precision, making them invaluable in population genetics, phylogenetic analyses, and gene mapping studies [6, 7].

21.2 Definition and significance of SCoT markers

Start codon targeted (SCoT) polymorphism markers are an innovative method developed to detect genetic polymorphisms by targeting the coding DNA sequences surrounding the translation start codons (ATG) Table 21.1. SCoT markers distinguish themselves from other molecular marker systems by focusing specifically on coding gene regions, providing a significant advantage in functional genomic studies. This characteristic not only facilitates the analysis of DNA sequence variations but also establishes a foundation for understanding the relationship between these variations and their biological functions in genetic diversity studies [8, 9].

Table 21.1: SCoT markers and applications of SCoT polymorphism markers in plant genetics.

SCoT name	Sequences (5′–3′)	Applications of SCoT and genomics
SCOT1	CAACAATGGCTACCACCA	Population genetics/genetic diversity
SCOT2	CAACAATGGCTACCACCC	Inter-specific/generic genetic relationship
SCOT3	CAACAATGGCTACCACCG	Cultivar/hybrid/species identification
SCOT4	CAACAATGGCTACCACCT	Gender discrimination
SCOT5	CAACAATGGCTACCACGA	Construction of linkage maps/association mapping
SCOT6	CAACAATGGCTACCACGC	Differential gene expression
SCOT7	CAACAATGGCTACCACGG	Genetic homogeneity testing in tissue-cultured plants
SCOT8	CAACAATGGCTACCACGT	
SCOT9	CAACAATGGCTACCAGCA	
SCOT10	CAACAATGGCTACCAGCC	
SCOT11	AAGCAATGGCTACCACCA	
SCOT12	ACGACATGGCGACCAACG	
SCOT13	ACGACATGGCGACCATCG	
SCOT14	ACGACATGGCGACCACGC	
SCOT15	ACGACATGGCGACCGCGA	
SCOT16	ACCATGGCTACCACCGAC	
SCOT17	ACCATGGCTACCACCGAG	
SCOT18	ACCATGGCTACCACCGCC	
SCOT19	ACCATGGCTACCACCGGC	
SCOT20	ACCATGGCTACCACCGCG	
SCOT21	ACGACATGGCGACCCACA	
SCOT22	AACCATGGCTACCACCAC	
SCOT23	CACCATGGCTACCACCAG	
SCOT24	CACCATGGCTACCACCAT	
SCOT25	ACCATGGCTACCACCGGG	
SCOT26	ACCATGGCTACCACCGTC	
SCOT27	ACCATGGCTACCACCGTG	
SCOT28	CCATGGCTACCACCGCCA	
SCOT29	CCATGGCTACCACCGGCC	
SCOT30	CCATGGCTACCACCGGCG	
SCOT31	CCATGGCTACCACCGCCT	
SCOT32	CCATGGCTACCACCGCAC	

Table 21.1 (continued)

SCoT name	Sequences (5′–3′)	Applications of SCoT and genomics
SCOT33	CCATGGCTACCACCGCAG	
SCOT34	ACCATGGCTACCACCGCA	
SCOT35	CATGGCTACCACCGGCCC	
SCOT36	GCAACAATGGCTACCACC	

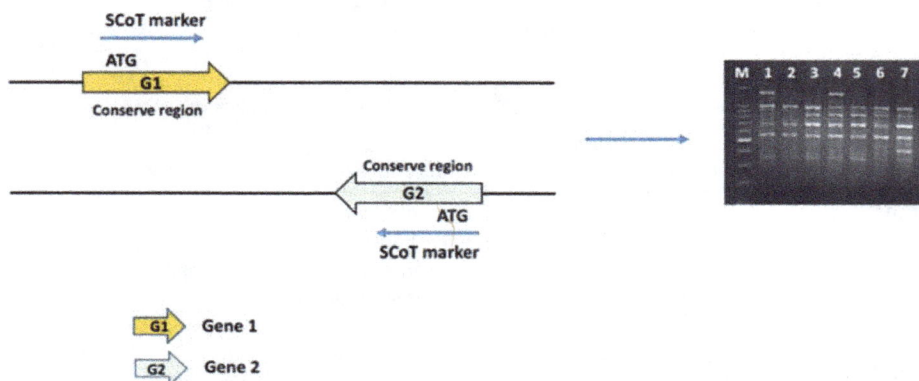

Figure 21.2: Schematic representation of SCoT marker-based PCR amplification [9].

PCR products obtained from SCoT analysis are easily visualized by standard agarose gel electrophoresis (Figure 21.2). SCoT polymorphism markers are dominant markers that target the regions flanking the start codon on both DNA strands (Figure 21.2). Due to their highly polymorphic nature, which reveals diversity at the level of genes and links to a specific trait, SCoT markers are advantageous in genetic and genomic studies as well as crop improvement programs for different groups of plants [9]. The significance of SCoT markers in genetic diversity analysis is supported by the following factors:

21.2.1 Targeted polymorphism analysis

SCoT markers enable the effective identification of genetic variations associated with gene functions by targeting polymorphisms around coding regions. This approach provides a unique opportunity to understand genetic diversity, particularly in relation to the regulation of gene expression and the control of biological processes [10].

21.2.2 High resolution and reproducibility

The PCR-based method's accuracy and primer design specificity lead to good repeatability rates for SCoT markers. This feature offers a significant advantage, particularly in the standardization of protocols across laboratories, and it provides reliable data for comparing different populations [10].

21.2.3 Cost-effective and practical application

The SCoT method stands out as a low-cost technique due to its lack of requirement for specific primer design and its ease of implementation with basic laboratory equipment. This makes the method accessible and preferable in both advanced and resource-limited research settings [11].

21.3 The use of SCoT markers in medicinal and aromatic plants

Understanding the genetic diversity of medicinal and aromatic plants is a critical step in conserving their natural populations and developing high-yielding varieties. SCoT markers, with their ability to detect genetic variations in both coding and noncoding regions, provide an effective tool for analyzing genetic diversity in these plants [12].

21.3.1 Use in functional genomic studies

SCoT markers can be utilized for the identification of genes involved in the synthesis of bioactive compounds and for analyzing the differentiation of these genes among populations [13].

21.3.2 Contribution to breeding programs

The association of genetic markers with target traits can be effectively utilized in the breeding of complex traits, such as stress tolerance, disease resistance, and secondary metabolite production [13] Table 21.2.

Table 21.2: Molecular markers used in plant breeding.

PCR-based markers	Hybridization-based markers
RAPD	RFLP
AFLP	
SRAP	
ISSR	
SSR	
SNP	
EST	
CAPS	
SCAR	

21.3.3 Population genetics and phylogenetic analyses

SCoT markers are used to identify genetic differences and phylogenetic relationships between populations from different geographic regions, thereby shedding light on the evolutionary history of medicinal and aromatic plants [13].

21.4 SCoT (start codon targeted) markers

SCoT markers are a modern molecular marker technique widely used in genetic diversity analysis. They are based on polymerase chain reaction (PCR)-based amplification of DNA sequences surrounding the translation start codon (ATG). The translation start codon (ATG) represents a sequence that plays a critical role in the initiation of protein synthesis and is a highly conserved feature of coding regions. The SCoT technique directly links genetic differences to biological functions by focusing on this conserved structure and the coding sequences that surround it. SCoT markers not only cover coding regions but also encompass regulatory sequences, which, although noncoding, play crucial roles in gene expression regulation and genetic adaptation. This broad analytical spectrum makes the SCoT technique a suitable tool for investigating the functional and adaptive consequences of genetic variations, going beyond merely detecting genetic polymorphisms [14, 15].

The SCoT technique is notable for providing high precision and specificity in genetic diversity analysis. The specific primers used target polymorphic regions surrounding the translation start codon, ensuring that the amplified DNA sequences are derived from biologically relevant regions. SCoT markers offer a more functional approach in functional genomic studies compared to traditional molecular marker methods due to their ability to examine the balance between coding and noncoding regions of genetic material. This versatile analytical capacity makes SCoT markers an indispensable tool for linking genetic diversity with phenotypic traits, adaptation mechanisms to environmental stress factors, and biological processes [16, 17].

21.5 Fundamental principles of SCoT

The principle behind SCoT markers is based on the amplification of DNA sequences surrounding the translation start codon (ATG) and its associated coding sequences through specific molecular primers. The translation start codon is an evolutionarily conserved DNA sequence that determines the initiation of gene expression, making the targeting of this region essential for studying genetic polymorphism in a biologically meaningful way. The primers used in the SCoT technique are designed to target highly polymorphic sequences located near the start codon. These regions contain critical regulatory and coding sequences that play a direct role in gene expression or contribute to environmental adaptation mechanisms. SCoT markers provide a tool for both detecting genetic variation and associating these variations with biological functions during the amplification process. Due to the high specificity of the primers, the amplified DNA sequences typically represent regions that are functionally or adaptively significant in genetic terms. This characteristic enables the SCoT technique to not only determine levels of genetic diversity but also analyze how this diversity impacts biological processes, such as phenotypic traits or sensitivity to environmental stress [18–20]. The strength of the SCoT method in genetic diversity analysis stems from the following features:

21.5.1 Targeted and functional polymorphism

SCoT markers have the ability to establish a direct and meaningful connection between genetic polymorphisms and biological functions due to the proximity of the translation start codon to biologically active and highly conserved coding DNA regions. The translation start codon (ATG) serves as the initiation point for translation, a critical process in which genetic information is transferred to protein synthesis. Therefore, SCoT markers not only detect genetic variations in polymorphism analysis but also serve as an ideal tool for studying the effects of these variations on gene expression, protein production, and genetic regulation processes. The SCoT method, by targeting the coding and regulatory regions surrounding the translation start codon, offers the capacity to evaluate polymorphisms within a functional genomic context. This provides significant advantages, particularly in understanding genetic adaptation mechanisms, identifying genes involved in stress tolerance, and studying the genetic basis of phenotypic traits. By targeting biologically active gene regions, SCoT markers become an indispensable tool for understanding gene expression regulation as well as the biochemical and physiological consequences of genetic variation. These features ensure that SCoT markers have a wide range of applications in adaptation processes, environmental stress responses, and functional genomic research [21, 22].

21.5.1.1 Primer structure specificity

The primers used in the SCoT technique are designed to provide high specificity in genetic diversity analysis, typically with a length of 18–20 bases. This specific length enables the primers to bind to target DNA sequences with high affinity, enhancing the sensitivity of the amplification process. In the design of SCoT primers, proximity to the translation start codon (ATG) and the surrounding coding sequences is a key criterion. This approach significantly increases the biological relevance of the amplified DNA segments, allowing for the analysis of genetic variations in a functional context. SCoT primers typically target biologically significant regions rather than random sequences. This improves the biological validity and dependability of the collected data, in addition to guaranteeing a high degree of reproducibility.

The structural features of the primers provide a strong foundation for examining the relationship between genetic variations, gene expression, adaptation, and phenotypic traits. Moreover, the specificity of these primer sequences reduces the risk of false-positive results while improving the accuracy of the amplification process. Through bioinformatics-assisted primer design processes, SCoT primers can be further optimized by expanding the method's applicability across different plant species and genetic diversity analyses. The targeted design of these primers toward coding regions makes the SCoT technique an indispensable tool in molecular biology and genetic studies for associating genetic polymorphism with biological functions [23, 24].

21.5.1.2 PCR-based economic technique

The SCoT method, as a PCR-based technique, offers a significant economic advantage in genetic diversity analysis. PCR is a widely used and cost-effective biotechnological method capable of amplifying genetic material using sequence-specific primers. SCoT utilizes this fundamental PCR principle, and since it does not require specialized enzymes or expensive reagents to detect genetic polymorphisms, it can be applied in laboratories at a low cost. This makes SCoT markers an accessible option, particularly in resource-limited laboratory and research environments. The SCoT technique significantly lowers the cost of genetic studies, and because PCR is so widely used, it can work with common PCR equipment found in many lab settings, thus eliminating the need for expensive chemical reagents or additional specialized equipment. As a result, SCoT markers provide a sustainable and cost-effective alternative for genetic diversity analysis, breeding programs, and biological research. Furthermore, since the SCoT technique offers high precision and specificity in genetic material analysis, it can be utilized without compromising the quality of the data obtained at a low cost. The low-cost nature of SCoT markers encourages their widespread use in large-scale genetic analyses, particularly in areas such as biodiversity studies, genetic resource conservation, and agricultural breeding strategies. This makes the SCoT method an economically sustainable and efficient tool for large-scale applications [25, 26].

21.6 Advantages and limitations of SCoT markers

SCoT markers are a powerful and effective molecular marker system for studying genetic diversity, genetic variation, and gene functions because they target translation start codons and nearby genetic regions. The fundamental working principle of SCoT markers is to identify polymorphisms in coding regions and link them to functional genes that are physiologically significant. With these characteristics, SCoT markers offer a wide range of applications, from evaluating genetic resources to genetic mapping and breeding studies [27].

However, like any molecular technique, SCoT markers come with specific advantages and limitations. The advantages of the method can be summarized as high reproducibility, ease of implementation, and the ability to obtain biologically meaningful genetic information. Since SCoT markers can detect polymorphisms in both coding and environmental genetic regions, they provide more specific and meaningful data compared to other random markers by focusing on the functional aspects of genetic variation. Moreover, the fact that these markers do not require specific genetic information or pre-existing sequence data for any species makes them easily applicable in new studies and across a wide range of plant species. However, there are several restrictions on SCoT markers. Due to the nature of the process, amplification products usually exhibit a dominant marker characteristic, which might restrict the resolution of genetic investigations and make it challenging to identify heterozygous individuals. Additionally, because SCoT markers target regions near the start codons, the data produced may not be comprehensive enough for phylogenetic or population genetics analyses when genetic variation is limited to this particular region. Furthermore, as the technique is PCR-based, technical factors such as DNA quality and reaction conditions, which might affect the results of the analysis, are critical for its effectiveness. In conclusion, while SCoT markers offer an innovative and functional tool for genetic research and molecular breeding studies, it is essential for researchers to consider the limitations of this method in relation to their specific research objectives and conditions. With a well-planned study design, SCoT markers can provide high-resolution and meaningful data, making valuable contributions to understanding genetic diversity, gene functions, and adaptive genetic variation [28–30]. Compared to other commonly used molecular marker techniques, such as random amplified polymorphic

DNA (RAPD) and inter-simple sequence repeat (ISSR), SCoT markers have several significant advantages.

21.6.1 Advantages

21.6.1.1 Sensitivity and relevance

SCoT markers offer a technique with high sensitivity and relevance in genetic diversity analyses. SCoT detects genetic polymorphisms by targeting DNA sequences near the coding regions surrounding the translation start codon (ATG). Since coding regions contain the parts of the genetic material that encode protein production, polymorphisms in these regions have the potential to reflect the direct biological impacts of genetic variations on functional processes. Therefore, SCoT markers not only detect genetic variation but also provide valuable information for understanding the relationship between these variations and biological processes, such as phenotypic traits and environmental adaptation. The sequences targeted by SCoT markers are located in critical regions where translation is initiated, which directly influences genetic expression. Analysis of polymorphisms in these domains allows for the generation of more direct and meaningful data on the functional consequences of genetic variation. This feature makes SCoT markers particularly advantageous for researchers investigating the biologically significant and adaptive effects of genetic diversity. In particular, SCoT markers are especially helpful for researching significant genetic processes like disease resistance, metabolic alterations, and environmental stress adaptation. The SCoT approach makes it possible to assess genetic variations in the context of their biological functional implications as well as genetic differences. Because of this, it is an essential tool in functional genomic research that aims to comprehend the genetic underpinnings of resilience and environmental adaptability. SCoT markers allow for a deeper understanding of the relationship between genetic variation and biological diversity and how an organism reacts to its surroundings [31–33].

21.6.1.2 High reproducibility

The SCoT method stands out as a molecular marker technique that offers high reproducibility. The specificity in the design of SCoT primers is a key factor that enhances the inter-laboratory consistency of the method. SCoT primers are designed with highly specific sequences found in the coding regions surrounding the translation start codon. This ensures that the primers amplify the same target DNA regions in each application, increasing the reproducibility of the results. Additionally, the PCR protocol used for the SCoT technique can be standardized and optimized, ensuring consistent and comparable results across different laboratories by following the same protocol. This feature significantly enhances the effectiveness of SCoT markers in large-scale genetic analyses and international research networks. In contrast, other molecular marker techniques, such as RAPD, often face challenges due to the random nature of primer sequences, which can negatively impact the accuracy and reproducibility of the analysis. In RAPD, the random selection of primers can result in varying degrees of primer-target DNA matching across different laboratories or time points, leading to erroneous or inconsistent results. However, the SCoT technique largely eliminates such issues by ensuring that primer sequences are highly specific and accurately reflect genetic diversity. As a result, SCoT markers provide high reproducibility and reliability, making them particularly valuable in genetic diversity studies, breeding programs, and biological research applications, where consistent data is critical [33–35].

21.6.1.3 Versatility

SCoT markers offer significant versatility in genetic analyses, enabling the examination of both coding (expression-related) and noncoding (regulatory and other nonfunctional) DNA regions. SCoT markers target the translation start codon (ATG) and the surrounding sequences, allowing not only the detection of genetic variations but also providing insights into the biological functional contexts of these variations. This approach encompasses not only polymorphisms in protein-coding genes but also variations in noncoding regions that may influence gene expression and genetic regulation. This versatile analytical capacity makes the SCoT technique useful across a broad range of applications. The versatility provided by SCoT markers is especially advantageous in fields like population genetics, phylogenetic analyses, genetic mapping, and breeding studies. SCoT markers are an excellent tool for population genetics research because they allow researchers to analyze genetic variation among populations and evaluate the functional impacts of that variability. It is easier to understand adaptation processes and environmental adaptability when genetic differences are understood in a functional context. In phylogenetic analyses, SCoT markers can be employed to identify shared ancestry and evolutionary relationships between species. The analysis of both coding and noncoding regions enables a more comprehensive evaluation of interspecies genetic differences. In genetic mapping and breeding studies, SCoT markers play a crucial role in mapping genetic markers associated with targeted traits. Such studies are especially valuable in agricultural plant breeding, where understanding the genetic foundations of key traits such as environmental stress tolerance, yield improvement, or disease resistance is essential. Ultimately, SCoT markers have broad application potential in genetic research, offering a powerful and versatile tool for both fundamental scientific research and practical applications [31–35].

21.6.1.4 High polymorphism rate

The high polymorphism rate provided by SCoT markers in genetic analyses offers a significant advantage, making studies of genetic diversity more in-depth, comprehensive, and reliable. Due to their high resolution and specificity in detecting genetic variation, SCoT markers stand out as an effective analytical method, particularly in genetically heterogeneous natural plant populations. These populations, which are frequently taken from the wild, usually exhibit a wide range of genetic variations that enable them to adapt to their surroundings. Correct instruments to evaluate population dynamics and ecosystem structure are necessary to comprehend this diversity. The ability of SCoT markers to clearly detect genetic variations in such populations is critical for uncovering the functional biological impacts of these variations.

This high polymorphism rate not only enables a more detailed understanding of genetic diversity but also contributes significantly to applied biological studies, such as ecosystem management and genetic resource conservation. Ecosystem management, particularly in ensuring sustainability and preserving biodiversity, requires the monitoring and assessment of genetic diversity within plant populations. The use of SCoT markers in this field allows for the tracking of changes in the genetic structures of ecosystems over time, facilitating the early detection of negative processes, such as genetic degradation or the decline in adaptive abilities of species. Similarly, the conservation of genetic resources is of vital importance in developing strategies to preserve the genetic diversity of endangered plant species. For these kinds of investigations, SCoT markers offer a formidable instrument that expands their potential in conservation biology and the long-term preservation of genetic variation. Therefore, the high polymorphism rate provided by SCoT markers is an important tool for applied biological investigations pertaining to biodiversity conservation and environmental management, as well as basic research. These markers promote the conservation of genetic resources and the sustainable management of ecosystems by offering a solid scientific basis for the precise identification and comprehension of genetic variation in plant populations [36–38].

21.6.1.5 Economic and easy applicability (same sequence issue)

The SCoT methodology is an economically low-cost, PCR-based molecular marker technology, similar to other PCR-based molecular marker methods such as RAPD and ISSR (Inter Simple Sequence Repeat). PCR-based technologies are often considered one of the most cost-effective options in genetic analyses because they offer a wide range of applications without requiring specialized enzymes or expensive reagents. While SCoT benefits from these advantages of PCR, it also provides data with higher biological relevance compared to other methods. SCoT primers' target-oriented design makes this method more successful and efficient. SCoT guarantees that the amplified regions are biologically significant by focusing on particular DNA sequences surrounding the translation initiation codon. This makes it possible for the data to show how genetic variation affects phenotypic features or environmental adaptability, in addition to detecting genetic variation. In this way, the SCoT technique adds value not just by detecting genetic differences in diversity analyses, but also by emphasizing the biological, functional, and adaptive significance of those differences. These advantages of SCoT not only enable the generation of biologically meaningful results but also make the analysis process more efficient. Like other PCR-based methods, SCoT is easily applicable in laboratory settings. Due to the widespread use of PCR, the SCoT technique can work with other standard equipment used in genetic analyses, ensuring high accuracy in laboratory applications. Therefore, SCoT offers an economically accessible option while providing in-depth, biologically relevant data, offering researchers a significant advantage in genetic analysis [36–38].

21.6.2 Restrictions

21.6.2.1 Polymorphism not fully restricted to coding regions

SCoT markers typically focus on DNA sequences surrounding coding regions to detect genetic diversity. However, the polymorphisms detected by these markers may not always be directly associated with coding genes. While the regions targeted by SCoT markers are adjacent to coding genes, polymorphisms in these regions may indicate genetic variation but may not always be related to phenotypic traits or gene expression. This could create some resolution limitations, especially for researchers trying to investigate the mechanisms of specific biological processes or phenotypic traits. Polymorphisms near coding genes have the potential to more clearly indicate the biological effects of genetic diversity, but some of these polymorphisms may only reflect fundamental differences in genetic structure and may struggle to establish a direct connection with functional biological processes. Another limitation is that SCoT markers can detect polymorphisms in regions that are not directly related to coding regions and do not directly influence gene expression or phenotypic traits. These types of polymorphisms can make data analysis more complex for researchers aiming to examine the effects of genetic variation on specific genetic networks or biological processes. To fully understand the biological significance of genetic diversity, it is necessary to consider not only polymorphisms in coding regions but also variations in regulatory and other noncoding areas. The ability of SCoT markers to fully resolve these complex relationships can sometimes make it difficult to interpret the biological relevance of polymorphism data from these regions. The applicability and resolution of SCoT markers may be limited by these restrictions; hence, it is crucial to carefully plan the analyses when employing this method. Additional techniques and analytical approaches could be needed in research to directly link the polymorphisms offered by SCoT markers with biological processes in order to comprehend the genetic basis of phenotypic features. Nevertheless, the genetic information provided by SCoT markers establishes a solid foundation for areas such as functional genomic research and plant breeding and, when used correctly, offers high-value insights. The requirement for bioinformatics analysis of data obtained using SCoT markers, especially in large-scale studies, necessitates the use of comprehensive bioinformatics tools. From primer design to the evaluation of genetic variations, this process may require a high level of technical knowledge and software infrastructure. This could pose a challenge for research teams with limited bioinformatics resources and training [39].

21.6.2.2 Lack of reference genome

SCoT markers provide a significant advantage in detecting and analyzing genetic diversity as a technique that does not require reference genome information. However, this characteristic may introduce some limitations, particularly for research aiming to conduct more detailed genomic analyses and extensive functional studies. While SCoT markers are effective in detecting genetic diversity by focusing directly on translation initiation codons, the absence of reference genome information can make it difficult to perform in-depth analyses of specific genetic variations. This issue can limit the applicability of SCoT markers, especially in species with limited genomic knowledge, such as medicinal and aromatic plants, where genetic databases are underdeveloped. Researchers seeking more detailed and comprehensive analyses in a genomic context may find that the lack of reference genomes requires additional reference sources to fully interpret the biological significance of data obtained with SCoT markers. In particular, when analyzing the genetic structure of a species from a broader perspective, the lack of a reference genome can hinder the accurate identification of different variants and the deeper exploration of functional genomic relationships. This deficiency can make it more challenging to conduct detailed analyses of genetic pathways involved in the biosynthesis of pharmacological compounds in medicinal plants. Additionally, the lack of a reference genome may make it more difficult to develop novel biotechnological applications as well as to gather fundamental data about genetic diversity. For instance, genetic studies aimed at identifying new drug candidates or plant components with high biological activity may be hindered by the lack of reference genomes, limiting the ability to identify some genetic variations and thus reducing marker effectiveness. Furthermore, in species like medicinal and aromatic plants, where there are gaps in genomic data, it can be more challenging to relate the results provided by SCoT markers to broader biological processes. As a result, while the independence of SCoT markers from reference genomes is advantageous in some cases, it can also limit their use in more detailed and comprehensive genomic analyses. It might be challenging for researchers to fully utilize the potential of SCoT markers in species without reference genomes, and in these situations, additional genomic data sources might be required. [38–40].

21.6.2.3 Standardization of the technique

The applicability and reliability of SCoT markers can vary due to differences in protocols and optimization methods used in different laboratories. In particular, the variety of primer sets and PCR settings can make it more difficult to compare the polymorphism data obtained in genetic investigations using SCoT markers. Since the PCR conditions used directly affect the amplification efficiency and accuracy of the targeted regions, such technical differences can hinder the standardization of results. This problem can be a serious obstacle, especially for research examining genetic variations among several populations and inter-species comparisons. The consistency of SCoT marker data across different laboratories depends on the standardization of the protocol and optimization of the methods. Without this, inconsistent results may be obtained between researchers, which could negatively impact the reliability and reproducibility of the overall findings. Despite being a powerful and practical tool for genetic diversity and functional genomics research, SCoT markers have limitations that should be carefully evaluated in light of the study's objectives and the genetic characteristics of the species under investigation. The genetic variation data provided by SCoT markers, especially targeting polymorphisms associated with coding regions and offering a relatively economical method, present an attractive option for laboratories with limited budgets. However, while this method is suitable for basic genetic analyses and genetic diversity detection, it may not be sufficient for deeper biological analyses or functional genomics studies that require high resolution. For such studies, integrating other molecular marker systems or bioinformatic tools that can provide higher resolution alongside SCoT markers will enable more comprehensive and accurate analysis of genetic variation. For example, additional omic data (e.g., transcriptomic and proteomic) and more sophisticated bioinformatic tools may be used for high-resolution genetic analyses or more detailed studies of genetic networks and biological processes. This approach will allow for deeper interpretation of the basic genetic data obtained from SCoT markers, facilitating more robust biological inferences. In conclusion, while SCoT markers are a powerful tool for genetic diversity analysis and functional genomics studies, the development of standard protocols and optimization processes is critical for methodological consistency and the comparability of results. The use of the method should be customized according to the targeted biological questions and the genetic features of the species under study. Additionally, considering the advantages and limitations of SCoT markers, it is recommended that they be supported by additional methods for studies requiring higher resolution and more in-depth examination of biological contexts. [41]

21.7 Conclusion and future perspectives

SCoT polymorphism markers have emerged as a significant innovation in recent years for studying genetic diversity in medicinal and aromatic plants. In molecular biology and genetic research, SCoT markers have become a powerful tool due to the unique advantages they provide, particularly in detecting genetic variation and conducting functional genomics studies. By focusing on translation initiation codons, SCoT markers not only help in understanding genetic diversity but also reveal its relationship with biological functions. This feature provides valuable insights not only into identifying genetic variation in plants but also into critical areas such as ecosystem management, the conservation of biological diversity, and the sustainable management of genetic resources [42]. In particular, in medicinal and aromatic plants, genetic analyses using SCoT markers allow for a deeper understanding of the plants' metabolic traits, stress tolerance, biological activity, and pharmacological potential. Identifying genetic diversity helps us understand how these plants adapt to various environmental conditions, the adaptation processes in their genetic structures, and the genetic variations involved in secondary metabolite production. In this context, SCoT markers are critical for the conservation and sustainable use of the genetic resources of plant species. Moreover, the availability of these markers contributes significantly to the design of breeding programs for biologically valuable species, such as medicinal and aromatic plants [42].

The economic and highly applicable nature of SCoT markers is a key factor that broadens their application in plant genetics. This strategy presents a compelling alternative for genetic research with constrained laboratory budgets, and its potential can be fully realized when paired with additional biotechnological instruments that offer higher resolution for more thorough functional investigations. The versatile use of SCoT markers extends beyond basic genetic analyses, holding considerable potential in biotechnological and agricultural applications as well. In the future, further development of SCoT markers and integration with new technologies, particularly advances in bioinformatics, will allow for more precise and comprehensive analysis of genetic diversity. Additionally, enhancing the effectiveness of SCoT markers in plant species lacking reference genomes will increase their applicability across a broader range of plant diversity. SCoT markers are expected to lay a solid foundation for future research in this field, paving the way for new applications in modern agriculture and plant biotechnology. Ultimately, SCoT markers will not only provide a solid scientific foundation for understanding the genetic makeup of medicinal and aromatic plants and their breeding but will also serve as a critical tool in the conservation of biological diversity, the management of genetic resources, and the optimization of environmental adaptation processes in the future [43].

21.8 Comparison of SCoT markers with other techniques

The comparison of SCoT markers with other molecular marker techniques highlights the unique advantages of this method more clearly.

21.8.1 Comparison with RAPD

RAPD markers are a method that amplifies randomly selected DNA regions in the genome without any specificity. This method has been a significant tool for screening genetic diversity and genome mapping, especially in the early stages. However, the low reproducibility of RAPD markers, random primer design, and the sensitivity of analysis results to environmental conditions limit the reliability of the method. In contrast, the SCoT marker system offers a more specific approach by targeting the translation start codon (ATG) and surrounding coding regions. This method allows for the study of regions that play a crucial role in gene regulation and expression levels, rather than biologically insignificant DNA regions. This specificity makes SCoT markers not only more reproducible but also more meaningful in understanding the functional and evolutionary characteristics of genetic material. The regions targeted by SCoT provide an advantage in determining the role of functional genes in genetic diversity studies within the genome. Therefore, the SCoT marker system stands out as a more reliable and biologically meaningful alternative compared to RAPD in genetic studies [44, 45].

21.8.2 Comparison with ISSR

ISSR markers are a molecular marker system frequently used for genetic diversity and polymorphism analysis, targeting regions between simple repetitive sequences known as microsatellites in the genome. The ISSR method works by binding primers to the complementary sequences of microsatellite repeats and is particularly suitable for detecting polymorphisms in noncoding DNA regions. While this feature makes ISSR an effective tool in genetic diversity analysis, its biological significance can be limited, as these repetitive regions are often located in nonfunctional or regulatory-independent areas. On the other hand, the SCoT marker system targets the translation start codon (ATG) and surrounding coding regions. This targeting strategy enables SCoT to analyze not only genetic diversity but also polymorphisms associated with gene expression. The examination of genetic variation in coding regions is crucial for determining gene functions and understanding biological processes. Therefore, SCoT markers are more advantageous than ISSR in functional genetic studies, particularly in the analysis of polymorphisms that can be linked to gene expression. This superiority of SCoT allows for a deeper understanding of the functional and evolutionary aspects of genetic material, while also providing more reliable results in applied fields such as gene resource development and plant breeding. While ISSR covers a broader region of the genome, SCoT's specific targeting of coding regions makes it particularly prominent in functional genomic research [44–46].

21.9 The role of SCoT markers in genetic diversity studies

The SCoT marker system is an innovative approach that broadens the scope of genetic studies by focusing on genetic diversity analysis and gene regions. This marker system, with its ability to detect polymorphisms surrounding coding regions that play crucial roles in gene expression and genetic regulation, provides solutions to fundamental issues encountered, particularly in the field of plant biology. SCoT markers go beyond traditional methods that target random DNA regions, enabling the identification of genetic variations that can be directly related to biological functions. The wide range of applications of this marker system, including population genetics, phylogenetic analysis, gene mapping, and breeding studies, demonstrates its versatility and effectiveness. For example, SCoT markers facilitate a detailed examination of genetic relationships between plant populations, leading to more accurate results in a phylogenetic context. Used as a valuable tool in gene mapping studies, this technique allows the identification of functional regions within genetic material, thereby contributing to the development of more effective strategies in breeding programs [46]. Another important aspect of SCoT markers is their critical role in the development of strategies for the conservation and sustainable use of genetic resources. Genetic diversity is essential for plants' adaptation to environmental stress conditions and agricultural productivity. SCoT markers provide the opportunity for detailed analysis of this diversity, leading to a better understanding of the genetic resources of plants with medicinal, aromatic, and agricultural significance. Furthermore, this marker system offers a reliable method for evaluating genetic variation and developing new genotypes in biotechnological applications. In conclusion, SCoT markers present a targeted, functional, and biologically meaningful approach in genetic diversity analysis, making them one of the innovative methods in molecular biology and genetics. This system has broad potential for both basic and applied research and stands out as an indispensable tool, especially in developing sustainable solutions for the management of genetic resources [47, 48].

21.10 Applicability of SCoT markers in medicinal and aromatic plants

Medicinal and aromatic plants are subjects of intensive research worldwide due to their high economic and biological value, both in the healthcare sector and various industrial applications. The secondary metabolites that confer therapeutic and aromatic properties to these plants are directly related to their genetic makeup, and their capacity for adaptation to environmental conditions is largely shaped by their levels of genetic diversity. Genetic diversity is also a critical parameter for the sustainable management of natural populations of these plants, enhancing their resistance to stress factors such as climate change and determining target traits in breeding programs. The SCoT marker system provides an effective and reliable tool for analyzing the genetic diversity and population dynamics of medicinal and aromatic plants in detail. The discovery of biologically relevant variations in genetic material is made possible by SCoT markers, which target translation initiation codons and the functional genomic sequences surrounding these sites. This feature provides a significant advantage, especially in identifying genetic regions responsible for secondary metabolite production and uncovering variations in these regions [49, 50]. The applicability of SCoT markers in medicinal and aromatic plants goes beyond merely screening for genetic diversity; it also allows for the functional analysis of genetic variations. This enables the study of specific genes' effects on biosynthetic pathways, the identification of genetic relationships among different ecotypes, and the development of strategies for the conservation of natural populations. Moreover, the high repeatability and specificity of SCoT markers make this method an ideal tool for both intraspecific and interspecific genetic studies. In conclusion, the SCoT marker system plays a significant role in modern molecular biology studies aimed at understanding the genetic structures, secondary metabolite profiles, and population dynamics of medicinal and aromatic plants. The detailed and functional analyses provided by this system hold great potential for the sustainable management of genetic resources and the development of innovative solutions in biotechnological applications [51–53].

21.11 Correlation with metabolite diversity

One of the key factors determining the high economic and biological value of medicinal and aromatic plants in the pharmaceutical, cosmetic, and food industries is the presence of secondary metabolites with biological activity. These metabolites not only support the plants' defense mechanisms against environmental stresses but also stand out due to their therapeutic effects on human health and their potential for industrial applications. Important secondary metabolites, such as terpenoids, alkaloids, and phenolic compounds, are synthesized by specific enzymes involved in biosynthetic pathways, and the genetic foundation of these pathways is directly linked to genetic diversity and polymorphisms [50–53]. The SCoT marker system stands out as an effective tool for detecting polymorphisms in genes related to metabolite biosynthesis, thereby uncovering the connection between these genetic variations and secondary metabolite profiles. SCoT markers, by targeting translation initiation codons and the surrounding genetic regions, provide high resolution for understanding the functional characteristics of genetic material. For instance, SCoT markers can be used to analyze genes encoding enzymes involved in the mevalonate pathway for terpenoid biosynthesis or phenylalanine ammonia-lyase (PAL), which plays a role in the production of phenylpropanoid derivatives. These analyses not only help in understanding the genetic foundations of biosynthetic pathways but also enable the identification of molecular-level targets aimed at enhancing the production capacity of bioactive compounds. The high specificity and repeatability offered by SCoT markers provide critical functional information at both genetic and biochemical levels, playing a key role in identifying genotypes with high bioactive compound content. Specifically, SCoT markers are a valuable method for investigating the effects of genetic variation on metabolite production across different plant species and selecting high-yielding genotypes based on these insights. This approach facilitates the effective evaluation and sustainable use of genetic resources, while also supporting the development of innovative solutions in plant breeding and biotechnological applications. In conclusion, the SCoT marker system offers a powerful tool for understanding genetic variations related to metabolite biosynthesis, contributing significantly to better understanding the genetic resources of medicinal and aromatic plants and developing strategies to enhance the economic value of these plants [54–56].

21.12 Population genetics and genetic diversity

The SCoT marker system is an effective and innovative tool for measuring genetic diversity and gaining a detailed understanding of the genetic structures of populations. This method, in particular, offers great potential for identifying genetic variation in medicinal and aromatic plant populations. Genetic diversity is a critical biological indicator that enhances the capacity of plant populations to adapt to environmental stress factors and supports the long-term sustainability of species. The analysis of genetic variation affecting secondary metabolite production and adaptive processes is of great importance for both the conservation of natural populations of these plants and the optimization of their agricultural and industrial uses [55–57]. SCoT markers target translation initiation codons and the surrounding functional genetic regions, providing analyses that highlight the biological significance and functional dimension of genetic variation. The high repeatability and specificity of these markers facilitate the comparison of genetic diversity across different populations. For example, comparing the genetic structures of medicinal and aromatic plant populations from various geographical regions enables the identification of genetic variants that confer adaptation to environmental stress factors. This information is valuable for understanding the adaptation capacities of ecotypes and developing strategies for conservation biology. In plant species with a narrow gene pool, the use of SCoT markers plays an important role in evaluating existing genetic resources and developing strategies to increase genetic diversity. In breeding programs, SCoT markers can contribute to the development of targeted breeding strategies between genotypes to broaden the genetic base. Additionally, this method can be applied in conservation programs aimed at preserving the genetic diversity of rare and endangered medicinal and aromatic plant species. In conclusion, the SCoT marker system provides a functional and biologically meaningful approach to genetic diversity analysis, making it an indispensable tool in both basic scientific research and applied plant biology and breeding studies. This technique significantly contributes to the development of modern approaches for the conservation, sustainable use, and enhancement of the adaptive capacities of the genetic resources of medicinal and aromatic plants [56–58].

21.13 Geographical distribution and genetic differences

SCoT markers are a highly effective and powerful molecular tool for investigating the genetic effects of geographic isolation and environmental factors on plant populations. Variations in the genetic structure of the same plant species growing in different ecological regions provide valuable insights into how genetic diversity is shaped by environmental adaptation and evolutionary processes. SCoT markers, by binding specifically to translation initiation codons and the surrounding genetic sequences, enable the detection of biologically meaningful genetic variations. This characteristic makes SCoT analysis an unparalleled method for assessing the genetic impacts of geographic and ecological differences in detail. Plant populations of the same species living in different ecological regions may exhibit genetic differences depending on habitat conditions. For example, the genetic structure of a population growing in a humid forested area may be distinctly different from that of the same species living in arid desert conditions due to the adaptive processes that help organisms cope with environmental stress factors. The use of SCoT markers allows for a detailed examination of these genetic differences and helps uncover adaptive genetic variation. This type of information not only aids in understanding the genetic structure of populations but also contributes to explaining the molecular-level selection processes occurring under environmental pressures. Genetic data obtained using SCoT markers form a crucial foundation for developing strategies for the conservation of natural populations of medicinal and aromatic plants. Preserving the genetic diversity of plant species that are threatened by habitat loss, climate change, and other anthropogenic effects is critical for their sustainable management. SCoT analyses provide detailed information on the genetic structure of plant populations, helping identify priority areas in conservation biology and contributing to the development of management strategies required to ensure the long-term sustainability of populations in suitable habitats. In conclusion, the SCoT marker system is an extremely effective method for studying the genetic variations of various plant species, particularly medicinal and aromatic plants, in response to geographical and environmental factors. This system plays an important role in the development of scientifically based and innovative approaches for the conservation of natural populations and the sustainable management of genetic resources [59].

21.14 Examples of applications

There are several studies that demonstrate the effectiveness of SCoT markers in analyzing genetic diversity and biological traits in medicinal and aromatic plants. These studies have significantly contributed to understanding the impact of genetic variation on secondary metabolite production and biological activities by examining the genetic structures of plants in detail.

Rosmarinus officinalis (Rosemary): SCoT markers have been successfully used to analyze the genetic diversity among natural populations of rosemary from different geographical regions Figure 21.3. These studies have revealed the effects of genetic differences between populations on secondary metabolite profiles. Specifically, significant correlations were found between genetic variation and biologically important traits, such as terpenoid content and antioxidant activity. For example, certain genotypes showed higher terpenoid content and stronger antioxidant activity, which were linked to genetic polymorphisms at the molecular level. These findings facilitate the expansion of the genetic base in breeding programs for high-value plants like rosemary, enabling the selection of genotypes with desired traits [60–62].

Ocimum basilicum (Basil): SCoT markers have also proven to be an effective tool in identifying genetic differences among basil species and varieties Figure 21.4. These analyses have facilitated the classification of different basil genotypes based on their genetic variation and uncovered relationships between genetic diversity and economically significant traits, such as essential oil content. In particular, the identification of genetic variations responsible for the production of volatile compounds like linalool and eugenol has provided a scientific basis for breeding programs aimed at increasing the commercial value of basil species. Such genetic analyses also contribute to the development of strategies for the conservation and sustainable management of the genetic resources of these species.

Figure 21.3: *Rosmarinus officinalis* (rosemary) plant [63].

Figure 21.4: *Ocimum basilicum* (basil) plant [64].

The applications of SCoT markers in these examples demonstrate that they are a powerful tool for better understanding the genetic diversity and biological traits of medicinal and aromatic plants. This technique not only reveals genetic variation but also elucidates how these variations affect biological activities and secondary metabolite production, making it crucial in both fundamental scientific research and applied breeding programs. Ultimately, SCoT markers provide an indispensable method for developing innovative approaches to the conservation, breeding, and sustainable use of genetic resources in medicinal and aromatic plants [65, 66].

21.15 The importance of SCoT markers in medicinal and aromatic plants

SCoT markers stand out as a scientifically robust and innovative tool for studying genetic diversity, population structures, and genetic variations related to metabolite biosynthesis in medicinal and aromatic plants. This technique, which focuses on translation initiation codons and the surrounding genetic regions, has the capacity to reveal biologically meaningful and functional aspects of genetic variation. SCoT markers enable the integration of genetic information with phenotypic traits, providing significant contributions to both theoretical and applied plant biology and genetic research. Medicinal and aromatic plants play a crucial role in agriculture, healthcare, and industry due to their ability to adapt to environmental stress factors and produce biologically active secondary metabolites. SCoT markers offer an effective method for understanding genetic diversity and population dynamics in these plants, allowing for the development of strategies for the conservation and sustainable use of natural resources. For example, the use of SCoT markers to study genetic variations in the genetic structures of the same species growing in different ecological regions provides critical data for understanding the effects of environmental factors on genetic diversity. Such analyses contribute to identifying priority areas for managing genetic resources and conserving threatened populations. The high reproducibility and specificity provided by SCoT markers allow for detailed correlation of genetic information with phenotypic traits. This characteristic enhances the effectiveness of the technique, especially in improving target traits such as stress tolerance, high yield, strong antioxidant capacity, and rich metabolite profiles. SCoT analyses uncover potential genetic links between stress tolerance genes and secondary metabolite biosynthesis, offering a scientific basis for molecular-level breeding studies. For instance, identifying genotypes with high essential oil content or detecting genetic variants resistant to stress conditions is possible through the application of SCoT markers. Additionally, SCoT markers play an important role in developing strategies to increase genetic diversity in species with a narrow genetic base. This contributes critically to hybridization and selection work aimed at improving target traits in modern breeding programs. This technique is of great importance in conservation biology programs aimed at preserving the genetic structures and biological diversity of natural populations. In conclusion, SCoT markers provide a comprehensive approach to understanding the genetic potential of medicinal and aromatic plants and integrating this knowledge into the development of phenotypic traits. This technique has become an indispensable tool for the conservation, sustainable management, and development of economically valuable new varieties of plants. Through the deepening of genetic knowledge and the enhancement of molecular breeding efforts, SCoT markers hold a pioneering place in plant biology and genetic research [67, 68].

21.16 The role of SCoT markers in genetic diversity analysis

SCoT markers are a powerful molecular tool that allows for the direct detection of genetic variations associated with coding regions. These features make SCoT markers not only important for identifying genetic diversity but also for studying the biological and functional outcomes of this diversity. SCoT markers provide in-depth insights into how genetic variations are related to metabolic pathways and biological processes in plants. In particular, the analysis of complex traits such as bioactive compound synthesis, stress tolerance, and environmental adaptation mechanisms is significantly strengthened by the genetic data provided by SCoT markers. The economic value of medicinal and aromatic plants largely depends on the diversity and abundance of secondary metabolites found in these plants. These metabolites carry significant medicinal and pharmaceutical value for humans, in addition to their biological functions in the plants. SCoT markers offer a valuable tool for understanding the genetic control mechanisms associated with secondary metabolite production in these species, helping to determine their biological potential. By targeting polymorphisms in coding regions, SCoT markers provide a significant advantage in identifying genetic variations related to metabolic pathways and understanding their biological consequences. The identification of genetic polymorphisms related to the production of bioactive compounds plays a critical role in the conservation of natural plant populations and the sustainable management of their genetic resources. The genetic information provided by SCoT markers is also useful in studying the population genetics and phylogenetic structures of these species, forming a foundation for the conservation of genetic diversity and the sustainable use of healthy genetic resources. Moreover, the data provided by SCoT markers contributes greatly to the development of high-yielding plant varieties with bioactive compounds in plant breeding programs. Identifying genetic polymorphisms and their relationship to secondary metabolite production can guide the selection of genetically superior genotypes and the development of more efficient plant varieties [69, 70].

The SCoT markers offer an extremely effective method for understanding genetic diversity and its functional outcomes in medicinal and aromatic plants. These markers go beyond merely detecting genetic variation; they uncover the relationship between these variations and metabolic pathways and biological functions, carrying significant potential for plant biology, breeding, and biotechnology. Therefore, the broader use of SCoT markers in these areas is crucial for the conservation, sustainable use, and enhancement of genetic diversity in medicinal and aromatic plants. SCoT markers are a powerful molecular tool that enables the direct detection of genetic variations associated with coding regions. This characteristic makes SCoT markers an essential platform not only for determining genetic diversity but also for examining the biological and functional consequences of this diversity. SCoT markers provide in-depth insights into how genetic variations are linked to metabolic pathways and biological processes in plants. Specifically, the analysis of complex traits such as bioactive compound synthesis, stress tolerance, and environmental adaptation mechanisms is significantly enhanced by the genetic data provided by SCoT markers. The economic value of medicinal and aromatic plants largely depends on the diversity and concentration of secondary metabolites in these plants. These metabolites contribute to the biological functions of plants and have considerable medicinal and pharmaceutical value for humans. SCoT markers offer a valuable tool for better understanding the genetic control mechanisms related to secondary metabolite production, helping to determine the biological potential of these plants. By targeting polymorphisms in coding regions, SCoT markers provide a distinct advantage in identifying genetic variations related to metabolic pathways and understanding their biological outcomes. In particular, identifying genetic polymorphisms involved in bioactive compound production plays a critical role in the conservation of natural plant populations and the sustainable management of their genetic resources. The genetic information provided by SCoT markers also facilitates the study of the population genetics and phylogenetic structures of these species, laying the foundation for the conservation of genetic diversity and the sustainable use of healthy genetic resources. Furthermore, the data generated by SCoT markers contribute significantly to plant breeding programs aimed at developing high-yielding and bioactive compound-rich plant varieties. The identification of genetic polymorphisms and their relationship to secondary metabolite production can guide the selection of genetically superior genotypes and the development of more efficient plant varieties. In conclusion, SCoT markers provide an extremely effective method for understanding the genetic diversity of medicinal and aromatic plants and the functional consequences of this diversity. Beyond detecting genetic variation, these markers reveal the relationship between these variations and metabolic pathways and biological functions, offering significant potential in the fields of plant biology, breeding, and biotechnology. Therefore, the broader use of SCoT markers in these areas is of critical importance for the conservation, sustainable use, and enhancement of genetic diversity in medicinal and aromatic plants [71–74].

SCoT markers play a major role in the analysis of the following:

21.16.1 Breeding studies

Breeding studies play a crucial role in optimizing plant species for desired traits. In this process, the use of genetic markers associated with phenotypic traits holds significant potential, particularly in improving targeted agronomic characteristics such as stress tolerance, disease resistance, yield, quality traits, and metabolite production. The use of genetic markers as molecular signals reflecting the genetic makeup of plants offers a faster and more precise approach to understanding their relationship with phenotypic traits. This provides a critical advantage for making breeding programs more targeted and achieving results more rapidly. In particular, the use of genetic markers to increase resistance to environmental factors (such as temperature, drought, and salinity) can help develop plant species with enhanced stress tolerance. Molecular markers like SCoT, which identify genetic variations associated with stress tolerance, enable the rapid selection of genotypes with better stress resilience. Similarly, identifying genetic variations related to disease resistance is an important step in developing plant varieties with greater resistance to pathogens. Metabolite production, especially in medicinal and aromatic plants, is a key phenotypic trait that determines the economic value of plants. SCoT markers, by identifying polymorphisms in genes involved in metabolite biosynthesis, provide an effective tool for identifying genetic structures that can efficiently produce bioactive compounds. This allows the selection of genotypes with desired metabolite profiles and enhances the efficient production of these compounds. Breeding studies become more efficient with the integration of genetic and phenotypic data. Correctly associating genetic markers with phenotypic traits enables breeders to direct plants more quickly and accurately. This becomes particularly important for achieving long-term breeding goals. The use of SCoT markers strengthens these associations while significantly improving the accuracy and speed of the breeding process. Additionally, it contributes to the development of more effective breeding strategies, opening the door to significant improvements in sustainable agriculture and agricultural production [75, 76].

21.16.2 Population genetics

Population genetics is a branch of science that studies genetic diversity, population structure, and evolutionary dynamics of a species. SCoT markers, as a significant tool in this field, allow for an in-depth examination of genetic diversity and population structure. The proximity of SCoT markers to coding gene regions offers a major advantage in identifying genetic variations and their phenotypic manifestations. These markers provide a robust solution for detecting genetic diversity and understanding the genetic relationships among populations, as they specifically target variations that are important in functional genomics contexts.

Genetic diversity analyses conducted with SCoT markers play a critical role in studying the genetic structure of a population. These analyses not only determine the genetic diversity within populations but also reveal how genetic differences among populations relate to environmental factors, adaptation processes, and evolutionary history. The conservation of genetic diversity is an important factor that directly influences a population's resilience to environmental changes and stress factors. Population genetics studies using SCoT markers also contribute to evaluating the risks of genetic erosion. Genetic erosion refers to the reduction of genetic diversity within species, thereby decreasing their adaptability and long-term survival prospects. The risk of genetic erosion becomes more pronounced in small and isolated populations, where a decline in genetic diversity can be more significant. SCoT markers, with their high reproducibility and ability to target genetic variations, provide a key tool in identifying genetic erosion in such populations. This genetic information plays a crucial role in the development of conservation strategies. Analyses conducted using SCoT markers allow for a better understanding of the structural characteristics of different populations in nature, the design of strategies for the conservation of genetic diversity, and the monitoring of the effectiveness of conservation measures. In particular, in the conservation of endangered species, SCoT markers are an important tool for assessing the genetic structures of local populations and sustainably preserving their diversity. In conclusion, SCoT markers not only enable a more comprehensive understanding of genetic diversity, evolutionary processes, and adaptation mechanisms in population genetics studies but also play a vital role in identifying genetic erosion risks and developing conservation strategies. In these respects, SCoT markers are an extremely valuable molecular tool for research focused on the preservation of ecosystems and biodiversity [77, 78].

21.16.3 Ecosystem dynamics

Ecosystem dynamics is a field that examines the effects of environmental factors on the interactions between populations and species of organisms. Understanding the role of medicinal and aromatic plants within ecosystems not only reveals their biological and economic value but also demonstrates how these plants adapt to environmental stress factors and ecosystem changes. Gaining insight into the genetic makeup of medicinal and aromatic plant populations from different geographical regions allows for a deeper examination of their ecological roles and environmental adaptation processes. Variations in the genetic makeup of these plants directly influence their response to environmental stresses, climate change, and alterations in habitat conditions. The use of SCoT markers enables a detailed analysis of the genetic diversity and structure of plant populations, revealing how the observed genetic differences among populations from different geographical regions are related to environmental factors. This analysis serves as a vital tool in understanding environmental adaptation processes, as genetic variation demonstrates how plants adjust to environmental changes and stresses, and what genetic mechanisms enable such adaptations. Environmental factors such as geographic isolation, habitat diversity, and microclimatic differences can influence the genetic diversity and adaptation capabilities of plants. Plant populations growing in different ecological regions may differ genetically, and these differences provide critical insights into how each population adapts to its environmental conditions. SCoT markers, while identifying genetic differences among these populations, also analyze how these differences reflect ecosystem characteristics and adaptation strategies. In particular, the genetic diversity of medicinal and aromatic plants is associated with phenotypic traits such as secondary metabolite profiles, stress tolerance, and adaptation capacity, which are key to their production. The genetic data provided by SCoT markers play a critical role in understanding the diversity of these plants in their natural environments, their responses to stress factors, and their resistance to global environmental threats such as climate change. In conclusion, a detailed study of the genetic structure of medicinal and aromatic plant populations provides an understanding of environmental adaptation processes and aids in the development of strategies for their sustainable use. SCoT markers are a powerful tool for understanding the dynamics of plant ecosystems and revealing how these populations adapt to environmental changes. This approach provides a scientifically rich foundation for understanding the ecological role and genetic adaptation potential of these plants within ecosystems [79].

21.17 Technological advancements and the future of SCOT markers

Technological advancements have led to revolutionary innovations in the applications of molecular biology and genetic research. These developments are expanding the scope of SCoT markers and enhancing the precision of these methods. The integration of high-throughput DNA sequencing technologies, particularly next-generation sequencing (NGS), has significantly increased the accuracy and resolution of SCoT markers in genetic analyses. NGS, with its ability to rapidly and cost-effectively generate large amounts of genetic data, enables deeper genetic analyses using SCoT markers. This development presents significant opportunities to detect more complex genetic structures and polymorphism varieties, as NGS expands the scope of SCoT markers by allowing simultaneous analysis of numerous genetic variations. Advances in bioinformatics provide a robust infrastructure for the effective use of SCoT markers in larger and more complex genomic projects. The development of automated data analysis and advanced bioinformatics tools allows for the rapid and accurate processing of genetic data. These technologies enable the high-efficiency application of SCoT markers and ensure that the results are interpreted more reliably. The integration of bioinformatics tools also facilitates the meaningful integration of SCoT data, automates analyses, and enables the processing of large datasets. As a result, SCoT markers are being incorporated into more extensive, multidimensional research in areas such as genetic diversity, phylogenetic analysis, population genetics, and functional genomics [79].

It is anticipated that SCoT markers will become more applicable to aromatic and medicinal plant species in the future. Genetic studies on these plants are often limited, and the integration of current technologies offers significant potential for a better understanding of their genetic makeup. The high efficiency of SCoT markers provides valuable opportunities to explore genetic variations, map metabolic pathways, and understand phenotypic traits such as stress tolerance and the production of bioactive compounds in these plants. As a result, continuous advancements in technology and the integration of SCoT markers with these innovations will expand the scope of genetic research and facilitate more sophisticated, precise, and large-scale analyses in medicinal and aromatic plants. These developments significantly enhance the future potential of SCoT markers in both basic scientific research and agricultural breeding programs [80].

21.18 Conclusion and application perspectives

SCoT markers have proven to be a significant molecular tool in a variety of biological and agricultural fields, including genetic diversity analysis, functional genomics, and plant breeding. These markers, particularly by targeting polymorphisms located within coding gene regions, allow for a deeper understanding of the genetic structures of plants. The versatile application capacity offered by SCoT markers will continue to expand, enabling more detailed genetic analyses in the future. In the future, the integration of SCoT markers with genomic, transcriptomic, and metabolomic data could create a significant paradigm shift in plant genetics and biology research. This integration will provide a deeper understanding of how genetic variations shape not only the DNA level but also gene expression and metabolic profiles. Especially, the genetic basis of complex phenotypic traits such as stress tolerance, secondary metabolite production, biochemical pathway regulation, and genetic adaptation will be better understood. The data provided by SCoT markers will allow for more detailed molecular examination of these traits, unveiling new insights into the biological processes of medicinal and aromatic plants. These advancements will contribute not only to theoretical scientific knowledge but also to important applied areas such as the sustainable management of medicinal and aromatic plants and the conservation of genetic diversity. SCoT markers will play a crucial role in the conservation of ecosystems and the management of natural resources, facilitating the conservation of genetic resources of medicinal plants, monitoring population structures, and assessing their ability to adapt to environmental stress conditions. Additionally, the integration of these markers into agricultural breeding programs will provide a strong foundation for developing new varieties with more efficient and higher bioactive compound production. The possibilities offered by SCoT markers in the field of molecular biology will make them an effective tool not only in basic research but also in commercial and practical applications. The data provided by these markers could lead to revolutionary changes in areas such as plant resistance to environmental stressors, the production and quality of active compounds used in medicinal plants, and the identification of genetic foundations for biotechnological applications. Therefore, SCoT markers will continue to be a vital tool in fields like molecular biology, genetic engineering, and agricultural biotechnology, finding additional research and application areas in the future.

References

[1]   Ma, S., Khayatnezhad, M. and Minaeifar, A. A. (2021). Genetic diversity and relationships among Hypericum L. species by ISSR markers: A high value medicinal plant from Northern of Iran. Caryologia, 74(1), 97–107.

[2]   Agrawal, L. and Kumar, M. (2021). Improvement in ornamental, medicinal, and aromatic plants through induced mutation. Journal of Applied Biology and Biotechnology, 9(4), 162–169.

[3]   Bi, D., Chen, D., Khayatnezhad, Hashjin, Z. S., Li, Z. and Ma, Y. (2021). Molecular identification and genetic diversity in Hypericum L.: A high value medicinal plant using RAPD markers markers. Genetika, 53(1), 393–405.

[4]   Anonymous, 2024a (Accessed December 5, 2024, at https://www.freepik.com/).

[5]   Swarup, S., Cargill, E. J., Crosby, K., Flagel, L., Kniskern, J. and Glenn, K. C. (2021). Genetic diversity is indispensable for plant breeding to improve crops. Crop Science, 61(2), 839–852.

[6]   Hoban, S., Paz-Vinas, I., Shaw, R. E., Castillo-Reina, L., Da Silva J. M., DeWoody, J. A. and Grueber, C. E. (2024). DNA-based studies and genetic diversity indicator assessments are complementary approaches to conserving evolutionary potential. Conservation Genetics, 1–7.

[7]   Omar, H. S., Elsayed, T. R., Reyad, N. E. H. A., Shamkh, I. M. and Sedeek, M. S. (2021). Gene-targeted molecular phylogeny, phytochemical analysis, antibacterial and antifungal activities of some medicinal plant species cultivated in Egypt. Phytochemical Analysis, 32(5), 724–739.

[8]   Nam, V. T., Hang, P. L. B., Linh, N. N., Ly, L. H., Hue, H. T. T., Ha, N. H. and Hien, L. T. T. (2020). Molecular markers for analysis of plant genetic diversity. Vietnam Journal of Biotechnology, 18(4), 589–608.

[9]   Rai, M. K. (2023). Start codon targeted (SCoT) polymorphism marker in plant genome analysis: Current status and prospects. Planta, 257(2), 34.

[10]   Chaturvedi, T., Gupta, A. K., Lal, R. K. and Tiwari, G. (2022). March of molecular breeding techniques in the genetic enhancement of herbal medicinal plants: Present and future prospects. The Nucleus, 65(3), 413–436.

[11]   Gupta, P., Mishra, A., Lal, R. K. and Dhawa, S. S. (2021). DNA fingerprinting and genetic relationships similarities among the accessions/species of *Ocimum* using SCoT and ISSR markers system. Molecular Biotechnology, 63, 446–457.

[12]   Liu, S., Wang, Y., Song, Y., Khayatnezhad, M. and Minaeifar, A. A. (2021). Genetic variations and interspecific relationships in Salvia (*Lamiaceae*) using SCoT molecular markers. Caryologia, 74(3), 77–89.

[13]   Ejaz, B., Mujib, A., Mamgain, J., Malik, M. Q., Syeed, R., Gulzar, B. and Bansal, Y. (2021). Comprehensive in vitro regeneration study with SCoT marker assisted clonal stability assessment and flow cytometric genome size analysis of Carthamus tinctorius L.: An important medicinal plant. Plant Cell, Tissue and Organ Culture (PCTOC), 1–16.

[14]   Hassan, F. A., Ismail, I. A., Mazrou, R. and Hassan, M. (2020). Applicability of inter-simple sequence repeat (ISSR), start codon targeted (SCoT) markers and ITS2 gene sequencing for genetic diversity assessment in Moringa oleifera Lam. Journal of Applied Research on Medicinal and Aromatic Plants, 18, 100256.

[15]   Borah, R., Bhattacharjee, A., Rao, S. R., Kumar, V., Sharma, P., Upadhaya, K. and Choudhury, H. (2021). Genetic diversity and population structure assessment using molecular markers and SPAR approach in Illicium griffithii, a medicinally important endangered species of Northeast India. Journal of Genetic Engineering and Biotechnology, 19(1), 118.

[16]   Arjmand, A., Ebrahimi, M. and Moradi, N. (2023). Genetic diversity of different ecotypes from three Althaea species using SCoT phytochemical and molecular markers. Iranian Journal of Medicinal and Aromatic Plants Research, 39(4), 622–637.

[17] Igwe, D. O., Ihearahu, O. C., Osano, A. A., Acquaah, G. and Ude, G. N. (2022). Assessment of genetic diversity of Musa species accessions with variable genomes using ISSR and SCoT markers. Genetic Resources and Crop Evolution, 69(1), 49–70.

[18] Pan, H., Deng, L., Zhu, K., Shi, D., Wang, F. and Cui, G. (2024). Evaluation of genetic diversity and population structure of Annamocarya sinensis using SCoT markers. PloS One, 19(9), e0309283.

[19] Yeken, M. Z., Emiralioğlu, O., Çiftçi, V., Bayraktar, H., Palacioğlu, G. and Özer, G. (2022). Analysis of genetic diversity among common bean germplasm by start codon targeted (SCoT) markers. Molecular Biology Reports, 49(5), 3839–3847.

[20] Bokaei, A. S., Sofalian, O., Sorkhilalehloo, B., Asghari, A. and Pour-Aboughadareh, A. (2023). Deciphering the level of genetic diversity in some aegilops species using CAAT box-derived polymorphism (CBDP) and start codon target polymorphism (SCoT) markers. Molecular Biology Reports, 50(7), 5791–5806.

[21] Hromadová, Z., Gálová, Z., Mikolášová, L., Balážová, Ž., Vivodík, M. and Chňapek, M. (2023). Efficiency of RAPD and SCoT markers in the genetic diversity assessment of the common bean. Plants, 12(15), 2763.

[22] Aydın, F., Özer, G., Alkan, M. and Çakır, İ. (2022). Start Codon Targeted (SCoT) markers for the assessment of genetic diversity in yeast isolated from Turkish sourdough. Food Microbiology, 107, 104081.

[23] Yilmaz, A. and Ciftci, V. (2021). Genetic relationships and diversity analysis in Turkish laurel (*Laurus nobilis* L.) germplasm using ISSR and SCoT markers. Molecular Biology Reports, 48(5), 4537–4547.

[24] Tereba, A. and Konecka, A. (2020). Comparison of microsatellites and SNP markers in genetic diversity level of two Scots pine stands. Environmental Sciences Proceedings, 3(1), 4.

[25] Çakır, E. (2023). Molecular markers from past to present and their application areas. Regenerative Agriculture, 19.

[26] El-Masry, S. S., Ibrahim, S. D., Abdel-Razek, F. S. and Sadik, A. S. (2024). Genetic variation of three BBTV-infected banana cultivars based on SCoT DNA marker technique. Journal of Agricultural Chemistry and Biotechnology, 15(1), 1–8.

[27] Chňapek, M., Mikolášova, L., Vivodík, M., Gálová, Z., Hromadová, Z., Ražná, K. and Balážová, Ž. (2021). Genetic diversity of oat genotypes using SCoT markers. Biology and Life Sciences Forum, 11, 1–29.

[28] Kocaman, B., Toy, S. and Maraklı, S. (2020). Application of different molecular markers in biotechnology. International Journal of Science Letters, 2(2), 98–113.

[29] Chňapek, M., Balážová, Ž., Špaleková, A., Gálová, Z., Hromadová, Z., Číšecká, L. and Vivodík, M. (2024). Genetic diversity of maize resources revealed by different molecular markers. Genetic Resources and Crop Evolution, 71(8), 1–19.

[30] Samarina, L. S., Malyarovskaya, V. I., Reim, S., Yakushina, L. G., Koninskaya, N. G., Klemeshova, K. V. and Ryndin, A. V. (2021). Transferability of ISSR, SCoT and SSR markers for Chrysanthemum× Morifolium Ramat and genetic relationships among commercial Russian cultivars. Plants, 10(7), 1302.

[31] Amiteye, S. (2021). Basic concepts and methodologies of DNA marker systems in plant molecular breeding. Heliyon, 7(10).

[32] Panchariya, D. C., Dutta, P., Ananya Mishra, A., Chawade, A., Nayee, N. and Kushwaha, S. K. (2024). Genetic marker: A genome mapping tool to decode genetic diversity of livestock animals. Frontiers in Genetics, 15, 1463474.

[33] Duta-Cornescu, G., Constantin, N., Pojoga, D. M., Nicuta, D. and Simon-Gruita, A. (2023). Somaclonal variation-advantage or disadvantage in micropropagation of the medicinal plants. International Journal of Molecular Sciences, 24(1), 838.

[34] Santoso, T. J., Husni, A., Nugroho, K., Ya'la, Z. R., Dewi, T., Marhawati, M. and Ndobe, S. (2024). Optimization of PCR analysis based on start codon targeted markers (SCoT Markers) for identification of genetic variation of seaweed from central sulawesi. Journal La Lifesci, 5(1), 37–48.

[35] Le, N. T. and Thai, T. B. (2022). Population genetic diversity of Camellia Dilinhensis on the Di Linh plateau of Vietnam revealed by ISSR and SCoT markers. Dalat University Journal of Science, 12(3), 43–55.

[36] Thilakarathna, M. K. S., Karunathilaka, R. I. S., Gunawardana, G. A. and Jayasooriya, R. G. P. T. (2022). Use of molecular biology techniques for animal identification and traceability. University of Colombo Review, 3(2).

[37] Hoban, S., Bruford, M. W., Funk, W. C., Galbusera, P., Griffith, M. P., Grueber, C. E. and Vernesi, C. (2021). Global commitments to conserving and monitoring genetic diversity are now necessary and feasible. Bioscience, 71(9), 964–976.

[38] Abdelhameed, A. A., Ali, M., Darwish, D. B. E., AlShaqhaa, M. A., Selim, D. A. F. H., Nagah, A. and Zayed, M. (2024). Induced genetic diversity through mutagenesis in wheat gene pool and significant use of SCoT markers to underpin key agronomic traits. BMC Plant Biology, 24(1), 673.

[39] Jamshidi, B., Etminan, A., Mehrabi, A. M., Pour-Aboughadareh, A., Shooshtari, L. and Ghorbanpour, M. (2024). Genetic diversity of Artemisia species based on CAAT-box derived polymorphism (CBDP) and start codon targeted (SCoT) markers. Genetic Resources and Crop Evolution, 71(7), 1–11.

[40] Wasi, A., Shahzad, A. and Tahseen, S. (2024). Synthesis of nonembryonic synseed, short term conservation, phytochemical evaluation and assessment of genetic stability through SCoT markers in Decalepis salicifolia. Plant Cell, Tissue and Organ Culture (PCTOC), 156(3), 101.

[41] Srivastava, A., Gupta, S., Shanker, K., Gupta, N., Gupta, A. K. and Lal, R. K. (2020). Genetic diversity in Indian poppy (*P. somniferum* L.) germplasm using multivariate and SCoT marker analyses. Industrial Crops and Products, 144, 112050.

[42] El-Hadi, A., Habiba, R. M., El-Leel, A., Omneya, F. and Ahmed, N. F. (2024). Genetic Diversity of some Basil Varieties Estimated using RAPD, SCoT and ISSR Techniques. Journal of Agricultural Chemistry and Biotechnology, 15(8), 93–101.

[43] Shayan, A., Shokrpour, M., Nazeri, V., Babalar, M. and Mehrabi, A. (2024). Assessment of diversity and genetic differentiation in polycross-derived populations of Thymus daenensis using ISSR and SCoT markers. Genetic Resources and Crop Evolution, 1–15.

[44] Shekhawat, J. K., Rai, M. K., Shekhawat, N. S. and Kataria, V. (2018). Start codon targeted (SCoT) polymorphism for evaluation of genetic diversity of wild population of Maytenus emarginata. Industrial Crops and Products, 122, 202–208.

[45] Vivodík, M., Balážová, Ž., Gálová, Z. and Petrovičová, L. (2017). Genetic diversity analysis of maize (Zea mays L.) using SCoT markers. The Journal of Microbiology, Biotechnology and Food Sciences, 6(5), 1170.

[46] Feng, S. G., He, R. F., Jiang, M. Y., Lu, J. J., Shen, X. X., Liu, J. J. and Wang, H. Z. (2016). Genetic diversity and relationships of medicinal Chrysanthemum morifolium revealed by start codon targeted (SCoT) markers. Scientia Horticulturae, 201, 118–123.

[47] Igwe, D. O., Afiukwa, C. A., Ubi, B. E., Ogbu, K. I., Ojuederie, O. B. and Ude, G. N. (2017). Assessment of genetic diversity in Vigna unguiculata L.(Walp) accessions using inter-simple sequence repeat (ISSR) and start codon targeted (SCoT) polymorphic markers. BMC Genetics, 18, 1–13.

[48] Mirzaei, K. and Mirzaghaderi, G. (2017). Genetic diversity analysis of Iranian *Nigella sativa* L. landraces using SCoT markers and evaluation of adjusted polymorphism information content. Plant Genetic Resources, 15(1), 64–71.

[49] Karaköy, T., Baloch, F. S., Toklu, F. and Özkan, H. (2014). Variation for selected morphological and quality-related traits among 178 faba bean landraces collected from Turkey. Plant Genetic Resources, 12(1), 5–13.

[50] Etminan, A., Pour-Aboughadareh, A., Mohammadi, R., Ahmadi-Rad, A., Noori, A., Mahdavian, Z. and Moradi, Z. (2016). Applicability of start codon targeted (SCoT) and inter-simple sequence repeat (ISSR) markers for genetic diversity analysis in durum wheat genotypes. Biotechnology & Biotechnological Equipment, 30(6), 1075–1081.

[51] Etminan, A., Pour-Aboughadareh, A., Noori, A., Ahmadi-Rad, A., Shooshtari, L., Mahdavian, Z. and Yousefiazar-Khanian, M. (2018). Genetic relationships and diversity among wild Salvia accessions revealed by ISSR and SCoT markers. Biotechnology & Biotechnological Equipment, 32(3), 610–617.

[52] Kobeissi, B., Saidi, A., Kobeissi, A. and Shafie, M. (2019). Applicability of SCoT and SSR molecular markers for genetic diversity analysis in Chrysanthemum morifolium genotypes. Proceedings of the National Academy of Sciences, India Section B: Biological Sciences, 89, 1067–1077.

[53] Nadeem, M. A., Yeken, M. Z., Shahid, M. Q., Habyarimana, E., Yılmaz, H., Alsaleh, A. and Baloch, F. S. (2021). Common bean as a potential crop for future food security: An overview of past, current and future contributions in genomics, transcriptomics, transgenics and proteomics. Biotechnology & Biotechnological Equipment, 35(1), 759–787.

[54] Sarrou, E., Ganopoulos, I., Xanthopoulou, A., Masuero, D., Martens, S., Madesis, P. and Chatzopoulou, P. (2017). Genetic diversity and metabolic profile of Salvia officinalis populations: Implications for advanced breeding strategies. Planta, 246, 201–215.

[55] Ramadan, W. A., Shoaib, R. M., Ali, R. T. and Abdel-Samea, N. S. (2019). Assessment of genetic diversity among some fennel cultivars (*Foeniculum vulgare* Mill.) by ISSR and SCoT Markers. African Journal of Biological Sciences, 15(1), 219–234.

[56] Nikkerdar, F., Farshadfar, M., Ebrahimi, M. A. and Shirvani, H. (2018). Genetic diversity among fennel (*Fueniculum vulgare* Mill.) landrace using Scot Markers. Journal of Crop Breeding, 9(24), 95–102.

[57] Ahmed, M. Z., Masoud, I. M. and Zedan, S. Z. (2018). Genetic diversity and relationship among nine cultivated flax genotypes (*Linum usitatissimum* L.) based on SCoT markers. Sciences, 8(04), 1480–1490.

[58] Yadav, C. and Malik, C. P. (2016). Molecular characterization of fennel (*Foeniculum vulgare* Mill.) accessions using Start Codon Targeted (SCoT) markers. Journal of Plant Science Research, 32(1).

[59] Shimira, F., Boyaci, H. F., Çilesiz, Y., Nadeem, M. A., Baloch, F. S. and Taşkin, H. (2021). Exploring the genetic diversity and population structure of scarlet eggplant germplasm from Rwanda through iPBS-retrotransposon markers. Molecular Biology Reports, 48(9), 6323–6333.

[60] Dhawan, S. S. (2016). Role of molecular markers in assessing genetic diversity in mentha: A review. Scientific Journal of Genetics and Gene Therapy, 2(1), 022–026.

[61] Hama Mostafa, K., Vafaee, Y., Khorshidi, J., Rastegar, A. and Morshedloo, M. R. (2024). Molecular and morphological characterization of wild Mentha langifolia L. accessions from Zagros Basin: Application for domestication and breeding. Genetic Resources and Crop Evolution, 71, 1–20.

[62] Foda, D. S., Sweelam, H. T. M. and Ibrahim, N. E. (2024). Role of rosmarinus officinalis aqueous extract in relieving the complications associated with ethylene glycol-induced urolithiasis in male rats. Current Bioactive Compounds, 20(10), E150224227008.

[63] Anonymous, 2024b (Accessed December 5, 2024, at https://www.freepik.com/free-photo/rosemary-plants-place-dark-floor_11995775.htm#fromView=search&page=1&position=11&uuid=3e29e9bb-4a74-45e7-9ad4-d68321369cb5).

[64] Anonymous, 2024c (Accessed December 5, 2024, at https://www.freepik.com/free-photo/vertical-high-angle-shot-beautiful-plant-white-vase-white-surface_10809663.htm#fromView=search&page=1&position=29&uuid=e68cdbe7-65eb-4148-8fcc-481b75cd079f).

[65] Ouahzizi, B., Lebkiri, N., Elbouny, H., Ouedrhiri, S., Saghir, K., Gaboun, F. and Diria, G. (2024). Morphological and genetic variability of thymus atlanticus (Ball) roussine; endemic plant of Morocco. Endemic Plant of Morocco.

[66] Elzaiat, M. A., Mandour, A. S., Youssef, M. A., Wafa, H. A., Aljahdali, S. M., Shakak, A. O. and Heakel, R. M. (2024). Biochemical and molecular characterization of five basil cultivars extract for enhancing

the antioxidant, antiviral, anticancer, antibacterial, and antifungal activities. *Pakistan Veterinary Journal*, 44(4).

[67] Urbanová, L., Farkasova, S., Speváková, I., Kyseľ, M., Šimora, V., Kacaniova, M. and Žiarovská, J. (2024). DNA-Based variability of length polymorphism of plant allergens coding genes homologs in selected *Lamiaceae* herbs. OBM Genetics, 8(3), 1–16.

[68] Heikrujam, M., Kumar, J. and Agrawal, V. (2015). Genetic diversity analysis among male and female Jojoba genotypes employing gene targeted molecular markers, start codon targeted (SCoT) polymorphism and CAAT box-derived polymorphism (CBDP) markers. Meta Gene, 5, 90–97.

[69] Hajibarat, Z., Saidi, A., Hajibarat, Z. and Talebi, R. (2015). Characterization of genetic diversity in chickpea using SSR markers, start codon targeted polymorphism (SCoT) and conserved DNA-derived polymorphism (CDDP). Physiology and Molecular Biology of Plants, 21, 365–373.

[70] Bisht, V., Rawat, J. M., Gaira, K. S., Purohit, S., Anand, J., Sinha, S. and Rawat, B. (2024). Assessment of genetic homogeneity of in-vitro propagated apple root stock MM 104 using ISSR and SCoT primers. BMC Plant Biology, 24(1), 240.

[71] Sanches, P. H. G., De Melo, N. C., Porcari, A. M. and De Carvalho, L. M. (2024). Integrating molecular perspectives: Strategies for comprehensive multi-omics integrative data analysis and machine learning applications in transcriptomics, proteomics, and metabolomics. Biology, 13(11), 848.

[72] Baloch, F. S., Guizado, S. J. V., Altaf, M. T., Yüce, I., Çilesiz, Y., Bedir, M. and Gómez, J. C. C. (2022). Applicability of inter-primer binding site iPBS-retrotransposon marker system for the assessment of genetic diversity and population structure of Peruvian rosewood (*Aniba rosaeodora* Ducke) germplasm. Molecular Biology Reports, 49, 1–12.

[73] Wang, M., Ma, Y., Yu, G., Zeng, B., Yang, W., Huang, C. and Wu, Z. (2024). Integration of microbiome, metabolomics and transcriptome for in-depth understanding of berberine attenuates AOM/DSS-induced colitis-associated colorectal cancer. Biomedicine & Pharmacotherapy, 179(47), 117292.

[74] Leal, K., Rojas, E., Madariaga, D., Contreras, M. J., Nuñez-Montero, K., Barrientos, L. and Iturrieta-González, I. (2024). Unlocking fungal potential: The crispr-cas system as a strategy for secondary metabolite discovery. Journal of Fungi, 10(11), 748.

[75] Yun, S., Noh, M., Yu, J., Kim, H. J., Hui, C. C., Lee, H. and Son, J. E. (2024). Unlocking biological mechanisms with integrative functional genomics approaches. Molecules and Cells, 100092.

[76] Karaköy, T., Toklu, F., Karagöl, E. T., Uncuer, D., Çilesiz, Y., Ali, A. and Özkan, H. (2024). Genome-wide association studies revealed DArTseq loci associated with agronomic traits in Turkish faba bean germplasm. Genetic Resources and Crop Evolution, 71(1), 181–198.

[77] Salami, M., Heidari, B., Batley, J., Wang, J., Tan, X. L., Richards, C. and Tan, H. (2024). Integration of genome-wide association studies, metabolomics, and transcriptomics reveals phenolic acid-and flavonoid-associated genes and their regulatory elements under drought stress in rapeseed flowers. Frontiers in Plant Science, 14, 1249142.

[78] Mansoor, S., Karunathilake, E. M. B. M., Tuan, T. T. and Chung, Y. S. (2024). Genomics, phenomics, and machine learning in transforming plant research: Advancements and challenges. Horticultural Plant Journal, 11(2), 486–503.

[79] Zhu, W., Li, W., Zhang, H. and Li, L. (2024). Big data and artificial intelligence-aided crop breeding: Progress and prospects. Journal of Integrative Plant Biology, 67, 722–739.

[80] Jiang, Y., Jin, Y., Shan, Y., Zhong, Q., Wang, H., Shen, C. and Feng, S. (2024). Advances in physalis molecular research: Applications in authentication, genetic diversity, phylogenetics, functional genes, and omics. Frontiers in Plant Science, 15, 1407625.

Index

https://doi.org/10.1515/9783111469713-022

www.ingramcontent.com/pod-product-compliance
Lightning Source LLC
Chambersburg PA
CBHW080338220326
41598CB00030B/4544